# Temporally and Spatially Resolved Molecular Science

Indian Institute of Science, Bangalore, India
12-14 January 2015

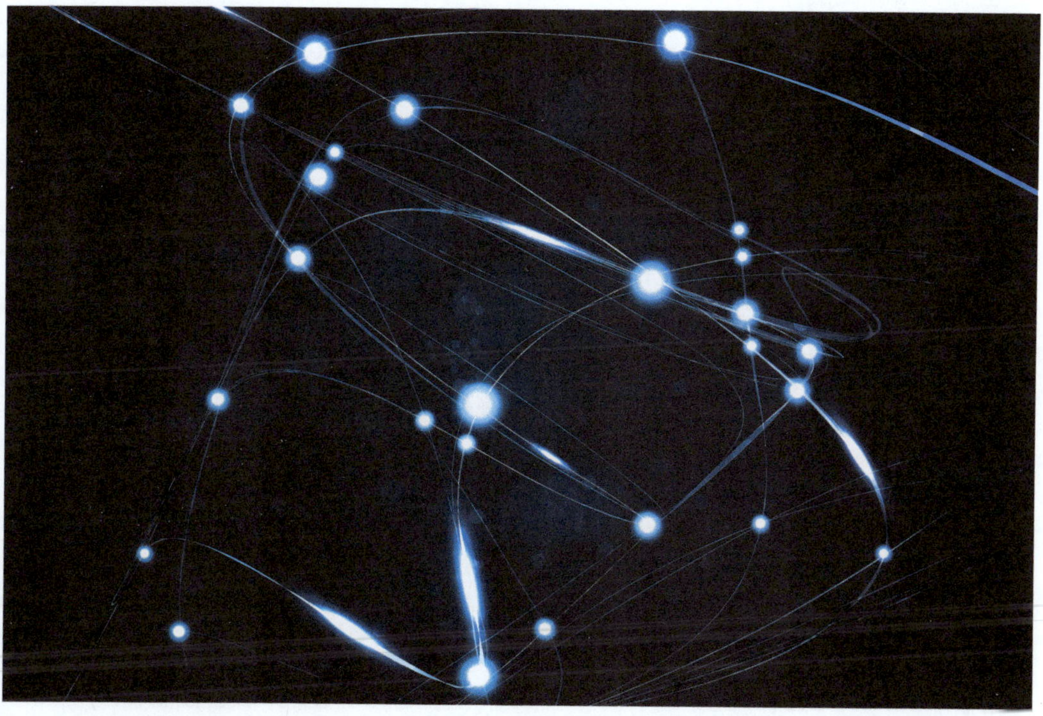

FARADAY DISCUSSIONS
Volume 177, 2015

ROYAL SOCIETY OF CHEMISTRY

The Faraday Division of the Royal Society of Chemistry, previously the Faraday Society, founded in 1903 to promote the study of sciences lying between Chemistry, Physics and Biology.




Faraday Discussions (Print ISSN 1359-6640, Electronic ISSN 1364-5498) is published 9 times a year by the Royal Society of Chemistry, Thomas Graham House, Science Park, Milton Road, Cambridge, UK CB4 0WF.

Volume 177 ISBN-13: 978-1-78262-177-5

2015 annual subscription price: print+electronic £877, US $1635; electronic only £833, US $1553. Customers in Canada will be subject to a surcharge to cover GST. Customers in the EU subscribing to the electronic version only will be charged VAT. All orders, with cheques made payable to the Royal Society of Chemistry, should be sent to RSC Order Department, Royal Society of Chemistry, Thomas Graham House, Science Park, Milton Road, Cambridge, CB4 0WF, UK.
Tel +44 (0) 1223 432398; E-mail www.orders@rsc.org

If you take an institutional subscription to any RSC journal you are entitled to free, site-wide web access to that journal. You can arrange access via Internet Protocol (IP) address at www.rsc.org/ip. Customers should make payments by cheque in sterling payable on a UK clearing bank or in US dollars payable on a US clearing bank.

Printed in the UK


MIX
Paper from responsible sources
FSC www.fsc.org FSC® C013604

Faraday Discussions are unique international discussion meetings that focus on rapidly developing areas of chemistry and its interfaces with other scientific disciplines.

# Temporally and Spatially Resolved Molecular Science

Faraday Discussions

www.rsc.org/faraday_d

A General Discussion on Temporally and Spatially Resolved Molecular Science was held in Bangalore, India on the 12th, 13th and 14th of January 2015.



## CONTENTS

ISSN 1359-6640; ISBN 978-1-78262-177-5

**Cover**
Front cover image courtesy of Shutterstock.

## INTRODUCTORY LECTURE

9 **Spatial resolution in Raman spectroscopy**
Volker Deckert, Tanja Deckert-Gaudig, Marco Diegel, Isabell Götz, Lucas Langelüddecke, Henrik Schneidewind, Gaurav Sharma, Prabha Singh, Pushkar Singh, Steffen Trautmann, Matthias Zeisberger and Zhenglong Zhang

## PAPERS AND DISCUSSIONS

21 **Breaking of a bond: when is it statistical?**
Pankaj Kumar Yadav and Srihari Keshavamurthy

33 **Negative hyperconjugation and red-, blue- or zero-shift in X–Z···Y complexes**
Jyothish Joy, Eluvathingal D. Jemmis and Kaipanchery Vidya

51 **Dynamics of the chemical bond: inter- and intra-molecular hydrogen bond**
Elangannan Arunan and Devendra Mani

SPONSOR

cogent | Skills for Science Based Industries

65 **Vibrational dynamics of the CO stretching of 9-fluorenone studied by visible-pump and infrared-probe spectroscopy**
Yuki Fukui, Kaoru Ohta and Keisuke Tominaga

77 **Dynamical, spectroscopic and computational imaging of bond breaking in photodissociation: roaming and role of conical intersections**
Masaaki Nakamura, Po-Yu Tsai, Toshio Kasai, King-Chuen Lin, Federico Palazzetti, Andrea Lombardi and Vincenzo Aquilanti

99 **The free-energy barrier to hydride transfer across a dipalladium complex**
C. R. Vanston, G. J. Kearley, A. J. Edwards, T. A. Darwish, N. R. de Souza, A. J. Ramirez-Cuesta and M. G. Gardiner

111 **Excited state dynamics of organic semi-conducting materials**
Kenneth P. Ghiggino, Andrew J. Tilley, Benjamin Robotham and Jonathan M. White

121 **Dynamics of chemical bond: general discussion**

155 **Pluronic copolymer encapsulated SCR7 as a potential anticancer agent**
Franklin John, Jinu George, Mrinal Srivastava, P. A. Hassan, V. K. Aswal, Subhas. S. Karki and Sathees. C. Raghavan

163 **The effect on structural and solvent water molecules of substrate binding to ferric horseradish peroxidase**
Niall Simpson, Katrin Adamczyk, Gordon Hithell, Daniel J. Shaw, Gregory M. Greetham, Michael Towrie, Anthony W. Parker and Neil T. Hunt

181 **Electronic excitations in molecular solids: bridging theory and experiment**
Jonathan M. Skelton, E. Lora da Silva, Rachel Crespo-Otero, Lauren E. Hatcher, Paul R. Raithby, Stephen C. Parker and Aron Walsh

203 **Resolution enhancement through microscopic spatiotemporal control**
Debabrata Goswami, Dhiman Das and Soumendra Nath Bandyopadhyay

213 **On the photophysics of butadiyne bridged pyrene–phenyl molecular conjugates: multiple emissive pathways through locally excited, intramolecular charge transfer and excimer states**
Avik Kumar Pati, Santosh J. Gharpure and Ashok K. Mishra

237 **Time resolved *in situ* X-ray diffraction study of crystallisation processes of large pore nanoporous aluminophosphate materials**
Kerry Simmance, Wouter van Beek and Gopinathan Sankar

249 ***In situ* characterization of Cu–Co oxides for catalytic application**
Z. Y. Tian, H. Vieker, P. Mountapmbeme Kouotou and A. Beyer

263 **Time and Space resolved Methods: general discussion**

293 **Electron transfer quenching in light adapted and mutant forms of the AppA BLUF domain**
Sergey P. Laptenok, Andras Lukacs, Richard Brust, Allison Haigney, Agnieszka Gil, Michael Towrie, Gregory M. Greetham, Peter J. Tonge and Stephen R. Meech

313 **Spatio-temporal correlations in aqueous systems: computational studies of static and dynamic heterogeneity by 2D-IR spectroscopy**
Rikhia Ghosh, Tuhin Samanta, Saikat Banaerjee, Rajib Biswas and Biman Bagchi

329 **Vibronic structure in the far-UV electronic circular dichroism spectra of proteins**
Zhuo Li, David Robinson and Jonathan D. Hirst

345 **Probing deactivation pathways of DNA nucleobases by two-dimensional electronic spectroscopy: first principles simulations**
Artur Nenov, Javier Segarra-Martí, Angelo Giussani, Irene Conti, Ivan Rivalta, Elise Dumont, Vishal K. Jaiswal, Salvatore Flavio Altavilla, Shaul Mukamel and Marco Garavelli

363 **Out-of-equilibrium dynamics of photoexcited spin-state concentration waves**
A. Marino, M. Buron-Le Cointe, M. Lorenc, L. Toupet, R. Henning, A. D. DiChiara, K. Moffat, N. Bréfuel and E. Collet

381 **Local and Global Dynamics: general discussion**

405 **Probing electronic and vibrational dynamics in molecules by time-resolved photoelectron, Auger-electron, and X-ray photon scattering spectroscopy**
Kochise Bennett, Markus Kowalewski and Shaul Mukamel

429 **X-ray diffraction in temporally and spatially resolved biomolecular science**
John R. Helliwell, Alice Brink, Surasak Kaenket, Victoria Laurina Starkey and Simon W. M. Tanley

443 **Disentangling detector data in XFEL studies of temporally resolved solution state chemistry**
Tim Brandt van Driel, Kasper Skov Kjær, Elisa Biasin, Kristoffer Haldrup, Henrik Till Lemke and Martin Meedom Nielsen

467 **Mapping atomic motions with ultrabright electrons: towards fundamental limits in space-time resolution**
Stephanie Manz, Albert Casandruc, Dongfang Zhang, Yinpeng Zhong, Rolf A. Loch, Alexander Marx, Taisuke Hasegawa, Lai Chung Liu, Shima Bayesteh, Hossein Delsim-Hashemi, Matthias Hoffmann, Matthias Felber, Max Hachmann, Frank Mayet, Julian Hirscht, Sercan Keskin, Masaki Hada, Sascha W. Epp, Klaus Flöttmann and R. J. Dwayne Miller

493 **Two-dimensional sum-frequency generation (2D SFG) spectroscopy: summary of principles and its application to amyloid fiber monolayers**
Ayanjeet Ghosh, Jia-Jung Ho, Arnaldo L. Serrano, David R. Skoff, Tianqi Zhang and Martin T. Zanni

507 **Phase modulation nanoscopy: a simple approach to enhanced optical resolution**
Robert Pal

517 **Future challenges: general discussion**

## CONCLUDING REMARKS

547 **Spatio-temporal resolution of primary processes of photosynthesis**
Wolfgang Junge

## ADDITIONAL INFORMATION

563 **Poster titles**
567 **List of participants**

Faraday Discussions

PAPER

# Spatial resolution in Raman spectroscopy

Volker Deckert,*[ab] Tanja Deckert-Gaudig,[a] Marco Diegel,[a] Isabell Götz,[a] Lucas Langelüddecke,[b] Henrik Schneidewind,[a] Gaurav Sharma,[b] Prabha Singh,[b] Pushkar Singh,[a] Steffen Trautmann,[a] Matthias Zeisberger[a] and Zhenglong Zhang[a]

Received 6th March 2015, Accepted 24th March 2015
DOI: 10.1039/c5fd90014j

This article is intended to set the scope of the meeting, in particular, the high spatial resolution section.

## 1 Introduction

A constant motivation to develop advanced technologies addressing temporal and spatial resolution is the investigation of structural changes of a single molecule while it is reacting. The dream is to follow a bond breakage or bond formation of a single selected molecule with a method that provides structural sensitivity, time resolution, and local specificity. Presently, such a combined technique is not yet available, however, regarding each single requirement of the mentioned list, tremendous progress has been made over the last few decades.

This overview will specifically discuss the spatial resolution aspects that are required to address, if not a single molecule, at least a sample volume that is small enough to distinct the target structure from the background. The 2014 Chemistry Nobel prize awarded to Moerner, Hell, and Betzig, highlighted the importance of high resolution techniques specifically for biochemical applications. While the awarded fluorescence approach is certainly extremely useful whenever dye labelling is possible and no interference with the function of the target system is expected, in particular for smaller molecular systems such labelling is not practical or even possible. In such cases vibrational spectroscopy techniques can be used, often at the cost of lower sensitivity, however, as structural information is obtained directly no labelling is required and sample preparation becomes easier. In this article we will solely present the high lateral resolution aspects of Raman spectroscopy, many of those aspects can be easily applied to infrared absorption spectroscopy if the specific factors of the particular

[a]*Leibniz Institute of Photonic Technology, Albert-Einstein-Str. 9, 07745 Jena, Germany. E-mail: Volker.deckert@ipht-jena.de; Fax: +49 3641 206 139; Tel: +49 3641 206 113*

[b]*Institute of Physical Chemistry and Abbe Center of Photonics, University of Jena, Helmholtzweg 4, 07743 Jena, Germany*

 

wavelength range are considered. As mentioned before, the sensitivity of fluorescence label techniques easily allows the investigation of single molecules. With normal Raman scattering this is not possible as the scattering cross sections are several orders of magnitude lower compared to fluorescence. Since the mid 1970s a method has existed to enhance the Raman signals considerably.[1–3] By using roughened noble metal surfaces or noble metal nanoparticles an enhancement of several orders of magnitude can be achieved mainly due to plasmonic effects. This bridges the gap between fluorescence and Raman spectroscopy and allows the investigation of single molecules with so called surface-enhanced Raman scattering (SERS).[4–6]

At present the only possible way to overcome the diffraction limit[7,8] with Raman spectroscopy requires the use of optical near-fields. Consequently, we will deal solely with the special properties of near-field Raman scattering,[9–11] which luckily is in many ways directly associated with the aforementioned plasmonic effects. Accordingly a combination of near-field optics and SERS provides access to a high lateral resolution, low detection limit, and high structural sensitivity. Furthermore, many aspects regarding the investigation of small sample volumes, and thus restricted number of sampled molecules, can be regarded as general aspects whenever such small dimensions are probed.

## 2 Experimental

### General setup

All investigations mentioned here were measured using a so-called tip-enhanced Raman scattering (TERS) setup. Fig. 1 shows the schematic sketch of our system. The main components are an atomic force microscope (AFM) and an inverted optical microscope equipped with an $x$, $y$, $z$ piezo stage for automated sample scanning. The whole system is designed for a precise positioning of the near-field optical probe with respect to laser focus and sample. These parts are shown in dark grey. Additionally, coarse sample and tip alignment standard $xy$ stages with $\sim$2 μm precision are also available, but not shown in detail. Further details about this particular setup can be found in the literature where a straightforward adaptation to a reflection setup giving access to opaque samples is also presented.[12–14] The illumination through substrate and sample on to the plasmonic tip seems to be demanding and prone to losses, however, this setup allows for the best collection efficiency and implementing proper polarisation allows an optimal excitation of the nanoparticles.

### TERS tips

**Theory.** The TERS tips are the integral element of the entire system. Their special properties are responsible for the combination of high lateral resolution and signal enhancement. When irradiating silver or gold (other metals like Cu or Al have also been tested for special purposes) nanoparticles with light of an appropriate wavelength, electrons in the conducting band are collectively excited and oscillate with the same frequency as the incoming electromagnetic field. This can be considered as an optical analogue of a receiving antenna[15] and leads to an enhancement and confinement of the fields depending on the shape and size of the actual nanoparticle. For details on the enhancement mechanism and a

 

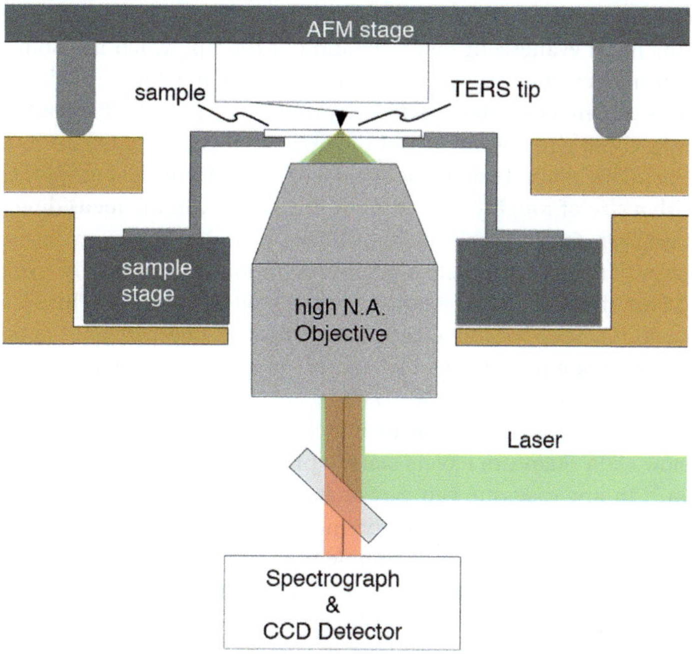


Fig. 1 Schematic view of a TERS epi-illumination setup. Tip and sample can be scanned independently using precision piezo stages. Illumination and collection for the Raman system is carried out *via* a high N. A. objective. The detection of the Raman signal is done after appropriate filtering using a standard multichromator/CCD detector arrangement.

thorough theoretical description see ref. 16–19. In essence, if the experimental conditions are matched properly the field is confined to a very small volume close to the tip. Fig. 2 shows model calculations of "ideal" tips, either with a single nanoparticle on an AFM tip apex or even with only a single metal sphere. In both

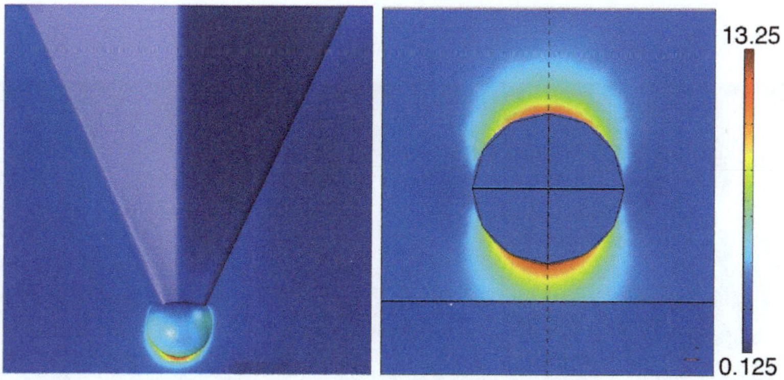


Fig. 2 Left panel: field distribution of an ideal TERS probe with one single nanoparticle attached to the apex of an AFM tip. Right panel: detail of the field distribution $E(x, z)$ obtained by FEM simulation for a silver sphere (similar to the dipol–mirror dipole model) of radius 10 nm and a distance from the surface of 5 nm, $\lambda = 413$ nm. Field distributions in both cases are comparable.

cases a total internal reflection illumination from the lower left was simulated. This results in an evanescent field excitation of the tip, which has some experimental advantages, but is not essential for such modelling. This geometry nevertheless explains the slight asymmetry of the electric field distribution. Most importantly, these simulations reveal the extreme light confinement in all three dimensions. The main lateral enhancement is concentrated underneath the sphere with a size of roughly the radius of the particle. In the focus direction the field decays even faster. As the Raman intensity scales approximately with the fourth power of the incident field a lateral resolution for realistic tip sizes (in our case 15–30 nm diameter) can be expected to be well below 5 nm. This value takes into account only the electromagnetic theory for the "macroscopic" appearance, specific atomic scale protrusions like corners or edges, as well as specific chemical interactions between the silver atoms and the sample, which could both potentially increase the resolution, are not considered. Interestingly, at very small scales below ~3 Å quantum effects start to play a role, finally limiting the lateral resolution.[20] In any case one can expect a spatial resolution that renders many samples as single crystals or at least oriented in a statistically unexpected direction, if special care is not taken in sample preparation.

**Tip preparation.** Fig. 3 shows typical SEM images of the tips mostly used in our group. These tips are obtained by physical vapour deposition of silver onto commercially available AFM tips[21] and the resulting silver nanoparticles have diameters of less than 30 nm. This approach aims to produce one more or less isolated small silver nanoparticle at the apex of the AFM tip and was initially developed for SERS substrates.[21–23] The nanoparticle at the tip apex serves as the active nano-antenna mentioned above. Inspecting more than a hundred tip batches, three main geometries of the outermost silver nanoparticles were found. Type (a) reveals a single silver nanoparticle at the top of the AFM tip in close vicinity with other particles. Most of the time the nanoparticles are slightly off-axis. A second type (b) shows two silver nanoparticles at the top. Usually, one of the two nanoparticles is located a few nanometers in front of the other one. Thus, we can expect that one acts as the plasmonically active device in the TERS experiment. About two thirds of our prepared tips fit these two classes. Last but

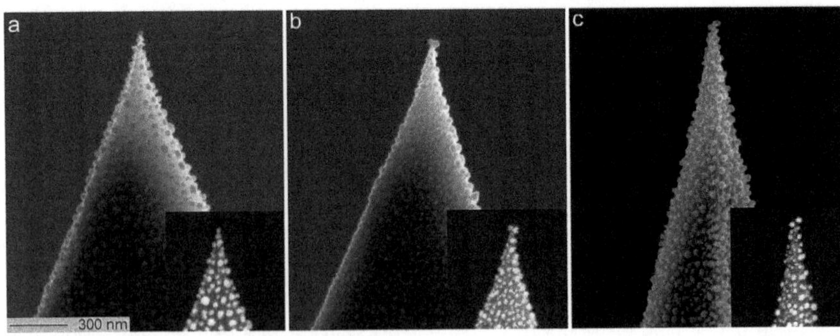


Fig. 3 SEM images of the main manifestations of the silver nanoparticles on top of the TERS tips: a single nanoparticle at the outermost top (a), two nanoparticles at the top (b), or a single nanoparticle at the edge of the AFM tip (c), respectively. The scale bar applies to all images. The insets show the same tips using the material contrast.

 

not least, a small number of tips showed one clearly isolated frontmost silver nanoparticle, again usually slightly off axis (c). As the practical yield, *e.g.* the ratio of tips in a batch that produce a considerable enhancement, of such tips is around 90% the actual geometry is influencing merely the details of the enhancement and, also quite importantly, the properties for the topographic sensing. Recently a more directed approach was successfully investigated for the particle attachment using dielectrophoresis.[24–28] This method allows single preselected nanoparticles to be placed onto an AFM cantilever, and hence, seems to be promising for dedicated applications with preformed particles. Furthermore, many different approaches to tip production have been presented and which type is most suited certainly depends on the specific experimental requirements.[23,24,26–28,29–31,32]

## Application examples

**Self-assembled monolayers.** In order to provide a reproducible method of sample preparation, self-assembly is a well-established tool. This way only a single layer of molecules can be investigated and the orientation is generally predetermined. Certainly this approach is not feasible for most realistic samples, nevertheless, the study of self-assembled monolayers (SAMs) is ideal to rule out specific concentration effects. In particular if working with an extremely distance dependent method like TERS, a layer with such thickness and orientation control is important to examine fundamental parameters. Several TERS studies have been performed on molecules adsorbed onto gold single crystals.[29,30] This provides reliable conditions, as any surface roughness effects can be ruled out. In particular for STM based reflection TERS systems this is an ideal arrangement. In order to utilize the advantages of single crystalline surfaces for epi-illumination setups like the one shown in Fig. 1, methods were developed to manufacture gold[32–35] or silver[36,37] single crystals thin enough for the transmission of light. Those were readily applied to study small molecules like thiophenols, amino acids and small peptides.[33,34,38,39] It became apparent that, in particular, the spectra of aromatic amino acid monolayers lacked some usually strong bands, but could still be identified by other marker signals. If the experimental results were compared with theoretical models of amino acids on gold (111) surfaces it became evident that, due to the most favoured flat orientation of the molecules *versus* the surface, the typical "ring breathing" modes were not efficiently excited in the TERS setup where the electric field vector is oriented orthogonally towards the surface and consequently orthogonal to the ring stretching vibrations.[36,40–43] Interestingly, the field components at the tip side (see Fig. 2) did not play a role, as these components should have excited the suppressed modes. This indirectly hints towards a high field confinement, *i.e.* a high lateral resolution, which is of course difficult to determine on a SAM. To further study that effect an even simpler system was chosen.[38] Thiophenol was immobilized on flat and rough gold and silver surfaces. Here, the "ring breathing" mode was excited, as the perpendicular orientation of the molecule with respect to the surface is favourable for TERS excitation. The main target of the work was to study differences between the SER and TER spectra of thiophenol monolayers. Often TERS signals tend to fluctuate slightly and the idea here was to control the orientation and layer thickness as well as possible. Fig. 4 shows the most intriguing features of the experiments. The two spectra (a) and (b) on the left side show the two selected SERS bands of a self-assembled

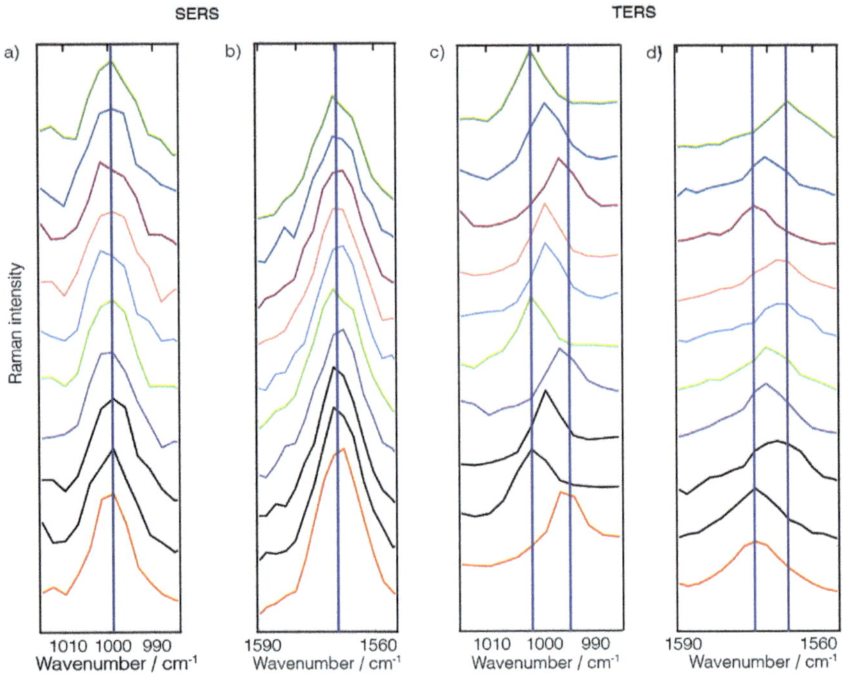


Fig. 4 Comparison between SERS (a and b) and TERS (c and d) spectra of thiophenol monolayers. Due to the small number of molecules detected in the TERS case, averaging effects like in the case of the SERS spectra are absent or not yet pronounced. This leads to stronger fluctuations in the TERS case.

thiophenol monolayer on a gold island film measured at arbitrary locations. The band parameters, width, intensity, and spectral position, hardly vary as can be expected if both the film and the substrate are homogeneous. The term "SERS substrate homogeneity" is a bit contradictory, as the substrates resemble more or less the silver island films on the AFM cantilevers shown in Fig. 3. On the other hand the laser spot is comparably large, and from the data it can be concluded that the inhomogeneities of the films were averaged out. In this respect, the corresponding TERS experiment should be even more reproducible: a ~30 nm thick gold flake, of diameter 2–3 μm, and a surface roughness of typically around 2 Å, is a much more regular sample support than any SERS substrate can intrinsically ever be. On top of this gold plate the TERS tip was positioned, whose enhancing properties hardly changed during the measurement. Still, the experimental data show much stronger band position fluctuations at different positions on the gold crystal compared to SERS (see Fig. 4c and d). A similar time dependent experiment over several minutes effectively showed the same behaviour for the SERS and TERS experiment, respectively. We attributed this behaviour to the fact that in the TERS experiment we probed such a small sample volume that averaging of minute molecule orientations was not observed, whereas in the SERS case many more molecules and also much more enhancing sites led to an effective averaging of the Raman signals. To support this hypothesis the band widths of the TERS and SERS experiments were also compared. Not surprisingly, the TERS data showed a broader distribution, however, the average width of the TERS experiments was always lower compared to the SERS spectra. As for both experiments the same physical constraints were applied, we hypothesise that the SERS spectra are a combination of many local single particles or single hot spot events, which would fundamentally resemble the TERS experiment. Moreover, these results raise the question how many molecules are actually probed in a TERS experiment? Again a direct estimation of the resolution is not possible from a monolayer measurement, but it is certainly interesting to speculate on the number of molecules required for averaging effects to dominate the spectra. For comparison, a hundred closely packed thiophenol molecules approximately fit on a $4 \times 4$ μm area.

### Protein structures and spatial resolution

As a kind of by-product of our TERS investigations of protein structure surfaces, namely amyloid fibrils, the lateral resolution aspect emerged again.[40–44] When investigating fibrils with TERS using a step size of 5 Å between subsequent positions, surprisingly pronounced spectral changes could be observed if a long enough acquisition time was chosen to allow a detailed band analysis. It should be mentioned here that Paulite *et al.* have presented TERS images of well-ordered amyloid nanotapes that appear homogeneous.[45] Unfortunately the signal-to-noise ratio in those experiments was chosen such that apart from the phenylalanine "ring breathing" mode no other signal was detected and a prediction regarding structural changes was hardly possible. Hence, we consider the signal variations detected in our experiments to be significant and related to the actual sample composition. If one considers furthermore the findings of the previous experimental and theoretical section concerning the lateral resolution, the choice of step size to be 5 Å was set quite conservatively to avoid accidental undersampling of the dataset. When investigating the data we typically find results like those shown in Fig. 5. Here we investigated an insulin amyloid protofilament and clearly saw structural changes occurring on a length scale of a few Ångströms. For instance, in the first four spectra one can observe an amide I band (at 1604 $cm^{-1}$) representing α-helix/unordered secondary structures. This is also observed, however slightly weaker, in the fifth spectrum, but then vanishes in the remaining two spectra. Similar patterns can be observed for the disappearance of the phenylalanine "ring breathing" mode and the appearance of a tyrosine marker band. The marker bands of cystine also change, however, as many different conformations around the disulfide bridge affect the exact spectral position and intensity ratios of the associated bands, a full assignment would be too complex here. Similarly, as detected in the self-assembled thiophenol monolayer case, one can observe similar spectral band position variations that are related to the immediate local environment of the sample. This can of course also involve the tip. These results can be simply understood as a way to investigate structural changes with high lateral resolution. But, in fact, this resolution is rather unexpected as we do not observe average peptide spectra, but rather locally dependent changes that reveal distinct amino acid contents.[44] Looking at the data of Fig. 4 and considering previous data on fibrils[41,42,44] and on DNA bases[46,47] a lateral resolution of about 1 nm can be estimated which is actually at the limit of sampling in the case presented. Bearing in mind the TERS SAM experiments

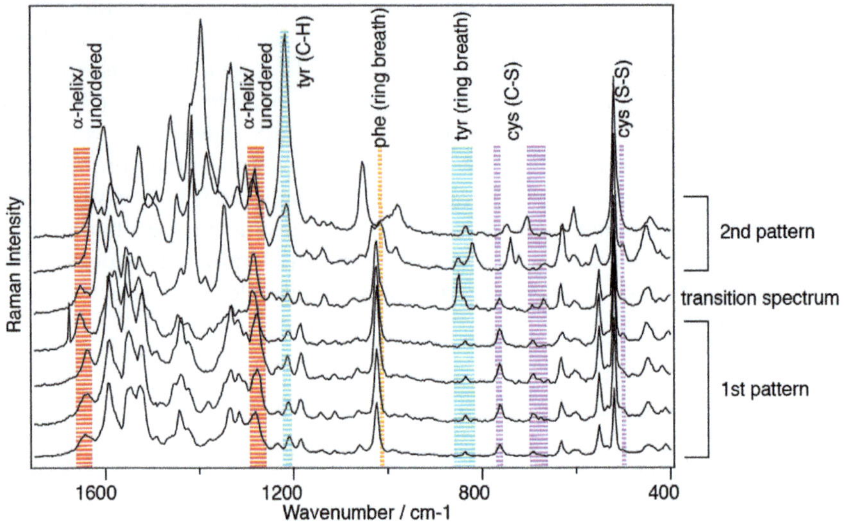


Fig. 5 TERS spectra detected on an insulin amyloid protofilament generated at pH 1.5, laser excitation 532 nm, acquisition time 10 s/spectrum, distance between positions 0.5 nm. Two major patterns can be clearly distinguished: only in the first 5 spectra α-helix/unordered structures can be clearly identified by amide I bands at 1640 $cm^{-1}$. A transition is also apparent in the synchronous appearance and disappearance of phenylalanine, tyrosine and cystine.

where a direct estimation of the resolution was not feasible, such a lateral resolution would relate to less than 10 molecules in the active region under the tip. For such a small number a "non-average" behavior becomes reasonable.

There is an important caveat to the generalization implied here. At the moment we are not aware of a full theoretical description that explains the lateral resolution of the experiments. It seems that either the size and shape parameters of the plasmonic tips have to be modeled differently, or that direct interactions of the tip and the sample provide an additional contrast that leads to the enhanced resolution observed. The latter aspect is often termed "chemical enhancement". While currently strong efforts are being made to unravel this issue, we think that any high resolution beyond 3–5 nm can be utilized to sequence virtually any material.

## General method for sequencing a single strand

For such sequencing we will provide the outline of this method and an estimation of the precision that is required for a successful application.[48] One can consider a tip as a simple point dipole, with the dipole in the $z$-direction, with a given field distribution that results in a TERS interaction volume as mentioned previously. This means for the following ideas we assume the theoretically predicted lateral resolution values and not even the values we experimentally determined. If, for instance a DNA single strand is aligned along the $x$-axis with bases at the positions $x_k$ ($k = 1, 2, \ldots, K$) and $z = 0$, and the tip moves along $x$ at height $z$, the signal at position $x_n$ for one particular Raman peak will be

 

$$S(x_n) = \sum_{k=1}^{K} b_k P(x_n - x_k, z), \quad n = 1 \ldots N$$

where $b_k$ equals either 1 or 0, depending on whether the monomer investigated (here a nucleobase) is present at $x_k$ (or not). $P(x,z)$ is related to the influence of the tip and is proportional to the 4th power of the field. For $N > K$, the coefficients $b_k$ can be calculated using a least square fitting method that requires the inversion of the matrix ($M$).

$$M_{lk} = \sum_{n=1}^{N} P(x_n - x_k, z) P(x_n - x_l, z), \quad k, l = 1 \ldots K$$

Such inverse problems are often ill conditioned, *i.e.* a small error in the input values ($S(x_n)$) can result in huge errors in the reconstructed solution ($b_k$). A useful method to evaluate this problem is the singular value decomposition of the matrix $M$ which provides the condition number $C$. This number allows the estimation of the error of the reconstructed $b$ values for a given error $\Delta S/S$ of the actually measured signal $S$.

$$\frac{\Delta b}{b} \leq C \frac{\Delta S}{S}$$

Fig. 6 shows the condition number of our problem as a function of the distance between tip dipole and base size normalized with respect to the base-to-base

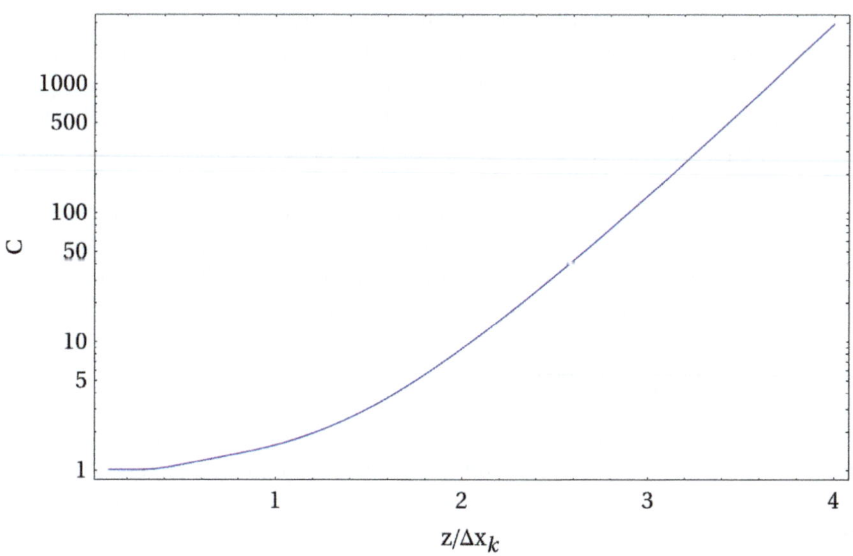


Fig. 6 Condition number $C$ of the inverse sequencing problem using the theoretically modeled enhancement profile of a TERS tip. The graph shows how quickly the method becomes unreliable if the distance $z$ between tip and sample increases. The $x$-axis values represent normalized distances normalized to a single unit of interest $\Delta x_k$ (for instance a base-to-base distance in DNA). Interestingly an AFM can achieve the small distances required.

distance. We assumed 5 samples per base, *i.e.* $N = 5K$, in other words, the polymer strand is scanned in such a way that each single base is sampled at five different subsequent TERS tip positions. The data show that a direct reconstruction of the $b_k$ values is only possible for extremely small distances between the bases in the DNA strand and the tip dipole which corresponds to the center of the Ag sphere in the experiment. If, for example, the error in the Raman signal is 1% this estimation results in a maximum distance $z$ of two base-to-base distances corresponding to roughly 1 nm. In addition to this general estimation we performed simulations where we started with several arbitrary base arrangements and calculated the corresponding TERS signal. After adding a certain noise to the signal we recalculated the base positions. The simulations show that slightly higher distances are allowed compared with the estimation given above, but the general trend was confirmed. While this renders the sequencing a challenging problem, it is also clear that the method's quality will improve dramatically with improving lateral resolution. Taking into account the experimental resolution value estimations and further improvements in the above mentioned method (for details see ref. 48) a direct sequencing as proposed already in the first TERS paper on DNA strands[49] is possible.

# 3 Conclusion

We presented the high resolution aspects of Raman spectroscopy using different approaches. By studying homogeneous samples under controlled conditions, a comparison of signal position fluctuations indicates that TERS probes very small sample volumes where individual orientation effects are not averaged. Complementary studies on proteins and protein fibrils reveal a surprising capability to distinguish amino acid distributions locally on a length scale of at least 1 nm, also confirming the high spatial resolution of TERS. Interestingly, no consistent theory is at hand to explain this resolution with the usual approach using "macroscopic" tip parameters. Last but not least we presented a general method that allows a direct sequencing of single polymer stands with TERS that would already work with the current theoretically predicted resolution.

The time resolution aspects of TERS have not been addressed here, but are currently being successfully investigated by the van Duyne group.[50] In this way the goal to combine ultimate temporal, spatial and spectral resolution as a general tool should be reached in the not too distant future.

# Acknowledgements

We gratefully acknowledge support from the Deutsche Forschungsgemeinschaft (FR 1348/19-1) the Alexander von Humboldt foundation, the Carl-Zeiss Foundation and the Thüringer Aufbaubank (FKZ: 2011 FE 9048; 2011 VF 0016) for their financial support.

# References

1 M. Fleischmann, P. Hendra and A. McQuillan, *Chem. Phys. Lett.*, 1974, **26**, 163–166.
2 M. G. Albrecht and J. A. Creighton, *J. Am. Chem. Soc.*, 1977, **99**, 5215–5217.

 

3 D. Jeanmaire and R. van Duyne, *J. Electroanal. Chem.*, 1977, **84**, 1–20.
4 S. Nie and S. Emory, *Science*, 1997, **275**, 1102–1106.
5 K. Kneipp, Y. Wang, H. Kneipp, L. Perelman, I. Itzkan, R. R. Dasari and M. S. Feld, *Phys. Rev. Lett.*, 1997, **78**, 1667–1670.
6 E. C. Le Ru, M. Meyer and P. G. Etchegoin, *J. Phys. Chem. B*, 2006, **110**, 1944–1948.
7 E. Abbe, *Arch. Mikrosk. Anat. Entwicklungsmech.*, 1873, **9**, 413–418.
8 Rayleigh, *London, Edinburgh Dublin Philos. Mag. J. Sci.*, 1879, **8**, 261–274.
9 D. A. Smith, S. Webster, M. Ayad, S. D. Evans, D. Fogherty and D. Batchelder, *Ultramicroscopy*, 1995, **61**, 247–252.
10 C. L. Jahncke, H. D. Hallen and M. A. Paesler, *J. Raman Spectrosc.*, 1996, **27**, 579–586.
11 D. Zeisel, B. Dutoit, V. Deckert, T. Roth and R. Zenobi, *Anal. Chem.*, 1997, **69**, 749–754.
12 A. Rasmussen and V. Deckert, *J. Raman Spectrosc.*, 2006, **37**, 311–317.
13 E. Bailo and V. Deckert, *Chem. Soc. Rev.*, 2008, **37**, 921–930.
14 T. Deckert-Gaudig, M. Richter, D. Knebel, T. Jähnke, T. Jankowski, E. Stock and V. Deckert, *Appl. Spectrosc.*, 2014, **68**, 916–919.
15 L. Novotny and B. Hecht, *Principles of Nano-Optics*, Cambridge University Press, 2012.
16 M. Moskovits, *Rev. Mod. Phys.*, 1985, **57**, 783–826.
17 M. Kerker, *SPIE Milestone Series*, 1990, **10**, 696.
18 A. Otto, I. Mrozek, H. Grabhorn and W. Akemann, *J. Phys.: Condens. Matter*, 1992, **4**, 1143–1212.
19 H. Kim, K. M. Kosuda, R. P. van Duyne and P. C. Stair, *Chem. Soc. Rev.*, 2010, **39**, 4820–4844.
20 K. J. Savage, M. M. Hawkeye, R. Esteban, A. G. Borisov, J. Aizpurua and J. J. Baumberg, *Nature*, 2012, **491**, 574–577.
21 R. Stockle, Y. Suh, V. Deckert and R. Zenobi, *Chem. Phys. Lett.*, 2000, **318**, 131–136.
22 R. Stockle, V. Deckert, C. Fokas and R. Zenobi, *Appl. Spectrosc.*, 2000, **54**, 1577–1583.
23 N. Hayazawa, Y. Inouye, Z. Sekkat and S. Kawata, *Opt. Commun.*, 2000, **183**, 333–336.
24 B. Ren, G. Picardi and B. Pettinger, *Rev. Sci. Instrum.*, 2004, **75**, 837–841.
25 C. Leiterer, T. Deckert-Gaudig, P. Singh, J. Wirth, V. Deckert and W. Fritzsche, *Electrophoresis*, DOI: 10.1002/elps.201400530.
26 G. Xu, Z. Liu, K. Xu, Y. Zhang, H. Zhong, Y. Fan and Z. Huang, *Rev. Sci. Instrum.*, 2012, **83**, 103708.
27 P. R. Brejna and P. R. Griffiths, *Appl. Spectrosc.*, 2010, **64**, 493–499.
28 T. Kim, K.-S. Jeon, K. Heo, H. M. Kim, J. Park, Y. D. Suh and S. Hong, *Analyst*, 2013, **138**, 5588–5593.
29 B. Pettinger, B. Ren, G. Picardi, R. Schuster and G. Ertl, *Phys. Rev. Lett.*, 2004, **92**, 096101.
30 B. Ren, G. Picardi, B. Pettinger, R. Schuster and G. Ertl, *Angew. Chem., Int. Ed.*, 2005, **44**, 139–142.
31 R. D. Rodriguez, E. Sheremet, S. Müller, O. D. Gordan, A. Villabona, S. Schulze, M. Hietschold and D. R. T. Zahn, *Rev. Sci. Instrum.*, 2012, **83**, 123708.
32 T. Deckert-Gaudig and V. Deckert, *Small*, 2009, **5**, 432–436.

 

33 T. Deckert-Gaudig and V. Deckert, *J. Raman Spectrosc.*, 2009, **40**, 1446–1451.
34 T. Deckert-Gaudig, E. Bailo and V. Deckert, *Phys. Chem. Chem. Phys.*, 2009, **11**, 7360–7362.
35 J.-S. Huang, V. Callegari, P. Geisler, C. Brüning, J. Kern, J. C. Prangsma, X. Wu, T. Feichtner, J. Ziegler, P. Weinmann, M. Kamp, A. Forchel, P. Biagioni, U. Sennhauser and B. Hecht, *Nat. Commun.*, 2010, **1**, 150.
36 T. Deckert-Gaudig, E. Rauls and V. Deckert, *J. Phys. Chem. C*, 2010, **114**, 7412–7420.
37 T. Deckert-Gaudig, F. Erver and V. Deckert, *Langmuir*, 2009, **25**, 6032–6034.
38 P. Singh, T. Deckert-Gaudig, H. Schneidewind, K. Kirsch, E. M. van Schrojenstein Lantman, B. M. Weckhuysen and V. Deckert, *Phys. Chem. Chem. Phys.*, 2015, **17**, 2991–2995.
39 E. M. van Schrojenstein Lantman, T. Deckert-Gaudig, A. J. G. Mank, V. Deckert and B. M. Weckhuysen, *Nat. Nanotechnol.*, 2012, **7**, 583–586.
40 A. V. Krasnoslobodtsev, A. M. Portillo, T. Deckert-Gaudig, V. Deckert and Y. L. Lyubchenko, *Prion*, 2010, **4**, 265–274.
41 D. Kurouski, T. Deckert-Gaudig, V. Deckert and I. K. Lednev, *J. Am. Chem. Soc.*, 2012, **134**, 13323–13329.
42 M. Schleeger, C. C. vandenAkker, T. Deckert-Gaudig, V. Deckert, K. P. Velikov, G. Koenderink and M. Bonn, *Polymer*, 2013, **54**, 2473–2488.
43 D. Kurouski, T. Deckert-Gaudig, V. Deckert, *et al.*, *Biophys. J.*, 2014, **106**, 263–271.
44 T. Deckert-Gaudig, E. Kämmer and V. Deckert, *J. Biophotonics*, 2012, **5**, 215–219.
45 M. Paulite, C. Blum, T. Schmid, L. Opilik, K. Eyer, G. C. Walker and R. Zenobi, *ACS Nano*, 2013, **7**, 911–920.
46 R. Treffer, X. Lin, E. Bailo, T. Deckert-Gaudig and V. Deckert, *Beilstein J. Nanotechnol.*, 2011, **2**, 628–637.
47 R. Treffer, R. Böhme, T. Deckert-Gaudig, K. Lau, S. Tiede, X. Lin and V. Deckert, *Biochem. Soc. Trans.*, 2012, **40**, 609–614.
48 V. Deckert and M. Zeisberger, register. dpma. de., 2013, DE102012024203A1.
49 E. Bailo and V. Deckert, *Angew. Chem., Int. Ed. Engl.*, 2008, **47**, 1658–1661.
50 J. M. Klingsporn, M. D. Sonntag, T. Seideman and R. P. van Duyne, *J. Phys. Chem. Lett.*, 2014, **5**, 106–110.

 

Faraday Discussions

PAPER

# Breaking of a bond: when is it statistical?

Pankaj Kumar Yadav[a] and Srihari Keshavamurthy*[b]

Received 23rd September 2014, Accepted 6th November 2014
DOI: 10.1039/c4fd00180j

Unimolecular dissociation dynamics of a model three degree of freedom triatomic molecule is studied in order to understand the mechanisms for deviations from statisticality. Performing a wavelet based time–frequency analysis of the dynamics allows for the dynamics to be followed on the network of nonlinear resonances, also called as the Arnold web. The results indicate that the long lifetime trajectories spend a considerable amount of time trapped near junctions in the web. It is argued that characterizing the dynamics near such junctions might lead to deeper insights into the origins of nonstatistical dynamics.

## 1 Introduction

A satisfactory answer to the title question is, perhaps surprisingly, still lacking despite several experimental and theoretical studies spanning more than half a century.[1] The term 'statistical' in statistical rate theories refers to a regime where dynamics can be safely ignored in calculating the rate constants and the celebrated Rice–Ramsperger–Kassel–Marcus (RRKM) theory[2–4] is a prime example. At the heart of RRKM theory is the crucial assumption that intramolecular vibrational energy redistribution (IVR) is sufficiently fast as compared to the reaction timescale. Provided this assumption is true, the microcanonical rate constant for a rotationless unimolecular reaction is given by,

$$k(E) = \frac{N^{\ddagger}(E)}{2\pi\hbar\rho(E)} \tag{1}$$

where $E$ is the total energy, $N^{\ddagger}(E)$ is the number of states at the transition state with energy less than or equal to $E$ and $\rho(E)$ is the total density of states. In addition, both the population of the unimolecular reactant:

$$N(t) = N(0)e^{-k(E)t} \tag{2}$$

and the so called lifetime distribution[5]

[a]Department of Chemistry, Indian Institute of Technology, Kanpur, UP, 208016, India
[b]Department of Chemistry, Indian Institute of Technology, Kanpur, UP, 208016, India. E-mail: srihari@iitk.ac.in; Fax: +91-512-2597436; Tel: +91-512-2597043

 

$$P(\tau) \equiv -\frac{1}{N(0)}\frac{\mathrm{d}N(\tau)}{\mathrm{d}\tau} = k(E)\mathrm{e}^{-k(E)\tau} \qquad (3)$$

are expected to decay exponentially. The lifetime distribution above is related to the gap time distribution and enormous efforts have gone into computing and analyzing such distributions. We refer the reader to the paper[6] by Ezra, Waalkens, and Wiggins for a clear exposition of the theory as well as for a detailed historical account.

It is important to note that the exponential decay is a necessary but not sufficient condition and one usually associates the notion of ergodicity with the validity of RRKM. The latter assumption, arising naturally given the classical dynamical origins of RRKM, is a notoriously difficult one to satisfy for Hamiltonians of isolated molecules since they typically exhibit a mixed regular-chaotic phase space even at energies close to the dissociation threshold. Therefore, it is hardly surprising that several studies have identified reactions which are intrinsically non-RRKM in nature.[7–14] Such studies are leading to much needed insights into the dynamics, and hence aiding in identifying novel mechanisms and raising hopes for targeted control of the reaction. Clearly, it is crucial to develop models for reaction rates that eschew the statistical assumption and provide insights into the mechanisms for nonstatistical behaviour.

Deviations from RRKM predictions can be analyzed from two perspectives. The first one[15] associates deviations from the exponential lifetime to the phase space being partitioned into two or more weakly coupled subregions and hence restricted IVR over the entire phase space. Kinetic modeling of such partitioned phase spaces is able to account for the nonexponential $P(\tau)$ in many cases. Nevertheless, highly correlated intramolecular dynamics can lead to the kinetic models being of doubtful utility. At the very least one would have to introduce the correlations into the kinetic scheme – a rather nontrivial task. The second perspective,[16,17] coming from the quantum state space model for IVR, has the correlated dynamics already built in and predicts power law scaling[18] for microcanonically averaged quantum survival probabilities. The exponent of the power law determines whether the system is in a regime of restricted or facile IVR. From the state space, and the associated[19,20] local random matrix theory (LRMT), perspective restricted IVR arises due to the state space being partitioned into weakly coupled subregions. The subregions themselves are a result of the various anharmonic resonant couplings leading to a sparse density of locally coupled states. In other words, restricted IVR is a result of the lack of a "percolating cluster" in the state space. Interestingly, a firm connection between the two perspectives is yet to be established. In essence, the issue here is to find the link between the partitioning of the classical phase space and the partitioning of the quantum state space. In other words, what is the quantum analog for the kinetic models? And, at the same time, what is the classical analog for the state space dynamics? Although, an answer to the latter question is starting to emerge,[21,22] answering the first question and linking the two perspectives together requires a detailed classical–quantum correspondence approach. Arguably, quantum effects like dynamical tunneling[23,24] will somewhat blunt the correspondence. Nevertheless, the failure of the classical–quantum correspondence approach will shed light on the relevance of quantum effects to the observed deviations from statisticality.

 

In order to address the issues mentioned above, it is crucial to study an appropriate model system from both the perspectives. Although it is desirable to study real molecular systems with large number of degrees of freedom (DOF), a disadvantage stems from the fact that several different phenomena in such systems can muddle the classical–quantum interpretations. In addition, quantum and classical studies for large systems at required levels of detail present daunting computational challenges. Thus, we choose a minimal model for our study which is still capable of connecting the two perspectives mentioned above. The model (described below) comes from the pioneering studies by Bunker[5,25] on the dissociation rates of model triatomic molecules. Based on a tour de force, at that time, calculation of classical rates and lifetime distribution Bunker provided a list of models that conform to the RRKM theory. A subsequent important paper[26] by Oxtoby and Rice provided insights into the Bunker results using tools, particularly the concept of nonlinear resonance overlap,[27] from the field of nonlinear dynamics. However, the work of Oxtoby and Rice replaces the three DOF ($s = 3$) Bunker models with an effective two DOF system. Consequently, as also realized[26] by Oxtoby and Rice, their study cannot reveal novel IVR pathways that are feasible in Bunker's models since there are significant differences between classical phase space transport for $s \geq 3$ systems as compared to the $s = 2$ systems.[28] In this study we undertake a detailed study of the classical dissociation dynamics of a $s = 3$ system, based on one of the Bunker models. We show that the mechanism of IVR and the subsequent dissociation dynamics can only be understood in terms of the network of nonlinear resonances, also known as the Arnold web.[29]

## 2 Model hamiltonian

The $s = 3$ Hamiltonian of interest is of the form,

$$H(\mathbf{p},\mathbf{q}) = \sum_{k=1}^{3}\left[\frac{1}{2}G_{kk}^{0}{p_k}^2 + V_k(q_k)\right] + \sum_{k<l=1}^{3} G_{kl}^{0}p_k p_l \tag{4}$$

with $k = 1, 2$ and $k = 3$ modeling the anharmonic stretching modes and a bending mode respectively of a nonlinear triatomic molecule. The potential functions $V_k(q_k)$ are chosen to be Morse oscillators for the stretching modes,

$$V_k(q_k) = D_k\left(1 - \mathrm{e}^{-a_k\left(q_k - q_k^0\right)}\right)^2, \tag{5}$$

and as a harmonic oscillator for the bending mode:

$$V_3(q_3) = \frac{1}{2G_{33}^0}{\omega_3}^2\left(q_3 - q_3^0\right)^2. \tag{6}$$

The various Hamiltonian parameters are determined such that $D_1 = D_2 = 24$ kcal mol$^{-1}$ with the resulting mode harmonic frequencies being $\nu_1 = 1112$ cm$^{-1}$, $\nu_2 = 1040$ cm$^{-1}$, and $\nu_3 = 632$ cm$^{-1}$, corresponding to Bunker's model 6 for ozone.[5] However, eqn (4) is a simplified version of the one used by Bunker in that the G-matrix elements have been replaced with their equilibrium values. The simplified version affords a couple of advantages. First, the action–angle representation of eqn (4) is easily obtained and allows one to extract the strengths of the various anharmonic resonances. Second, gauging the influence of the bending mode on the dissociation dynamics is straightforward without losing the essential three DOF nature of the system. Thus, setting $G_{13}^0 = G_{23}^0 = 0$ yields a (2 + 1) DOF system with the bend mode decoupled from the coupled stretching modes subsystem. Note that the subsystem itself has been the subject of detailed dynamical studies[30] starting with the classic work[31] of Thiele and Wilson.† Hence, the minimal model of eqn (4) is ideally suited for highlighting the central issues of interest to the present work.

### 2.1 Action–angle representation: location and strengths of the resonances

The Hamiltonian in eqn (4) can be conveniently transformed to the action–angle $(\boldsymbol{J}, \boldsymbol{\theta})$ representation since the exact action–angle variables for both the Morse[32] and the harmonic oscillators are known. The resulting Hamiltonian $H(\boldsymbol{J}, \boldsymbol{\theta}) \equiv H_0(\boldsymbol{J}) + V(\boldsymbol{J}, \boldsymbol{\theta})$ has a integrable zeroth-order part,

$$H_0(\mathbf{J}) = \omega_3 J_3 + \sum_{k=1,2} \omega_k \left[ J_k - \frac{\omega_k}{4D_k} J_k^{\,2} \right], \tag{7}$$

and the coupling term which can be explicitly written as a Fourier expansion,

$$V(\boldsymbol{J}, \boldsymbol{\theta}) = \sum_{\alpha,\beta=1}^{\infty} v_{\alpha\beta}(J_1, J_2)[\cos(\alpha\theta_1 - \beta\theta_2) - \cos(\alpha\theta_1 + \beta\theta_2)] + \sum_{k=1,2} \sum_{\alpha=1}^{\infty} u_\alpha(J_k, J_3) \times [\sin(\alpha\theta_k - \theta_3) + \sin(\alpha\theta_k + \theta_3)]. \tag{8}$$

In the above, the main terms responsible for IVR and dissociation are the various resonances‡ of the form $\cos(\alpha\theta_i - \beta\theta_j)$ or $\sin(\alpha\theta_i - \beta\theta_j)$ with strengths $v_{\alpha\beta}(\mathbf{J})$ or $u_\alpha(\mathbf{J})$. A specific resonance occurs when,

$$\frac{\mathrm{d}}{\mathrm{d}t}(\alpha\theta_i - \beta\theta_j) = \alpha\dot{\theta}_i - \beta\dot{\theta}_j \equiv \alpha\Omega_i - \beta\Omega_j = 0, \tag{9}$$

with $\Omega_i$ being the nonlinear frequency associated with the $i$th mode. The above resonance condition, satisfied for certain resonant action values $\mathbf{J}^r$, can be compactly written as $\mathbf{r}\cdot\boldsymbol{\Omega}^r = 0$ with $\mathbf{r} = (\alpha, -\beta)$ and $\boldsymbol{\Omega}^r \equiv \boldsymbol{\Omega}(\boldsymbol{J}^r) = (\Omega_i^r, \Omega_j^r)$. The zone of influence of a resonance scales as the square root of the strength of the resonance which decrease rapidly with the so called order of a resonance $O = |\alpha| + |\beta|$.

The role of such resonances in the dissociation dynamics comes from the Chirikov overlap criterion[27] whereby overlap of various resonances lead to stochasticity in the phase space and hence, presumably, rapid IVR followed by dissociation of the critical bond. Oxtoby and Rice hypothesized[26] that the validity of RRKM at a specific energy is linked to the extensive resonance overlap that occurs at that energy. However, as mentioned in the introduction, the hypothesis has a very strong two DOF flavor to it and the key assumption of the existence of isolated resonance zones needs to be scrutinized carefully.

† It is interesting that the referee to this classic work did point out the "danger of generalizing from triatomic to complex polyatomic molecules". The referee's concerns are true even regarding generalization from a linear triatomic (2 DOF) to a bent triatomic (3 DOF).

‡ Quantum analogs of the classical nonlinear resonances are precisely the Fermi resonances.

 

### 2.2 Two *versus* three DOF: isolated resonances *versus* the Arnold web

The key difference between two and three DOF classical dynamics for our model Hamiltonian can be illustrated rather dramatically by first considering the dynamics with the bending mode decoupled from the stretching modes. Moreover, the total energy of the 2 DOF stretching subsystem is fixed at 15 kcal mol$^{-1}$, far below the dissociation energy of the oscillators. In Fig. 1A all possible resonances up to order five are shown within the Chririkov approximation for the subsystem. Clearly, at 15 kcal mol$^{-1}$ there is no extensive overlap of the resonances and one does not expect any dissociation to occur. The corresponding phase space in Fig. 1B qualitatively agrees with the state space picture in Fig. 1A and shows the existence of the 1:1 resonance along with a substantial degree of chaos. As expected, none of the four initial conditions shown in Fig. 1B dissociate. In this 2 DOF case, initial conditions inside the 1:1 island are trapped forever classically.

The above brief description gives the essence of the IVR dynamics in 2 DOF and emphasizes a crucial feature – at a given energy, trapping in isolated resonance zones will lead to non-RRKM behavior. However, adding the third bending DOF to the subsystem leads to a qualitative and fundamental change of this picture. Addition of the third mode, and hence for the full Hamiltonian in eqn (4), the general resonance condition involves three frequencies and can be written down as,

$$\alpha\Omega_1 + \beta\Omega_2 + \gamma\Omega_3 = 0. \tag{10}$$

The above condition represents a resonance plane in the full $(J_1, J_2, J_3)$ state space. These resonance planes intersect the constant energy surface in resonance lines. The collection of all such resonance lines on a given energy surface is known as the Arnold web.[29] The crucial difference between two and three DOF is

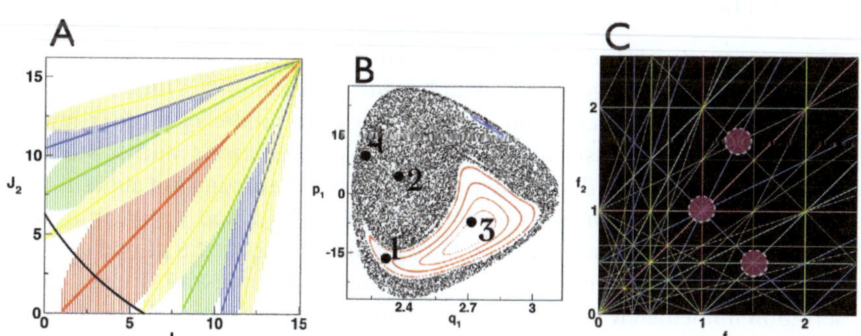


Fig. 1 (A) Resonances up to a maximum order of five and their widths in the $(J_1, J_2)$ state space for the coupled Morse stretch subsystem. The black line is the $H_0(J_1, J_2) = 15$ kcal mol$^{-1}$ isoenergy line. (B) Poincaré surface of section $(q_1, p_1)$ at $q_2 = q_2^0$ for the subsystem at 15 kcal mol$^{-1}$ shows a large 1:1 resonance zone embedded in the chaotic sea. Black dots are specific initial conditions of interest. (C) Arnold web represented in the frequency ratio space $(f_1, f_2) = (\Omega_1/\Omega_3, \Omega_2/\Omega_3)$ showing the possible resonances (without the widths) up to order five. A few example resonance junctions are highlighted with circles. In (A) and (C) the resonances are color coded with red, green, blue and yellow representing orders of 2, 3, 4, and 5 respectively.

now apparent; in contrast to the former case, resonances in the latter case form an intricate network and are not isolated on the energy surface. In Fig. 1C we show the Arnold web in the so called frequency ratio space (FRS)[33] wherein eqn (10) can be expressed in terms of $(f_1, f_2) = (\Omega_1/\Omega_3, \Omega_2/\Omega_3)$ as,

$$f_2 = -\frac{\alpha}{\beta}f_1 - \frac{\gamma}{\beta} \tag{11}$$

A distinct feature of systems with DOF > 2 is the occurrence of junctions wherein several independent resonances on the energy surface intersect. For example, Fig. 1C shows a junction at $(f_1, f_2) = (1, 1)$ formed by two independent resonances $(\alpha, \beta, \gamma) = (1, 0, -1)$ and $(0, 1, -1)$. The number of independent resonances forming the junction determine the multiplicity of the junction. Thus, (1, 1) in Fig. 1C is a multiplicity two junction. Two important points are worth noting. First, an infinity of resonances of various orders emanate from a junction. Second, large scale transport can occur on the Arnold web with the junctions playing a central role in the dynamics. Thus, although a trajectory can be indefinitely trapped in the 1:1 zone in Fig. 1A and B, coupling in the bending mode can detrap the trajectory. Consequently, arguments based on the Chirikov overlap criterion alone are not sufficient to predict the statisticality of the bond breaking process. Understanding the dynamics from the Arnold web perspective is necessary. Is it possible, however, to quantify the extent of non-RRKM behaviour from such a perspective? What role do the junctions in Fig. 1C play in the dissociation dynamics of our model Hamiltonian? Is there a connection, both qualitative and quantitative, between the lifetime distribution of eqn (3) and the dynamics on the FRS? In the next few sections we attempt to answer these questions.

# 3 Dissociation dynamics: role of the resonance junctions

We start this section by noting that the initial conditions shown in Fig. 1B, corresponding to a total energy of 25 kcal mol$^{-1}$ with 15 kcal mol$^{-1}$ in the subsystem, do not dissociate. However, upon coupling the bend modes of all four initial conditions do lead to dissociation. Interestingly, trajectories 1 and 3, starting from a regular region of the subsystem phase space, have lifetimes of about 1.4 ps and 8.6 ps respectively. On the other hand, trajectories 2 and 4 start from the chaotic region of the phase space and have lifetimes of about 2.8 ps and 18.8 ps respectively. Clearly, trajectories 3 and 4 belong to the non-RRKM class. This observation immediately shows that any attempt to *a priori* associate trajectories originating from chaotic regions in the subsystem phase space with RRKM behaviour is incorrect. Nevertheless, it is crucial to understand the mechanism which leads to such long lifetime trajectories. As argued in the previous section, detailed insights in this regard are expected[33] to come from the dynamics in the FRS. However, this requires extracting the mode frequencies as a function of time and several studies have shown[21,22,34–39] that a wavelet based time–frequency analysis[34,35] is ideally suited for this purpose. In particular, for irregular trajectories, the variations of the nonlinear frequencies with time is capable of providing deep insights into the transport occurring in the multidimensional phase space. Note that the static Arnold web in Fig. 1C only indicates the possible resonances that can play a role and it is absolutely important to extract the time dependent frequencies in order to identify the key resonances and the dynamically relevant regions of the web. In this sense, Fig. 1C can be thought of as a template to understand the dynamics in the FRS.

We give here a brief description of the time–frequency analysis since details can be found in previous works.[22,34,35] The time series $z_k(t) = q_k(t) + ip_k(t)$, obtained from the dynamics, for the three modes ($k = 1, 2, 3$) are subjected to a continuous wavelet transform,[34]

$$L_\psi z_k(a,b) = a^{-1/2}\int_{-\infty}^{\infty} z_k(t)\psi^*\left(\frac{t-b}{a}\right)\mathrm{d}t, \tag{12}$$

for $a > 0$ and $b$ real and,

$$\psi(t) = \frac{1}{\sigma\sqrt{2\pi}}\mathrm{e}^{2\pi i\lambda t}\mathrm{e}^{-t^2/2\sigma^2}, \tag{13}$$

with $\lambda = 1$ and $\sigma = 2$ is the Morlet–Grossman wavelet. Eqn (12) gives the local frequency of $z_k(t)$ over a small interval of time around $t = b$ and inverse of the scale factor $a$ is proportional to the frequency. There are two ways of analyzing the transform. First is to extract the dominant frequency at a given time by computing the maximum of the modulus of the wavelet transform of $z(t)$ *i.e.*, $\max_a|L_\psi z(a, b)|$. The ratios of the dominant frequencies can then be used to analyze the dynamics in the FRS. A second more comprehensive way[35] is to compute the full scalogram $|L_\psi z_k(a, b)|^2$ to look at the ridge maxima and dominant frequency variations over time.

The results of our computation are summarized in Fig. 2 for two of the initial conditions labeled 1 and 3 in the subsystem phase space shown in Fig. 1B. The total energy is fixed at 25 kcal mol$^{-1}$, just above the dissociation energy of the Morse oscillators. Breaking of either mode 1 or mode 2 was identified as dissociation in accordance with the criteria used by Bunker. Clearly, Fig. 2 shows a distinct difference between the FRS dynamics of the two trajectories. The short lifetime trajectory visits very few of the numerous resonance junctions whereas trajectory 3 with $\tau \approx 8.6$ ps spends an appreciable amount of time near (2, 1), (2, 2), and (1, 2) junctions. Moreover, as shown by a blue arrow in Fig. 2 (upper right panel), there are indications that the dynamics is influenced by resonance junctions formed by resonances of orders higher then five. Note that the FRS range shown in Fig. 2 is energetically accessible and hence a lack of density in certain regions suggests the existence of dynamical barriers.

In Fig. 3 we summarize the FRS dynamics for the initial conditions 2 and 4 as shown in Fig. 1C along with the scalograms. Note that both trajectories start in the chaotic part of the subsystem phase space and yet their lifetimes differ by nearly a factor of six. The reasons for this difference becomes clear from the FRS shown in Fig. 3 wherein the long lifetime trajectory is seen accessing several resonance junctions. Apart from the main (1, 1) junction, seen in the previous cases as well, the trajectory is clearly spending sufficient amount of time near the (3/2, 1), (4/3, 2/3), and the (1/2, 1/2) junctions. The main point here is that the FRS shows significant anisotropy despite the considerable delocalization with the anisotropy originating from the trajectory "sticking" around the various junctions. In contrast, the shorter lifetime trajectory is accessing far fewer junctions

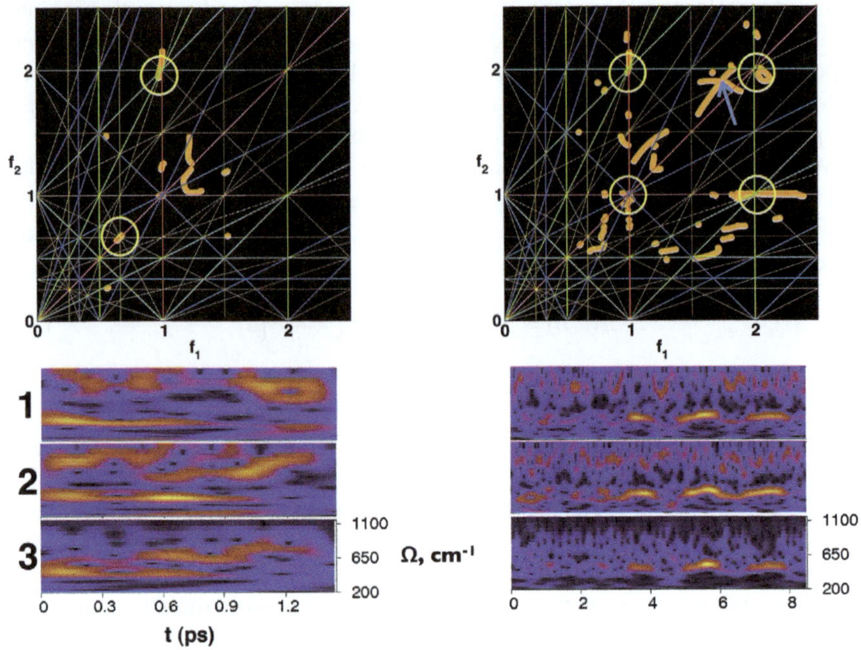


Fig. 2 The top left and right panels show the evolution of the dynamics on the frequency ratio space for initial conditions 1 and 3 shown in Fig. 1B. Circled regions highlight the resonance junctions near which significant activity is observed. The corresponding full scalograms (modes indicated) for initial condition 1 and 3 are shown in the bottom left and right multipanels respectively. Scalograms for each case have the same axis range as shown for mode 3. In the top tight FRS, the blue arrow indicates high density possibly near junctions of much higher order than shown in the plot. Total energy at 25 kcal mol$^{-1}$ for both trajectories.

with almost no density near junctions formed by high order resonances such as the (4/3, 2/3) junction.

As a contrast to the cases discussed above, Fig. 4 shows the FRS dynamics for two trajectories. The first one is the trajectory 3 as before except that the coupling of the bend mode to the stretches has been turned off. As expected from the surface of section shown in Fig. 1B, and in strong contrast to Fig. 2 (top right panel), the trajectory is localized in the 1:1 stretch–stretch resonance for all time. The second case pertains to a trajectory with a total energy of 40 kcal mol$^{-1}$ which is far above the dissociation energy. The lifetime of this particular trajectory is about 1.2 ps and is relatively large compared to the typical lifetimes for most dissociating trajectories at this energy. The reason for this becomes clear from Fig. 4 wherein trapping near (1/2, 3/2), (1/2, 1), and (1, 1/2) junctions can be seen.

The results shown in this section clearly indicate a strong correlation between long lifetime trajectories and enhanced density around various multiplicity two resonance junctions. Such junctions invariably involve the bending mode and the scalograms show that the bending modes actively participate in the IVR even at early times. We stress that the examples shown are representative and several other trajectories exhibit very similar characteristics in the FRS. It is also worth pointing out that significant differences can be seen between Fig. 2 and 3

 

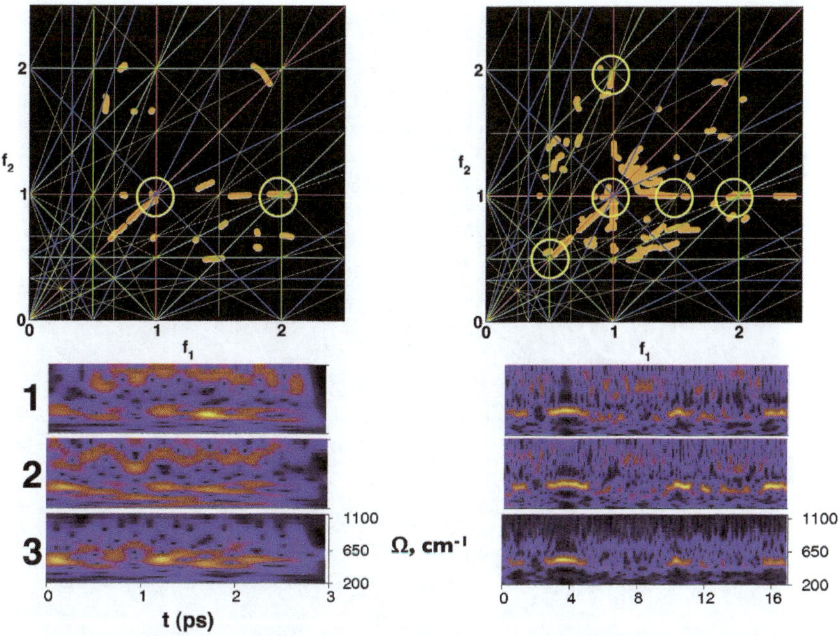


Fig. 3 Same as in Fig. 2 for initial conditions 2 and 4 respectively. Total energy at 25 kcal $mol^{-1}$ for both trajectories.

corresponding to the dynamics of the two long lifetime trajectories. Note that existence of such junctions and their influence on the dissociation dynamics is a genuine three DOF feature and cannot be uncovered by effective models which have lesser than three DOF.

# 4 Discussion and open problems

The results shown in this work provide an important clue to the title question. Dissociating trajectories which have very small sojourn times near the resonance junctions are expected to behave statistically. This observation is reminiscent of an earlier work[40] by Hamilton and Brumer in that trapping near junctions lead to long time dynamical correlations and hence would violate the so called zero relevance condition. The emphasis of this work on junctions is also in line with the early observations[41] of Engel and Levine who argued that large scale IVR requires the participation of at least three modes. Interestingly, our results indicate that the junctions might slow down IVR just enough to violate statisticality. Studies by Bach, Hostettler and Chen on the unimolecular dissociation of the ethyl radical, using the approach of time-frequency analysis, hint towards episodes of regularity in the dynamics even at energies far above the dissociation threshold.[36] It is not known whether there is any role of junctions in their studies.

This work is inspired by the seminal work of Martens, Davis, and Ezra[33] where the first attempts to use the FRS to understand the IVR dynamics of planar OCS were undertaken. More recently, Paskauskas, Chandre, and Uzer provided[42] a detailed study of the phase space of planar OCS and identified one possible

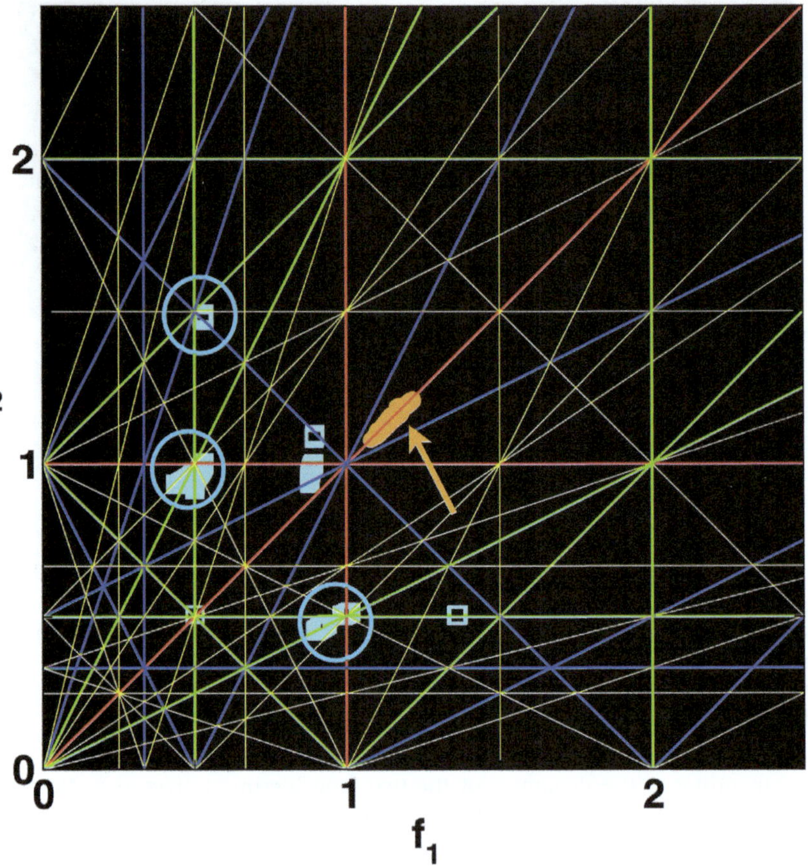


Fig. 4 FRS dynamics for two specific trajectories. The first one (orange) corresponds to trajectory 3 in Fig. 1C with the bend mode decoupled and total energy of 25 kcal mol$^{-1}$. The dynamics is clearly localized (arrow) around the 1:1 resonance. The second trajectory (cyan) is at a total energy of 40 kcal mol$^{-1}$ and has a lifetime of ~1.2 ps. Trapping near resonance junctions are highlighted with circles.

mechanism of trapping and escape involving a class of two-dimensional invariant tori. The results presented in this work point towards a connection between these studies which, however, is not yet established. Of course, the clue regarding links between non-RRKM and trapping near junctions raises several questions and we list a few of them below for further work and discussions.

(1) Which features of the junction determine the extent of trapping of a trajectory? This requires detailed studies comparing, for instance, the nature of the dynamics near a junction involving a strong low order resonance and a weak high order resonance with dynamics near a junction formed by two high order resonances. Although methods do exist[43–46] to study such problems, it remains to be seen if quantitative measures for deviations from RRKM can be obtained using such approaches.

(2) In this study of a three DOF system, one can only have junctions with multiplicity two. In higher DOF systems one can have junctions of different

 

multiplicities and such a scenario would be generic to many realistic molecular systems. However, not much is known about the dynamics near junctions with less then maximal multiplicity in such cases. A puzzling observation here is worth mentioning. For systems with very large number of DOF it appears that the Arnold web will be densely populated with resonance junctions. In turn, given that trajectories can get trapped near junctions, this implies that nonstatisticality may be the rule rather than exception.

(3) Recent work[47] suggests that local dynamical traps near junctions can slow down the quantum IVR dynamics. However, the issue of whether quantum dynamical tunneling can also help in escaping from the trap is largely unexplored. If dynamical tunneling dominates, as is expected[48] for polyatomic systems with a large density of states, then that might resolve the puzzle mentioned above.

(4) Finally, is it possible to relate the number of dominant junctions with the number of partitions in the kinetic schemes mentioned earlier? Clearly, in order to establish such a connection one would have to relate the trapping or residence times[46] near the junctions with the weighting factors $\lambda_i$ in the multi exponential fit[15] $P(\tau) = \sum_i \lambda_i \exp(-k_i\tau)$ to the lifetime distribution. Since several studies have already established the close correspondence between the FRS and the quantum state space models of IVR, progress in this regard will help in relating the two perspectives mentioned in the introduction.

## Acknowledgements

PY gratefully acknowledges the University Grants Commission of India for a graduate fellowship.

## References

1 W. L. Hase and T. Baer, *Unimolecular Reaction Dynamics – Theory and Experiment*, Oxford University Press, New York, 1st edn, 1996.
2 O. K. Rice and H. C. Ramsperger, *J. Am. Chem. Soc.*, 1927, **49**, 1617–1629.
3 L. S. Kassel, *J. Phys. Chem.*, 1928, **32**, 225–242.
4 R. A. Marcus, *J. Am. Chem. Soc.*, 1952, **20**, 359–364.
5 D. L. Bunker, *J. Chem. Phys.*, 1962, **37**, 393–403.
6 G. S. Ezra, H. Waalkens and S. Wiggins, *J. Chem. Phys.*, 2009, **130**, 164118.
7 U. Lourderaj and W. L. Hase, *J. Phys. Chem. A*, 2009, **113**, 2236–2253.
8 J. Rehbein and B. K. Carpenter, *Phys. Chem. Chem. Phys.*, 2011, **13**, 20906–20922.
9 M. Paranjothy, R. Sun, A. K. Paul and W. L. Hase, *Z. Phys. Chem.*, 2013, **227**, 1361–1379.
10 R. A. Marcus, *Proc. Natl. Acad. Sci. U. S. A.*, 2013, **110**, 17703–17707.
11 M. Kryvohuz and R. A. Marcus, *J. Chem. Phys.*, 2010, **132**, 224304.
12 M. Berblinger and C. Schlier, *J. Chem. Phys.*, 1994, **101**, 4750–4758.
13 B. C. Dian, G. E. Brown, K. O. Douglass and B. H. Pate, *Science*, 2008, **320**, 924–928.
14 B. C. Dian, G. G. Brown, K. O. Douglass, F. S. Rees, J. E. Johns, P. Nair, R. D. Suenram and B. H. Pate, *Proc. Natl. Acad. Sci. U. S. A.*, 2008, **105**, 12696–12700.

15 R. A. Marcus, W. L. Hase and K. N. Swamy, *J. Phys. Chem.*, 1984, **88**, 6717–6720.
16 M. Gruebele, *Adv. Chem. Phys.*, 2000, **114**, 193–261.
17 M. Gruebele and P. G. Wolynes, *Acc. Chem. Res.*, 2004, **37**, 261–267.
18 S. Schofield and P. G. Wolynes, *J. Chem. Phys.*, 1993, **98**, 1123–1131.
19 D. E. Logan and P. G. Wolynes, *J. Chem. Phys.*, 1990, **93**, 4994–5012.
20 D. M. Leitner and P. G. Wolynes, *J. Phys. Chem. A*, 1997, **101**, 541–548.
21 A. Semparithi and S. Keshavamurthy, *J. Chem. Phys.*, 2006, **125**, 141101.
22 S. Keshavamurthy, *Adv. Chem. Phys.*, 2013, **153**, 43–110.
23 M. J. Davis and E. J. Heller, *J. Chem. Phys.*, 1981, **75**, 246–254.
24 S. Keshavamurthy, *Int. Rev. Phys. Chem.*, 2007, **26**, 521–584.
25 D. L. Bunker, *J. Chem. Phys.*, 1964, **40**, 1946–1957.
26 D. W. Oxtoby and S. A. Rice, *J. Chem. Phys.*, 1976, **65**, 1676–1683.
27 B. V. Chirikov, *Phys. Rep.*, 1979, **52**, 263–379.
28 S. Wiggins, *Chaotic Transport in Dynamical Systems*, Springer-Verlag, New York, 1st edn, 1992.
29 A. L. Lichtenberg and M. A. Lieberman, *Regular and Chaotic Dynamics*, Springer, New York, 2nd edn, 1992.
30 E. Thiele, *J. Chem. Phys.*, 1963, **38**, 1959–1966.
31 E. Thiele and D. J. Wilson, *J. Chem. Phys.*, 1961, **35**, 1256–1263.
32 R. B. Shirts, *J. Phys. Chem.*, 1987, **91**, 2258–2267.
33 C. C. Martens, M. J. Davis and G. S. Ezra, *Chem. Phys. Lett.*, 1987, **142**, 519–528.
34 L. V. Vela-Arevalo and S. Wiggins, *Int. J. Bifurcation Chaos Appl. Sci. Eng.*, 2001, **11**, 1359–1380.
35 C. Chandre, S. Wiggins and T. Uzer, *Phys. D*, 2003, **181**, 171–196.
36 A. Bach, J. M. Hostettler and P. Chen, *J. Chem. Phys.*, 2005, **123**, 021101.
37 A. Shojiguchi, C. B. Li, T. Komatsuzaki and M. Toda, *Phys. Rev. E: Stat., Nonlinear, Soft Matter Phys.*, 2007, **76**, 056205.
38 A. Sethi and S. Keshavamurthy, *Mol. Phys.*, 2012, **110**, 717–727.
39 J. C. Losada, R. M. Benito and F. Borondo, *Eur. Phys. J.: Spec. Top.*, 2008, **165**, 183–193.
40 I. Hamilton and P. Brumer, *J. Chem. Phys.*, 1985, **82**, 1937–1946.
41 Y. M. Engel and R. D. Levine, *Chem. Phys. Lett.*, 1989, **164**, 270–278.
42 R. Paskauskas, C. Chandre and T. Uzer, *J. Chem. Phys.*, 2009, **130**, 164105.
43 C. Efthymiopoulous, *Celestial Mech. Dyn. Astron.*, 2008, **102**, 49–68.
44 M. Guzzo and E. Lega, *Chaos*, 2013, **23**, 023124.
45 G. Haller, *Phys. Lett. A*, 1995, **200**, 34–42.
46 S. Honjo and K. Kaneko, *Adv. Chem. Phys.*, 2005, **130B**, 437–463.
47 P. Manikandan and S. Keshavamurthy, *Proc. Natl. Acad. Sci. U. S. A.*, 2014, **111**, 14354–14359.
48 D. M. Leitner and P. G. Wolynes, *Phys. Rev. Lett.*, 1996, **76**, 216–219.

 

Faraday Discussions

PAPER

# Negative hyperconjugation and red-, blue- or zero-shift in X–Z···Y complexes†

Jyothish Joy,[a] Eluvathingal D. Jemmis*[b] and Kaipanchery Vidya[a]

Received 26th September 2014, Accepted 11th November 2014
DOI: 10.1039/c4fd00183d

A generalized explanation is provided for the existence of the red- and blue-shifting nature of X–Z bonds (Z = H, halogens, chalcogens, pnicogens, *etc.*) in X–Z···Y complexes based on computational studies on a selected set of weakly bonded complexes and analysis of existing literature data. The additional electrons and orbitals available on Z in comparison to H make for dramatic differences between the H-bond and the rest of the Z-bonds. The nature of the X-group and its influence on the X–Z bond length in the parent X–Z molecule largely controls the change in the X–Z bond length on X–Z···Y bond formation; the Y-group usually influences only the magnitude of the effects controlled by X. The major factors which control the X–Z bond length change are: (a) negative hyperconjugative donation of electron density from X-group to X–Z $\sigma^*$ antibonding molecular orbital (ABMO) in the parent X–Z, (b) induced negative hyperconjugation from the lone pair of electrons on Z to the antibonding orbitals of the X-group, and (c) charge transfer (CT) from the Y-group to the X–Z $\sigma^*$ orbital. The exchange repulsion from the Y-group that shifts partial electron density at the X–Z $\sigma^*$ ABMO back to X leads to blue-shifting and the CT from the Y-group to the $\sigma^*$ ABMO of X–Z leads to red-shifting. The balance between these two opposing forces decides red-, zero- or blue-shifting. A continuum of behaviour of X–Z bond length variation is inevitable in X–Z···Y complexes.

## 1. Introduction

The early stages of the studies of weak interactions in chemistry were dominated by the H-bond due to its obvious importance in biology.[1,2] Detailed understanding of the nature, variety and mechanism of X–H···Y interactions, where X is an atom or molecular fragment and Y is an atom or molecule, has led to many novel ideas in weak interactions. The inevitability of the H-bond in all branches of chemistry

[a]*School of Chemistry, Indian Institute of Science Education and Research-Thiruvananthapuram, Thiruvananthapuram, Kerala, 695016, India*

[b]*Department of Inorganic and Physical Chemistry, Indian Institute of Science, Bangalore, 560012, India. E-mail: jemmis@ipc.iisc.ernet.in; Fax: +91-2360-1552; Tel: +91-80-2293-3347*

† Electronic supplementary information (ESI) available: Coordinates of the model complexes and evidence for proper and pro-improper nature of model systems. See DOI: 10.1039/c4fd00183d

 

is now known and its detailed understanding has helped to bring new perceptions to research in material science and crystal engineering.[3,4] Weak interactions akin to the H-bond must exist for any element in the periodic table in appropriate molecular combinations. Indeed, several examples have appeared in the past for weak interactions involving elements other than hydrogen.[5] While the formulation of X–Z···Y for these interactions and the name Z-bond where Z is any atom, with or without further substituents, brings similarity to the H-bond and the X–H···Y interaction, major additional factors are expected to play an important role in Z-bonds. Halogen bonds,[6–8] chalcogen bonds,[9,10] pnicogen bonds,[11,12] tetral bonds[13–15] and others have far more flexibility than the H-bond as the H atom with its one electron and one valence orbital is unique. Additionally the unit mass of H which is invariably several times lighter than X makes the X–H bond tractable by variations in the X–H stretching frequency. We are particularly interested in understanding blue- and red- shifting in the X–Z bond and contrast this with H-bonds. On the other hand, pure X–Z stretching modes are rare.[6,16–18] This forces us to depend on the X–Z bond length rather than its stretching frequency.

Various explanations have been given by different groups of investigators for the red- and blue-shifting nature of the hydrogen bond.[19–25] A few years ago we examined these aspects in detail and proposed a unified explanation for the nature of X–H bond length variation based on the change in electron density.[26] We showed that there is nothing "improper" or "anti" about the blue-shifting H-bonds and that all are a part of a continuum of behaviour.[27,28] Using the terminology prevalent at that time we classified the H-bonds as proper, pro-improper and improper based on the X–H bond length and vibrational frequency change during the complex formation. Proper H-bonds show an increase in X–H bond length and a decrease in vibrational frequency (red-shifting) when X–H and Y form the X–H···Y complex. Pro-improper H-bonds may be red- or blue-shifting, but show an initial X–H bond length contraction and increase in vibrational frequency (blue-shifting) at large donor–acceptor separation and, finally red-shift or blue-shift at equilibrium geometry. Improper H-bonds blue-shift in the X–H stretching frequency from larger intermolecular distances as X–H and Y are brought together. A similar understanding is possible for Z-bonds in the X–Z···Y interaction. Red- and blue-shifting X–Z bonds in the X–Z···Y interaction, especially halogen bonds (Z = halogen), have been studied by many groups in recent years, but detailed analysis and general explanations that can be transferred to other examples are not available. Apart from a charge analysis, previous studies have not concentrated on the influence of the starting X–Z distance (brought in by the X-group in the monomer X–Z) on the nature of blue- and red- shifting in the X–Z···Y complex.[7] Studies on weak X–Z···Y bonds where Z = chalcogen,[9,10] pnicogen,[11,12] and Carbon[13–15] are evolving, but none has yet focussed on the continuum nature of red- to blue-shifting X–Z bonds in X–Z···Y. The X-group plays an important role in deciding the nature of X–Z bond length variation during complex formation, much more than what is possible for the H-bond and hence a detailed analysis of the mechanism is necessary.

The present study focuses on the detailed explanation of the continuum nature of red- to blue-shifting X–Z bonds in any X–Z···Y complex, using the model complexes FCl, $CF_3Cl$ and $NO_2Cl$ as X–Z and $NH_3$ as Y. A generalized concept of X–Z bond length change in any weakly interacting complex based on the relative competition between negative hyperconjugation in the donor molecule and charge transfer (CT) from the incoming acceptor molecule evolves from this study. The additional electrons and orbitals available on Z in comparison to H make for dramatic differences between the H-bond and the rest of the Z-bonds.

# 2. Computational details

The three model systems under detailed investigation are calculated at *ab initio* second order Møller–Plesset theory (MP2). All the geometries are optimized at MP2(full)/aug-cc-pVTZ level of theory without any constrains and are verified to be ground state by frequency calculations using Gaussian09 quantum chemistry package.[29] This level of theory and basis set are reported to give relatively good results for weak interactions.[30,31] Double hybrid functional with Grimmes third order dispersion correction,[32,33] B2PLYP-D3 was also used to check the reliability of calculations and the results were found to be in good agreement with the MP2 results. Errors due to basis set superposition (BSSE) are corrected using the standard Counter Poise (CP) correction method of Boys and Bernardi.[34] Electron Density analysis was performed using the ADF2013.01 DFT package[35] and AIM2000 suite of programs.[36] Deformation density plots, which give information about the extent of electron density reorganization during the complex formation, are obtained using the ETS-NOCV[37,38] method implemented in ADF2013.01. Wavefunction files required for AIM analysis were generated from MP2 calculations. Population analysis, partial atomic charges and charge transfer analysis on MP2 optimized geometries are done by the standalone NBO6 program[39] at M06-2x/aug-cc-pVTZ level of theory using Gaussian09 as the ESS host package. Most of the complexes analysed to establish the continuum nature are well known and are available in the literature. We have re-computed these geometries at MP2(full)/6-311+g(d,p) level of theory to substantiate the proposed hypothesis (ESI materials†).

While going from the H-bond to the Z-bond, we realize that the variation in stretching frequency is no longer the most suitable parameter to represent bond strength for the following reason. In the X–H bond, it is almost always possible to get a pure X–H stretching frequency. However, heavier Z atoms have several vibrational modes mixing with the X–Z stretching mode, so that an accurate estimate of blue- and red-shifting is not easily possible.[6,16–18] Therefore we rely largely on bond length variation as an indicator of bond strength and red-, zero- or blue-shift in the X–Z stretch.

# 3. Results and discussions

We have performed detailed theoretical analysis on three selected halogen bonded systems, F–Cl···$NH_3$ (red-shifted), $F_3C$–Cl···$NH_3$ (moderately blue-shifted) and $O_2N$–Cl···$NH_3$ (strongly blue-shifted) (Table 1). These structures are optimized by keeping the Z···Y distance at several selected values by scanning the potential energy surface. The well-established idea of charge transfer (CT) from the acceptor group (Y) to the X–Z $\sigma^*$ anti-bonding molecular orbital (ABMO) of the donor as the reason for bond length elongation (red-shift), prompted us to speculate the situation where X–Z $\sigma^*$ ABMO is already occupied at the monomer level. What will happen if an electron rich Y-group approaches such an X–Z molecule? Our analysis depends on the variation of several parameters as a function of the Z···Y distance, through a series of scan calculations. The following parameters are especially analysed as a function of Z···Y distance: (a) total energy; (b) X–Z bond length; (c) variation of the occupancy of the X–Z σ* ABMO; (d) charge transfer from Y to X–Z; (e) negative hyperconjugation from X to Z and from Z to X; (f) atomic population density; (g) electron density at bond critical points for X–Z and Z···Y bonds; and (h) deformation density at equilibrium geometry.

### 3.1 Geometrical parameters and properties of the studied molecules

The scan calculations along the Cl···N distance between the donor and acceptor molecules ranged from 6.0 Å to 2.0 Å with a step size of 0.2 Å. Except for this distance all other parameters are completely optimized for all separate values of Cl···N. The plots of X–Cl bond length and total energy against the scan coordinate are shown in Fig. 1. Complex F–Cl···$NH_3$ (Fig. 1A) shows red-shift, and complexes $F_3C$–Cl···$NH_3$ (Fig. 1B) and $O_2N$–Cl···$NH_3$ (Fig. 1C) show blue-shift compared to the free monomers. The highly blue-shifted $O_2N$–Cl···$NH_3$ complex shows a drastic change in N–Cl bond length as compared to the other two systems and this behaviour seems to be highly abnormal. While the attractive interaction between monomers is justified in terms of the long range electrostatic attraction between the sigma-hole on the halogen and the electron density in the Y ($NH_3$) group,[40–42] the details of the electron reorganization that causes this have to be analysed.

Electrostatic attraction between the σ-hole on the donor and electron density on the acceptor nitrogen is the primary reason for the interaction between monomers at larger separations. Nevertheless, complex (A) becomes red shifted and complexes (B) and (C) become blue-shifted. Different authors give different explanations for this unusual behaviour. We believe that the factors that control both red and blue-shifting in X–Z bond length exist in all complexes, except that these factors contribute in different magnitudes in different complexes. It is clear from the available literature that F–Cl belongs to the proper category (pure red-shifting) of Z-bond donors and there is no example of a blue-shifted F–Cl···Y complex. $F_3C$–Cl and $O_2N$–Cl belong to the pro-improper category of weak interactions, where red- or blue-shifting results depending on the nature of the incoming Y-group. We propose that it is the modulation of the already existing negative hyperconjugative donation of electron density from the X-group to the X–Z σ* ABMO in the monomer by the incoming Z-bond acceptor that causes the pro-improper behaviour of the X–Z bond. Negative hyperconjugation[43,44] is defined as the donation of electron density from the filled lone pair or p-orbital to the neighbouring σ* orbital which is anti-periplanar to the donor orbital. The

Table 1 Geometric parameters of the studied complexes at the equilibrium geometry. B.E = Binding Energy. ZPE = Zero Point Energy. BSSE = Basis Set Superposition Error. Calculations are done at MP2(full)/aug-cc-pVTZ level of theory

| Complex | *d*(Hal···Y) (Å) | Δ(X–Hal) (Å) | ∠(X–Hal···Y) (degree) | $B.E_{ZPE}$ (kcal $mol^{-1}$) | $B.E_{ZPE-BSSE}$ (kcal $mol^{-1}$) |
|---|---|---|---|---|---|
| F–Cl···$NH_3$ | 2.207 | +0.0786 | 180.00 | 10.44 | 8.52 |
| $F_3C$–Cl···$NH_3$ | 3.041 | −0.0064 | 179.96 | 2.31 | 1.78 |
| $O_2N$–Cl···$NH_3$ | 2.730 | −0.0497 | 179.94 | 3.06 | 2.20 |

 

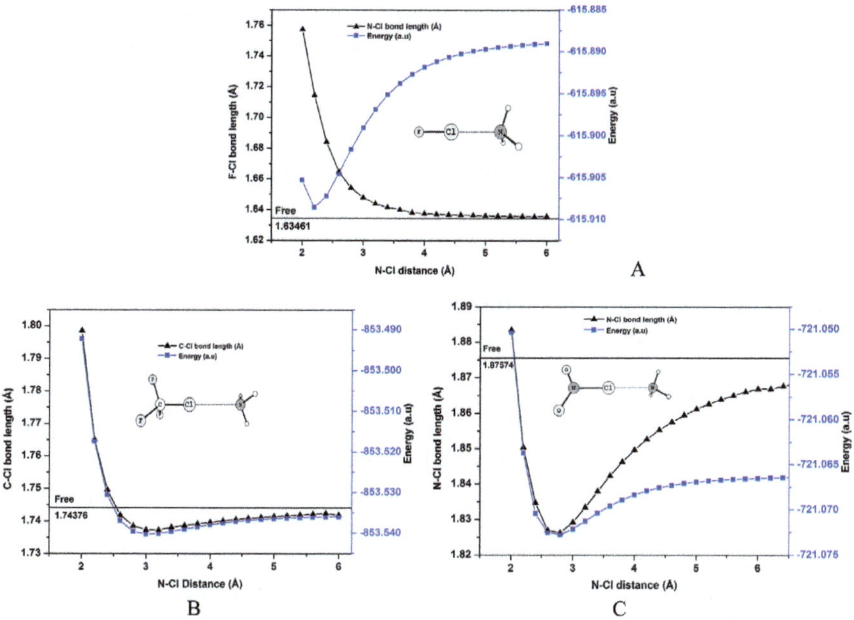


Fig. 1 Variation of X–Cl bond length and total energy during the potential energy scan: (A) F–Cl···$NH_3$, (B) $F_3C$–Cl···$NH_3$ and (C) $O_2N$–Cl···$NH_3$. Calculations are done at MP2(full)/aug-cc-pVTZ level of theory.

immediate consequence of this effect is an elongation of the neighbouring bond which receives the electron density to its antibonding molecular orbital.

The negative hyperconjugative donation of electron density from the F atom to the C–Cl σ* ABMO of the $F_3C$–Cl molecule is shown in Fig. 2. The same interaction is possible for the lone pair on the Cl atom to the C–F σ* ABMO also. We do not anticipate any fluorine-bonds in the $F_3CCl$ molecule because the positive

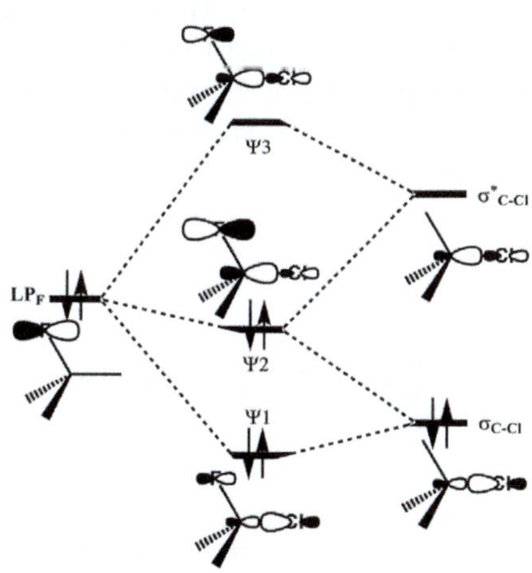


Fig. 2 Interaction diagram showing the negative hyperconjugative donation of the F-lone pair ($LP_F$) to the C–Cl σ* bond orbital in the $F_3C$–Cl molecule.

electrostatic potential over the Fluorine atom is not strong enough to interact with the $NH_3$ molecule. Hence we only consider the donation of electron density from F atoms of the $-CF_3$ group to the σ* C–Cl bond. The extra electron density at the ABMO makes the C–Cl bond elongated and it also decreases the electron density around the Cl atom because of its synergic donation to the C–F σ* orbital. Bond orbitals participating in negative hyperconjugation and hyperconjugation (CT) in the $F_3C$–Cl···$NH_3$ complex are shown in Fig. 3.

In proper X–Z bonds, where negative hyperconjugation from the X-group is absent, the σ* X–Hal orbital is vacant in the monomer and increases its occupancy during complex formation through charge transfer from the Y-group. This increases the X–Hal bond length (red shifting). The X–Hal bond will be highly elongated at the monomer level itself in systems where negative hyperconjugation is strong due to the presence of electrons in the σ* ABMO. Any incoming Y-group can perturb these extra electrons through long range exchange repulsion. This leads to the back flow of electron density from the partially occupied σ* ABMO of the X–Hal bond to other parts of the molecule *i.e.*, towards the X-group. This effect of back flow has two consequences: (i) extra electron density flows back from the anti-bonding molecular orbital leading to bond length contraction and (ii) electron density shift to the X-group makes it more negatively charged and the connected halogen atom becomes more positively charged. The combined effect of electron density decrease at the σ* orbital, and an increase in opposite charges on the X-group and Halogen atom makes the X–Hal bond shorter (blue-shifting). The possibility for charge transfer interactions in blue-shifted systems cannot be neglected. This is the reason why some of the systems which initially show preference towards blue-shifting become red-shifting at equilibrium geometry (pro-improper). Hence the preference towards resultant red- or blue-shift in geometry is decided by the balance between negative hyperconjugation and charge transfer. Observations made by Hobza *et al.*[19,21] about blue-shift in hydrogen bonding are in line with our arguments, where they observed the flow of electron density through charge transfer to various parts of the donor molecule, apart from the X–H σ* ABMO orbital. The source of electron density involved in this process has to be evaluated in each case. As the shift of electron density begins even when the Y-group is at 5.0 Å (Fig. 1B and C), this is not caused by the additional electron density resulting from CT from the lone pair on Y. This is the long range effect of the Y-group on the electron density at the X–Z bond. Careful analysis of dimer geometries obtained at each scan step gives evidence for the existence of the proposed mechanism.

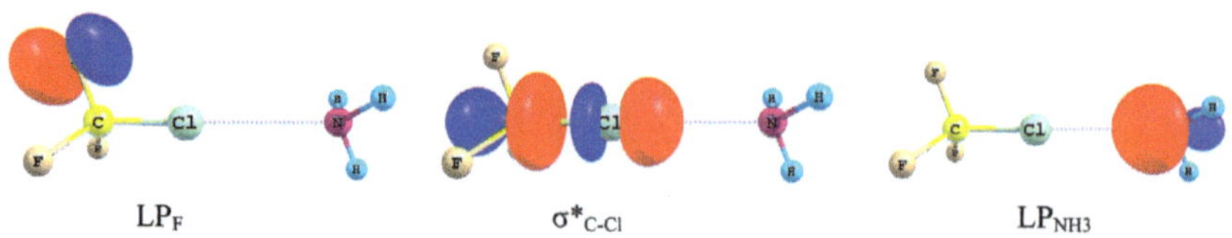


Fig. 3 NBO bond orbitals of the $F_3C$–Cl···$NH_3$ complex: LP$_F$ donates to σ*$_{C-Cl}$ (negative hyperconjugation). The CT at short distances takes place from LP$_{NH3}$ to σ*$_{C-Cl}$, often termed as hyperconjugation (this is not strictly hyperconjugation, which represents a delocalization involving pi-type and sigma type orbitals, but frequently used in this context). Calculations are done at MP2(full)/aug-cc-pVTZ level of theory.

 

### 3.2 Negative hyperconjugation and σ* occupancy of the X–Hal bond

The effect of negative hyperconjugation on the X–Hal bond length can be demonstrated well by analysing the σ* ABMO occupancy. Let us consider the examples of $O_2N$–Cl and $O_2NO$–Cl[45] as halogen bond donors. Here $O_2N$–Cl has reasonably high electron occupancy at the N–Cl σ* ABMO but O–Cl in $O_2NO$–Cl has no σ* occupancy from negative hyperconjugation. This makes $O_2N$–Cl blue-shifting (−0.0497 Å) as electron density can be shifted away from the σ* ABMO and $O_2NO$–Cl red shifting (+0.0454 Å) as the major contribution from CT is to the σ* ABMO during complex formation with $NH_3$ (Fig. 4).

Plots of the X–Hal σ* orbital occupancy and total energy of the studied complexes against the scan coordinates (Fig. 5) support these ideas. The occupancy of σ* F–Cl increases as $NH_3$ approaches (Fig. 5A). The absence of negative hyperconjugation in the F–Cl molecule leaves the F–Cl σ* orbital vacant and CT from $NH_3$ leads to bond length elongation (red shift). The complex $O_2N$–Cl···$NH_3$ (Fig. 5C) shows pure blue-shifting where the σ* occupancy decreases drastically during the approach of $NH_3$ through long range exchange repulsion and it shows minimum occupancy at equilibrium geometry and hence bond length contraction. Complex (B) is a mixture of both the effects operating in complexes (A) and (C), where σ* occupancy decreases to a minimum and the bond contracts at the first stage but at equilibrium geometry (3.04 Å) occupancy slightly increases from the minimum due to the presence of CT from the $NH_3$ lone pair, Table 2. Hence complex (B) shows only a slight blue-shift because of the balance between CT and negative hyperconjugation. The balance between these two opposing forces and the equilibrium distance is to be contrasted with the thick horizontal line, which shows the bond occupancy in the monomer (Fig. 5). This, in turn, reflects the extent of the occupancy of the X–Hal σ* orbital. This is a handle for designing red- and blue-shifting bonds in any weakly interacting complexes. It is possible to tune the Y-group to force $F_3C$–Cl to form a red-shifting complex. Hence there is a strong correlation between σ* occupancy and the nature of the X–Z Bond in deciding the bond length variations in the X–Z···Y complex. The higher the σ* occupancy in the monomer X–Z bond, the greater will be the chance for blue-shifting.

NBO second order perturbation analysis (Table 3) gives information about CT from the Y-group and strength of negative hyperconjugation from the X-group to the X–Hal σ* ABMO. For pure red shifting cases, there is no negative hyperconjugation and hence the X–Z bond length always elongates. In the case of complexes (B) and (C), both negative hyperconjugation and CT are active. Negative hyperconjugation dominates at larger monomer separations and decreases as

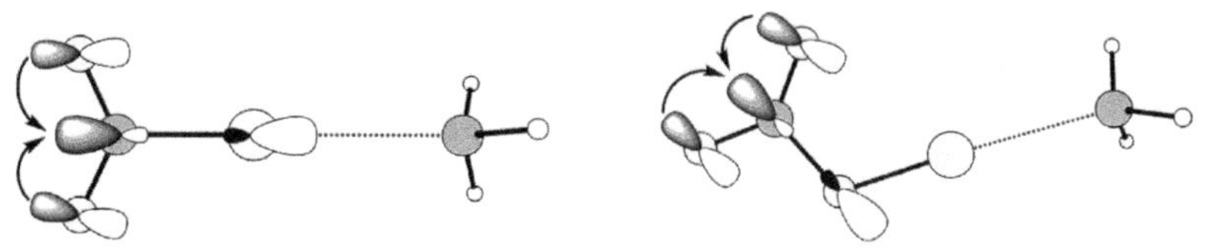


Fig. 4 Presence of negative hyperconjugative donation to N–Cl σ* orbital in $O_2N$–Cl···$NH_3$ and its absence in O–Cl σ* orbital in $O_2NO$–Cl···$NH_3$.

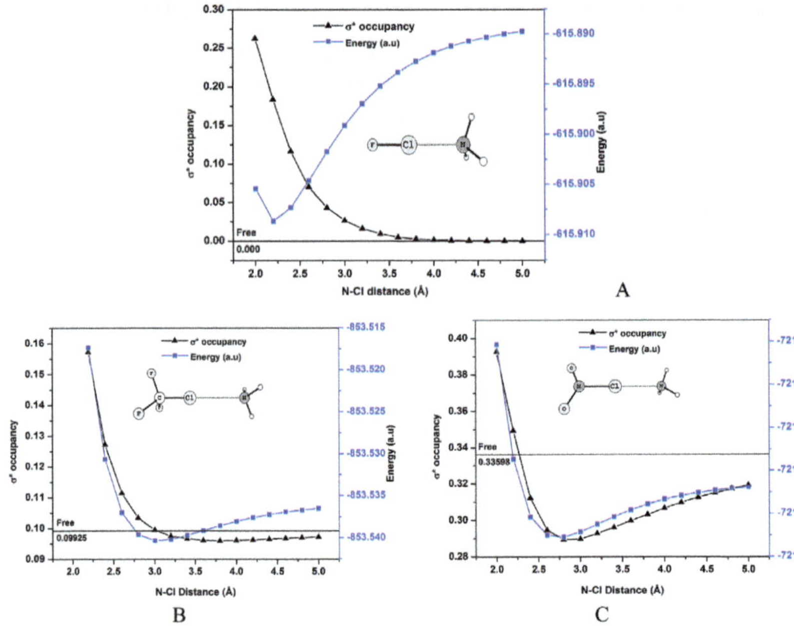


Fig. 5 Variation of X–Cl σ* bond occupancy and total energy during the potential energy scan: (A) F– Cl···$NH_3$, (B) $F_3C$–Cl···$NH_3$ and (C) $O_2N$–Cl···$NH_3$. Calculations are done at M06-2x/aug-cc-pVTZ level of theory.

the Y-group approaches and CT shows the opposite trend. CT starts only below 3.5 Å of Hal···Y separation and hence we can conclude that any bond length change that occurs beyond 3.5 Å is purely due to the effect of long range exchange repulsion of the Y-group on σ* ABMO electron density. The decrease in negative hyperconjugation during the complex formation supports our idea of back flow of electron density to the σ* ABMOs of the substituents of the X-group, *i.e.*, towards C–F σ* of the –$CF_3$ group and N–O σ* of the –$NO_2$ group.

### 3.3 Population and charge analysis

Population density analysis at each of the scanned geometries gave information on the influence of the long range exchange repulsion from the Y-group to the electron density at the X-Hal σ* ABMO. The results of the population analysis are shown in Fig. 6.

Table 2 X–Cl σ* occupancy of monomer, dimer and the resultant charge transfer (CT) from $NH_3$. Calculations are done at M06-2x/aug-cc-pVTZ level of theory

| Complex | Monomer σ* occupancy | Dimer σ* occupancy | CT | σ* occupancy in the absence of CT |
|---|---|---|---|---|
| FCl···$NH_3$ | 0.0000 | 0.1809 | 0.1772 | 0.0037 |
| $F_3CCl$···$NH_3$ | 0.0992 | 0.0990 | 0.0080 | 0.0910 |
| $O_2NCl$···$NH_3$ | 0.3359 | 0.2903 | 0.0267 | 0.2636 |

 

Table 3 NBO second order perturbation analysis (kcal mol$^{-1}$) showing hyperconjugation (CT) and negative hyperconjugation (NHC) at each of the scan geometries along with change in X–Hal bond length during the formation of X–Hal···Y. Values at equilibrium geometries are shown in bold numbers. Calculations are done at M06-2x/aug-cc-pVTZ level of theory

| Hal···Y (Å) | $FCl\cdots NH_3$ | | | $F_3C–Cl\cdots NH_3$ | | | | $O_2N–Cl\cdots NH_3$ | | | |
|---|---|---|---|---|---|---|---|---|---|---|---|
| | CT[a] | NHC[b] | ΔF–Cl | CT[c] | NHC[d] | NHC[e] | ΔC–Cl | CT[f] | NHC[g] | NHC[h] | ΔN–Cl |
| 2.2 | **53.69** | **0.00** | **+0.079** | 31.13 | 11.70 | 9.66 | +0.021 | 53.93 | 42.25 | 14.11 | −0.025 |
| 2.4 | 27.71 | 0.00 | +0.049 | 16.20 | 12.29 | 10.46 | +0.006 | 29.12 | 41.89 | 16.71 | −0.041 |
| 2.6 | 14.43 | 0.00 | +0.030 | 8.78 | 12.71 | 11.03 | −0.002 | 15.83 | 44.73 | 17.98 | −0.049 |
| 2.8 | 7.89 | 0.00 | +0.019 | 4.88 | 12.98 | 11.28 | −0.005 | **8.97** | **47.31** | **18.46** | **−0.049** |
| 3.0 | 4.39 | 0.00 | +0.013 | **2.79** | **13.29** | **11.23** | **−0.006** | 5.20 | 49.52 | 18.48 | −0.047 |
| 3.2 | 2.42 | 0.00 | +0.009 | 1.56 | 13.44 | 10.81 | −0.006 | 2.99 | 51.44 | 18.31 | −0.042 |
| 3.4 | 1.30 | 0.00 | +0.007 | 0.86 | 13.63 | 10.15 | −0.005 | 1.68 | 52.96 | 18.04 | −0.038 |
| 3.6 | 0.67 | 0.00 | +0.005 | 0.45 | 13.72 | 10.03 | −0.005 | 0.92 | 54.24 | 17.74 | −0.033 |
| 3.8 | 0.34 | 0.00 | +0.004 | 0.23 | 13.77 | 9.00 | −0.005 | 0.49 | 55.24 | 17.46 | −0.029 |
| 4.0 | 0.17 | 0.00 | +0.003 | 0.12 | 13.88 | 8.37 | −0.004 | 0.25 | 56.12 | 17.21 | −0.026 |

[a] [$L.P_{NH3} \rightarrow \sigma^*_{F-Cl}$]. [b] No NHC. [c] [$L.P_{NH3} \rightarrow \sigma^*_{C-Cl}$], Fig. 3. [d] [$L.P_F \rightarrow \sigma^*_{C-Cl}$], Fig. 3. [e] [$L.P_{Cl} \rightarrow \sigma^*_{C-F}$]. [f] [$L.P_{NH3} \rightarrow \sigma^*_{N-Cl}$]. [g] [$L.P_O \rightarrow \sigma^*_{N-Cl}$]. [h] [$L.P_{Cl} \rightarrow \pi^*_{N-O}$].

The population of fluorine and chlorine atoms in the $F–Cl\cdots NH_3$ complex is almost invariant at longer distances and increases as the donor–acceptor distance decreases (Fig. 6A). Both the donor and acceptor groups start reorganizing their electron population below 3.0 Å. This systematic change in population can be accounted for by CT, which is largely a short range interaction. In contrast both Fig. 6B and C show that the population of atoms in the halogen donor molecule changes even at longer distances where the $NH_3$ population remains almost unchanged. This shows that even before $NH_3$ starts a CT interaction with the donor molecule, the long range interactions begin. This reorganizes the electron density distribution of the halogen donor molecule. Analysis evidently shows that the halogen atom in the donor molecule loses electron density and the remaining portion of the molecule gains it. It can be observed from Table 3 that the incoming Y group can reduce the magnitude of negative hyperconjugation that existed in the parent X–Z system and in addition force a negative hyperconjugative flow of electron density from the p-type lone pair on the chlorine atom to the ABMOs' of the X-group. Electron density from the chlorine atom is shifted to the anti-bonding N–O orbital in the $O_2N–Cl\cdots NH_3$ complex and towards the anti-bonding C–F bond orbital in the $F_3C–Cl\cdots NH_3$ complex. This type of intramolecular stabilization is absent in the X–H bond (no extra orbitals and electrons on H). Therefore the magnitude of blue-shift in H-bonded complexes will be much smaller than those in halogen and similar Z-bonds. Justification for the small blue-shift observed in the $CH_3Cl\cdots NH_3$[46] complex can also be explained using the above arguments, where negative hyperconjugation to the C–Cl σ* ABMO in $CH_3Cl$ is absent. The incoming $NH_3$ increases the extent of negative hyperconjugation from chlorine to C–H σ* ABMO and hence increases the intra molecular interaction which leads to C–Cl bond length contraction. Thus in general, back flow of electron density leads to the generation of positive charge on

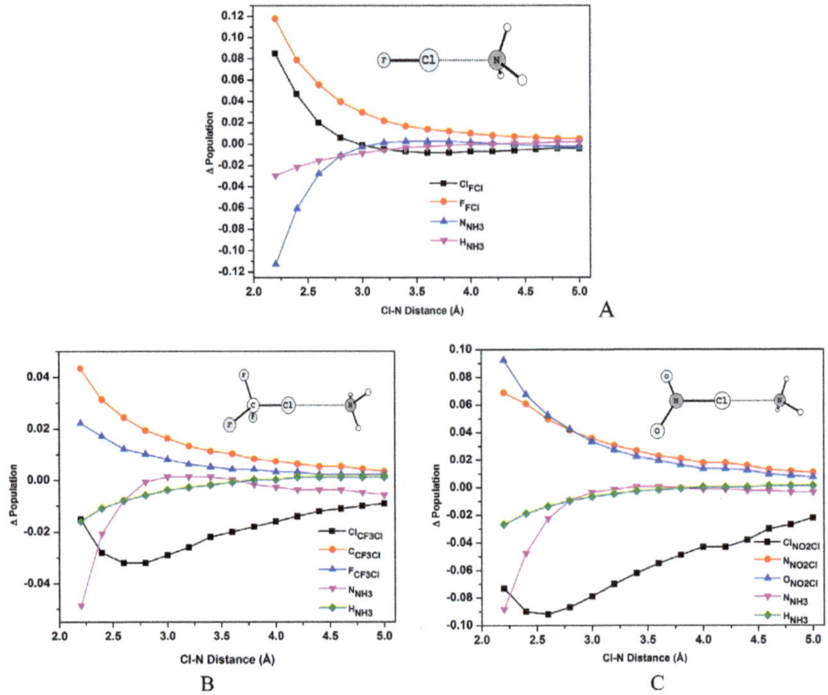


Fig. 6 Change in partial atomic population of the complex compared to the corresponding monomers during the scan: (A) $FCl\cdots NH_3$, (B) $F_3CCl\cdots NH_3$ and (C) $O_2NCl\cdots NH_3$. Calculations are done at M06-2x/aug-cc-pVTZ level of theory.

the halogen atom which was slightly negatively charged in the monomer. The initial decrease in population density at the chlorine atom in blue-shifted systems is a manifestation of the extent of the negative hyperconjugation that existed in the parent X–Z. This analysis shows the existence of a long range exchange repulsive interaction which gets activated in pro-improper Z-bonded systems. An immediate consequence of this is the redistribution of charge density in the X–Z molecule. In the case of the $F–Cl\cdots NH_3$ (red shifted) complex both F and Cl gain electron density during the approach of $NH_3$. In $F_3C–Cl\cdots NH_3$ and $O_2N–Cl\cdots NH_3$ (blue-shift), the Cl atom loses electron density and the $–CF_3$ and $–NO_2$ groups gain electron density. This also shows that $NH_3$ does not lose any charge before it reaches a minimum distance ($\approx 3.0$ Å). Charge analysis also supports our arguments of long range exchange repulsion from Y and its consequence of intra molecular electron density reorganization in the Z-bond donor molecule, without accepting any electron density from Y.

There exists a strong correlation between the nature of the X-group, mode of σ-hole generation over Z atom and negative hyperconjugation from the X-group. Politzer *et al.*[47–49] proposed that the presence of a positive electrostatic potential region (σ hole) at the Z atom is the reason for electrostatic attraction between monomers. The electronegativity of fluorine compared to chlorine in FCl makes the σ-hole in the F–Cl molecule concentrated on the chlorine end. The σ hole in $CH_3Cl$ is expected on the carbon end. However in $CF_3Cl$, the highly electron withdrawing fluorine substituents make carbon more electron demanding from chlorine, effectively increasing the electronegativity of carbon as suggested by Mulliken and Jaffe.[50–52] Increased positive charge and hence increased electronegativity of carbon pulls electrons from the chlorine atom and creates a σ-hole at the chlorine atom of the C–Cl bond in $CF_3Cl$. This is reflected in the orbital coefficients of the σ* MO of the C–Cl bond (Fig. 3). The extent of the stretching of the C–Cl bond from negative hyperconjugation in $F_3CCl$ is more than anticipated due to the larger size of the lobe of the σ* C–Cl MO at the carbon atom. Incoming $NH_3$ decreases the negative hyperconjugation and hence there is more possibility for blue shifting of the C–Cl bond. The same effects can be observed in the case of the $O_2N$–Cl molecule also. Thus the larger the electronegativity difference between the Z atom and the connected atom of the X-group, the larger will be the magnitude of negative hyperconjugation and consequently blue-shifting. H-bonded complexes show a lower magnitude in blue shifting for the same reason.

### 3.4 Analysis of variation in electron density at bond critical points during complex formation

We analysed the evolution of the donor–acceptor system from monomers by carefully studying the position and properties of electron density at the bond critical points (BCP) during each scan step. Analysing the position and topological properties[53] of the BCPs on either side of the chlorine atom gives some defining parameters for characterizing red and blue-shifting bonds. The two bond critical points (red triangle and green dot) which are directly connected to the halogen atom are shown in Fig. 7.

The red triangle represents the BCP at the X–Hal bond and the green dot represents the BCP connecting the donor–acceptor system. Koch and Poplier[54] proposed eight criteria for characterizing H-bonds which are based on electron density at the BCP connecting donor and acceptor (green dot). Many studies showed that halogen bonds also follow some of those criteria. The systems studied here also agree with the density at the BCP criteria (0.002 a.u to 0.04 a.u) but the Laplacian of density at the BCP is slightly less than the predicted range for Hydrogen bonds (−0.15 a.u to −0.02 a.u). Analysis of electron density at both BCPs for the three systems under our study is given in Fig. 8.

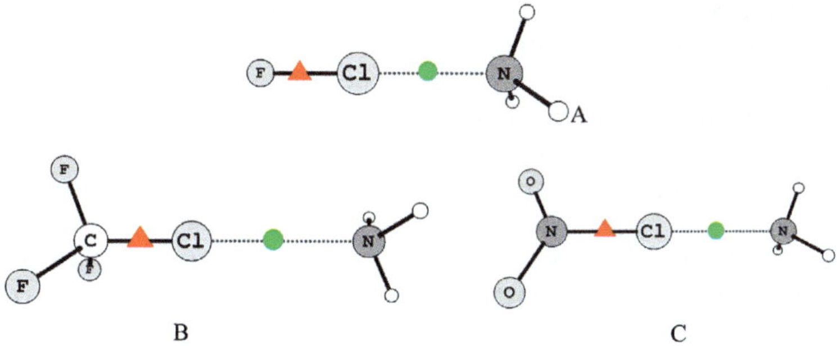


Fig. 7 Bond critical points (BCPs) under investigation: (A) $FCl\cdots NH_3$, (B) $F_3CCl\cdots NH_3$ and (C) $O_2NCl\cdots NH_3$.

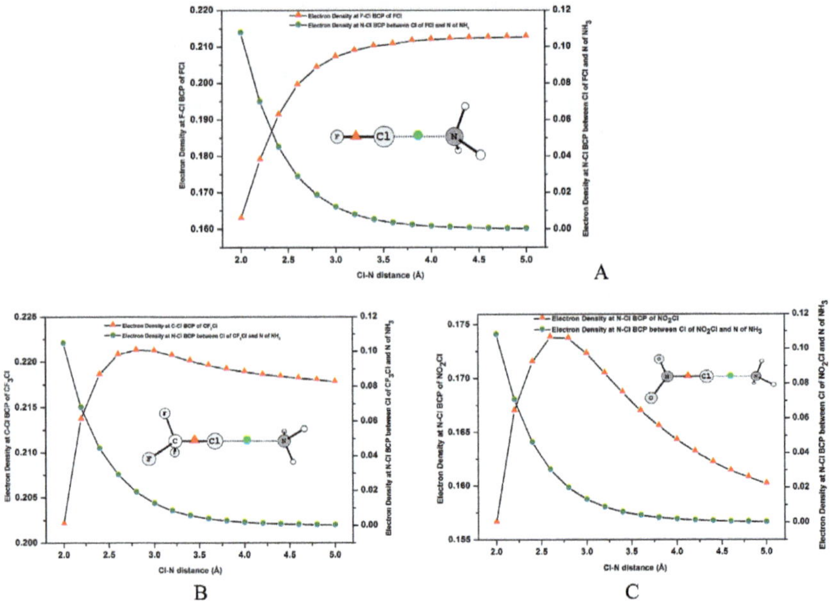


Fig. 8 Variation of electron density at BCPs on either side of Cl atom during scan: (A) FCl···$NH_3$, (B) $F_3C$–Cl···$NH_3$ and (C) $O_2N$–Cl···$NH_3$. Calculations are done at MP2 optimized geometries.

Fig. 8A shows the characteristic electron density change in a normal red shifted system. Electron densities at both the BCPs do not change at longer distances but start changing when they reach the distance where CT begins. The decrease in electron density at the F–Cl bond during the approach of $NH_3$ shows the nature of bond length elongation and extra electron occupancy at the F–Cl σ* orbital. Fig. 8B and C show the characteristics of electron density change in blue-shifting bonds, where the approach of the Y-group removes extra electron density from the σ* X–Z bond orbital and hence electron density at the BCP increases. In the case of the $F_3C$–Cl···$NH_3$ system, change in density at the C–Cl BCP is not so large and can be correlated to the already explained population analysis and σ* occupation. The drastic decrease in N–Cl bond length during formation of the $O_2N$–Cl···$NH_3$ complex leads to the removal of extra electron density at the N–Cl σ* ABMO. Hence, the density curve sharply rises till equilibrium geometry and then starts decreasing due to the larger CT from the $NH_3$ lone pair. It can be seen that irrespective of the X–Z bond length change (as red- or blue-), the density at the donor–acceptor interaction region keeps on increasing. This observation prompts us to propose that the weakly interacting complexes form not only because of the electrostatic attraction between the donor and acceptor but also due to the reorganization of electrons which binds the donor and acceptor molecule. Similar arguments were proposed by Grabowski about the covalent nature of H-bonded complexes based on QTAIM analysis.[55] Very recently Arunan *et al.* verified the covalency and ionicity of weakly interacting X–Z···Y complexes at the Z···Y region for hydrogen, halogen and lithium bonds.[31]

Analysis of computational deformation density 2D contour maps also supports our argument of electron density reorganization as the reason for X–Z bond length variation during complex formation. Deformation density gives information about the extent of electron density reorganization during the formation of the dimer from two separate monomers. Fig. 9 shows the deformation density contour maps at 0.0003 a.u isosurface. Red contours represent the electron density depleted region and the blue contours represent the electron density concentrated region.

Fig. 9A shows the characteristic donation of a lone pair on the Y-group to the X–Z σ* ABMO, where both the Z atom and X-group/atom receive electron density through charge transfer. Fig. 9B and C show the electron density reorganized state of the blue-shifting complexes, where electron density flows away from the Z atom and concentrates over the X-group due to the long range exchange repulsion from the Y-group on to the extra electron density at the σ* ABMO. The electron density concentration in between the monomers represents the glue of electrons which holds the X–Z and Y molecules. Though we have not discussed this explicitly, the variation in the position of the BCPs also follows from these arguments.

Now it is clear that negative hyperconjugation in the X–Z molecule and extra electron occupancy at the X–Z σ* orbital are the primary reasons for the occurrence of pro-improper bonds in weakly interacting complexes and their absence leads to red shifting bonds. A few exceptions are known for this observation, *e.g.* $CH_4\cdots NH_3$,[22] where a blue-shift in C–H bond length is observed even in the absence of negative hyperconjugation. Such situations can be treated as special cases of the generalized electron density reorganization mechanism. The negligible electronegativity difference between Hydrogen and Carbon atoms, where part of the electron density localized over the Hydrogen can easily be shifted towards the C–H bonding region by the incoming Y-group, provides an exception. The applicability of the proposed concept and the inevitability of the continuum nature of red- to blue-shifting bonds can be proved by analysing some well-studied weak interactions such as hydrogen, halogen, chalcogen and pnicogen bonds.

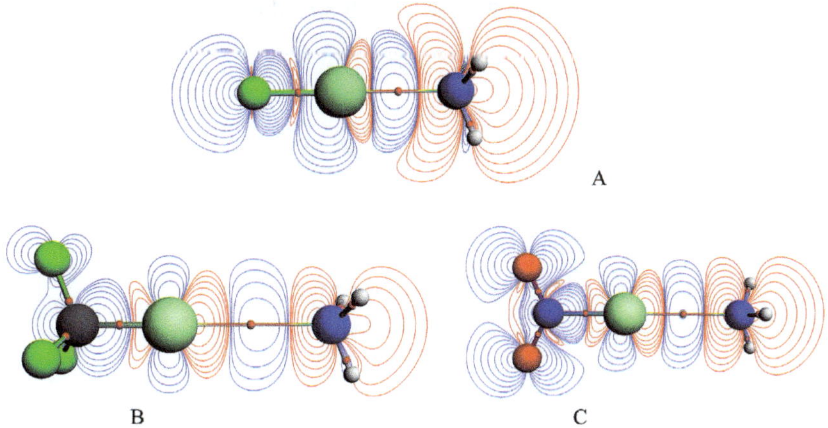


Fig. 9 2D Contour maps of the deformation density at 0.0003 a.u isosurface using equilibrium geometry: (A) $FCl\cdots NH_3$, (B) $F_3C-Cl\cdots NH_3$ and (C) $O_2N-Cl\cdots NH_3$. Calculations are done at M06-2x/TP2Z level of theory.

# 4. On the continuum nature of weak interactions

The halogen bonded complexes discussed above are used as prototypical systems for understanding the interactions that govern the nature of weak X–Z···Y complexes. If the X–Z molecule does not have negative hyperconjugation from Z to X, the X–Z σ* is likely to be more concentrated on Z. This helps the flow of electrons from Y to the X–Z σ*. Hence we conclude that the absence of negative hyperconjugation from the X-group to the X–Z σ* orbital in the X–Z molecule leads to proper Z-bonds (red shift). On the other hand, the presence of negative hyperconjugation from the X-group to the X–Z σ*-orbital leads to red- or blue-shifting depending on the magnitude of hyperconjugation in X–Z. Thus, the electron donating capacity of Y decides the blue or red shift in the X–Z bond in X–Z···Y. These are called pro-improper Z-bonds. Computations on selected sets of complexes and literature data support these arguments (Table S1†). A literature survey of four series of weakly interacting complexes: the hydrogen-bond, halogen-bond, chalcogen-bond, and pnicogen-bond, supports the above arguments. This, along with the arguments developed in this article, allows the prediction of a continuum in weak interactions (Z-bonds) from red-, zero- to blue-shifting. This can be achieved by properly tuning the X- and Y-groups of the X–Z···Y complex. Fig. 10 shows the relative ordering of weakly interacting complexes based on the magnitude of X–Z bond length change and interaction energy. Halogen bonded complexes are found to be largely blue-shifted with lesser interaction energies, followed by chalcogens, hydrogens and pnicogens.

There exists a direct correlation between interaction energy and equilibrium monomer separation. Studies showed that the major factors which decide the equilibrium geometry and hence interaction energy are the electrostatic attraction and the Pauli repulsion between monomers along with polarization and dispersion components.[12,56] The electron density around the Z atom of the X–Z molecule decides the balance between these opposing forces. In the case of the F–Cl molecule, the higher electron withdrawing nature of F makes the Cl atom electron deficient in the σ-hole direction. Consequent higher electrostatic attraction and lower Pauli repulsion lead to higher interaction energy and smaller monomer separation. This is not the case for the $CF_3Cl$ molecule. Here, the $CF_3$–

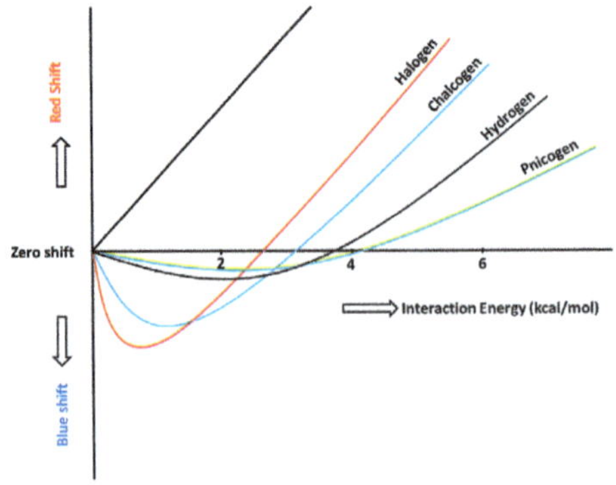


Fig. 10 A qualitative relationship between interaction energy and X–Z bond length change during the formation of different sets of weakly interacting complexes.

group electronegativity is only sufficient to make a weak σ-hole over the Cl atom. Thus, larger electron density around the Cl atom creates higher Pauli repulsion and lower electrostatic attraction. This leads to lower interaction energy and larger monomer separations. Hence in general, blue-shifting halogen bonded complexes show lower interaction energy and higher monomer separation due to the higher electronegativity of the Z atom compared to the connected atom of the X-group. Similar arguments can also be extended to chalcogen, pnicogen and hydrogen bonded complexes (Table S2†). Thus, based on the above premise we suggest that equilibrium monomer separation, and hence total interaction energy, changes with variations in the effective charge density around the Z atom.

Halogen bonded complexes show a higher magnitude of X–Z bond length compression (blue-shifting). This can be inferred from the magnitude of negative hyperconjugation observed in these complexes. As explained above, the higher electronegativity of the Z atom leads to larger MO coefficients at the X atom which is directly connected to the Z atom at the σ* ABMO. This helps to increase the magnitude of electron density donation from its substituents through negative hyperconjugation. This increases the σ* ABMO occupancy of the X–Z bond and hence the bond will be highly elongated in the monomer itself. The extra electron density at the σ* orbital can easily be perturbed by the incoming Y-group which results in the shift of electron density from the anti-bonding MOs to where it came from (reversal of negative hyperconjugations). This leads to X–Z bond compression and hence blue-shift. Thus, it can be concluded that, the larger the electronegativity difference between the X- and Z-atoms, the higher will be the extent of blue-shifting. This is the reason for lower magnitudes of blue-shifting in hydrogen and pnicogen bonded complexes compared to that of halogen and chalcogen bonded complexes, as implied in Fig. 10.

In general, red shifting bonds of varying interaction energies and bond lengths can be observed as shown in the upper portion of Fig. 10, for all Z atoms, but the magnitude of blue-shifting and interaction energy are shown schematically for special characteristic features for each class of interactions. This essentially proves that the proper selection of X- and Y-group for a particular class of Z- will eventually lead to X–Z···Y complexes with zero-shift in X–Z bond length and with non-zero interaction energy. This confirms the inevitability of a continuum of X–Z bond length change from red-, zero- to blue-shifting.

## 5. Conclusions

The extent of negative hyperconjugation and consequent lengthening of the X–Z bond in a parent Z-bond donor differentiate the nature of proper and pro-improper in X–Z···Y complexes. The continuum nature of weak interactions from red-, zero- to blue-shifting has been analysed and correlated to the nature of X- and Y-groups along with the electron density around the Z atom. Our understanding about the nature of X–Z···Y interactions and red- and blue-shifting X–Z bonds can be summarized as follows:

(1) A new description for proper and pro-improper bonds has been given; monomers with no negative hyperconjugation mostly fall in the category of proper (red-shifting) and those with negative hyperconjugation fall in the category of pro-improper (red- or blue-shifting).

(2) A direct correlation exists between total interaction energy and monomer separation at equilibrium geometry, which is decided by the electron density around the Z atom.

(3) The magnitude of blue-shifting depends on the electronegativity difference between the Z atom and the X-group atom connected to Z. The larger the electronegativity difference, the larger will be the negative hyperconjugation and hence larger blue-shifting.

(4) The continuum nature of weak interactions from red-, zero- to blue-shifting can be applied to any X–Z···Y complex. It is decided by the composite of the magnitudes of σ-hole and negative hyperconjugation from the X-group, and the exchange repulsion and charge transfer from the Y-group, the variation of the individual magnitudes estimated as mentioned above.

## Acknowledgements

The authors thank IISER-TVM, SERC-IISc-Bangalore and CMSD-Hyderabad for computational facilities. EDJ thanks the DST for funding through the J. C. Bose Fellowship. JJ and KV thank UGC-India for Senior Research Fellowship.

## References

1 G. R. Desiraju and T. Steiner, *The weak hydrogen bond: In structural chemistry and biology*, Oxford University Press, Oxford, New York, 1999.
2 G. A. Jeffrey and W. Saenger, *Hydrogen bonding in biological structures*, Student edn, Springer-Verlag, Berlin, New York, 1994.
3 S. Srivastava, *Recent development in material science*, Research India Press, New Delhi, 1st edn, 2011.
4 E. R. T. Tiekink, J. J. Vittal and M. Zaworotko, *Organic crystal engineering: frontiers in crystal engineering*, Wiley, Chichester, U.K, 2010.
5 E. Arunan, *Curr. Sci.*, 2013, **105**, 892–894.
6 W. Wang, N.-B. Wong, W. Zheng and A. Tian, *J. Phys. Chem. A*, 2004, **108**, 1799–1805.
7 W. Wang and P. Hobza, *J. Phys. Chem. A*, 2008, **112**, 4114–4119.
8 C. Wang, D. Danovich, Y. Mo and S. Shaik, *J. Chem. Theory Comput.*, 2014, **10**, 3726–3737.
9 J. Fanfrlík, A. Přáda, Z. Padělková, A. Pecina, J. Macháček, M. Lepšík, J. Holub, A. Růžička, D. Hnyk and P. Hobza, *Angew. Chem., Int. Ed.*, 2014, **53**, 10139–10142.
10 R. Gleiter and G. Haberhauer, *J. Org. Chem.*, 2014, **79**, 7543–7552.
11 S. Zahn, R. Frank, E. Hey-Hawkins and B. Kirchner, *Chem.–Eur. J.*, 2011, **17**, 6034–6038.
12 S. Scheiner, *Int. J. Quantum Chem.*, 2013, **113**, 1609–1620.
13 A. Bauzá, T. J. Mooibroek and A. Frontera, *Angew. Chem., Int. Ed.*, 2013, **52**, 12317–12321.
14 D. Mani and E. Arunan, *Phys. Chem. Chem. Phys.*, 2013, **15**, 14377–14383.
15 S. J. Grabowski, *Phys. Chem. Chem. Phys.*, 2014, **16**, 1824–1834.
16 W. Wang, Y. Zhang, B. Ji and A. Tian, *J. Chem. Phys.*, 2011, **134**, 224303–224308.
17 B. Ji, Y. Zhang, D. Deng and W. Wang, *CrystEngComm*, 2013, **15**, 3093–3096.
18 B. Raghavendra and E. Arunan, *J. Phys. Chem. A*, 2007, **111**, 9699–9706.

 

19 W. Zierkiewicz, D. Michalska, Z. Havlas and P. Hobza, *ChemPhysChem*, 2002, **3**, 511–518.
20 S. Scheiner, S. J. Grabowski and T. Kar, *J. Phys. Chem. A*, 2001, **105**, 10607–10612.
21 B. J. van der Veken, W. A. Herrebout, R. Szostak, D. N. Shchepkin, Z. Havlas and P. Hobza, *J. Am. Chem. Soc.*, 2001, **123**, 12290–12293.
22 Y. Gu, T. Kar and S. Scheiner, *J. Am. Chem. Soc.*, 1999, **121**, 9411–9422.
23 W. Qian and S. Krimm, *J. Phys. Chem. A*, 2002, **106**, 6628–6636.
24 I. V. Alabugin, M. Manoharan, S. Peabody and F. Weinhold, *J. Am. Chem. Soc.*, 2003, **125**, 5973–5987.
25 I. V. Alabugin, S. Bresch and M. Manoharan, *J. Phys. Chem. A*, 2014, **118**, 3663–3677.
26 J. Joseph and E. D. Jemmis, *J. Am. Chem. Soc.*, 2007, **129**, 4620–4632.
27 P. Hobza, V. Špirko, H. L. Selzle and E. W. Schlag, *J. Phys. Chem. A*, 1998, **102**, 2501–2504.
28 P. Hobza and Z. Havlas, *Chem. Rev.*, 2000, **100**, 4253–4264.
29 M. J. Frisch, G. W. Trucks, H. B. Schlegel, G. E. Scuseria, M. A. Robb, J. R. Cheeseman, G. Scalmani, V. Barone, B. Mennucci, G. A. Petersson, H. Nakatsuji, M. Caricato, X. Li, H. P. Hratchian, A. F. Izmaylov, J. Bloino, G. Zheng, J. L. Sonnenberg, M. Hada, M. Ehara, K. Toyota, R. Fukuda, J. Hasegawa, M. Ishida, T. Nakajima, Y. Honda, O. Kitao, H. Nakai, T. Vreven, J. A. Montgomery, J. E. Peralta, F. Ogliaro, M. Bearpark, J. J. Heyd, E. Brothers, K. N. Kudin, V. N. Staroverov, R. Kobayashi, J. Normand, K. Raghavachari, A. Rendell, J. C. Burant, S. S. Iyengar, J. Tomasi, M. Cossi, N. Rega, J. M. Millam, M. Klene, J. E. Knox, J. B. Cross, V. Bakken, C. Adamo, J. Jaramillo, R. Gomperts, R. E. Stratmann, O. Yazyev, A. J. Austin, R. Cammi, C. Pomelli, J. W. Ochterski, R. L. Martin, K. Morokuma, V. G. Zakrzewski, G. A. Voth, P. Salvador, J. J. Dannenberg, S. Dapprich, A. D. Daniels, Ö. Farkas, J. B. Foresman, J. V. Ortiz, J. Cioslowski and D. J. Fox, *Gaussian, Inc.*, Wallingford CT, 2010.
30 C. Bleiholder, D. B. Werz, H. Köppel and R. Gleiter, *J. Am. Chem. Soc.*, 2006, **128**, 2666–2674.
31 A. Shahi and E. Arunan, *Phys. Chem. Chem. Phys.*, 2014, **16**, 22935–22952.
32 S. Grimme, J. Antony, S. Ehrlich and H. Krieg, *J. Chem. Phys.*, 2010, **132**, 1–19.
33 L. Goerigk and S. Grimme, *J. Chem. Theory Comput.*, 2010, 7, 291–309.
34 S. F. Boys and F. Bernardi, *Mol. Phys.*, 1970, **19**, 553–566.
35 G. te. Velde, F. M. Bickelhaupt, E. J. Baerends, C. Fonseca Guerra, S. J. A. van Gisbergen, J. G. Snijders and T. Ziegler, *J. Comput. Chem.*, 2001, **22**, 931–967.
36 J. B. Biegler-König and F. Schönbohm, *J. Comput. Chem.*, 2001, **22**, 545–559.
37 M. Mitoraj and A. Michalak, *J. Mol. Model.*, 2007, **13**, 347–355.
38 T. Ziegler and A. Rauk, *Inorg. Chem.*, 1979, **18**, 1755–1759.
39 E. D. Glendening, C. R. Landis and F. Weinhold, *J. Comput. Chem.*, 2013, **34**, 1429–1437.
40 S. M. Huber, J. D. Scanlon, E. Jimenez-Izal, J. M. Ugalde and I. Infante, *Phys. Chem. Chem. Phys.*, 2013, **15**, 10350–10357.
41 P. Politzer, J. S. Murray and T. Clark, *Phys. Chem. Chem. Phys.*, 2010, **12**, 7748–7757.
42 M. D. Esrafili and B. Ahmadi, *Comput. Theor. Chem.*, 2012, **997**, 77–82.

43 I. Fleming, *Molecular orbitals and organic chemical reactions*, Wiley, Chichester, West Sussex, U.K, Student edn, 2009.
44 A. D. McNaught and A. Wilkinson, *Compendium of chemical terminology: IUPAC recommendations*, Blackwell Science, Oxford England, Malden, MA, USA, 2nd edn, 1997.
45 A. Obermeyer, H. Borrmann and A. Simon, *J. Am. Chem. Soc.*, 1995, **117**, 7887–7890.
46 J.-W. Zou, Y.-J. Jiang, M. Guo, G.-X. Hu, B. Zhang, H.-C. Liu and Q.-S. Yu, *Chem.–Eur. J.*, 2005, **11**, 740–751.
47 T. Clark, M. Hennemann, J. Murray and P. Politzer, *J. Mol. Model.*, 2007, **13**, 291–296.
48 P. Politzer, J. S. Murray and P. Lane, *Int. J. Quantum Chem.*, 2007, **107**, 3046–3052.
49 P. Politzer, J. S. Murray and T. Clark, *Phys. Chem. Chem. Phys.*, 2013, **15**, 11178–11189.
50 R. S. Mulliken, *J. Chem. Phys.*, 1934, **2**, 782–793.
51 J. Hinze and H. H. Jaffe, *J. Am. Chem. Soc.*, 1962, **84**, 540–546.
52 J. Hinze, M. A. Whitehead and H. H. Jaffe, *J. Am. Chem. Soc.*, 1963, **85**, 148–154.
53 R. F. W. Bader, *Atoms in molecules: a quantum theory*, Clarendon Press, Oxford, New York, 1990.
54 U. Koch and P. L. A. Popelier, *J. Phys. Chem.*, 1995, **99**, 9747–9754.
55 S. J. Grabowski, *Chem. Rev.*, 2011, **111**, 2597–2625.
56 D. R. Duarte, G. Sosa and N. Peruchena, *J. Mol. Model.*, 2013, **19**, 2035–2041.

 

Faraday Discussions

PAPER

# Dynamics of the chemical bond: inter- and intra-molecular hydrogen bond†

Elangannan Arunan*[a] and Devendra Mani[ab]

Received 5th September 2014, Accepted 12th November 2014
DOI: 10.1039/c4fd00167b

In this discussion, we show that a static definition of a 'bond' is not viable by looking at a few examples for both inter- and intra-molecular hydrogen bonding. This follows from our earlier work (Goswami and Arunan, *Phys. Chem. Chem. Phys.* 2009, **11**, 8974) which showed a practical way to differentiate 'hydrogen bonding' from 'van der Waals interaction'. We report results from *ab initio* and atoms in molecules theoretical calculations for a series of Rg···HX complexes (Rg = He/Ne/Ar and X = F/Cl/Br) and ethane-1,2-diol. Results for the Rg···HX/DX complexes show that Rg···DX could have a 'deuterium bond' even when Rg···HX is not 'hydrogen bonded', according to the practical criterion given by Goswami and Arunan. Results for ethane-1,2-diol show that an 'intra-molecular hydrogen bond' can appear during a normal mode vibration which is dominated by the O···O stretching, though a 'bond' is not found in the equilibrium structure. This dynamical 'bond' formation may nevertheless be important in ensuring the continuity of electron density across a molecule. In the former case, a vibration 'breaks' an existing bond and in the later case, a vibration leads to 'bond' formation. In both cases, the molecule/complex stays bound irrespective of what happens to this 'hydrogen bond'. Both these cases push the borders on the recent IUPAC recommendation on hydrogen bonding (Arunan *et al. Pure. Appl. Chem.* 2011, **83** 1637) and justify the inclusive nature of the definition.

## Introduction

In September 2013, *Chemical and Engineering News* turned 90 and it chose 9 discoveries in chemistry that 'transformed the planet'.[1] Topmost among them was the 'chemical bond'. One would expect that nearly 75 years after Pauling's classic book[2] on *The Nature of the Chemical Bond* was first published, chemists, if not the world, would understand the chemical bond. However, Ritter mentions in

[a]*Department of Inorganic and Physical Chemistry, Indian Institute of Science, Bangalore, India 560012. E-mail: arunan@ipc.iisc.ernet.in; Fax: +91 8023600282; Tel: +91 8022932828*

[b]*Department of Physical Chemistry II, Ruhr-University Bochum, 44780, Bochum, Germany. E-mail: devendra.mani@rub.de*

† Electronic supplementary information (ESI) available: A powerpoint presentation showing torsional vibration in ethane. See DOI: 10.1039/c4fd00167b

 

the lead article of this special issue "*...chemists still struggle to come up with an apt description of a chemical bond*".[3] Hoffmann was quoted in Ritter's article: "*Any rigorous definition of a chemical bond is bound to be impoverishing*".

In this discussion, we address a more specific, and perhaps one of the most important chemical bonds discovered by chemists, the hydrogen bond. If Hoffmann felt a rigorous definition of 'a chemical bond would be impoverishing', a rigorous definition of 'a hydrogen bond' was thought to be impossible. It took 14 experts and about 25 referees (and numerous others interested in the outcome) before a modern definition of a 'hydrogen bond' was accepted by the International Union of Pure and Applied Chemistry in 2011.[4] A short definition had to be followed by 3 pages of caveats listing six criteria and six characteristics, with a warning that 'these are not exhaustive'. A technical report running to 18 pages accompanied the definition detailing the rationale behind the new definition.[5]

Why was defining the hydrogen bond such an arduous task? To understand this, we need to go back to the ideal gas law and van der Waals equation.[6] The ideal gas law was derived with two assumptions: gaseous molecules can be treated as point masses occupying no space and there is no attractive interaction between them. van der Waals introduced two parameters, aptly denoting them $a$ and $b$, and corrected the ideal gas law. The parameter $a$ accounted for the attractive forces between molecules and the parameter $b$ accounted for the finite volume of gaseous molecules, which in effect reduces the volume of the container leading to the term 'excluded volume'. This led to all intermolecular forces being called 'van der Waals forces' and the intermolecular interaction being noted as 'van der Waals interaction'. However, in the early part of 20th century, the advent of X-ray diffractometers among other things, produced accurate structural information in the solid phase. It was discovered that each $H_2O$ molecule is surrounded by four other $H_2O$ molecules in ice but each $H_2S$ molecule is surrounded by twelve other $H_2S$ molecules in solid $H_2S$.[1] They were as different as apples and oranges, see Fig. 1. Only a sphere could accommodate 12 neighbours and $H_2S$ is effectively spherical. The difference was stunning and it clearly warranted different names for the intermolecular interaction found between $H_2O$ molecules and that found between $H_2S$ molecules. The anisotropic interaction between $H_2O$ molecules was christened 'hydrogen bonding' while the isotropic interaction between $H_2S$ molecules was left as 'van der Waals interaction'.

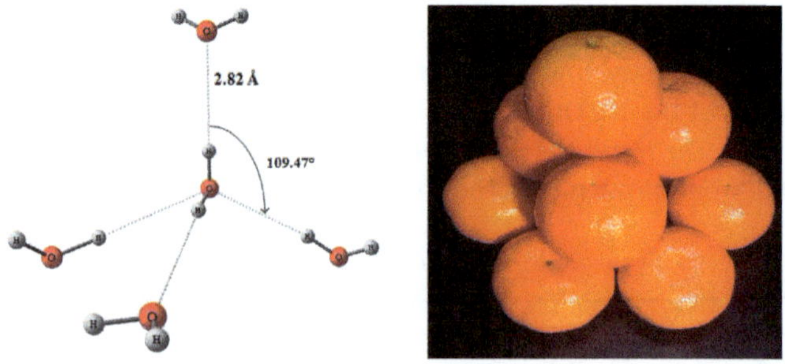


Fig. 1 Comparison of the structures of solid $H_2O$ (left, close to real) and solid $H_2S$ (right, cartoon). Packing of $H_2O$ is anisotropic influenced by the O–H···O 'hydrogen bonds' and packing of $H_2S$ is isotropic having 'van der Waals interaction'.

 

It is interesting to learn that the concept of 'hydrogen bonding' was being discussed even before Lewis/Langmuir[7,8] defined a 'covalent bond'. The reader is referred to the IUPAC technical report[5] and references therein for historical accounts. It is important to realize that van der Waals was not aware of the physical forces involved in the interaction between molecules and did not spell them out. The several authors who invoked hydrogen bonding to explain the difference between the macroscopic properties of different molecules could not spell out what 'hydrogen bonding' was made of either. Pauling included a chapter titled 'Hydrogen bond' in his book.[2] He concluded that the hydrogen which was already bonded to an atom could not form another bond and hence 'hydrogen bonding' must be a simple electrostatic dipole–dipole interaction. We know today that intermolecular interactions between $H_2O$ molecules as well as those between $H_2S$ molecules arise from electrostatic forces between multipoles, inductive forces, exchange correlation, charge transfer/covalency and London dispersive forces to varying extents.

In terms of the physical forces involved, there is really no difference between 'hydrogen bonding' and 'van der Waals interactions'. It is not clear to the authors of the present discussion, who equated 'van der Waals forces' to London dispersive forces.[9] That dispersion is attractive at all orientations was taken to mean that dispersion is 'isotropic'. Clearly, earlier results such as that evident from Fig. 1 could have led to such (mis)understanding. Equating of van der Waals forces to London dispersive forces is indeed unfortunate as it has taken the due credit for dispersive forces away from London. One hopes that dispersive forces are referred to only as London forces from now on.

In the next section, we refer to the recent IUPAC definition[4] and a previous criterion[10] for hydrogen bonding and provide the motive for the present work. This is followed by a brief description of the theoretical methods used. Results on several test cases, which push the borders of these definitions, are discussed next followed by some concluding remarks.

## Definitions of hydrogen bonding

### $H_2O$ and $H_2S$: dimer and solid

The recent IUPAC recommendation[4] defines hydrogen bonding as *an attractive interaction between a hydrogen atom from a molecule or a molecular fragment X–H in which X is more electronegative than H, and an atom or a group of atoms in the same or a different molecule, in which there is evidence of bond formation*. During the discussion of this definition, one of the suggestions was to have 'when' instead of 'in which' before the 'evidence of bond formation' in the definition given above.

If one looks at the structures of ice and solid $H_2S$ when they freeze, they are stunningly different, see Fig. 1. Water is hydrogen bonded and $H_2S$ has van der Waals interaction. On the other hand, if one looks at the optimized geometry of the $H_2O$ dimer and $H_2S$ dimer, see Fig. 2, they look very similar and both are hydrogen bonded. These seemingly contradicting observations were resolved by Goswami and Arunan[10] who suggested that the barriers to torsional motion which can break hydrogen bonds are more important than the barrier along the stretching coordinate which leads to the dissociation of the dimers. Nearly all prior work on hydrogen bonding[2,11–13] concentrated on the barrier along the stretching coordinate and decided on some arbitrary cut off on the dissociation

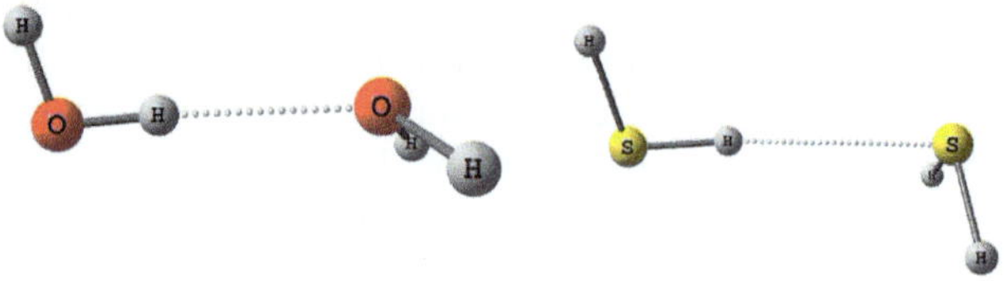


Fig. 2 Optimized structures of $H_2O$ dimer and $H_2S$ dimer, with hydrogen bonds shown by dotted lines.

energies to distinguish between hydrogen bonding and the van der Waals interaction. Goswami and Arunan[10] suggested that, *for a 'hydrogen bonded complex', the zero point energy along any large amplitude vibrational coordinate that destroys the orientational preference for the hydrogen bond should be significantly below the barrier along that coordinate so that there is at least one bound level.*

Goswami and Arunan[10] realized that the thermal motions in solid $H_2S$ lead to an average 'spherical shape' and predicted that $H_2S$ solid at lower temperature should exhibit anisotropic packing and hydrogen bonding. They did not realize that it had already been proved by Loveday *et al.*[14] who reported the crystal structure of $D_2S$ at lower $T$ and higher $P$. Loveday *et al.* had indeed concluded that $H_2S$ can be a good model system where we can tune the hydrogen bond strength from "absent or very weak to structurally significant". The choice of $D_2S$ is interesting as the heavier isotopologue would have shorter amplitude vibrations compared to the lighter isotopologue.

## Rare gas···HX complexes

The crystal structure of $D_2S$ discussed above pegs a question. Could we have a complex in which XD···Y has one bound level below the torsional barrier and XH···Y does not? Could we have a 'deuterium bond' even when a 'hydrogen bond' is not possible? The complexes formed between rare gas atoms and hydrogen halides provide such a possibility. The IUPAC report[5] points out that the He···HF complexes have HF as a nearly free rotor whereas in Kr···HF complexes, this motion remains a bending vibration. It was decided to investigate several Rg···HX complexes and determine the bending frequency and the barrier for internal rotation of HX within the complex. Results from these detailed calculations on a series of Rg···HX (Rg = He/Ne/Ar and X = F/Cl/Br) show that it is indeed possible to have a 'deuterium bond' when a 'hydrogen bond' does not exist according the criterion given by Goswami and Arunan.[10]

## Ethane-1,2-diol

In the previous section, we considered the cases where a large amplitude vibration breaks a hydrogen bond existing at an equilibrium structure. Could we have a molecular system in which a bond between two atoms is not apparent at the equilibrium structure but forms during the course of a normal mode vibration? Ethane-1,2-diol exhibits such a possibility. Whether an intra-molecular hydrogen bond is present in this molecule, commonly known as ethylene glycol, has been hotly debated. This molecule has been a textbook example for an 'intra-molecular hydrogen bond' finding a mention in Eliel's classic book.[15] Microwave spectroscopy reveals that the *gauche* conformer is more stable than the *trans* conformer.[16] This was attributed to the intra-molecular hydrogen bond though microwave spectroscopy cannot provide direct evidence for the bond. After careful NMR experiments on this molecule, Roberts and coworkers[17] concluded that "*the bulk of the NMR evidence indicates that intra-molecular hydrogen bonding between the hydroxyl groups is unlikely to be a significant factor in determining that preference, except possibly in fairly non-polar solvents*". Though they did not find any evidence, they were cautious in their conclusion. Atoms in molecules (AIM) theoretical analysis on this molecule by various authors[18–20] did not show any evidence of a bond critical point between the H of one OH and the O of another OH. The $\angle$O–H···O angle is about 108°, below even the generously given lower limit for a hydrogen bond, $\angle$X–H···Y, an angle of 110°, by the IUPAC recommendation.[4] The distance between H and O is less than the sum of their van der Waals radii but more than the sum of hydrogen bond radii for OH and O.[21–23] However, both the van der Waals 'radii' and hydrogen bond 'radii' do not account for the effect of anisotropy and conclusions from them are suspect. The IUPAC recommendation specifically discourages the use of van der Waals radii in deciding the hydrogen bond. An X-ray study including charge density analysis[23] of the solid sample did not find any evidence for the hydrogen bond as well, though the distance between two OH groups was slightly larger than that in the gas phase, due to the strong inter-molecular hydrogen bonds. As of now, the only experimental evidence for the intra-molecular hydrogen bond is a small red-shift of about 30 $cm^{-1}$ in the O–H stretching frequency.[24] However, this is smaller than the intercept in the frequency shift *versus* bond energy correlation plot for O–H···O hydrogen bonds, 40 $cm^{-1}$, implying a binding energy of zero or no bond.[25] While the experimental lack of evidence remains uncontested, the lack of a bond critical point in AIM theoretical analysis, as evidence against a bond, has been criticized severely.[26] A recently introduced 'non covalent index (NCI) plot'[27] does reveal some evidence for the intra-molecular hydrogen bond.

Are all the discussions mentioned above only semantic, as everyone agrees that the *gauche* conformer is more stable than the *trans* conformer? This debate is indeed important, as ethane-1,2-diol is part of the training set for parametrizing molecular mechanics program.[28] It is a ubiquitous synthon in sugars[29] and many pharmacologically active materials[30] and it is commonly used in step-growth polymerizations for the synthesis of polyurethanes and polyesters.[31] It is important to understand the properties of the monomer as thoroughly as one could. Moreover, the presence of a bond, *i.e.* a local concentration of electrons in between two atoms, as revealed by a bond critical point in AIM analysis is important in the concept of an 'electron wire'[32] enabling the formation of a cooperative network as in liquid water. What if thermal vibrations bring two atoms close enough to be 'bonded'? This could still help in the concept of an 'electron wire' and in some respect it would make more sense when such thermal motions can lead to this 'bond formation' leading to direct, continuous electron density between the two atoms. Atoms in molecules theoretical analysis has recently been employed along the reaction coordinate to understand bond formation and breaking during a reaction.[33] We report the AIM theoretical analysis as a function of the O···O stretching vibration in ethane-1,2-diol which indeed shows evidence of intra-molecular hydrogen bond formation as the two OH groups come closer.

# Computational methods

All calculations reported in this work have been carried out at MP2 level with a correlation consistent triple zeta basis set, aug-cc-pvtz (aug-cc-pvdz in the case of ethane-1,2-diol), using the Gaussian 09 programme.[34] Our main objective in carrying out these calculations is conceptual rather than an accurate determination of structure and energetics and in our experience this method provides reasonable results. All optimized structures were confirmed to be true minima using frequency calculations. Energies were corrected for basis set superposition errors using the counterpoise method[35] inherent in Gaussian09. Atoms in molecules theoretical analysis was carried out using the AIMALL program.[36]

# Results and discussion

### Hydrogen *vs.* deuterium bond: Rg···HX complexes

The Rg···HX complexes have been studied in detail both experimentally and theoretically. The reader is referred to a recent comprehensive study by Murdachaew *et al.*[37] for a detailed discussion of the multidimensional potential surface and all the associated energy levels. Our objective in this discussion is more conceptual and limited to determining the barrier for the internal rotation of HX which can lead to the isomerization between the Rg···HX and Rg···XH minima. This is compared to the zero point energy along this bending/internal rotation coordinate for both Rg···HX and Rg···DX to determine if the interaction is strong enough to preserve the orientation. We use one-half of the normal mode frequency as the zero point energy in this conceptual exercise. Accurate calculation of the energy levels for this vibrational motion which can change from torsion to hindered rotation to free rotation is quite involved[38] and is beyond the scope of this discussion.

Table 1 lists our results on Rg···HX complexes for Rg = He/Ne/Ar and X = F/Cl/Br. The results are not surprising. The barrier for internal rotation decreases as Rg···HF > Rg···HCl > Rg···HBr. For a given HX, the barrier increases as He···HX < Ne···HX < Ar···HX. The results given in Table 1 may not be the most accurate estimates but they are reasonable. For example, the lowest bound level for Ne···DF has been experimentally determined to be 4–6 $cm^{-1}$ above the barrier for internal rotation[39] and Table 1 gives this to be about 8 $cm^{-1}$.

Table 1 Barrier for bending/internal rotation of HX/DX in Rg···HX/DX complexes and zero point energies (ZPE) (in $cm^{-1}$) along this coordinate

| Complex | Barrier | ZPE (HX) | ZPE (DX) |
|---|---|---|---|
| He···HF | 20.7 | 48.7 | 35.4 |
| He···HCl | 12.2 | 20.5 | 14.9 |
| He···HBr | 9.7 | 41.6 | 30.1 |
| Ne···HF | 39.9 | 66.6 | 47.8 |
| Ne···HCl | 21.8 | 29.4 | 21.0 |
| Ne···HBr | 18.7 | 49.9 | 35.5 |
| Ar···HF | 130.8 | 98.8 | 70.5 |
| Ar···HCl | 76.2 | 48.9 | 34.2 |
| Ar···HBr | 63.9 | 56.5 | 40.2 |

 

All the Ar···HX complexes are 'hydrogen bonded' according to the criterion given by Goswami and Arunan.[10] They do exhibit a bond critical point between Ar and H and hence they satisfy the criteria for a bond by Bader[40] and also by Koch and Popelier.[41] According to Bader, the presence of a bond critical point between two atoms is enough to conclude the presence of a 'bond'. Koch and Popelier were trying to distinguish a 'van der Waals interaction' from a C–H···O hydrogen bond and listed eight criteria based on the electron density topology for C–H···O hydrogen bonds.

The 'hydrogen bonds' in these cases are strong enough to preferentially orient the H of HX towards Ar in the minimum. Not surprisingly, all Ar···DX complexes are more localized than the Ar···HX complexes and the deuterium bond is stronger than the hydrogen bond, a well known observation. The calculations were not carried out for Kr···HX and Xe···HX, for which the barriers will be significantly higher resulting in a stronger hydrogen bond. This observation is not new and has been pointed out in many earlier publications, including in a crossed beam scattering study[42] with rare gases and $H_2O$, aptly titled as the 'birth of a hydrogen bond'.

Clearly, none of the He···HX complexes are 'hydrogen bonded', even though all of them exhibit a bond critical point between He and H in their optimized structure.[43] Hence, the He···HX complexes satisfy the criterion for a bond by Bader, but these do not satisfy either the criterion given by Koch and Popelier for C–H···O hydrogen bonds[41] or that given by Goswami and Arunan.[10] For He···HX (X = F/Cl/Br), the electron densities at the BCP are 0.0017/0.0012/0.0009, respectively,[43] below the limit of 0.002 recommended by Koch and Popelier.[41] Even for He···DX, the zero point energies are higher than the barrier. While for He···DF and He···DBr, the zero point energies are significantly higher, for He···DCl, the zero point energy is only 2 $cm^{-1}$ above the barrier. This implies that the DCl in He···DCl will be more dominantly oriented as 'deuterium bonded' compared to He···HCl as 'hydrogen bonded'. It is important to note that all the He···HX complexes as a whole are bound even though the nearly free internal rotation can break the 'hydrogen/deuterium bond'.

The results from the Ne···HX complexes offer an interesting middle ground. All Ne···HX complexes have a zero point energy above the barrier. The Ne···DF and Ne···DBr complexes also have zero point energies above the barrier. However, Ne···DCl has a zero point energy just below the barrier. It does have one bound level below the barrier making it 'deuterium bonded' according to the criterion by Goswami and Arunan,[10] even when Ne···HCl is not 'hydrogen bonded' according to the same criterion. The potential energy scans along the bending/internal rotation coordinate for He/Ne/Ar···HCl are shown in Fig. 3, 4, and 5, respectively and these provide a nice summary of the three possibilities.

We note that the bending frequencies for the Rg···HCl complexes and the corresponding zero point energies are lower than those of both the Rg···HF and Rg···HBr complexes. The barriers for internal rotation reduce by close to 50% from Rg···HF to Rg···HCl, but drop by only 10–20% from Rg···HCl to Rg···HBr. Clearly, these barriers cannot explain the unusual trend in frequencies, which indicate that the bending motion is stiffer in Rg···HBr compared to Rg···HCl. The reduced masses for this normal coordinate do not vary much and the changes in frequencies are dominated by the changes in force constants. For example, the reduced masses for the Ne···HF/HCl/HBr complexes for the bending vibration are

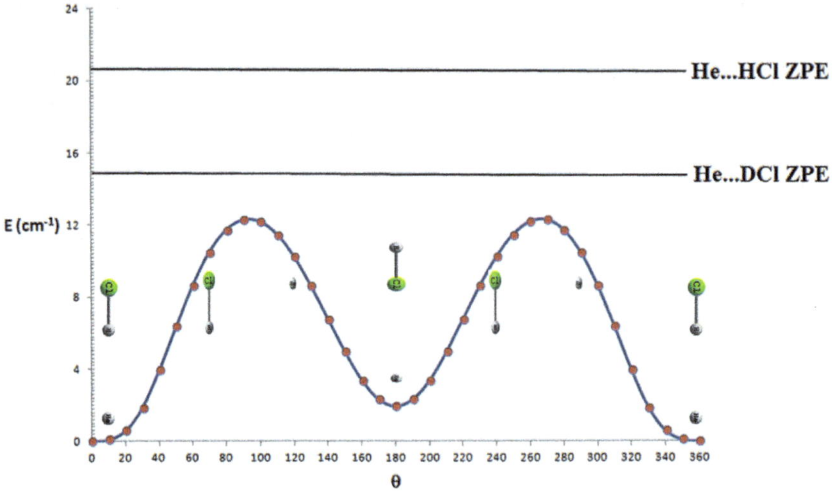


Fig. 3 Potential energy curve along the bending/internal rotation coordinate for He–HCl, with the zero point energy marked for He···HCl and He···DCl. The 'hydrogen/deuterium bond' is not strong enough to restrict internal rotation. It may be noted that the frequencies were calculated at the hydrogen bonded minimum.

1.0374/1.0257/1.0193 amu, respectively. The force constants for these are 0.0108/0.0021/0.0060 mdyn Å$^{-1}$, respectively. In any case, these are floppy complexes that are very weakly bound and the one-dimensional analysis we report here is not intended for accurate description of the inter-molecular potentials.

We reiterate that these results are more conceptual than quantitative, and they can explain some experimental observations. The crystal structure of $D_2S$ showed

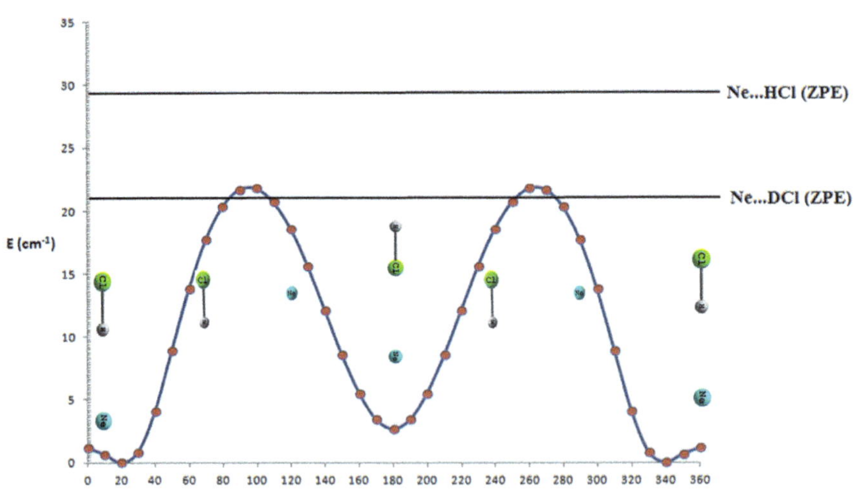


Fig. 4 Potential energy curve along the bending/internal rotation coordinate for Ne–HCl, with the zero point energy marked for Ne···HCl and Ne···DCl. The 'hydrogen bond' is not strong enough to restrict internal rotation but 'deuterium bond' is. It may be noted that the frequencies were calculated at the hydrogen bonded minimum.

 

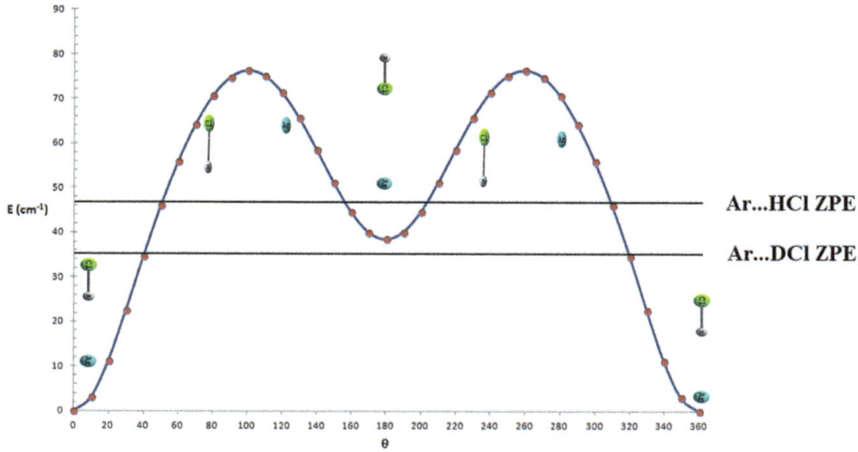


Fig. 5 Potential energy curve along the bending/internal rotation coordinate for Ar–HCl, with the zero point energy marked for Ar···HCl and Ar···DCl. The 'hydrogen/deuterium bond' is strong enough to restrict the internal rotation. It may be noted that the frequencies were calculated at the hydrogen bonded minimum.

evidence for 'hydrogen bonding' more readily than that of $H_2S$.[14] Another manifestation of significant difference in vibrational averaging was seen in the example of the Ar···$H_2S$ complex. The experimental rotational constant for the Ar···$D_2S$ complex was higher than that of the Ar···$H_2S$ complex.[44] This is the result of a significantly shorter intermolecular distance in the Ar···$D_2S$ complex as compared to that in the Ar···$H_2S$ complex, outweighing the increase in mass. Similarly, the gas phase experimental (vibrationally averaged) structure of the HI dimer and DI dimer show significant differences, which can be explained by the dramatic differences in zero point vibrations.[45]

In the earlier days of the hydrogen bond,[2] it was considered that only H attached to F, O, and N atoms could form a hydrogen bond. These molecules typically have a larger dipole moment due to the larger electronegativity difference. Clearly, the intermolecular potential for interaction of these molecules with hydrogen bond acceptors has significant anisotropy which could be detected readily by macroscopic experiments. On the contrary, for molecules such as $H_2S$, HCl, and HBr, the intermolecular potential is not isotropic, but it is much less anisotropic in comparison and their experimental manifestations could not be detected earlier. These examples show that a hydrogen bond that exists in the equilibrium structure could be broken even at zero Kelvin due to zero point vibrations. Depending on the actual difference between the barrier and zero point energy, the intermolecular potential may effectively become isotropic, leading to a structure for $H_2S$ solid at −82 °C in which $H_2S$ looks spherical. At even lower temperatures, the thermal motion is restricted pushing the molecular system into potential wells that are no longer isotropic.

## Ethane-1,2-diol

As discussed earlier, the presence of an intra-molecular hydrogen bond in ethane-1,2-diol has been hotly debated in recent years. Previous calculations by various authors[18–20] have concluded that there is no bond critical point between the H of one O–H group and the O of the other O–H group. Having seen the effect of even zero point vibration breaking a hydrogen bond, one wonders if a large amplitude vibration in ethane-1,2-diol could lead to a hydrogen bond. Much of the earlier work on this molecule focused on the O–H stretching frequency and the small red shift observed. An inspection of its normal mode vibrations in our calculations revealed that the nominal O···O stretching mode appears at 173 $cm^{-1}$. We carried out AIM theoretical analysis along this vibrational coordinate, starting from the equilibrium structure. During this vibration, the distance between the two atoms would increase and decrease compared to that in the equilibrium distance. However, when it increases, a BCP is unlikely to appear and that was indeed the case. When the distance between the two atoms decreases, a BCP does indeed appear. The potential energy curve for this vibration along with the AIM molecular graphs are shown in Fig. 6. The energy levels for $\nu = 0$ and 1 are shown for comparison, assuming harmonic oscillator energies.

As is evident from Fig. 6, there is no BCP corresponding to an intra-molecular hydrogen bond at the equilibrium structure. A BCP does not appear during zero point vibration either. However, when the value of this normal coordinate increases to just below the outer turning point of $\nu = 1$, a BCP starts appearing corresponding to an intramolecular hydrogen bond. This being a low frequency vibration, the thermal population at 300 K is close to 44% and clearly thermal vibrations can bring the two OH groups close enough to be 'bonded', even according to the AIM analysis. Interestingly, when an intra-molecular hydrogen bond is formed, a five member ring forms and a ring critical point has to appear to

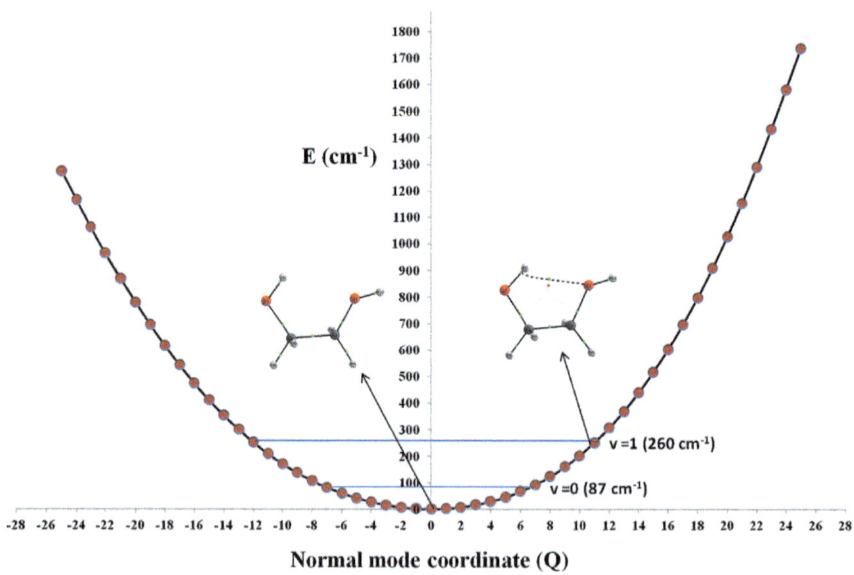


Fig. 6 The potential energy along the nominal O···O stretching coordinate. At the equilibrium structure, the AIM molecular graph shows no BCP between the H of one OH and the O of the other OH. As the O···O distance reduces, a bond critical point corresponding to the intra-molecular hydrogen bond appears along with a ring critical point, starting from step 11 in this coordinate. This is below the outer turning point for the $v = 1$ level.

 

satisfy the topological requirement, and it does appear. If the newly formed bond is weaker, the ring critical point will be very close to the bond critical point. A ring critical point does indeed appear and at larger values of the normal coordinate, it moves towards the centre of the ring, as expected.

Clearly, as the ethane-1,2-diol molecule vibrates, an intra-molecular hydrogen bond does appear satisfying the requirements of the proponents of AIM theory. One natural question that arises now is whether an isotopic substitution with a heavy atom would restrict the amplitude of vibration to an extent that no BCP is formed during the vibration. To test this, the calculations were repeated with $^{18}O$ and D substitutions. In every case, a BCP does appear during the vibration. As mentioned earlier, Roberts and co-workers[17] could not find any evidence for spin–spin coupling through the intramolecular hydrogen bond in their NMR experiments. It would be interesting to see the temperature effect on the NMR spectrum of a very dilute solution of ethane-1,2-diol in a non-polar solvent.

## What then is a hydrogen bond?

This analysis of ethane-1,2-diol has led us to a contrasting conclusion. A careful analysis of the structure of the $H_2S$ dimer and crystal led us to propose a practical criterion to differentiate hydrogen bonding and van der Waals interaction. This does make sense. An analogy with internal rotation in ethane would convince a reader, if (s)he is not yet convinced. Ethane, and substituted ethanes can exist in different minima accessible by internal rotation of C–C. These minima are characterized as *trans*, *gauche*, staggered and eclipsed depending on the dihedral angle. The barrier for this internal rotation is typically about 3–5 kcal $mol^{-1}$, very similar to the hydrogen bond energy. At ambient conditions, the vibrational energy along this coordinate, usually called a torsion, is limited and the molecule could exist as *trans* and *gauche*, for substituted ethanes. However, if this vibrational energy is increased, this could become a hindered rotation and eventually a free rotation. Three movie files given in the ESI† show the three motions, torsion, hindered rotation and free rotation. If an ethane molecule has free internal rotation, it would not have *trans*, *gauche*, staggered or eclipsed conformers, but it would remain ethane. For it to become two methyl radicals, there has to be sufficient energy given along the C–C stretching coordinate, which is typically much larger than the barrier along the internal rotation coordinate.

The difference between barriers along torsional coordinates and the stretching coordinate for a hydrogen bond would typically be much less than that of ethane. This has led to the arbitrary classification of hydrogen bonding as different from van der Waals interaction. However, with the advent of experimental and theoretical methods, even small differences in the barriers can lead to experimentally observable manifestations. Clearly, hydrogen bonding is a specific subset of the van der Waals interaction.

The results for ethane-1,2-diol show that an intra-molecular hydrogen bond, which does not exist in an equilibrium structure, could be formed during a vibration. While it would also be broken in this vibration, this would indeed be important in ensuring that a bond path is created for electron density to flow from O to H, through this transient bond.

## Conclusions

We have considered two examples of hydrogen bonds in this discussion. The first one is formed between a rare gas and HX in its equilibrium structure. However, depending on the rare gas and X, this may not survive even the zero point vibration. We could have a strong 'hydrogen bond' which can ensure the directionality of a hydrogen bond in one extreme. We could have a weak hydrogen bond that does not hinder a free rotation of HX leading to an apparent isotropic interaction, often called van der Waals interaction. The second example has the potential of an intra-molecular hydrogen bond. The equilibrium structure does not have a corresponding BCP, but during thermal vibration this bond could be formed. Indeed, the recent IUPAC definition of the hydrogen bond insisting on evidence could end with 'when there is evidence of a bond'. The dynamics due to molecular vibrations cannot be ignored. Could an experiment be carried out in which zero point vibrations can be viewed?

## Acknowledgements

Authors thank Abhishek Shahi for discussion and also for making the power point presentation given in the supporting document. EA thanks Prof. S. R. Gadre for discussion about the ethane-1,2-diol vibrations leading to bond formation and IUPAC for supporting the project to redefine hydrogen bonds.

## Notes and references


1 A. Maureen Rouhi, *Chem. Eng. News*, 2013, **91**, 5.
2 L. Pauling, *The Nature of the Chemical Bond*, Cornell University Press, 1939.
3 S. K. Ritter, *Chem. Eng. News*, 2013, **91**, 28–33.
4 E. Arunan, G. R. Desiraju, R. A. Klein, J. Sadlej, S. Scheiner, I. Alkorta, D. C. Clary, R. H. Crabtree, J. J. Dannenberg, P. Hobza, H. G. Kjaergaard, A. C. Legon, B. Mennucci and D. J. Nesbitt, *Pure Appl. Chem.*, 2011, **83**, 1637–1642.
5 E. Arunan, G. R. Desiraju, R. A. Klein, J. Sadlej, S. Scheiner, I. Alkorta, D. C. Clary, R. H. Crabtree, J. J. Dannenberg, P. Hobza, H. G. Kjaergaard, A. C. Legon, B. Mennucci and D. J. Nesbitt, *Pure Appl. Chem.*, 2011, **83**, 1619–1636.
6 J. D. van der Waals, Ph.D. Thesis, Univ. Leiden, 1873.
7 G. N. Lewis, *J. Am. Chem. Soc.*, 1916, **38**, 762–785.
8 I. Langmuir, *J. Am. Chem. Soc.*, 1919, **41**, 868–1543.
9 F. London, *Z. Phys.*, 1930, **63**, 240.
10 M. Goswami and E. Arunan, *Phys. Chem. Chem. Phys.*, 2009, **11**, 8974–8983.
11 G. R. Desiraju and T. Steiner, *The Weak Hydrogen Bond: In Structural Chemistry in Biology*, Oxford University Press, Oxford, 1999.
12 G. A. Jeffery and W. Saenger, *Hydrogen bonding in biological structure*, Springer Verlag, Berlin, 1991.
13 S. Scheiner, *Hydrogen Bonding: A Theoretical Perspective*, Oxford University Press, Oxford, 1997.
14 J. S. Loveday, R. J. Nelmes, S. Klotz, J. M. Besson and G. Hamel, *Phys. Rev. Lett.*, 2000, **85**, 1024–1027.


 

15 E. L. Eliel, *Stereochemistry of Carbon Compounds*, McGraw-Hill, New York, 1962, p. 131.
16 D. Christen, L. H. Coudert, J. A. Larsson and D. Cremer, *J. Mol. Spectrosc.*, 2001, **205**, 185–196.
17 K. A. Petterson, R. S. Stein, M. D. Drake and J. D. Roberts, *Magn. Reson. Chem.*, 2005, **43**, 225–230.
18 R. A. Klein, *J. Comput. Chem.*, 2002, **23**, 585–599.
19 C. Trindle, P. Crum and K. Douglass, *J. Phys. Chem. A*, 2003, **107**, 6236–6242.
20 M. Mandado, A. M. Grana and R. A. Mosquera, *Phys. Chem. Chem. Phys.*, 2004, **6**, 4391–4396.
21 B. Lakshmi, A. G. Samuelson, K. V. Jovan Jose, S. R. Gadre and E. Arunan, *New J. Chem.*, 2005, **29**, 371–377.
22 B. Raghavendra, P. K. Mandal and E. Arunan, *Phys. Chem. Chem. Phys.*, 2006, **8**, 5276–5286.
23 D. Chopra, T. N. Guru Row, E. Arunan and R. A. Klein, *J. Mol. Struct.*, 2010, **964**, 126–133.
24 D. L. Howard and H. G. Kjaergaard, *J. Phys. Chem. A*, 2006, **110**, 10245–10250.
25 A. V. Iogansen, *Spectrochim. Acta, Part A*, 1999, **55**, 1585–1612.
26 J. R. Lane, J. Contreras-Garica, J. P. Piquemal, B. J. Miller and H. G. Kjaergaard, *J. Chem. Theory Comput.*, 2013, **9**, 3263–3266.
27 J. Contreras-Garcia, E. R. Johnson, S. Keinan, R. Chaudret, J. P. Piquemal, D. N. Beraton and W. T. Yang, *J. Chem. Theory Comput.*, 2011, **7**, 625–632.
28 S. O. Jonsdottir and R. A. Klein, *Fluid Phase Equilib.*, 1997, **132**, 117–127.
29 Y. L. Li, X. L. Sun and Y. L. Wu, *Tetrahedron*, 1994, **50**, 10727–10738.
30 R. N. Patel, A. Banerjee, Y. R. Pendri, J. Liang, C. P. Chen and R. Mueller, *Tetrahedron: Asymmetry*, 2006, **17**, 175–178.
31 X. Wang, L. Wang, H. Li, X. Tang and F. C. Chang, *J. Appl. Polym. Sci.*, 2000, **77**, 184–188.
32 C. Manca, C. Tanner and S. Leutwyler, *Int. Rev. Phys. Chem.*, 2005, **24**, 457–488.
33 C. L. Parworth, M. K. Tucker, B. E. Holmes and G. L. Heard, *J. Phys. Chem. A*, 2011, **115**, 13133–13138.
34 J. A. Frisch, M. J. Trucks, G. W. Schlegel, H. B. Scuseria, G. E. Robb, M. A. Cheeseman, J. R. Montgomery Jr., J. A. Vreven, T. Kudin, K. N. Burant, J. C. Millam, J. M. Iyengar, S. S. Tomasi, J. Barone, V. Mennucci, B. Cossi and M. Scalmani, *Gaussian 09, Revision C.01*, Gaussian, Inc., Wallingford CT, 2009.
35 S. F. Boys and F. Bernardi, *Mol. Phys.*, 1970, **19**, 553–566.
36 T. A. Keith, *AIMAll (Version 13.05.06)*, TK Gristmill Software, Overland Park KS, USA, 2013, http://www.aim.tkgristmill.com.
37 G. Murdachaew, K. Szalawicz, H. Jiang and Z. Bačić, *J. Chem. Phys.*, 2004, **121**, 11839–11855.
38 Y.-Y. Chuang and D. G. Truhlar, *J. Chem. Phys.*, 2000, **112**, 1221–1228.
39 C. M. Lovejoy and D. J. Nesbitt, *J. Chem. Phys.*, 1991, **94**, 208–223.
40 R. F. W. Bader, *Atoms in Molecules: A Quantum Theory*, Oxford University Press, Oxford, 1990.
41 U. Koch and P. L. A. Popelier, *J. Phys. Chem.*, 1995, **99**, 9747–9754.
42 V. Aquilanti, E. Cornicchi, M. M. Teixidor, N. Saendig, F. Pirani and D. Cappelletti, *Angew. Chem., Int. Ed.*, 2005, **44**, 2356–2360.
43 B. Raghavendra, Ph.D. Thesis, Indian Institute of Science, 2008, p.106.

44 H. S. Gutowsky, T. Emilsson and E. Arunan, *J. Chem. Phys.*, 1997, **106**, 5309–5315.

45 Z. Wang, B. A. McElmury, R. R. Lucchese, J. W. Bevan and L. H. Coudert, *J. Chem. Phys.*, 2011, **134**, 064317.

 

Faraday Discussions

PAPER

# Vibrational dynamics of the CO stretching of 9-fluorenone studied by visible-pump and infrared-probe spectroscopy

Yuki Fukui, Kaoru Ohta and Keisuke Tominaga*

Received 8th September 2014, Accepted 19th November 2014
DOI: 10.1039/c4fd00169a

We studied the effects of hydrogen bonds on the vibrational structures and vibrational dynamics of the CO stretching mode of 9-fluorenone (FL) in the electronically excited state in aprotic and protic solvents using sub-picosecond visible-pump and IR-probe spectroscopy. The transient IR spectrum of the CO stretching band in methanol-$d_4$ has two bands at 1529.9 cm$^{-1}$ and 1543.4 cm$^{-1}$, which are assigned to an FL-solvent complex and free FL, respectively. In the aprotic solvents, the CO stretching bands show blue-shifts in time. This shift is due to vibrational cooling, which is derived from anharmonic couplings with some low-frequency modes. Interestingly, a red-shift is observed at later delay time for the band at 1529.9 cm$^{-1}$ in methanol-$d_4$. A possible mechanism of this spectral shift is related to the hydrogen bond dynamics between the solute and solvent.

## Introduction

In solution, a solute molecule relaxes to the most stable state in the electronically excited state ($S_1$) after electronic excitation to the Franck–Condon state. Examples of the relaxation processes are vibrational energy relaxation, cooling of the locally heated environment around the solute, and solvation dynamics.[1–8] These processes influence chemical reactions initiated by the electronic excitation. In a protic solvent, a solute molecule forms intermolecular hydrogen bonds with solvent molecules if the solute has hydrogen bonding sites such as a carbonyl group. The hydrogen bonds influence various properties of the solute, such as its electronic states and the relaxation processes.

9-Fluorenone (FL) has a carbonyl group which can form hydrogen bonds, and spectroscopic studies[9–14] as well as theoretical calculations[15] have been reported on the nature of the excited state of FL. The vibrational dynamics of FL in alcohol solutions have been reported in the electronic ground state[12] and the excited state.[10,13,14] Using sub-picosecond UV-pump infrared (IR)-probe spectroscopy we

*Department of Chemistry, Graduate School of Science, Kobe University, Molecular Photoscience Research Center, Kobe University, Nada, Kobe, 657-8501, Japan. E-mail: tominaga@kobe-u.ac.jp*

have observed transient IR absorption spectra of FL in acetotrnile-$d_3$ and methanol-$d_4$.[13] By comparison with theoretical calculations of the vibrational structure in the $S_1$ state,[15] we assigned the vibrational bands observed in the transient spectra. In this study, we used sub-picosecond visible-pump and IR-probe spectroscopy to study the effects of hydrogen bonding on the vibrational structures and vibrational dynamics of the CO stretching mode of the excited state FL in various solvents, including aprotic solvents.

## Experimental

Details of the experimental setup are as follows. The pump pulse was generated by frequency doubling of the output of a Ti:sapphire regenerative amplifier. A tunable IR pulse was generated by difference frequency mixing of the signal and idler outputs of an optical parametric amplifier. The probe and reference pulses were obtained using the reflection of a $CaF_2$ wedged window. The probe and reference pulses were dispersed by a monochromator and the absorption changes were measured with a 32-channel MCT (HgCdTe) array detector. The experiments were conducted using 100 μm-thick liquid samples in a rotating cell. The concentration of all FL solutions was 45 mM.

FL was purchased from Sigma-Aldrich and recrystallized from ethanol once before use. Aprotic solvents (cyclohexane, acetonitrile-$d_3$, tetrahydrofurane (THF), and dimethyl sulfoxide (DMSO) from Sigma-Aldrich) and a protic solvent (methanol-$d_4$ from Merck) were used as received.

We obtained optimized geometries and frequencies of normal mode coordinates of the electronic ground state of FL in the gas phase by performing density functional theory (DFT) calculations with the 6-31++G(d, p) basis set using the B3LYP functional in the Gaussian 03 package. Calculations for the hydrogen-bonded complex were also conducted. For the $S_1$ state time-dependent (TD) DFT calculation was performed with the B3LYP functional and the 6-31G(d, p) basis set. The solvent effect was taken into account by self-consistent reaction field theory for a polarizable continuum model with the specified dielectric constant of the solvent.

## Results and discussion

Fig. 1 shows steady-state IR absorption spectra of FL in acetonitrile-$d_3$ and methanol-$d_4$. In aprotic solvents, such as acetonitrile-$d_3$, a sharp single band due to the CO stretching mode of FL is observed, which is reproduced well with a single Lorentzian function. For protic solvents such as methanol-$d_4$, three Lorentzian functions are required to simulate the CO stretching band. By comparison with normal modes obtained from DFT calculations Hirai and coworkers showed that the highest wavenumber band (band *a* in Fig. 1) is assigned to the CO stretching band of FL which does not form a hydrogen bond with the solvent (abbreviated as [FL] hereafter). The next highest band (band *b*) is assigned to the band of FL that is hydrogen-bonded with one alcohol molecule ([FL:ROH]), and the lowest wavenumber band (band *c*) is assigned to the band of FL bonded with two alcohol molecules ([FL:$(ROH)_2$]).[12] In this work we also used Gaussian functions to fit the spectrum, and found that the Lorentzian function gave better spectral simulation results. Of course, it is possible to simulate the spectra with

 

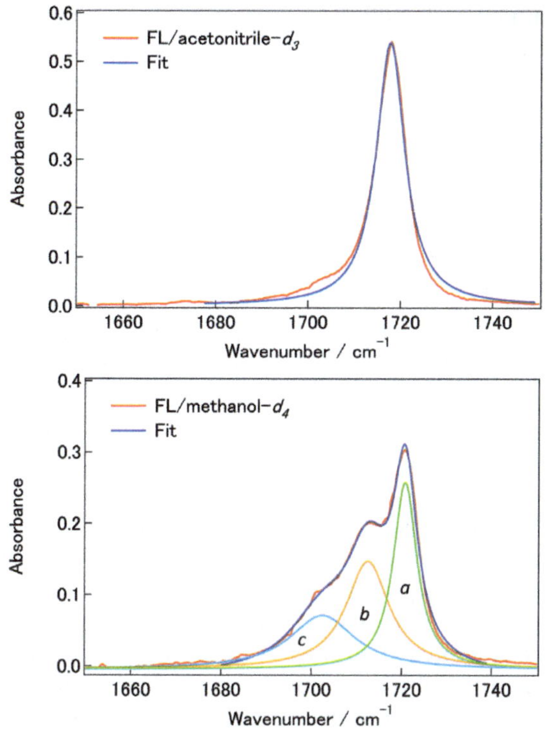


Fig. 1 FTIR spectra of FL in acetonitrile-$d_3$ (upper) and methanol-$d_4$ (lower). Results of the spectral simulation are also shown.

Gaussian functions and the final conclusion will differ from that obtained using Lorentzian functions.

In Table 1 we summarize the peak wavenumbers of the CO stretching band of FL in various solvents. It can be seen that in more polar solvents the peak wavenumber shifts toward the lower-frequency side. This is because the vibrational excited state is more energetically stable in polar solvents. Levinson *et al.* explained the solvatochromism of the CN stretching band of benzonitrile and 4-aminobenzonitrile in terms of the Onsager factor ($F_{onsager}$), a parameter indicating the polarity of solvent.[16] In the model a solute is assumed to be a spherical

Table 1 Peak wavenumbers of the vibrational bands in the IR absorption spectrum of FL in various solvents

| Solvents | Peak wavenumber (cm$^{-1}$) | | |
|---|---|---|---|
| **Aprotic solvents** | | | |
| Cyclohexane | 1725.3 ± 0.1 | | |
| Acetonitrile-$d_3$ | 1717.9 ± 0.1 | | |
| Tetrahydrofurane | 1719.7 ± 0.1 | | |
| Dimethyl sulfoxide | 1714.3 ± 0.1 | | |
| **Protic solvent** | Band *a* | Band *b* | Band *c* |
| Methanol-$d_4$ | 1720.9 ± 0.1 | 1712.5 ± 0.1 | 1702.6 ± 0.5 |

 

cavity in a continuum with dielectric constant $\varepsilon$ and refractive index $n$. The Onsager factor can be expressed as,

$$F_{\text{onsager}} = \frac{2(\varepsilon - 1)(n^2 + 2)}{3(2\varepsilon + n^2)}. \tag{1}$$

In Fig. 2 the peak frequencies of the CO stretching band of FL in aprotic solvents and free FL in methanol-$d_4$ (band *a* in Fig. 1) is plotted as a function of $F_{\text{onsager}}$. A good correlation is observed in the relation between the peak frequency and $F_{\text{onsager}}$.

Fig. 3 shows the time evolution of the transient IR absorption spectra of FL in the electronically excited state in acetonitrile-$d_3$ and methanol-$d_4$. The observed wavenumber region is 1492–1576 $cm^{-1}$. We first discuss the results for the acetonitrile-$d_3$ case. We show a transient spectrum at 48 ps together with a spectral simulation using three Lorentzian functions in Fig. 4. These three bands are named bands *a*, *a'* and *d* from the higher wavenumber side. In Table 2 the peak frequencies of the bands in several aprotic solvents are summarized. The band position of band *d* depends on the solvent, and in cyclohexane and THF band *d* is not observed; it is probably located below 1492 $cm^{-1}$. Using nanosecond time-resolved IR spectroscopy Tanaka *et al.* found that the CO stretching band of FL in the $S_1$ state is located at 1544 $cm^{-1}$ in acetonitrile-$d_3$.[10] They observed an isotope shift caused by substitution of $^{16}O$ to $^{18}O$ in the carbonyl group of FL. Therefore, band *a* is assigned to the CO stretching mode of FL in the $S_1$ state. Furthermore, band *d* is located at 1495.7 $cm^{-1}$, which corresponds to the band at 1496 $cm^{-1}$ that Tanaka *et al.* observed previously. From TD-DFT calculations we obtain the normal modes and vibrational frequencies in the $S_1$ state in acetonitrile. The CO stretching band is predicted to be at 1644.5 $cm^{-1}$. Other bands in this wavenumber region are a band at 1637.4 $cm^{-1}$ in which the CO stretching mode and a fluorene ring mode are coupled, and a band at 1604.7 $cm^{-1}$ corresponding to a vibrational mode localized in the fluorene ring. It is suggested that these modes correspond to the experimentally observed bands at 1533.2 $cm^{-1}$ (band *a'*) and 1495.7 $cm^{-1}$ (band *d*) in acetonitrile-$d_3$, respectively, although the agreement between the calculation and experiment is not perfect. We summarize

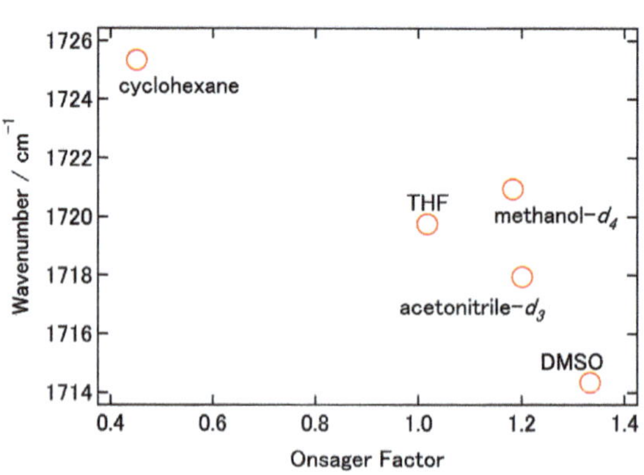


Fig. 2 Peak wavenumbers of free FL (band *a*) in the $S_0$ state as a function of solvent polarity.

 

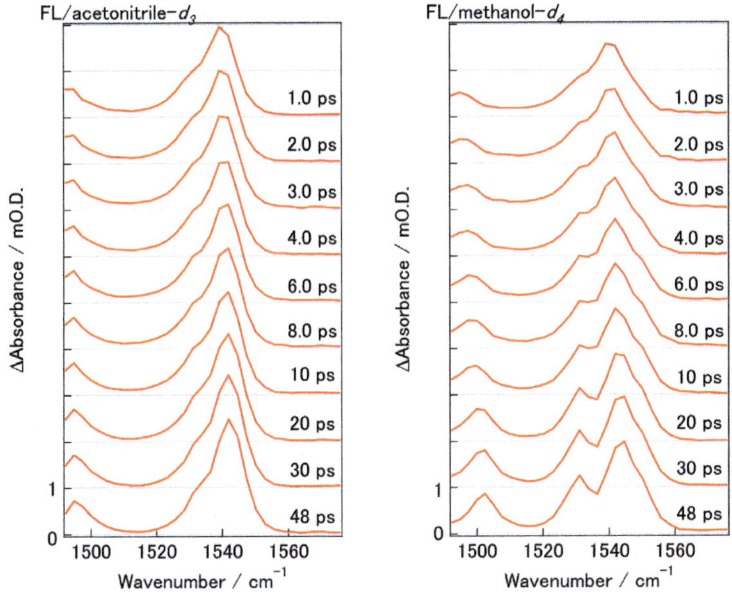


Fig. 3 Time evolution of transient IR spectra of FL in acetonitrile-$d_3$ (left) and methanol-$d_4$ (right).

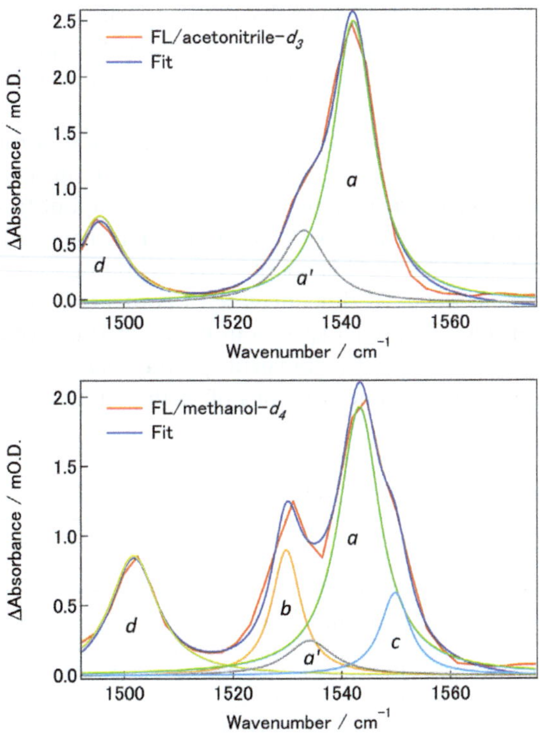


Fig. 4 Transient IR spectra of FL in acetonitrile-$d_3$ (upper) and methanol-$d_4$ (down) at a delay time of 48 ps. Spectral simulation results are shown.

Table 2 Peak wavenumbers of the vibrational bands in the transient IR absorption spectrum of FL in aprotic solvents

| | Band $a$/cm$^{-1}$ | Band $a'$/cm$^{-1}$ | Band $d$/cm$^{-1}$ |
|---|---|---|---|
| Cyclohexane | 1540.0 ± 0.5 | 1533.7 ± 2.9 | — |
| Acetonitrile-$d_3$ | 1542.4 ± 0.2 | 1533.2 ± 0.8 | 1495.7 ± 0.5 |
| Tetrahydrofurane | 1539.1 ± 0.3 | 1529.8 ± 2.2 | — |
| Dimethyl sulfoxide | 1539.7 ± 0.4 | 1531.8 ± 1.3 | 1494.1 ± 0.8 |

the computed frequencies in Table 3, and normal mode pictures are shown in Fig. 5.

In Fig. 6 the time evolution of the peak wavenumber of band $a$ is shown in acetonitrile-$d_3$. The band shows a blue-shift as a function of delay time. The time dependence is simulated well in terms of a bi-exponential function with time constants of 1.1 ± 0.2 ps and 10.5 ± 0.4 ps. The other two bands, bands $a'$ and $d$, also show similar blue-shifts. It has been suggested that this blue-shift is due to vibrational cooling.[17] The high-frequency intramolecular modes of FL are anharmonically coupled to the low-frequency bath modes. After photoexcitation, the vibrational energy relaxation takes place from the Franck–Condon state to the vibrational ground state in the $S_1$ state. The solute molecule in the excited state is locally heated because of excess energy which is dissipated into the intramolecular degrees of freedom and local environment surrounding the solute. The vibrationally hot $S_1$ state is cooled down due to thermal diffusion, which causes a blue-shift of the vibrational band, because of anharmonic coupling with the bath modes. This vibrational cooling has been observed in other solution systems, which show a picosecond time scale for this process.[17,18]

In the right panel of Fig. 3 we display the time evolution of the transient IR absorption spectra of FL in the electronically excited state in methanol-$d_4$. The band at 1544.7 cm$^{-1}$ is assigned as the CO stretching band of free FL from the similarity of the band shape and peak wavenumber to band $a$ in the aprotic solvents. The band at 1532.2 cm$^{-1}$ is not observed in aprotic solvents. This band becomes prominent as the delay time increases. In order to assign this band we measured the transient vibrational spectra in binary mixtures of dimethyl sulfoxide (DMSO) and methanol-$d_4$. The transient spectra at a delay time of 48 ps are shown in Fig. 7 for different concentrations of methanol-$d_4$. In pure DMSO the

Table 3 Wavenumbers of FL in the $S_1$ state computed by TD-DFT[a]

| SCRF | | Number of hydrogen bonds | | Wavenumbers/cm$^{-1}$ | |
|---|---|---|---|---|---|
| | $\varepsilon$ | | CO stretching mode | CO stretching and fluorene ring mode | Fluorene ring mode |
| Acetonitrile | 36.64 | | 1644.5 | 1637.4 | 1604.7 |
| Methanol | 24.55 | 0 | 1644.6 | 1637.4 | 1604.7 |
| | | 1 | 1624.5 | 1644.3 | 1605.3 |
| | | 2 | 1611.5 | 1645.2 | 1604.8 |

[a] Normal mode pictures are shown in Fig. 5.

 

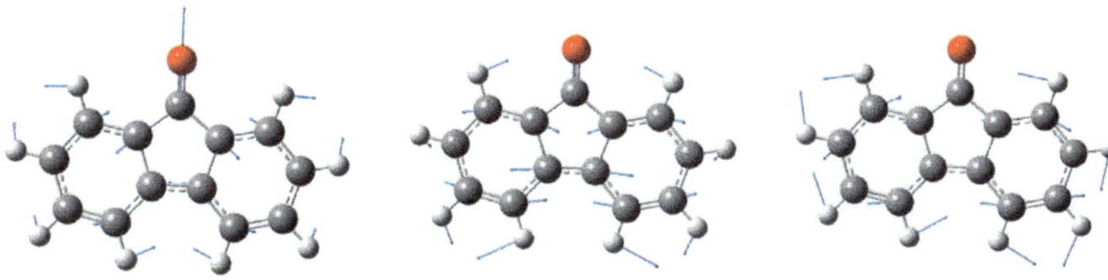

Fig. 5 Normal modes of FL in the $S_1$ state computed by TD-DFT. CO stretching mode (left), mixed mode of CO stretching and fluorene ring modes (middle), fluorene ring mode (right).

transient spectrum is similar to that in acetonitrile-$d_3$. With increasing concentrations of methanol-$d_4$ this band (named band *b*) emerges as seen from the figure. Band *b* is characteristic to protic solvents; therefore, we conclude that band *b* is the CO stretching mode of FL which is hydrogen-bonded with methanol-$d_4$, [FL:$CD_3OD$]. It should be noted that at $t = 1.0$ ps in Fig. 3 band *b* in methanol-$d_4$ is not clearly observed, and the spectral shape looks like that in acetonitrile-$d_3$. This fact suggests that just after photoexcitation FL, which forms hydrogen bonds with the solvent in the $S_0$ state, experiences hydrogen bond breakage. As the increase of the delay time, hydrogen bonds are formed in the excited $S_1$ state again. At $t = 3.0$ ps a shoulder peak at around 1530 $cm^{-1}$ is already observed, suggesting that hydrogen bond formation takes place within this time scale.

Since band *b* is located in the lower wavenumber than band *a*, the CO stretching band of FL in the $S_1$ state is red-shifted by the formation of hydrogen bonds with the solvent, as in the case of the $S_0$ state. This trend is also reproduced by TD-DFT calculations. In Table 3 the calculation results for methanol solvent are also summarized. When the carbonyl group is hydrogen-bonded with one methanol molecule, the vibrational wavenumber is shifted to 1624.5 $cm^{-1}$, about 20 $cm^{-1}$ lower than the free FL case.

For the $S_0$ state in methanol-$d_4$, a hydrogen-bonded complex [FL:$(CD_3OD)_2$] is observed in the IR absorption spectrum. In the $S_1$ state, band *d* was previously assigned to the CO stretching mode of [FL:$(CD_3OD)_2$] by sub-picosecond visible-pump IR-probe spectroscopy.[14] However, we observe band *d* even in aproptic solvents, such as DMSO. Therefore, we conclude that band *b* is not due to

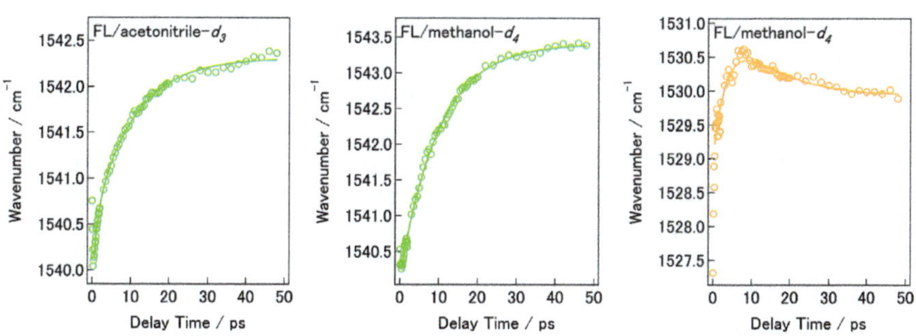


Fig. 6 Time evolution of the peak wavenumbers of the transient IR spectra of FL. Band *a* in acetonitrile-$d_3$ (left), band *a* in methanol-$d_4$ (middle), and band *b* in methanol-$d_4$ (right).

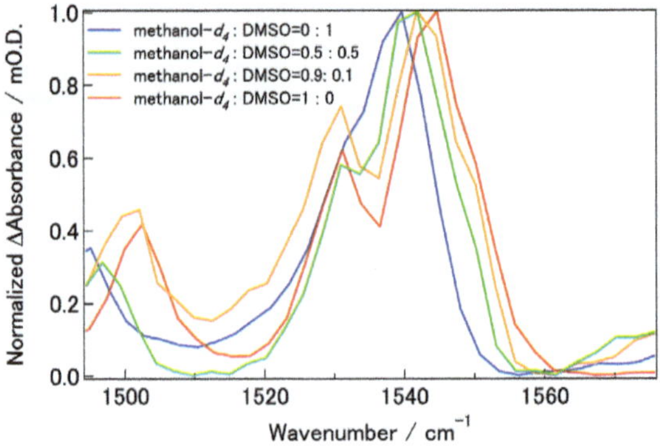


Fig. 7 Transient IR spectra of FL in a binary solvent of methanol-$d_4$ and DMSO at a delay time of 48 ps. The concentrations are indicated in the figure as a molar ratio.

[FL:$(CD_3OD)_2$], but due to free FL. Since TD-DFT calculations predict that the CO stretching mode is more red-shifted when FL forms more hydrogen bonds with solvents, it is deduced that the CO stretching band of [FL:$(CD_3OD)_2$], if it exists, will be located at a lower frequency than bands *a* or *b*. However, we do not observe such a band in that frequency region beside band *d*. Thus, it is suggested that hydrogen-bonded complex [FL:$(CD_3OD)_2$] does not exist in the $S_1$ state, probably due to the hydrogen bond with the solvent molecule being weaker for FL in the $S_1$ state than in the $S_0$ state.

In aprotic solvents we observed band *a*′, which is formed by mixing of the CO stretching mode and fluorene ring vibrational mode. Similarly, in protic solvents it is natural to assume that this mixed mode exists for free FL. Actually, from TD-DFT calculations it is predicted that band *a*′ is located at 1637.4 cm$^{-1}$ and the fluorene ring mode is at 1640.7 cm$^{-1}$ for free FL in methanol. Furthermore, a shoulder band is observed at 1550.2 cm$^{-1}$ in methanol, which is not observed in aprotic solvents and is named band *c*. TD-DFT calculations show that the mixed mode (band *a*′) is blue-shifted from 1637.4 cm$^{-1}$ to 1644.3 cm$^{-1}$ by the formation of hydrogen-bonds with the solvent, in contrast to the case of the CO stretching band. In our previous work we assigned bands *a* and *b* to the CO stretching modes of free FL and the complex [FL:$CD_3OD$], respectively, which is opposite to the present conclusion, by comparison with TD-DFT calculations with B3LYP/TZVP level.[15] Zhao and Han concluded that the bands at 1592 cm$^{-1}$ and 1595 cm$^{-1}$ are the CO stretching modes of free FL and the complex [FL:$CD_3OD$], respectively. We performed TD-DFT calculations with the same basis set for free FL and the hydrogen-bonded complex and obtained similar results in terms of frequency. However, from the normal mode picture, it is found that the bands at 1590 cm$^{-1}$ and 1592 cm$^{-1}$ are not CO stretching modes but rather the fluorene ring modes of free FL and the hydrogen-bonded complex, respectively. The CO stretching bands of free FL and the complex are located at 1626 cm$^{-1}$ and 1619 cm$^{-1}$, respectively. This suggests that the fluorene ring mode might have been confused with the CO stretching mode.

To summarize, there are five bands to simulate the transient spectra of FL in protic solvent, such as methanol. Three of them (bands *a*, *a*′ and *d*) are observed in aprotic solvents also, and the remaining two (bands *b* and *c*) are observed only

 

Table 4 Peak wavenumbers of the vibrational bands in the transient IR absorption spectrum of FL in methanol-$d_4$

| | Band *c* | Band *a* | Band *a*′ | Band *b* | Band *d* |
|---|---|---|---|---|---|
| Peak wavenumber (cm$^{-1}$) | 1550.0 ± 0.7 | 1543.4 ± 0.3 | 1534.2 ± 0.3 | 1529.9 ± 0.3 | 1501.8 ± 0.3 |

in protic solvents. To simulate the spectra with five Loretzian functions we make the following assumption. It is expected that the wavenumbers of bands *a*′ and *b* are very close to each other, so that it is difficult to separate them in the spectral simulation. We assume that the wavenumber difference and the relative intensities between bands *a* and *a*′ for the methanol solvent case are the same as those for acetonitrile, since the polarities of the two solvents are similar. We first obtain time evolutions of the wavenumber difference and the relative intensities between bands *a* and *a*′ for acetonitrile from the spectral analysis and use these results for the analysis of the methanol case. The result of the spectral analysis using five Loretzian functions is shown in Fig. 4, and the peak wavenumbers of the five bands are summarized in Table 4.

In Fig. 6 the time evolutions of the peak wavenumbers of bands *a* and *b* are displayed. Band *a* shows a blue-shift, which is reproduced well in terms of a single exponential function with a time constant of 10.0 ± 0.2 ps. As in the aprotic solvent cases, this blue-shift is due to vibrational cooling. Interestingly, band *b* first shows a blue-shift which is followed by a red-shift. The time dependence is simulated by a bi-exponential function with time constants of 3.9 ± 1.4 ps (blue-shift) and 11.1 ± 6.9 ps (red-shift). It seems two different dynamics cause the vibrational frequency shift in opposite ways. The blue-shift is probably due to vibrational cooling. However, it should be noted that hydrogen bond formation also occurs during this early time scale ($t < 3$ ps) as mentioned previously. On the other hand, the red-shift may be related to the hydrogen bond between the solute and solvent, since only the hydrogen-bonded complex shows this red-shift. A possible dynamics for this is the structural rearrangement of the hydrogen-bonded complex, and there are two conceivable mechanisms of this

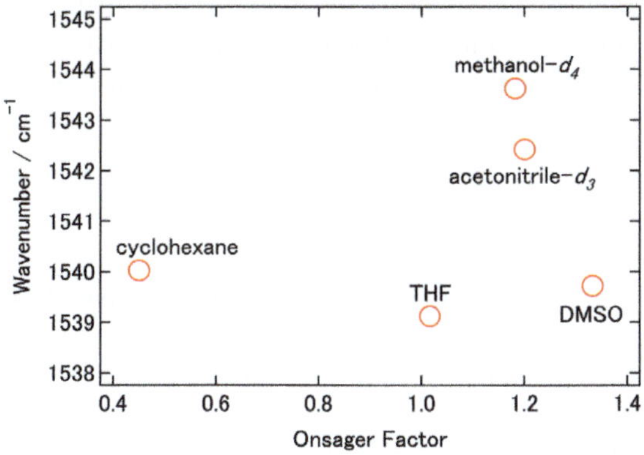


Fig. 8 Peak wavenumbers of free FL (band *a*) in the $S_1$ state as a function of solvent polarity.

rearrangement: one is that the hydrogen-bonded complex relaxes to the most stable state by cooling in the surrounding environment around the complex; the other is that, as a result of polarization relaxation in the solvent (solvation dynamics), reorganization of the solvent molecules takes place, which causes the rearrangement of the structure of the hydrogen-bonded complex. At present it is difficult to distinguish which mechanism cooling or solvation dynamics, is the dominant cause of this red-shift.

In Fig. 8 we plot the peak wavenumbers of band *a* as a function of solvent polarity $F_{\text{onsager}}$. In the $S_0$ state we observe good correlation between the peak wavenumber and $F_{\text{onsager}}$; the peak frequency of the CO stretching mode shifts to lower frequencies as a function of solvent polarity (see Fig. 2). However, in the $S_1$ state such a good correlation is not observed (see Fig. 7). This is probably because the electronic structure of the $S_1$ state depends on solvent polarity so the vibrational structure of the $S_1$ state is also influenced by the solvent. It is well known that the $n\pi^*$ and $\pi\pi^*$ states are mixed in the electronically excited state of FL and the degree of the mixing depends on solvent polarity.

## Conclusion

The transient vibrational spectra of FL were obtained in both aprotic and protic solvents by sub-picosecond visible-pump and IR-probe spectroscopy. The transient IR spectrum of the CO stretching band in methanol-$d_4$ has peaks at 1529.9 $cm^{-1}$ and 1543.4 $cm^{-1}$, which are assigned to bands of hydrogen-bonded complexes with the solvent and free FL, respectively. In the aprotic solvents, the CO stretching bands show a blue-shift over time, which is due to vibrational cooling. The free FL in methanol-$d_4$ also shows the same blue-shift; however, a red-shift is observed at a later time for the band at 1529.9 $cm^{-1}$. The hydrogen bond dynamics between FL and solvent molecules may be related to this red-shift.

## Acknowledgements

We thank Dr Masahiro Higashi of the University of Ryukyus and Prof. Shinji Saito of Institute for Molecular Science for helpful discussions, especially on theoretical calculation. Theoretical calculations were performed using the facilities of Research Center for Computational Sciences, Okazaki, Japan. The authors thank Prof. Dipak K. Palit of the Bhabha Atomic Research Centre for valuable discussions.

## References

1 H. Hamaguchi and T. L. Gustafson, *Annu. Rev. Phys. Chem.*, 1994, **45**, 593–622.
2 P. Hamm and W. Zinth, *J. Phys. Chem.*, 1995, **99**, 13537–13544.
3 C. Chudoba, E. T. J. Nibbering and T. Elsaesser, *J. Phys. Chem. A*, 1999, **103**, 5625–5628.
4 E. T. J. Nibbering, F. Tschirschwitz, C. Chudoba and T. Elsaesser, *J. Phys. Chem. A*, 2000, **104**, 4236–4246.
5 J. B. Asbury, Y. Wang and T. Lian, *Bull. Chem. Soc. Jpn.*, 2002, **75**, 973–983.

 

6 W.-M. Kwok, M. W. George, D. C. Grills, C. Ma, P. Matousek, A. W. Parker, D. Phillips, W. T. Toner and M. Towrie, *Angew. Chem., Int. Ed.*, 2003, **42**, 1826–1830.
7 E. T. J. Nibbering, H. Fidder and E. Pines, *Annu. Rev. Phys. Chem.*, 2005, **56**, 337–367.
8 P. Kukura, D. W. McCamant and R. A. Mathies, *Annu. Rev. Phys. Chem.*, 2007, **58**, 461–488.
9 K. Yoshihara and D. R. Kearns, *J. Chem. Phys.*, 1966, **45**, 1991–1999.
10 S. Tanaka, C. Kato, K. Horie and H. Hamaguchi, *Chem. Phys. Lett.*, 2003, **381**, 385–391.
11 V. Samant, A. K. Singh, G. Ramakrishna, H. N. Ghosh, T. K. Ghanty and D. K. Palit, *J. Phys. Chem. A*, 2005, **109**, 8693–8704.
12 S. Hirai, M. Banno, K. Ohta, D. K. Palit and K. Tominaga, *Chem. Phys. Lett.*, 2007, **450**, 44–48.
13 S. Hirai, M. Banno, K. Ohta, D. K. Palit and K. Tominaga, *Chem. Lett.*, 2010, **39**, 932–934.
14 H. N. Ghosh, K. Adamczyk, S. Verma, J. Dreyer and E. T. J. Nibbering, *Chem. - Eur. J.*, 2012, **18**, 4930–4937.
15 G.-J. Zhao and K.-L. Han, *J. Phys. Chem. A*, 2007, **111**, 9218–9223.
16 N. M. Levinson, S. D. Fried and S. G. Boxer, *J. Phys. Chem. B*, 2012, **116**, 10470–10476.
17 P. Hamm, S. M. Ohline and W. Zinth, *J. Chem. Phys.*, 1997, **106**, 519–529.
18 K. Iwata and H. Hamaguchi, *J. Phys. Chem. A*, 1997, **101**, 632–637.

 

Faraday Discussions

PAPER

# Dynamical, spectroscopic and computational imaging of bond breaking in photodissociation: roaming and role of conical intersections

Masaaki Nakamura,[a] Po-Yu Tsai,[ab] Toshio Kasai,[ac] King-Chuen Lin,*[ab] Federico Palazzetti,[de] Andrea Lombardi[e] and Vincenzo Aquilanti[efg]

Received 12th September 2014, Accepted 27th October 2014
DOI: 10.1039/c4fd00174e

Recent experimental and theoretical advances in the study of the dissociation of excited molecules are revealing unexpected mechanisms, when their outcomes are tackled by combining (i) space-time ion imaging of translational features, with (ii) spectroscopic probing of rotational and vibrational distributions; crucial is the assistance of (iii) the quantum chemistry of structural investigations of rearrangements of chemical bonds, and of (iv) the simulations of molecular dynamics to follow the evolution of selective bond stretching and breaking. Here we present results of such an integrated approach to methyl formate, $HCOOCH_3$, the simplest of esters; the main focus is on the rotovibrationally excited CO ($v = 1$) product and in general on the energy distribution in the fragments. Previous laser studies of dissociation into CO and $CH_3OH$ at a sequence of various wavelengths discovered signatures of a roaming mechanism by the late arrival of CO ($v = 0$) products in time-of-flight ion imaging. Subsequent detailed investigations as a function of excitation energy provided the assessment of the threshold, which opens for triple breakdown into CO and further fragments H and $CH_3O$, as spectroscopically characterized by ion imaging and FTIR respectively. Accompanying quantum mechanical electronic structure calculations and classical molecular dynamics simulations clarify the origin of these fragments through "roaming" pathways involving incipient radical intermediates at energies below the triple fragmentation threshold: a specific role is played by nonadiabatic transitions at a conical intersection between ground and excited states; alternative pathways focalize our

[a]*Department of Chemistry, National Taiwan University, Taipei 106, Taiwan. E-mail: kclin@ntu.edu.tw; Fax: +886 2 23621483; Tel: +886 2 33661162*
[b]*Institute of Atomic and Molecular Sciences, Academia Sinica, Taipei 106, Taiwan*
[c]*Graduate School of Science, Department of Chemistry, Osaka University, Toyonaka, 560-0043 Osaka, Japan*
[d]*Scuola Normale Superiore, 56126 Pisa, Italy*
[e]*Dipartimento di Chimica, Biologia e Biotecnologie, Università di Perugia, 06123 Perugia, Italy. E-mail: vincenzoaquilanti@yahoo.it; Fax: +39 075 5855606; Tel: +39 075 5855512*
[f]*Consiglio Nazionale delle Ricerche, Istituto di Struttura della Materia, 00016 Roma, Italy*
[g]*Instituto de Física, Universidade Federal da Bahia, 40210 Salvador, Brazil*

 

attention to regions of the potential energy surfaces other than those in the neighbourhoods of saddle points along minimum energy paths: eventually this leads us to look for avenues in reaction kinetics beyond those of venerable transition state theories.

# 1. Introduction

## 1.A. Motivation

Dynamically and spectroscopically following the fragments of photodissociated molecules is a technique of choice, whereby in chemistry one can investigate reaction pathways and probe the main features of the involved energy landscapes. According to the Franck–Condon principle applied to molecular photodissociation, after absorption of photons, molecules access an excited state or a manifold of states, along which they evolve to deliver products carrying the translational and internal energy features due to their experiencing interactions governed by the relevant potential energy surfaces: often systems undergo nonadiabatic transitions back to the ground state without emitting light; the localization of the intersections and the nonadiabatic couplings are key factors to be studied in order to understand the dissociation mechanisms and the effective pathways of molecular fragmentation. An observed effect that can be attributed to the occurrence of nonadiabatic transitions is the evidence of alternative routes to molecular dissociation that appear to circumvent the minimum energy reaction path: the presence of these alternative pathways can be inferred by the properties of the energy distribution of the product fragments.

A striking feature is provided by roaming pathways that comprise a new class of reaction mechanisms, whose experimental fingerprints are slow and/or internally cold photodissociation products. The roaming atom mechanism has been found to mainly occur in the photodissociation of simple organic molecules (for a review see for example ref. 1–3 and references therein), *e.g.* formaldehyde,[4] acetaldehyde,[5] propionaldehyde,[6] methyl formate,[7,8] and nitrobenzene.[9]

The origin of roaming as a kind of dynamical isomerization or intramolecular rearrangement typically between a weakly bound pair of incipient radicals, is connected to high-lying regions of the electronic ground state surface, corresponding to molecular configurations far from minimum energy paths and transition state geometries. A typical condition for roaming to occur is a pronounced bond elongation up to a partial bond-cleavage in the direction of the dissociation into two radicals (*e.g.* the C–OCH3 bond of methyl formate), which corresponds to evolution of the system along a high-lying, small gradient region of the potential energy surface. This incipient radical dissociation can become the radical product channel, but when the excitation energy is sufficiently high to lead to further dissociation the result is multiple fragmentation (*e.g.* triple fragmentation for methyl formate). This common root of roaming and triple fragmentation pathways leads to interchanges of signatures as a function of the excitation energy, a fact amply documented for methyl formate[8] and acetaldehyde[11,12] (see also the discussion on this topic reported in *Faraday Discuss.* 157 (ref. 10)).

Previous work[8] has emphasized the role of conical intersections of potential energy surfaces as providing a general support for nonadiabatic transitions, channelling the evolution of systems in configurations characterized by highly stretched bonds.

 

This emerging picture of the roaming phenomenon, in accord with the view of measurable delay in the arrival of product fragment in the molecular channel and its unusual internal rotational energy distribution in CO ($\nu = 0$), is corroborated by quantum chemistry and molecular dynamics simulations, exemplified in comparisons extensively sampling initial conditions centred around the transition state and the tip of conical intersection.[8] The experimentally more demanding probes on CO ($\nu = 1$), reported and discussed in the following, provide additional insight regarding the vibrational degrees of freedom.

### 1.B. Plan of the paper

In this paper, the extensive set of novel measurements on methyl formate photodissociation has been specifically focused at the laser excitation wavelength value of 248 nm, which has been demonstrated as slightly below the threshold for the triple dissociation into H, CO, and $CH_3O$. These measurements of the internal energy, and of the images in space and time of the translational features of CO are accompanied by spectroscopic probing of the vibrational and rotational energy deposited in the HCO radical fragment from excited molecules, without however sufficient energy to further decompose, so to provide clear evidence of the roaming phenomenon, and in general of the internal energy distribution of excited molecules. Previously, the photodissociation of methyl formate $HCOOCH_3$ by laser excitation at 234 nm discovered signatures of a roaming mechanism by imaging of the CO products:[7] such a mechanism is inferred by the late arrival of products in time-of-flight measurements. The subsequent detailed investigations,[8] as a function of the excitation energy using a sequence of various laser wavelengths and also spectroscopic characterization of the fragments, H by ion imaging and HCO by Fourier transform infrared spectroscopy (FTIR, see Appendix 1), provided systematic data on the pathways followed by the system for producing radicals and specifically on the threshold for the opening of the breakdown in three fragments. Quantum mechanical electronic structure calculations and classical molecular dynamics simulations clarified the role of non-adiabatic transitions at a conical intersection of ground state and excited potential energy surfaces.

Focusing on CO ($\nu = 1$) from methyl formate dissociation, a clearer signature of roaming is obtained as the bimodal character of the kinetic energy distribution of fragments with respect to the ($\nu = 0$) case, where a definite characterization was prevented by the presence of another important mechanism that generates slow fragments, the CO production by triple fragmentation.[10,13]

In this combined experimental and theoretical study of the photodissociation of methyl formate in a subthreshold regime (see ref. 8) with respect to triple fragmentation, using laser excitation at 248 nm wavelength, fragmentation into CO and $CH_3OH$ has been characterized by an ion imaging technique, described in Sec. 2. The technique permits vibrational state-specific detection of CO ($\nu$): we report probes of population for $\nu = 1$ vibrational levels for $J = 2$, 7, and 11 rotational levels.

Theoretical approaches are described in Sec. 3. Available quantum chemical information at the conical intersection and main molecular configurations, particularly at the minimum energy path across the transition state and in the exit channels (molecular and radical, see ref. 8) are in Sec. 3.A. Sec. 3.B describes how we constructed the ground potential energy surface of methyl formate and its analytical representation suitable for classical trajectory simulations. Methods for the latter and for the phase space assessment of triple fragmentation features are given in Sec. 3.B. The results and their discussion are presented in Sec. 4, conclusions follow in Sec. 5. Two appendices provide complementary information on an experimental set up and on theoretical tools: in Appendix A.1, we describe the Fourier transform infrared apparatus; Appendix A.2 provides details on the representation of the ground potential energy surface.

## 2. Experimental: ion imaging

The kinetic energy distribution of rotationally and vibrationally selected photofragments of methyl formate (in Sec. 3.A we report a description of the dissociative channels) has been measured through an ion imaging technique (see for example ref. 5, 7 and 14–16 and references therein). The recorded images are two-dimensional projections of three-dimensional distributions with cylindrical symmetry around the photolysis laser polarization axis. The three-dimensional spatial distributions of the fragments can thus be reconstructed from the obtained two-dimensional images by efficient deconvolution methods. The speed distributions of the CO fragment were then converted into their corresponding translational energy distributions in the centre-of-mass frame. The imaging results were treated by deconvolution into separate contributions from the various dissociative channels. In this section, we give a detailed description of the experiment performed to probe the vibrationally excited CO ($\nu = 1$) fragment (Fig. 1) and show new acquired data for $J = 2$, 7, and 11 compared with classical trajectory simulation results. The measurements of higher $J$ and CO ($\nu = 2$) were attempted, but limited by a poor signal-to-noise ratio. We also recall a previous work on the detection of CO ($\nu = 0$) and H fragments. The FTIR technique, employed in accompanying work for the detection of photofragments, is described in Appendix 1.

The photofragment ion images were obtained using a velocity mapping ion imaging apparatus. The methyl formate vapor was premixed with He gas in 10% concentration. The sample gas was ejected at 900–1000 Torr stagnation pressure through a pulsed valve (General Valve Co.) with a 0.6 mm-diameter orifice operating synchronously with the laser pulses at 10 Hz and expanded into the source chamber. After passing through a 1 mm-diameter skimmer and a collimator, the molecular beam was intersected perpendicularly by a linearly polarized laser beam in a two-stage ion lens region. The skimmer was mounted 50 mm downstream from the nozzle to divide the source chamber from the main chamber. A KrF excimer laser (Compex 102, Lambda Physik) emitting at 248 nm with 300–350 μJ per pulse energy was used as the photolysis source. The laser radiation through an iris was linearly polarized perpendicular to the flight tube and parallel to the detection surface, and then softly focused to the leading edge of the skimmed beam to minimize cluster formation. A 308 nm XeCl excimer laser (LPX 200, Lambda Physik) pumped dye laser beam (PD 3000, Lambda Physik) was guided in the opposite direction to probe the CO fragments. Its output was frequency-doubled to emit at ~230 nm with ~180 μJ per pulse energy. Following the ester photolysis, the focused probe beam was applied in a 20 ns delay to detect CO ($\nu = 1$) by (2 + 1) resonance-enhanced multiphoton ionization (REMPI) method *via*

 

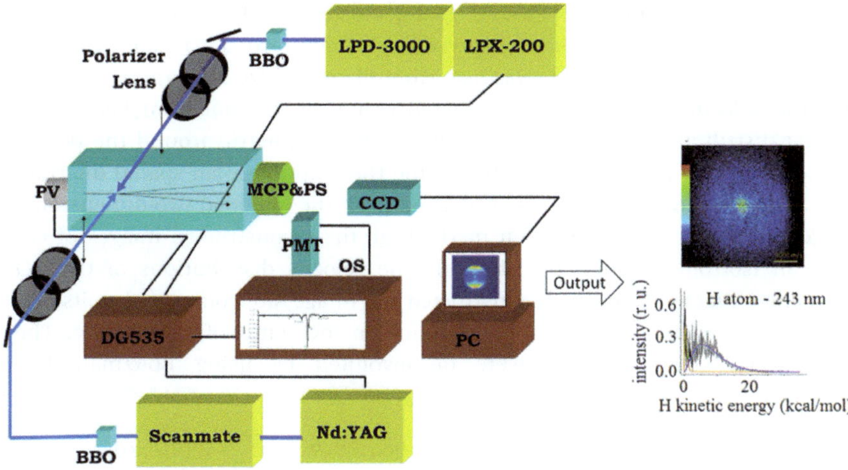


Fig. 1 Experimental apparatus for measurements of velocity mapping ion imaging. A Nd:YAG laser pumped dye laser operating with different laser dyes is used as the photolysis source, its output frequency is doubled by a BBO (β-barium borate) crystal. A 308 nm XeCl excimer laser (Lambda Physics LPX-200) pumping dye laser beam (LPD-3000) is guided in the opposite direction to probe the fragments (in the present case CO). In this case, the output is also frequency-doubled to emit at the requested wavelength and to ionize the product. The reactant precursor (in this case methyl formate) is prepared in a helium-seeded free-jet molecular beam (PV = pulse valve), the photofragments are then produced and ionized in the laser-molecular beam intersection region downstream. The ion-cloud expansion is mapped onto a two-stage microchannel plate (MCP) and a phosphorous screen (PS), then recorded by a charge coupled device (CCD) camera. All the ion signals without gate restriction are acquired by a photomultiplier tube (PMT) and transferred to a transient digitizer to display the time-of-flight mass spectrum. Results are displayed by an oscilloscope (OS) and a personal computer (PC). The 2-Dimensional image of the H atom and its deconvoluted plot, which enables the kinetic energy distribution of the photofragment, are shown in the panels on the right adapted from ref. 7.

a B $^1\Sigma^+ \leftarrow$ X $^1\Sigma^+$ two photon transition.[17] Laser pulse powers were adjusted to optimize the two-laser ion signal while minimizing the single-laser ion signal. The resulting CO ions were extracted and accelerated into a 36 cm-length field-free drift tube along the molecular beam direction. The ion-cloud expansion was mapped onto a two-stage microchannel plate and phosphor screen (FM3040, Galileo). The microchannel plate could be gated within a minimum duration of 90 ns for mass selection. The ion imaging on the phosphor screen was recorded by a charge coupled device camera (200XL4078, Pixelfly). The ion signal intensity was acquired by a photomultiplier tube and a transient digitizer was used to display the time-of-flight mass signal during the experiments.

Methyl formate was photolyzed at 248 nm and the CO ($\nu = 1$) fragment was probed at the selected rotational levels of $J = 2$, 7, and 11. Images at higher $J$ could not be acquired because of the poor signal-to-noise ratio; because of the low resolution especially at low rotational levels, the image at $J = 2$ might be contributed significantly to by the adjacent levels. Each image was an accumulation of 30 000–50 000 laser shots. The background images obtained with either photolyzing or ionizing pulses were both subtracted from raw images to eliminate the contributions from background gases, clusters, and the single-laser ion signal. The production of clusters, present as a non-resonant signal in a wide range of ionizing wavelengths, was also carefully minimized in sample preparation and selection of the laser-intercepted region of the molecular beam. The angular distribution of fragments is cylindrically symmetric around the polarization axis of the photolysis laser beam; the three-dimensional spatial distributions of the fragments can be reconstructed from the obtained two-dimensional images by the basis-set expansion method: all the reconstructed images of CO show an isotropic angular distribution. The velocity distributions of the CO fragment, extracted from the reconstructed ion image, are convoluted to its corresponding translational energy distribution in the centre-of-mass frame. The translational energy profiles were deconvoluted by using Boltzmann-like functions:

$$I_{\text{total}} = I_{\text{slow}} + I_{\text{fast}} = A(E_{\text{CO}})^{a_1}[\exp(a_2 E_{\text{CO}})] + B(E_{\text{CO}})^{b_1}[\exp(b_2 E_{\text{CO}})], \quad (1)$$

where the total intensity $I_{\text{total}}$ is supposed to be a sum of the intensity of a slow contribution $I_{\text{slow}}$ and the intensity of a fast contribution $I_{\text{fast}}$; the explicit form of $I_{\text{total}}$ (see eqn (1)) depends on the kinetic energy of CO, $E_{\text{CO}}$, on the weighting factors $A$ and $B$ and on the parameters $a_i$ and $b_i$ ($i = 1, 2$) that define the profile shape and are also connected to the rotational temperatures of the two distributions.

# 3. Theoretical tools

In Fig. 2, we report a schematic view of the potential energy landscape for methyl formate ground state $S_0$: a sketch is given of the minimum energy path, which joins the equilibrium geometry to the molecular products $CH_3OH + CO$, through a transition state; the figure also gives the feature of a conical intersection that connects the ground state to the excited state surface $S_1$ (details on the excited state $S_1$, on the conical intersection and on the modes excited as the molecule switches from $S_0$ to $S_1$ are given in ref. 18).

### 3.A. Quantum chemistry and mechanism

Qualitative features and quantitative assessments of the energy and structure of relevant configurations are based on ref. 18 (and private communications by its author), on ref. 19 and 20 and on our previous work.[7]

We introduce the following dissociative channels:

$$HCOOCH_3 + h\nu \rightarrow CH_3OH + CO\text{, molecular product channel} \quad (2)$$

$$HCOOCH_3 + h\nu \rightarrow CH_3O + HCO \rightarrow CH_3O + H + CO\text{, formyl radical channel} \quad (3)$$

$$HCOOCH_3 + h\nu \rightarrow COCH_3O + H\text{, methoxycarbonyl radical channel} \quad (4)$$

The choice to take into account only channels (2), (3), and (4) relies on experimental information and quantum chemical calculations within the range of excitation energies. The molecular channel (2) ($\Delta H = ca.$ 9 kcal mol$^{-1}$, taking into account the zero point energies of reagent and products) leads to the

 

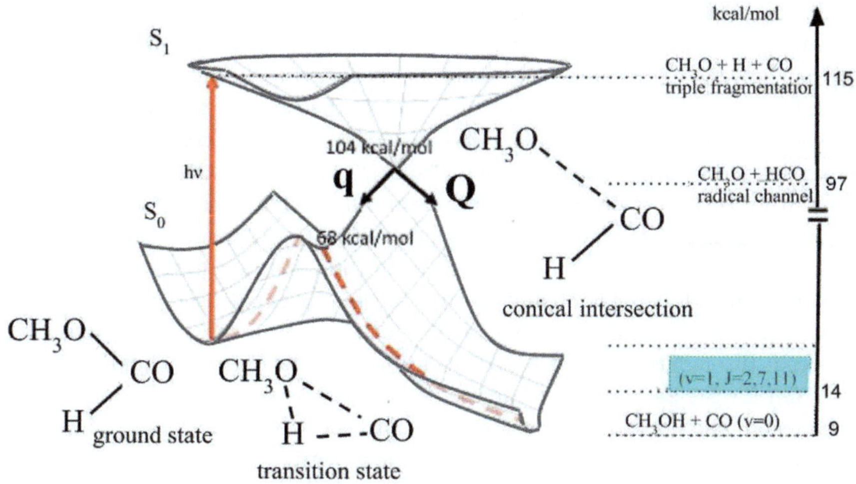


Fig. 2 A schematic view of the photodissociation pathways of methyl formate. The minimum energy path, from the ground state to the dissociated products $CH_3OH$ + CO passing through the transition state, is represented by the red-dashed line. From the ground state ($S_0$), the molecule is promoted to the excited state ($S_1$) through laser excitation at 248 nm. The very large vibronic coupling induces the transition to the ground state through the conical intersection giving $CH_3OH$ + CO products, while the triple dissociation channel, which yields $CH_3O$ + CO + H, is not accessible because its threshold is slightly higher than the energy provided by the laser. For example, vectors **q** and **Q** are shown pointing in the direction of increasing angle bending, and of the $CH_3O$-CO bond stretching, respectively. While the H − CO distance is nearly constant ($r = 0.11$ nm ca.) at the ground state, transition state and conical intersection, the CH3O − CO distance is $R = 0.13$, 0.18 and 0.23 nm ca., respectively The latter values corroborate the view that the main radical path responsible for the observed roaming originates from the elongation of the $CH_3O$–CO bond.

formation of $CH_3OH$ and CO. It can proceed directly through the minimum energy path (crossing a transition state, $\Delta H = ca.$ 68 kcal mol$^{-1}$) or through the roaming mechanism.

The formyl (3) and methoxycarbonyl (4) radical channels can be considered as two step processes, which finally yield the products: CO + H + $CH_3O$ ($COCH_3O$ can further dissociate to $CH_3O$ + CO). The threshold of these triple fragmentation processes has been estimated by quantum chemical calculations: in channel (3) the first dissociation (formation of $CH_3O$ + HCO products) requires a variation of enthalpy of *ca.* 97 kcal mol$^{-1}$, while for the final step (dissociation of HCO), $\Delta H$ is *ca.* 18 kcal mol$^{-1}$; in the case of the methoxycarbonyl radical channel, for the first step $\Delta H$ is *ca.* 100 kcal mol$^{-1}$ and for the second step $\Delta H$ is *ca.* 15 kcal mol$^{-1}$; hence, the threshold for both channel (3) and (4) is *ca.* 115 kcal mol$^{-1}$. Kinematic arguments attribute to channels (3) and (4) rotationally and translationally cold CO products and slow H fragments. It has been argued[8] that channel (4) can be excluded in the probed energy range not only because of an exit energy barrier but also because in configuration space, it lies far away from the configurations considered operative in the present situation. In addition, our simulations of the photodissociation of CO ($\nu = 0$) indicate that less than 3% of methyl formate dissociates generating methoxycarbonyl radicals at 248 nm.

 

Experimental and theoretical studies on aldehydes[4–6] revealed signatures of roaming in the CO product. Evidence is being accumulated that this mechanism typically involves elongated C–R bonds (where R is H for formaldehyde, $CH_3$ for acetaldehyde and $CH_3O$ for methyl formate) so that molecules become close to dissociated into radicals; the system is in a wide region of the configuration space characterized by a flat potential energy surface, where R abstracts the hydrogen atom from HCO and yields the molecular products (channel (2)).

In Fig. 2 and 3, we show two vectors **q** and **Q** (Fig. 2), centred at the conical intersection[21] placed close to the product region and pointing along the directions of stretching $CH_3O$–CO bond and opening the bending angle H–C–$OCH_3$. In such a way, the dissociation process can be illustrated as a three-body event that proceeds through the elongation along the **Q** vector: the $CH_3O$–CO bond length is 0.13 nm for the equilibrium structure of methyl formate, 0.18 nm for the transition state and 0.23 nm for the structure corresponding to the conical intersection, the **r** vector is about 0.10 nm and is almost constant for the three structures, while the H–C–$OCH_3$ angle is 109° for the equilibrium structure and 49° at the transition state. The simplification to a three-body problem is only for illustration, since in our model we also explicitly consider the internal rotovibrational degrees of freedom of CO.

The presence of a conical intersection near the product region at low enough energy, accessible in our experiment, suggests that the formyl radical channel should play a role in the dissociation of methyl formate. A highlight of the

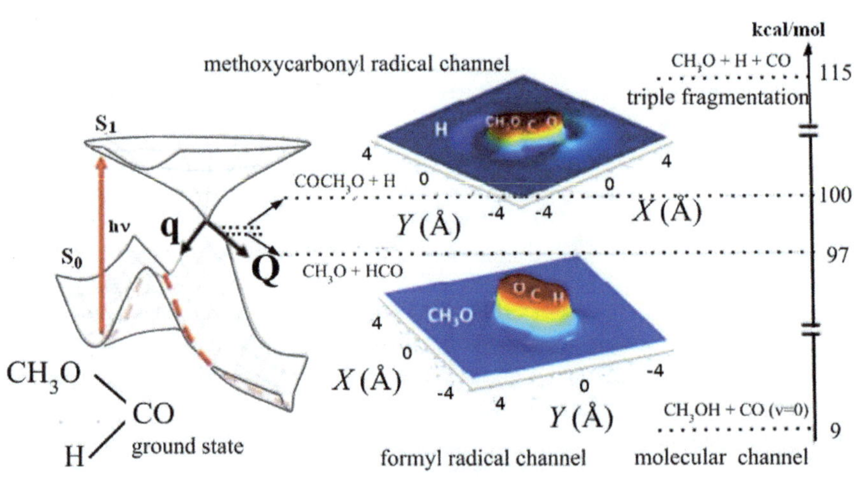


Fig. 3 We report a detail of the methoxycarbonyl and formyl radical channels. The illustrative tridimensional plots (on the right-hand side) show the interaction of H + $COCH_3O$ and $CH_3O$ + HCO as reduced three-body problems (the internal degrees of freedom of CO and of $CH_3O$ being considered as frozen); the origin of the Cartesian coordinates (*X*,*Y*) in each plot is the center of mass of the triatomic fragment, and the CO vector points in the positive direction of the *X*-axis. The geometry of each triatomic fragment is fixed at the equilibrium structure of $HCOOCH_3$, and the interaction energy (in kcal $mol^{-1}$) is taken with respect to the four-body dissociation (H + C + O + $CH_3O$) as zero. In the methoxycarbonyl radical channel, the **q** vector varies and the **Q** vector is fixed (see Fig. 2), while *vice versa* in the formyl radical channel case. This latter case corresponds to the elongation of the CO–$CH_3O$ bond, a situation which occurs in the structure of methyl formate at the conical intersection.

 

potential energy surface referred to the radical channels (channel (3) and channel (4)) after being funnelled through the conical intersection is shown in Fig. 3. As mentioned previously, the role of channel (4) can be considered negligible, because its configuration is far from the region of the conical intersection; on the contrary, channel (3) leads to the formation of radical HCO and $CH_3O$. Remarkably, the H–C–O angles in the methyl formate ground state minimum and transition state (as obtained from our mode potential energy surface, see Sec. 3.B) do not differ much, being 136° and 134° respectively, while no such information is available for the conical intersection geometry (the angle is missing in ref. 18, the sole thorough study of methyl formate ground and excited states). If the available energy is enough to overcome the triple dissociation threshold HCO dissociates to yield $CH_3O$ + H + CO, while if the available energy does not allow the system to reach a complete dissociation, $CH_3O$ remains in a flat weakly bound region to finally yield the molecular products $CH_3OH$ + CO (roaming mechanism). Our simulations indicate a fraction of *ca.* 0.15 for the methyl formate molecules that finally dissociate as HCO + $CH_30$ at 248 nm.

### 3.B. The ground state potential energy surface of methyl formate

Available full-dimensional *ab initio* quantum chemical information on methyl formate[18,19] is the starting point to build a reduced four-center ground state potential energy surface suitable for classical trajectory simulations. The reduction consists in taking the $CH_3O$ group as "frozen"; considering it as a spectator (although possibly available for energy disposal, with the internal energy equally distributed among its degrees of freedom). The potential energy surface of this effective four-center system has been generated in parametric form according to the many-body expansion procedure (see for example ref. 22), previously applied to several few-body molecules, including formaldehyde, of which the structure HCOH can be thought of as a prototype of the present system (for details of the application of many-body expansion on methyl formate see ref. 23). The potential energy function (available on request) has been constructed with the aid of additional *ab initio* calculations.[24] The same model potential energy surface has been used in simulations presented in ref. 7. The physically motivated many-body expansion method, used to build the potential energy surface presents definite advantages with respect to the numerical fitting of a huge amount of *ab initio* points calculated on grids. The calculation of an accurate single energy point requires a higher level of calculation when compared to the optimization of geometries, adopted in the present method: the calculation of the energy in the long-range region and for bond breaking and formation is in fact critical and requires a very high computational level. The ground state energy surface might also undergo strong coupling with various excited states even at the asymptotic limit. The advantage of this approach is thus to provide an efficient way to generate the global potential energy function with acceptable accuracy for this truncated model system.

The interaction potential of this effective four-body system is given by the following expansion as a sum of $n$-body contributions ($n = 1, 2, 3, 4$):

$$V_{\text{HCOM}} = V^{(2)}_{\text{CH}} + V^{(2)}_{\text{CO}} + V^{(2)}_{\text{OH}} + V^{(2)}_{\text{OM}} + V^{(2)}_{\text{HM}} + V^{(2)}_{\text{CM}} + V^{(3)}_{\text{HCO}} + V^{(3)}_{\text{MCO}} + V^{(3)}_{\text{HCM}} + V^{(3)}_{\text{HOM}} + V^{(4)}_{\text{HCOM}}, \quad (5)$$

where the superscripts in parenthesis identify the $n$-body interaction terms and subscripts denote the interacting atoms, or the "frozen" $CH_3O$ that here is referred to as M. The interaction energy at the asymptotic limit of the four single-body terms, $V_H^{(1)}$, $V_C^{(1)}$, $V_O^{(1)}$ and $V_M^{(1)}$, is set to zero. The force constants for each $n$-body term have been obtained by DFT calculations with a B3LYP functional and a 6-311++g(2df,2p) basis set.[24]

The interaction potentials of the 2-body terms $V^{(2)}$, are expressed as Morse potentials whose parameters can be readily determined *via* force constants, equilibrium geometries and potential well depths:

$$V^{(2)} = D_e[\exp(-2a(r - r_e)) - 2\exp(-a(r - r_e))], \tag{6}$$

where $a = \sqrt{\frac{k_e}{\mu}}$, $D_e$ is the dissociation energy, $r$ is the interatomic distance, $r_e$ is the interatomic distance at the equilibrium, $k_e$ is the force constant at the minimum of the potential well and $\mu$ is the reduced mass of the two-body term (see Table 1).

The interaction potential of the 3-body terms, $V^{(3)}$ is given in an analytical form:

$$V^{(3)}(R_1, R_2, R_3) = V^0\left(1 + \sum_i C_i\rho_i + \sum_{i\le j} C_{ij}\rho_i\rho_j + \ldots\right) \times \prod_i [1 - \tanh(\gamma_i\rho_i/2)], \tag{7}$$

where $\rho_i = R_i - R_i^{eq}$ and $i = 1$, 2, and 3. Given the coordinate at equilibrium geometry $R_i^{eq}$ of triatomic fragment ABC, the term $V_0$ can be calculated by the following expression:

$$V = V_{ABC}(R_1^{eq}, R_2^{eq}, R_3^{eq}) - V_C^{(1)} - V_B^{(1)} - V_A^{(1)} - V_{AC}^{(2)}(R_3^{eq}) - V_{AB}^{(2)}(R_1^{eq}) - V_{BC}^{(2)}(R_2^{eq}) \tag{8}$$

where $V_{ABC}(R_1^{eq}, R_2^{eq}, R_3^{eq})$ is the potential of the isolated system ABC and can be obtained, as in the case, by quantum chemical calculations; $V_{AC}^{(2)}(R_3^{eq})$, $V_{AB}^{(2)}(R_1^{eq})$, and $V_{BC}^{(2)}(R_2^{eq})$ are the two-body terms introduced in eqn (6) at the distance of equilibrium of the ABC system. For the specific case of methyl formate (HCOM), ABC corresponds to any of HCO, HOM, OCM, or HCM.

As done in ref. 22, we denote the first term of the right hand side of eqn (7) as the expansion function and the second term as the switch function. The expression coefficients $C_i$ of the expansion function are correlated to the derivatives of the potential energy (see Appendix 2), $R_i$ are the normal coordinates, $R_i^{eq}$ is the equilibrium position of the $n$-body term, and $\gamma_I$, are adjustable parameters in the

Table 1 Morse parameters for 2-body terms

| | C–O | C–M | O–M | C–H | O–H | M–H |
|---|---|---|---|---|---|---|
| $D_e$ (eV) | 11.224 | 3.959 | 2.231 | 3.648 | 4.632 | 4.477 |
| $r_e$ (nm) | 0.113 | 0.134 | 0.132 | 0.112 | 0.097 | 0.095 |
| $k_e$ (a. u.) | 2.4649 | 0.233 | 0.451 | 0.330 | 0.555 | 0.567 |

 

hyperbolic tangent functions (second term). The divergence of the expansion function at large distance can be compensated by the switch function, which behaves like an exponential decay along the coordinates but gives a finite value at the origin.

The interaction potential of the 4-body term is also expressed in analytic form, not given explicitly here. An example is given in ref. 22. Further details on the method to obtain the potential energy surface are reported in Appendix 2.

### 3.C. Classical trajectory simulations

The population of the CO products in the $\nu = 1$ vibrational state is expected to be much smaller than the population of CO in the vibrational ground state.

Indeed, in the experiments, the imaging of CO ($\nu = 1$) required much longer accumulation times than CO ($\nu = 0$) to obtain reliable results and, in the same manner, a consistently large number of trajectories were necessary to obtain an accurate statistical sampling for the CO ($\nu = 1$) dynamics. For the theoretical modeling of the photodissociation experiment on CO ($\nu = 1$), whose results are reported here, around $3 \times 10^6$ trajectories were run corresponding to a set of given initial geometries of methyl formate. A set of preliminary calculations was performed to establish the best initial geometries, in which the initial configurations were generated according to the $S_1/S_0$ conical intersection configuration of the methyl formate as it is known from ref. 18, but altering, by a random sampling, the HCO–$OCH_3$ bond distance to make it vary between 0.21 nm and 0.30 nm. This led us to retain the initial HCO–$OCH_3$ distance of *ca.* 0.23 nm as the best initial value for the trajectories. The rest of the initial conditions were generated according to the following scheme: the total internal energy $E$ of the molecule was given a fixed value consistent with the experimental excitation wavelength (115.29 kcal mol$^{-1}$ for 248 nm); the initial configuration corresponding to a given potential energy $V$ was assigned to the molecule, and the energy difference ($E - V$) randomly distributed among the internal degrees of freedom. Trajectories were subjected to the constraint of zero total angular and linear momentum, consistent with the rotationally cold distribution in the beam of molecules to be photoexcited. Alternative initial geometries were also tested, in particular around the minimum energy configuration and the transition state configuration but results were largely not satisfactory, confirming the role of the conical intersection region in the mechanism of the methyl formate photodissociation.

## 4. Results and discussion

In Fig. 4, we report the imaging results of CO ($\nu = 1; J = 2, 7, 11$). The CO ($\nu = 1$) kinetic energy distributions (green curve) are deconvoluted by fitting data into a bimodal Boltzmann function in such a way that deconvoluted different contributions can be disentangled: the minimum energy path and roaming. These data were compared with those obtained by classical trajectory simulations (see right-hand side panels in the figure).

The sharp cold component peaking at 2 kcal mol$^{-1}$ (blue curve) is attributed to the roaming process, while the broad component peaking at 5–6 kcal mol$^{-1}$ (red curve) is ascribed to the minimum energy path mechanism. The normalized yields of the two mechanisms were evaluated by means of the integrated area of

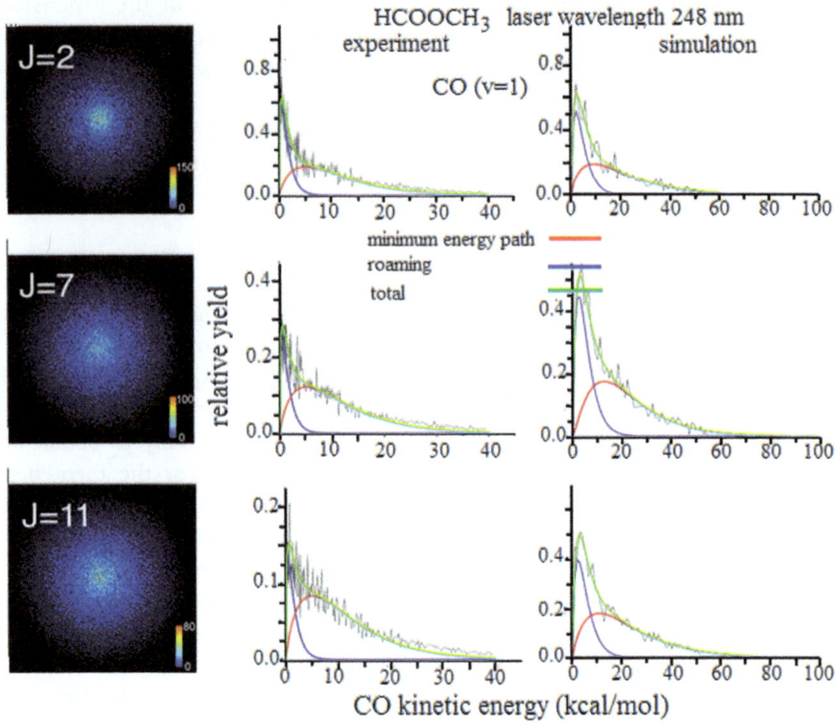


Fig. 4 The figure shows the image (left panels) and the extracted kinetic energy distributions of vibrational excited CO ($v = 1$) from photodissociation of methyl formate at 248 nm, for the selected rotational levels $J = 2$, 7, and 11. In the central panels, fits of the experimental data (green curve) and deconvolution of the slow (in blue) and fast (in red) components are reported. Panels on the right-hand side show the corresponding results as obtained by classical trajectory simulations. The normalized yields by the integration of the area of the deconvoluted profiles (central panels) are: for the roaming mechanism; $0.29 \pm 0.07$ for $J = 2$, $0.20 \pm 0.09$ for $J = 7$, and $0.17 \pm 0.04$ for $J = 11$. For the minimum energy path mechanism; $0.71 \pm 0.12$ for $J = 2$; $0.80 \pm 0.20$ for $J = 7$, and $0.83 \pm 0.08$ for $J = 11$.

the deconvoluted profiles (see Fig. 4); the roaming fraction tends to decrease from 0.29 to 0.17, for $J$ increasing from 2 to 11, while the minimum energy path fraction increases from 0.71 to 0.83. Under the assumption of a Boltzmann distribution, the rotational temperatures of the CO products from each pathway, determined from the $J$-dependence of the branching fractions, may be estimated to be 200 K for roaming and 420 K for the molecular channel. The total roaming fraction is on average $0.2 \pm 0.06$, as evaluated from the available ratios for $J = 2, 7, 11$, with the roaming CO product appearing rotationally cold and with a small translational energy. A previous measurement with time-resolved FTIR[25] emission spectroscopy, gave for roaming a rotational temperature of 470 K and for the minimum energy path, 1050 K, and a roaming fraction of $0.3 \pm 0.04$ consistent with the imaging result. The discrepancy of the rotational temperatures could be due to a difference in the experimental conditions: the parent molecules are cooled down in the molecular beam experiment, while the FTIR measurements are carried out at room temperature. In addition, the system might not completely reach thermal equilibrium in the molecular beam, although we assumed the Boltzmann

 

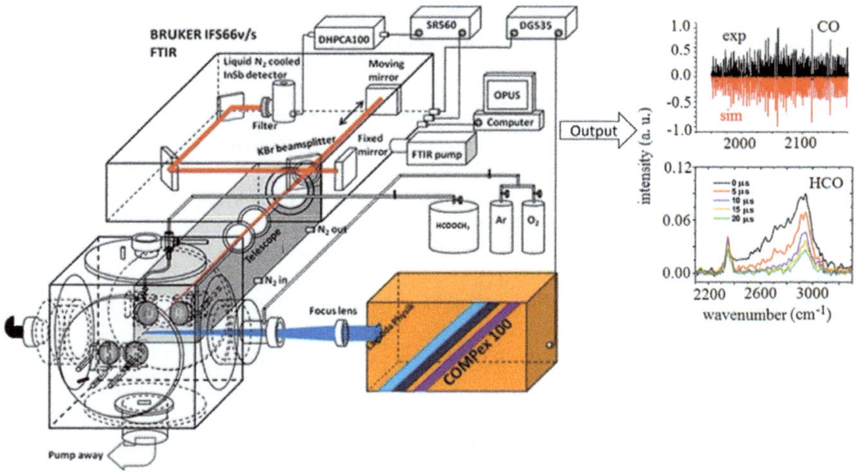


Fig. 5 Experimental apparatus for FTIR emission measurements. The photodissociation source is a KrF excimer laser (Lambda Physik COMPex 100); the incident UV laser beam is focused into a stainless steel gas chamber with a spherical lens. The reactant precursor (in this case methyl formate) flows into the chamber along an inlet intruding from the top of the chamber, while the Ar or $O_2$ collision partners are injected close to the rear window of the chamber. The fragment emission signals are collected in a perpendicular direction to the laser beam *via* a nitrogen-purged light path which is composed of a pair of focal lenses (telescope). The IR emission light is introduced into the FT-IR spectrometer (IFS 66v/S, Bruker). The transmitted signal is monitored with an InSb detector cooled at 77 K, the obtained interferograms can be Fourier transformed to give rise to a series of time-resolved spectra *via* OPUSTM software. DHPCA100: DHPCA100 current preamplifier. SR560: voltage preamplifier (Stanford Research Systems, SR560). DG535: time-delay generator (Stanford Research Systems, DG535). As an example of the output, we report (right panels) the spectra with different resolution of the CO and HCO fragments.

Table 2 The expansion coefficients of MCO 3-body term (distances are in nm, energies in eV, force constants of internal coordinate in atomic units)

| MCO | $V_0$ | $C_1$ | $C_2$ | $C_3$ |
|---|---|---|---|---|
| | 3.033 | 1.978804 | 5.689299 | 1.341308 |
| | $C_0$ | $C_{11}$ | $C_{22}$ | $C_{33}$ |
| | 1.0 | 5.840956 | 42.290240 | 7.390537 |
| $R_{eq1}$(CM) | $\gamma_1$ | $C_{12}$ | $C_{13}$ | $C_{23}$ |
| 1.32747 | 3.796384 | 14.924260 | 4.628371 | 8.436104 |
| $R_{eq2}$(CO) | $\gamma_2$ | $F_{11}$ | $F_{22}$ | $F_{aa}$ |
| 1.17985 | 17.742849 | 0.375867 | 0.921841 | 0.307491 |
| $R_{eq3}$(OM) | $\gamma_3$ | $F_{12}$ | $F_{1a}$ | $F_{2a}$ |
| 2.25006 | 3.151990 | 0.106342 | 0.0610 | 0.048397 |

equation to estimate the rotational distribution: CO fragments are ionized in a short time, *ca.* 20 ns, right after the photolysis laser radiation, while the response time of FTIR measurement is of the order of $10^{-6}$ s.

The results of the simulations show a very good qualitative agreement, reproducing bimodality in the kinetic energy distribution of both excited ($\nu = 1$) and ground state CO products. Disagreements occur in the energy range covered

Table 3 The expansion coefficients of the HOM 3-body term (distances are in nm, energies in eV, force constants of internal coordinate in atomic units)

| HOM | $V_0$ | $C_1$ | $C_2$ | $C_3$ |
|---|---|---|---|---|
| | 0.1733 | −4.307835 | 1.369019 | 2.123089 |
| | $C_0$ | $C_{11}$ | $C_{22}$ | $C_{33}$ |
| | 1.0 | 9.491546 | −13.384507 | 22.562432 |
| $R_{eq1}$(OM) | $\gamma_1$ | $C_{12}$ | $C_{13}$ | $C_{23}$ |
| 2.20174 | 0.854314 | −2.215224 | −35.588642 | 3.303192 |
| $R_{eq2}$(OH) | $\gamma_2$ | $F_{11}$ | $F_{22}$ | $F_{aa}$ |
| 0.96919 | 1.602842 | 0.015698 | 0.5108 | 0.01715 |
| $R_{eq3}$(HM) | $\gamma_3$ | $F_{12}$ | $F_{1a}$ | $F_{2a}$ |
| 2.3398 | 12.218217 | 0.00177 | −1.34873 | −0.003325 |

by the CO products, and appear to be larger for the simulated results. This is probably due to the increased amount of energy available for rotations and translations in our reduced model, where the internal degrees of freedom of the $CH_3O$ group (frozen) are not available for energy disposal.

The $\nu = 1$ results can be further discussed with reference to the CO ($\nu = 0$) results. In ref. 8, CO fragments ($J = 10$, 14, 19, and 24) in their ground vibrational state were obtained by photodissociation at 225, 234, 250, and 255 nm, and probed by ion imaging. The obtained kinetic energy distribution curve appears to be consistent with a bimodal distribution, made up of a slower sharp component and a broad and faster one. Phase space theory calculations and dynamics simulations indicate that at energies above the triple fragmentation threshold (estimated at 248 nm), this channel plays a significant role in contributing to the slow peak in the deconvolution of the kinetic energy distribution.

For completeness, we cite results from measurements at higher energies where triple dissociation is open: the kinetic energy distribution of the H atom (see ref. 8) was measured through ion imaging, for the photodissociation of methyl formate at 243 nm (above the estimated threshold for triple dissociation). In Fig. 7 of ref. 8, we reported a fit of the experimental data and the corresponding deconvolution into slow and fast components. The slow component must originate from HCO spontaneous decomposition. According to calculations in ref. 7, the radical channel lies 97.1 kcal mol$^{-1}$ above the methyl formate ground state global

Table 4 The expansion coefficients of HCM 3-body term (distances are in nm, energies in eV, force constants of internal coordinate in atomic units)

| HCM | $V_0$ | $C_1$ | $C_2$ | $C_3$ |
|---|---|---|---|---|
| | 0.2223 | 3.793969 | 7.558600 | −3.578366 |
| | $C_0$ | $C_{11}$ | $C_{22}$ | $C_{33}$ |
| | 1.0 | 51.922756 | −11.434842 | 42.887273 |
| $R_{eq1}$(CM) | $\gamma_1$ | $C_{12}$ | $C_{13}$ | $C_{23}$ |
| 1.3197 | 3.691497 | 77.886641 | −66.760355 | −60.012112 |
| $R_{eq2}$(CH) | $\gamma_2$ | $F_{11}$ | $F_{22}$ | $F_{aa}$ |
| 1.08078 | 2.315571 | 0.46092 | 0.3270 | 0.1952 |
| $R_{eq3}$(HM) | $\gamma_3$ | $F_{12}$ | $F_{1a}$ | $F_{2a}$ |
| 2.12 | 3.298011 | 0.141 | −0.14488 | −0.15186 |

 

Table 5 The expansion coefficients of HCOM 4-body interaction term (distances are in nm, energies in eV, force constants of internal coordinate in atomic units)

| HCOM | $V_0$ | $C_1$ | $C_2$ | $C_3$ |
|---|---|---|---|---|
| | −0.606778 | 9.945154 | 17.127433 | 55.888183 |
| | $C_0$ | $C_4$ | $C_5$ | $C_6$ |
| | 1.0 | −20.358639 | 2.354528 | −8.842246 |
| $R_{eq1}$(CH) | $\gamma_1$ | $C_{11}$ | $C_{22}$ | $C_{33}$ |
| 1.096720 | 17.187900 | 59.811865 | 147.350834 | 1199.089874 |
| $R_{eq2}$(CM) | $\gamma_2$ | $C_{44}$ | $C_{55}$ | $C_{66}$ |
| 1.33720 | 18.029200 | 197.839527 | −14.560888 | 22.429016 |
| $R_{eq3}$(CO) | $\gamma_3$ | $C_{12}$ | $C_{13}$ | $C_{14}$ |
| 1.199260 | 36.923900 | 132.489306 | 454.087857 | −139.150904 |
| $R_{eq4}$(CM) | $\gamma_4$ | $C_{15}$ | $C_{16}$ | $C_{23}$ |
| 2.036620 | −22.909700 | 18.663936 | −59.772061 | 619.211472 |
| $R_{eq5}$(OM) | $\gamma_5$ | $C_{24}$ | $C_{25}$ | $C_{26}$ |
| 2.259870 | −8.012290 | −264.525530 | 36.965650 | −70.453209 |
| $R_{eq6}$(HM) | $\gamma_6$ | $C_{34}$ | $C_{35}$ | $C_{36}$ |
| 1.988260 | −2.651190 | −761.097790 | −70.045682 | −202.258401 |
| | | $C_{45}$ | $C_{46}$ | $C_{56}$ |
| | | −14.482735 | −87.086552 | 11.589886 |
| $F_{11}$ | $F_{22}$ | $F_{33}$ | $F_{44}$ | $F_{55}$ |
| 0.33613 | 0.16776 | 0.31924 | −0.32005 | $-2.3977 \times 10^{-2}$ |
| $F_{66}$ | $F_{12}$ | $F_{13}$ | $F_{14}$ | $F_{15}$ |
| −0.11238 | 0.49588 | 0.21452 | $-7.445 \times 10^{-2}$ | −0.30104 |
| $F_{16}$ | $F_{23}$ | $F_{24}$ | $F_{25}$ | $F_{26}$ |
| −0.12487 | 0.70676 | −0.27129 | −0.22727 | $-6.6139 \times 10^{-2}$ |
| $F_{34}$ | $F_{35}$ | $F_{36}$ | $F_{45}$ | $F_{46}$ |
| 0.36027 | $1.4466 \times 10^{-1}$ | $2.89012 \times 10^{-1}$ | 0.33168 | $9.61107 \times 10^{-2}$ |
| $F_{56}$ | | | | |
| 0.17254 | | | | |

minimum energy, while the energy required for dissociation of HCO at its ground electronic state is *ca.* 17.9 kcal mol$^{-1}$, so we estimate a total available energy of *ca.* 2.6 kcal mol$^{-1}$, which consistently falls within the experimentally probed range. The broad component is assigned to the two body fragmentation H + $COCH_3O$ (channel 3, see Sec. 3.A).

# 5. Conclusions

The exploration of alternative roads that open to molecules with sufficient energy to yield different products is a widely debated subject in the scientific community. Ample discussion on this topic is reported in the proceedings of *Faraday Discuss.* 157, held in Assisi (Perugia, Italy), June 25–27 2012,[10] where it was specifically argued by a kinematic approach that slow product fragments can be due to multiple dissociation processes (specifically, triple fragmentation).

In this paper, we reported an overview on the study of photodissociative channels of methyl formate. We presented extensive combined experimental and theoretical studies of methyl formate photodissociation in the 225–255 nm range, where the presence of slow and fast products, which are manifestations of minimum energy path and roaming-mediated mechanisms, established the role of a roaming pathway at these energies. The importance of the triple fragmentation mechanism, whose possible role in the photodissociation of methyl formate was just suggested in the *Faraday Discuss.* 157, has been assessed, a threshold for the opening of the corresponding channel established, and it was demonstrated that roaming and triple fragmentation are alternative mechanisms responsible for rotationally and translationally cold CO products.

We also reported new acquired data on the probe of vibrationally excited CO ($\nu = 1$) for excitation energy at 248 nm, which was slightly lower than the triple fragmentation threshold.

In perspective, a complete study on the photodissociation of methyl formate requires a characterization of the mechanisms also for vibrationally excited molecules. Work in progress using an FTIR technique (see Appendix 1) accompanied by computational tools should permit us to determine branching ratios and rotational distributions for the various vibrational levels.[25]

# Appendix

## 1. Time-resolved Fourier-transform infrared emission spectroscopy

In this appendix, we report alternative techniques, not employed in the present work to observe evidences of the roaming mechanism.

Houston and Kable[26] employed laser-induced fluorescence in the comparable case of the photodissociation of acetaldehyde to CO and $CH_4$, to resolve the rotational states of the vibrationally selected CO products.

In the photodissociation of methyl formate,[8] Fourier transform infrared (FTIR) emission spectroscopy[27–30] (see the apparatus in Fig. 5) was used to probe the formation of HCO fragments. As discussed in Sec. 4, the presence of a slow peak ascribable to triple fragmentation at 243 nm in the H atom images (see Sec. 3.A for the arguments to exclude channel (4)) demonstrates that this channel is open at the shorter wavelengths considered in the images of the CO fragments (225 and 240 nm), indicating its contribution to the very sharp and slow peak of the deconvoluted kinetic energy distributions. The threshold for the opening of the triple fragmentation channel (*ca.* 115 kcal mol$^{-1}$) was then indirectly confirmed by comparison of the time-resolved FTIR emission spectra of the HCO fragments after photodissociation of methyl formate at 248 nm. The spectrum was compared with the HCO spectrum from the photodissociation of the acetaldehyde at 308 nm and both yielded a similar and consistent decay time, between 5–10 μs, under otherwise identical conditions. In contrast, the acetaldehyde photolysis gives, in the FTIR experiment at 248 nm, an HCO fragment lifetime lower than 5 μs, indicating efficient decomposition. Since the HCO fragment from acetaldehyde photodissociation at 308 nm does not have enough energy to decompose (see ref. 7, 11 and 12), we conclude that the HCO obtained from methyl formate at 248 nm should not undergo significant decomposition, taking the corresponding irradiation energy value as a threshold for this molecule (115.3 kcal mol$^{-1}$).

## 2. Representation of the ground state reduced-dimension potential energy surface of methyl formate

In this appendix, we report details on the method used to build the reduced-dimension potential energy surface of methyl formate, needed for the molecular dynamics simulations.

 

*Force constants*. As mentioned in the main text, the required geometric factors and spectroscopic data of isolated 3- and 4-body terms, in internal coordinates (see Tables 2–5) have been obtained by their equilibrium geometries, optimized *via* a DFT method (B3LYP/6-311g(2df,2p)) by using Gaussian 03 software. The relationship between the internal coordinates ($r_1$, $r_2$, $\alpha$) and normal coordinates ($R_1$, $R_2$, $R_3$) and the formulas to calculate the force constants are the following:

The calculated harmonic force constant can be expressed in interatomic displacements:

$$\tilde{V}^{(3)}_{\mathrm{MCO}}(r_1, r_2, a) \equiv V^{(3)}_{\mathrm{MCO}}(R_1, R_2, R_3) = V_{\mathrm{MCO}}(R_1, R_2, R_3) - V^{(2)}_{\mathrm{CM}}(R_1) - V^{(2)}_{\mathrm{CO}}(R_2) - V^{(2)}_{\mathrm{OM}}(R_3)$$

$$f_{ij} = \left(\frac{\partial^2 V}{\partial R_i \partial R_j}\right)_{\mathrm{eq}}, \quad F_{ij} = \left(\frac{\partial^2 \tilde{V}}{\partial r_i \partial r_j}\right)_{\mathrm{eq}}$$

$$F_{11} = f_{11} + 2f_{13}\left(\frac{\partial R_3}{\partial r_1}\right) + f_{33}\left(\frac{\partial R_3}{\partial r_1}\right)^2 + \left(\frac{\partial V}{\partial R_3}\right)_{\mathrm{eq}}\left(\frac{\partial^2 R_3}{\partial r_1^2}\right)$$

$$F_{22} = f_{22} + 2f_{23}\left(\frac{\partial R_3}{\partial r_2}\right) + f_{33}\left(\frac{\partial R_3}{\partial r_2}\right)^2 + \left(\frac{\partial V}{\partial R_3}\right)_{\mathrm{eq}}\left(\frac{\partial^2 R_3}{\partial r_2^2}\right)$$

$$F_{aa} = f_{33}\left(\frac{\partial^2 R_3}{\partial a^2}\right) + \left(\frac{\partial V}{\partial R_3}\right)_{\mathrm{eq}}^2\left(\frac{\partial^2 R_3}{\partial a^2}\right)$$

$$F_{12} = f_{12} + f_{13}\left(\frac{\partial R_3}{\partial r_2}\right) + f_{23}\left(\frac{\partial R_3}{\partial r_1}\right) + f_{33}\left(\frac{\partial R_3}{\partial r_1}\right)\left(\frac{\partial R_3}{\partial r_2}\right) + \left(\frac{\partial V}{\partial R_3}\right)_{\mathrm{eq}}\left(\frac{\partial^2 R_3}{\partial r_1 \partial r_2}\right)$$

$$F_{1a} = f_{13}\left(\frac{\partial R_3}{\partial a}\right) + f_{33}\left(\frac{\partial R_3}{\partial r_1}\right)\left(\frac{\partial R_3}{\partial a}\right) + \left(\frac{\partial V}{\partial R_3}\right)_{\mathrm{eq}}\left(\frac{\partial^2 R_3}{\partial r_1 \partial a}\right)$$

$$F_{2a} = f_{23}\left(\frac{\partial R_3}{\partial a}\right) + f_{33}\left(\frac{\partial R_3}{\partial r_2}\right)\left(\frac{\partial R_3}{\partial a}\right) + \left(\frac{\partial V}{\partial R_3}\right)_{\mathrm{eq}}\left(\frac{\partial^2 R_3}{\partial r_2 \partial a}\right)$$

$$f_{11} = F_{11}$$

$$f_{22} = F_{22}$$

$$f_{33} = \left(F_{33} - \left(\frac{\partial V}{\partial R_3}\right)_{\text{eq}}\left(\frac{\partial^2 R_3}{\partial a^2}\right)\right) \Big/ \left(\frac{\partial R_3}{\partial a}\right)^2$$

$$f_{13} = \left(F_{1a} - \left(\frac{\partial V}{\partial R_3}\right)_{\text{eq}}\left(\frac{\partial^2 R_3}{\partial r_1 \partial a}\right)\right) \Big/ \left(\frac{\partial R_3}{\partial a}\right) - f_{33}\left(\frac{\partial R_3}{\partial r_1}\right)$$

$$f_{23} = \left(F_{2a} - \left(\frac{\partial V}{\partial R_3}\right)_{\text{eq}}\left(\frac{\partial^2 R_3}{\partial r_2 \partial a}\right)\right) \Big/ \left(\frac{\partial R_3}{\partial a}\right) - f_{33}\left(\frac{\partial R_3}{\partial r_2}\right)$$

$$f_{12} = F_{12} - f_{13}\left(\frac{\partial R_3}{\partial r_2}\right) - f_{23}\left(\frac{\partial R_3}{\partial r_1}\right) - f_{33}\left(\frac{\partial R_3}{\partial r_1}\right)\left(\frac{\partial R_3}{\partial r_2}\right) - \left(\frac{\partial V}{\partial R_3}\right)_{\text{eq}}\left(\frac{\partial^2 R_3}{\partial r_1 \partial r_2}\right)$$

The expansion coefficients of the analytical function $P$ (see eqn (7) and Sec. 3.B) can be evaluated, up to quadratic terms at this moment, *via* the relationship between the coefficients and derivatives of potential. The derivatives of the range function $T$ (see eqn (7) and Sec. 3.B) can be expressed as:

$$T = [1 - \tanh(\gamma S_i/2)] = 1 - \tanh(\gamma(R_i - R_i^{\text{eq}})/2)$$

$$= 1 - \left(\frac{e^{\gamma S_i} - 1}{e^{\gamma S_i} + 1}\right) = \left(\frac{2}{e^{\gamma S_i} + 1}\right)$$

$$\left(\frac{\partial T^{-1}}{\partial R_i}\right)_{\text{eq}} = \left(\frac{\gamma\, e^{\gamma S_i}}{2}\left(\frac{\partial(R_i - R_i^{\text{eq}})}{\partial R_i}\right)\right)_{\text{eq}} = \frac{\gamma}{2}$$

The following equation can be treated as the starting point for obtaining the expansion coefficients:

$$\left(\frac{\partial P}{\partial R_i}\right)_{\text{eq}} = \left(V_{\text{ABC}}{}^{(3)}\frac{\partial T^{-1}}{\partial R_i} + T^{-1}\frac{\partial V_{\text{ABC}}^{(3)}}{\partial R_i}\right)_{\text{eq}} = V_0\frac{\gamma}{2\sqrt{3}} + \left(\frac{\partial V_{\text{ABC}}^{(3)}}{\partial R_i}\right)_{\text{eq}}$$

Analytical forms of the coefficients are listed below (up to quadratic terms):

$$g_i = \frac{\gamma}{2}$$

$$G_{ij} = V_0{}^{-1}\left(\frac{\partial^2 V_{\text{ABC}}^{(3)}}{\partial R_i \partial R_j}\right)_{\text{eq}}$$

and so on

 

$$\left(\frac{\partial V_{\mathrm{ABC}}^{(3)}}{\partial R_i}\right)_{\mathrm{eq}} = \left(\frac{\partial V_{\mathrm{ABC}}}{\partial R_i}\right)_{\mathrm{eq}} - \left(\frac{\partial \sum_j V^{(2)}(R_j)}{\partial R_j}\right)_{\mathrm{eq}} = -\left(\frac{\partial V^{(2)} R_j}{\partial R_i}\right)_{\mathrm{eq}}$$

Linear terms:

$$C_i = g_i + G_i$$

$$C_1 = \frac{\gamma_1}{2} - \frac{1}{V_0}\left(\frac{\mathrm{d}V_{\mathrm{AB}}^{(2)}}{\mathrm{d}R_1}\right)_{\mathrm{eq}}, \quad C_2 = \frac{\gamma_2}{2} - \frac{1}{V_0}\left(\frac{\mathrm{d}V_{\mathrm{BC}}^{(2)}}{\mathrm{d}R_2}\right)_{\mathrm{eq}}, \quad C_3 = \frac{\gamma_3}{2} - \frac{1}{V_0}\left(\frac{\mathrm{d}V_{\mathrm{AC}}^{(2)}}{\mathrm{d}R_3}\right)_{\mathrm{eq}}$$

Quadratic terms:

$$C_{ii} = g_i^2 + g_i G_i + G_{ii}/2$$

$$C_{ij} = g_i g_j + g_i G_j + g_j G_i + G_{ij}$$

$$C_{11} = \frac{{\gamma_1}^2}{4} - \left(\frac{\gamma_1}{2V_0}\right)\left(\frac{\partial V_{\mathrm{CM}}^{(2)}}{\partial R_1}\right)_{\mathrm{eq}} + \left(f_{11} - \left(\frac{\partial^2 V_{\mathrm{CM}}^{(2)}}{\partial {R_1}^2}\right)_{\mathrm{eq}}\right)\Big/ 2V_0$$

$$C_{22} = \frac{{\gamma_2}^2}{4} - \left(\frac{\gamma_2}{2V_0}\right)\left(\frac{\partial V_{\mathrm{CO}}^{(2)}}{\partial R_2}\right)_{\mathrm{eq}} + \left(f_{22} - \left(\frac{\partial^2 V_{\mathrm{CO}}^{(2)}}{\partial {R_2}^2}\right)_{\mathrm{eq}}\right)\Big/ 2V_0$$

$$C_{33} = \frac{{\gamma_3}^2}{4} - \left(\frac{\gamma_3}{2V_0}\right)\left(\frac{\partial V_{\mathrm{OM}}^{(2)}}{\partial R_3}\right)_{\mathrm{eq}} + \left(f_{33} - \left(\frac{\partial^2 V_{\mathrm{OM}}^{(2)}}{\partial {R_3}^2}\right)_{\mathrm{eq}}\right)\Big/ 2V_0$$

$$C_{12} = \frac{\gamma_1\gamma_2}{4} - \left(\frac{1}{2V_0}\right)\left[\gamma_1\left(\frac{\partial V_{\mathrm{CM}}^{(2)}}{\partial R_1}\right)_{\mathrm{eq}} + \gamma_2\left(\frac{\partial V_{\mathrm{CO}}^{(2)}}{\partial R_2}\right)_{\mathrm{eq}}\right] + \frac{f_{12}}{V_0}$$

$$C_{13} = \frac{\gamma_1\gamma_3}{4} - \left(\frac{1}{2V_0}\right)\left[\gamma_1\left(\frac{\partial V_{\mathrm{CM}}^{(2)}}{\partial R_1}\right)_{\mathrm{eq}} + \gamma_3\left(\frac{\partial V_{\mathrm{OM}}^{(2)}}{\partial R_3}\right)_{\mathrm{eq}}\right] + \frac{f_{13}}{V_0}$$

$$C_{23} = \frac{\gamma_2\gamma_3}{4} - \left(\frac{1}{2V_0}\right)\left[\gamma_2\left(\frac{\partial V_{\mathrm{CO}}^{(2)}}{\partial R_2}\right)_{\mathrm{eq}} + \gamma_3\left(\frac{\partial V_{\mathrm{OM}}^{(2)}}{\partial R_3}\right)_{\mathrm{eq}}\right] + \frac{f_{23}}{V_0}$$

The evaluated coefficients are listed in Tables 2–5. Note that in all the 3-body terms, none has more than one minimum except the term for HCO. H can attach to either the O-atom end or C-atom end. Due to the limited amount of data, the expansion coefficients are evaluated up to quadratic terms.

After obtaining the 3-body and 4-body force constants from *ab initio* calculations or literature data, the undetermined parameters ($\gamma$) can be optimized by following the steps reported below:

(i) Initial values of $\gamma$ in the 3-body terms are chosen according to *ab initio* 3-body interaction energies and geometries;

(ii) The $\gamma$ parameters of the 3-body terms are adjusted manually according to the *ab initio* 4-body geometry and behaviour of the radical dissociation channels HCOM $\rightarrow$ HCO + M and HCOM $\rightarrow$ H + COM;

(iii) The $\gamma$ parameters of the 3-body terms are optimized according to the *ab initio* 4-body geometry (without the 4-body term);

(iv) $\gamma$ of the 4-body term is optimized according to the *ab initio* 4-body interaction energy and geometry (with $\gamma$ of 3-body terms fixed).

According to the *ab initio* calculations, the four atoms are placed within a plane at the equilibrium geometry and such a geometrical constraint has been considered during the optimization of the adjustable parameters. Besides, note that although the expansion coefficients of the HCO 3-body term have already been refined by optimizing the adjustable parameter $\gamma$, these are treated as adjustable variables again without changing the expansion coefficients themselves, in order to obtain a better optimization for the HCOM geometry and the global minimum. The refined $\gamma$ of the HCO 3-body terms are $\gamma_1$ (CH) = 1.718887, $\gamma_2$ (CO) = 1.255612, and $\gamma_3$ (OH) = 1.843658.

The relative displacements (in nm) of the optimized geometry of the potential energy surface and the corresponding *ab initio* values (in parenthesis) are as follows: CH = 0.10693 (0.10967); CM = 0.13361 (0.13372); CO = 0.12495 (0.11993); OH = 0.20451 (0.20366); OM = 0.23045 (0.22599); HM = 0.1981 (0.19883).

The averaged error is around 0.0025 nm and corresponds to a mean relative error of 1.6%. Angles (in degrees) of the OCH, HCM and OCM bonds are 123.5753 (124.9), 110.395 (109.2) and 126.03 (125.9), respectively. The global minimum energy related to the asymptotic limit of the four bodies is −16.5418 eV, while the *ab initio* value is −16.51 eV.

## Acknowledgements

The work carried out in Taiwan was supported by the Ministry of Science and Technology, Taiwan, Republic of China under contract no. NSC 102-2113-M-002-009-MY3. T. Kasai is grateful to the Ministry of Science and Technology, for supporting his stay at the Department of Chemistry, National Taiwan University. A. Lombardi acknowledges financial support from MIUR PRIN 2010–2011 (contract 2010ERFKXL 002) and EGI Inspire. Thanks are also due to IGI (Italian Grid Infrastructure) and COMPCHEM Virtual Organization for the allocated computing time. V. A. acknowledges CAPES (Brazil) for the position of Pesquisador Visitante Especial at UFBA.

## References

1 J. M. Bowman, *Mol. Phys.*, 2014, **112**, 2516–2528.

 

2 T. Kasai, D.-C. Che, M. Okada, P.-Y. Tsai, K.-C. Lin, F. Palazzetti and V. Aquilanti, *Phys. Chem. Chem. Phys.*, 2014, **16**, 9776–9790.
3 A. G. Suits, *Acc. Chem. Res.*, 2008, **41**, 873.
4 D. Townsend, S. A. Lahankar, S. K. Lee, S. D. Chambreau, A. G. Suits, X. Zhang, J. Rheinecker, L. B. Harding and J. M. Bowman, *Science*, 2004, **306**, 1158.
5 K.-C. Hung, P.-Y. Tsai, H.-K. Li and K.-C. Lin, *J. Chem. Phys.*, 2014, **140**, 064313.
6 P.-Y. Tsai, K.-C. Hung, H.-K. Li and K.-C. Lin, *J. Phys. Chem. Lett.*, 2014, **5**, 190–195.
7 M. H. Chao, P. Y. Tsai and K. C. Lin, *Phys. Chem. Chem. Phys.*, 2011, **13**, 7154.
8 P.-Y. Tsai, M.-H. Chao, T. Kasai, K.-C. Lin, A. Lombardi, F. Palazzetti and V. Aquilanti, *Phys. Chem. Chem. Phys.*, 2014, **16**, 2854–2865.
9 M. L. Hause, N. Herath, R. Zhu, M. C. Linand and A. G. Suits, *Nat. Chem.*, 2011, **3**, 932.
10 *Molecular Reaction Dynamics in Gases, Liquids and Interfaces, Faraday Discuss.*, Royal Society of Chemistry Publishing, Cambridge, United Kingdom, 2012, vol. 157.
11 L. Rubio-Lago, G. A. Amaral, A. Arregui, G. A. Amaral, A. Arregui, L. Rubio-Lago, J. D. Rodríguez and L. Bañares, *Phys. Chem. Chem. Phys.*, 2010, **133**, 64303.
12 L. Rubio-Lago, G. A. Amaral, A. Arregui, J. Gonzalez-Vazquez and L. Banares, *Phys. Chem. Chem. Phys.*, 2012, **14**, 6067.
13 G. de Wit, B. R. Heazlewood, M. S. Quinn, A. T. Maccarone, K. Nauta, S. A. Reid, M. J. T. Jordan and S. H. Kable, *Faraday Discuss.*, 2012, **157**, 227.
14 T. J. B. Eppink and D. H. Parker, *Rev. Sci. Instrum.*, 1997, **68**, 3477.
15 Y. Tang, W. B. Lee, Z. F. Hu, B. Zhang and K. C. Lin, *J. Chem. Phys.*, 2007, **126**, 064302.
16 X. P. Zhang, W. B. Lee, D. F. Zhao, M. K. Hsiao, Y. L. Chen and K. C. Lin, *J. Chem. Phys.*, 2009, **130**, 214305.
17 M. Castillejo, S. Couris, E. Lane, M. Martin and J. Ruiz, *Chem. Phys.*, 1998, **232**, 353.
18 G. Cui, F. Zhang and W. Fang, *J. Chem. Phys.*, 2010, **132**, 034306.
19 J. S. Francisco, *J. Am. Chem. Soc.*, 2003, **125**, 10475.
20 S. H. Lee, *J. Chem. Phys.*, 2008, **129**, 194304.
21 M. Baer, *Beyond Born–Oppenheimer: Electronic Nonadiabatic Coupling Terms and Conical Intersections*, John Wiley & Sons, Hoboken, NJ, USA, 2006.
22 J. N. Murrell, S. Carter, S. C. Farantos, P. Huxley and A. J. C. Varandas, *Molecular potential energy functions*, John Wiley & Sons, London, 1984.
23 A. Lombardi, F. Palazzetti, K.-C. Lin and P.-Y. Tsai, *Lecture Notes in Computer Science (including subseries Lecture Notes in Artificial Intelligence and Lecture Notes in Bioinformatics)*, 2014, **vol. 8579**(1), 452–467.
24 *GAUSSIAN 03, Revision C.02*, Gaussian, Inc., Wallingford, CT, 2004.
25 M. Nakamura, P.-Y. Tsai, T. Kasai, K.-C. Lin, F. Palazzetti, A. Lombardi and V. Aquilanti, in preparation.
26 P. L. Houston and S. H. Kable, *Proc. Natl. Acad. Sci. U. S. A.*, 2006, **103**, 16079.
27 Y. T. Liu, M. T. Tsai, C. Y. Liu, P. Y. Tsai, K. C. Lin, Y. H. Shih and A. H. Chang, *J. Phys. Chem. A*, 2010, **114**, 7275.
28 Y. Y. Yeh, M. H. Chao, P. Y. Tsai, Y. B. Chang, M. T. Tsai and K. C. Lin, *J. Chem. Phys.*, 2012, **136**, 044302.

29 R. A. Palmer, J. L. Chao, R. M. Dittmar, V. G. Gregoriou and S. E. Plunkett, *Appl. Spectrosc.*, 1993, **47**, 1297.
30 G. V. Hartland, W. Xie, H. L. Dai, A. Simon and M. J. Anderson, *Rev. Sci. Instrum.*, 1992, **63**, 3261.

 

Faraday Discussions

PAPER

# The free-energy barrier to hydride transfer across a dipalladium complex

C. R. Vanston,[a] G. J. Kearley,[b] A. J. Edwards,†[b] T. A. Darwish,[b] N. R. de Souza,[b] A. J. Ramirez-Cuesta[c] and M. G. Gardiner*[a]

Received 26th September 2014, Accepted 4th November 2014
DOI: 10.1039/c4fd00182f

We use density-functional theory molecular dynamics (DFT-MD) simulations to determine the hydride transfer coordinate between palladium centres of the crystallographically observed terminal hydride locations, Pd–Pd–H, originally postulated for the solution dynamics of the complex bis-NHC dipalladium hydride $[\{(MesIm)_2CH_2\}_2Pd_2H][PF_6]$, and then calculate the free-energy along this coordinate. We estimate the transfer barrier-height to be about 20 kcal $mol^{-1}$ with a hydride transfer rate in the order of seconds at room temperature. We validate our DFT-MD modelling using inelastic neutron scattering which reveals anharmonicity of the hydride environment that is so pronounced that there is complete failure of the harmonic model for the hydride ligand. The simulations are extended to high temperature to bring the H-transfer to a rate that is accessible to the simulation technique.

## Introduction

Transition metal hydrides are used or invoked as intermediates in a wide range of roles in catalysis and molecular hydrogen activation and storage, with compounds ranging from mononuclear complexes with single hydride ligands to large catalyst clusters or surfaces designed for hydrogen capture.[1–9] As well as providing an alternative to traditional catalytic cycles based on mononuclear complexes, some clusters are known to act as reservoirs for active mononuclear species.[10] The compound studied in the present work, $[\{(MesIm)_2CH_2\}_2Pd_2H]$-$[PF_6]$, denoted MesPdH throughout this paper, is of interest in relation to systems where partially reduced precatalysts/catalysts function either as reservoirs for active mononuclear catalysts, as noted above, and/or intermediates prior to larger

[a]*School of Physical Sciences (Chemistry), University of Tasmania, Private Bag 75, Hobart, Tasmania 7001, Australia. E-mail: Michael.Gardiner@utas.edu.au*
[b]*Bragg Institute, Australian Nuclear Science and Technology Organization, Locked Bag 2001, Kirrawee DC, NSW 2232, Australia*
[c]*Oakridge National Laboratory-Chemical and Engineering Materials Division, PO Box 2008 MS6473, Oakridge, TN 37831-6473, USA*
† Presenting Author.

 

cluster formation that eventually lead to inactive metal deposits in ethylene/carbon monoxide polymerisation.[11,12]

One important aspect of metal hydride chemistry is the ability of hydride ligands to undergo fluxional processes while remaining coordinated to the metal centre, a property that is useful to some catalyst activity and selectivity. Catalysts of this type may range from long established systems such as Wilkinson's catalyst for alkene hydrogenation[13] to multinuclear clusters with known or potential use for the catalysis of hydrogen transfer reactions.[14,15]

Solution $^1$H NMR spectroscopic studies of MesPdH suggested the occurrence of a fluxional process, which was thought to involve the hydride ligand exchanging between the palladium centres (Fig. 1). A major objective of the current study is to examine if the same process could occur in the solid state.

Previous experimental studies on palladium hydride clusters have shown/suggested that hydride transfer processes between metal centres can occur either with the Pd–Pd bond remaining intact, or with cleavage of the Pd–Pd *via* a Pd–H–Pd geometry.[14,15] These studies are mostly based on the interpretation of solution phase fluxionality investigated by variable temperature NMR spectroscopy. The presence of good NMR probes in the clusters, phosphine ligands, can be invaluable in discriminating between mechanistic pathways. In the present case, we have no experimental evidence for such detail of the hydride ligand transfer process. We note, however, that in this investigation of the H-transfer in the solid state, the more significant changes to the cluster core expected to accompany cleavage of the Pd–Pd bond in a linearly bridged Pd–H–Pd structure are likely to be disfavoured in the solid state due to crystal packing effects. In initial calculations on H-transfer in MesPdH it was revealed that it would be too slow to observe the H-transfer process directly, using for example quasielastic neutron scattering, so simulation at high temperatures seemed to be a possible way forward.

It is straightforward using DFT-MD to show that the dynamics of the H-ligand in MesPdH are largely determined by the time-dependent free-energy landscape of neighbouring molecular groups. These dynamics can be scaled up by increasing the temperature, and it turns out that at 1500 K a H-transfer event can just be seen, within 8 ps, but that higher temperatures and longer times lead to destruction of the model. The central question is: is it only the amplitude of the fluctuations in the H-transfer free-energy surface that increase with temperature, or is a new process activated? We will show that the former case applies here, and a validation of the model at low temperature should still apply reasonably well at high temperature.

There are three difficulties to overcome with this strategy. Firstly, MesPdH, particularly in the full crystal structure, has a formidable number of degrees of

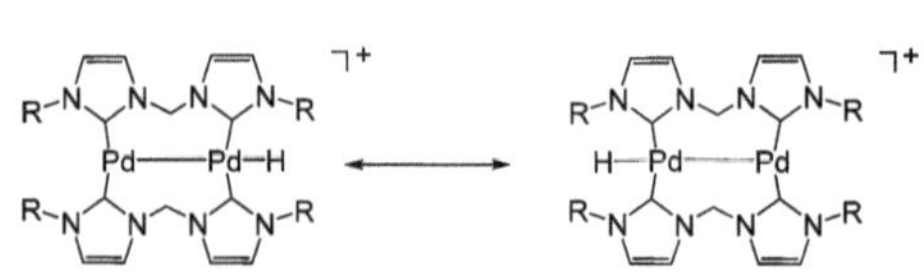


Fig. 1 Proposed solution phase dynamics of MesPdH hydride transfer in $[\{(MesIm)_2CH_2\}_2Pd_2H][PF_6]$, $R$ = 2,4,6-trimethylphenyl.

freedom, but we are only interested in those that involve the H-ligand. The associated vibrational dynamics of this ligand can be conveniently studied almost in isolation by the use of inelastic neutron scattering (INS). Because the H-nucleus has a uniquely high incoherent neutron scattering cross-section, selective deuteration of all H-atoms, except the H-ligand, makes this ligand's vibrations more pronounced in the INS spectrum. Secondly, because the transfer process involves bond-making and breaking, and due to the general lack of a suitable force-field for MesPdH, we need to use DFT-MD. This method has the advantage of removing ambiguity, but is vastly more computationally expensive than empirical force-field methods, and consequently the size of the model and the time-scale covered will be rather limited. Finally, the unit-cell of the crystal is large and it is important to establish how the crystal forces may distort the gas-phase molecular conformation. It turns out that these solid-state effects are rather small and that the calculated vibrational dynamics of the isolated molecule and the full periodic crystal are very similar.

This paper is therefore primarily concerned with the dynamics of the isolated MesPdH molecules and two of its selectively deuterated isotopomers illustrated in Fig. 2. We retain MesPdH to denote the isotopically normal material, and use DMesPdD and DMesPdH for the fully deuterated and protonated hydride, respectively.

# Experimental

The complex MesPdH was synthesised using literature methods.[11] Handling of air-sensitive chemicals was carried out in a dry glove box (Innovative Technologies) under a nitrogen atmosphere with Schlenk type glassware.

### Preparation of selectively deuterated samples for TOSCA

20 g of the ligand precursor *N*-2,4,6-trimethylphenyl imidazole (denoted MesIm) was provided to the National Deuteration Facility at ANSTO for chemical deuteration. This was achieved by heating the compound at 150 °C for 3 days under pressure in deuterium oxide in the presence of a mixture of Pt/C and Pd/C catalysts and using 20% v/v THF as a co-solvent.[16] The hydrothermal reaction mixture was purified using silica column chromatography to give deuterated MesIm (96% yield) where the two phenyl protons remained essentially un-

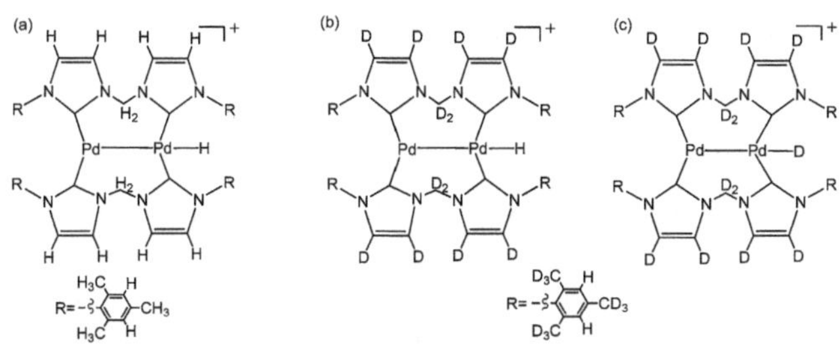

Fig. 2 Deuterated isotopomers of MesPdH: (a) MesPdH (b) DMesPdH (c) DMesPdD. The ring H-atoms of the R groups could not be deuterated, but these are the same in (b) and (c) and so will contribute equally to the INS spectra.

exchanged. Conversion to the deuterated diimidazolium salt using literature procedures[17] with deuterated dibromomethane (99% D) resulted in an overall molecule deuteration of 84 ± 2%. The deuteration percentages at each location of the molecule were calculated by $^{1}H$ and $^{13}C$ {$^{1}H$, $^{2}H$} NMR spectroscopy, along with mass spectrometry and isotopic distribution analysis to confirm the specific site, as well as the overall isotopic purity, respectively.

The deuterated imidazolium salt was then converted through previously reported methods[11] to the cationic dipalladium hydride DMesPdH; exchange of the hydride ligand to deuteride was accomplished by dissolving the relevant dipalladium hydride complex in deuterated methanol and stirring at 60 °C for 2 days. Removal of the methanol under vacuum and subsequent redissolving and stirring in THF-$d_8$ produced the deuterated THF solvate DMesPdD.

### Instrumentation

Inelastic neutron scattering spectroscopy was performed using the time-of-flight spectrometer TOSCA at the ISIS facility in Chilton, UK.[18] Approximately 1 g of the samples of the isotopomers shown in Fig. 2 was available and in each case was placed in an aluminium sachet, which was cooled in a standard closed-cycle cryostat. Spectral data were accumulated until an acceptable statistical quality was obtained at sample temperatures of 80 K and at 7 K. Data correction and transformation to the energy spectrum were made using standard local algorithms.

### DFT calculations

All DFT calculations were carried out using plane-wave code, Vienna *ab initio* Simulation Package (VASP).[19–21] Isolated (or gas-phase) molecule calculations were performed on an isolated $\{(MesIm)_2CH_2\}_2Pd_2H^+$ cation composed of 117 atoms and contained in a periodic cell of dimensions 25 × 18 × 18 Å. This cell is large enough to avoid fictive intermolecular interactions. Geometry optimization, vibrational dynamics, and MD simulations were made using the Projector Augmented Wave (PAW) potential[22] and the Perdew–Burke–Ernzerhof (PBE) exchange-correlation functional.[23] In all cases a single *k*-point was used ($\Gamma$-point) and for accurate calculations a 400 eV energy cut-off, with an electronic-energy convergence of $1.0 \times 10^{-5}$ eV. For all MD simulations the energy cutoff was reduced to 300 eV and the electronic convergence to $1.0 \times 10^{-4}$ eV, with a time step of 1 fs being used. The crystallographic unit-cell volume is almost 7000 Å$^3$ and contains over 700 atoms so only a limited number of calculations were possible at low precision to obtain an estimate of the solid-state effects on the H-ligand dynamics.

### Geometry optimization

Geometry optimization was achieved by allowing the atoms to relax to their minimum energy positions. For the crystal simulation the values of the cell constants were constrained to the experimental values and the full symmetry of the crystal structure was maintained during this optimization. For the isolated molecule optimization the initial conformation was taken from the crystal and relaxed with no symmetry constraints. The final molecular structure was very similar to the initial structure illustrating that the crystal forces only play a minor role in determining the molecular conformation. An energy-convergence criterion of $1.0 \times 10^{-4}$ eV was used for the isolated molecule and $1.0 \times 10^{-3}$ for the crystal.

### Matrix method molecular vibrations

Normal-mode vibrational frequencies and amplitudes were calculated for the molecule by obtaining the Hessian matrix using VASP with the same criteria as for the geometry optimization. Atoms were displaced in positive and negative Cartesian directions from the minimum-energy positions by 0.03 Å to determine the second derivatives of the energy. The INS spectrum was calculated from the frequencies and amplitudes using CLIMAX[24] applying the instrumental parameters from the TOSCA spectrometer. Spectra for all isotopomers were calculated using the same force-constants, but with the appropriate atomic masses being used in the corresponding inverse kinetic-energy matrix.

### DFT-MD simulations

The matrix method uses the harmonic approximation, which turned out to be poor for vibrations of the H-ligand. *Ab initio* MD allows exploring the form and origin of anharmonicity taking finite temperature of the sample into account, which is required not only for spectral agreement, but more importantly for the study of the H-transfer process. For all simulations a 4 ps equilibration on the energy-minimized structure in the isokinetic ensemble at the target temperature was performed to distribute the kinetic energy within the system. For the periodic crystal model this was followed by 4.2 ps production run in the microcanonical ensemble. For the isolated molecule models, an additional 4 ps equilibration in the microcanonical ensemble was made followed by an 8.5 ps production run. The MD trajectories from the low temperature runs were used to calculate the INS spectrum using nMoldyn.[25] This was achieved *via* the incoherent intermediate scattering function $F(Q,t)$ that was calculated on a regular grid in both momentum transfer, $Q$, and time, $t$, which was then Fourier transformed to obtain the scattering function $S(Q,\omega)$. The appropriate cut in $Q,\omega$ was taken to coincide with the experimental INS spectrum on TOSCA. This spectrum was then convoluted with the instrumental resolution function to obtain the calculated INS spectrum.

## Results and discussion

We will first describe the vibrational spectra of the MesPdH analogues and how these are interpreted using two different standard approaches. This provides a validation of the MD approach, and we then provide some justification for being able to extend this to much higher temperatures. This gives access to the H-transfer process that is seen a single time in the high-temperature DFT-MD simulation. The accessible free-energy profile is then analysed and extended to include the H-transfer.

### Comparison of observed and calculated INS spectra

The experimental INS spectra of DMesPdD and DMesPdH are remarkably similar, as illustrated in Fig. 3. The expectation from the matrix-method calculation, is that the H-ligand present in DMesPdH would produce a strong feature in the

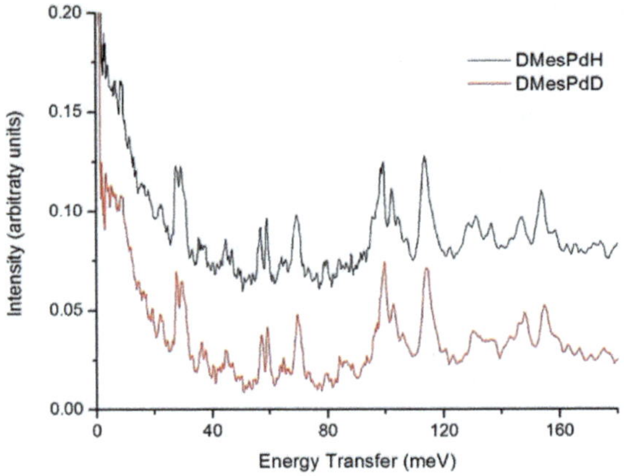


Fig. 3 Comparison of INS spectra for DMesPdH and DMesPdD obtained at TOSCA. Relative intensity shifted for clarity. Note the strong similarity between these spectra despite the isotopic difference of the H(D) ligand.

50 meV region corresponding to an out-of-plane wagging (H–Pd–Pd deformation), that would be absent in the spectrum of DMesPdD, as shown in the lower part of Fig. 4. It is worth noting that in both analogues there are 2 un-exchanged H-atoms on each trimethylphenyl unit (see Fig. 2) and these will contribute extra peaks from vibrations of the organic ligands, but these are the same for both analogues.

The absence of the strong peak at around 50 meV predicted by the matrix method is due to a breakdown of the harmonic approximation for modes of the H-ligand. This is easily demonstrated by using the alternative MD method to produce the calculated INS spectrum. Whilst DFT-MD lacks proper treatment of overtones and combinations, it has the advantage that it makes no assumption about the form of the vibrational (or other) potential. The result shown in the upper part of Fig. 4 is that the strong peak at 50 meV is indeed absent in the calculated spectrum, and even a casual examination of the MD trajectories for the

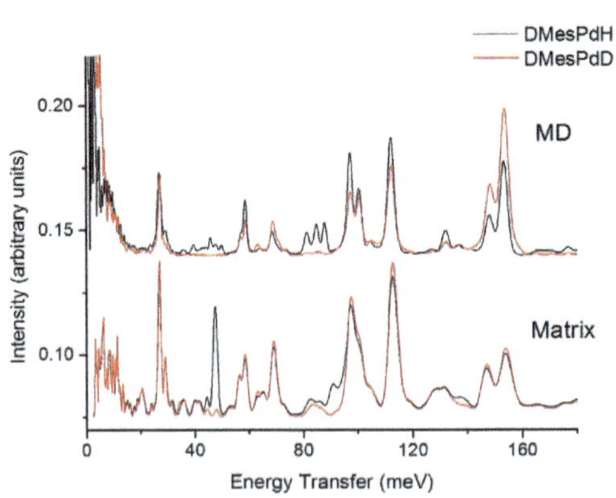


Fig. 4 Comparison of the calculated spectra of DMesPdH and DMesPdD by MD and Matrix simulation methods. These spectra should also be compared with the experimental spectra in Fig. 3. Relative intensities shifted for clarity.

 

H-ligand show the dynamics to be locally diffusive rather than oscillatory. Normally the difference between the INS spectra calculated by the two methods is small, and even the differences in Fig. 4 between 80 and 100 meV that are in part due to the lack of overtones and combinations in the MD, would be considered as significant. To our knowledge, the difference at around 50 meV in the current example is much larger than any reported so far.

The main aim of the current work is to study the H-transfer and the purpose of the INS is to validate that DFT-MD can at least reproduce the dynamics measured on the low-frequency part of the vibrational spectrum. We will take the general spectral agreement, and in particular the correct prediction of the non-oscillatory H-ligand dynamics, as a validation of the method.

## MD at high temperature

Computationally based energy-barrier calculations are generally performed by mapping the potential energy of the system as a function of some pre-defined reaction coordinate, taking no account of thermal motion. Even in better defined systems such as linear hydrogen bonds the involvement of many degrees of freedom can play a crucial role[26] so in MesPdH, where the H-atom is weakly bound and surrounded by molecular groups, each with many degrees of freedom, the choice of reaction coordinate is bewildering. The simple solution is to use MD to sample the free energy surface in a systematic and physically meaningful way, albeit at lower precision and greater computational expense. The difficulty is that the amplitudes sampled are limited by the available kinetic energy (simulation temperature). Increasing the MD simulation temperature should be more realistic than mapping potential-energies along pre-determined coordinates (at zero K), but it is still useful to provide additional justification.

We increased the simulation temperature until a transfer process occurred once within the timescale of the simulations (8 ps). The crucial coordinate for the transfer process turned out to be the dihedral angle between the plane of a heterocyclic ring and the Pd–H vector, which approximates to an extension of the mode that was incorrectly predicted by the harmonic approximation at 50 meV in the INS. The actual definition used is shown in Fig. 5a, but for convenience we will refer to this as Pd–Pd–H, to which it approximates. Clearly, this coordinate is not a normal coordinate of the system and so has a distributed spectral profile, which is shown (from the MD simulations) in Fig. 5b. The crucial point illustrated in this figure is that this frequency distribution does not change significantly between 50 K, where this is validated against experiment, and 1500 K where the transfer was seen in the MD, although there is a large increase in amplitude. A limited MD simulation for the periodic crystal at 300 K shows a very similar profile in Fig. 5b to those of the isolated molecule. This indicates that not only are the vibrational dynamics dominated by intra-molecular interactions, but also the H-transfer mechanism.

## Hydride transfer energy barrier

The actual transfer process is almost certainly from terminal H at one Pd-atom, to bridging-H between the two Pd-atoms to terminal H at the second Pd-atom. We never observed this whole process, or the competing "bridge and return" process. The resources for the longer simulations required for this are not available to us

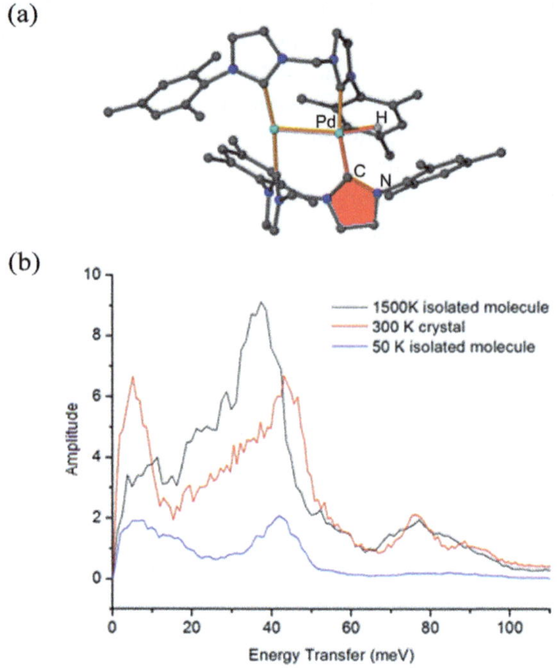


Fig. 5 (a) Structure of the cation of MesPdH indicating plane used to define the dihedral torsion coordinate. We do not actually use Pd–Pd–H, although we denote it this way in the text, because that coordinate includes motion of both Pd atoms. (b) Frequency spectra of the hydride torsion showing comparison between the isolated molecule at 1500 K and 50 K, as well as with the periodic structure calculation. Note that other than the lattice modes below 10 meV, there is little difference between the crystal and isolated models.

and increasing the temperature destroys the model before any transfer process is seen. Consequently we base our analysis on the dihedral angle shown in Fig. 5a. The temporal evolution of this Pd–Pd–H coordinate contains the resultant of the dynamics of the neighbouring ligands to which it is strongly coupled. To estimate the barrier to the transfer process we calculated the free energy from the frequency distribution of this angle during the 8 ps simulation at 1500 K as:

$$\mathrm{FE}_{(\mathrm{Pd-Pd-H})} = kT \ln\left(\frac{P_{(\mathrm{Pd-Pd-H})}}{P_{(\mathrm{max})}}\right) \tag{1}$$

where $T$ is the temperature of the simulation, $P_{(\mathrm{Pd-Pd-H})}$ is the time average population of a particular value of the angle, and $P_{(\mathrm{max})}$ is the maximum value of $P_{(\mathrm{Pd-Pd-H})}$. This energy function is shown in Fig. 6. Because only a single transfer was seen during the simulation, there are no data around the peak of the barrier, and it would take an extremely long simulation to obtain these. In the absence of a known functional form we fit these angle-dependent data with a sinusoidal function resulting in an estimate for the energy barrier between 19 and 23 kcal mol$^{-1}$. By using the Arrhenius equation:

$$k = A\mathrm{e}^{\frac{-E_\mathrm{a}}{RT}} \tag{2}$$

 

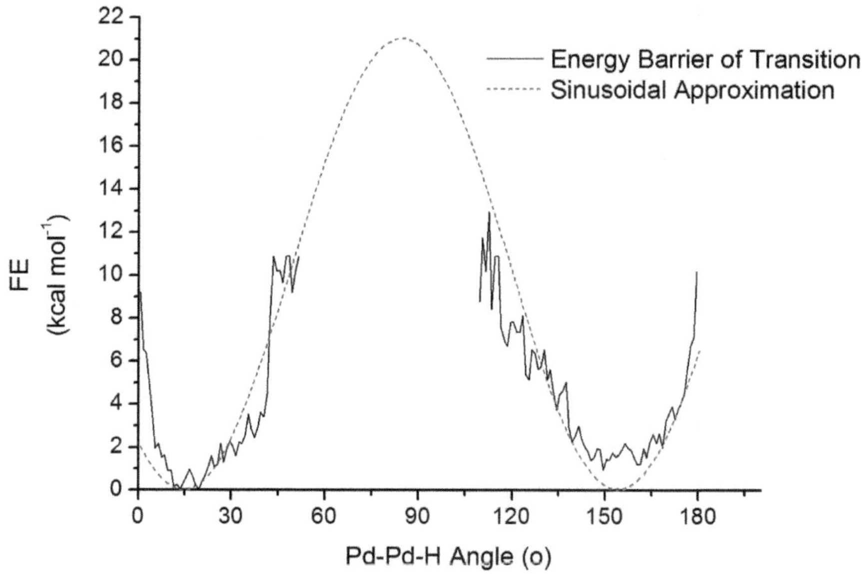


Fig. 6 Free energy from the frequency distribution of the Pd–Pd–H angle approximated with a sinusoidal function to estimate the energy barrier.

where $k$ is the rate constant, $A$ is the rate pre-factor determined using the observed single transfer in the MD as the rate at 1500 K, $E_a$ is the activation energy, $R$ is the universal gas constant $1.986 \times 10^{-3}$ kcal mol$^{-1}$ K$^{-1}$ and $T$ is the temperature, a room-temperature transfer rate of one transfer per 5.3 seconds was calculated. However, because the error in the barrier transforms exponentially to the rate, the limiting values are 360 ms and 77 s.

We have made no attempt at this stage to analyse the correlations between torsional motions of the methyl groups, whole-body torsions of the ligands and the H-transfer process. These are clearly central to the H-transfer process and could be chemically tailored to either increase or decrease the rate, or indeed to trap the H-ligand in a bridging or terminal position.

# Conclusions

Many studies have successfully used potential-energy mapping to obtain calculated dynamics that agree well with experimental results. This method ignores entropy, which enters in the current example through the highly flexible local environment of the H-ligand, and as we have shown, the mapping methods are then not appropriate. From a solid-state structural view the H-ligand is bonded as part of the molecular solid, but from a dynamical view the H-atom experiences a more clathrate-like confinement, in which host–guest interactions inevitably dominate. The H-transfer process is correspondingly more diffusive in nature, broadly analogous to transport between different pores. The H-atom is held as much within the confines of the neighbouring ligands as by the closest metal atom, and our calculations show that the crystal forces have little effect on this environment, or the overall molecular conformation.

It is difficult to determine barriers to slow transfer processes in complex systems that have many degrees of freedom so we have attempted to circumvent this using approximations that may not be general, but are suitable for the system under study. In particular, we have used an MD simulation at a temperature that is high enough to see the H-transfer process within our resource-limited time, but even though this is above the decomposition temperature, it is low enough that this does not occur within the simulation period. In this way the MD probes physically meaningful coordinates and takes account of the dynamics of the whole system, albeit somewhat limited close to the actual transfer barrier. Additional MD simulations show that it is not the amplitude of the dynamics that lead to H-transfer, but rather a change in their geometry, or the onset of a new process. At least for these transfer-coordinates it is reasonable to extrapolate the low-temperature INS validation to the higher temperature dynamics.

It would be difficult, but very interesting to measure this transfer process in the solid state experimentally. This would provide a strong incentive to put further effort into the detailed understanding of the transfer process and the role of slow dynamics of the hydride environment.

## Acknowledgements

We would like to acknowledge beamtime on TOSCA at the ISIS facility, Oxford and the National Deuteration Facility, ANSTO for access *via* proposal NDF1196. Thanks to the Access to Major Research Facilities Program (AMRFP) for travel funding. Thanks to AINSE (for Research Awards and Postgraduate Research Award to CRV), the University of Tasmania (scholarship support for CRV) and the Australian Research Council for major project funding (Discovery Grant).

## Notes and references

1 N. K. Nag, *J. Phys. Chem. B*, 2001, **105**, 5945.

2 C. M. Hagen, L. Vieille-Petit, G. Laurenczy, G. Suss-Fink and R. G. Finke, *Organometallics*, 2005, **24**, 1819.

3 S. K. Brayshaw, A. Harrison, J. S. McIndoe, F. Marken, P. R. Raithby, J. E. Warren and A. S. Weller, *J. Am. Chem. Soc.*, 2007, **129**, 1793.

4 S. Chaudhuri, J. Graetz, A. Ignatov, J. J. Reilly and J. T. Muckerman, *J. Am. Chem. Soc.*, 2006, **128**, 11404.

5 M. Pérez Jigato, B. Coussens and D. A. King, *J. Chem. Phys.*, 2003, **118**, 5623.

6 F. Ozawa, H. Okamoto, S. Kawagishi, S. Yamamoto, T. Minami and M. Yoshifuji, *J. Am. Chem. Soc.*, 2002, **124**, 10968.

7 M. S. Viciu, G. A. Grasa and S. P. Nolan, *Organometallics*, 2001, **20**, 3607.

8 M. A. Zuideveld, P. C. J. Kamer, P. W. N. M. van Leeuwen, P. A. A. Klusener, H. A. Stil and C. F. Roobeek, *J. Am. Chem. Soc.*, 1998, **120**, 7977.

9 Y. Duan, L. Li, M. Chen, C. Yu, H. Fan and Y. Zhou, *J. Am. Chem. Soc.*, 2014, **136**, 7688.

10 M. Baya, J. Houghton, D. Konya, Y. Champouret, J. Daran, K. Q. A. Leñero, L. Schoon, W. P. Mul, A. B. van Oort, N. Meijboom, E. Drent, A. G. Orpen and R. Poli, *J. Am. Chem. Soc.*, 2008, **130**, 10612.

 

11 P. D. W. Boyd, A. J. Edwards, M. G. Gardiner, C. C. Ho, M. Lemee-Cailleau, D. S. McGuinness, A. Riapanitra, J. W. Steed, D. N. Stringer and B. F. Yates, *Angew. Chem., Int. Ed.*, 2010, **49**, 6315.
12 M. G. Gardiner, C. C. Ho, F. M. McKay, D. S. McGuinness and M. Tucker, *Dalton Trans.*, 2013, **42**, 7447.
13 S. B. Duckett, C. L. Newell and R. Eisenberg, *J. Am. Chem. Soc.*, 1994, **116**, 10548.
14 D. Evrard, K. Groison, Y. Mugnier and P. D. Harvey, *Inorg. Chem.*, 2004, **43**, 790.
15 C. Cugnet, D. Lucas, E. Collange, B. Hanquet, A. Vallat, Y. Mugnier, A. Soldera and P. D. Harvey, *Chem.–Eur. J.*, 2007, **13**, 5338.
16 T. A. Darwish, A. R. G. Smith, I. R. Gentle, P. L. Burn, E. Luks, G. Moraes, M. Gillon, P. J. Holden and M. James, *Tetrahedron Lett.*, 2012, **53**, 931.
17 M. G. Gardiner, W. A. Herrmann, C. Reisinger, J. Schwarz and M. Spiegler, *J. Organomet. Chem.*, 1999, **572**, 239.
18 A. J. Ramirez-Cuesta, S. F. Parker and J. Tomkinson, in *TOSCA User Guide*, 2003.
19 G. Kresse and J. Furthmüller, *Comput. Mater. Sci.*, 1996, **6**, 15.
20 G. Kresse and J. Hafner, *Phys. Rev. B: Condens. Matter Mater. Phys.*, 1993, **47**, 558.
21 G. Kresse and D. Joubert, *Phys. Rev. B: Condens. Matter Mater. Phys.*, 1999, **59**, 1758.
22 P. E. Blochl, *Phys. Rev. B: Condens. Matter Mater. Phys.*, 1994, **50**, 17953.
23 J. P. Perdew, K. Burke and M. Ernzerhof, *Phys. Rev. Lett.*, 1996, **77**, 3865.
24 G. J. Kearley, *Nucl. Instrum. Methods*, 1995, **354**, 53.
25 G. R. Kneller, V. Keiner, M. Kneller and M. Schiller, *Comput. Phys. Commun.*, 1995, **91**, 191.
26 M. Plazanet, N. Fukushima, M. R. Johnson, A. J. Horsewill and H. P. Trommsdorff, *J. Chem. Phys.*, 2001, **115**, 3241.

Faraday Discussions

PAPER

# Excited state dynamics of organic semi-conducting materials†

Kenneth P. Ghiggino,*[a] Andrew J. Tilley,[ab] Benjamin Robotham[a] and Jonathan M. White[ab]

Received 10th September 2014, Accepted 27th October 2014
DOI: 10.1039/c4fd00171k

Time-resolved absorption and emission spectroscopy has been applied to investigate the dynamics of excited state processes in oligomer models for semi-conducting organic materials. Following the photo-excitation of a pentamer oligomer that is a model for the conjugated polymer MEH-PPV, an ultrafast component of a few picoseconds is observed for the decay of the initially formed transient species. Variable temperature absorption and emission spectra combined with X-ray crystallography and calculations support the assignment of this rapid relaxation process to an excited state conformational rearrangement from non-planar to more planar molecular configurations. The implications of the results for the overall photophysics of conjugated polymers are considered.

## Introduction

Conjugated polymers are finding widespread application in light-emitting diodes,[1–5] photovoltaic cells[6–10] and other organic electronic devices.[11–15] Since the discovery of efficient electroluminescence in conjugated polymers during the 1990s,[16] research has focused on improving optoelectronic performance to bring conjugated polymers in line with existing inorganic technologies. Materials based on poly(*p*-phenylenevinylene) (PPV) have received considerable attention owing to their favourable luminescent and semi-conducting properties.[17–20] In recent years, there has been a considerable research effort toward understanding the fundamental photophysical and light harvesting properties of this important polymer; in particular, the more useful solubilised analogues such as poly[2-methoxy-5-(2′-

[a]*School of Chemistry, University of Melbourne, Melbourne, Victoria, Australia. E-mail: ghiggino@unimelb.edu.au; Fax: +61 3 93475180; Tel: +61 3 83448939*

[b]*Bio21 Molecular Science and Biotechnology Institute, University of Melbourne, Parkville, Victoria, 3010, Australia*

† Crystallographic information file (CIF) for the bromo PPV dimer ((*E*)-1-bromo-2-hexyloxy-4-(2′-hexyloxy-5′-methoxystyryl)-5-methoxy benzene). CCDC 1023608. For crystallographic data in CIF or other electronic format see DOI: 10.1039/c4fd00171k.

ethylhexyloxy)-1,4-phenylenevinylene] (MEH-PPV) have been of particular interest.[21–25]

In conjugated polymers, a number of competing exciton relaxation pathways are accessible following photoexcitation, often leading to complex photophysical behavior. In MEH-PPV, it is thought that chain defects cause polymer backbone conjugation to be segmented into smaller sub-units, leading to broadened absorption spectra.[26,27] Following photoexcitation, energy transfer from higher energy, shorter segments to lower energy, longer segments has been proposed to lead to emission from a restricted subset of conjugated sequences, resulting in the observed structured solution emission spectrum. However, it is now known that this picture is too simplistic, and that excited states of conjugated polymers are likely governed by complex energy transfer and structural relaxation processes, and that the degree of torsional freedom in individual conjugated segments can influence energy transfer dynamics.[28–30] In addition, quantum coherence has been proposed to contribute to the polymer energy transfer manifold.[21,22]

Recently, we reported the synthesis and photophysics of structurally well-defined MEH-PPV oligomers.[31] It was noted in that work that the spectral profiles of the oligomers were quite similar to those of the bulk polymer, suggesting that the observed asymmetry between absorption and fluorescence spectra is due to

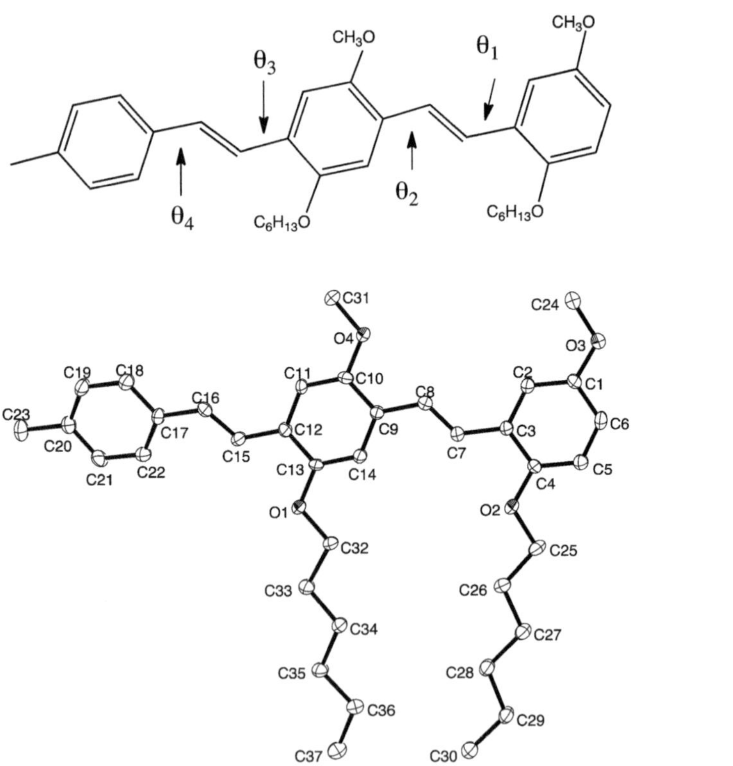


Fig. 1 Conformations of a PPV trimer where and $\theta_1$–$\theta_4$ are the twist dihedral angles (upper figure). X-ray structure of the trimer (for R = $C_6H_{13}$) where $\theta_1 = -8.7(7)°$, $\theta_2 = 7.4(7)°$, $\theta_3 = -15.0(7)°$, $\theta_4 = -6.9(7)°$.

 

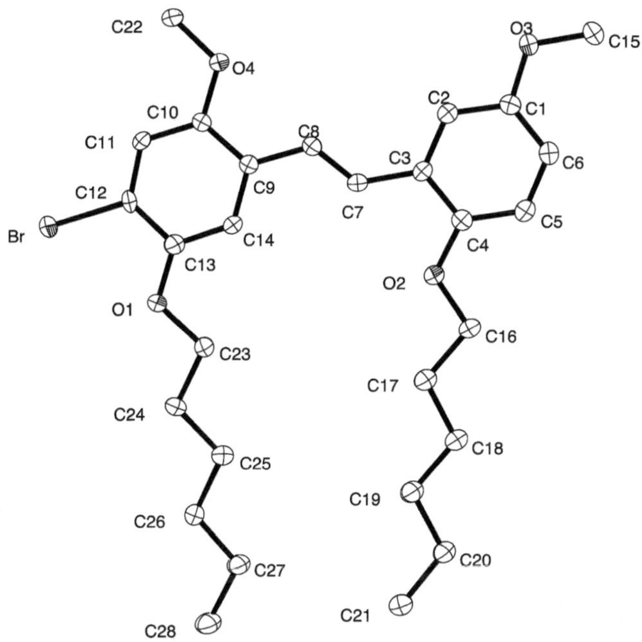


Fig. 2 X-ray structure of a bromo PPV dimer with 30% ellipsoids where $\theta_3 = 11.9(4)°$, $\theta_4 = -9.4(4)°$ and $\theta_3$ = C14–C9–C8–C7 and $\theta_4$ = C7–C3–C4–O2 dihedral angles.

properties inherent to the individual chromophores that make up the polymer backbone. It was proposed that the ground state likely consists of a population of non-degenerate torsional conformers, each possessing slightly different transition energies, leading to broadened absorption spectra. The possible conformations of a model PPV trimer oligomer are illustrated for example in Fig. 1. Molecular mechanics calculations show that the trimer structure is not planar but consists of a family of up to eight twisted conformations that have similar energies (within 0.84 kJ mol$^{-1}$) where the four dihedral angles $\theta_1$ to $\theta_4$ deviate from planarity by angles from 10° to 15°. These qualitative predictions are supported by the crystal structure of the trimer (also see Fig. 1) which adopts a twisted conformation with dihedral angles $\theta_1 = -8.7(7)°$, $\theta_2 = 7.4(7)°$, $\theta_3 = -15.0(7)°$, $\theta_4 = -6.9(7)°$.[31] We also report here that the crystal structure of the bromo PPV dimer ((*E*)-1-bromo-2-hexyloxy-4-(2′-hexyloxy-5′-methoxystyryl)-5-methoxy benzene, Fig. 2) similarly shows a significant deviation from molecular planarity due to non-zero dihedral angles.

In order to explain the reason structured fluorescence from the oligomers is observed, in contrast to the broad absorption spectra profiles, it is proposed that following photo-excitation rapid conformational relaxation of the twisted conformations to more planar structures occurs. These proposals are consistent with several other theoretical and experimental studies on related systems.[32–35]

In the present paper the excited state dynamics of a MEH-PPV oligomer have been investigated using ultrafast transient absorption spectroscopy and other photophysical methods. Earlier work by others on a related PPV trimer using lower resolution time-correlated single photon counting experiments has found evidence for a <50 ps fluorescence decay component following photoexcitation that was attributed to conformational relaxation.[34] Owing to the ultrafast nature of this relaxation process, picosecond transient absorption spectroscopy has been applied here to resolve the time-scale of the initial relaxation processes.

# Experimental

The synthesis of pentamer **1** (Fig. 3) is reported elsewhere.[31] UV-Vis absorption spectra were recorded on a Varian Cary 50 Bio spectrophotometer. All solution absorption spectra were recorded against solvent blanks in matched 1.0 cm path length quartz cells. Fluorescence spectra were recorded on a Varian Eclipse fluorescence spectrofluorometer. The optical densities of the solutions were kept below 0.20 at the excitation wavelength to minimize reabsorption effects. Low temperature studies were carried out in an Oxford Instruments Optistat liquid nitrogen cryostat fitted with a temperature controller.

The transient absorption instrumentation has been described elsewhere.[36] In summary, it consists of a Ti:sapphire laser (Coherent, Mira) combined with a regenerative amplifier (Coherent, RegA 9050) operating at a 94 kHz pulse repetition rate and with an average output power of ~500 mW at 830 nm. Half of the amplified output was directed to an OPA (Coherent, OPA 9400) and the second harmonic output at 415 nm was used as the pump beam that passes through a light chopper to provide modulation synchronised to one twentieth of the pulse repetition rate (4.7 kHz). The other half of the amplified laser output was focused into a 3 mm sapphire crystal (Crystalsystems) to provide a continuum probe beam. The relative polarization of the pump and probe beams are set to 54.7 degrees (the magic angle) to remove any distortion effects due to molecular rotation. After passing through the sample contained in a 2 mm path length cell the probe beam is delivered to a CMOS detector (Ultrafast Systems) by a 200 μm fiber (Ocean Optics, P200-2-UV-VIS). The time resolution of the system is approximately 200 fs. The kinetics scheme modelling the pentamer transient decay was solved within MATLAB (Mathworks) software using the ode23s ordinary differential equation solver. Optimisation of the fit was achieved by minimisation of the chi-squared fitting parameter. The goodness of fit was determined by the $R^2$ regression parameter. The concentration of pentamer **1** was approximately $1 \times 10^{-6}$ M and ~10% of the sample was excited by the excitation laser pump pulse as assessed by the ground state depletion. The solutions were stirred continuously throughout the measurement by a magnetic stirrer bar.

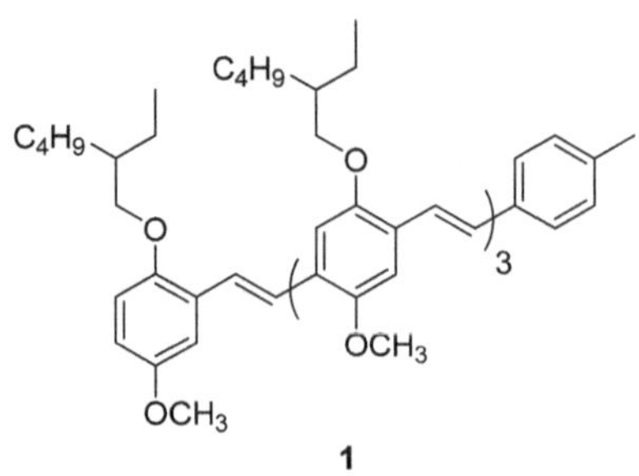


**1**

Fig. 3 Structure of pentamer 1 studied in this work.

# Results and discussion

### Steady state photophysics

The variable temperature absorption and emission spectra for pentamer **1** are shown in Fig. 4 and 5. These spectra were recorded in 2-methyltetrahydrofuran, which provided an optically clear matrix as the solution was cooled.

At room temperature the absorption spectrum of **1** is broad and featureless while, in contrast, the fluorescence spectrum shows vibrational structure. The variable temperature absorption spectra presented in Fig. 4 are similar to those for the related tetramer reported elsewhere,[31] although the spectra for the pentamer are more red-shifted which is consistent with the additional backbone π-conjugation. The absorption spectra show a marked increase in vibrational structure as the solution is cooled, coupled with a red-shift in the absorption maximum. This behaviour is consistent with absorption by a restricted set of torsional configurations in the low temperature solution environment whereas at higher temperatures additional configurations are accessible. The emission

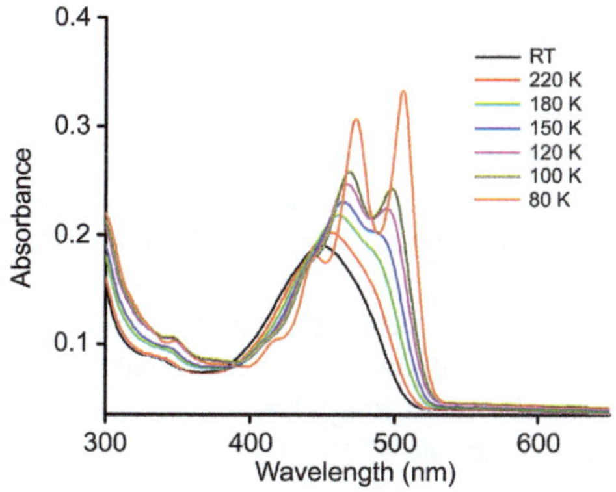


Fig. 4 Absorption spectra of pentamer **1** in 2-methyltetrahydrofuran as a function of temperature.

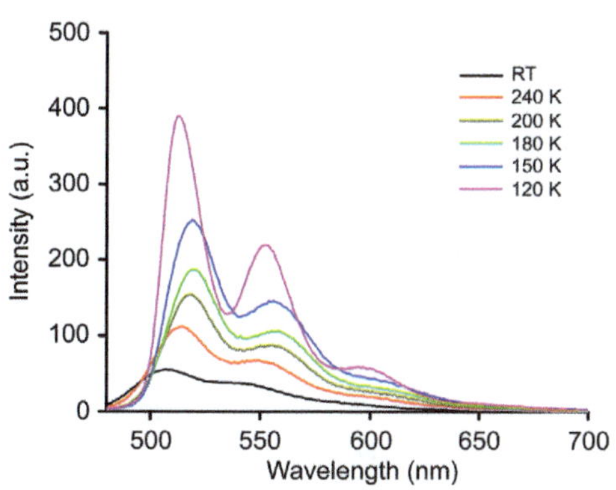


Fig. 5 Emission spectra of pentamer **1** in 2-methytetrahydrofuran as a function of temperature. $\lambda_{exc} = 470$ nm.

spectrum becomes more structured and increases in intensity as the temperature is lowered but does not show the large red-shift that is observed in absorption. At low temperature a strong mirror image relationship is apparent between absorption and emission profiles and there is a reduced Stokes shift compared with the room temperature spectra.

The similarity (in spectral position) between the 120 K and room temperature emission spectra suggests that the emitting conformations present at low temperature are similar to those populated at room temperature. This supports a model of conformational relaxation to low energy structures prior to emission in room temperature solutions, and suggests that the large Stokes shift observed in the room temperature spectra arises from planarization of the oligomer backbone following excitation. These findings are also consistent with previous theoretical and experimental studies,[37,38] which have found evidence for a large degree of torsional flexibility in the ground state of PPV oligomers.

## Transient absorption spectroscopy

To investigate the dynamics of the conformational relaxation process, ultrafast transient absorption spectroscopy was employed. The transient absorption spectrum of pentamer **1** was recorded in a dilute chloroform solution at room temperature and is shown in Fig. 6.

A clear transient absorption signal can be seen with the band maximum centred at 770 nm, while both ground state bleach and stimulated emission signals are evident in the shorter wavelength region (<600 nm). There is a concomitant recovery of the ground state bleach signal with the decay of the transient, indicating that the transient signal observed corresponds to an $S_1$–$S_n$ transition and that the decay reflects the depopulation of the $S_1$ state. Of particular interest is the initial rapid decay in transient signal intensity observed within the first 20 ps following the excitation pulse.

The transient decay trace at 770 nm (Fig. 7) shows an initial ultrafast component, coupled with a longer-lived decay. The latter component decays on a similar time scale to the fluorescence from the pentamer measured by the time-correlated single photon counting (TCSPC) technique that has a major component lifetime of 800 ps.[31] The time resolution of ~50 ps of the TCSPC

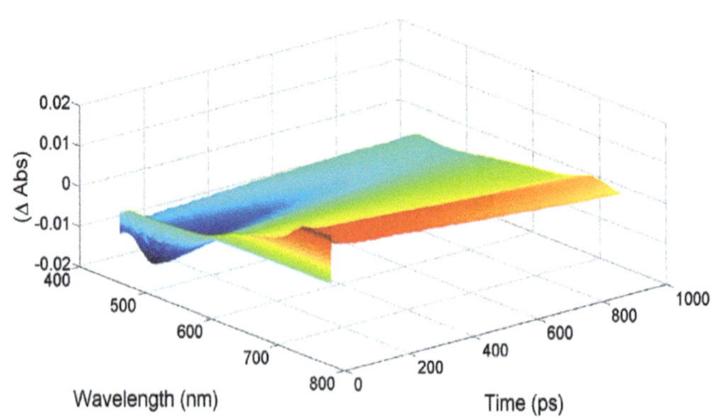


Fig. 6 Transient absorption spectrum of pentamer 1 recorded in chloroform solution at room temperature. An excitation wavelength of 415 nm was used.

measurements would not have revealed the shorter decaying components observed in the transient studies reported here.

In order to quantify the ultrafast component from the transient decay, the decay profile was fitted to the simple kinetic model shown in the inset of Fig. 7. In this model $k_1$ is the rate constant for conformational relaxation of the excited population to give the planar species ($S_1^{planar}$), $k_2$ is the competing rate of the initially prepared configuration directly to the ground state (including both radiative and non-radiative processes), while $k_3$ is the total decay rate from the conformationally relaxed planar configurations.

This kinetic model gave an adequate fit to the data ($R^2$ fitting parameter of 0.996) yielding a value for the conformational relaxation rate ($k_1$) of $1.47 \times 10^{11}$ $s^{-1}$ (see Fig. 7). The fitted values for $k_2$ and $k_3$ are $3.23 \times 10^{10}$ $s^{-1}$ and $1.10 \times 10^9$ $s^{-1}$ respectively. Comparison of the fitted $k_1$ and $k_2$ rate constants indicates that conformational relaxation is the dominant decay channel for the excited PPV oligomer, with approximately 80% of the initially excited population relaxing to planar states prior to emission. The inverse of $k_3$ gives the lifetime of the $S_1^{planar}$ state (910 ps) that compares well with the fluorescence lifetime recorded from time-correlated single photon counting (800 ps).

In light of the steady state temperature dependent absorption and fluorescence results, it seems reasonable to assign the observed ultrafast decay component to rapid conformational relaxation of the initially excited non-planar configurations of the oligomer to form the more planar relaxed structures from which the majority of fluorescence arises. Further evidence for this conclusion comes from the fluorescence spectrum of the pentamer oligomer when it is dispersed in a solid poly(methyl methacrylate) (PMMA) film at low concentration to avoid aggregation effects. Under these conditions a broad structureless emission spectrum is observed with a maximum at 498 nm compared to the structured emission observed with a maximum at 518 nm (*cf.*, Fig. 5). This blue shift in the fluorescence maximum and absence of spectral structure in the solid film is

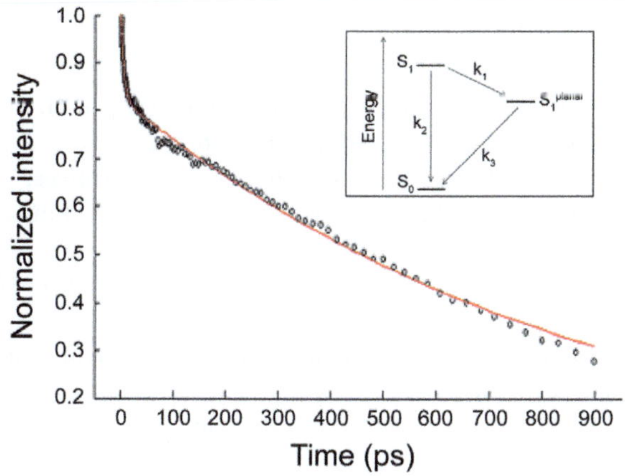


Fig. 7 Pentamer transient decay recorded at 770 nm. Points: experimental data; solid line: fit to a kinetic model that includes excited state conformational relaxation with $k_1 = 1.47 \times 10^{11}$ $s^{-1}$, $k_2 = 3.23 \times 10^{10}$ $s^{-1}$, $k_3 = 1.10 \times 10^9$ $s^{-1}$ (see text). Inset: kinetic model describing excited state processes in the PPV oligomer defining $k_1$, $k_2$ and $k_3$.

consistent with emission originating from the broad set of unrelaxed configurations.

Pentamer **1** can be considered a model for the chromophore subunits of the extended conjugated polymer MEH-PPV. The results of this work indicate that the observed broad absorption but structured emission of MEH-PPV in solution, often attributed to energy transfer processes from a range of absorbing polymer segments to low energy emitting segments, is more likely the result of rapid conformational relaxation processes within the segments themselves. Nevertheless exciton transfer within a MEH-PPV polymer chain has been demonstrated by a number of bulk and single molecule measurements, *e.g.* ref. 21, 28, 39. The short time relaxation behaviour in the pentamer suggests that, in MEH-PPV solutions, conformational relaxation likely competes with intramolecular energy transfer and needs to be considered in any complete theoretical description of the polymer energy transfer dynamics.

# Conclusion

Conformational relaxation dynamics of a MEH-PPV oligomer have been investigated using variable temperature steady state measurements and ultrafast TAS. The steady state measurements, combined with X-ray studies and theoretical calculations, provided evidence for a ground state populated by a family of torsional isomers that upon photo-excitation relaxes to more planar arrangements. Transient spectroscopy of the pentamer in room temperature solution showed a clear ultrafast component ($\sim$7 ps) in the transient decay that can be attributed to this conformational relaxation process.

# Acknowledgements

AJT thanks the David Hay Memorial Fund for financial assistance during the preparation of this paper. This research was supported through an Australian Research Council Discovery Project (DP 0986166).

# References

1 K. Brunner, A. Dijken, H. Borner, J. J. A. M. Bastiaansen, N. M. M. Kiggen and B. M. W. Langeveld, *J. Am. Chem. Soc.*, 2004, **126**, 6035.

2 M. A. Reddy, G. Mallesham, A. Thomas, K. Srinivas, V. J. Rao, K. Bhanuprakash, L. Giribabu, R. Grover, A. Kumar, M. N. Kamalasanan and R. Srivastava, *Synth. Met.*, 2011, **161**, 869.

3 Q. Li, J. Li, H. Ren, Y. Duan, Z. Gao and D. Liu, *Macromol. Rapid Commun.*, 2011, **32**, 736.

4 W. Lu, J. Kuwabara, T. Iijima, H. Higashimura, H. Hayashi and T. Kanbara, *Macromolecules*, 2012, **45**, 4128.

5 T. W. Yoo, C. Park, T. M. Nguyen, D. U. Kim and L. S. Park, *Mol. Cryst. Liq. Cryst.*, 2011, **551**, 69.

6 Y. Zhang, H. L. Yip, O. Acton, S. K. Hau, F. Huang and A. K. Y. Jen, *Chem. Mater.*, 2009, **21**, 2598.

7 D. M. Huang, S. A. Mauger, S. Friedrich, S. J. George, D. Dumitriu-LaGrange, S. Yoon and A. J. Moule, *Adv. Funct. Mater.*, 2011, **21**, 1657.

 

8 A. Gupta, S. E. Watkins, A. D. Scully, T. B. Singh, G. J. Wilson, L. J. Rozanski and R. A. Evans, *Synth. Met.*, 2011, **161**, 856.
9 H. Zhou, L. Yang and W. You, *Macromolecules*, 2012, **45**, 607.
10 H. J. Son, B. Carsten, I. H. Jung and L. Yu, *Energy Environ. Sci.*, 2012, **5**, 8158.
11 T. Kojima, D. Kumaki, J.-i. Nishida, S. Tokito and Y. Yamashita, *J. Mater. Chem.*, 2011, **21**, 6607.
12 T. Lei, Y. Cao, Y. Fan, C.-J. Liu, S.-C. Yuan and J. Pei, *J. Am. Chem. Soc.*, 2011, **133**, 6099.
13 H. Bronstein, Z. Chen, R. S. Ashraf, W. Zhang, J. Du, J. R. Durrant, P. Shakya Tuladhar, K. Song, S. E. Watkins, Y. Geerts, M. M. Wienk, R. A. J. Janssen, T. Anthopoulos, H. Sirringhaus, M. Heeney and I. McCulloch, *J. Am. Chem. Soc.*, 2011, **133**, 3272.
14 A. Facchetti, *Chem. Mater.*, 2011, **23**, 733.
15 Z. Li, J. Lu, S.-C. Tse, J. Zhou, X. Du, Y. Tao and J. Ding, *J. Mater. Chem.*, 2011, **21**, 3226.
16 J. H. Burroughes, D. D. C. Bradley, A. R. Brown, R. N. Marks, K. Mackay, R. H. Friend, P. L. Burn and A. B. Holmes, *Nature*, 1990, **347**, 539.
17 D. Braun and A. J. Heeger, *Appl. Phys. Lett.*, 1991, **58**, 1982.
18 C. Liedenbaum, Y. Croonen, P. van de Weijer, J. Vleggaar and H. Schoo, *Synth. Met.*, 1997, **91**, 109.
19 G. Yu, *Synth. Met.*, 1996, **80**, 143.
20 S. L. M. van Mensfoort, M. Carvelli, M. Megens, D. Wehenkel, M. Bartyzel, H. Greiner, R. A. J. Janssen and R. Coehoorn, *Nat. Photonics*, 2010, **4**, 329.
21 E. Collini and G. D. Scholes, *Science*, 2009, **323**, 369.
22 E. Collini and G. D. Scholes, *J. Phys. Chem. A*, 2009, **113**, 4223.
23 S. Habuchi, S. Onda and M. Vacha, *Chem. Commun.*, 2009, 4868.
24 S. Habuchi, S. Onda and M. Vacha, *Phys. Chem. Chem. Phys.*, 2011, **13**, 1743.
25 H. Lin, Y. Tian, K. Zapadka, G. Persson, D. Thomsson, O. Mirzov, P.-O. Larsson, J. Widengren and I. G. Scheblykin, *Nano Lett.*, 2009, **9**, 4456.
26 B. J. Schwartz, *Annu. Rev. Phys. Chem.*, 2003, **54**, 141.
27 S. Saini and B. Bagchi, *Phys. Chem. Chem. Phys.*, 2010, **12**, 7427.
28 I. Hwang and G. D. Scholes, *Chem. Mater.*, 2011, **23**, 610.
29 Y. Ebihara and M. Vacha, *J. Phys. Chem. B*, 2008, **112**, 12575.
30 J.-S. Yang, J.-L. Yan, C.-K. Lin, C.-Y. Chen, Z.-Y. Xie and C.-H. Chen, *Angew. Chem., Int. Ed.*, 2009, **48**, 9936.
31 A. J. Tilley, S. M. Danczak, C. Browne, T. Young, T. Tan, K. P. Ghiggino, T. A. Smith and J. White, *J. Org. Chem.*, 2011, **76**, 3372.
32 J. Gierschner, H. G. Mack, L. Lüer and D. Oelkrug, *J. Chem. Phys.*, 2002, **116**, 8596.
33 S. T. Hoffman, H. Bässler and A. Köhler, *J. Phys. Chem. B*, 2010, **114**, 17037.
34 R. E. Di Paolo, J. Seixas de Melo, J. Pina, H. D. Burrows, J. Morgado and A. L. Maçanita, *ChemPhysChem*, 2007, **8**, 2657.
35 J. Seixas de Melo, J. Pina, H. D. Burrows, R. E. Di Paolo and A. L. Macanita, *Chem. Phys.*, 2006, **330**, 449.
36 S. K. Sugunan, B. Robotham, R. P. Sloan, J. D. Szmytkowski, K. P. Ghiggino, M. F. Paige and R. P. Steer, *J. Phys. Chem. A*, 2011, **115**, 12217.
37 F. Sterpone and P. J. Rossky, *J. Phys. Chem. B*, 2008, **112**, 4983.
38 J. B. Lagowski, *J. Mol. Struct.: THEOCHEM*, 2002, **589–590**, 125.
39 E. N. Hooley, A. J. Tilley, J. M. White, K. P. Ghiggino and T. D. M. Bell, *Phys. Chem. Chem. Phys.*, 2014, **16**, 7108.

Faraday Discussions



DISCUSSIONS

# Dynamics of chemical bond: general discussion

Kunal Dhoke, Martin Zanni, Upendra Harbola, Ravi Kumar Venkatraman, Elangannan Arunan, King-Chuen Lin, Artur Nenov, Jonathan Skelton, R J Dwayne Miller, Jonathan D. Hirst, Vincenzo Aquilanti, John R. Helliwell, Srihari Keshavamurthy, Sai Ramesh, Mike Ashfold, Anuradha Pallipurath, Priyadarshi Roy Chowdhury, Sanghamitra Mukhopadhyay, Jemmis E D, Himani Medhi, Debabrata Goswami, Prasenjit Halder, Wolfgang Junge, Mahesh Hariharan, Santosh Kumar Singh, Siva Umapathy, Adithya Lakshmannam, Martin Meedom Nielsen, Sankarampadi Aravamudhan, Volker Deckert, Kenneth Ghiggino, Keisuke Tominaga and Alison Edwards

DOI: 10.1039/C5FD90016F

**Wolfgang Junge** began the discussion by asking **Volker Deckert:** If your AFM-tip is optimally placed on a given target, what would be the highest time resolution for a Raman-transient under a single turnover?

**Volker Deckert** answered: We have not done such experiments with our tips yet, however, the results from the van Duyne group report Raman spectra with pulse durations of 1 ps.[1,2]

1 E. A. Pozzi *et al.*, Ultrahigh Vacuum Tip-Enhanced Raman Spectroscopy with Picosecond Excitation, *The Journal of Physical Chemistry Letters*, 2014, **5**, 2657–2661.
2 J. M. Klingsporn, M. D. Sonntag, T. Seideman and R. P. van Duyne, Tip-Enhanced Raman Spectroscopy with Picosecond Pulses, *The Journal of Physical Chemistry Letters*, 2014, **5**, 106–110.

**Siva Umapathy** followed with the question: When you compare the SERS *vs.* the TERS spectrum do you expect the spectral intensities of various Raman peaks to remain the same, considering the molecular conformation of the adsorbed molecule may be different?

**Volker Deckert** responded: The way we compare SERS and TERS spectra, we do not consider and compare the intensities for exactly this reason. However, the SERS spectra, in particular the ones measured on the silver island films, show only minute intensity variations. So, macroscopically different enhancement sites, hot-spots and various conformation changes are being averaged in the focus. The intensity comparison of TERS is a two-fold issue. Comparison of the same tip at different sites results in demonstrated peak shifts and possibly also intensity shifts if the structures are varying on the length scale of the tip sample interaction. Apart from that, on an ideally flat and homogeneous sample, neither intensity variation nor peak shifts are expected. If two tips are compared, however, the different tip geometries will almost certainly result in different enhancement factors and therefore different intensities.

**Martin Zanni** addressed **Volker Deckert:** Single molecules measured with IR do not necessarily need to be lorentzian because their lineshapes will be given by Kubo lineshape theory so it depends on the timescales of motions.

**Volker Deckert** answered: I was discussing Raman line shapes, but the same principles apply of course. Two issues must be considered, firstly the observed line narrowing was not associated with a transition from gaussian to lorentzian. As correctly stated, single molecules do not have to be lorentzian as diffusion and rotation can definitely blur the line shapes. This is definitely the case under the presented conditions (room temperature). However, if when cooling down a change in band shape can be detected, this has been associated with cancelling out motion effects.

**Adithya Lakshmannam** continued the discussion by asking **Volker Deckert** : Tip enhanced Raman scattering (TERS) provides the Raman signature of a species at a very high spatial resolution. This helps in understanding the sequence in DNA. How does TERS help in understanding mitochondria which itself is micron size in dimension?

**Volker Deckert** responded: In general TERS is a surface sensitive technique. Relating to mitochondria specifically, this means, depending on the scattering cross sections of the molecules in question, structural information down to 10-15 nm can be gained. This has been tested by comparing the TERS intensity of lipid signals from the very mitochondrial surface to cytochrome c signals which are known to be in the inter membrane space.[1] So TERS can help in understanding the structure and potentially also the dynamics at the surface with nanometer spatial resolution. If information from deeper layers is required, other nanoscale techniques must be used. At present this would mean sacrificing the possibility to directly obtain structure information or "cutting" the mitochondrium open with all the related consequences and investigate it layer by layer with TERS.

1 R. Böhme *et al.*, Characterizing cytochrome c states – TERS studies of whole mitochondria, *Chem. Commun.*, 2011, **47**, 11453–11455.

**R J Dwayne Miller** opened the discussion of the paper by **Srihari Keshavamurthy:** It is quite a tour de force to map out the different nuclear configurations and resonant terms in the interaction leading to nonstatistical dynamics for dissociation. Even in the 3 degree of freedom problem, you see the need for a bending coordinate to assist dissociation. The question is how to scale this approach to complex systems. We have been able to observe a number of formally chemical processes at the atomic level with sufficient time resolution to map out spatial correlations in the reaction dynamics.[1,2] The systems studied involve many 10's to hundreds of degrees of freedom and yet involve only a few key modes. You make the observation (point 2 in the discussion section) that for high density of state systems the Arnold web is densely populated with resonant junctions and trapping of trajectories that necessarily lead to nonstatistical behaviour – likely as a general rule. This indeed seems to be the case. Can you comment on how to extend your approach to more complex systems to test this idea? It would be a great simplification if we could cast chemistry for even complex systems in terms of key reaction modes.

1 Gao *et al.*, *Nature*, 2013, **496**, 343–346.
2 Jean-Ruel *et al.*, *J. Phys. Chem. B*, 2013, **117**, 15894–15902.

**Srihari Keshavamurthy** replied: A very interesting and challenging question indeed! Currently, our approach has been used to study the IVR dynamics in systems with three degrees of freedom only. The motivation is to finally address the key differences and new dynamical features that arise in going from two to three degrees of freedom. In principle, features like the Arnold web, trapping near resonance junctions and transport on the web will manifest in more complex systems and our model studies highlight the importance of such features. However, scaling our approach to complex systems with several degrees of freedom requires some effort since the Arnold web itself will be high dimensional. One possible approach involves slicing the web at conditions corresponding to specific modes being locked in the resonances of interest and studying the dynamics on the projected web. This can get cumbersome and there may be more intelligent/efficient ways to extract the projected dynamics. On the other hand, we would like to point out that one can use our approach to extract the important junctions for a complex system based on the trapping times near them. This information can be obtained by simply processing the data from our time-frequency analysis and does not require visualization of the web itself. Presumably, the identity of the key resonance junctions can provide insights into the identity of the key modes that control the reaction dynamics. However, this remains an open problem.

**Siva Umapathy** addressed **Srihari Keshavamurthy**: The paper is interesting, however, for a large polyatomic system would it be possible to have such a well defined energy relaxation path way considering (a) the density of states to which we are exciting and (b) the lack of *a priori* knowledge of the vibrational normal mode to which energy is being deposited.

**Srihari Keshavamurthy** responded: Indeed, extending our technique to complex systems with a large density of states is quite challenging. However, a large molecule does not necessarily mean that the IVR state space is also large – it is quite possible for large systems to have only a few modes that actively participate in the IVR dynamics. In this regard, see also the remarks in the previous question by **Dwayne Miller** on our paper. For example, a large body of work by Wolynes, Gruebele, Leitner, and coworkers[1] describes fairly large molecules at significant levels of excitation showing a fairly small effective IVR dimensionality in the state space. What one thus hopes to uncover is the relevant set of modes and then that yields a plausible dominant pathway(s) for IVR. Regarding the second part of the question, our technique is precisely trying to identify the modes where energy is being, perhaps preferentially, deposited.

1 R. Bigwood, M. Gruebele, D. M. Leitner and P. G. Wolynes, *Proc. Natl. Acad. Sci (USA)*, 1998, **95**, 5960.

**Debabrata Goswami** enquired: The Arnold web dynamics is based on the statistical distribution of states wherefrom the excitation energy can be redistributed within the density of states. Could you comment on the change of such an Arnold web structure in trying to get to the right dynamics and control under the effect of shaped pulses? Can we generalize this picture to even NMR principles?

**Srihari Keshavamurthy** answered: We would like to make it clear that the Arnold web, or frequency ratio space, does not presuppose statistical dynamics. In fact, we propose to establish the statisticality of a unimolecular reaction based on certain features of the web – the resonance junctions. Having said that, indeed one of the interests that we have is to use the Arnold web structure to infer the crucial anharmonic resonances and junctions they are forming, leading to IVR. Therefore, any perturbation of the key junction(s) is bound to affect the subsequent IVR dynamics. Such perturbations can come in the form of an external field coupled to the modes and creating field-matter resonances. We believe that one can locally influence the dynamics in this fashion.[1] The challenge however is to come up with some guiding principles to gauge the influence on other weaker parts of the web. We do not think that our approach can be applied in the NMR setting.

1 A. Sethi and S. Keshavamurthy, *Mol. Phys.*, 2012, **110**, 717-727.

**Kunal Dhoke** asked: You have a molecular system with 2 and 3 degrees of freedom, is there any difference between the two systems near the junctions?

**Srihari Keshavamurthy** responded: Yes, there is a fundamental difference in classical dynamics between systems with two and more than two degrees of freedom. Resonance junctions do not exist for two degrees of freedom and are ubiquitous in cases of three and higher degrees of freedom.[1]

1 A. J. Lichtenberg and M. A. Lieberman, *Regular and Chaotic Dynamics*, Applied Mathematical Sciences, Springer Science+Business Media, 1992, ch. 6, pp. 373–456.

**Vincenzo Aquilanti**, on behalf of **S Cavalli** and **D De Fazio**, commented:

One of the several interesting issues stimulated by this paper concerns the question of the validity of the statistical treatment of intramolecular modes involved in reaction rates and in the decompositions of excited molecules. This question, particularly relevant for the venerable TST and RRKM theories and their modern applications, has been standing at the very foundations of theoretical chemical kinetics for almost a century. Approaches have typically relied on frameworks provided by classical mechanics, upon which quantum features are added by semiclassical — at times highly sophisticated — techniques. Recent progress has been greatly helped by advances in the quantum mechanics of structure and dynamics, and interestingly by the experimentally motivated shift of attention towards the striking low temperature ranges attainable in the laboratory and occurring in astrochemical environments.

For definiteness, we refer here to a triatomic case, as in this paper, and more specifically to the archetypal reaction $F{+}H_2 \rightarrow HF + H$ [for earlier work, see our review,[1] and related statistics of quantum levels[2]]. Benchmark calculations for this reaction[3,4] and the isotopic variant F + HD[5–7] are now feasible all the way from the potential energy surface to cross sections (to be compared with molecular beam experiments) and to rate constants (to be compared with gas kinetics data). The system $HeH_2^+$, prototypical of ion–molecule reactions, is of particular interest in plasmas and in astronomical and astrophysical environments.[8–10] The role of quantum effects is expected to be of limited relevance at high temperature, but its relevance increases in the subthermal range as we go down to the cold and ultracold regimes, which have recently been of growing interest. There, theory has predicted the results being currently obtained from demanding experiments, anticipating and interpreting trends thanks to proper accounting of tunnel and resonance effects. Systematics of narrow and overlapping resonances, repulsion between overlapping series, and relationships with metastable trappings in the entrance and exit channels, or in the intermediate complex, have been documented both by lifetime scattering matrix[11,12] and Regge pole[13–14] analysis. An intriguing question arises regarding the Arnold nets of resonances and their junctions, central to the paper by Yadav and Keshavamurthy: do they bear similarities to the quantum resonance phenomenology, regarding sensitivity to the local topology of the potential energy function, and thus provide information on specific features relevant to understanding their role for intermolecular and reactive dynamics?

Finally, low temperature deviations in reaction rates from Arrhenius behavior in the cold regime have been accounted for by heuristically deforming Boltzmann probability distributions.[16-19] Can it be argued that techniques discussed in the paper by Yadav and Keshavamurthy may be relevant to describing deviations from Arrhenius law?


1 V. Aquilanti, S. Cavalli, D. De Fazio, A. Volpi, A. Aguilar, X. Giménez and J. M. Lucas, *Phys. Chem. Chem. Phys.*, 2002, **4**, 401–415.
2 L. Peroncelli, G. Grossi and V. Aquilanti, *Mol. Phys.*, 2004, **102**, 2345–2359.
3 V. Aquilanti, S. Cavalli, D. De Fazio, A. Volpi, A. Aguilar, X. Giménez and J. M. Lucas, *Chem. Phys. Lett.*, 2003, **371**, 504–509
4 V. Aquilanti, S. Cavalli, D. De Fazio, A. Volpi, A. Aguilar and J. M. Lucas, *Chem. Phys.*, 2005, **308**, 237–253.
5 D. De Fazio, V. Aquilanti, S. Cavalli, A. Aguilar and J. M. Lucas, *J. Chem. Phys.*, 2006, **125**, 133109.
6 D. De Fazio, S.Cavalli, V. Aquilanti, A. Buchachenko and T. V. Tscherbul, *J. Phys. Chem. A*, 2007, **111**, 12538–12549.
7 D. De Fazio, V. Aquilanti, S. Cavalli, A. Aguilar and J.M. Lucas, *J. Chem. Phys.*, 2008, **129**, 064303.
8 V. Aquilanti, G. Capecchi, S. Cavalli, D. De Fazio, P. Palmieri, C. Puzzarini, A. Aguilar, X. Gimenez and J. M. Lucas, *Chem. Phys. Lett.*, 2000, **318**, 619–628

9 D. De Fazio, M. de Castro, A. Aguado, V. Aquilanti and S. Cavalli, *J. Chem. Phys.*, 2012, **137**, 244306–244320.
10 D. De Fazio, *Phys. Chem. Chem. Phys.*, 2014, **16**, 11662–11672.
11 V. Aquilanti, S. Cavalli, A. Simoni, A. Aguilar, J. M. Lucas and D. De Fazio, *J. Chem. Phys.*, 2004, **121**, 11675–11690.
12 V. Aquilanti, S. Cavalli, D. De Fazio, A. Simoni and T. V. Tscherbul, *J. Chem. Phys.*, 2005, **123**, 054314.
13 D. Sokolovski, S.K. Sen, V. Aquilanti, S. Cavalli and D. De Fazio, *J. Chem. Phys.*, 2007, **126**, 084305.
14 D. Sokolovski, D. De Fazio, S. Cavalli, and V. Aquilanti, *J. Chem. Phys.*, 2007, **126**, 121101.
15 D. Sokolovski, D. De Fazio, S. Cavalli and V. Aquilanti, *Phys. Chem. Chem. Phys.*, 2007, **9**, 5664–5671.
16 D. De Fazio, J. M. Lucas, V. Aquilanti and S. Cavalli, *Phys. Chem. Chem. Phys.*, 2011, **13**, 8571–8582.
17 V. Aquilanti, K. C. Mundim, S. Cavalli, D. De Fazio, A. Aguilar and J. M. Lucas, *Chem. Phys.*, 2012, **398**, 186–191.
18 V. H. C. Silva, V. Aquilanti, H. C. B. de Oliveira and K. C. Mundim, *Chem. Phys. Lett.*, 2013, **590**, 201–207.
19 S. Cavalli, V. Aquilanti, K. C. Mundim and D. De Fazio, *J. Phys. Chem. A*, 2014, **118**, 6632–6641.

**Srihari Keshavamurthy** responded: Thank you for the references and summary of the detailed work done by your group and others. If one does have a situation wherein a long lived intermediate complex is formed then the intramolecular dynamics of the complex can be studied using the techniques used in our paper. In a way, one can imagine this to be the unimolecular dynamics of a complex intermediate. For the reactive scattering problem, assuming one has a bottleneck region, it would be very interesting to study the Arnold web structure in the transition region and correlate it with the lifetime scattering matrix and possibly the local topology of the potential surface. However, very little efforts have been made in this direction. Regarding the non-Arrhenius behavior, it is not immediately apparent to us as to whether our work can shed any light on this issue. Partly, this is due to the fact that our current work is at zero temperature and on isolated molecule dynamics. However, if there are deviations due to tunneling, then we do believe that the resonance network can be helpful. Particularly, dynamical tunneling has been shown to intricately depend on the network of resonances.[1] How one may, if at all, connect to the heuristic deformation of the Boltzmann distribution is not yet apparent to us.

1 S. Keshavamurthy, *Int. Rev. Phys. Chem.*, 2007, **26**, 521–584, and references therein.

**Upendra Harbola** addressed **Srihari Keshavamurthy**: In this work, the bond breaking process in a three atom molecule is studied using the classical Hamiltonian dynamics. My question is, since it is an isolated system, there is no "real" bond breaking process due to recurrences. So, what are the criteria used to define the bond breaking? Also, is there strictly no bond-breaking if only two stretching modes are coupled or is it much less likely as compared to the case when the bending mode is connected? How does the Arnold web change with time, do the number of junctions and their order change as the system evolves in time? How crucial is the approximation of taking the G-matrices (coupling matrix) as the equilibrium G-matrix?

**Srihari Keshavamurthy** replied: This is an isolated system. However, the bond does break (as it happens in any gas phase unimolecular reaction) and we use the criteria used by Bunker in his original studies involving the specific distance exceeding a certain threshold value. Moreover, monitoring the trajectory gives a clear indication of the bond breaking. Bending certainly plays an important role, but the stretch mode sub-system also leads to bond breaking provided there is sufficient energy for dissociation. The Arnold web, strictly speaking, provides a template of the various types of nonlinear resonances that can manifest at the energy of interest. The web has several junctions formed by independent resonances of various orders. The dynamical evolution of the system on the web determines the regions on the Arnold web that are crucial in a given time range. So, the number/order of the junctions does not change but the regions that are sampled by the dynamics do change and reflect the key resonances that are influencing the energy flow dynamics.

The approximation of taking the G-matrix at equilibrium geometry means that certain other couplings are being ignored. They may change the IVR dynamics but one does not *a priori* know whether it will lead to statistical decay or not. The Arnold web picture and the ensuing interpretation, however, are not limited by this approximation. Note that Bunker in his original work did not make this approximation. However, the subsequent analysis by Oxtoby and Rice[1] does invoke this approximation and they find quantitative agreement with Bunker's predictions.

1 D. W. Oxtoby and S. A. Rice, *J. Chem. Phys.*, 1976, **65**, 1676.

**Sai Ramesh** asked: The inter-mode couplings in the model Hamiltonian of your paper are quadratic couplings that are kinetic in origin. Indeed, these kinds of couplings are important and hold the key to the essential dynamics in the system. However, to what extent would the bilinear potential couplings affect the findings of the paper? Depending on the signs of the potential and kinetic contributions, the effective coupling constants in equation (8) would be modified, perhaps quite significantly in some chemical systems. While the potential terms wouldn't alter the resonance conditions, they would likely alter the dynamics. Would it be possible to get a qualitative sense of the magnitudes of the changes in terms of, say, the dissociation timescales that the paper discusses?

**Srihari Keshavamurthy** answered: The model that we used is more appropriately labeled as a local mode picture. In the normal mode representation one would have potential couplings. Nevertheless, having both in our model will, as you rightly point out, change certain effective coupling constants without changing the nature of the resonance. It is crucial though to realize that even weak resonances can have nontrivial consequences for long lifetime IVR. The key here, as established by Leitner and Wolynes,[1] is that weak high order resonances establish the transition to facile IVR but the rate is controlled by low order resonances. In any case, our analysis will still hold and we believe that our conjecture is general enough – it depends only on the dynamics near the junctions. On the other hand, it is not easy to even qualitatively predict the impact of such additional couplings on the lifetime distributions.

1 D. M. Leitner and P. G. Wolynes, *J. Chem. Phys.*, 1996, **105**, 11226.

**Elangannan Arunan** queried:

(i) Would the conclusions about the lack of IVR alter if rotations were included in theory?

(ii) As your model considers only vibrational motions, would the results be more applicable for reactions in solutions rather than isolated molecules in the gas phase? In the solutions, the inter-molecular vibrations may play a significant role.

**Srihari Keshavamurthy** replied: Certainly, inclusion of rotations can change the nature of IVR. However, certain forms of rotation–vibration coupling may lead even to restricted IVR and certain others can accelerate IVR. For instance, in the work by Parmenter and Stone,[1] the torsional modes of a methyl rotor have been implicated in enhanced IVR in *para*-fluoro Toluene in comparison to *para*-difluoro benzene. A beautiful analysis by Perry *et al.* highlighted the complexities of IVR in the presence of rotations.[2] In their systems, but-1-yne and ethanol, the rotational coupling still does not lead to a completely statistical IVR. Our analysis will be useful for short lifetime IVR in a molecule in the condensed phase with solvents that are not H-bonded. However, once the intermolecular couplings take over, the longer lifetime IVR will need some more effort to understand. Moreover, the issue of slow+fast modes undergoing IVR has not yet been analyzed from the perspective of our work.

1 C. S. Parmenter and B. M. Stone, *J. Chem. Phys.*, 1986, **84**, 4710.
2 David S. Perry, G. A. Bethardy, Michael J. Davis and J. Go, *Faraday Discuss.*, 1995, **102**, 215-226.

**Anuradha Pallipurath** asked: Amongst all the resonance junctions why is the 1,1 considered the most important?

**Srihari Keshavamurthy** replied: This particular junction is formed by the intersection of two low order anharmonic (nonlinear) resonances. Low order resonances, apart from being the leading order perturbations that arise in a molecule, have larger strengths when compared to high order resonances. Thus, the junctions formed by low order resonances are more crucial then the ones formed by the intersection of high order resonances.

**Anuradha Pallipurath** said: Perhaps a long shot but could your work be applied to non-equilibrium processes in the future – like ball milling for instance?

**Srihari Keshavamurthy** responded: Well, we do have a dynamical system here and we do not assume equilibrium. So, our approach can be used for any dynamical system. But, we do not have any idea as to whether there is any relation or use to ball milling or any such process.

**Himani Medhi** opened the discussion on the paper by **Eluvathingal Jemmis**:

(i) The Z atom has the ability to form Z-bonding like H-bonding when the Z atom is electron sufficient in the X–Z species. Will you please elaborate on this statement?

 

(ii) Can we explain all types of coordination bonds as like Z-bonds? Can the bond $BF_3$–$NH_3$ be referred to as a B-bond?

**Jemmis E D** replied:

We consider the X–H···Y interaction as a H-Bond if it follows the various criteria defined for a H-Bond. One factor that is taken for granted is that the H has already satisfied the traditional valence of one. The H-Bond is in addition to this. This aspect of the definition is obvious when we consider halogen bonds or chalcogen bonds. Thus X–Cl···Y corresponds to the chlorine bond. There is no room for confusion here as the chlorine is also monovalent and has fulfilled its octet of electrons in X–Cl. It is important to note that in representing the sulphur bond as X–S···Y, we assume that X–S represents a molecule where the divalence of sulphur is already satisfied. Thus in MeHS···$NH_3$, XS corresponds to MeSH and S has already satisfied its octet. Similarly the carbon bond given by E. Arunan is an XC···Y interaction where the carbon in XC has already satisfied its tetravalence.[1]

The second question requires more explanation. $H_3B$–$NH_3$ or any other obvious coordination bond cannot be defined as a Z-Bond as it stands now. However, as the Table below (MP2=full/6-311+g(d,p)) shows, the distinction fades away gradually with appropriate Lewis acids and bases. The interaction of $BF_3$ with NCH is not strong and with $N_2$ as the base it is very weak.[2] It is not possible to treat either of these as a coordinate bond as we understand it. The geometry of $XF_3$ is not affected much in the complex. The B···N distance is very long. The term weak interaction is appropriate here, represented by XB···Y. According to me this is better termed as a B-Bond (boron-bond), even though formally speaking the boron in the XB molecule has not saturated its valence shell. This brings in the larger question of a bonding continuum. In the presentation I emphasized the continuum nature of blue- and red-shifts in X–Z···Y interactions. It seems artificial that we need to differentiate the interaction of the lone pairs of Y with the vacant sigma* level of the X–H bond in X–H···Y as a weak interaction and the weak donation of the lone pair from NCH to the p orbital of boron (F3B···NCH) in $F_3B$ as something else. This is perhaps the right time to see entire bond formation as a continuum, going from the 2c–2e C–C bond to the weak hydrogen-, carbon- and other Z-bonds as a part of a continuum. The idea lends itself to obvious extrapolation.

| Complex $X_3B\cdots NH_3$ | D(B···N) (Å) | $\Delta E_{ZPE\text{-}BSSE}$ (kcal/mol) | Δ(XBX) (degrees) | Δd(B–X) (Å) |
|---|---|---|---|---|
| $H_3B\cdots NH_3$ | 1.653 | 21.71 | −6.27 | +0.01830 |
| $F_3B\cdots NH_3$ | 1.665 | 13.96 | −5.80 | +0.05957 |
| $F_3B\cdots NCH$ | 2.434 | 2.97 | −3.43 | +0.00594 |
| $F_3B\cdots N_2$ | 2.791 | 0.97 | −0.85 | +0.00097 |

1 D. Mani and E. Arunan, *Phys. Chem. Chem. Phys.*, 2013, **15**, 14377–14383.
2 S. J. Grabowski, *ChemPhysChem*, 2014, **15**, 2985–2993.

**Artur Nenov** asked: X–Z···Y bonds are generally weak. Can you comment on the stability of these bonds in the spirit of the definition of a hydrogen bond given by Goswami and Arunan that "for a hydrogen bonded complex the zero point energy along any large amplitude vibration coordinate that destroys the orientational preference for the hydrogen bond should be significantly below the barrier along that coordinate so that there is at least one bound level"? Are the X–Z···Y bonds strong enough to impose a linear arrangement and if not how do you expect this to affect the negative hyperconjugation?

**Jemmis E D** answered: In most examples the cited "Arunan" criterion is met. The occurrence of negative hyperconjugation at several parts of the X–Z molecule is important in the context of blue- or red-shifting of the X–Z bond after the Z-Bond (*i.e.* X–Z···Y) is formed. This can happen even in very weak Z-Bonds which may not be experimentally realizable Z-Bonds. In other words a decrease or increase in hyperconjugation is possible in weak and strong Z-Bonds.

**Himani Medhi** queried **Jemmis E D**: To form a Z-bond a Z atom should be electron sufficient in the X–Z species. But, in the case of $F_3B\cdots NH_3$, the B atom is not electron sufficient yet still B has the ability to form Z-like bonds. What is your opinion?

**Jemmis E D** replied: $F_3B\cdots NH_3$ is a weak regular coordinate bond. The point I want to make is that there is a gradual change from the regular to the weak bond. If $NH_3$ is replaced by $N_2$, the bonding in $F_3B\cdots N_2$ becomes closer to the regular weak bond (*i.e.* the Z-Bond, in this case the B-Bond). Please also see the answer to the earlier question raised by Himani Medhi.

**Kunal Dhoke** opened the discussion of the paper by **Elangannan Arunan:** Did you observe any RCP (Ring Critical Point) during the AIM (Atoms in Molecule) calculation?

**Elangannan Arunan** responded: Yes of course. When the intra-molecular hydrogen bond critical point (BCP) appears between the H from one OH and the oxygen from the other OH, a ring is formed and an RCP does appear. In Fig. 6 of our paper, the RCP is shown in red and the BCPs are shown in green.

**John R. Helliwell** commented: In your paper page 4 lines 45 to 47 you ask: "Could we have a molecular system in which a bond between two atoms is not apparent at the equilibrium structure but forms during the course of a normal mode vibration?" I can offer as an example the study "X-ray and molecular dynamics studies of concanavalin A glucoside and mannoside complexes: Relating structure to thermodynamics of binding".[1]

In this study the equilibrium structure is that from X-ray crystallography. We sought to explain the differing energetics of binding to the protein of these two sugars, as measured by calorimetry. Molecular dynamics simulations based on the X-ray crystal structure revealed fleeting interactions which gave an improved fit to the energetics data.

1 G. M. Bradbrook, T. Gleichmann, S. J. Harrop, J. Habash, J. Raftery, A.J. Kalb (Gilboa), J. Yariv, I. H. Hillier and J. R. Helliwell, *Faraday Trans.*, 1998, **94**(11), 1603–1611.

 

**Elangannan Arunan** answered: Thank you very much for bringing this to our attention. We are pleased to learn of examples already existing in the literature showing that a bond that does not exist in an equilibrium structure could be formed due to vibrational dynamics. Indeed, a static definition of a bond is inadequate.

**Jonathan Skelton** commented: Linking in with other comments on this work, it is fast becoming a topical question in theoretical chemistry/materials science whether or not calculations should take into account the effects of structural dynamics (*e.g.* through lattice-dynamics or molecular-dynamics calculations) to go beyond results obtained from the 0 K equilibrium structure. Examples in the solid-state include thermal expansion[1] and renormalisation[2] of the electronic structure due to electron–phonon coupling.

1 J. M. Skelton, S. C. Parker, A. Togo, I. Tanaka and A. Walsh, *Phys. Rev. B*, 2014, **89**, 205203, DOI: 10.1103/PhysRevB.89.205203.
2 Z. M. Gibbs, H. Kim, H. Wang, R. L. White, F. Drymiotis, M. Kaviany and G. J. Snyder, *Appl. Phys. Lett.*, 2013, **103**, 262109, DOI: 10.1063/1.4858195.

**Elangannan Arunan** responded: Thanks for your comment and references. Indeed, the dynamics is important and the advent of experimental and theoretical techniques in the last few decades has allowed us to worry about them. A static definition of a chemical bond is inadequate. The example discussed in our work, hydrogen sulphide, is a classic case. It shows isotropic interactions when it is frozen at −82 degrees C and as it is cooled to lower temperatures at higher pressures, the structure becomes orderly resembling ice and exhibits anisotropy.

**Santosh Kumar Singh** queried:

(i) Was the potential energy of ethane-1,2-diol calculated along the O–O coordinate at room temperature?

(ii) If at a higher temperature there is a possibility of intramolecular hydrogen bonding, then is it possible to see the spectroscopic signature of the hydrogen bond (O–H stretching frequency) through any spectroscopic methods?

**Elangannan Arunan** responded: The potential energy is at zero Kelvin. Depending on the temperature, the vibrational distribution along this mode will change. Even at zero Kelvin, the molecule will have the O–O bond vibrating with zero point energy. Within this oscillation, the calculations do not show any bond formation according to the Atoms in Molecules (AIM) theory. If the $\nu = 1$ is populated, the distance between the two O atoms can reduce to an extent that AIM analysis does reveal the formation of an intramolecular hydrogen bond. At room temperature, assuming a Boltzmann distribution, there will be a significant population (more than 35%) in this level and the intramolecular hydrogen bond should be there. This should lead to observable changes in the IR and NMR spectrum as a function of temperature.

**R J Dwayne Miller** commented: I appreciate that the motion along the normal mode O–O coordinate (ethane-1,2-diol) can affect the local electron density and by that the instantaneous frequency of the OH. However, there is no well defined minimum in the potential that would give this configuration a finite lifetime that could be ascribed to a bond. Shouldn't the definition of H-bonding need to also include the interaction in question having a finite lifetime, *i.e.* the state can be described for some time as a population? To put this question in another way, how can we distinguish this effect from mode–mode coupling between the O–O stretching coordinate and the OH?

**Elangannan Arunan** replied: This is an important question and indeed we do have contributions from O–H stretching to the normal mode dominated by the O···O stretching mode. Our calculations show that a red-shift of about 150 $cm^{-1}$ is seen in the O–H stretching frequency on both sides of the minimum shown in Fig. 6 of our paper and so this frequency shift cannot be used as 'evidence' for a hydrogen bond. We decided to calculate the O–H stretching frequency after identifying a bond critical point when the O···O distance reduced and were pleased to see a large red-shift. To be certain, we repeated the frequency calculations as the O···O distance increased and this also led to similar shifts in O–H stretching. Hence, this experiment cannot give evidence for a 'hydrogen bond'. In any case, it will be interesting to see the oscillations in O–H stretching frequency as the nominal O···O stretching vibration happens.

We do agree that a finite lifetime is important to have any 'stable/bound' species. However, a transient 'bond' can still be important in controlling the properties of a molecule. We were pleasantly surprised to find that Prof. John R. Helliwell has already observed such transient 'bonds' in crystals. Dr Jonathan Skelton has pointed out that similar discussions are happening among solid state physicists as well. I think the time is ripe for a 'transient bond' and I refer to an article on 'vibrational bonding' that was just published towards the end of 2014.[1]

1 D. G. Fleming, J. Manz, K. Sato and T. Takayanagi, *Angew. Chem., Int. Ed.*, 2014, **53**, 13706–13709.

**Priyadarshi Roy Chowdhury** asked:

(i) For the case of ethane-1,2 diol and its potential energy diagram (Fig. 6 of your article); beyond the potential energy limit is the deformation restricted to the O···O bond only or does it extend to the entire molecule?

(ii) In ethane 1,2 diol, do the O···Hstr and O···O vibrations take place simultaneously or not?

**Elangannan Arunan** responded:

(i) The nominal O···O stretching discussed in the manuscript is one of the normal modes of vibration for ethane-1,2-diol dominated by O···O stretching. It does involve O···H stretching. As mentioned during the talk, we noted that the O–H stretching frequency red-shifts to 150 $cm^{-1}$ as the O···O distance increases. This was an exciting observation as we thought that this was due to the formation of a hydrogen bond. However, similar shifts happen even when the O···O distance reduces. It was found that the O···H distance changes along this coordinate as the molecule vibrates away from the equilibrium value on both sides, see Fig. 6 in the manuscript. A bond critical point, however appears only when the O···O distance reduces from the equilibrium value.

(ii) Ethane-1,2-diol, with 10 atoms, has 24 normal modes of vibration and all of them occur simultaneously. If one were to really look at the way the atoms dance somewhat independently and arbitrarily, it would be difficult to observe any normal mode vibrations. Normal modes are obtained by normal mode analysis by finding 3N-6 orthogonal vibrational coordinates. If only one mode, such as O···H stretching, is excited to a high overtone state, it may be possible to observe that the vibration is largely localized along this coordinate. It is useful to think about translational degrees of freedom. Whether a cricket ball or atoms, they do not need to move along the *X*, *Y* or *Z* coordinates and they may actually move any which way. However any translational motion can be considered as the sum of motion along the *X*, *Y* and *Z* coordinates, which are the normal modes for translational degrees of freedom. When one looks at a molecule dancing at a given temperature, it includes translational, rotational and vibrational motions, all occurring simultaneously.

**Priyadarshi Roy Chowdhury** asked a further question: Are there any other systems beyond the $H_2S$ dimer that could easily demarcate between the H-bond and van der Waals interactions for ethane-1,2 diol systems? What would happen if the molecule exists in an intermediate conformation between the staggered and the eclipsed?

**Elangannan Arunan** replied: we do not see a relation between the $H_2S$ dimer and ethane-1,2-diol. We used the $H_2S$ dimer just to illustrate that the characterization of 'hydrogen bonding' as different from 'van der Waals interaction' is arbitrary and condition dependent.

Ethane-1,2-diol is unlikely to 'exist' in an intermediate conformation as there are no other minima. The molecule could explore these regions as it vibrates and one might say it can have a 'transient existence' in intermediate conformations. It should be remembered that there are 10 unique conformers for this molecule due to torsions about the two C–O bonds and one C–C bond and the conformers are named gGg, gTt, *etc.*, to indicate that it has a *gauche*/*trans* form along each of these bonds. The middle letter is for the C–C bond. More information on the conformers may be found in the literature.

**Vincenzo Aquilanti**, on behalf of **F. Pirani** and **D. Cappelletti**, remarked:

The results reported in the paper by Arunan and Mani are exemplary of *ab initio* evaluations of the energetics and structure of systems involving inter- and intra-molecular hydrogen bonds. Progress in the understanding of the nature of these bonds comes from experimental systematic characterizations of the ranges and strengths of the involved forces, possibly accounting for their anisotropies and for the role of the various contributions to the global interactions: welcome are analyses based on state-of-the-art quantum chemical calculations.

Over the years, in our laboratory molecular beam scattering experiments have been carried out on a variety of systems: relevant here are those including water, ammonia, hydrogen sulfide (some of these systems are referred to in the paper by Arunan and Mani). High angular and velocity resolution conditions permitted measurements of the uncertainty-principle limited quantum mechanical integral cross section for collisions of these and other molecules with atomic and molecular targets.[1–6] The observation of interference effects (the "glory") was crucial for the determination of the absolute scale of the interaction in a wide intermolecular distance range. The obtained accuracy demands proper analysis of the interaction components such that the accompanying quantum chemical *ab initio* calculations[7–9] are carried out with a highly extended basis set – often requiring extrapolation to an infinite order. The emerging picture for the interaction in these weakly bound systems, prototypical of the insurgence of the hydrogen bond, albeit in embryonic form,[1] has its focus on the height and location of potential energy barriers, associated with hindered internal motions, and especially on their anisotropy. Insight comes from considering how the basic components scale differently both with the intermolecular distance $R$ and with the angles of mutual orientation: (i) the ubiquitous "physical" van der Waals component $V_{vdW}$, (which comprehends the molecular size repulsion and the dispersion attraction) at large $R$ scales as $R^{-6}$ and has a much weaker dependence on orientation than that of (ii) the "chemical" charge transfer contribution $V_{ct}$ , which decays exponentially with $R$ and is strongly orientation dependent.

In the $H_2O$–noble-gas systems[7,10] the absolute interaction anisotropy increases along the series, going from He to Xe due to changes both of the absolute and relative contributions of $V_{vdW}$ and $V_{ct}$. Similar trends have also been quantitatively established for the series involving $NH_3$ and $H_2S$. Accordingly, the number and type of quantum levels, confined in the intermolecular interaction potential wells, vary regularly along each series, depending also on the ratio between the force constant and the reduced mass of each system. This establishes the basic features of the dynamics of weak chemical bonds, of interest to Arunan and Mani, also beyond these elementary systems.

Indeed, recently[11,12] we investigated in detail experimentally the stereo-dynamics of water–water encounters. Molecular orientation effects, appearing as anomalously large measured cross sections, are interpreted as arising from the selective role of the long-range anisotropy associated to the dipole–dipole interaction. These long-range effects originate from intermolecular electric field gradients sufficient to promote the formation of pendular states even at intermolecular distances of around 1 nm, *i.e.* when the interaction anisotropy becomes comparable or even exceeds the separation of the molecular rotational levels.

The picture of stereo-dynamically driven collisions between interacting partners with possible orientation/alignment effects promoted by intermolecular interactions, can also be of relevance for the HX–Ng systems discussed in the paper by Arunan and Mani (X and Ng are halogen and noble gas atoms, respectively). We point out similarities with the X–Ng systems on account that both the HX molecule and the open shell halogen atom X exhibit an internal angular momentum (rotational for HX and electronic for $X(2P_J)$). At large $R$ the interaction is substantially isotropic in both cases.[13] At short $R$ the effective potential energy depends on the balance of isotropic, anisotropic and centrifugal interaction components with the spin–orbit splitting in $X(2P_J)$–Ng and with the roto-vibrational quantum levels in HX–Ng . A transition from atomic to diatomic and possibly molecular states occurs during X–Ng encounters.[14–19] Correspondingly, in the dynamics involving diatomic molecules[20] the transition is operative from free rotation to hindered rotation and finally pendular states. We would be curious to have Professor Arunan's opinion on these remarks.

1 V. Aquilanti, E. Cornicchi, M. Moix-Teixidor, N. Saendig, F. Pirani and D. Cappelletti, *Angew. Chem., Int. Ed.*, 2005, **44**, 2356.
2 D. Cappelletti, A. F. A. Vilela, P. R. P. Barreto, R. Gargano, F. Pirani and V. Aquilanti, *J. Chem. Phys.*, 2006, **125**, 133111.
3 V. Aquilanti, D. Cappelletti, F. Pirani, L. F. Roncaratti, *Int. J. Mass Spectrometry*, 2009, **280**, 72.
4 F. Pirani, L. F. Roncaratti, L. Belpassi, F. Tarantelli and D. Cappelletti, *J. Chem. Phys.*, 2011, **135**, 194301.
5 F. Pirani, D. Cappelletti, L. Belpassi and F. Tarantelli , *J. Phys. Chem. A*, 2013, **117**, 12601.
6 A. Bartocci, D. Cappelletti, F. Pirani, F. Tarantelli and L. Belpassi, *J. Phys. Chem. A*, 2014, **118**, 6440.
7 L. F. Roncaratti, L. Belpassi, D. Cappelletti, F. Pirani and F. Tarantelli, *J. Phys. Chem. A*, 2009, **113**, 15223.
8 L. Belpassi, M. L. Reca, F. Tarantelli, L. F. Roncaratti, F. Pirani, D. Cappelletti, A. Faure and Y. Scribano, *J. Am. Chem. Soc.*, 2010, **132**, 13046.
9 G. Bistoni, L. Belpassi, F. Tarantelli, F. Pirani and D. Cappelletti, *J. Phys. Chem. A*, 2011, **115**, 14657.
10 D. Cappelletti, E. Ronca, L. Belpassi, F. Tarantelli and F. Pirani, *Acc. Chem Res.*, 2012, **45**, 1571.
11 L. F. Roncaratti, D. Cappelletti and F. Pirani, *J. Chem. Phys.*, 2014, **140**, 124318.
12 L. F. Roncaratti, D. Cappelletti, P. Candori and F. Pirani, *Phys. Rev. A*, 2014, **90**, 012705.
13 F. Pirani, G. S. Maciel, D. Cappelletti and V. Aquilanti, *Int. Rev. Phys. Chem.*, 2006, **25**, 165, and references therein.
14 F. Pirani, A. Giulivi, D. Cappelletti and V. Aquilanti, *Mol. Phys.*, 2000, **98**, 1749.
15 V. Aquilanti, D. Cappelletti and F. Pirani, *Chem. Phys. Lett.*, 1997, **271**, 216.
16 V. Aquilanti, D. Cappelletti, V. Lorent and F. Pirani , *J. Phys. Chem.*, 1993, **97**, 2063.
17 V. Aquilanti, D. Cappelletti and F. Pirani , *J. Chem. Soc. Faraday Trans.*, 1993, **89**, 1467.
18 V. Aquilanti, S. Cavalli, F. Pirani, A. Volpi and D. Cappelletti, *J. Phys. Chem. A*, 2001, **105**, 2401.

**Elangannan Arunan** answered: Thank you for your comment and pointing out that even at 1 nm distance, anisotropy in intermolecular interactions can be found.

I am aware of the beautiful series of experiments using the crossed beam apparatus at Perugia which explored the intermolecular potential surface accurately. The only difficulty I have is the way the van der Waals interaction is defined in this series of papers. They are all consistent within themselves. However, I have seen the van der Waals interaction equated to (i) only dispersion;[1] (ii) exchange repulsion and dispersion;[2] (iii) size repulsion, induction and dispersion in your Angew Chem paper, but induction missing in your query; (iv) a long-range weak attraction between permanent and induced electric dipole (and higher) moments as well as instantaneous asymmetric charge distributions in atoms and molecules (dispersion interaction);[3] and (v) the forces of attraction and repulsion between molecules are called van der Waals forces.[4] We have a recent perspective in PCCP where we discuss all intermolecular 'bonding' in terms of ionicity and covalency as for intramolecular bonding.[5] Interaction energy is one and there are numerous ways it can be decomposed and in my view all of them give an incomplete picture. Directionality is the key for hydrogen bonding as beautifully demonstrated by the structure of ice and solid $H_2S$.

1 E. R. Bernstein, *Annu. Rev. Phys. Chem.*, 1995, **46**, 197.
2 Desiraju and Steiner, *The Weak Hydrogen bond*, Oxford University press, 1999, pp. 17.
3 Koperski, *van der Waals complexes in supersonic beams*, Wiley VCH, 2003.
4 A. J. Stone, *The Theory of Intermolecular Forces*, Clarendon Press, Oxford, 1996.

5 A. Shahi and E. Arunan, *Phys. Chem. Chem. Phys.*, 2014, **16**, 22935–22952.

**Srihari Keshavamurthy** continued the discussion of the paper by **Elangannan Arunan**: According to the definition proposed, there needs to be at least one bound state below the barrier in order to have a hydrogen bond. Nevertheless, from the quantum mechanical point of view, even a state above the barrier has a finite reflection probability. Thus, is it possible to give a more refined definition from this perspective? Also, does the definition take into account the multidimensionality of the barrier due to coupling between the relevant modes?

**Elangannan Arunan** replied: We agree that the definition insisting on one bound level is somewhat classical and there can be a finite reflection probability above the classical barrier. Moreover, if one were to calculate the wave function for the energy level just above the barrier, it is likely that it is largely localized within the potential well. It may be a worthwhile effort to calculate the wave functions for energy levels above the barrier and see when they are spread all over the torsional coordinate (dihedral angle). One might then say that the probability of finding the hydrogen atom within the well should be above 50% to be called a hydrogen bond. While these considerations are important and would appeal to the physical chemistry community, for most chemists and biologists, a simpler definition would be more appealing.

**Prasenjit Halder** asked: Does the 'transient bond' concept (molecular vibration leads to hydrogen bond) explain why the *gauche* conformer is more stable than the *trans* conformer? As the 'transient bond' is dynamic in nature, how does it stabilize the *gauche* conformer?

**Elangannan Arunan** explained: The preference for *gauche* conformer can be justified as the molecules are never static, not even at zero Kelvin. The 'structure' is dynamic and the molecule does explore all regions available with the thermal energy it has. Accurate equilibrium calculations considering all the energy levels of various conformers should yield the relative population correctly and this will be influenced by the transient bonds.

**R J Dwayne Miller** opened a general discussion of the paper by **Keisuke Tominaga**: I would like to make a connection to the work presented by Prof. Jemmis where he discusses how hydrogen bonds can lead to either red or blue shifts in the vibrational frequency depending on how the electronic density is affected at the donor group by the changes in H-bond length. This paper was quite enlightening. Photoexcitation necessarily changes the electron density at the carbonyl group that would then change any electrostatic interaction with the surrounding solvent for both aprotic and protic solvents. Is it possible that the blue shift in the carbonyl frequency you observe is due to an increase in H-bonding for methanol and the dynamics are related to solvent rearrangement to accommodate this? This question is pertinent as you observe both a blue shift followed by a red shift for methanol (b mode), which is rather unusual if the dynamics were solely due to thermal effects.

 

Did you do a temperature dependence study on the CO stretch frequency in acetonitrile and/or methanol to see if the magnitude of the blue shift corresponds to that expected for the lattice temperature reached after vibrational cooling? From your quantum calculations, it should also be possible to determine if electronic factors are possible in explaining both the transient blue and red shifts.

**Keisuke Tominaga** responded: The temperature dependence for optical-pump IR-probe spectroscopy is very interesting. We are planning to do that, but not yet. We believe that the blue shift is just due to vibrational cooling, which is observed for many other molecules in the excited state. The non-hydrogen bonded molecule shows only a blue shift, and the hydrogen-bonded one shows both blue and red shifts. So, we consider that the red shift observed only for the hydrogen-bonded molecule is related to the hydrogen bonding dynamics.

**Ravi Kumar Venkatraman** addressed **R J Dwayne Miller** and **Keisuke Tominaga**: I would like to answer **R J Dwayne Miller**'s question regarding the electron density at the oxygen atom of the carbonyl site for the 9-fluorenone in the excited state. Ke-li Han and co-workers have done TD/DFT studies on this molecule and showed that on excitation to the $S_1$ state, the electron density drifts from the aromatic ring to the carbonyl site.[1] Further they have also proposed that in the $S_1$ state the hydrogen bond is strengthened relative to the $S_0$ state. Therefore upon photoexcitation to the $S_1$ state, the increased hydrogen bonding red shifts the carbonyl frequency of the hydrogen bonded complex relative to the free carbonyl. Similar work has been done by us recently using steady-state absorption and resonance Raman spectroscopy in conjunction with TD/DFT calculations on 9,10-Phenanthrenequinone and we have proposed that on photoexcitation to the $S_2$ state, the hydrogen bond is strengthened.[2]

So could we associate this hydrogen bond strengthening of the hydrogen bonded complex in the $S_1$ state to your observed initial blue shift and the latter red shift of the 1529.9 cm$^{-1}$ band? The carbonyl stretching mode being a Franck-Condon active mode would populate the higher vibrational levels of the $S_1$ state upon photoexcitation, which then relaxes to the lowest vibrational level by vibrational cooling and could account for the early blue-shift of the 1529.9 cm$^{-1}$ band. While the hydrogen bond strengthening of the $S_1$ state is demonstrated by the latter red-shift of the band.

1 G. J. Zhao and K. Han, *Accounts Chem. Res.*, 2012, **45**, 404.
2 R. K. Venkatraman, N. Rajkumar, S. Umapathy, *J. Chem. Phys.*, 2015, **142**, 024305.

**Keisuke Tominaga** responded: Generally speaking, the vibrational energy relaxation in the $S_0$ state is in a picosecond scale for polyatomic molecules, whereas the relaxation from the higher vibrational states in the electronically excited state is very rapid, faster than a picosecond. So, the blue shift occurring in a few picoseconds to tens of picoseconds cannot be vibrational relaxation from the higher vibrational state.

**Siva Umapathy** asked: This is regarding the lifetime of the breaking of the hydrogen bond and the differences in the dynamics in methanol and ethanol. H-bonding must last much longer in ethanol than methanol which is now what you have. Is this right? Since methanol is known as a fast solvent, the H-bond must break much faster in methanol than ethanol. Therefore when you excite them the reorganisation of the hydrogen bond should be reflected in the frequencies and in the lifetime? Would that be correct?

**Debabrata Goswami** further asked: Hydrogen bonding strength in methanol is much higher than its higher homologues. The hydrogen bonding network is much larger and stronger for methanol as compared to ethanol, propanol, butanol, *etc.* The structure of a solvent is highly dependent on its hydrogen bonding network. We have shown this distinctive nature of solvent character very effectively through ultrafast thermal lens spectroscopy of the alcohol series.[1] The results presented here might follow from the fact that ethanol has a lesser strength of hydrogen bonding.

1 P. Kumar, S. Dinda, A. Chakraborty, D. Goswami, *Phys. Chem. Chem. Phys.*, 2014, **16**(24), 12291–12298.

**Keisuke Tominaga** replied to both **Siva Umapathy** and **Debabrata Goswami**: In the $S_0$ state we measured the FTIR spectra of FL in different linear alcohols, and found that the peak shift of the CO stretch due to hydrogen bond formation is almost the same for all the alcohols we investigated. We also performed optical-pump IR-probe measurements for ethanol solvents; we did not observe a major change between the methanol and ethanol solvent cases. I do not think that the hydrogen bond lifetime of methanol is so different from that of ethanol. For example, the time scale of the solvation dynamics of ethanol is longer than that of methanol, but not so significant.

**Jemmis E D** commented: You have mentioned that with time the excited state C=O stretching frequency of the ketone blue shifts. Could this be a result of the departure of one or two of the solvent molecules so that the CO–H bond is broken?

**Keisuke Tominaga** responded: At every delay time, the two species, hydrogen-bonded and non-hydrogen-bonded fluorenone, can be separately observed. So, this blue shift is due to time-evolution of the carbonyl group of the non-HB FL.

**Siva Umapathy** remarked: In your ethanol spectra, there is a shoulder on the 1550 $cm^{-1}$ band. If we look carefully it looks to me that there is a very interesting time dependant change in the ratio of 1450 to 1550 $cm^{-1}$, an example of blue shift. Is that correct?

**Keisuke Tominaga** answered: The spectral range I showed was 1490 to 1580 $cm^{-1}$. I did not show a spectrum at 1450 $cm^{-1}$. I guess you are confused with the band at 1490 $cm^{-1}$. As you said, the band ratio between 1490 $cm^{-1}$ and 1550 $cm^{-1}$ is time-dependent, and this time dependence is different for ethanol and methanol. We suggested this band is due to a vibrational mode localized in the fluorene ring. We do not know if the intensity of this band should be correlated with that of the CO stretch. Moreover, it is difficult to discuss the intensity of the vibrational band since the width also changes with the delay time. In order to discuss the band intensity more quantitatively we need an area of the vibrational band, which is difficult to obtain accurately.

**Jonathan D. Hirst** asked: Regarding the statement that the nπ* and ππ* states are mixed, can you comment on the mechanism by which this occurs? One might anticipate that the mixing of states of different symmetry types would be limited. And would you comment on how well time-dependent DFT would be expected to describe a multi-configurational electronic excited state?

**Keisuke Tominaga** answered: The mixing of nπ* and ππ* states in the organic molecules has been extensively discussed for a long time. The nature of the excited state of fluorenone has also been studied for many years.[1–3] We are not so familiar with the methodology of the quantum chemical computations. Please refer to the article by Zhao and Han for calculations.[4]

1 A. Kuboyama, *Bull. Chem. Soc. Jpn.*, 1964, **37**, 1540.
2 Yoshihara and D. R. Kearns, *J. Chem. Phys.*, 1966, *45*, 1991.
3 T. Kobayashi and S. Nagakura, *Chem. Phys. Lett.*, 1976, **43**, 429.
4 G.-J. Zhao and K-L. Han, *J. Phys. Chem. A*, 2007, **111**, 9218–9223.

**Artur Nenov** queried: In the experimental section you state that "Calculations for the hydrogen-bonded complex were also conducted". Could you explain what methodology was utilized?

**Keisuke Tominaga** replied: We just put a methanol molecule near fluorenone so that the carbonyl group can form a hydrogen bond with the hydroxyl group of methanol, and performed DFT calculations to optimize the geometry. The basis set and functional we used are mentioned in the text.

**King-Chuen Lin** enquired: Vibrational cooling by relaxation does not seem to occur as fast as the picosecond period in the gas phase, since the lifetime of the vibrational excited state is usually in the ms regime. However, solvent effects may play an important role in speeding up the relaxation in the liquid phase. Please describe how the solvent acts in the vibrational cooling and how this phenomenon proceeds.

**Keisuke Tominaga** responded: Generally speaking, the vibrational energy relaxation takes place in sub-picosecond and picosecond time scales for polyatomic molecules.[1] By IR pulse excitation, particular vibrational modes, which should be IR active, are excited, and vibrational population relaxation as well as intramolecular vibrational redistribution occurs. The excitation energy is dissipated into the surrounding solvent molecules. Finally, the energy will be diffused as heat into the bulk solvent, which can be described by the thermal diffusion equation with a reported thermal conductivity. See, for example, Iwata and Hamaguchi.[2]

1 J. C. Owrutsky, D. Raftery, and R. M. Hochstrasser, *Annu. Rev. Phys. Chem.*, 1994, **45**, 519-555.
2 K. Iwata and H. Hamaguchi, *J. Phys. Chem. A*, 1997, **101**, 632.

**Srihari Keshavamurthy** asked: In the paper it is mentioned that the $S_1$ excited state does not show a good correlation with the Onsager reaction field parameter. In this hydrogen bonding system, should one really expect a coarse parameter like the reaction field to give the right trends?

**Keisuke Tominaga** answered: Since a good correlation can be observed in the $S_0$ state, the Onsager parameter is not a bad parameter to describe a peak shift of the carbonyl group due to solvent polarity.

**Jonathan Skelton** remarked: In this work, the authors are able to measure the vibrational frequencies of 9-fluorenone in the first excited state. This sort of experimental data would be useful to quantitatively evaluate the performance of different theoretical methods in reproducing the properties of electronically-excited molecules. Have the authors explored this? Would their experimental protocol be applicable to other small molecules, which could be used as simple test cases for benchmarking excited-state DFT?

**Keisuke Tominaga** responded: We believe so. In this article we investigated only fluorenone, and optical-pump IR-probe spectroscopy can be applied to any molecule if an excitation pulse has enough energy to pump the molecule.

**Anuradha Pallipurath** noted: The band a' has been assigned to a combination band, more often than not there should be a complimentary band in the near-IR, have you found any evidence for such a band ?

**Keisuke Tominaga** answered: We have not done any measurements in the near-IR region. So, I have no comment on this.

**Anuradha Pallipurath** enquired: What was the resolution with which the spectra were acquired?

**Keisuke Tominaga** responded: The FTIR measurements were conducted at every 1.0 $cm^{-1}$, and approximately every 2.5 $cm^{-1}$ for the time-resolved measurements.

**Martin Zanni** opened the discussion of the paper by **King-Chuen Lin**: The roaming mechanism is conceptually very interesting. The roaming atom mechanism occurs in simple organic molecules – a very broad statement – is there something conceptual/fundamental about that?

 

**Mike Ashfold** expanded on this question: The Introduction to your paper[1] highlights the number of organic molecules for which photodissociation has now been shown to involve a roaming component, but we should not overlook a celebrated non-organic example – the formation of NO following visible excitation of the $NO_3$ radical.[2–4]

Fig. 4 of your paper[1] illustrates the clear similarities between the measured and calculated kinetic energy distributions of the CO($v$=1, $J$) products from the 248 nm photolysis of methyl formate, and the decomposition of these distributions into 'fast' and 'slow' components. These are viewed as manifestations of dissociations *via* the minimum energy path and roaming mediated mechanisms, respectively. Have the individual trajectories falling within the 'slow' component been studied in detail and, if so, is it possible to identify and define characteristics of trajectories that correspond to the roaming dissociation mechanism in this molecule?

1 M. Nakamura *et al.*, *Faraday Discuss.*, 2015, **177**, DOI 10.1039/c4fd00174e.
2 M. P. Grubb, M. L. Warter, A. G. Suits and S. W. North, *J. Phys. Chem. Lett.*, 2010, **1**, 2455.
3 M. P. Grubb, M. L. Warter and S. W. North, *Phys. Chem. Chem. Phys.*, 2012, **14**, 6733.
4 M. P. Grubb, M. L. Warter, H. Y. Xiao, S. Maeda, K. Morokuma and S. W. North, *Science*, 2012, **335**, 1075.

**King-Chuen Lin** and **Vincenzo Aquilanti** answered the questions raised by **Martin Zanni** and **Mike Ashfold** regarding the roaming mechanism: The observation that increasing phenomenology regarding the characterization and the understanding of the roaming mechanism is being accumulated with reference to organic molecules, is certainly correct, but by no means exclusive. Early and current examples, including those in our paper (see refs therein), indeed involve the photolysis of several carbonyl compounds and signatures of roaming came by the late arrival of a rotationally cold CO product in ion imaging experiments. See our paper in this discussion for a *résumé* of some milestones. Often, accompanying classical trajectory simulations can contribute to attaching meaning to what can be considered as the prevailing significance of the name "roaming", contrasting with either "direct" mechanisms, or transition state mediated pathways. Clearly the ten year old name should not be given anytime as alternative roads show up in a supposedly elementary (mostly unimolecular) chemical process, but is fully justified only within a definition of the area of application of the concept, in order to avoid excessive vagueness and to exclude unwarranted generalizations. Let us try to formulate two distinct paradigms, which in our opinion are to be distinguished in applying unambiguously the nomenclature:

(i) The simplest, yet perhaps the most genuine picture of the origin, nature and manifestation of roaming, is associated with pathways on the potential energy surface of a given elementary chemical process, exhibiting what in the established chemical kinetics literature has the characteristics of "loose" transition states, or even of asymptotically flat exit channels. These pathways have features intrinsically alternative to the "camino real", namely the direct road along the minimum energy path through a transition state, technically distinguished by the adjective "tight". In the case of aldehydes, recent papers[1–3] have included extensions to higher members of the series, showing that, as the molecular size increased, the roaming pathway increased in importance with respect to the direct one.

(ii) Often, processes involve more than a single potential surface, and are mediated by nonadiabatic effects in specific neighborhoods where surfaces approach, and even where their conical intersections occur (these features are ubiquitous in modern organic photochemistry). In the case of methyl formate $HCOOCH_3$, fragmented by photolysis into $CH_3OH$ and CO, the latter serves as the "scout" carrying information from explorations of vast territories of the potential energy surface possibly far away from the minimum energy path. In spite of an extensive search, we failed to reproduce computationally the experimental bimodal images of time-of-flight arrival of fast and slow CO for the system dissipating the photoexcitation energy on the ground $S_0$ surface, while in the experiments at various wavelengths we even observed the opening of the triple fragmentation channel in $CH_3O$, H and CO. Accordingly, exploiting the theoretically available information on the existence of a conical intersection available in the energy range of the experiments provided support for the mechanism of photoexcitation as raising the system to an upper branch of the surface, designated $S_1$. There the delivery back to $S_0$ takes place, in a far away region of configuration space corresponding to the conical intersection. And there the molecule is very distorted with respect to both the ground state and the lower transition state, in particular along the C–O bond connecting the incipiently separating HCO and $CH_3O$ radicals (Fig. 2 and 3 in our paper). It is in that region where we initiate the classical trajectories of our dynamical simulations.

1 K.-C. Hung, P.-Y. Tsai, H.-K. Li and K.-C. Lin, Photodissociation of $CH_3CHO$ at 248 nm by time-resolved Fourier-transform infrared emission spectroscopy: Verification of roaming and triple fragmentation, *J. Chem. Phys.*, 2014, **140**,064313.
2 P.-Y. Tsai, K.-C. Hung, H.-K. Li and K.-C. Lin, Photodissociation of Propionaldehyde at 248 nm: Roaming Pathway as an Increasingly Important Role in Large Aliphatic Aldehyde, *J. Phys. Chem. Lett.*, 2014, **5**,190–195.
3 H.-K. Li, P.-Y. Tsai, K.-C. Hung, T. Kasai, and K.-C. Lin, Photodissociation of $CH_3CHO$ at 308 nm: Observation of H-roaming, $CH_3$-roaming, and transition state pathways together along the ground state surface, *J. Chem. Phys.*, 2015, **142**, 041101.

**Elangannan Arunan** addressed **Vincenzo Aquilanti** and **King-Chuen Lin**: Is roaming limited to unimolecular reactions? In a bimolecular reaction, the transition state would not have any 'lifetime' by definition and it would not support roaming by an atom or a group of atoms. If that were to happen, there needs to be a long-lived intermediate and then one can think of it again as a unimolecular reaction of this 'complex'.

**King-Chuen Lin** and **Vincenzo Aquilanti** answered: The question posted by **Elangannan Arunan** on whether there is documentation on roaming other than in photofragmentations and isomerizations (*i.e.* whether the phenomenon is essentially limited to the decomposition of highly excited molecules and to suprathermal unimolecular processes), has been circulated widely since the nomenclature was introduced a decade ago. Even previous extensive chemical reactions experimentally investigated over decades of chemical kinetics and theoretically tackled by quantum mechanics or simulated by classical trajectories, have been carefully scrutinized to possibly discover previously overlooked manifestations of roaming, specifically in bimolecular elementary chemical processes. The search has so far given no convincing evidence, although the updated review by

 

Bowman[1] cites as candidates three recent papers on bimolecular reactions by classical trajectories and a couple of quantum investigations of the $MgH_2$ system. Here we add similar behavior found in $SN_2$ bimolecular reactions,[2] although in reference to this important subject no use is normally made of the term "roaming" to characterize a possibly similar dynamical phenomenon. Inspired by crossed molecular beams studies with oriented reactants,[3] in a recent quantum mechanical first-principles study of the OH + HBr → $H_2O$ + Br reaction,[4] which shows a striking anti Arrhenius temperature dependence, we believe it possible to label as roaming the effect occurring in the search for the severely restricted mutual orientation of reactants, necessary for the favorable H-atom exchange. This represents an example of a stereodynamical origin of roaming in a bimolecular chemical reaction, for which the phenomenology is still limited. Work on extending transition-state based theories to cope with these situations continues.[1]

1 J. M. Bowman, Roaming, *Mol. Phys.*, 2014, **112**, 2516–2528.
2 J. Mikosch, S. Trippel, C. Eichhorn, R. Otto, U. Lourderaj, J. X. Zhang, W. L. Hase, M. Weidemuller, R. Wester, Imaging Nucleophilic Substitution Dynamics, *Science*, 2008, **319**,183–186.
3 T. Kasai, D.-C. Che, M. Okada, P.-Y. Tsai, K.C. Lin, F. Palazzetti, V. Aquilanti, Directions of chemical change: experimental characterization of the stereodynamics of photodissociation and reactive processes, *Phys. Chem. Chem. Phys.*, 2014, **16**, 9776–9790.
4 N. D. Coutinho, V. H. C. Silva, H. C. B. de Oliveira, A. J. Camargo, K. C. Mundim, V. Aquilanti, Sterodynamical origin of anti-Arrhenius kinetics: negative activation energy and roaming for a four-atom reaction, to be submitted, 2015.

**Srihari Keshavamurthy** asked **Vincenzo Aquilanti** and **King-Chuen Lin**: Given that the conical intersection is an important feature in the system of interest, certain quantum effects (for example, surface hopping due to non adiabaticity) are not taken into account in the classical dynamical investigations presented in this work. Would inclusion of such quantum effects potentially change the extent of roaming?

A second question is regarding the treatment of the $CH_3CO$ moiety as frozen. As indicated by you, this does alter the nature and extent of the intramolecular vibrational energy flow in the molecule. Again, how crucial is this approximation to the observed relative proportions of the direct and roaming trajectories?

**Artur Nenov** further asked **King-Chuen Lin** and **Vincenzo Aquilanti**: You perform classical MD simulations with initial conditions generated by sampling around the conical intersection. Do you expect the momentum accumulated in the HCO–$OCH_3$ dissociation direction, as well as non-adiabatic effects in the vicinity of the conical intersection to affect the outcome of the dynamics?

**King-Chuen Lin** and **Vincenzo Aquilanti** answered: The relevant questions asked by **Srihari Keshavamurthy** and by **Artur Nenov** give us the opportunity to enter into greater detail regarding our proposed mechanism involving nonadiabatic effects in the various outcomes of methyl formate photolysis and specifically the role of a conical intersection in mediating the path leading to delayed delivery of decomposition products. The probe of this observed delay of a fraction of CO produced also in low rotational states comes from ion imaging and is considered a "signature" of the roaming phenomenon. The energy and features of the conical intersection seam,[1] as well additional information on the lower branch of the potential energy surface, denoted $S_0$,[2] come from full dimensional eight-atom quantum chemistry calculations. All channels considered in the paper, up to triple breakdown, can be comprised within an effective reduced three-center model, but since we probed CO translational and roto-vibrational energies[3] we included CO fully as two particles and set up a four-body $S_0$ surface for dynamics, keeping only the $CH_3O$ adiabatically frozen. The concerns of Srihari Keshavamurthy were the exclusion of quantum effects in our model and consequences of the freezing of $CH_3O$ internal modes. We failed to find slow CO products from large samplings of trajectories originating from the ground state and evolving on $S_0$ along the minimum energy path through the transition state saddle. Therefore, we systematically studied the effect of trajectories on $S_0$ originating from the neighborhood of the tip of the lower cone looking for possible branching ratios of exit channels, independently of the mechanism for the transition from the upper $S_1$ surface, due to nonadiabatic coupling. In the comparison between experiment and simulation, we attributed the somewhat underestimated yield of $CH_3O$ to the assumed freezing, because the artificial truncation of the degrees of freedom may be responsible for the limitation of the phase space available for energy and momentum disposal. Finally, we agree with Artur Nenov's remark that accumulation of linear momentum along the dissociating bond direction may not be negligible as a bias affecting the outcome of our classical molecular dynamics simulations. However, we cannot evaluate at present the role of such a bias in influencing branching ratios of alternative channels, and its phenomenological account would require additional parameters with respect to our assumption of initial isotropic distribution as the system moves away from the conical intersection.

1 G. Cui, F. Zhang and W. Fang, Insights into the mechanistic photodissociation of methyl formate, *J. Chem. Phys.*, 2010, **132**, 034306.
2 A. Lombardi, F. Palazzetti, K.-C. Lin and P.-Y. Tsai, *Lecture Notes in Computer Science (including subseries Lecture Notes in Artificial Intelligence and Lecture Notes in Bioinformatics)*, 2014, **8579**, 452–467.
3 P.-Y. Tsai, M.-H. Chao, T. Kasai, K.-C. Lin, A. Lombardi, F. Palazzetti and V. Aquilanti, Roads leading to roam. Role of triple fragmentation and of conical intersections in photochemical reactions: experiments and theory on methyl formate, *Phys. Chem. Chem. Phys.*, 2014, **16**, 2854–2865.

**Mike Ashfold** commented: Much of our recent work has centred on exploring similarities and differences between molecular photofragmentation dynamics in the gas and condensed phase.[1] Studies of photoinduced O–H (S–H) bond fission in gas phase phenol (thiophenol) encourage the view that such results can provide a very useful guide to the early time dynamics following photoexcitation of these same molecules in solution in weakly interacting solvents.[2–5] This, in turn, opens the way to using solution phase experiments as a route to studying photofragmentation processes that are traditionally viewed as hard to study in the gas phase. Photoinduced ring opening reactions constitute one such example. Our recent study of the 267 nm photoexcitation of thiophenone in solution in acetonitrile, which transient infrared absorption spectroscopy shows leads to prompt ring opening and formation of a ketene isomer, provides one such example.[6] The ketene products are formed with high levels of internal (vibrational) relaxation, but then relax on a ps timescale through interaction with solvent molecules – as evidenced by the narrowing and blue shifting of the antisymmetric C=C=O stretch band shown in Fig. 1 below.

Companion *ab initio* calculations show that ring opening involves radiationless transfer through a conical intersection with the ground ($S_0$) state potential energy surface (PES) at extended C–S bond lengths. Having accessed the $S_0$ PES, the molecules have a choice – to complete the ring opening process or reform the cyclic parent molecule – with the eventual products in each case stabilised by collisional relaxation. Parent bleach recovery measurements show that ~60% of the initially photoexcited thiophenone molecules reform the parent molecule under the prevailing experimental conditions.[6]

Investigating such photoinduced ring opening processes in the gas phase is challenging. The parent and product species are isomers, which limits the utility of mass-selective ionisation detection as a means of following the reaction. The highly vibrationally excited nature of the ring opened products complicates their identification and characterisation by frequency resolved spectroscopies. As shown below, photofragment translational spectroscopy can offer (indirect) insights into such processes, and ultrafast time-resolved studies employing X-ray or electron probe pulses such as previewed in this discussion[7,8] offer the prospect of observing such processes directly in the not too distant future.

Fig. 2 shows results for the near UV photolysis of 2-bromothiophene, wherein the Br atom serves as a reporter which we monitor by velocity map ion imaging.[9] The ground state Br atom products formed when exciting at long wavelengths (*e.g.* 266.6 nm, fig. 2(a)) display near limiting parallel recoil anisotropy, consistent with prompt C–Br bond fission following electron promotion to a $\sigma^*_{C-Br}$ orbital – behaviour reminiscent of that seen following similar long wavelength excitation of alkyl and aryl bromides. Upon tuning to shorter wavelengths (*i.e.* higher photon

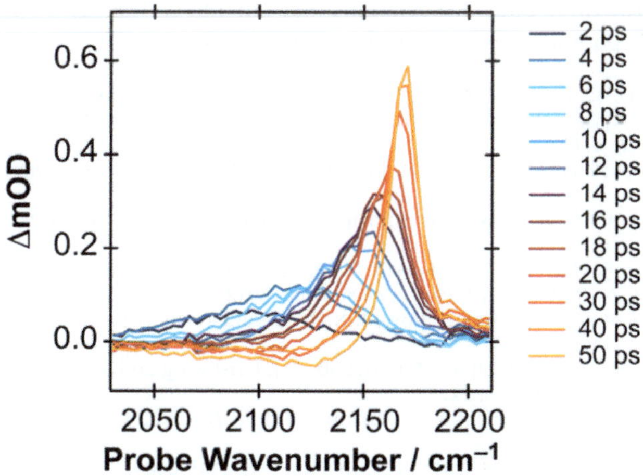


Fig. 1 Transient IR spectra obtained at a range of time delays following 267 nm excitation of a 23 mM solution of thiophenone in acetonitrile showing the narrowing and blue shifting of the antisymmetric C=C=O stretch absorption band of the ring opened ketene product.

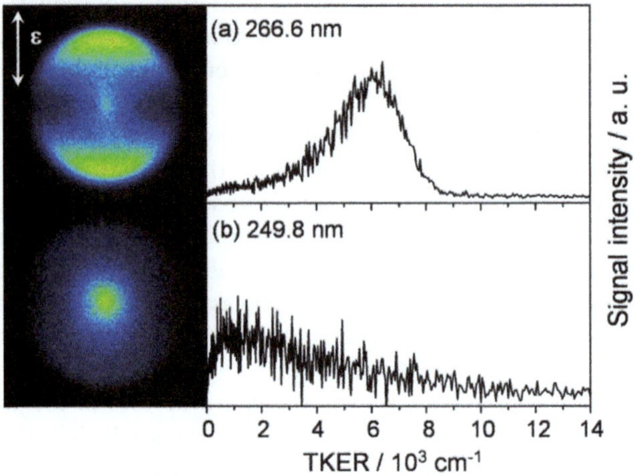


Fig. 2 Images of the ground state Br products resulting from photolysis of jet-cooled 2-bromothiophene at (a) 266.6 nm and (b) 249.8 nm, with the electric (ε) vector of the photolysis laser radiation aligned vertically in the plane of the image (as shown in panel (a)). The total kinetic energy release (TKER) spectrum derived from analysis of each image is shown to the right.

energies), this fast component fades and is progressively replaced by a slow, isotropic recoil distribution (see fig. 2(b)). Again, complementary electronic structure calculations suggest that this switch in fragmentation behaviour reflects the opening of a rival excited state C–S bond extension pathway to a conical intersection with the ground state PES. The resulting biradicals are formed with more than sufficient internal energy to sample the configuration space associated with several parent isomers and to dissociate to yield Br atoms in tandem with both cyclic and ring-opened partner fragments.[9] In the context of (hetero)cyclic dissociations, the Br atom loss subsequent to ring-opening may be viewed as the analogue of the three body dissociations that Lin and colleagues worked hard to avoid in their photodissociation studies of methyl formate.[10]

1 S. J. Harris, D. Murdock, Y. Zhang, T. A. A. Oliver, M. P. Grubb, A. J. Orr-Ewing, G. M. Greetham, I. P. Clark, M. Towrie, S. E. Bradforth and M. N. R. Ashfold, *Phys. Chem. Chem. Phys.*, 2013, **15**, 6567.
2 T. A. A. Oliver, Y. Zhang, M. N. R. Ashfold and S. E. Bradforth, *Farad. Disc. Chem. Soc.*, 2011, **150**, 439.
3 Y. Zhang, T. A. A. Oliver, M. N. R. Ashfold and S. E. Bradforth, *Farad. Disc. Chem. Soc.*, 2012, **157**, 141.
4 D. Murdock, S. J. Harris, T. N. V. Karsili, G. M. Greetham, I. P. Clark, M. Towrie, A. J. Orr-Ewing and M. N. R. Ashfold, *J. Phys. Chem. Letts.*, 2012, **3.**, 3715.
5 Y. Zhang, T. A .A. Oliver, S. Das, A. Roy, M. N. R. Ashfold and S .E. Bradforth, *J. Phys. Chem. A.*, 2013, **117**, 12125.
6 D. Murdock, S. J. Harris, J. Luke, M .P. Grubb, A. J. Orr-Ewing and M. N. R. Ashfold, *Phys. Chem. Chem. Phys.*, 2014, 16, 21271.
7 K. Bennett, M. Kowalewski and S. Mukamel, *Farad. Discuss.*, 2014, **177**, DOI 10.1039/c4fd00178h.
8 S. Manz *et al.*, *Farad. Discuss.* 2014, **177**, DOI 10.1039/c4fd00204k.
9 B. Marchetti, T. N. V. Karsili and M. N. R. Ashfold, unpublished results.
10 M. Nakamura, P.-Y. Tsai, T. Kasai, K.-C. Lin, F. Palazzetti, A. Lombardi and V. Aquilanti, *Farad. Discuss.*, 2014, **177**, DOI 10.1039/c4fd00174e.

 

**King-Chuen Lin** and **Vincenzo Aquilanti** answered: We find interesting **Mike Ashfold**'s observation of a possible connection between his novel results on a change in mechanism in 2-bromothiophene photolysis at two wavelengths and the opening of the triple fragmentation channel in the case of methyl formate[1] as a function of the increasing energy of the photolysing light. We note a further similarity in the involvement of an unusually large extension of a C–S bond as a pathway to a conical intersection with the ground state potential energy surface in his case, and a similar stretch of the C–O bond between the HCO and $OCH_3$ departing radicals invoked in the simulations of our experiments. For the disentangling of the triple fragmentation signatures in the TKER we used a phase-space model,[2] valid for barrier-less processes. An impulsive model of more general validity is available.[3]

1 P.-Y. Tsai, M.-H. Chao, T. Kasai, K.-C. Lin, A. Lombardi, F. Palazzetti and V. Aquilanti, Roads leading to roam. Role of triple fragmentation and of conical intersections in photochemical reactions: experiments and theory on methyl formate, *Phys. Chem. Chem. Phys.*, 2014, **16**, 2854–2865.
2 G. de Wit, B. R. Heazlewood, M. S. Quinn, A. T. Maccarone, K. Nauta, S. A. Reid, M. J. T. Jordan and S. H. Kable, Product state and speed distributions in photochemical triple fragmentations, *Faraday Discuss.*, 2012, **157**, 227-241.
3 P.-Y. Tsai and K.-C. Lin, Insight into the Photodissociation Dynamical Feature of Conventional Transition State and Roaming Pathways by an Impulsive Model, *J. Phys. Chem. A*, 2015, **119**, 29–38.

**Anuradha Pallipurath** queried: Why were the rotational states $J = 2,7,\ 11$ of interest over others?

**King-Chuen Lin** and **Vincenzo Aquilanti** answered: The values of the CO rotational states from ion imaging experiments on methyl formate photofragmentation, chosen as illustrative in the vibrational level $\nu = 1$ in the present paper (and $\nu = 0$ previously[1]), correspond to ranges where the experimental signal-to-noise ratio was favorable with respect to unambiguous deconvolutions. The FTIR technique, briefly alluded to in the paper (see also further work[2]), permitted us to obtain full rotational bands for $\nu = 1, 2, 3, 4$, and bimodality for $\nu = 1$ and 2, from which energy disposal in vibrational and rotational degrees of freedom could be characterized. The results, supported by accompanying classical trajectory simulations and additional[2] to those presented in this paper, demonstrate that when molecules are excited as much as possible but below the threshold for triple dissociation, the energy deposited in the CO fragment is essentially found in vibrations, while rotation modes are colder. Roaming persists as a substantial phenomenon also for the production of highly vibrational CO states.[2]

1 P.-Y. Tsai, M.-H. Chao, T. Kasai, K.-C. Lin, A. Lombardi, F. Palazzetti and V. Aquilanti, Roads leading to roam. Role of triple fragmentation and of conical intersections in photochemical reactions: experiments and theory on methyl formate, *Phys. Chem. Chem. Phys.*, 2014, **16**, 2854–2865.
2 A. Lombardi, F. Palazzetti, V. Aquilanti, H.-K. Li, P.-Y. Tsai, T. Kasai, K.-C. Lin, Rotovibrationally excited molecules excited on the verge of a triple breakdown, to be submitted.

**Martin Zanni** opened the discussion of the paper by **Alison Edwards**: In the introduction to your paper, one rationale of the work is that catalysts can act as a reservoir for mono nuclear catalysis. What do you mean by reservoir?

**Alison Edwards** responded: The classification of Pd(I) complexes acting as a "reservoir" for active catalysts relates to a range of systems where equilibria exist between Pd(I) dimers and Pd(II)+Pd(0) complexes in which one or both of the Pd(II) or Pd(0) are active. In other cases the Pd(I) dimers have been shown to be the active catalyst.

**Jonathan Skelton** remarked: In this paper, a mechanism for hydride transfer across a dipalladium complex was proposed based on molecular-dynamics simulations.

The authors appear to dismiss transition-state modelling in favour of molecular-dynamics simulations, on the basis of the fact that there are too many degrees of freedom to consider in the first instance. However, as the authors obtained a good idea of what the transition state was – and of the major degrees of freedom involved in the reaction coordinate – from their MD simulations, they may be able to investigate their mechanism further using other transition-state modelling techniques, for example the nudged elastic-band method.[1] This might give them a complete energy profile for the reaction, rather than the partial profile + sinusoidal fit they show in Fig. 6 in the paper.

A comment here is that transition-state modelling and (constrained) MD can be good complementary techniques – this approach was used for example by J. M. Skelton *et al.*[2]

Also, it is sometimes the case that molecular-dynamics simulations using the classical (Newtonian) equations of motion can fail to describe the behaviour of hydrogen atoms, because, being light particles, these can sometimes display quantum phenomena (*e.g.* tunnel through energy barriers). In such systems, path-integral molecular dynamics may be required.[3]

Do the authors have any feel for whether non-classical dynamics may play any role in the hydride transfer in this system?

1 G. Mills, H. Jonsson and G. K. Schenter, *Surface Science*, 1995, **324**(2), 305–337, DOI: 10.1016/0039-6028(94)00731-4.
2 J. M. Skelton *et al.*, *CrystEngComm*, 2015, **17**(2), 383–394, DOI: 10.1039/C4CE01411A.
3 B. J. Berne, *Ann. Rev. Phys. Chem.*, 1986, **37**, 401–424.

**Alison Edwards** replied: The term "dismiss" is perhaps too strong, we did consider it. Both the experimental and MD data relate the PE profile around the minima, which is far from the transition state, so we lack data here. We could indeed make some estimate of what the transition-state may be from the single transfer to the transition state observed in the MD, but how to characterise the ligand conformations at the transition state? We considered that overall, an estimate of the reaction coordinate based on the profile around the minima, assuming the functional form of the PE, is less of an approximation estimating the transition state. This is open to discussion, and it would indeed be interesting to compare the results of the two approaches.

 

**Jemmis E D** asked: Is it possible to make the chemical environment of the two metals unequal by changing the nature of the ligands? If this is possible in a systematic way, it would help to see the origin of the barrier (or lack there of) for the H-shift. This would give hints to decrease or increase the barrier. Calculation of the molecule by removing the proton and analyzing the electron distribution is another way to gather information on the preferred position of the proton, as done by Prof Roald Hoffmann in many organometallic complexes.

**Alison Edwards** answered: We are currently studying dicarbene ligands with lowered symmetry (altering the *N*-substituents using the methods published in reference 12 of our paper (*Dalton Trans.*, 2013, **42**, 7447) for this very reason and hope to report soon.

**Elangannan Arunan** enquired: With the beautiful symmetry in the molecule, how does one know if a reaction has taken place at all unless one can watch the hydride moving from one side to another? I am reminded of the $H + H_2 \rightarrow H_2 + H$ reaction. In this case, an isotopic variant is studied experimentally unlike the 'reaction' discussed by you.

**Alison Edwards** responded: This is a purely classical molecular dynamics simulation based on the assumption that the height and width of the barrier for proton-transfer from one end of the molecule to the other are so large that the tunnelling probability is negligible. We know this from the rate. Thus, we can identify the classical location of the H-atom as being at one end or the other (or between) of the molecule. Simply consider the distances involved.

**R J Dwayne Miller** commented: In terms of trying to observe the bridged hydride intermediate, are there any estimates of the lifetime of this intermediate. It may be too short lived to observe. Is there any evidence of dynamics related to hydride transitions in the diffraction, regarding linewidths, diffuse scatterring?

**Alison Edwards** replied: Unfortunately not. There are no relevant experimental data and in the simulation only a single transfer to the intermediate was observed. Longer simulations would give some estimate, but given the limited stability of the system at the temperatures used in our simulation, a lower temperature, and thus even longer simulations would be required.

**Sanghamitra Mukhopadhyay** said: In your submitted paper for Faraday discussions 177 titled "The free energy barrier to hydride transfer across a dipalladium complex", it is mentioned in the Experimental section and DFT calculations subsection (line number 227, page 4), that "Isolated (gas phase) molecule calculations were performed on an isolated $\{(MesIm)_2CH_2\}2Pd_2H^+$ cation composed of 117 atoms and contained in a periodic cell of dimensions 25x18x18 Å. This cell is large enough to avoid fictive intermolecular interactions". This statement indicates that all lattice dynamics (static) calculations are done on isolated molecules. Moreover as mentioned in the same section these calculations are done on a "single $k$-point. (Gamma point)", but there is no mention about how many phonon wave-vector transfers (q) were used in the calculations. As read from the text (line 39 on the same page as above), only MD calculations are done on periodic solid state samples with "over 700 atoms so only a limited number of calculations were possible at low precision". Simulations of inelastic neutron scattering (INS), particularly for low temperatures, requires detailed solid state calculations (not gas phase) with reasonable $k$-point and $q$-points. Can the author comment on how these low precision and molecular level calculations are comparable to INS experiments? Some more details such as unit cell size and number of molecules in a unit cell used for lattice dynamics calculations will be useful.

**Alison Edwards** responded: Unfortunately not. There are no relevant experimental data and in the simulation only a single transfer to the intermediate was observed. Longer simulations would give some estimate, but given the limited stability of the system at the temperatures used in our simulation, a lower temperature, and thus even longer simulations would be required.

**Martin Zanni** opened the discussion of the paper by **Kenneth Ghiggino**: You have done a study on the oligomers. When people study energy transfer in the system they look at large polymers. Are the thousamers soluble or insoluble? Regarding the absorption spectra, does the polymer have a broader absorption band for one molecule or a film? Does it get broader for a film?

**Kenneth Ghiggino** replied: The oligomers reported in this study represent short chain models for the conjugated polymer poly[2-methoxy-5-(2'-ethylhexyloxy)-1,4-phenylenevinylene] (MEH-PPV). High molecular weight MEH-PPV is soluble in a number of solvents (*e.g.* chloroform, tetrahydofuran, chlorobenzene). The absorption spectrum of MEH-PPV in solution and in films has been reported by others.[1] There are shifts and broadening of the absorption spectrum of MEH-PPV in films compared to solution and these have been attributed to aggregation of the polymer chains and/or conformational differences depending on the solvent from which the solid films are cast.

1 B. J. Schwartz, *Annu. Rev. Phys. Chem.*, 2003, **54**, 141–172, for example.

**Mahesh Hariharan** questioned **Ken Ghiggino**: The change in electronic absorption and fluorescence. How do you allow for the possibility of aggregation. Can it be both or fully excluded?

**Kenneth Ghiggino** answered: We report in our paper changes in the absorption and fluorescence spectra with temperature of a dilute 2-MTHF solution (approx. 2 micromolar) of the MEH-PPV oligomer. At this concentration aggregation is very unlikely. We have in fact studied neat solid films of the oligomer where aggregation does occur and this results in broadening and a small red-shift of the absorption spectrum rather than the marked increase in structure and large red shift we report in this paper as temperature is varied so we are reasonably confident that aggregation is not playing a role in the spectral changes reported in this paper.

**Martin Meedom Nielsen** commented: While you have been looking at isolated oligomers, the performance of a device built from oligomeric or polymeric systems depends to a large extent on charge transport between oligomers or polymer chains, which again is governed by overlap in π-orbitals in most models. The change in planarity of molecules could potentially modify such π-overlaps. Could you comment on the perspective of your studies of planarity in this context?

**Kenneth Ghiggino** responded: In opto-electronic devices charge transport in the semiconducting material can indeed play an important role in determining the operating efficiency. In the case of bulk heterojunction photovoltaic devices the degree of orbital delocalisation in the material can also assist charge separation and reduce geminate recombination losses. The twisted conformations reported in our paper for the MEHPPV oligomer model compound have only a relatively small degree of non-planarity (~10–15 degrees dihedral angle deviation) and, while affecting the electronic energy levels sufficiently to broaden the absorption spectrum at room temperature, would still maintain a significant degree of π-orbital delocalisation. Considerably higher twist angles (>40 degrees) are normally considered necessary before the π conjugation is considered to be disrupted. Nevertheless the non-planarity of the ground state oligomers and polymers probably has some impact on charge transport. As discussed in our paper, the conformational relaxation that occurs in the excited state likely plays a role in energy transfer processes following photoexcitation. In other work we have demonstrated that by disrupting the extended π-orbital conjugation completely in confined conjugation length MEHPPV type polymers, the yields of polarons formed following photoexcitation are significantly reduced, confirming the importance of orbital delocalisation in charge separation and transport processes.[1]

1 T. A. T. Tan *et al.*, *Polym. Chem.*, 2013, **4**, 5305.

**Martin Zanni** enquired: What is the delocalization length of MEH-PPV when initially excited? When it is collapsed?

**Kenneth Ghiggino** replied: An MEH-PPV polymer chain can be considered to consist of sequences of light absorbing chromophores of different lengths and conformational orientations. These chromophores have been estimated to have a delocalization length of between 2 to 12 repeat units.[1] These estimates are based on theoretical simulations and the observed absorption and fluorescence spectra. The electronic couplings of these conformational subunits play a role in exciton transport and ultimately the photochemical properties of these semi-conducting materials. When the chains adopt collapsed conformations in poor solvents there are shifts in absorption spectra and solution and single molecule studies demonstrate that the exciton transport properties are significantly affected.[2,3]

1 T. E. Dykstra *et al.*, *J. Phys. Chem. B*, 2009, **113**, 656-667, for example.
2 E. Collini and G. D. Scholes, *Science*, 2009, **323**, 369–373.
3 E. N. Hooley *et al.*, *Phys. Chem. Chem. Phys.*, 2014, **16**, 7108–7114.

**John R. Helliwell** addressed **Ken Ghiggino**: In the materials world (or in the living cell) you do not have the ultra-dilute solution conditions preferred in your measurements but a complex mixture. Can you please comment on your spectrocopic results and how you extrapolate from them to the more complex situations to which I refer?

**Kenneth Ghiggino** responded: We are studying very dilute solutions in our paper to characterize and understand intramolecular excited state processes in the model oligomer compound. For higher molecular weight polymers, concentrated solutions and in solid films it is certainly true that additional interactions (such as intermolecular energy transfer) can occur. However results on these systems will be difficult to fully understand unless the behavior of the isolated units is resolved. This is the motivation for this work. As an example, theoretical and spectroscopic studies of exciton diffusion in semi-conducting polymers often invokes a role for the dynamics of torsional relaxation in the component polymer segments that is the topic of our investigation.[1]

1 T. E. Dykstra *et al.*, *J. Phys. Chem. B*, 2009, **113**, 656–667, for example.

**Srihari Keshavamurthy** asked: A kinetic model was presented in this work wherein the non planar conformers relax to planar forms which then undergo a slow relaxation. It was also commented in the paper that this represents a case wherein intramolecular vibrational energy flow competes on the timescales of interest. I would like to point out that the ~1 ns planar lifetime seems in accordance with an earlier work by Engelhardt *et al.*[1] wherein they observed very slow IVR for molecules in planar conformations.

1 Engelhardt *et al.*, *J. Phys. Chem. A*, 2001, **105**, 6800.

**Kenneth Ghiggino** replied: The lifetime of the relaxed planar singlet excited state that occurs on a nanosecond time-scale will be determined by the competing rates of fluorescence and other non-radiative electronic relaxation processes.

**Martin Zanni** remarked: I presented a technique from a published paper of mine[1] that allows one to measure the absolute transition dipoles of transitions by taking the ratio of 1D and 2D spectra. The absolute transition dipole strength is a measure of exciton delocalization. Would it be useful to apply this approach to polymers like those presented by Ken Ghiggino?

1 M Gecko and M Zanni, *J. Chem. Phys.*, 2012, **137**, 184202.

**Kenneth Ghiggino** replied: This approach would certainly be worth looking into to see if it can provide additional information.

 

**Debabrata Goswami** commented: The ring opening class of reactions is important and we have worked a lot on the ring opening of dicyclopentadiene in the gas phase with ultrafast laser pulses.[1] You make an interesting attempt to connect liquid to gas phase work. However, we have shown ring opening in the gas phase to be immensely effected by laser properties. In fact it is possible to control such reactions based on how you change your laser character which can completely change the laser interaction and the ensuing chemical process. Simply going between the gas phase and liquid phase without laser property consideration is a highly oversimplified picture.

1 T. Goswami, D.K. Das, D. Goswami, *Chem. Phys. Lett.*, 2013, **558**, 1–7.

**Mike Ashfold** answered: **Debabrata Goswami**'s point is well made. Clearly, experiments seeking to explore similarities (and possible differences) between the photofragmentation dynamics of any given molecule in the gas and solution phases should start from 'as level a playing field' as possible – as in our comparative studies of the one (UV) photon induced X–H/$CH_3$ (X = S, O) bond fissions in thiophenols, thioanisoles and phenol itself,[1] which have served to inform our more recent pump-probe studies of UV photoinduced ring-opening (in thiophenone and furanone)[2] and the photoisomerism of N-methyl-2-pyridone[3] in weakly interacting solvents like cyclohexane or acetonitrile. The photo-induced ring opening study to which Debabrata Goswami refers is a very different beast, involving intense laser fields, absorption of multiple 800 nm photons and eventual monitoring of fragment ions with m/z 66 – which has little connection with any photochemistry that would be induced by single (267 nm) photon absorption by dicyclopentadiene either in the gas phase or when dissolved in a weakly interacting solvent.

1 S. J. Harris, D. Murdock, Y. Zhang, T. A. A. Oliver, M. P. Grubb, A. J. Orr-Ewing, G. M. Greetham, I. P. Clark, M. Towrie, S. E. Bradforth and M. N. R. Ashfold, *Phys. Chem. Chem. Phys.*, 2013, **15**, 6567.
2 D. Murdock, S. J. Harris, J. Luke, M. P. Grubb, A. J. Orr-Ewing and M. N. R. Ashfold, *Phys. Chem. Chem. Phys.*, 2014, **16**, 21271.
3 D. Murdock, S. J. Harris, I. P. Clark, G. M. Greetham, M. Towrie, A. J. Orr-Ewing and M. N. R. Ashfold, *J. Phys. Chem. A*, 2015, **119**, 88.

**Debabrata Goswami** addressed **Kenneth Ghiggino**: The aspect of torsional properties discussed in the context of this work seems to be at least an order of magnitude slower in comparison to free torsional motion which is in the order of femtoseconds. Is this due to the extreme viscous nature of the polymeric sample being studied?

**Kenneth Ghiggino** answered: The torsional relaxation we are observing involves a significant geometry change to achieve planarization in the excited state following excitation of ground state twisted configurations and there is also likely solvent involvement. A timescale for this process in the picosecond regime does not seem unreasonable. There have been various theoretical and experimental reports of time-scales for similar torsional relaxation in related polymers and oligomers from >100 fs[1] to <50 picoseconds.[2]

1 K. S. Wong *et al.*, *Chem. Phys. Lett.*, 1998, **288**, 59–64.
2 R. E. Di Paolo *et al.*, *ChemPhysChem*, 2007, **8**, 265–2664.

**Anuradha Pallipurath** queried: Have you observed any evidence for your torsional isomers in the ORTEP plots obtained from single crystal studies?

**Kenneth Ghiggino** answered: The ORTEP plots indicate the oligomers studied adopt a non-planar twisted conformation in the crystal rather than being planar. While there is just one conformation observed in the crystal form, the observation that a non-planar configuration is adopted in the crystal lends support to the results of theoretical calculations that indicate a range of twisted conformations of quite similar energy are likely to be present in fluid solution that should all be populated at room temperature.

**Artur Nenov** questioned: The absorption spectra show a pronounced vibrational structure with cooling. Could a similar effect be achieved by immobilizing the PPV pentamer through interchain couplings of the aliphatic chains?

**Kenneth Ghiggino** responded: We attribute the increased vibrational structure and the red-shift of our pentamer absorption spectrum with cooling to the favoring of lower energy conformations of the molecule at the lower temperatures (*i.e.* higher energy rotational isomers are not populated at the low temperatures). If we could find a way to "fix" the PPV pentamer molecules in a particular configuration at room temperature we might expect to see a similar effect.

 

Faraday Discussions

PAPER

# Pluronic copolymer encapsulated SCR7 as a potential anticancer agent

Franklin John,*[a] Jinu George,[a] Mrinal Srivastava,[b] P. A. Hassan,[c] V. K. Aswal,[d] Subhas. S. Karki[e] and Sathees. C. Raghavan[b]

Received 15th September 2014, Accepted 4th November 2014
DOI: 10.1039/c4fd00176a

Nonhomologous end joining (NHEJ) of DNA double strand breaks (DSBs) inside cells can be selectively inhibited by 5,6-bis-(benzylideneamino)-2-mercaptopyrimidin-4-ol (SCR7) which possesses anticancer properties. The hydrophobicity of SCR7 decreases its bioavailability which is a major setback in the utilization of this compound as a therapeutic agent. In order to circumvent the drawback of SCR7, we prepared a polymer encapsulated form of SCR7. The physical interaction of SCR7 and Pluronic® copolymer is evident from different analytical techniques. The *in vitro* cytotoxicity of the drug formulations is established using the MTT assay.

## Introduction

Cancer drug targeting is challenged by problems like multi-drug resistance and low solubility of potential drug candidates. There are two general classes of resistance to anti-cancer drugs: those that impair delivery of anti-cancer drugs to tumor cells, and those that arise in the cancer cell itself due to genetic and epigenetic alterations that affect drug sensitivity. Impaired drug delivery can result from poor absorption of orally administered drugs, increased drug metabolism or increased excretion, resulting in lower levels of drug in the blood and reduced diffusion of drugs from the blood into the tumor.[1,2]

Cytotoxic therapy offers one prominent approach towards cancer treatment. Traditional cytotoxic therapy includes radiation and chemotherapeutic compounds such as platinum-based drugs.[3–6] Among the genetic damages, DNA double strand breaks (DSBs) are considered as the most lethal as they affect the integrity and continuity of the genome.[7–10] Inappropriate repair of DSBs may result in deletions, inversions, duplications and chromosomal translocations.[11–14]

[a]*Biotechnology Laboratory, PG and Research Department of Chemistry, Sacred Heart College, Kochi 682 013, India. E-mail: franklinshc@gmail.com*
[b]*Department of Biochemistry, Indian Institute of Science, Bangalore 560 012, India*
[c]*Chemistry Division, Bhabha Atomic Research Centre, Mumbai 400 085, India*
[d]*Solid State Physics Division, Bhabha Atomic Research Centre, Mumbai 400 085, India*
[e]*Department of Pharmaceutical Chemistry, KLE University, Bangalore 560 010, India*

Nonhomologous DNA end joining (NHEJ) is one of the major DNA DSB repair pathways.[15] Recently, a novel inhibitor of NHEJ, 5,6-bis-(benzylideneamino)-2-mercaptopyrimidin-4-ol (SCR7) has been reported.[16] Inhibition of NHEJ by SCR7 in cancer cells results in the accumulation of unrepaired DNA double-strand breaks. Despite promising results of SCR7 as a good anticancer agent, it showed high $IC_{50}$ which could be attributed to its high hydrophobicity.

## Results and discussion

Although SCR 7 has shown promising results as a potential anticancer drug, its limited solubility in water may lead to poor bioavailability and requires the search for novel drug delivery systems. Traditional methods to improve the bioavailability of poorly soluble drugs include encapsulating them in nanosized carriers such as liposomes,[17] emulsions, polymer micelles, niosomes,[18] lipid particles *etc.* Poly Ethylene Glycol (PEG)[19] based block copolymers have the distinct advantage as compared to other delivery systems due to its ability to encapsulate large amounts of drug. One of the most widely studied classes of amphiphilic copolymers in this field is the Pluronic triblock copolymers. Thus, we investigated the role of Pluronic P123 in encapsulating and delivering SCR7 *via* hydrophobic interactions.

Fluorescence emission spectra of SCR7 showed a blue shift upon encapsulation with P123 copolymer in DMSO as solvent as shown in Fig. 1(a). The fluorescence spectra of SCR7 in DMSO exhibits an emission peak at ~474 nm while SCR7 in the presence of P123 show a maximum at ~442 nm with the same solvent (DMSO due to the high solubility of SCR7).[20] The shift in the $\lambda_{max}$ of fluorescence spectrum is an indication of interaction between SCR7 and the copolymer P123. Such shift in the fluorescence emission is reported for doxorubicin encapsulated with chitosan nano particles and curcumin casein micelles.[21,22]

Fig. 1(b) shows the UV-Vis absorption spectra of SCR7-P123 dissolved in a minimum amount of water. The hydrophobic PPO core of the copolymer effectively encapsulates the drug molecule and makes it dispersible in the aqueous layer. To reiterate the encapsulation and release of SCR7 from P123, the aqueous

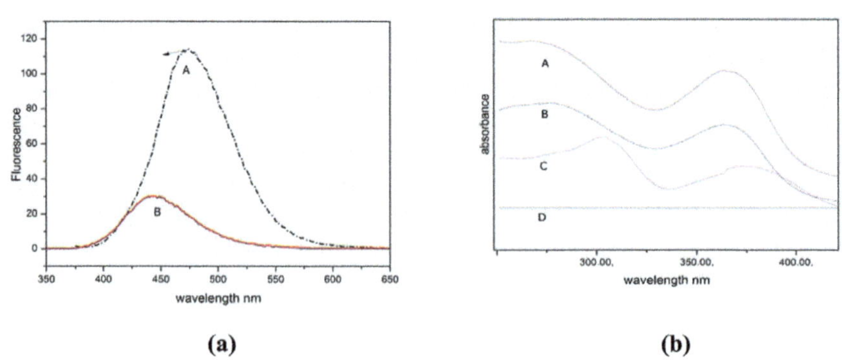


Fig. 1 (a) (A) Fluorescence emission spectra of SCR7 in DMSO (B) SCR7 loaded P123 in DMSO. (b) (A) UV absorption spectra of SCR7 loaded P123 in water (B) chloroform extract (C) SCR7 alone in water (D) P123 in water.

 

layer was extracted with chloroform (3×). The chloroform layer is dried with $Na_2SO_4$. UV absorption spectra of the chloroform layer show absorption maxima at 267 nm and 365 nm. The optical absorption spectrum of the chloroform layer is compared with SCR7 in pure water and SCR7 encapsulated P123 in water (A and C of Fig. 1(b)). This indicates the ability of P123 in encapsulating the drug and its release upon treatment with hydrophobic solvents like chloroform.

The FTIR spectra of SCR7 and SCR7 encapsulated with P123 are depicted in Fig. 2. The C–S stretching vibrations occur at 3271 $cm^{-1}$, while the C–S bending occurs at 1047 $cm^{-1}$ and 1082 $cm^{-1}$. The C–N stretching vibrations occur at 1026 $cm^{-1}$, whereas the N–H bending vibrations occur at 924 $cm^{-1}$. The absorptions at 1250 $cm^{-1}$, 1117.6 $cm^{-1}$ and 1031.1 $cm^{-1}$ are due to C–O stretching vibrations from the P123 matrix. The absorptions at 957 $cm^{-1}$ and 817 $cm^{-1}$ are arising from the aromatic ring of SCR7. As seen, the spectrum of SCR7-P123 did show the characteristic absorption band of SCR7 at 2965 $cm^{-1}$ and 1107 $cm^{-1}$. FT-IR spectral analysis reveals that there is no appreciable shift in the IR signal of SCR7 when it changes from the free state to the polymer bound state. It can also be concluded that there is no chemical interaction taking place between P123 and SCR7. The result confirmed the presence of SCR7 in the encapsulated formulation.

Successful incorporation of SCR7 in P123 and its release in different solvents was elucidated by $^1H$ NMR spectroscopy. As is evident from Fig. 3, the $^1H$ NMR spectra of SCR7-P123 employing DMSO-d6 as the solvent showed characteristic aromatic signals from SCR7 at 7.1–7.5 ppm. In all the spectra, the aromatic proton signals corresponding to SCR7 are seen with lower intensity than compared with the Pluronic counterpart. The small signals from SCR7 are indicated by * in Fig. 3.

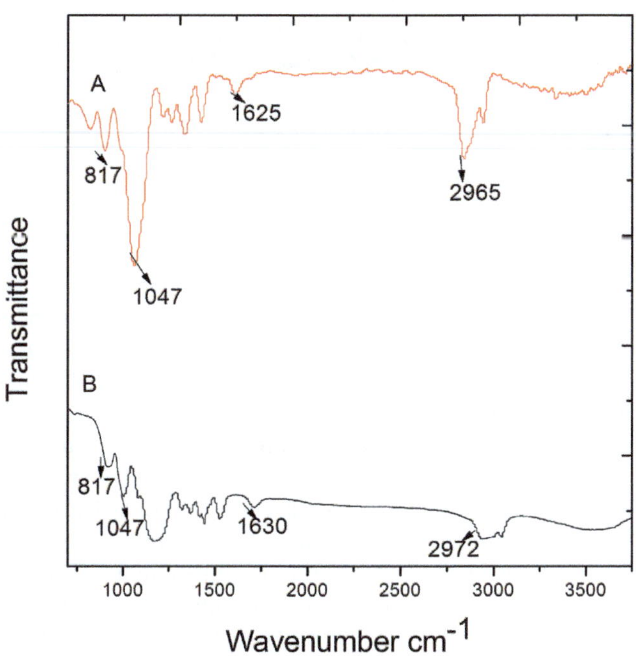


Fig. 2 FT-IR spectra of (A) SCR7 and (B) SCR7-P123.

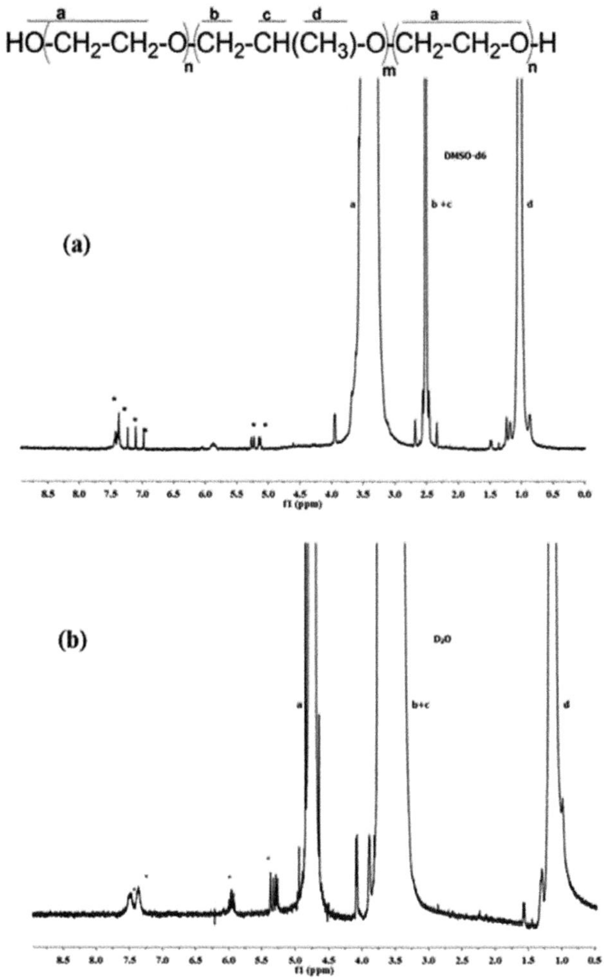


Fig. 3 $^1$H NMR spectra of SCR7P123 in (a) DMSO-d6 (b) $D_2O$.

### *In vitro* cytotoxic assay

A cytotoxic assay was performed to measure the metabolic activity of cells. This was carried out by reducing yellow coloured tetrazolium salt 3-(4,5-dimethyl thiazol-2-yl)-2,5-diphenyltetrazolium bromide to purple coloured formazan. The test sample was prepared by dissolving 200 mg of SCR7-P123 in 0.5 mL of DMSO and made up to 5.8 mL with culture medium containing serum and sterilized filters. This was diluted with culture medium to 50%, 25% and 12.5%. Cells cultured in normal medium were considered as the cell control. An equal volume (100 μL) of various dilutions of test samples, extract of negative control, cell control and positive control were placed on subconfluent monolayer of L929 cells. After incubation of the cells with various concentrations of test sample and controls at 37 °C for 24 ± 2 h, the extract and control medium was replaced with 50 μL MTT solution (1 mg $mL^{-1}$ in medium without supplements), wrapped with

 

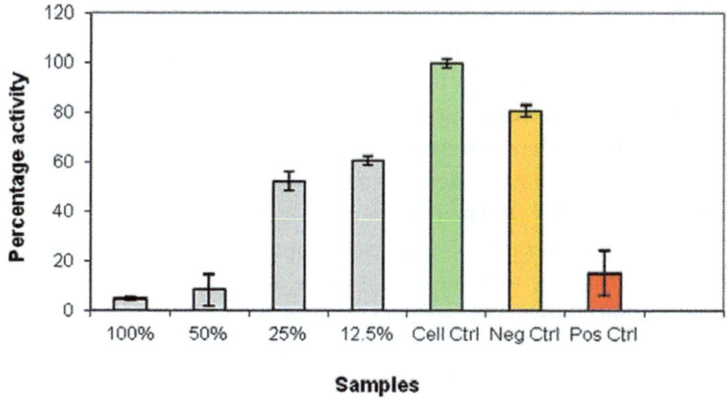


Fig. 4 MTT assay profile of SCR7P123.

aluminium foil and incubated at 37 °C for 2 hours. After discarding the MTT solution 100 μL of isopropanol was added to all wells and the plates were shaken. The developed colour was quantified by measuring absorbance at 570 nm using a spectrophotometer. The data obtained for the test sample were compared with the cell control. As shown in Fig. 4, the MTT assay of L929 cells after 24 hour contact with samples 100%, 50%, 25% and 12.5% of test material SCR7P123 showed 4.72%, 8.23%, 52.29%, and 60.69% metabolic activity respectively. The positive control (diluted phenol with culture medium containing serum) showed 15.14% and the negative control (ultra high molecular weight poly ethylene with culture medium containing serum) showed 80.56% metabolic activity.

# Experimental section

Pluronic P123 and $D_2O$ were purchased from Aldrich (Bangalore, India). All the salts and solvents used in the study were purchased from Merck (Mumbai, India). All other reagents and buffer solution components were analytical grade. Distilled and deionized water was used in all experiments.

### Preparation of SCR7-loaded polymeric micelles

SCR7 loaded micelles were prepared by the thin-film hydration method. 8 mg of SCR7 and 440 mg of Pluronic block copolymer were dissolved in 10 mL acetonitrile in a round-bottom flask. The solvent was evaporated by rotary evaporation at 50 °C for about 1 h to obtain a solid SCR7-P123 matrix. Residual acetonitrile remaining in the film was removed under vacuum overnight at room temperature. Then, the resultant thin film was hydrated with different amounts of water, while the hydration temperature was varied according to the experimental design. The mixture was stirred at 700 rpm for 30 min to obtain a micelle solution, which was then filtrated through 0.2 μm filter membrane to remove the unincorporated drug aggregates, followed by lyophilization.

### Characterization

Fluorescence emissions were recorded on a F-2500 fluorescence spectrometer (Hitachi, Japan). UV measurements were performed on a Shimadzu (UV-2450) UV-visible double beam spectrophotometer with a matched pair of stoppered fused silica cells of 1 cm optical path length. Fourier transformed infrared (FT-IR) measurements were obtained using a SHIMADZU 8400 FT IR spectrometer. The drug loading and release characteristics of P123 encapsulated SCR7 was analyzed by $^1$H NMR spectra which was recorded on a Varian 400 MHz spectrometer (Varian, Palo Alto, CA, USA) in deuterated dimethyl sulfoxide (DMSO-d6), chloroform ($CDCl_3$) and water ($D_2O$) at room temperature.

Thin layer chromatography (TLC) was adopted to study the physical interaction of SCR7 and P123. 10% methanol in chloroform was used as the mobile phase and silica gel G was used as the stationary phase. The spots were detected under UV light and stained using alkaline $KMnO_4$ solution and the $R_f$ values were determined.[23] To evaluate the cell cytotoxic potential of SCR7-P123, a MTT assay was performed. The cell line was obtained from the biotechnology wing of SCTIMST Thiruvananthapuram. The source of the cell line is the ATCC strain and L-929. L-929 is an established and well characterized mammalian cell line that has demonstrated reproducible results. The cells were cultured in 10% FBS. Ultra high molecular weight poly ethylene was used as the negative control and it was prepared by incubating 3 cm$^2$ ultra high molecular weight poly ethylene with culture medium containing serum at 37 ± 1 °C for 24 ± 2 h. Diluted phenol was used as the positive control and was prepared by diluting phenol stock solution (13 mg mL$^{-1}$) to 1.3 mg mL$^{-1}$ with culture medium containing serum.

# Conclusions

Anti cancer agent SCR7 has the capability to inhibit NHEJ in a Ligase IV dependent manner within cells. Hydrophobicity of SCR7 is a major setback in the utilization of this compound. Polymer encapsulation of small molecules to enhance their physicochemical properties represents an alternative approach in recent drug discovery research. The major focus of the reported study is to formulate an aqueous soluble matrix for drug delivery. The thin film hydration method is confirmed to be one of the most suitable methods for the encapsulation of SCR7 within the copolymer. $^1$H NMR spectra confirmed the entrapment of the drug within the Pluronic polymeric core, along with fluorescence, UV-Vis and FT-IR spectroscopic techniques. SCR7 is efficiently diffused from the polymeric core as evidenced from the chloroform extract. *In vitro* cytotoxic assays showed that the encapsulated form of SCR7 induced cytotoxicity in a concentration dependent manner. Biocompatibility of the polymeric matrix can serve as an ideal formulation for controlled drug delivery applications.

# Acknowledgements

The authors thank Department of Biotechnology, Govt. of India (DBT, no. BT/PR7703/27/493/2013) for financial support. We are grateful to Sophisticated Instrumentation Facility, CUSAT, Kochi for the analytical instrumentation service. Fluorescence and TOF MS/MS spectra were recorded at School of Environmental Sciences, M.G University, Kottayam.

 

## Notes and references


1 M. M. Gottesman, T. Fojo and S. E. Bates, *Nat. Rev. Cancer*, 2002, **2**, 48.
2 D. M. Molina, R. Jafari, M. Ignatushchenko, T. Seki, E. A. Larsson, C. Dan, L. Sreekumar, Y. Cao and P. Nordlund, *Science*, 2013, **341**, 84.
3 S. Aggarwal, *Nat. Rev. Drug Discovery*, 2010, **9**, 427.
4 H. A. Shih, J. S. Loeffler and N. J. Tarbell, Effects of CNS Radiation Therapy, *Late Effects of Treatment for Brain Tumors*, Springer, US, 2009, p. 23.
5 S. P. Jackson and J. Bartek, *Nature*, 2009, **461**, 1071.
6 T. Helleday, E. Petermann, C. Lundin, B. Hodgson and R. A. Sharma, *Nat. Rev. Cancer*, 2008, **8**, 193.
7 S. Sharma and S. C. Raghavan, *J. Nucleic Acids*, 2010, **1**, 2010.
8 M. R. Lieber, K. Yu and S. C. Raghavan, *DNA Repair*, 2006, **5**, 1234.
9 C. Wyman and R. Kanaar, *Annu. Rev. Genet.*, 2006, **40**, 363.
10 V. Gopalakrishnan and S. C. Raghavan, *Future Oncol.*, 2012, **8**, 1121.
11 M. Nambiar and S. C. Raghavan, *Cell. Mol. Life Sci.*, 2013, 1.
12 M. Nambiar and S. C. Raghavan, *Nucleic Acids Res.*, 2011, **39**, 5813.
13 S. C. Raghavan, P. C. Swanson, X. Wu, C. L. Hsieh and M. R. Lieber, *Nature*, 2004, **428**, 88.
14 F. W. Alt, Y. Zhang, F. L. Meng, C. Guo and B. Schwer, *Cell*, 2013, **152**, 417.
15 M. L. Hefferin and A. E. Tomkinson, *DNA Repair*, 2005, **4**, 639.
16 M. Srivastava, M. Nambiar, S. Sharma, S. S. Karki, G. Goldsmith, M. Hegde, S. Kumar, M. Pandey, R. K. Singh, P. Ray, R. Natarajan, M. Kelkar, A. De, B. Choudhary and S. C. Raghavan, *Cell*, 2012, **151**, 1474.
17 J. W. Lee, E. D. Jeong, E. J. Cho, J. A. Gardella Jr, W. Hicks Jr, R. Hard and F. V. Bright, *Appl. Surf. Sci.*, 2008, **255**, 2360.
18 W. Chen, Y. Shi, H. Feng, M. Du, J. Z. Zhang, J. Hu and D. J. Yang, *J. Phys. Chem. B*, 2012, **116**, 9231.
19 D. Qiu, C. Flood and T. A. Cosgrove, *Langmuir*, 2008, **24**, 2983.
20 F. John, J. George, S. V. Vartak, M. Srivastava, P. A. Hassan, V. K. Aswal, S. S. Karki and S. C. Raghavan, *Macromol. Biosci.*, 2014, DOI: 10.1002/mabi.201400480.
21 S. Alexander, T. Cosgrove, T. C. Castle, I. Grillo and S. W. Prescott, *J. Phys. Chem. B*, 2012, **116**, 11545.
22 P. Alexandridis and L. Yang, *Macromolecules*, 2000, **33**, 5574.
23 M. J. Ansari, S. Ahmad, K. Kholi, J. Ali and R. K. Khar, *J. Pharm. Biomed. Anal.*, 2005, **39**, 132.

Faraday Discussions

PAPER

# The effect on structural and solvent water molecules of substrate binding to ferric horseradish peroxidase

Niall Simpson,[a] Katrin Adamczyk,†[a] Gordon Hithell,[a] Daniel J. Shaw,[a] Gregory M. Greetham,[b] Michael Towrie,[b] Anthony W. Parker[b] and Neil T. Hunt*[a]

Received 22nd August 2014, Accepted 27th October 2014
DOI: 10.1039/c4fd00161c

Ultrafast, multi-dimensional infrared spectroscopy, in the form of 2D-IR and pump–probe measurements, has been employed to investigate the effect of substrate binding on the structural dynamics of the horseradish peroxidase (HRP) enzyme. Using nitric oxide bound to the ferric haem of HRP as a sensitive probe of local dynamics, we report measurements of the frequency fluctuations (spectral diffusion) and vibrational lifetime of the NO stretching mode with benzohydroxamic acid (BHA) located in the substrate-binding position at the periphery of the haem pocket, in both $D_2O$ and $H_2O$ solvents. The results reveal that, with BHA bound to the enzyme, the local structural dynamics are insensitive to H/D exchange. These results are in stark contrast to those found in studies of the substrate-free enzyme, which demonstrated that the local chemical and dynamic environment of the haem ligand is influenced by water molecules. In light of the large changes in solvent accessibility caused by substrate binding, we discuss the potential for varying roles for the solvent in the haem pocket of HRP at different stages along the reaction coordinate of the enzymatic mechanism.

## Introduction

The peroxidase enzymes are commonplace in biological systems, where they are responsible for catalysing the oxidation of substrates ranging from proteins, as in the case of cytochrome c peroxidase, to a variety of small organic molecules.[1] Despite the large number of different peroxidase enzymes, a remarkably high degree of similarity is seen across the family in terms of both the haem pocket structure and the mechanism by which they operate.

[a]*Department of Physics, University of Strathclyde, SUPA, 107 Rottenrow East, Glasgow, G4 0NG, UK. E-mail: neil.hunt@strath.ac.uk*

[b]*Central Laser Facility, Research Complex at Harwell, STFC Rutherford Appleton Laboratory, Harwell Oxford, Didcot, Oxon, OX11 0QX, UK*

† Present address: University of Hamburg & MPI for the Structure and Dynamics of Matter, Center for Free Electron Laser Science (CFEL), Luruper Chaussee 149, 22761 Hamburg, Germany.

Horseradish peroxidase (HRP) is widely used as a typical exemplar of this category of enzymes, and the mechanism of HRP can be described as:

$$\text{PorFe(III)} + H_2O_2 \Rightarrow \text{O=Fe(IV)Por}^{+\bullet}\ \text{(CpdI)} + H_2O \quad (1)$$

$$\text{O=Fe(IV)Por}^{+\bullet} + \text{HS} \Rightarrow \text{O=Fe(IV)Por} + \text{S}^{\bullet} \quad (2)$$

$$\text{O=Fe(IV)Por} + \text{HS} \Rightarrow \text{PorFe(III)} + \text{S}^{\bullet} + H_2O \quad (3)$$

where Por represents the haem propionate moiety and S the substrate.

Although the physiological substrate of HRP has not been determined thus far,[1] the structure of HRP and the involvement of the various elements of the active site in this mechanism have been the subject of much study.[2–9] It has been well-established from crystallography that the haem pocket of HRP features the histidine and arginine amino acid residues, located on the distal side of the haem, that are conserved in most peroxidases (His42 and Arg38 in HRP, see Fig. 1). Furthermore, molecular methods have shown these residues to be key to the formation of the oxo-ferryl intermediate species denoted Compound I (CpdI, see reaction (1)), which is one of the central aspects of the peroxidase mechanism. These residues are potentially also important in the onward reaction steps from CpdI.[3]

Despite this progress, the intimate molecular events that contribute to the HRP mechanism are yet to be fully established and important questions remain. One such issue relates to the mechanistic role of water in the active site of HRP. The steps leading to CpdI formation are believed to begin with the binding of a molecule of hydrogen peroxide to the 5-coordinate ferric haem of the enzyme and

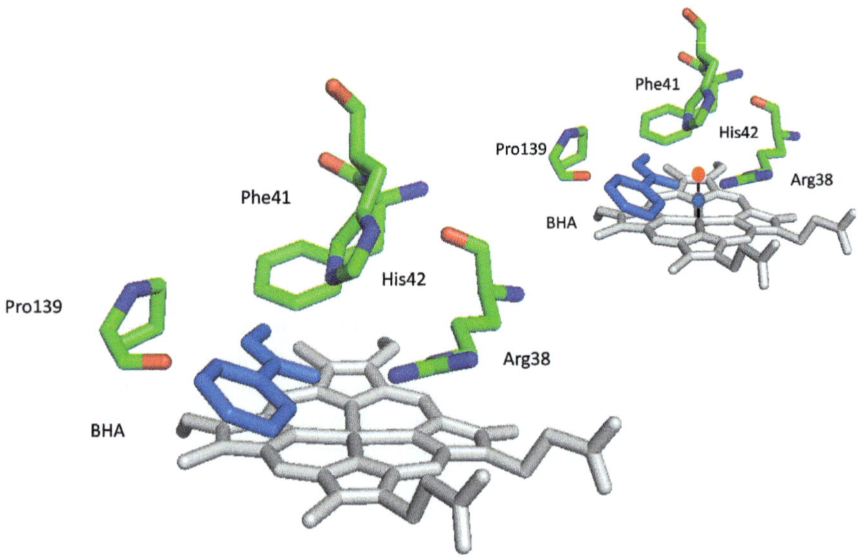


Fig. 1 Structure of the haem pocket of HRP–BHA showing residues with side chains located in close proximity to the haem (PDB:2ATJ).[5] The resolution of the structure was 2 Å. The inset shows a schematic representation of the same structure with the NO binding position indicated; N is shown as a blue sphere; O, red.

to progress to a precursor species similar to the Compound 0 (Cpd0) that has been observed in catalases, which takes the form Fe(III)–OOH. The formation of Cpd0 requires a proton transfer step from the bound $H_2O_2$ to the distal histidine side chain but the distance between the haem ligand and His42 has been shown by computational simulations to be too great to be spanned by a single H-bond interaction. Moreover, molecular dynamics simulations have suggested that proton transfer is likely to be energetically more favourable if mediated by the presence of disordered water molecules in the active site.[10,11] These findings are also consistent with experimental investigations suggesting a mechanistic role for water,[1,12,13] though some of these conclusions have been the subject of debate.[10] Finally, recent studies using ultrafast IR spectroscopy have demonstrated a clear dependence of the vibrational and structural dynamics of the haem pocket of HRP upon H/D exchange of the solvent, inferring the presence of water molecules in close contact with the haem ligand.[14]

Water molecules are also cited as being influential in the later stages of the peroxidase mechanism during which reduction of CpdI occurs *via* oxidation of larger substrates in two one-electron steps (steps (2) and (3) above). Here, the reason for the divergence of the peroxidase mechanism from that of the closely-related catalase enzyme, in which CpdI reacts *via* a single two-electron step involving a second molecule of $H_2O_2$, is widely attributed to the removal of water, or otherwise, from the vicinity of CpdI. This has led to the introduction of the concepts of a 'wet' CpdI in peroxidase and a 'dry' variant in catalase.[15] In apparent support of this, it is notable the that the active site of HRP is readily accessible by solvent molecules,[9] in contrast to what is believed to be the case for catalase. More recently it has been suggested that such concepts may have a more subtle, dynamic, interpretation following spectroscopic studies indicating that both enzymes feature interactions between the haem ligand and water molecules and so are likely to undergo similar water-mediated processes leading to CpdI.[14] This study did however report significant dynamic differences between the active sites of catalase and peroxidase that may contribute to the selection of different onward steps from CpdI. Specifically, HRP shows a much more flexible active site than catalase and it is plausible that this promotes the ability to bind a multitude of substrates, while the more restricted catalase pocket featuring well-defined low-frequency vibrational modes that couple to the haem ligand stretching vibration, may tailor functionality specifically toward subsequent reactions only with hydrogen peroxide.

A clear need thus exists to fully characterise water involvement in the HRP active site at all stages of the enzyme mechanism. In doing so, insights into the fast structural dynamics that accompany the presence of water and hydrogen bond interactions, which could influence individual mechanistic steps, will be of considerable value. As a result, the question we seek to address here is: what is the effect on the structural and solvation dynamics of the haem pocket of binding the model substrate molecule benzohydroxamic acid (BHA) to HRP and does this shed any light on our molecular-level understanding of the later stages of the peroxidase mechanism?

Although not a physiological substrate of HRP, BHA has been shown by co-crystallisation studies to bind strongly near the δ-edge of the haem group as a result of both hydrophobic interactions between the phenyl ring and the protein architecture and specific H-bond interactions involving the hydrophilic BHA head group and the His42 and Arg38 residues near the haem Fe (Fig. 1).[5,16] Under conditions where the haem is not ligated by a diatomic molecule, BHA binding promotes the formation of a 6-coordinate aqua complex of the ferric haem[17,18] but, otherwise, BHA binding seems to exclude water from the normally solvent-accessible pocket. In part, this is apparently achieved through competitive replacement of hydrogen bonding links near the haem but the binding of BHA also leads to a reorientation of the F68 residue, which serves to occlude the solvent access channel to the haem pocket.[5,17]

It is noted that crystallography methods are not normally sensitive to mobile water and so we employ a complementary approach, using ultrafast infrared spectroscopy to determine the spectral diffusion (2D-IR) and vibrational lifetime (pump–probe) of NO bound to the haem Fe of the HRP–BHA complex under both aqueous ($H_2O$) and deuterated ($D_2O$) solvent conditions in order to establish the effect of substrate binding on access of mobile solvent molecules to the haem centre. These measurements will further support crystal-derived data by reporting on changes in the dynamic nature of the haem site caused by substrate interactions and also on the relationship between crystal structure and the solution-phase environment.

Despite its prominent place in biological research, HRP has not been widely examined using ultrafast or dynamic spectroscopy methods. The nature of the haem pocket of HRP has been investigated using ligand rebinding studies, where NO recombination with the ferric haem following photolysis was reported to show the straightforward dynamics consistent with an active site that is readily solvent-accessible.[19] The equilibrium structural dynamics of the non-physiological ferrous state of HRP have been investigated using 2D-IR methods, though the analysis pre-dated molecular dynamics investigations into the mechanistic role of water and as such this was not considered.[20] Despite this, a general restriction of the dynamic nature of the haem cavity was observed on BHA binding.

## Experimental

For all FTIR, IR pump–probe and 2D-IR experiments, HRP (Sigma) was concentrated to the level of ~7.5 mM. MAHMA NONOate was used to nitrosylate the ferric HRP as has been described previously.[21] For experiments carried out under aqueous ($H_2O$) conditions, a phosphate buffer solution was employed at pH 7. For deuterated samples ($D_2O$), repeated buffer-exchange was performed using pD7 deuterated phosphate buffer solution; care was taken to ensure deuteration of all readily exchangeable protons. BHA (Sigma) was added to the sample at concentrations of 12 mM and binding was confirmed *via* detection of a single, narrow UV-Vis absorption band at 407 nm.[22] For all IR experiments, the samples were held between two $CaF_2$ windows separated by a PTFE spacer of 100 μm thickness.

The method for obtaining IR pump–probe and 2D-IR spectra has been described previously.[21,23] Briefly, 2D-IR spectra were acquired using the FT-2D-IR method described in ref. 21 using a sequence of three mid-infrared (IR) laser pulses arranged in a pseudo pump–probe beam geometry with parallel relative polarization directions (the effects of molecular rotation on the signals was found to be negligible on the timescales studied).[24,25] For IR pump–probe measurements, the same spectrometer was used but only two of the three pulses used for 2D-IR spectroscopy were employed. The pulses were generated by a Ti:sapphire laser pumping a white-light seeded optical parametric amplifier (OPA) equipped with difference frequency mixing of the signal and idler outputs. Mid IR pulses with a temporal duration of ~100 fs; a central frequency of 1900 cm$^{-1}$ with a bandwidth of ~200 cm$^{-1}$ were employed.

# Results

### FTIR spectroscopy

Introduction of BHA and NO into solutions of HRP resulted in the observation of a single new peak in the infrared absorption spectrum near 1900 cm$^{-1}$ that is attributed to the NO stretching vibration ($\nu_{NO}$) of the nitrosylated ferric haem.[21,23,26] Representative spectra are shown in Fig. 2A. This was observed to be the case irrespective of the order of addition of BHA and NO. When an aqueous solvent was employed, this peak was well-represented by a single Gaussian line-shape with a central frequency of 1912.1 cm$^{-1}$ and a full width half maximum (FWHM) value of 7.7 cm$^{-1}$. In deuterated solvent, this peak yielded values of 1911.5 cm$^{-1}$ and 8.8 cm$^{-1}$ for the central frequency and FWHM respectively, which do not represent a significant isotope-dependence of these parameters. The results of the data fitting are given in Table 1, where the values for the BHA-free enzyme are also quoted for reference. It is noted that the binding of BHA to HRP leads to a shift of around 8 cm$^{-1}$ in the frequency of the NO stretching vibration relative to the substrate-free enzyme as well as a significant narrowing of the peak by around 6 cm$^{-1}$.[14] The absence of signals due to BHA-free HRP in the FTIR data clearly shows that the vast majority of HRP molecules in the ensemble were in the form of a BHA complex.

### IR pump–probe spectroscopy

IR pump–probe spectroscopy was used to recover the vibrational dynamics of the $\nu_{NO}$ mode of nitrosylated HRP–BHA (simply denoted HRP–BHA hereafter as all samples discussed are nitrosylated); example spectra are shown in Fig. 2B. In both solvents, a negative peak was detected at a frequency corresponding to the single absorption observed *via* FTIR spectroscopy, which is assigned to bleaching and stimulated emission processes featuring the $v = 0$–1 transition of the $\nu_{NO}$ mode of HRP–BHA. In addition, a positive peak due to the $v = 1$–2 transient absorption was located at 1891 cm$^{-1}$. The lower frequency of this transition relative to the fundamental indicates the anharmonic shift of the $\nu_{NO}$ mode of HRP–BHA.

Both the $v = 0$–1 and $v = 1$–2 peaks were observed to decrease in amplitude in a similar manner as a function of pump–probe delay time (Fig. 2B and C). This effect is attributable to vibrational relaxation of the $\nu_{NO}$ $v = 1$ level and the data could be fitted effectively to a single exponential decay function (Fig. 2C). In the case of aqueous solvents, a vibrational lifetime of 16.5 ps was extracted while in deuterated solvents a value of 16.3 ps was recovered. The results of the fitting process are given in Table 1.

### 2D-IR spectroscopy

Representative 2D-IR spectra of HRP–BHA in aqueous and deuterated solvents are shown in Fig. 3A–C and 4A–C respectively for a range of experimental waiting times (equivalent to a pump–probe delay time). In both solvents, the spectra each

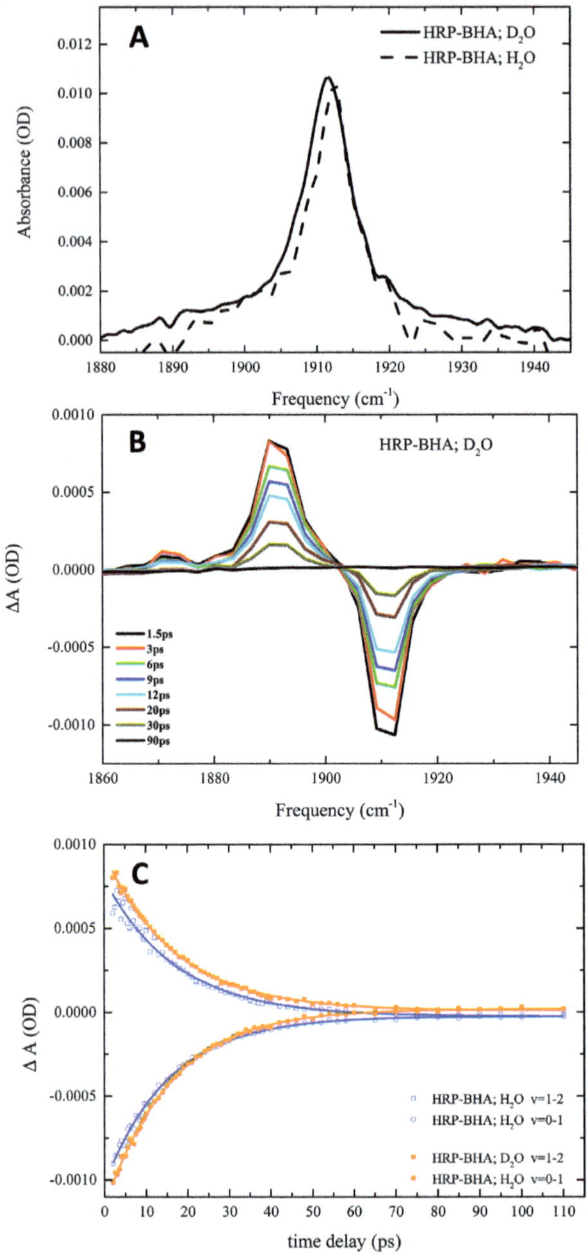


Fig. 2 (A) FTIR spectra of HRP–BHA in deuterated (solid line) and aqueous (dashed line) solvents. (B) Ultrafast IR pump–probe spectra of HRP–BHA in deuterated solvent at a range of pump–probe delay times. (C) Temporal decay profiles of the peaks observed in pump–probe spectra for HRP–BHA in aqueous (blue) and deuterated (orange) solvents. Solid lines indicate the results of fitting to exponential functions as described in the text.

show two peaks: a negative peak is located on the spectrum diagonal, with values of pump and probe frequency equal to the fundamental $\nu_{NO}$ absorption band in the FTIR spectrum. In addition, a second, positive feature is present that is shifted

 

Table 1 Results of fitting experimental data from FTIR, IR pump–probe and 2D-IR spectroscopy experiments to functions as described in the text

| | HRP + BHA | | Substrate-free HRP[a] | |
|---|---|---|---|---|
| | $D_2O$ | $H_2O$ | $D_2O$ | $H_2O$ |
| **FTIR** | | | | |
| Frequency ($cm^{-1}$) | 1911.5 | 1912.1 | 1902.7 | 1903.9 |
| Linewidth ($cm^{-1}$) | 8.8 | 7.7 | 14.7 | 13.9 |
| $\mathbf{IR_{pump-probe}}$ | | | | |
| $\nu = 1$ lifetime (ps) | 16.3 | 16.5 | 18.5 | 14.8 |
| **2D IR** | | | | |
| Spectral diffusion (ps) | 3.8 | 6.3 | 7.4 | 3.3 |
| Static offset | 0.35 | 0.3 | 0.27 | 0.45 |

[a] Data from ref. 14.

26 $cm^{-1}$ to a lower probe frequency. These peaks are assigned to the $\nu = 0$–1 and $\nu = 1$–2 transitions of the $\nu_{NO}$ mode respectively, consistent with the peaks observed in the pump-probe data.

Irrespective of the solvent used, the lineshapes of the 2D-IR spectral features were observed to evolve as the experimental waiting time increased; becoming more circular in profile and losing the diagonal elongation of the peaks present at

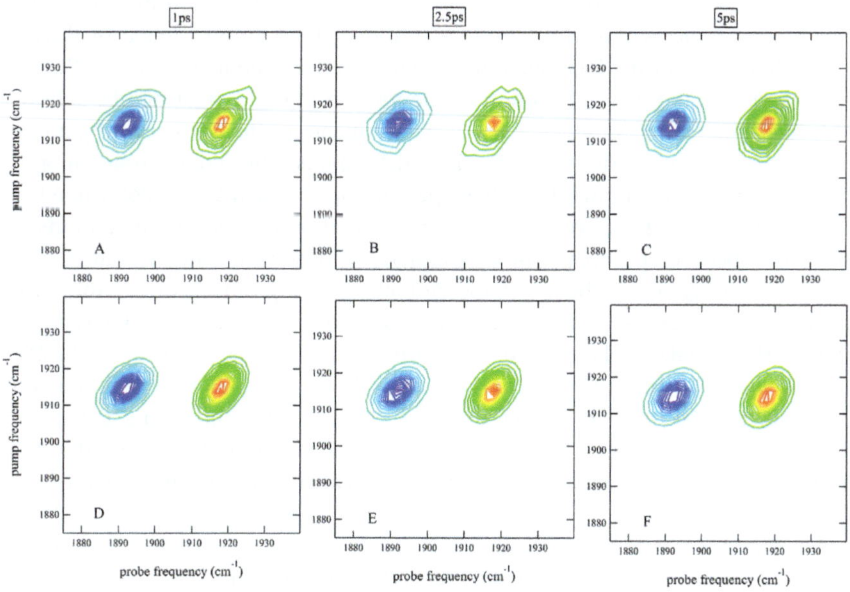


Fig. 3 (A–C) 2D-IR spectra of HRP–BHA in aqueous solvents at a range of experimental waiting times. (D–F) Fits of the data in (A–C) to a two-dimensional Gaussian function as discussed in the text.

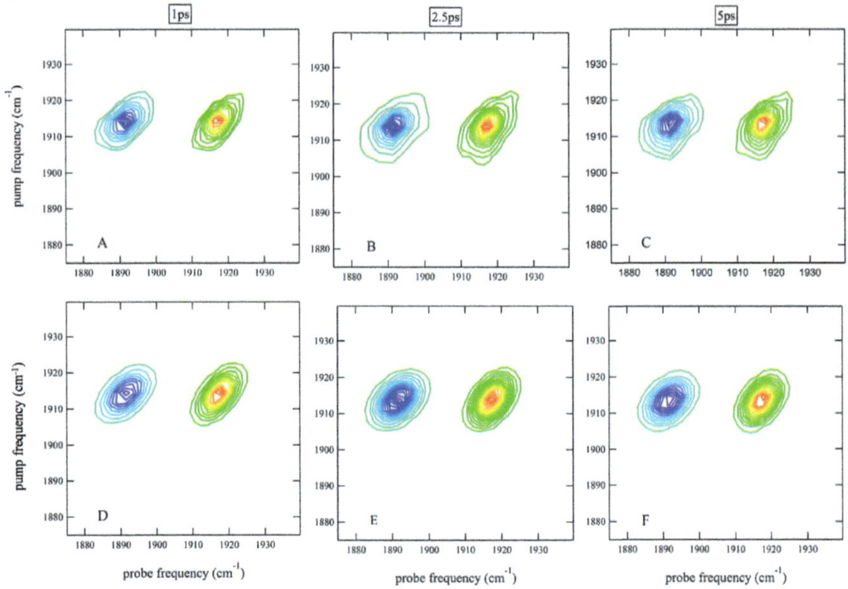


Fig. 4 (A–C) 2D-IR spectra of HRP–BHA in deuterated solvents at a range of experimental waiting times. (D–F) Fits of the data in (A–C) to a two-dimensional Gaussian function as discussed in the text.

short waiting times. This effect is well established for diatomic ligands bound to haem proteins and is assigned to spectral diffusion of an inhomogeneously-broadened transition.[20,23,27–34]

When located in the protein haem site the ligand vibrational absorption of the ensemble of molecules in the sample is inhomogeneously broadened by fluctuations that occur in the local chemical environment of the ligand. These fluctuations can take the form of motion of the protein scaffold or of nearby solvent molecules and lead to small changes of the frequency of the NO stretching mode *via* electrostatic interactions with the ligand.[35] This is reflected as broadening of the absorption band obtained *via* FTIR spectroscopy but in a 2D-IR plot, when the waiting time is short with respect to the dynamic timescale of the fluctuations, a correlation between pump and probe frequency is observed, which is manifest as a diagonal elongation of the lineshape. At longer waiting times, comparable with the dynamic processes occurring, the HRP–BHA ensemble has evolved leading to a loss of the initial correlation and so a more circular lineshape. Fitting of the 2D-IR lineshapes and quantification of the temporal evolution, referred to as spectral diffusion, can be used to extract the timescales for the underlying dynamics present in the Frequency–Frequency Correlation Function (FFCF) of the NO mode, which in turn provides insight into the local chemical environment of the NO probe.[36,37]

For HRP–BHA in aqueous and deuterated solvents, the spectra were fit to a 2D-Gaussian lineshape featuring a cross-correlation parameter $C_{2D}$. It has been shown previously that the temporal dependence of $C_{2D}$ accurately reflects the FFCF dynamics of NO in the protein environment. Fig. 3D–F and 4D–F show the fitting results for the experimental data displayed in Fig. 3A–C and 4A–C

 

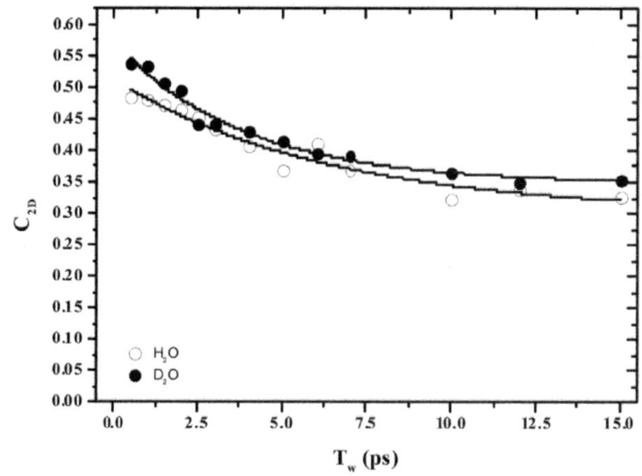


Fig. 5 Spectral diffusion dynamics for HRP–BHA in both aqueous (open circles) and deuterated (filled circles) solvents. Solid lines indicate fits to a single exponential decay as described in the text.

respectively.[21,23] The variation of $C_{2D}$ with waiting time (Fig. 5) was fit to an exponential decay function featuring a static offset parameter; the results of this are also given in Table 1.

For aqueous samples, the $C_{2D}$ data shows a decay with a timescale of $6.3 \pm 1.9$ ps and a substantial static offset (~0.30) while, in deuterated samples, the decay timescale was found to be $3.8 \pm 0.6$ ps with a static offset of ~0.35. In both cases, the initial value of $C_{2D}$ is reduced from unity (as might be expected for a fully-inhomogeneously-broadened system) by a combination of fast dynamics beyond the time resolution of the spectrometer (~200 fs) and the homogeneous linewidth of the transition.

# Discussion

Each of the three sets of experimental data presented above allow inferences to be drawn regarding the chemical or dynamic nature of the environment in which the NO ligand is located in the HRP–BHA complex.

### FTIR spectroscopy

In the case of the FTIR spectra, binding of BHA to HRP leads to a significant narrowing and blue-shifting of the NO stretching vibrational mode in comparison to the substrate-free enzyme in both solvents. For ease of reference the corresponding data for the substrate-free HRP enzyme are also given in Table 1.[14]

It has been shown that the absorption frequency of the $\nu_{NO}$ mode in ferric haem systems is sensitive to local contacts with the protein scaffold.[21,23,35] Of particular relevance here, studies on myoglobin and its mutants have indicated that the presence of a strong interaction between distal components of the haem pocket, such as the nearby histidine side chain, leads to an increase in the NO stretching vibrational frequency.[23,38] This effect has been attributed to the fact that, when bound to ferric haems, the electronic ground state of the Fe(III)–NO moiety is somewhat better-described by the Fe(II)–NO+ formalism.[39] This engenders interactions between the NO and the distal histidine in myoglobin that occur through a lone pair donation mechanism rather than a hydrogen bond leading to an upshift of the frequency.

In the case of substrate-free HRP, although a distal histidine residue is present in the active site, it has been shown to be more distant from the haem than would be required for direct interaction.[10,11] Additionally, previous work employing ultrafast infrared spectroscopy in both aqueous and deuterated solvents has shown results that are more consistent with the primary interaction of the NO being with water molecules in the haem pocket.[14] When BHA binds to HRP however, X-ray crystallography indicates that the phenyl ring is located near the δ-edge of the haem while the hydrophilic head group is located close to the Fe centre and the nearby His42 and Arg38 residues, to which the oxygen and nitrogen atoms of BHA form H-bonds (see Fig. 1). In the un-liganded form of the protein, a single water molecule is observable by crystallography located in a position 2.6 Å above the haem iron that is most consistent with a weak interaction with the Fe centre.[5,16,17] In the case of a haem centre with the diatomic CN ligand bound to it, the binding of BHA has been shown to be similar but there are indications of a rotated BHA head group in order to better accommodate interactions with the haem ligand.[5]

In light of this, the FTIR data are consistent with a major change in the local environment of the NO ligand occurring upon substrate coordination. Further, the spectral changes appear to be in agreement with the picture derived from the crystal structures, with the upshift of the $\nu_{NO}$ frequency supporting a close interaction between either the O or N atom of the BHA head group and the NO ligand. Alongside the frequency upshift, the marked narrowing of the $\nu_{NO}$ lineshape suggests a less inhomogeneously-broadened environment with the substrate in place, which is consistent with the replacement of water with more defined, directional interactions from BHA. Similar conclusions have been reported using IR absorption spectroscopy in concert with molecular dynamics simulations on the ferrous carbonmonoxy variant of HRP.[40] Thus, the FTIR data would seem to support the removal of water from the vicinity of the haem ligand upon binding of BHA but confirmation of this requires consideration of the time-resolved experiments.

### IR pump–probe spectroscopy

The vibrational lifetime of the $\nu_{NO}$ mode obtained from pump–probe spectroscopy provides a further useful indicator of the changes that are caused upon substrate binding to HRP. No solvent isotope dependence of the NO stretching mode vibrational lifetime is detected with the substrate in place; the value derived is essentially identical in both aqueous and deuterated solvents (Table 1).

This observation contrasts sharply with the situation found in the substrate-free enzyme in which the $\nu_{NO}$ mode lifetime became longer by a factor of ~25% upon deuteration of the solvent (Table 1).[14] This effect was attributed to the significant increase in the solvent-related density of vibrational states in the region of the IR spectrum near the NO stretching mode frequency in aqueous *versus* deuterated solvents. As a result of the greater overlap between solvent modes and the $\nu_{NO}$ mode in aqueous water, more rapid energy transfer and vibrational relaxation is facilitated in relation to deuterated conditions; this effect is not limited to proteins and has been observed for various solvent–solute systems.[41–43] This isotope-dependent relaxation mechanism appears to be absent in the HRP–BHA data.

Overall, the vibrational lifetime data for HRP–BHA points to the fact that the mechanism by which the NO mode relaxes is different to that of the substrate-free enzyme in that this no longer occurs through direct involvement of the solvent. This observation is also important in that it would seem to clearly separate the impact on $\nu_{NO}$ relaxation of deuteration of the protein backbone from that of the isotopic content of the solvent. This question arises because, upon transfer into a deuterated solvent, care is taken to ensure replacement of all readily-exchangeable protons within the enzyme structure. Thus, deuteration will necessarily alter the vibrational density of states of the protein and could, in principle, affect $\nu_{NO}$ mode relaxation dynamics. However, if this were the mechanism by which H/D exchange impacted upon the vibrational lifetime of the NO probe, this would have been observed in the substrate-bound enzyme also. That it is not confirms the original assignment of the isotope-dependent dynamics to water involvement in the haem pocket but also adds weight to the argument that this water is removed from the environment of the NO ligand upon substrate binding.

It is interesting also to note that the vibrational relaxation of the NO stretching vibration of ferric myoglobin is similarly insensitive to solvent isotope exchange.[14] As myoglobin is well-known to have no direct interaction between water and the haem ligand, rather a strong H-bond to the distal histidine residue occurs, this is also consistent with the observations for HRP–BHA.

The magnitude of the vibrational lifetime observed for HRP–BHA is not substantially different to either the aqueous or deuterated forms of the substrate-free enzyme (Table 1), indeed the values for the HRP–BHA complex fall between those of substrate-free HRP in aqueous and deuterated solvents.[14] Overall, this suggests that BHA binding does not lead to the creation of a markedly more efficient route for vibrational relaxation than the water-mediated environment of HRP. A possible reason for this is that the ultimate destination of the vibrational energy imparted by the pump pulse is likely to be the same low-frequency modes of the protein architecture. The change from solvent-sensitive to solvent-independent relaxation dynamics merely seems to indicate that the initial step in this process has changed from interaction with water-related vibrational modes to that with modes of BHA. It is relevant here that BHA also possesses a range of overtone and combination band transitions near 1900 $cm^{-1}$ arising from ring vibrations and those of the head group. These modes could potentially allow NO relaxation in place of those of the solvent for the substrate-free enzyme. Notably, these BHA-derived modes will be largely unaffected by deuteration of the solvent as BHA possesses only one or two exchangeable protons, unlike those of water. As such, large changes in the relaxation rate upon substrate binding should perhaps not be expected.

### 2D-IR spectroscopy

The picture emerging from the FTIR and vibrational lifetime data is further supported by the 2D-IR measurements of spectral diffusion processes affecting the $\nu_{NO}$ mode. The dynamic timescales recovered from this experiment provide information relating to the rate of fluctuation of the $\nu_{NO}$ frequency in response to local changes in the electrostatic environment of the ligand. These in turn have been shown to derive from motions within the local haem pocket.[35]

When BHA is bound to HRP these dynamics show two main components; a picosecond-timescale decay on the order of ~4–6 ps and a static offset. Studies applying 2D-IR to a range of haem proteins have indicated that these fluctuations can have a variety of origins. Fast, few ps, dynamics typically arise from highly-specific motion of distal residues in close proximity to the ligand while slower, more global motions of the haem pocket structure contribute to the longer dynamic components.[20,21,23,27–33,44–46] The latter tend to appear as a static offset in the data because they are too slow to fall within the temporal range of the 2D-IR experiment, which is limited by the vibrational lifetime of the probe group. More recently, it has been shown that contributions to fast spectral diffusion processes can also occur from the presence of water near the haem ligand. These have been differentiated from protein-derived motions *via* a solvent isotope-dependence.[14]

On binding a substrate analogue to HRP the results do not appear consistent with water interacting with the haem ligand. When considering the uncertainty in the spectral diffusion timescales recovered for HRP–BHA (Fig. 5), it is hard to establish a clear solvent-dependent separation between the picosecond-timescale dynamics observed in the two solvents. Perhaps the most important aspect of the data however is that the timescale recovered under aqueous conditions is longer than that in deuterated solvents, which is opposite to what one would expect for typical solvent-dependent dynamics.

As with the vibrational relaxation dynamics, it is interesting that the absolute magnitude of the spectral diffusion timescales observed in the picosecond time range for the NO stretching mode of HRP–BHA does not differ markedly from that of substrate-free HRP. The value of ~4–6 ps, falls between the limits of 3.3 ($H_2O$) and 7 ps ($D_2O$) observed for the substrate-free enzyme. The data from HRP experiments were explained by a model in which the water in the active site of HRP was dynamically-constrained by the nature of the haem pocket, leading to the slowing of the water motion from the 1–2 ps dynamics observed for Fe–NO units in bulk solution.[47] This was attributed to the small free volume present in the distal part of the pocket such that, rather than being free to undergo reorientational motion as in the bulk, water molecules are subject to interactions with polar, hydrophilic residues near the haem, potentially restricting motion and slowing the observed dynamics. Such an effect has been observed in reverse micelles.[48] This assignment was supported by quantitative analysis of the free-volume in the haem pocket of the substrate-free enzyme obtained from X-ray crystallography, which showed sufficient volume in the cavity for only a very small pool of <10 water molecules. While it has been shown that such a pool can support fast dynamic processes (<2 ps),[48] these simulations also suggested that the majority of the non-structural water was displaced from the centre of the haem in HRP, such that the haem-bonded NO would only interact with the extremities of the pool, where the interactions with protein residues are likely to become more dominant in restricting water motion.

In the case of the substrate-bound enzyme, the evidence suggests that, rather than there being an interaction between NO and the water pool, BHA displaces the water from near the haem, disrupting the pool and making H-bonding interactions with the NO ligand, consistent with the structures revealed in the crystalline phase.[5,16] This would explain the removal of the isotope dependence of the data. Given this scenario, it may be considered that interactions between the larger head-group of BHA and the NO ligand might be expected to yield slower spectral diffusion timescales than those found when smaller, more mobile water molecules are responsible for the dynamics. The fact that the timescales are not significantly different however suggests that the fluctuations of the BHA head group, being H-bonded to the protein architecture as they are, will be influenced to a certain extent by the same motions of the active site as water molecules were in the substrate-free enzyme, leading to similar observed dynamic timescales, albeit without the same sensitivity to solvent isotope which are superimposed on these motions in the presence of water.

In the case of the static offset parameter, this too does not vary noticeably with solvent isotopic composition in the substrate-bound system, in contrast to the data for substrate-free HRP, which showed a noticeable increase in amplitude on progressing from $D_2O$ to $H_2O$ solvents.[14] This also points to a reduction in the influence of water on the observed dynamics with the substrate present.

The persistence of a significant static component to the data in HRP–BHA, similar to that of the substrate-free enzyme, suggests that the overall flexibility of the haem pocket observed for HRP does not change greatly on binding BHA. It is interesting to note that previous studies of substrate binding to enzymes have determined that a significant shortening of spectral diffusion timescales are observed upon binding the correct combination of substrate and transition state analogue inhibitor molecules.[49] This was exemplified by a significant reduction in the amplitude of the pseudo static component of the spectral diffusion dynamics observed.[17,49] In the case of HRP–BHA however, such a shortening is not observed but it is noted that NO is not a precise structural analogue of CpdI, furthermore, BHA is not a natural substrate of HRP, merely one of a range of molecules that can bind to the substrate binding position.[1] As such, although it is instructive to employ BHA to shed light on water behaviour in substrate-bound HRP, this result suggests that it does not lead to the kind of large-scale dynamic modification of the pocket that has been indicated for some systems that are perhaps closer to the physiological situation.

Related to this, the first report of the use of 2D-IR spectroscopy to interrogate the structural dynamics of HRP, including changes to these dynamics upon BHA binding, noted that an increase in the rate of the picosecond fluctuations was observed upon substrate binding.[20] In that case, the ferrous form of HRP was studied with CO used as the spectroscopic probe. In addition, a general slowing of the longer timescale dynamics was attributed to a restriction of the structure following substrate biding. The change in the fast dynamics is replicated in this study of the ferric system also, though only for deuterated systems. In contrast, the reverse situation occurs for aqueous solvents with the fast timescale slowing on addition of BHA.

Finally, even upon substrate binding, the type of heavily restricted dynamics and coherent vibrational motions observed in catalase are not replicated, maintaining the view that these enzymes have dynamically very different active sites and that it is this that determines the differences in reactions catalysed.[14,21]

### Mechanistic implications

It is interesting to consider these results in light of the mechanism of action of HRP in which substrate molecules interact with the reactive intermediate, CpdI, undergoing oxidation. These results suggest that, although water molecules are clearly important in the formation of CpdI in HRP, as they are in catalase,[14] there is a likelihood that binding of substrate molecules leads to exclusion of water from the vicinity of the haem, such that they are unlikely to play any role in later steps. Such a scenario has also been proposed following simulations.[10,11] This is perhaps consistent with the scenario in which water molecules are needed to bridge distances between the haem ligand and the nearby residue side chains prior to CpdI formation but the insertion of a substrate into the haem pocket, which will establish its own H-bonds to these residues, removes this need allowing subsequent steps to proceed directly.

This conclusion would seem to be at odds with the view of a 'wet' CpdI, as is widely believed to occur in HRP. However, ultrafast dynamic measurements now suggest that the concept of a wet intermediate needs to be viewed as a subtle construct related to the fact that the dynamics near the HRP haem ligand during CpdI formation are solvent-mediated but rather more flexible, and thus slightly closer to bulk-like water, than those of catalase. These dynamics appear to specifically relate to CpdI attainment rather than the onward steps and so these results are not inconsistent with the established view.

It is noteworthy that no difference in the ability of HRP to bind NO was noted upon prior binding of BHA. This suggests that, although restricted by the motion of F68, it is still possible to gain access from solvent to haem with BHA in place. Rather, it is the close matching of H-bonding components of BHA to distal residues in the pocket that leads to exclusion of water from near the haem. Interestingly, this poses the possibility that $H_2O_2$ could also gain access to the haem with a substrate in place though it is not clear whether CpdI can be achieved under these conditions.

## Conclusions

Ultrafast, multi-dimensional infrared spectroscopy methods have been employed to investigate the effect of substrate binding on the structural dynamics of the Horseradish peroxidase (HRP) enzyme. Measurements of the frequency fluctuations (spectral diffusion) and vibrational lifetime of the stretching vibrational mode of NO bound to ferric HRP with benzohydroxamic acid (BHA) located in the substrate-binding position at the periphery of the haem pocket, in both $D_2O$ and $H_2O$ solvents reveal no sensitivity of these parameters to H/D exchange. Comparison of these results with those for the substrate-free enzyme suggests that large changes in solvent residence in the haem pocket are caused by substrate binding. Overall, these observations imply that, although water molecules are vital for the formation of CpdI, they are likely to play a less important role in onward reaction processes of peroxidase and that it is the overall dynamics and relative flexibility of the respective pockets that determines the differences in behaviour between peroxidases and catalases. In turn, we suggest that the concept of the 'wet' and 'dry' CpdI intermediates proposed for these enzymes[9,15] can be modified to a more nuanced picture upon consideration of ultrafast dynamics.

## Acknowledgements

NTH, KA, GH and NS would like to acknowledge funding for this work from the EPSRC and support from STFC for access to the Central Laser Facility. Useful discussions with Dr A Gumiero are also gratefully acknowledged.

## References

1 T. L. Poulos, *Chem. Rev.*, 2014, **114**, 3919–3962.
2 B. Zelent, A. D. Kaposi, N. V. Nucci, K. A. Sharp, S. D. Dalosto, W. W. Wright and J. M. Vanderkooi, *J. Phys. Chem. B*, 2004, **108**, 10317–10324.
3 G. I. Berglund, G. H. Carlsson, A. T. Smith, H. Szoke, A. Henriksen and J. Hajdu, *Nature*, 2002, **417**, 463–468.
4 H. K. Baek and H. E. Vanwart, *J. Am. Chem. Soc.*, 1992, **114**, 718–725.
5 A. Henriksen, D. J. Schuller, K. Meno, K. G. Welinder, A. T. Smith and M. Gajhede, *Biochemistry*, 1998, **37**, 8054–8060.
6 A. Henriksen, A. T. Smith and M. Gajhede, *J. Biol. Chem.*, 1999, **274**, 35005–35011.
7 A. Gumiero, S. K. Badyal, T. Leeks, P. C. E. Moody and E. L. Raven, *Dalton Trans.*, 2013, 16972.
8 G. H. Carlsson, P. Nicholls, D. Svistunenko, G. I. Berglund and J. Hajdu, *Biochemistry*, 2005, **44**, 635–642.
9 P. Jones and H. B. Dunford, *J. Inorg. Biochem.*, 2005, **99**, 2292–2298.
10 P. Vidossich, G. Florin, M. Alfonso-Prieto, E. Derat, S. Shaik and C. Rovira, *J. Phys. Chem. B*, 2010, **114**, 5161–5169.
11 E. Derat, S. Shaik, C. Rovira, P. Vidossich and M. Alfonso-Prito, *J. Am. Chem. Soc.*, 2007, **129**(20), 6346–6347.
12 I. Efimov, S. K. Badyal, C. L. Metcalfe, I. Macdonald, A. Gumiero, E. L. Raven and P. C. E. Moody, *J. Am. Chem. Soc.*, 2011, **133**, 15376–15383.
13 J. P. Roth and C. J. Cramer, *J. Am. Chem. Soc.*, 2008, **130**(25), 7802–7803.
14 K. Adamczyk, N. Simpson, G. M. Greetham, A. Gumiero, M. A. Walsh, M. Towrie, A. W. Parker and N. T. Hunt, *Chem. Sci.*, 2015, **6**, 505–516.
15 P. Jones, *J. Biol. Chem.*, 2001, **276**, 13791–13796.
16 A. Gumiero, E. J. Murphy, C. L. Metcalfe, P. C. E. Moody and E. L. Raven, *Arch. Biochem. Biophys.*, 2010, **500**, 13–20.
17 B. D. Howes, J. N. RodriguezLopez, A. T. Smith and G. Smulevich, *Biochemistry*, 1997, **36**, 1532–1543.
18 C. Indiani, E. Santoni, M. Becucci, A. Boffi, K. Fukuyama and G. Smulevich, *Biochemistry*, 2003, **42**, 14066–14074.
19 X. Ye, A. Yu and P. M. Champion, *J. Am. Chem. Soc.*, 2006, **128**, 1444–1445.
20 I. J. Finkelstein, H. Ishikawa, S. Kim, A. M. Massari and M. D. Fayer, *Proc. Natl. Acad. Sci. U. S. A.*, 2007, **104**, 2637–2642.
21 M. Candelaresi, A. Gumiero, K. Adamczyk, K. Robb, C. Bellota-Antón, V. Sangul, J. T. Munnoch, G. M. Greetham, M. Towrie, P. A. Hoskisson, A. W. Parker, N. P. Tucker, M. A. Walsh and N. T. Hunt, *Org. Biomol. Chem.*, 2013, **11**, 7778–7788.

22 H. A. Heering, A. T. Smith and G. Smulevich, *Biochem. J.*, 2002, **363**, 571–579.
23 K. Adamczyk, M. Candelaresi, R. Kania, K. Robb, C. Bellota-Antón, G. M. Greetham, M. R. Pollard, M. Towrie, A. W. Parker, P. A. Hoskisson, N. P. Tucker and N. T. Hunt, *Phys. Chem. Chem. Phys.*, 2012, **14**, 7411–7419.
24 L. P. DeFlores, R. A. Nicodemus and A. Tokmakoff, *Opt. Lett.*, 2007, **32**, 2966–2968.
25 S.-H. Shim and M. T. Zanni, *Phys. Chem. Chem. Phys.*, 2009, **11**, 748.
26 N. T. Hunt, G. M. Greetham, M. Towrie, A. W. Parker and N. P. Tucker, *Biochem. J.*, 2011, **433**, 459–468.
27 K. Adamczyk, M. Candelaresi, K. Robb, A. Gumiero, M. A. Walsh, A. W. Parker, P. A. Hoskisson, N. P. Tucker and N. T. Hunt, *Meas. Sci. Technol.*, 2012, **23**, 062001.
28 N. T. Hunt, *Chem. Soc. Rev.*, 2009, **38**, 1837–1848.
29 M. C. Thielges, J. K. Chung and M. D. Fayer, *J. Am. Chem. Soc.*, 2011, **133**, 3995–4004.
30 M. C. Thielges, J. K. Chung, J. Y. Axup and M. D. Fayer, *Biochemistry*, 2011, **50**, 5799–5805.
31 S. Bagchi, B. T. Nebgen, R. F. Loring and M. D. Fayer, *J. Am. Chem. Soc.*, 2010, **132**, 18367–18376.
32 S. Kim, J. K. Chung, K. Kwak, S. E. J. Bowman, K. L. Bren, B. Bagchi and M. D. Fayer, *J. Phys. Chem. B*, 2008, **112**, 10054–10063.
33 H. Ishikawa, I. J. Finkelstein, S. Kim, K. Kwak, J. K. Chung, K. Wakasugi, A. M. Massari and M. D. Fayer, *Proc. Natl. Acad. Sci. U. S. A.*, 2007, **104**, 16116–16121.
34 M. Cheng, J. F. Brookes, W. R. Montfort and M. Khalil, *J. Phys. Chem. B*, 2013, **117**, 15804–15811.
35 E. S. Park, M. R. Thomas and S. G. Boxer, *J. Am. Chem. Soc.*, 2000, **2000**, 12297–12303.
36 K. Kwak, S. Park, I. J. Finkelstein and M. D. Fayer, *J. Chem. Phys.*, 2007, **127**, 124503.
37 S. T. Roberts, J. J. Loparo and A. Tokmakoff, *J. Chem. Phys.*, 2006, **125**, 084502.
38 A. V. Soldatova, M. Ibrahim, J. S. Olson, R. S. Czernuszewicz and T. G. Spiro, *J. Am. Chem. Soc.*, 2010, **132**, 4614–4625.
39 D. R. Nutt and M. Meuwly, *ChemPhysChem*, 2007, **8**, 527–536.
40 A. D. Kaposi, N. V. Prabhu, S. D. Dalosto, K. A. Sharp, W. W. Wright, S. S. Stavrov and J. M. Vanderkooi, *Biophys. Chem.*, 2003, **106**, 1–14.
41 P. Hamm, M. Lim and R. M. Hochstrasser, *J. Chem. Phys.*, 1997, **107**, 10523–10531.
42 K. Ohta, H. Maekawa and K. Tominaga, *Chem. Phys. Lett.*, 2004, **386**, 32–37.
43 J. T. King, M. R. Ross and K. J. Kubarych, *J. Phys. Chem. B*, 2012, **116**, 3754–3759.
44 S. Bagchi, D. G. Thorpe, I. F. Thorpe, G. A. Voth and M. D. Fayer, *J. Phys. Chem. B*, 2010, **114**, 17187–17193.
45 H. Ishikawa, K. Kwak, J. K. Chung, S. Kim and M. D. Fayer, *Proc. Natl. Acad. Sci. U. S. A.*, 2008, **105**, 8619–8624.
46 H. Ishikawa, S. Kim, K. Kwak, K. Wakasugi and M. D. Fayer, *Proc. Natl. Acad. Sci. U. S. A.*, 2007, **104**, 19309–19314.
47 J. F. Brookes, K. M. Slenkamp, M. S. Lynch and M. Khalil, *J. Phys. Chem. A*, 2013, **117**, 6234–6243.

 

48 E. E. Fenn, D. B. Wong, C. H. Giammanco and M. D. Fayer, *J. Phys. Chem. B*, 2011, **115**, 11658–11670.

49 J. N. Bandaria, S. Dutta, M. W. Nydegger, W. Rock, A. Kohen and C. M. Cheatum, *Proc. Natl. Acad. Sci. U. S. A.*, 2010, **107**, 17974–17979.

Faraday Discussions

PAPER

# Electronic excitations in molecular solids: bridging theory and experiment†

Jonathan M. Skelton,* E. Lora da Silva, Rachel Crespo-Otero, Lauren E. Hatcher, Paul R. Raithby, Stephen C. Parker and Aron Walsh

Received 5th September 2014, Accepted 22nd October 2014
DOI: 10.1039/c4fd00168k

As the spatial and temporal resolution accessible to experiment and theory converge, computational chemistry is an increasingly powerful tool for modelling and interpreting spectroscopic data. However, the study of molecular processes, in particular those related to electronic excitations (*e.g.* photochemistry), frequently pushes quantum-chemical techniques to their limit. The disparity in the level of theory accessible to periodic and molecular calculations presents a significant challenge when modelling molecular crystals, since accurate calculations require a high level of theory to describe the molecular species, but must also take into account the influence of the crystalline environment on their properties. In this article, we briefly review the different classes of quantum-chemical techniques, and present an overview of methods that account for environmental influences with varying levels of approximation. Using a combination of solid-state and molecular calculations, we quantitatively evaluate the performance of implicit-solvent models for the $[Ni(Et_4dien)(\eta^2\text{-}O,ON)(\eta^1\text{-}NO_2)]$ linkage-isomer system as a test case. We focus particularly on the accurate reproduction of the energetics of the isomerisation, and on predicting spectroscopic properties to compare with experimental results. This work illustrates how the synergy between periodic and molecular calculations can be exploited for the study of molecular crystals, and forms a basis for the investigation of more challenging phenomena, such as excited-state dynamics, and for further methodological developments.

## 1 Introduction

Continuous improvements in spectroscopic techniques, such as the advent of next-generation X-ray free-electron laser (XFEL) facilities, are allowing the structural dynamics of molecular crystals to be studied in unprecedented detail. These

*Department of Chemistry, University of Bath, Claverton Down, Bath, BA2 7AY, UK. E-mail: j.m.skelton@bath.ac.uk*

† Electronic supplementary information (ESI) available: includes additional data from the parameter testing in Section 3, including the effect of different DFT functionals on molecular geometry and extended basis-set convergence results, plus the full dataset from the molecular TD-DFT calculation. See DOI: 10.1039/c4fd00168k

advances bring with them, however, a number of significant challenges in interpreting the data in terms of processes occurring at the molecular level. The length and timescales accessible to these advanced experimental techniques are rapidly converging on those which are accessible to theoretical study, and as such computational chemistry, *e.g.* within the ubiquitous density-functional theory (DFT) framework, can be a powerful complement to experiment.

A particular problem arises with the study of molecular solids, namely that the calculations need to capture both the chemistry of the molecular species, and the influence of the crystalline environment. At present, there exists a large disparity between the level of theory which is accessible to molecular and solid-state (periodic) calculations, due to the higher complexity of the latter. However, accurate descriptions of molecular properties, particularly those related to electronic excitations, *e.g.* absorption spectra and excited-state dynamics, frequently require high-level methods. Through continual advances in computing power and software efficiency, the gap between molecular and solid-state calculations is closing, but at present molecular crystals still frequently stretch the limits of what is possible in periodic calculations, mainly due to their large unit cells, and to the spectrum of weak non-bonding interactions which play an important role in defining their structure and properties.

To study electronic excitations in molecular solids, therefore, approximate methods of accounting for the effect of the crystalline environment in molecular calculations are required. In this work, we review several different approaches, and evaluate quantitatively the performance of the simplest one for computing various properties of the well-studied $[Ni(Et_4dien)(\eta^2\text{-O,ON})(\eta^1\text{-}NO_2)]$ linkage-isomerisation system. The key focus of our work is to explore how best to exploit the synergy between solid-state and molecular calculations, and to provide new theoretical insight to complement ongoing experimental work on this and related systems.

## Towards chemical accuracy: the "Jacob's ladder" of approximations

Quantum chemistry aims to model the properties of quantum-mechanical systems from first principles by solving the Schrödinger equation, typically within the Born–Oppenheimer (clamped-nuclei) approximation. The ultimate goal is to be able to predict properties with "chemical accuracy", that is, accuracy on the same scale as state-of-the-art experimental techniques. For example, historically, chemically-accurate energies are usually taken as implying an uncertainty of less than 1 kcal mol$^{-1}$ (4.18 kJ mol$^{-1}$), although tighter criterion are often required in practice. Due to Perdew and coworkers,[1] quantum chemistry techniques can be classified according to a "Jacob's ladder" of approximations, extending to the "heaven" of chemical accuracy.

DFT, which recasts the many-body Hamiltonian in the electronic Schrödinger equation into a functional of the spatial electron density $n(\mathbf{r})$,[2,3] is perhaps the most widely-used theoretical "workhorse" at present. Leaving aside implementation details, such as the mathematical basis used to express the electron orbitals, the key parameter in DFT is the form of the exchange–correlation (XC) functional used to calculate the contributions to the total energy from quantum-mechanical electron exchange and correlation. The simplest functional form is the local-density approximation (LDA), in which the exchange and correlation energy for a given $n(\mathbf{r})$ are obtained using a homogenous electron gas as a model system. For some systems, the LDA benefits from a fortuitous cancellation of errors and performs far better than expected, but, in general, it is too big an approximation to model subtle properties accurately.

The LDA can be improved upon substantially by also including the gradient of the electron density, $\nabla n(\mathbf{r})$, which forms the basis for semi-local generalised-gradient approximation (GGA) functionals. The logical extension to these are the *meta*-GGA functionals, which include the second derivative of the electron density, usually in the form of the electron kinetic-energy density $\tau(\mathbf{r})$. At the next level, more accurate functionals can be obtained by replacing a fraction of the DFT exchange energy with the exact Hartree–Fock exchange, and such "hybrid" functionals then require the electron orbitals, as well as $n(\mathbf{r})$, as input. Hybrid functionals are standard in molecular quantum-chemistry, and are used routinely for accurate electronic-structure calculations on periodic systems.

Finally, there exist a growing number of "beyond DFT" and "post Hartree–Fock" methods, which are becoming increasingly popular and accessible with advances in computing power. Examples of this include GW theory,[4] a perturbative approach to treating many-body physics, and second-order Møller–Plesset perturbation theory[5] (MP2) and coupled-cluster theory,[6] which both aim at a more accurate description of electron correlation.

In practice, (*meta*-)GGA functionals typically offer a good balance between computational cost and accuracy for a number of properties, including energetics and forces, and hence equilibrium geometries and vibrational frequencies.[7] Hybrids usually give more accurate electronic structures (*i.e.* orbital energies in molecular systems) than semi-local functionals, and so are often a prerequisite for the computation of optical properties and for studying electronic excitations in time-dependent DFT (TD-DFT) calculations. Post-DFT methods are useful in cases where even higher accuracy is required, *e.g.* to construct benchmark datasets against which functionals can be tested, and are sometimes necessary for systems where many-body effects are prominent, for which the more approximate approaches frequently fail.

## Environmental effects in molecular calculations: a spectrum of approaches

The most common methods for accounting for the solid-state environment in calculations on the component species in molecular crystals may be divided into three classes, with differing computational cost and accuracy.

At one end of the scale, full periodic calculations can be performed. The molecular environment is treated explicitly, but the computational cost of using high-level theories is prohibitive for large systems, and lower-level theories may be insufficient to describe certain properties with the required level of accuracy. At the other end, continuum models[8] (*e.g.* the polarisable-continuum model (PCM)[8] or COnductor-like Screening MOdel (COSMO)[9] schemes) assume that the most important environmental effects can be captured by a simple dielectric screening. This is likely to be a good approximation when the intermolecular interactions are minimal, and is commonly used to implicitly model the effect of a solvent. Forming a "middle ground" between these techniques are embedding methods, in which a large system is separated spatially into different regions which are then treated at different levels of theory.

## Explicit periodic calculations

With current hardware and software algorithms, it is feasible to perform periodic calculations on medium-sized molecular crystals (up to a few hundred atoms) using DFT with semi-local (*meta-*)GGA functionals, or, in some cases, with hybrids.[10–13] This is usually sufficient for studying equilibrium geometry and lattice dynamics (*e.g.* vibrational frequencies), but semi-local functionals are often not able to describe electronic structure quantitatively, as is required, for example, for accurate prediction of optical properties.

A particular issue with molecular crystals, as opposed to many simple bulk materials, is that weak interactions (*e.g.* van der Waals' dispersion forces) often play an important role in defining the structure.[14–16] Dispersion forces are, in principle, a non-local electron-correlation effect, and thus a first-principles quantum-chemical description requires a non-local functional. To circumvent this, several approximate methods have been developed to correct GGA energies and forces, *e.g.* the DFT-D2[17]/D3[18] and DFT-TS[19]/TS + SCS[20] methods, and non-local correlation functionals such as vdW-DF[21]/DF2.[22] However, these approximate forms may not account for more complex many-body dispersion interactions, which appear to be significant in some molecular systems.[16]

A good compromise to obtain accurate electronic properties is to optimise structures at a moderate level of theory, and to then perform more accurate single-shot calculations with hybrid functionals. In many cases, this approach gives good results, but for systems where many-body effects are prominent, *e.g.* excitonic effects in optical spectra,[23] the required higher levels of theory sometimes cannot be used due to computational limitations. Also, since geometry relaxation with non-local functionals is not feasible, it is not realistically possible to study excited-state dynamics with periodic calculations, which precludes the investigation of photochemical reactions.

## Embedding

In typical embedding methods, a large bulk system is divided into regions, and the core (*e.g.* a single molecule, or a dimer, *etc.*) is treated with a quantum-mechanical (QM) method, while a surrounding "shell" is treated with an empirical molecular-mechanics (MM) method such as a parameterised force field, reverting to a continuum model at larger distances. The layers are then interfaced together to account for the interaction energy between them. The MM atoms are treated as point charges or multipoles, which can polarise the wavefunction in the QM region, while the QM atoms can likewise exert forces on the MM atoms. This allows for an explicit atomistic treatment of large systems, with high accuracy in a region of interest, at a manageable computational cost. This QM/MM approach is commonly used to model biochemical systems, for example to look at redox processes at the active centres of enzymes.

The idea behind QM/MM embedding schemes can be extended to analogous QM/QM methods, where the core region is treated with a high-level quantum-chemical method (*e.g.* MP2 or coupled-cluster) and the shell with a lower-level one such as DFT with a GGA functional.[24–27] This class of methods also encompasses techniques in which the periodic wavefunctions obtained from an extended-system calculation are converted to localised orbitals, and selected local states then treated with higher-level theories.[28,29]

 

Embedding schemes have successfully been applied to a variety of molecular crystals,[30–32] and the particularly ambitious study by Kochman *et al.*[32] utilised a scheme where a single molecule in a periodic DFT calculation was treated with a molecular code, allowing exploration of its excited-state potential energy surface using TD-DFT.

### Continuum models

At the lowest level of approximation, environmental effects can be included using continuum models, which assume that the main influence of surrounding molecules is a dielectric-screening effect. The key parameter in these models is the static dielectric constant of the medium, $\varepsilon_{\text{static}}$, which captures its ability to screen an applied electric field. These methods are routinely used to model solvent effects in molecular calculations, but could also feasibly be used for molecular crystals, provided that the interaction between molecular units is negligible.

The molecule is treated as being a solute within a cavity, surrounded by a dielectric continuum of the solvent. The charge–density cloud of the molecule polarises the medium, which responds by generating screening charges on the cavity surface. In the COSMO model, the screening charges are obtained by modelling the response of an ideal solvent with $\varepsilon_{\text{static}} = \infty$, which is then scaled to account for the $\varepsilon_{\text{static}}$ of a non-ideal medium. Some continuum models additionally contain the shape of the solvent molecule as a parameter, which is then taken into account when defining the cavity around the solute.

Continuum models are widely used in the study of solution chemistry (*e.g.* solution thermodynamics[33] and reactions[34,35]), and to model biochemical systems.[36] They have also been used to model environmental effects on electronic excitations,[37,38] *e.g.* within a PCM/TD-DFT formalism.[38] Continuum models are generally used in quantum chemistry to model solvents, although there are a few cases where they have been employed to model the dielectric environment of a crystal in a molecular calculation.[39,40]

### Test system: linkage isomerisation in [Ni(Et$_4$dien)($\eta^2$-O,ON)($\eta^1$-NO$_2$)]

Solid-state linkage isomerisation is a topical example of a reversible phase transition generating one or more metastable structural isomers in response to external stimuli, *e.g.* illumination or temperature changes.[41] Since the initial observation by Coppens *et al.* of linkage isomerisation in sodium nitroprusside,[42] several families of linkage-isomer systems have been discovered, among which Ni–NO$_2$ (ref. 39 and 43–47) and Ru–SO$_2$ (ref. 48–51) complexes are perhaps the most well-known.

Linkage-isomer systems are typically studied using photocrystallography, where a sample, usually a single crystal, is irradiated *in situ* on the diffractometer, allowing the steady-state populations of the different isomers to be quantified as a function of temperature and of the wavelength of radiation used.[52] By performing pseudo-steady-state experiments, where the crystal is continuously irradiated during the data-collection process,[41] additional short-lived metastable species can also be identified.[46]

The [Ni(Et$_4$dien)($\eta^2$-O,ON)($\eta^1$-NO$_2$)] linkage-isomer system is an ideal test case for continuum models. The isomerisation takes place within a large reaction cavity, which allows the ligand binding to be switched without inducing a large stress on the crystal.[39,40,43,51] Furthermore, previous computational modelling[40] and experimental measurements of the transition kinetics[46] both suggest that, to a good approximation, the molecules in the solid behave as isolated units, whose properties and behaviour are influenced by the dielectric environment of the crystal.

This system exists as three isomers, in which the $\eta^1$-bound $NO_2$ ligand coordinates to the Ni centre *via* either N or O (Fig. 1). The N-bound ground-state (GS) isomer is obtained on cooling the crystal in the absence of illumination, and is the thermodynamically-stable form. The GS can be converted to the O-bound MS1 isomer with excellent yield on photoactivation, and is also formed thermally in significant population at moderate temperatures. A second O-bound isomer, MS2, was recently observed as a transient species close to the so-called metastable limit, the temperature at which MS1 begins to decay thermally, and can interconvert with MS1 with a relatively small energy barrier.[40]

A number of open questions remain on linkage isomerisation. Firstly, although the isomers themselves have been characterised, understanding the reaction paths which connect them is a work in progress.[53–55] Numerical modelling has allowed mechanisms to be proposed for some systems,[39,40] although explicit study of the excited-state potential energy surfaces, and establishing the photochemical-isomerisation pathways, remains an important undertaking. A second key challenge is to understand the role that the crystalline environment plays in controlling the kinetics and energetics of the process, which would, in principle, provide valuable input to crystal-engineering approaches for tuning the properties of different systems for specific applications.

In the present work, we focus on the GS and MS1 isomers, since these have been well characterised experimentally,[44,46] providing a good set of reference data

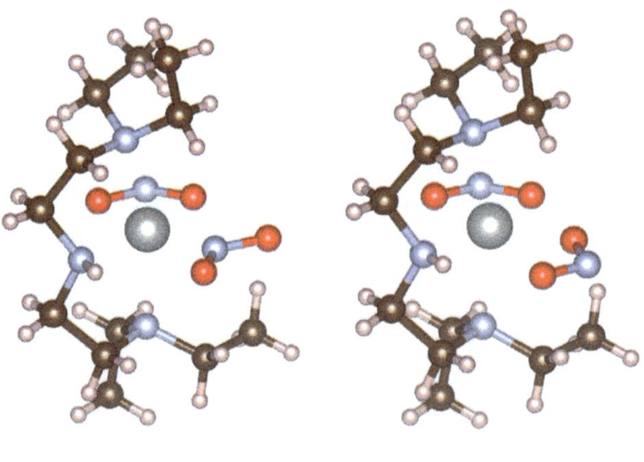


Fig. 1 Geometries of the [Ni(Et$_4$dien)($\eta^2$-O,ON)($\eta^1$-NO$_2$)] isomers. The left-hand image shows the ground-state (GS) N-bound isomer, and the right-hand image shows the metastable O-bound MS1 isomer. These structures were taken from the data in Hatcher *et al.*[44]

 

against which to optimise computational parameters and to evaluate the performance of continuum models.

## 2 Methods

All computational work was performed within the Kohn-Sham DFT formalism.[3] Different codes were used for the periodic and molecular calculations, as outlined below.

### Periodic calculations

Periodic plane-wave pseudopotential calculations were performed using the Vienna *Ab initio* Simulation Package (VASP) code.[56] Projector augmented-wave (PAW) pseudopotentials[57,58] were used to treat core electrons, with the outer s and p electrons of H, N and O, and the 4s, 3d and semicore 3p states of Ni, being treated as valence states. Where available, we used "hard" pseudopotentials, with a smaller core region to allow for more flexibility in the description of the valence wavefunctions. A kinetic-energy cutoff of 944.5 eV was used for the plane-wave basis, and the electronic wavefunctions were evaluated at the $\Gamma$ point; these parameters were found to be sufficient to converge the total energies to within less than 1 meV per atom. Spin-polarisation was used in all calculations, with each Ni complex constrained to have a triplet magnetic configuration, as was established to be the lowest-energy state in previous work.[40]

A selection of DFT exchange–correlation (XC) functionals were used in these calculations, as described in the text: we tested the PBE[59] and PBEsol[60] GGA functionals, the TPSS[61] *meta*-GGA, the dispersion-corrected PBE-D2,[17] PBE-D3 [18] and TPSS-D2 functionals, and the PBE0 hybrid.[62]

Convergence criteria of $10^{-6}$ and $10^{-5}$ eV were employed during electronic minimisation and atom position/unit-cell parameter optimisations, respectively. During the calculation of the dielectric functions, as described in the text, the number of electronic bands was increased to 1136, triple the default, to ensure convergence of the sum over empty electronic states. To obtain a high resolution for the calculated function, the number of grid points used to evaluate the electronic density of states (DOS) was increased to 5000 or 10 000, depending on whether or not the resulting function was to be used to compute optical properties.

### Molecular calculations

Molecular calculations were carried out using the NWChem[63] and Gaussian 09 [64] codes. As described in the text, several Pople split-valence[65] and Dunning[66] basis sets were tested for the light atoms, together with the Los Alamos effective-core pseudopotential and corresponding double-zeta basis set (LANL2DZ) for the Ni atom.[67] As in the VASP calculations, the molecular complex was constrained to have a triplet magnetic configuration. For comparison with the periodic calculations, we tested PBE, PBE-D2 and TPSS, and we also used the M06 [68] and B3LYP[69] functionals, which are both popular choices for molecular calculations. Convergence criteria of $10^{-6}$ and $10^{-5}$ were used for the total energy and density, respectively, while geometry optimisations were performed to a force tolerance of $5 \times 10^{-4}$ Hartree per Bohr. For the continuum calculations, we tested the

COnductor-like Screening MOdel (COSMO) method,[9] as implemented in NWChem, and the polarisable-continuum model (PCM) in Gaussian.

### Vibrational spectroscopy

Room-temperature infrared (IR) and Raman spectra of crystalline [Ni(Et$_4$dien)($\eta^2$-O,ON)($\eta^1$-NO$_2$)] were collected for comparison with calculated vibrational spectra. IR spectra were obtained from a single crystal using a Perkin Elmer Spectrum 100 FT-IR spectrometer with the ATR accessory. A spectral resolution of 1 cm$^{-1}$ between 4000 and 600 cm$^{-1}$ was available with this instrument, and acquisition and processing was performed using the Perkin Elmer Spectrum software.

Raman spectra were collected from a single crystal using a Renishaw inVia Raman spectrometer with the Renishaw WiRE 4.1 software. A 532 nm excitation laser (Renishaw diode laser, 380 mW at source/1.9 mW after attenuation) was used with a 20× objective lens (LEICA), and the instrument was calibrated to the 520 cm$^{-1}$ line of Si. Spectra were obtained between ~100 and 4000 cm$^{-1}$ at a resolution of 1.2 cm$^{-1}$.

It is worth noting that, since the GS-to-MS1 transition is thermally activated, at ambient temperature, and in particular under laser irradiation at 532 nm during the Raman experiments, the spectra are expected to contain bands due to both the GS and MS1 forms.

## 3 Results

### Energetics

Among the most straightforward property to compute with DFT is (relative) total energy. The per-molecule enthalpy difference between the GS and MS1 crystals has been determined from photocrystallographic measurements of the temperature dependence of the isomer populations,[44] providing a benchmark against which to quantitatively compare calculated energy differences.

In our recent work,[40] we found that the energetics were sensitive to the computational parameters employed, in particular the DFT functional used and, for molecular calculations, the choice of basis set. As a foundation for testing continuum models, in this section we systematically investigate these dependencies, and quantify the effect of the crystalline environment on the energy differences between the isomers.

To test the effect of the choice of functional on the predicted energetics, we carried out plane-wave pseudopotential calculations on the reported GS and MS1 crystal structures[44] with six commonly-used functionals, *viz.* PBE, PBEsol, PBE + D2, PBE + D3, TPSS and TPSS + D2. For each, the atomic positions and unit-cell parameters were fully relaxed, and the per-molecule GS-MS1 energy differences were then calculated from these optimised models; some structural parameters from the models (*e.g.* lattice parameters, cell volume and bond lengths) are given in the ESI.† To investigate the effect of the crystalline environment on the energetics, we also carried out an equivalent set of calculations on the individual GS/MS1 complexes in the gas phase. These were performed by placing single molecules from the crystal structures in periodic cells, with an (initial) vacuum gap of 10 Å between images, and optimising the geometries.

 

Of the selection of functionals we employed, PBE represents a typical choice for many solid-state problems, although PBEsol, a variant of PBE revised to better reproduce the properties of solids, is also used routinely. PBE + D2 and PBE + D3 both add a semi-empirical correction to the PBE energies to account for dispersion forces, and are frequently used for systems where weak interactions are expected to be significant. TPSS is a typical *meta*-GGA functional, and perhaps the most routinely used functional of this type, and we also tested it in combination with the D2 dispersion correction (TPSS + D2).

Fig. 2 compares the per-molecule energy differences in the crystal and in the gas phase. Compared to the experimental enthalpy difference of 9.69 kJ mol$^{-1}$, all of the functionals apart from TPSS significantly overestimate the energy difference between the GS and MS1 crystals. The PBE energy difference of 14.57 kJ mol$^{-1}$ is around 1.5× larger than the experimental one, while PBEsol and the three dispersion-corrected functionals overestimate the difference by a factor of two. On the other hand, the TPSS value of 7.75 kJ mol$^{-1}$ is in remarkably good agreement, suggesting that the improved accuracy of *meta*-GGA functionals provides a good description of the energetics of this system.

Comparing the gas-phase energy differences to those in the solid, all six functionals predict that the difference between the isomers is substantially reduced. However, in all cases it remains positive, indicating that the GS isomer is the ground state in the gas phase as well as in the molecular crystal.

Having found that TPSS gives improved energetics compared to the GGA functionals, we performed single-point energy calculations on our six optimised GS and MS1 structures using the PBE0 hybrid functional, and recalculated the energy differences (see ESI†). We observed some small variation between the different starting geometries, although the calculated values appear to be relatively insensitive to this. The best PBE0 value (7.05 kJ mol$^{-1}$) is obtained when the atomic positions, but not the cell parameters, are optimised with TPSS as a

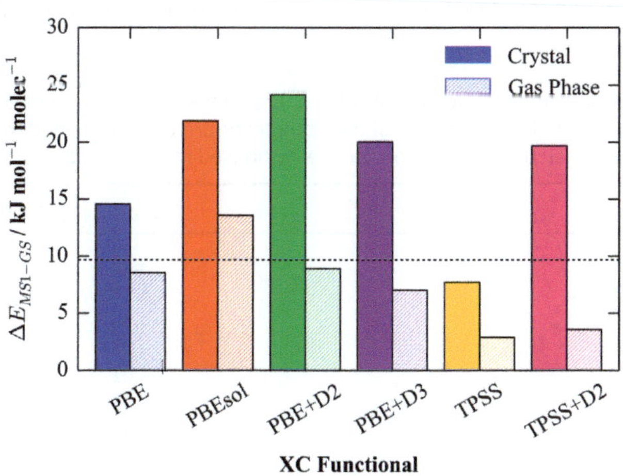


Fig. 2 Calculated per-molecule GS-MS1 energy differences in the molecular crystal (solid) and in vacuum (hatched) for a series of different exchange–correlation (XC) functionals. The experimentally-measured enthalpy difference is overlaid as a dashed black line.

starting point; it is possible that the agreement might improve further if a geometry optimisation could be performed with the hybrid.

The tight convergence criteria employed in the plane-wave calculations on the molecular species means that the calculated energy differences should be close to the basis-set convergence limit for the molecular calculations. We therefore used these values as a reference to optimise the basis set for these simulations. We carried out single-point energy calculations on the optimised gas-phase GS and MS1 structures with a number of different basis sets, and compared the calculated energy differences to the plane-wave values. Table 1 shows the results for a subset of the basis sets tested and the PBE functional; additional data for PBE, and a corresponding set of data for PBE + D2, are given in the ESI.†

In general, the double-zeta 6-31G family of basis sets predict a qualitatively incorrect energy ordering. Among the triple-zeta basis sets, the accuracy of the predicted energy differences improves when diffuse and polarisation functions are added (6-311G+/++ and 6-311G(d)/(d,p), respectively). This appears to be more important for the heavier atoms, with the 6-311+G(d) basis, which includes diffuse functions and d-type polarisation functions on the heavy atoms, giving very similar results to 6-311++G(2d,2p), which includes diffuse functions and two d/p polarisation functions on both the light and heavy atoms.

We note that the results in Table 1 were all obtained using the LANL2DZ pseudopotential and corresponding double-zeta basis set to treat the Ni atom; during testing, we found that varying the Ni basis made relatively little difference to the energetics when the better-converged basis sets were used for the other atoms (see ESI†), but that, without the pseudopotential, the computational cost was significantly increased.

In contrast to the Pople basis sets, the Dunning bases are designed to give more systematic convergence with respect to the number of basis functions, although they are typically more computationally expensive than the Pople ones. Like the equivalent Pople double-zeta basis sets, the double-zeta cc-pVDZ basis predicts an incorrect energy ordering, while the augmented cc-pVDZ (aug-cc-pVDZ) basis set and the bare and augmented triple-zeta (aug-)cc-pVTZ sets give the best results overall. For the augmented cc-pVDZ and the two cc-pVTZ basis

Table 1 Calculated single-molecule GS-MS1 energy differences with the PBE functional and various basis sets. The structures of the complexes are those obtained from the gas-phase plane-wave calculations, and are not re-optimised in these tests

| Basis set | $\Delta E_{\text{MS1-GS}}$ kJ$^{-1}$ mol$^{-1}$ molec$^{-1}$ |
|---|---|
| 6-31G | −2.82 |
| 6-31G(d) | −3.28 |
| 6-31G(d,p) | −3.46 |
| 6-311G(d) | 3.02 |
| 6-311G(d,p) | 3.01 |
| 6-311+G(d) | 5.35 |
| 6-311++G(d,p) | 4.93 |
| 6-311++G(2d,2p) | 5.58 |
| cc-pVDZ | −1.47 |
| aug-cc-pVDZ | 7.31 |
| cc-pVTZ | 7.20 |
| aug-cc-pVTZ | 6.51 |

 

sets, we observed near-quantitative agreement with the plane-wave values for energetics calculations with PBE, PBE + D2 and TPSS (Fig. 3), suggesting that the energy difference with these basis sets is close to the convergence limit.

We note that these basis sets are computationally expensive to work with, and we found them unwieldy for geometry optimisation; however, the results in Table 1 suggest that a good compromise would be to optimise with the more efficient 6-311+G(d) basis, and to then perform further calculations with one of the triple-zeta Dunning bases where possible, which is the approach taken in the following sections.

## Dielectric properties

In this section, we calculate the macroscopic dielectric constants of the molecular crystals from solid-state calculations, and evaluate the performance of the implicit-solvent COSMO model in including the effect of the crystalline environment in the molecular calculations.

As discussed in Section 1, the static dielectric constant (permittivity), $\varepsilon_{\text{static}}$, of a material measures its ability to screen an applied electric field, and as such can be used to model approximately the effect of a dielectric continuum (*e.g.* a solvent) on various properties of a molecular system. $\varepsilon_{\text{static}}$ can be decomposed into the sum of two contributions, *viz.* electronic polarisation, and ionic relaxation:[70]

$$\varepsilon_{\text{static}} = \varepsilon_{\text{polarisation}} + \varepsilon_{\text{ionic}} \tag{1}$$

$\varepsilon_{\text{polarisation}}$ can be calculated either from the response of the system to a finite electric field (*e.g.* using density-functional perturbation theory (DFPT)), or obtained as the zero-frequency component of the real part of the frequency-dependent dielectric function ($\varepsilon(E) = \varepsilon_{\text{r}}(E) + i\varepsilon_{\text{i}}(E)$). Calculating $\varepsilon_{\text{ionic}}$ requires additionally the vibrational modes of the system to be evaluated, which for large systems can be very time consuming.

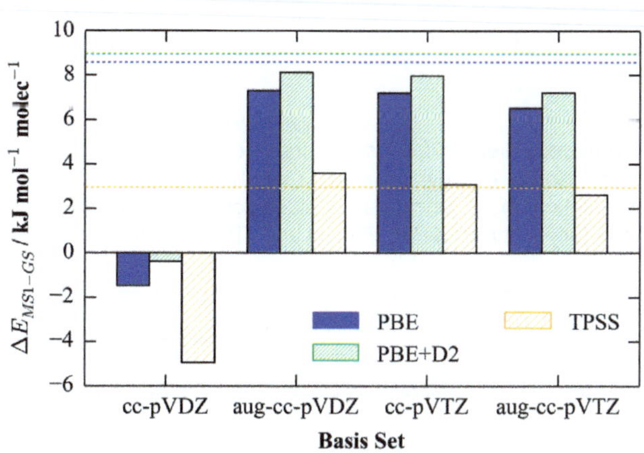


Fig. 3 Calculated single-molecule GS-MS1 energy differences with four Dunning basis sets and the PBE, PBE + D2 and TPSS functionals. As for the values in Table 1, the structures of the complexes are those obtained from the corresponding gas-phase plane-wave calculations, and are not re-optimised. For comparison, the reference plane-wave values are overlaid as dashed lines.

In the current version of the VASP code, $\varepsilon_{polarisation}$ can be computed using DFPT, or by calculating the dielectric function *via* the linear optical response, while both components of $\varepsilon_{static}$ can be calculated by combining a DFPT calculation of $\varepsilon_{polarisation}$ with a DFPT phonon calculation. However, VASP currently does not support DFPT calculations with semi-empirical and *meta*-GGA functionals, and thus we were only able to calculate both components of $\varepsilon_{static}$ with PBE and PBEsol. For the other functionals, we obtained the value of $\varepsilon_{polarisation}$ by calculating the dielectric function. Table 2 lists the values of $\varepsilon_{polarisation}$, $\varepsilon_{ionic}$ and $\varepsilon_{static}$ for the GS and MS1 crystals obtained with PBE and PBEsol, and the corresponding values of $\varepsilon_{polarisation}$ computed with a selection of other functionals are listed in Table 3.

There is a clear overall trend visible in this data. For both crystals, $\varepsilon_{static}$ is around 3–4, with the major contribution being from electronic polarisation rather than ionic relaxation. This value is comparable to a solvent such as benzene, toluene or diethylamine (2.27, 2.38 and 3.58, respectively), rather than a more polar solvent such as ethanol or water (24.5/80.1). On this scale, the variation between the functionals considered here is relatively small, which, in the absence of experimental measurements, lends a reasonable degree of confidence to the computed range.

Having obtained a reliable estimate of the dielectric constants of the GS and MS1 crystals, we then attempted to reproduce the solid-state energetics in molecular calculations using COSMO. We calculated the GS-MS1 energy differences for molecular complexes optimised with a selection of functionals, *viz.* PBE,

Table 2 Dielectric constants of the GS and MS1 crystal structures, computed with PBE and PBEsol. Each row gives the calculated values of $\varepsilon_{polarisation}$, $\varepsilon_{ionic}$, and the sum, $\varepsilon_{static}$

| System | Functional | $\varepsilon_{polarisation}$ | $\varepsilon_{ionic}$ | $\varepsilon_{static}$ |
|---|---|---|---|---|
| GS | PBE | 2.414 | 1.144 | 3.558 |
| | PBEsol | 2.652 | 0.971 | 3.623 |
| MS1 | PBE | 2.369 | 0.715 | 3.084 |
| | PBEsol | 2.592 | 0.714 | 3.305 |

Table 3 High-frequency dielectric constants (real components of the optical dielectric functions at zero photon energy) ($\varepsilon_{polarisation}$) of the GS and MS1 crystal structures, computed with a selection of functionals. Calculations with the PBE0 functional were carried out on the fully-optimised PBE, PBEsol and TPSS structures, which is denoted by "PBE0@XC"

| | $\varepsilon_{polarisation}$ | |
|---|---|---|
| Functional | GS | MS1 |
| TPSS | 2.582 | 2.535 |
| PBE-D2 | 3.142 | 3.070 |
| PBE-D3 | 3.042 | 2.986 |
| TPSS-D2 | 3.033 | 2.966 |
| PBE0@PBE | 2.134 | 2.099 |
| PBE0@PBEsol | 2.261 | 2.217 |
| PBE0@TPSS | 2.134 | 2.105 |

 

PBE-D2, TPSS, B3LYP and M06, and with dielectric constants of $\varepsilon_{static} = 3$, 3.5 and 4; as discussed at the end of the previous section, the geometry optimisations were performed with the 6-311+G(d) basis set, and the final energy differences were obtained from single-point energy calculations with the cc-pVTZ basis.

The calculated energy differences are compared in Fig. 4. PBE, PBE-D2 and TPSS were also used in the periodic calculations, allowing for a quantitative comparison. For all three functionals, the presence of a continuum increases the energy difference relative to the gas phase, mirroring the plane-wave calculations, and in general the correspondence between the energy differences obtained from the full-periodic and continuum calculations is very good. The only exceptions to this are the PBE-D2 results, the reasons for which are not clear.

B3LYP and M06 are hybrid and *meta*-hybrid functionals, respectively, and thus should in principle yield more accurate energetics than the others. M06 performs similarly well to TPSS in these calculations, yielding energy differences fairly close to the experimental values. On the other hand, B3LYP significantly underestimates the energy differences, yielding values lower than the TPSS gas-phase results, which suggests that this functional does not provide a good description of this system. We note in passing that we encountered problems with spin contamination in the B3LYP/$\varepsilon_{static} = 3.5$ single-point calculation with the cc-pVTZ basis set, and for the same reason we were not able to obtain energy differences with PBE0. Since this problem did not appear to affect the gas-phase calculations with the same functionals, and the small dielectric constant of the continuum is not expected to lead to large perturbations to the orbital structure, this is most likely an issue with the current implementation of COSMO in NWChem.

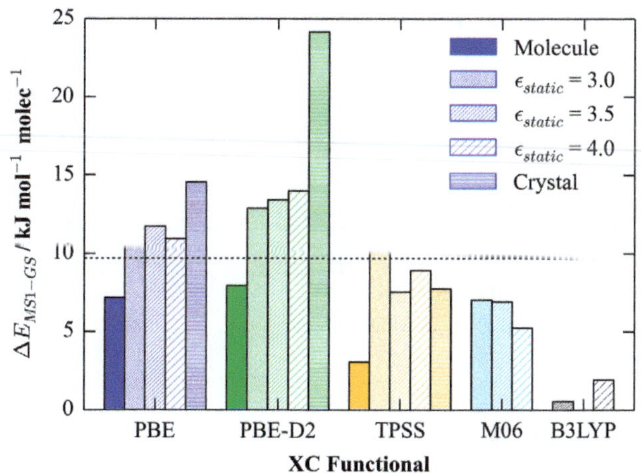


Fig. 4 Calculated single-molecule GS-MS1 energy differences, computed using the COSMO model with three dielectric constants ($\varepsilon_{static}$), *viz.* 3, 3.5 and 4, and a selection of DFT functionals. For PBE, PBE-D2 and TPSS, the plane-wave values, computed for the molecule and the crystal, are included for comparison, and the experimentally-determined energy difference is overlaid as a dashed black line. As discussed at the end of the section on energetics, for each calculation, the geometries of the GS and MS1 molecules were optimised with the 6-311+G(d) basis set, and the energies then recalculated with the cc-pVTZ basis set. There is no B3LYP value for $\varepsilon_{static} = 3.5$, due to issues with this particular calculation (see text).

Overall, the results suggest that a polarisable-continuum model, using the dielectric constant obtained from periodic calculations, is an effective way to combine the strengths of solid-state and molecular calculations for this system. In the following sections, we use our optimised parameters to model various spectroscopic properties, in particular IR/Raman frequencies and UV-visible absorption profiles.

### Vibrational spectra

Since (*meta*-)GGA functionals typically yield good forces, it is possible to perform periodic vibrational-frequency calculations, although this can be expensive for large systems. However, while calculating the IR intensity for a vibrational mode is fairly straightforward, Raman intensities are more computationally demanding to model, and so for molecular crystals the spectra are ideally best obtained from molecular, rather than periodic, calculations.

The IR spectra of the molecular crystals were obtained with the PBE and PBEsol functionals as a by-product of the calculations of $\varepsilon_{\text{ionic}}$ in the previous section. To further quantify the performance of the polarisable-continuum model, we computed the spectra with PBE and a 6-311+G(d) basis set using the Gaussian 09 code, using the polarisable-continuum model[8] with diethylamine (DEA) as a solvent ($\varepsilon_{\text{static}} = 3.6$). The spectra are compared in Fig. 5.

The overall correspondence between the two sets of spectra is very good. Although there are minor variations in the calculated intensities, the spectra have similar overall forms, and most of the calculated peak positions are likewise very similar. The only notable exception is in the position of a pair of bands around

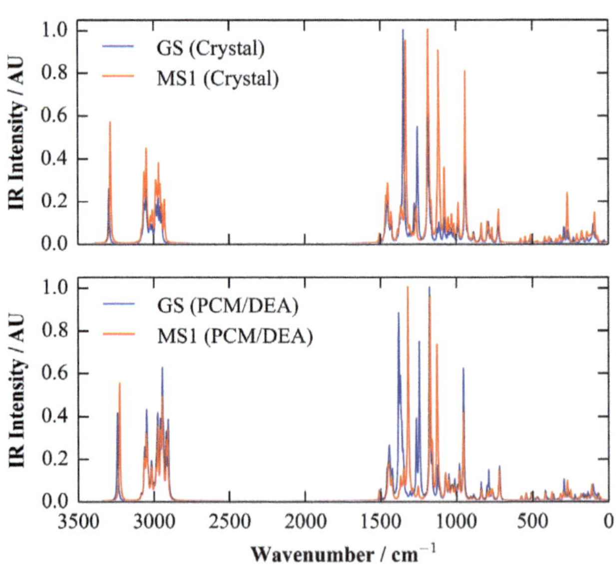


Fig. 5 Simulated IR spectra for the GS and MS1 systems, calculated with the PBE functional. The top panel shows the spectra obtained from a periodic calculation on the two fully-optimised molecular crystals, while the lower panel shows the spectra from molecular calculations with a polarisable-continuum model using diethylamine as a solvent (DEA; $\varepsilon_{\text{static}} = 3.6$). The spectra were all broadened using a Lorentzian function with a full-width at half-maximum (FWHM) of 7.5 cm$^{-1}$.

 

~1400 cm$^{-1}$, which correspond to a mixture of vibrations of the isomerisable ligands and parts of the Et$_4$dien backbone, and which lie on top of each other in the periodic calculations, but are visibly separated in the continuum ones. This suggests a slight difference in the bonding of the ligand in the two sets of calculations, which may well be related to the small energy differences between the periodic and continuum calculations evident in Fig. 4.

It is also worth noting that, despite the use of the continuum model, as only a single molecule is present, the molecular calculations would not be able to reproduce phonon modes corresponding to collective motions of two or more of the four molecules in the crystallographic unit cell, which could potentially lead to discrepancies, in particular in the low-wavenumber ("fingerprint") region of the spectra. The good correspondence between the periodic and molecular spectra in Fig. 5 therefore indicates that, to a very good approximation, the vibrational modes of the molecular crystal correlate with those of the molecular species, lending further support to the treatment of the molecules in the solid as independent entities.

Given the good performance of the continuum model with the calculated dielectric constant, we opted to compute IR and Raman spectra for both isomers with the more accurate M06 functional, which, as a hybrid functional, is in particular more likely to predict Raman polarisabilities more accurately than (*meta-*)GGAs. The calculated spectra are compared against room-temperature IR and Raman data collected from single crystals in Fig. 6.

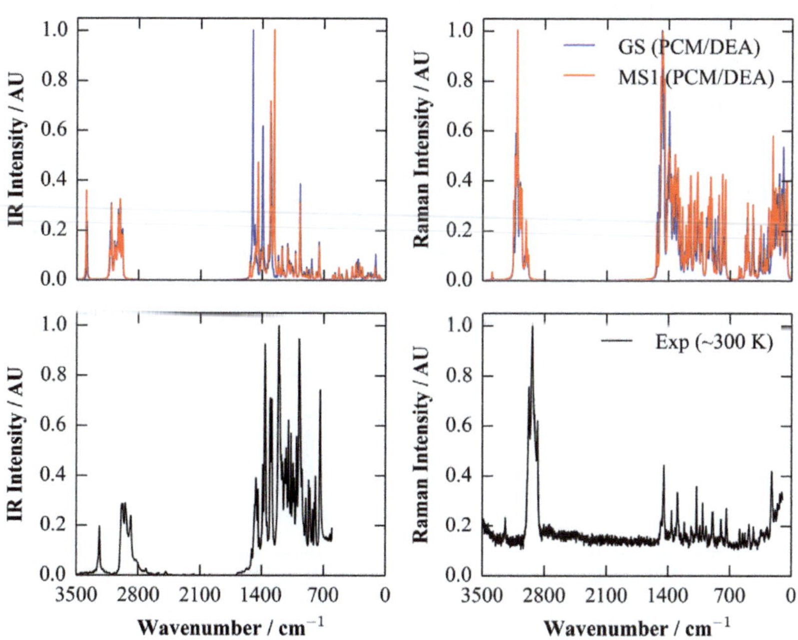


Fig. 6 Simulated IR and Raman spectra of the GS (blue) and MS1 (red) molecules, compared to room-temperature (~300 K) spectra taken from single crystals (black). The simulated spectra were obtained using the M06 functional and a polarisable continuum of diethylamine (DEA; $\varepsilon_{static} = 3.6$). As in Fig. 5, these spectra were broadened using a Lorentzian function with a full-width at half-maximum (FWHM) of 7.5 cm$^{-1}$.

There is a generally good correspondence between the simulated and experimental spectra. There are some notable mismatches in the predicted intensities, particularly in the case of the Raman spectra, although this could be due in part to the (arbitrary) choice of the broadening width for the calculated spectra. However, the positions of the peaks appear to be fairly well modelled, enabling assignment of some of the IR bands between ~1000 and 1500 cm$^{-1}$ to the GS and MS1 isomers. This agreement nicely illustrates a potential utility of computation for supporting characterisation.

## Optical properties

Optical properties arise from electronic excitations, and are a response property of the system. Modelling them formally requires solution of the time-dependent Schrödinger equation, or, more practically, modelling a perturbation to the ground-state electronic wavefunctions in response to an external field (*e.g.* as in the TD-DFT formalism).

The method implemented in VASP uses a perturbative linear-response method to obtain the frequency-dependent dielectric function, $\varepsilon(E)$, which, in essence, entails an enumeration over transitions between filled and empty electronic states.[71] From $\varepsilon(E)$, various optical properties can be calculated, including refractive indices ($n$), extinction coefficients ($\kappa$), and absorption coefficients ($\alpha$), the latter of which can be compared to experimentally-recorded UV-visible spectra. Fig. 7

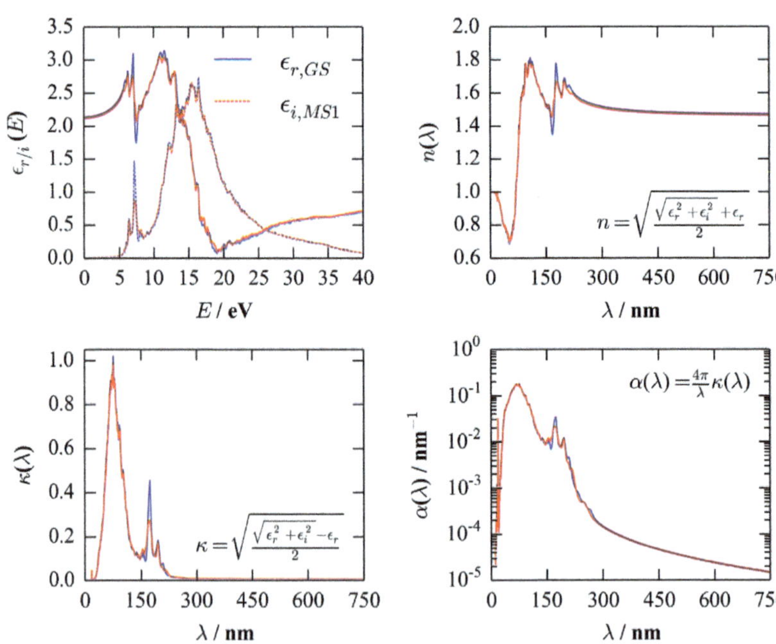


Fig. 7 Optical properties of the GS (blue) and MS1 (red) isomers, computed from the TPSS-optimised crystal structures using a PBE0 ground-state reference. The plots compare the real (solid lines) and imaginary (dashed lines) parts of the frequency-dependent dielectric function $\varepsilon(E)$, $\varepsilon_r/\varepsilon_i$, the refractive index, $n$, the extinction coefficient, $\kappa$, and the absorption coefficient, $\alpha$. Each of the latter three plots are annotated with the equations used to calculate the quantities from $\varepsilon(E)$.

 

compares these quantities for the GS and MS1 structures, computed from the TPSS-optimised crystal structures with a PBE0 ground-state reference.

The UV-visible spectrum of the GS molecular crystal published previously[46] does not extend into the deep UV, but the key features in the visible region are a long absorption tail around 400 nm, plus a second, broad isolated peak at ~650 nm. The most likely comparison between this and the spectrum in Fig. 7 is that the absorption tail is blue-shifted by ~200 nm, while the peak at 650 nm is not reproduced.

For comparison, we calculated spectra for the GS and MS1 molecules using the linear-response TD-DFT implemented in Gaussian. As for the calculations in the previous section, we used the M06 functional, together with a 6-311+G(d) basis set and a polarisable continuum of diethylamine. To compute the spectra, we analysed the 30 lowest-energy triplet transitions. The calculated spectra are shown in Fig. 8.

The overall structure of both spectra is similar to the 150–250 nm region of the solid-state one, consisting of primary high- and low-intensity peaks with some fine structure. The red shift of ~200 nm compared to the periodic spectra suggests that the more sophisticated TD-DFT method implemented in the molecular code gives a better description of the optical-absorption profile, and the position of the absorption tail agrees much better with the experimental data.

Based on our previous calculations,[40] the majority of the transitions correspond to delocalised metal-to-ligand transitions, with large components on the Ni d orbitals, while the majority of the states from ~450–350 nm can be assigned as metal-to-ligand charge-transfer bands. We note that, while the computational parameters used in these and in the present calculations are very similar, the continuum solvent used in the previous study ($H_2O$) has a considerably larger dielectric constant ($\varepsilon_{static} = 80.1$). However, we found that, while the positions of the absorption bands were sensitive to the dielectric constant, their nature and relative ordering were not, with the assignments being the same both in the gas phase and in a continuum of $H_2O$.

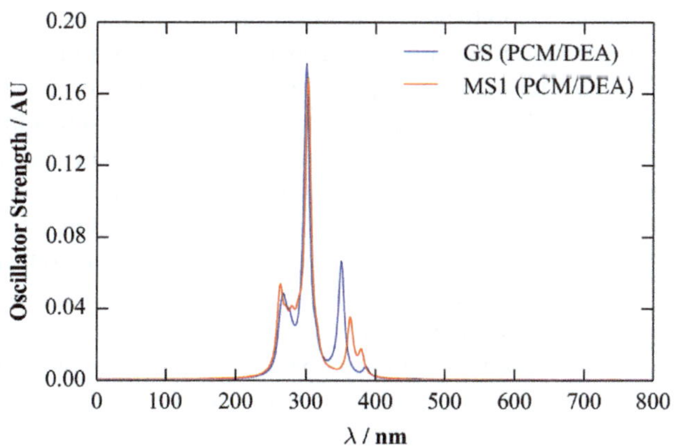


Fig. 8 UV-visible spectra of the GS and MS1 molecules, calculated using linear-response TD-DFT with M06 and a polarisable continuum of diethylamine (DEA; $\varepsilon_{static} = 3.6$). The spectra have been broadened using a Lorentzian function with a full-width at half-maximum (FWHM) of 10 nm.

The broad peak at 650 nm is again not reproduced in these spectra. However, considering the individual calculated transitions (see ESI†), there are several states with lower excitation energies, but with negligible oscillator strengths. The orbitals involved have a significant Ni d component, suggesting them to be (symmetry forbidden) d–d transitions, which perhaps become weakly allowed when coupled to vibrations. This could, in principle, be modelled by performing TD-DFT calculations on structures displaced along the vibrational modes, although this is much more computationally demanding.

These results, and also those in our previous work,[40] suggest that further optimisation of the computational parameters may be required to reproduce accurately the experimental UV-visible spectra for this system. However, the superior performance of the continuum TD-DFT calculation over the periodic linear-response method highlights the potential benefits of being able to use molecular calculations, and their excited-state functionality, while accounting approximately for the influence of the crystalline environment.

# 4 Discussion and conclusions

The results in the previous section demonstrate quantitatively the performance of polarisable-continuum models in approximately accounting for the influence of the molecular crystal environment in molecular calculations on the component species. With well-converged basis sets, continuum calculations, in conjunction with dielectric constants obtained from periodic calculations, were able to reproduce a broad range of properties of the [Ni(Et$_4$dien)($\eta^2$-O,ON)($\eta^1$-NO$_2$)] system, including energetics, vibrational frequencies, and, to a reasonable extent, optical properties. The correspondence between the continuum and periodic calculations was very good, and the comparison with available experimental data was also generally favourable. Moreover, and particularly with regard to modelling electronic excitations, the benefits of the higher levels of theory accessible in molecular calculations were clearly demonstrated.

The method explored in this work is fairly straightforward, and effectively combines some of the strengths of periodic and molecular calculations to allow for a more complete theoretical characterisation of molecular solids. For its broader applicability, the most important proviso is the assumption inherent in the polarisable-continuum approach, *i.e.* that the interactions between the molecular units in the solid are well represented by a dielectric-screening effect. It remains to be discussed, however, for which sorts of system this is a good approximation, and for which systems it should be expected to fail, *e.g.* crystals with longer-range intermolecular interactions such as π-stacking. For systems where this method works well, however, we anticipate that theoretical calculations should be a valuable tool for modelling and understanding a broad spectrum of state-of-the-art experimental work.

While the main focus of this study has been on the methodology, the calculations also provide some interesting insight into the chemistry of [Ni(Et$_4$dien)($\eta^2$-O,ON)($\eta^1$-NO$_2$)], reinforcing the findings of our previous study.[40] It is quite clear that the crystalline environment significantly influences the relative stabilities of the different linkage isomers, in this case increasing the energy difference between the GS and MS1 forms relative to the gas phase. To explore this further, as a theoretical exercise we calculated the GS-MS1 energy difference using

 

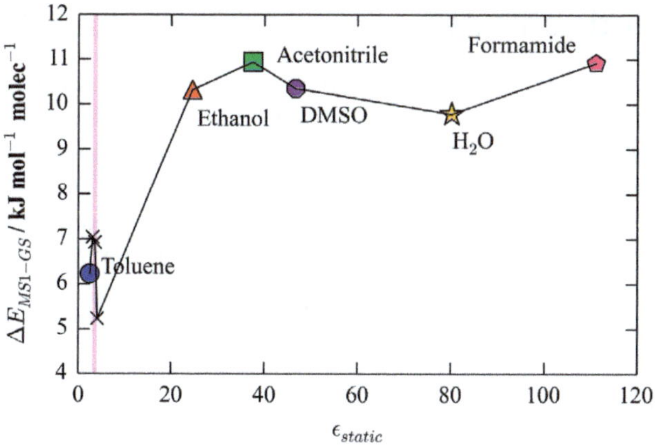


Fig. 9 Dependence of the GS-MS1 energy difference on the dielectric constant, $\varepsilon_{static}$, of the medium. The shaded region corresponds to values between 3 and 4, which is the range computed for [Ni(Et$_4$dien)($\eta^2$-O,ON)($\eta^1$-NO$_2$)]. Points outside of this range are annotated with solvents with comparable values of $\varepsilon_{static}$. All values were computed with the M06 functional, according to the same procedure used to obtain the values in Fig. 4.

COSMO for a range of dielectric constants, spanning the two-orders-of-magnitude range covered by common solvents (Fig. 9).

This plot clearly illustrates that, by hypothetically tuning the dielectric constant of the crystal, the energy difference between the two isomers can be adjusted over quite a large range. In general, a stronger dielectric screening appears to increase the energy difference, with the effect saturating beyond values of $\varepsilon_{static}$ around 30. Up to this limit, however, there is potentially scope for controlling the kinetics of the isomerisation process through the chemistry of the molecular species. This adds a new perspective to the established practice of using crystal engineering to create a "reaction cavity" for the isomerisation to occur within.[39,40,43,51]

To investigate this finding in more detail, it would be interesting to calculate the dielectric constants of other known Ni–$NO_2$ linkage-isomer systems, and to relate these to the photoconversion yield and/or the presence or absence of thermal–isomerisation pathways. It may perhaps also be of interest to relate the dielectric constant to other properties, such as the optical-absorption profile, and the energy difference between magnetic configurations.

To conclude, polarisable-continuum models represent an effective means of including the influence of the crystalline environment in molecular calculations on the [Ni(Et$_4$dien)($\eta^2$-O,ON)($\eta^1$-NO$_2$)] system. We have compared quantitatively a number of properties, and observed good overall agreement both with periodic calculations and experimental data, as well as obtaining new insight into the properties of this system, in particular the effect of the dielectric environment on the linkage isomerism. This work represents a step towards developing more general approaches to the theoretical characterisation of molecular solids, and, ultimately, to working on more complex topics, such as performing high-level electronic-structure calculations and exploring excited-state potential-energy surfaces.

## Acknowledgements

The authors would like to thank Mr Richard Driscoll for his assistance with collecting Raman spectra of [Ni(Et$_4$dien)($\eta^2$-O,ON)($\eta^1$-NO$_2$)]. The authors gratefully acknowledge financial support from an EPSRC programme grant (Grant no. EP/K004956/1) and the European Research Council (Grant no. 277757). We acknowledge use of the HECToR and ARCHER supercomputers through membership of the UK HPC Materials Chemistry Consortium, which is funded by EPSRC Grant no. EP/F067496, in the completion of this work.

## References

1 J. P. Perdew and K. Schmidt, *Jacob's ladder of density functional approximations for the exchange-correlation energy*, Antwerp, Belgium, 2001.
2 P. Hohenberg and W. Kohn, *Phys. Rev. [Sect.] B*, 1964, **136**, 864–871.
3 W. Kohn and L. J. Sham, *Phys. Rev.*, 1965, **140**, 1133–1138.
4 L. Hedin, *Phys. Rev.*, 1965, **139**, A796–A823.
5 C. Moller and M. S. Plesset, *Phys. Rev.*, 1934, **46**, 0618–0622.
6 J. Čížek, *J. Chem. Phys.*, 1966, **45**, 4256–4266.
7 J. M. Skelton, S. C. Parker, A. Togo, I. Tanaka and A. Walsh, *Phys. Rev. B: Condens. Matter Mater. Phys.*, 2014, **89**, 205203.
8 J. Tomasi, B. Mennucci and R. Cammi, *Chem. Rev.*, 2005, **105**, 2999–3093.
9 A. Klamt and G. Schuurmann, *J. Chem. Soc., Perkin Trans. 2*, 1993, 799–805.
10 A. M. Reilly, D. A. Wann, M. J. Gutmann, M. Jura, C. A. Morrison and D. W. H. Rankin, *J. Appl. Crystallogr.*, 2013, **46**, 656–662.
11 D. M. S. Martins, D. S. Middlemiss, C. R. Pulham, C. C. Wilson, M. T. Weller, P. F. Henry, N. Shankland, K. Shankland, W. G. Marshall, R. M. Ibberson, K. Knight, S. Moggach, M. Brunelli and C. A. Morrison, *J. Am. Chem. Soc.*, 2009, **131**, 3884–3893.
12 S. Hunter, T. Sutinen, S. F. Parker, C. A. Morrison, D. M. Williamson, S. Thompson, P. J. Gould and C. R. Pulham, *J. Phys. Chem. C*, 2013, **117**, 8062–8071.
13 J. Binns, M. R. Healy, S. Parsons and C. A. Morrison, *Acta Crystallogr., Sect. B: Struct. Sci., Cryst. Eng. Mater.*, 2014, **70**, 259–267.
14 S. Hirata, K. Gilliard, X. He, J. Li and O. Sode, *Acc. Chem. Res.*, 2014.
15 R. A. DiStasio, O. A. von Lilienfeld and A. Tkatchenko, *Proc. Natl. Acad. Sci. U. S. A.*, 2012, **109**, 14791–14795.
16 R. A. DiStasio, V. V. Gobre and A. Tkatchenko, *J. Phys.: Condens. Matter*, 2014, 26.
17 S. Grimme, *J. Comput. Chem.*, 2006, **27**, 1787–1799.
18 S. Grimme, J. Antony, S. Ehrlich and H. Krieg, *J. Chem. Phys.*, 2010, 132.
19 A. Tkatchenko and M. Scheffler, *Phys. Rev. Lett.*, 2009, 102.
20 A. Tkatchenko, R. A. DiStasio, R. Car and M. Scheffler, *Phys. Rev. Lett.*, 2012, 108.
21 M. Dion, H. Rydberg, E. Schroder, D. C. Langreth and B. I. Lundqvist, *Phys. Rev. Lett.*, 2004, 92.
22 K. Lee, E. D. Murray, L. Z. Kong, B. I. Lundqvist and D. C. Langreth, *Phys. Rev. B: Condens. Matter Mater. Phys.*, 2010, 82.

 

23 P. Cudazzo, M. Gatti and A. Rubio, *Phys. Rev. B: Condens. Matter Mater. Phys.*, 2012, 86.

24 T. A. Wesolowski and A. Warshel, *J. Phys. Chem.*, 1993, **97**, 8050–8053.

25 T. Wesolowski, R. P. Muller and A. Warshel, *J. Phys. Chem.*, 1996, **100**, 15444–15449.

26 H. P. Hratchian, P. V. Parandekar, K. Raghavachari, M. J. Frisch and T. Vreven, *J. Chem. Phys.*, 2008, 128.

27 P. V. Parandekar, H. P. Hratchian and K. Raghavachari, *J. Chem. Phys.*, 2008, 129.

28 V. Bezugly and U. Birkenheuer, *Chem. Phys. Lett.*, 2004, **399**, 57–61.

29 C. Willnauer and U. Birkenheuer, *J. Chem. Phys.*, 2004, **120**, 11910–11918.

30 S. H. Wen, K. Nanda, Y. H. Huang and G. J. O. Beran, *Phys. Chem. Chem. Phys.*, 2012, **14**, 7578–7590.

31 O. Bludsky, M. Rubes and P. Soldan, *Phys. Rev. B: Condens. Matter Mater. Phys.*, 2008, 77.

32 M. A. Kochman, A. Bil and C. A. Morrison, *Phys. Chem. Chem. Phys.*, 2013, **15**, 10803–10816.

33 C. P. Kelly, C. J. Cramer and D. G. Truhlar, *J. Chem. Theory Comput.*, 2005, **1**, 1133–1152.

34 T. N. Truong, *J. Phys. Chem. B*, 1998, **102**, 7877–7881.

35 F. P. Cossio, G. Roa, B. Lecea and J. M. Ugalde, *J. Am. Chem. Soc.*, 1995, **117**, 12306–12313.

36 R. M. Jackson and M. J. E. Sternberg, *J. Mol. Biol.*, 1995, **250**, 258–275.

37 R. Cammi, B. Mennucci and J. Tomasi, *J. Phys. Chem. A*, 2000, **104**, 5631–5637.

38 G. Scalmani, M. J. Frisch, B. Mennucci, J. Tomasi, R. Cammi and V. Barone, *J. Chem. Phys.*, 2006, 124.

39 M. R. Warren, T. L. Easun, S. K. Brayshaw, R. J. Deeth, M. W. George, A. L. Johnson, S. Schiffers, S. J. Teat, A. J. Warren, J. E. Warren, C. C. Wilson, C. H. Woodall and P. R. Raithby, *Chem.–Eur. J.*, 2014, **20**, 5468–5477.

40 J. M. Skelton, R. Crespo-Otero, L. E. Hatcher, S. Parker, P. Raithby and A. Walsh, *CrystEngComm*, 2015, **17**, 383–394.

41 L. E. Hatcher and P. R. Raithby, *Coord. Chem. Rev.*, 2014, **277–278**, 69–79.

42 M. D. Carducci, M. R. Pressprich and P. Coppens, *J. Am. Chem. Soc.*, 1997, **119**, 2669–2678.

43 M. R. Warren, S. K. Brayshaw, A. L. Johnson, S. Schiffers, P. R. Raithby, T. L. Easun, M. W. George, J. E. Warren and S. J. Teat, *Angew. Chem.*, 2009, **121**, 5821–5824.

44 L. E. Hatcher, M. R. Warren, D. R. Allan, S. K. Brayshaw, A. L. Johnson, S. Fuertes, S. Schiffers, A. J. Stevenson, S. J. Teat, C. H. Woodall and P. R. Raithby, *Angew. Chem., Int. Ed.*, 2011, **50**, 8371–8374.

45 S. K. Brayshaw, T. L. Easun, M. W. George, A. M. E. Griffin, A. L. Johnson, P. R. Raithby, T. L. Savarese, S. Schiffers, J. E. Warren, M. R. Warren and S. J. Teat, *Dalton Trans.*, 2012, **41**, 90–97.

46 L. E. Hatcher, J. Christensen, M. L. Hamilton, J. Trincao, D. R. Allan, M. R. Warren, I. P. Clarke, M. Towrie, D. S. Fuertes, C. C. Wilson, C. H. Woodall and P. R. Raithby, *Chem.–Eur. J.*, 2014, **20**(11), 3128–3134.

47 P. Coppens, I. Novozhilova and A. Kovalevsky, *Chem. Rev.*, 2002, **102**, 861–883.

48 D. A. Johnson and V. C. Dew, *Inorg. Chem.*, 1979, **18**, 3273–3274.

49 A. Y. Kovalevsky, K. A. Bagley and P. Coppens, *J. Am. Chem. Soc.*, 2002, **124**, 9241–9248.
50 A. Y. Kovalevsky, K. A. Bagley, J. M. Cole and P. Coppens, *Inorg. Chem.*, 2003, **42**, 140–147.
51 A. E. Phillips, J. M. Cole, T. d'Almeida and K. S. Low, *Phys. Rev. B: Condens. Matter Mater. Phys.*, 2010, **82**, 155118.
52 S. K. Brayshaw, J. W. Knight, P. R. Raithby, T. L. Savarese, S. Schiffers, S. J. Teat, J. E. Warren and M. R. Warren, *J. Appl. Crystallogr.*, 2010, **43**, 337–340.
53 D. Schaniel, N. Mockus, T. Woike, A. Klein, D. Sheptyakov, T. Todorova and B. Delley, *Phys. Chem. Chem. Phys.*, 2010, **12**, 6171–6178.
54 Z. Tahri, R. Lepski, K.-Y. Hsieh, E.-E. Bendeif, S. Pillet, P. Durand, T. Woike and D. Schaniel, *Phys. Chem. Chem. Phys.*, 2012, **14**, 3775–3781.
55 B. Cormary, S. Ladeira, K. Jacob, P. G. Lacroix, T. Woike, D. Schaniel and I. Malfant, *Inorg. Chem.*, 2012, **51**, 7492–7501.
56 G. Kresse and J. Hafner, *Phys. Rev. B: Condens. Matter Mater. Phys.*, 1993, 47.
57 P. E. Blochl, *Phys. Rev. B: Condens. Matter Mater. Phys.*, 1994, **50**, 17953–17979.
58 G. Kresse and D. Joubert, *Phys. Rev. B: Condens. Matter Mater. Phys.*, 1999, 59.
59 J. P. Perdew, K. Burke and M. Ernzerhof, *Phys. Rev. Lett.*, 1996, **77**, 3865–3868.
60 L. A. Constantin, J. M. Pitarke, J. F. Dobson, A. Garcia-Lekue and J. P. Perdew, *Phys. Rev. Lett.*, 2008, 100.
61 J. M. Tao, J. P. Perdew, V. N. Staroverov and G. E. Scuseria, *Phys. Rev. Lett.*, 2003, 91.
62 C. Adamo and V. Barone, *J. Chem. Phys.*, 1999, **110**, 6158–6170.
63 M. Valiev, E. J. Bylaska, N. Govind, K. Kowalski, T. P. Straatsma, H. J. J. Van Dam, D. Wang, J. Nieplocha, E. Apra, T. L. Windus and W. de Jong, *Comput. Phys. Commun.*, 2010, **181**, 1477–1489.
64 M. J. Frisch, G. W. Trucks, H. B. Schlegel, G. E. Scuseria, M. A. Robb, J. R. Cheeseman, G. Scalmani, V. Barone, B. Mennucci, G. A. Petersson, H. Nakatsuji, M. Caricato, X. Li, H. P. Hratchian, A. F. Izmaylov, J. Bloino, G. Zheng, J. L. Sonnenberg, M. Hada, M. Ehara, K. Toyota, R. Fukuda, J. Hasegawa, M. Ishida, T. Nakajima, Y. Honda, O. Kitao, H. Nakai, T. Vreven, J. A. Montgomery Jr, J. E. Peralta, F. Ogliaro, M. J. Bearpark, J. Heyd, E. N. Brothers, K. N. Kudin, V. N. Staroverov, R. Kobayashi, J. Normand, K. Raghavachari, A. P. Rendell, J. C. Burant, S. S. Iyengar, J. Tomasi, M. Cossi, N. Rega, N. J. Millam, M. Klene, J. E. Knox, J. B. Cross, V. Bakken, C. Adamo, J. Jaramillo, R. Gomperts, R. E. Stratmann, O. Yazyev, A. J. Austin, R. Cammi, C. Pomelli, J. W. Ochterski, R. L. Martin, K. Morokuma, V. G. Zakrzewski, G. A. Voth, P. Salvador, J. J. Dannenberg, S. Dapprich, A. D. Daniels, Ö. Farkas, J. B. Foresman, J. V. Ortiz, J. Cioslowski and D. J. Fox, Gaussian, Inc., Wallingford, CT, USA, 2009.
65 R. Ditchfield, W. J. Hehre and J. A. Pople, *J. Chem. Phys.*, 1971, **54**, 724–728.
66 T. H. Dunning, *J. Chem. Phys.*, 1989, **90**, 1007–1023.
67 P. J. Hay and W. R. Wadt, *J. Chem. Phys.*, 1985, **82**, 270–283.
68 Y. Zhao and D. G. Truhlar, *Theor. Chem. Acc.*, 2008, **120**, 215–241.
69 A. D. Becke, *J. Chem. Phys.*, 1993, **98**, 1372–1377.
70 R. A. Lunt, A. J. Jackson and A. Walsh, *Chem. Phys. Lett.*, 2013, **586**, 67–69.
71 M. Gajdos, K. Hummer, G. Kresse, J. Furthmuller and F. Bechstedt, *Phys. Rev. B: Condens. Matter Mater. Phys.*, 2006, 73.

 

Faraday Discussions

PAPER

# Resolution enhancement through microscopic spatiotemporal control

Debabrata Goswami,* Dhiman Das and Soumendra Nath Bandyopadhyay

Received 15th September 2014, Accepted 24th November 2014
DOI: 10.1039/c4fd00177j

Operating at biologically benign conditions, multi-photon fluorescence imaging microscopy has benefitted immensely from recent developments in microscopic resolution enhancement. Fluorescence microscopy continues to be the best choice for experiments on live specimens, however, multi-photon fluorescence imaging often suffers from overlapping fluorescence of typical dyes used in microscopy, limiting its scope. This limitation has been the focus of our research where we show that by making simple modifications to the laser pulse structure, it is possible to resolve these overlapping fluorescence complications. Specifically, by using pairs of femtosecond pulses with variable delay in place of single pulse excitation, we show controlled fluorescence excitation or suppression of one of the fluorophores over the other through wave-packet interferometry. Such an effect prevails even after the fluorophore coherence timescale, which effectively results in a higher spatial resolution. Here we extend the effect of our pulse-pair technique to microscopic axial resolution experiments and show that such pairs of pulses can also 'enhance' axial resolution.

## 1 Introduction

Fluorescence microscopy continues to be the best choice for monitoring live specimens in spite of the resolution advantage of electron microscopes as the energy deposited in electron microscopy adversely affects the viability of live specimens. This practical compromise implicitly sets resolution enhancement as one of the most important developmental aspects in fluorescence microscopy.[1] Demonstrating better depth resolution through multi-photon excitation in laser scanning microscopy is thus an important achievement. However, the broad spectral window of an ultrafast laser pulse required for two-photon fluorescence (TPF) microscopy and the overlapping multi-photon absorption spectra of common fluorophores lead to the simultaneous excitation of many different fluorophores. Consequently, selective enhancement or suppression of fluorescence is important for developing better resolution in TPF microscopy. In this context, ultrafast laser pulse-shaping[2]

*Department of Chemistry, Indian Institute of Technology Kanpur, Kanpur, India. E-mail: dgoswami@iitk.ac.in; Fax: +91 512 259 7554; Tel: +91 512 259 7187*

approaches to quantum control that have shown promise in discriminating nearly identical fluorophores[3–5] can be very effectively applied to microscopy.[4–9] Manipulation of excited state populations and the corresponding spontaneous emission (and/or fluorescence) through coherent quantum interference is possible through precise control over the inter-pulse delay and phase in a pulse pair[10,11] or pulse train[12] excitation. Previously, this approach of fluorophore discrimination was demonstrated in the solution phase.[13] On the other hand, pulse-pair fluorophore excitation can also occur by the manipulation of excited state photo-physics, *i.e.*, incoherent population dynamics, where the only "control factor" is the time delay between the pulse pairs. We have earlier reported selective TPF suppression using a pulse-pair excitation scheme, which was explained on the basis of selective stimulated emission by a time delayed second pulse following the excitation pulse.[14] In our approach, though simultaneous two-photon absorption (TPA) of the two different fluorophores occurred, the ensuing one-photon stimulated emission was selective to only one fluorophore, which resulted in the observed discrimination.[15] Thereafter, we also demonstrated selective fluorescence suppression by a train of excitation pulses, where, instead of two time-delayed identical pulses, many pulses with gradually decreasing pulse intensities (in geometric progression) and controllable delay between the successive pulses, were employed.[16] In the multi-pulse scheme, the first pulse induced TPA while the subsequent successive pulses resulted in one-photon stimulated emission in preference over TPA. The effect of varying polarization of the individual pulses used under such incoherent schemes was also explored, and, as expected, was found to be only minimal unlike the coherent case.[1] Thus, using either a simple pair of femtosecond pulses with variable delay or a train of laser pulses at 20–50 gigahertz excitation, we developed controlled fluorescence excitation or suppression of one of the fluorophores with respect to the other. Since this is an incoherent process, the effect prevails even after the coherence lifetime of the fluorophore and can be used in both single-photon as well as in multi-photon excitation conditions to result in an effective higher image contrast resolution. Higher image contrast is achieved as fluorescence from one of the two fluorophores with respect to the other is enhanced as a function of the relative time delay between the pulse pair. Such high image contrast with pulsed broadband excitation can be of immense benefit to multi-photon microscopy.

Use of pairs of pulses in place of a single pulse in microscopy begets the natural question of whether such modified pulses would affect the axial resolution of the microscope as well. Here, we address this particular aspect of the use of femtosecond pulse pairs with variable delay on microscopic axial resolution for the first time and relate it to our contrast resolution enhancement results in the image plane. In order to make a valid correlation, we compare the enhanced contrast resolution situation in the image plane with that of the axial plane under identical experimental conditions. Comparisons have been drawn from several experiments performed under appropriate conditions with different laser and/or sample characteristics.

## 2 Experimental

Fig. 1 shows our experimental setup. The laser system used in our experiments is a mode-locked Ti:Sapphire oscillator (Mira900-F pumped by Verdi V5, Coherent Inc., USA) producing femtosecond pulses (pulse-width ranging 120–180 fs

 

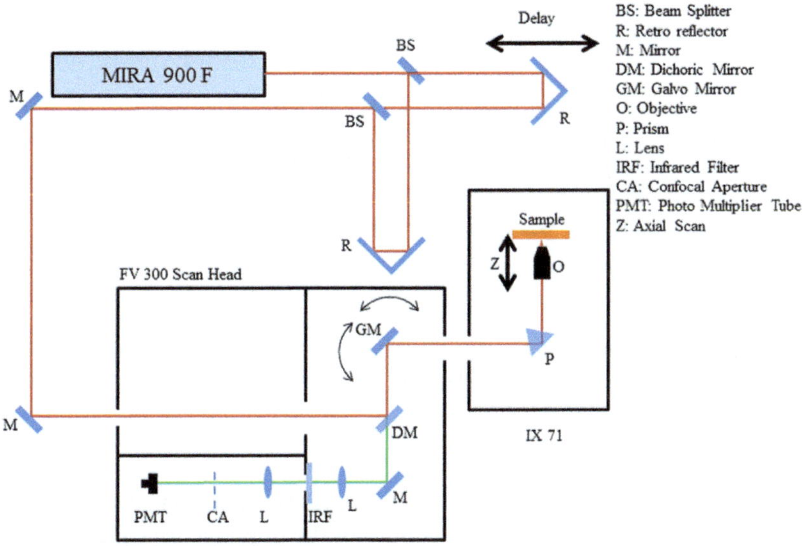


Fig. 1 Overall experimental setup used in our experiments. The abbreviations used are expanded in the legend on the side.

depending on the center wavelength) at 76 MHz repetition rate and wavelength tunable within 720–980 nm. The pulse pair was generated using a collinear Mach-Zehnder-type interferometer where the laser beam was separated into two parts using a beam-splitter and recombined using another beam-splitter. One of the split beams from the first beam-splitter was passed through a movable retro-reflecting mirror mounted on a mechanical stage (UE1724SR driven by ESP300, Newport) while the other one was routed through a fixed retro-reflecting mirror. The delay stage was interfaced with a computer controlled GPIB card (National Instruments) to provide precise delay steps. The separation between the pulse pair depended on the precise delay of the movable arm in the Mach-Zehnder interferometer. The laser pulse pair thus generated from the Mach-Zehnder interferometer was measured just before it entered the microscope with a commercial autocorrelator (Model: Pulse Check, A.P.E GmbH). Representative spectra measured with a fiber coupled monochromator (HR2000) as well as the autocorrelation trace for such pulse pairs are shown in Fig. 2. Depending on the sample used in the specific experiment, the center wavelength of the laser as well the corresponding pulse-width varied. The collinearly propagating beams were sent into our confocal-ready multi-photon microscope system (FV300 scan-head coupled with IX71 inverted microscope, Olympus, Japan). The laser beam was focused on the sample using a high numerical aperture (we used two different cases: either the UPlanApoN 40× 1.30NA or the UPlanApoN 60× 1.42NA, both from Olympus) objective lens. We used an immersion oil of refractive index 1.518 between the objective lens and coverslip. Two mirrors before the scan head were used for fine alignment of the laser beam into the scan-head and for maximizing the signal. For imaging, as well as for the axial resolution measurements, the total average power of the laser beam entering the scan-head was ~30 mW. A pair of lenses was used in the detection path to collimate the beam for getting the

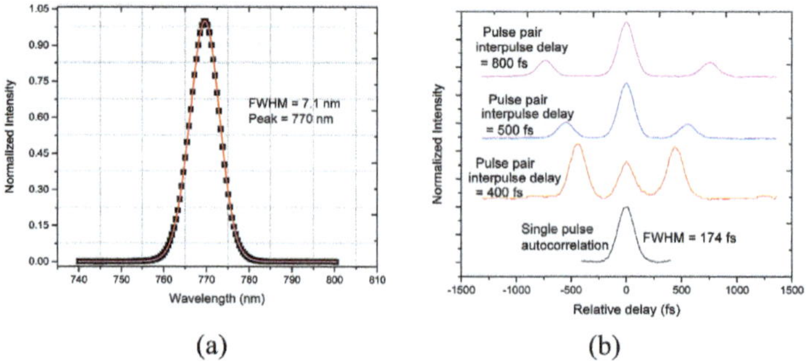


Fig. 2 (a) Laser pulse spectrum as well as (b) representative autocorrelation trace for pulse pairs at a few different inter-pulse delays are shown. These measurements are made before the pulses enter the microscope.

optimum fluorescence signal from the detector (PMT, Hamamatsu Inc.). Finally an infra-red (IR) cut-off filter was introduced in the fluorescence collection path for blocking any un-wanted back-scattered IR light.

### 2.1 Enhanced contrast fluorescence imaging microscopy

For pulse pair imaging microscopy, slides of bovine pulmonary artery endothelial (BPAE) cells having nuclei stained with DAPI (blue fluorophore staining nuclei) and Texas Red-X phalloidin (red fluorophore staining mitochondria) were used and are commercially available (Molecular Probes, Invitrogen, USA). At ~730 nm (pulse width ~130 fs) both these dyes are simultaneously excited through two-photon excitation, even at 10 mW average power before the microscope, which makes these fluorophore labeled samples a perfect choice for our selective excitation experiments. However the fluorescence from Texas Red extends up to ~750 nm, which suggests the possibility of its fluorescence suppression through stimulated emission in contrast to the fluorescence from DAPI that almost dies out at ~650 nm. The image acquisition and intensity counts were performed using FLUOVIEW 300 software (Olympus, Japan). The image resolution for each image collected at different time delays between the pulse-pair was set to 512 × 512 pixels.

### 2.2 Measuring the axial resolution in fluorescence microscopy

For measuring axial resolution, we first used a millimolar ($10^{-3}$M) methanolic solution of Rhodamine-6G (R6G) in a culture dish (35 mm Glass Bottom Dish No. 0 (MatTek Co., USA)). In these experiments, the center wavelength of the Ti:Sapphire excitation laser (Mira900-F) was tuned to 770 nm (pulse-width ~120 fs). The confocal aperture was kept fully open at 300 μm. We performed the axial scan (along $z$-axis) of the R6G solution to generate the axial edge response as shown in Fig. 3a. The $z$-axis piezo-stage of the microscope objective was moved at a step-size of 0.25 μm during this data acquisition. This axial edge response data is approximated to a sigmoidal (shown by a dotted line in Fig. 3a), whose derivative was fitted to a Gaussian function (point spread function, Fig. 3b).[20–24] The full-width-half-maximum (FWHM) of this Gaussian fit was taken as the measure of the axial resolution of the microscope for a specific pulse pair delay. We

 

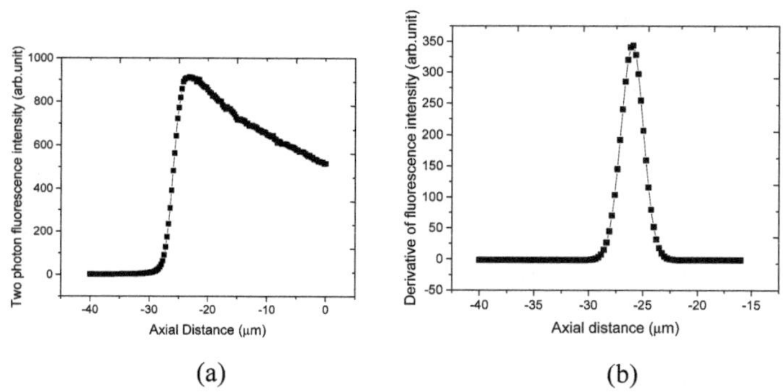


Fig. 3 (a) Axial edge response function corresponding to the fluorescence intensity as a function of objective position along the optic axis during axial scan and (b) its derivative fitted to a Gaussian function.

collected such axial resolution data at different inter-pulse delays for the pulse pair and compared them to the single pulse data. We used two different focussing conditions by using the two objective lenses of the microscope. Next we collected axial intensity profiles for the BPAE fixed cells stained with DAPI at specific pulse pair delays at 750 nm (pulse-width ~130 fs) using the UPlanApoN 40× 1.30NA objective. FWHM of the Gaussian fit of each axial intensity profile collected is the measure of axial resolution of the BPAE cells. A point to note is that all our experiments are performed with spatially overlapped laser beams; the changing parameter is only the temporal delay between pulses. Such axial resolution data from the scanned images of BPAE at different inter-pulse delays for the pulse pairs were collected and compared to that of the single pulse data. Thus we were able to generate three different cases of series of axial resolution measurements for various pulse pair delays that were compared to the single pulse data.

# 3 Results and discussion

As discussed at the very outset, each of our experiments was performed at one specific center wavelength and as such is a one-color experiment. This is different from typical pump–probe experiments[17,18] where two different colors are used: one for the pump and the second color for the probe. The pump pulse excites a part of the ground state population to the excited state where it rapidly relaxes to the lowest vibrational level of the excited state from where the fluorescence emission occurs. However, if a time-delayed probe pulse that is wavelength-tuned to the red edge of fluorescence sends the population back to the ground electronic state by stimulated emission in competition with the fluorescence process, fluorescence suppression is observed. Fluorescence, being an incoherent emission, is omnidirectional. Stimulated emission, on the other hand, is a coherent process and, as such, photons generated by stimulated emission travel in the direction of the stimulating beam. So, if we collect only the back-scattered fluorescence (epi-fluorescence as in typical confocal microscopy) there is suppression in the fluorescence signal as discussed previously.[17] However, under the same experimental condition, if one collects the forward scattered photons, a gain in total measured signal is observed due to the additional stimulated emission which could even transform weakly fluorescent molecules into suitable candidates for microscopy applications.[18]

Our pulse-pair microscopy, however, focuses on experiments that are of practical relevance to two photon fluorescence (TPF) laser scanning microscopy and only restricted to a one-color excitation scheme.[19] It is important that when the identical replica pulse is delayed with respect to the original, it is beyond the FWHM of the pulse otherwise the interference between the pulses is too overwhelming. For image contrast enhancements based on fluorophore discrimination, in our version of one color pulse-pair excitation we found interesting control strategies that materialized even within a time window of 100 fs, as shown in the region of 0.9 ps to 1.0 ps at a step-size of 5 fs, comparing the suppression of Texas Red fluorescence to that of DAPI fluorescence in the BPAE cell (Fig. 4a and b). In this case, the first pulse causes a two photon absorption for both DAPI and Texas Red. The time-delayed second pulse causes a similar two-photon absorption as in case of the first pulse but, in addition, dumps the population by (one-photon) stimulated emission if the red edge of fluorescence coincides with the excitation wavelength (as is the case for Texas Red) (Fig. 4c). Thus a distinction is seen between DAPI and Texas-Red in the two-photon fluorescence image. This contrast between the fluorophores at a ~1 ps time delay between the pulse pair can be critical in distinguishing the overlapping fluorophores.

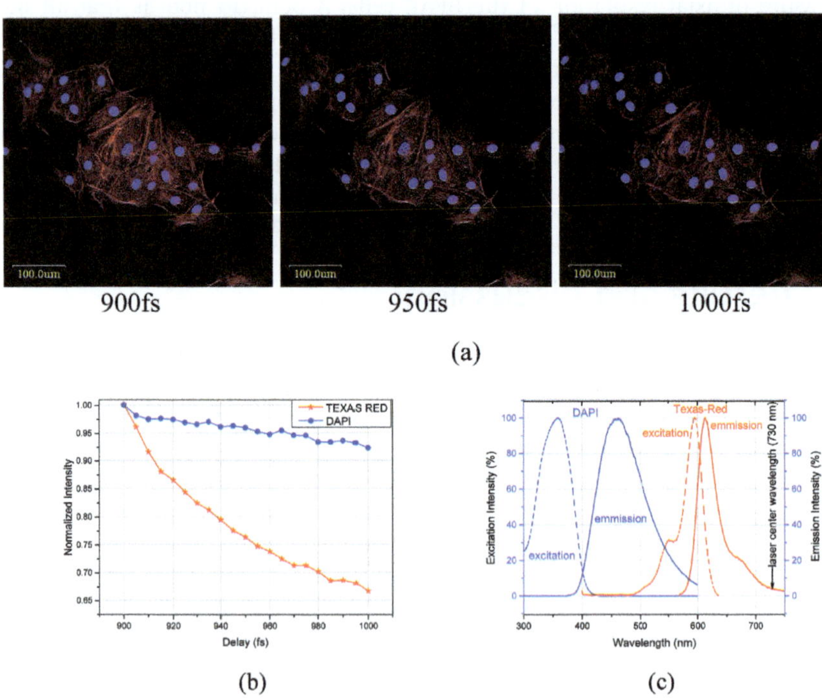


Fig. 4 (a) Three representative fluorescence suppression images of Texas Red relative to DAPI as a function of time delay between pulses under one-color (730 nm) pulse-pair excitation, (b) change in the relative fluorescence intensity of DAPI versus Texas Red, and (c) corresponding excitation (broken line) and emission (solid line) spectra of Texas-Red and DAPI relative to the excitation laser wavelength (730 nm) position.

In our axial resolution studies with R6G solution, the data at every 50 fs inter-pulse delay for the pulse pair are shown in Fig. 5, and are compared to the single pulse data for two different objective lenses having different numerical apertures (NA). As discussed earlier, each data point corresponds to the width of the point spread function (Fig. 3b). Several measurements at each inter-pulse delay were collected, which enabled used to report the error bars for each of the plotted data points. Very near to the overlapped condition between the pairs of pulses (*i.e.*, close to zero delay), the axial resolution (as defined by the FWHM of Gaussian fit to the derivative of the edge response function) shows an oscillatory nature with a large error bar. The oscillatory nature could be attributed to the interference

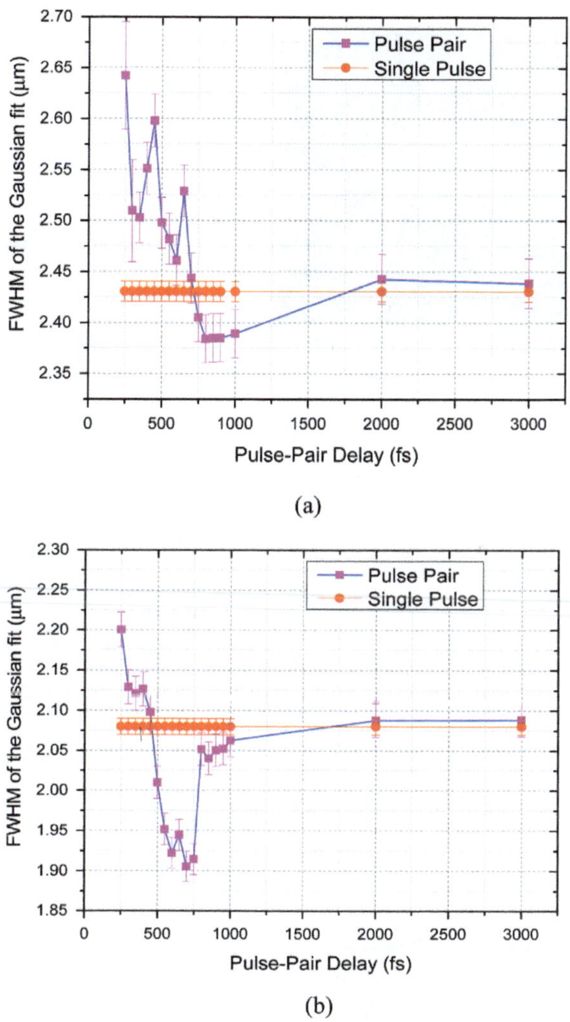


Fig. 5 Change in axial resolution as a function of the relative delay between the pulses in a pulse pair excitation scheme for Rh-6G dye solution in methanol. This change is compared to the traditional microscopy case of a single pulse excitation condition. Error-bars for each experimental data point is also provided. Two cases with different NA objectives have been used: (a) 40× 1.30NA objective and (b) 60× 1.42NA objective.

pattern of the nearly time-overlapped pulse pair. However, it is important to note that the best possible resolution under such a closely overlapping pulse pair is always lower than the single pulse case. As the pulse separation increases between the pulse pair for both the microscope objective cases (Fig. 5a and b), until about 1 ps, the axial resolution as per our definition keeps getting better and better with smaller and smaller error-bars (Fig. 5). The effect of NA is also quite obvious between these two cases. Interestingly enough, this resolution enhancement starts to fall off for inter-pulse separations beyond 1000 fs to reach the axial resolution comparable to that of a single pulse case.

The TPF intensity profile of a DAPI stained BPAE cell at different inter-pulse delays during an axial scan along the *z*-axis of the XY-image (Fig. 6) was fitted with

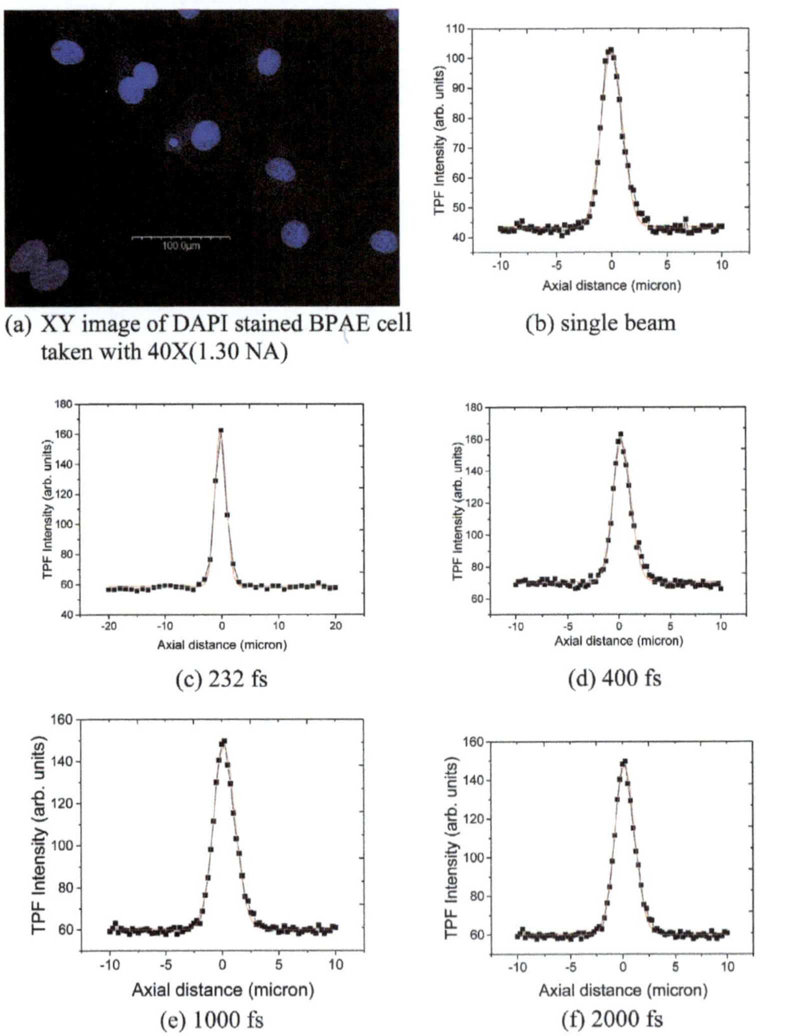


Fig. 6 Axial two photon fluorescence intensity profile along with Gaussian fit of DAPI stained BPAE cell at different inter-pulse delays (b–f) during axial scans of the XY-image (shown in (a)) along the *Z*-axis. The FWHM of Gaussian fit is taken as a measure of axial resolution.

 

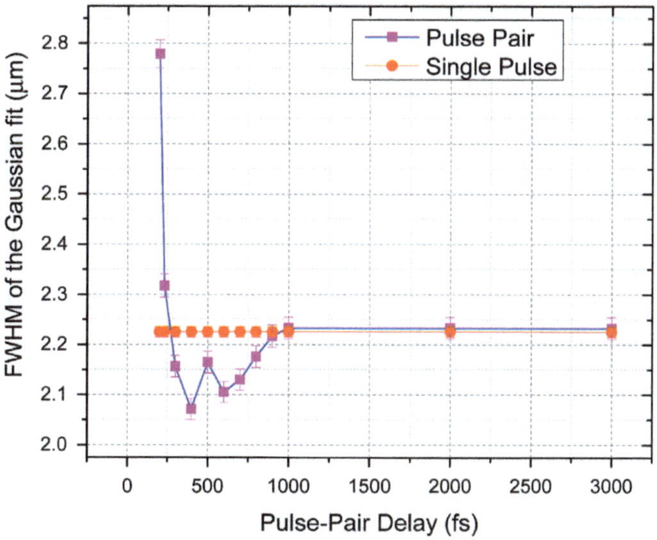


Fig. 7 Change in axial resolution as a function of the relative delay between the pulses in a pulse pair excitation scheme for a DAPI stained BPAE cell. The change is compared to the traditional microscopy case of a single pulse excitation condition.

the Gaussian function. The FWHM of the Gaussian fit was taken as a measure of the axial resolution for these images. Fig. 7 shows the axial resolution of the BPAE cell at different inter-pulse delays for the pulse pair as well as for the single pulse data. Comparing Fig. 5 and 7, we find that typically, for a pulse separation of ~2 ps and above, only single pulse character is observed in the axial resolution, while for a pulse separation of less than ~1 ps there is a major effect of one pulse on the resolution of the other resulting in the best possible axial resolution. The increased fluorophore image contrast at ~1 ps (Fig. 4b) for the BPAE cells also correlates well with this study showing the coupled impact of both the pulses in the pulse pair.

## 4 Conclusion

We have shown how the innovative use of relative delay between pulses in pulsed laser illumination can be utilized for resolution enhancement in fluorescence microscopy. Our experiments on pulse-pair control mediated by stimulated emission have interesting applications in microscopy. The control method is based on competition between fluorescence and stimulated emission and has been demonstrated for fluorophore discrimination. We extended the effect of our pulse-pair technique to microscopic axial resolution experiments. Such experiments show that the pairs of pulses can also 'enhance' the axial resolution as long as one pulse can feel the impact of the other beyond the pulse temporal interference zone. Within errors of the experiments performed on different dyes, there is a strong indication of axial resolution enhancement at ~1 ps delay between the femtosecond pulse pair. This observed axial resolution enhancement might have an important effect on 3D microscopic development.

# References


1 A. K. De and D. Goswami, *Int. Rev. Phys. Chem.*, 2011, **30**(3), 275–299.

2 D. Goswami, *Phys. Rep.*, 2003, **374**, 385–481.

3 T. Brixner, N. H. Damrauer, P. Niklaus and G. Gerber, *Nature*, 2001, **414**, 57–60.

4 M. Roth, L. Guyon, J. Roslund, V. Boutou, F. Courvoisier, J. P. Wolf, J. Roslund and H. Rabitz, *Phys. Rev. Lett.*, 2010, **102**, 253001.

5 M. Roth, L. Guyon, J. Roslund, V. Boutou, F. Courvoisier, J. P. Wolf, J. Roslund and H. Rabitz, *Phys. Rev. Lett.*, 2010, **105**, 073003.

6 I. Pastirk, J. M. Dela, J. M. Cruz, K. A. Walowicz, V. V. Lozovoy and M. Dantus, *Opt. Express*, 2003, **11**, 1695–1701.

7 J. M. Dela Cruz, I. Pastirk, M. Comstock, V. V. Lozovoy and M. Dantus, *Proc. Natl. Acad. Sci. U. S. A.*, 2004, **101**, 16996–17001.

8 J. Ogilvie, D. Debarre, X. Solinas, J. L. Martin, E. Beaurepaire and M. Joffre, *Opt. Express*, 2006, **14**, 759–766.

9 K. Isobe, A. Suda, M. Tanaka, F. Kannari, H. Kawano, H. Mizuno, A. Miyawaki and K. Midorikawa, *Opt. Express*, 2009, **17**, 13737–13746.

10 N. F. Scherer, A. J. Ruggiero, M. Du and G. R. Fleming, *J. Chem. Phys.*, 1990, **93**, 856–857.

11 N. F. Scherer, R. J. Carlson, A. Matro, M. Du, A. J. Ruggiero, V. Romero-Rochin, J. A. Cina, G. R. Fleming and S. A. Rice, *J. Chem. Phys.*, 1991, **95**, 1487–1511.

12 A. Rebane, C. Sigel and M. Drobizhev, *Chem. Phys. Lett.*, 2000, **322**, 287–292.

13 A. K. De, D. Roy and D. Goswami, *Phys. Rev. A*, 2011, **83**, 015402.

14 A. K. De and D. Goswami, *J. Biomed. Opt.*, 2009, **14**, 064018.

15 A. K. De, D. Roy and D. Goswami, *Proc. SPIE*, 2010, **7569**, 756929.

16 A. K. De, D. Roy and D. Goswami, *J. Biomed. Opt.*, 2010, **16**, 060502.

17 M. Dyba, T. A. Klar, S. Jakobs and S. W. Hell, *Appl. Phys. Lett.*, 2000, **77**, 597.

18 W. Min, S. Lu, S. Chong, R. Roy, G. R. Holtom and X. S. Xie, *Nature*, 2009, **461**, 1105.

19 A. K. De, D. Roy and D. Goswami, *J. Biomed. Opt. Lett.*, 2011, **16**(10), 100505.

20 S. W. Hell, A. R. Utz, M. Schrader, P. E. Hänninen and E. Soini, *J. Opt. Soc. Am. A*, 1995, **12**, 9.

21 J. Bewersdorf and S. W. Hell, *Journal of Microscopy*, 1998, **191**, 28–38.

22 N. Martini, J. Bewersdorf and S. W. Hell, *Journal of Microscopy*, 2002, **206**, 146–151.

23 M. Dyba and S. W. Hell, *Phys. Rev. Lett.*, 2002, **88**, 163901–163904.

24 M. C. Lang, T. Staudt, J. Engelhardt and S. W. Hell, *New Journal of Physics*, 2008, **10**, 043041.


 

Faraday Discussions

PAPER

# On the photophysics of butadiyne bridged pyrene–phenyl molecular conjugates: multiple emissive pathways through locally excited, intramolecular charge transfer and excimer states†

Avik Kumar Pati,[a] Santosh J. Gharpure*[b] and Ashok K. Mishra*[a]

Received 8th September 2014, Accepted 27th October 2014
DOI: 10.1039/c4fd00170b

The present work describes the photophysical properties of a group of butadiyne bridged pyrene–phenyl molecular hybrids having different substitutions with varying donor and acceptor abilities. In addition to emission from the locally excited (LE) state originating from the pyrene moiety, intramolecular charge transfer (ICT) emissions were observed in molecules with donor–acceptor character. The positions of the ICT emission maxima varied over a wide range of wavelengths (475–600 nm). Pyrene behaved as a donor when a strong acceptor group (–CN) was attached to the phenyl ring and it behaved as an acceptor when the phenyl group contained a strong donor group ($-NMe_2$). In mixed aqueous solvents at higher percentages of water (80–99%), the derivatives showed emissions from the aggregate state in addition to the LE and ICT states. Emissions from the aggregate states of the derivatives were centred in the range 510–560 nm. The aggregate state emissions were found to originate from static excimers involving pyrene moieties. A detailed structure–property relationship of the butadiynyl derivatives was revealed in this study.

## Introduction

The quest for the development of new organic fluorophores with efficient intramolecular charge transfer (ICT) character has been an important area of research in recent years. Organic derivatives with donor and acceptor units bridged

[a]Department of Chemistry, Indian Institute of Technology Madras, Chennai 600036, Tamil Nadu, India. E-mail: mishra@iitm.ac.in; Fax: +91-44-22574202; Tel: +91-44-22574207

[b]Department of Chemistry, Indian Institute of Technology Bombay, Powai, Mumbai 400076, India. E-mail: sjgharpure@iitb.ac.in; Fax: +91-22-2576 7152; Tel: +91-22-2576 7171

† Electronic supplementary information (ESI) available: General experimental section for synthesis, measurement of quantum yield, IR and NMR spectral data, emission spectra in DCM–$CH_3CN$ and THF–$CH_3CN$ mixtures, Cartesian coordinates of optimized ground state geometries. See DOI: 10.1039/c4fd00170b

through $\pi$ conjugation often exhibit ICT character. Such molecules find applications in different fields such as light emitting diodes, field effect diodes, sensors *etc.*[1–4] ICT also plays an important role in several biological and chemical phenomena.[5–13] The bridge between the donor and acceptor moieties in the ICT process generally comprises of various aromatic moieties such as benzene, tetrathiafulvalene, thiophene *etc.* and olefinic moieties or alkyne units.[14–25] However, the role of conjugated diyne, that is the butadiyne moiety, as a bridge in the ICT process between the donor and acceptor moieties has not been studied in great detail. In this context, recently, we studied the photophysics of substituted diphenyl butadiynes.[26] The diphenyl butadiynyl derivatives were found to exhibit ICT character.[26] Butadiyne moieties being present in several biologically important molecules[27] and in various systems such as molecular rotors,[28] liquid crystals[29–32] *etc.* attract special attention. Pyrene has been used as an acceptor unit in ICT processes.[24,25] The distinct photophysical behaviors of pyrene such as monomer and excimer emissions, long excited state lifetime, and solvent sensitive electronic transitions have made it a prototype fluorophore for standard photophysical investigations.[33–35] However, a detailed study on the photophysics of pyrene attached butadiynyl fluorophores is scarcely found in the literature.[36]

As a part of our ongoing research directed at the understanding of photophysics of butadiynyl fluorophores,[26,37] herein, we report the photophysical properties of a group of butadiyne bridged pyrene–phenyl molecular hybrids.

# Experimental methods

### Solution preparation for photophysical studies

Spectroscopic grade solvents were used for all the photophysical studies. A stock solution of $10^{-3}$ M concentration in $CH_2Cl_2$ was prepared for all the derivatives individually. The solutions in other solvents of different polarities were prepared by evaporating $CH_2Cl_2$ from the desired amount of the stock solution by purging nitrogen gas. Then the desired solvent was added to it.

### Absorption and steady state fluorescence experiments

Absorption spectra of all the derivatives were recorded with a Jasco V-650 UV-vis spectrometer. Fluorescence experiments were carried out with a Horiba Jobin-Yvon FluoroMax-4 spectrofluorometer, with a 450 W xenon lamp as the light source. The emission spectra of the derivatives were collected with slit widths of either 3/3 nm or 5/5 nm.

### Time resolved fluorescence measurements

Fluorescence lifetime experiments were done using a Horiba Jobin Yvon TCSPC lifetime instrument in a time-correlated, single-photon counting arrangement. A 370 nm nano-LED was used as the light source for all the experiments. The pulse repetition rate was set to 1 MHz and the instrumental full width half maximum of the 370 nm LED, including the detector response was around 800 ps. The instrument response function was collected using a scatterer (Ludox AS40 colloidal silica). The decay data were analyzed using IBH software. A value of $\chi^2$, $0.95 \leq \chi^2 \leq 1.3$, was considered as a good fit which was further verified by the symmetrical distribution of the residuals.

 

### Computational methodology

All computational studies were performed using Gaussian 09 computational package.[38] Optimization of the ground state of all the derivatives was carried out using density functional theory (DFT) with the B3LYP[39] hybrid functional and 6-311G(d,p) basis set without any symmetry constrains. Vibrational analyses were carried out to ascertain the absence of imaginary frequencies. Vertical absorption energies of all the derivatives were calculated using time dependent density functional theory (TD-DFT) at the CAM-B3LYP[40]/6-311+G(d,p) level of methodology. The effect of solvent in the vertical absorption energy calculations of the derivatives was included through the self-consistent reaction field (SCRF) using the polarizable continuum model (PCM).[41] The standard parameters ($\alpha = 0.19$, $\beta = 0.46$, and $\mu = 0.33$) were used for the CAM-B3LYP functional.

## Results and discussion

The butadiynyl derivatives **PyBP**, **PyBPOMe**, **PyBPCN**, and **PyBPNMe$_2$** were synthesized using Eglinton coupling of terminal alkynes in good yield (Fig. 1).[42] The identities of the derivatives were confirmed through NMR spectroscopy and mass spectrometry techniques. All the coupling reactions produced about 10–15% homo-coupling products of the terminal alkynes in addition to the desired hetero-coupling product. However, no homo-coupling product was observed from

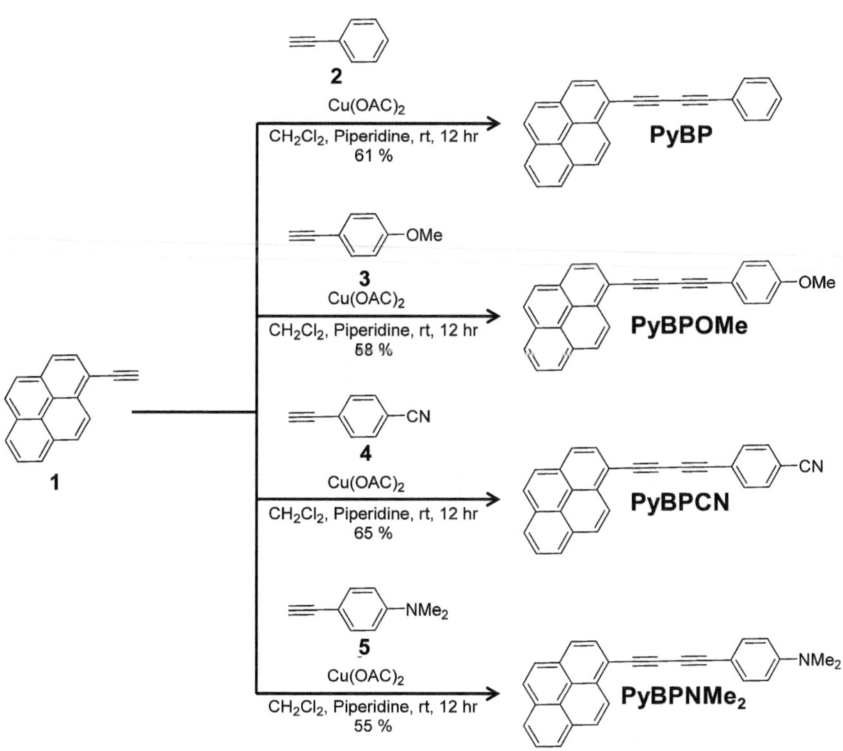


Fig. 1 Butadiyne bridged pyrene–phenyl molecular conjugates (**Py**, **B**, and **P** notations stand for pyrene, butadiyne bridge, and phenyl moiety respectively).

the terminal alkyne **4**. The desired hetero-coupling derivatives were purified through extensive column chromatography to make sure of the absence of any impurities before carrying out photophysical investigations. The electronic substitution at the *para* position of the phenyl ring of the parent derivative **PyBP** was varied by weak donor (–OMe), strong acceptor (–CN), and strong donor (–$NMe_2$) groups to understand the structure–property relationship of the butadiynyl derivatives.

## Steady state absorption and emission properties

Absorption spectra of the derivatives **PyBP**, **PyBPOMe**, **PyBPCN**, and **$PyBPNMe_2$** were recorded in solvents of different polarities such as heptane, cyclohexane, dioxane, tetrahydrofuran (THF), dichloromethane (DCM), isopropanol, methanol (MeOH), acetonitrile ($CH_3CN$), and *N*,*N*-dimethylformamide (DMF). The longest wavelength absorption maxima ($\lambda_{max}$, 0–0) of all the butadiynyl derivatives remained almost unchanged with solvent polarity, suggesting insensitivity of the derivatives to solvent environment within a short femto second time scale of absorption phenomenon. **PyBP** and **PyBPOMe** exhibited a $\lambda_{max}$ at 396 (Fig. 2a) and 398 (Fig. 2b) nm in heptane respectively. The $\lambda_{max}$ of the derivative **PyBPCN** having a strong acceptor cyano group and the derivative **$PyBPNMe_2$** having a strong donor *N*,*N*-dimethylaniline group was found at 410 (Fig. 2c) and 416 (Fig. 2d) nm in heptane respectively. The onset of absorption spectra of **PyBPCN** and **$PyBPNMe_2$** got red shifted in polar solvents when compared with the non-polar solvents such as heptane and cyclohexane. The red shift of onset of **PyBPCN** and

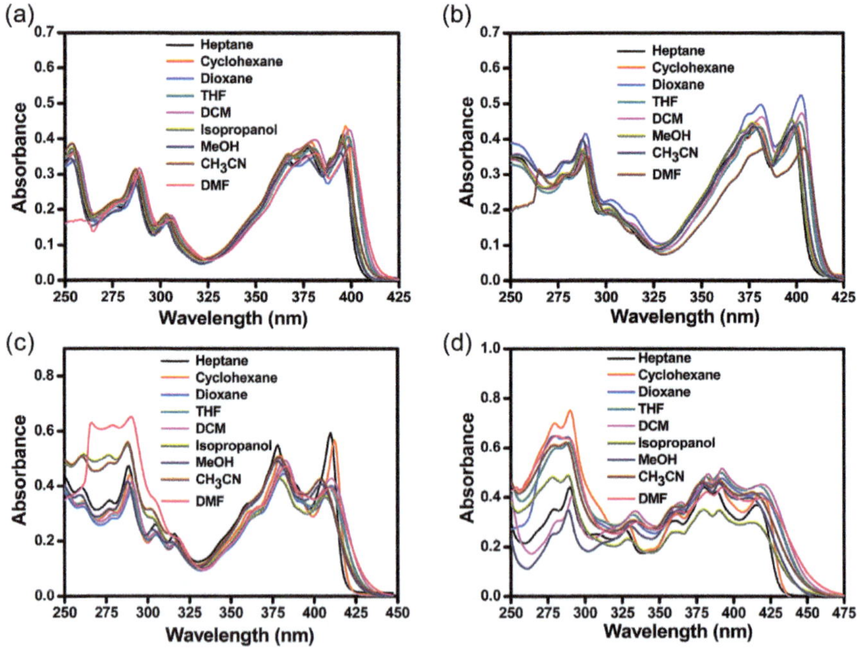


Fig. 2 UV-visible absorption spectra of (a) PyBP, (b) PyBPOMe, (c) PyBPCN, and (d) $PyBPNMe_2$ in different solvents ($C = 10^{-5}$ M, instrumental band width = 2 nm).

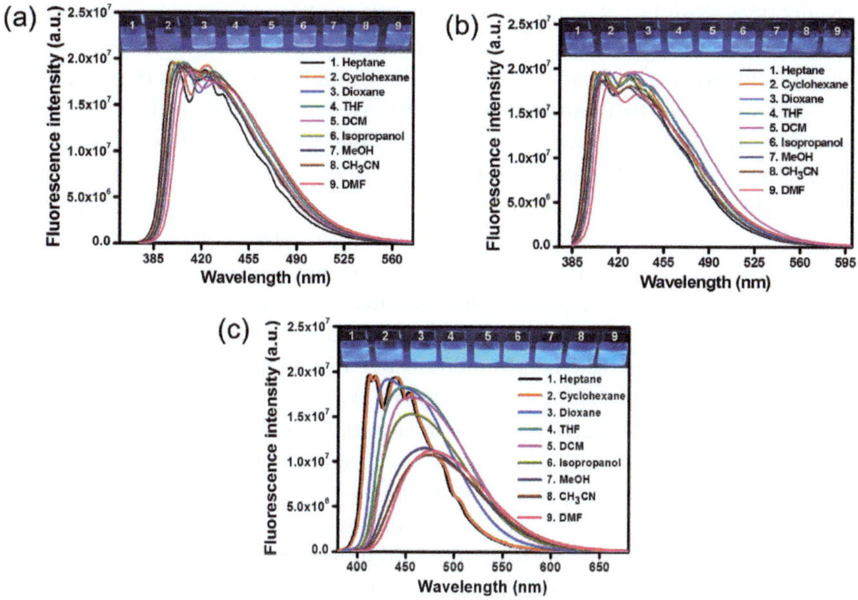


Fig. 3 Emission spectra of (a) **PyBP** ($\lambda_{ex} = 360$ nm), (b) **PyBPOMe** ($\lambda_{ex} = 370$ nm), and (c) **PyBPCN** ($\lambda_{ex} = 360$ nm for heptane, cyclohexane, dioxane, and $\lambda_{ex} = 380$ nm for others) in different solvents ($C = 1 \times 10^{-5}$ M, instrumental slit width = 3 nm).

**PyBPNMe$_2$** in polar solvents indicated the ground state intramolecular charge transfer (ICT) character of the derivatives. The derivatives showed structured absorption spectra in all the solvents of varying polarities.

The emission spectra of the derivatives **PyBP**, **PyBPOMe**, and **PyBPCN** are shown in Fig. 3. The derivatives exhibited blue emission under illumination of the 365 nm UV lamp. **PyBP** and **PyBPOMe** showed structured emissions in all the

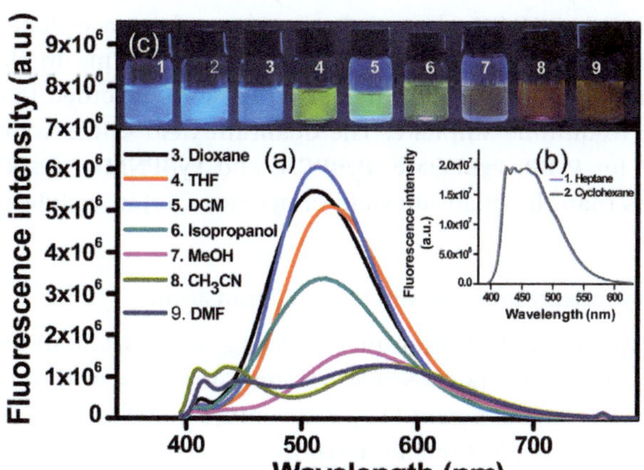


Fig. 4 (a) and (b) Emission spectra of the fluorophore **PyBPNMe$_2$** in different solvents ($C = 10^{-5}$ M, $\lambda_{ex} = 380$ nm, instrumental slit width = 3 nm) and (c) photograph of the fluorophore **PyBPNMe$_2$** under 365 nm UV light.

solvents (Fig. 3a and b). For the derivative **PyBPCN**, although structured emission was found in non-polar heptane and cyclohexane, the vibronic emission bands were not observed in other solvents of moderate and high polarities (Fig. 3c). The emission spectra of **PyBP** and **PyBPOMe** showed less spectral changes towards solvent polarity. The emission maxima of **PyBPCN** in polar solvents such as MeOH, $CH_3CN$ and DMF were found to be considerably red shifted compared to those of the non-polar solvents such as heptane, cyclohexane and dioxane. For example, the emission maximum of the derivative **PyBPCN** was at 413 nm in heptane (considering the shortest wavelength vibronic band maximum); while it was at 476 nm in DMF. The fluorescence intensity of **PyBPCN** also decreased considerably in polar solvents with the concomitant red shift of the emission maxima. This suggested the involvement of ICT in the polar solvents.

The ICT character of the fluorophore **PyBPNMe$_2$** having a strong donor moiety ($-NMe_2$) was also investigated through steady state fluorescence experiments in solvents of different polarities. The emission spectra of the derivative **PyBPNMe$_2$** exhibited dual emissions in polar solvents (Fig. 4a). The shorter wavelength band at around 400–450 nm was assigned to a locally excited (LE) emission whereas the longer wavelength band was set to an ICT emission. **PyBPNMe$_2$** mainly showed an LE emission in non-polar solvents like cyclohexane and heptane (Fig. 4b). In particular, the ICT emission band largely red shifted in highly polar solvents like $CH_3CN$ and DMF with a concomitant decrease of fluorescence intensity compared to the other solvents of moderate polarities. The ICT character of the derivative **PyBPNMe$_2$** was much stronger than that of the derivative **PyBPCN. PyBPNMe$_2$** was found to emit at 424 nm in heptane and 576 nm in $CH_3CN$, thus exhibiting a large emission window. This was also evident from the colour of **PyBPNMe$_2$** in solvents of different polarities under illumination of UV light at 365 nm (Fig. 4c). The Stokes' shift was high in $CH_3CN$ (6911 $cm^{-1}$), suggesting a strong ICT character of the derivative **PyBPNMe$_2$**. A detailed list of absorption, emission, Stokes' shift and molar extinction co-efficient ($\varepsilon$) values of all the derivatives **PyBP**, **PyBPOMe**, **PyBPCN**, and **PyBPNMe$_2$** is given in Table 1. It has been observed earlier that the presence of one acetylene moiety between pyrene and *N*,*N*-dimethylaniline in *N*,*N*-dimethyl-4-(pyren-1-yl)benzenamine (PyDMA)[43] produced a 15 nm red shift of emission maxima in $CH_3CN$ whereas the presence of two acetylene moieties in PyDMA afforded a 44 nm red shift of the emission maxima in the derivative **PyBPNMe$_2$**. The quantum yields of the derivatives were recorded in cyclohexane with reference to quinine sulphate. The quantum yield values were 0.44, 0.40, 0.45, and 0.37 for **PyBP**, **PyBPOMe**, **PyBPCN**, and **PyBPNMe$_2$**, respectively. The values were less than the quantum yield of pyrene (0.58) in cyclohexane.

### Mirror image symmetry and allowed $S_0 \rightarrow S_1$ transition

The longest wavelength absorption maxima ($\lambda_{0-0}$, $S_0 \rightarrow S_1$) of all the butadiyne bridged pyrene derivatives ($\lambda_{max} \sim$ 394–419 nm) were observed to be much red shifted compared to that in pyrene ($\lambda_{max} \sim 355$ nm, $\varepsilon \sim 200\ M^{-1}\ cm^{-1}$). Generally, the strongly allowed $S_0 \rightarrow S_1$ transition in polycyclic aromatic compounds leads to (a) mirror symmetry in the absorption and emission spectra, (b) insensitivity of vibronic emission towards solvent polarity and (c) a shorter fluorescence lifetime.[44] For pyrene, the $S_0 \rightarrow S_1$ transition is weakly allowed and hence the molar extinction co-efficient for the longest wavelength transition is very low

 

Table 1 Photophysical parameters of the butadiynyl derivatives in different solvents

| | PyBP | | | | PyBPOMe | | | | PyBPCN | | | | $PyBPNMe_2$ | | | |
|---|---|---|---|---|---|---|---|---|---|---|---|---|---|---|---|---|
| Solvents | $\lambda_{abs}$ (nm) | $\lambda_{em}$ (nm) | Stokes' shift ($cm^{-1}$) | $\varepsilon$ ($\times 10^4$) $M^{-1}\ cm^{-1}$ | $\lambda_{abs}$ (nm) | $\lambda_{em}$ (nm) | Stokes' shift ($cm^{-1}$) | $\varepsilon$ ($\times 10^4$) $M^{-1}\ cm^{-1}$ | $\lambda_{abs}$ (nm) | $\lambda_{em}$ (nm) | Stokes' shift ($cm^{-1}$) | $\varepsilon$ ($\times 10^4$) $M^{-1}\ cm^{-1}$ | $\lambda_{abs}$ (nm) | $\lambda_{em}$ (nm) | Stokes' shift ($cm^{-1}$) | $\varepsilon$ ($\times 10^4$) $M^{-1}\ cm^{-1}$ |
| Heptane | 396 | 398 | 127 | 4.1 | 398 | 402 | 250 | 4.3 | 410 | 413 | 177 | 5.9 | 416 | 424 | 453 | 3.7 |
| Cyclohexane | 397 | 399 | 126 | 4.3 | 399 | 403 | 248 | 4.3 | 410 | 413 | 177 | 5.6 | 418 | 426 | 449 | 4.1 |
| Dioxane | 399 | 406 | 432 | 3.8 | 402 | 409 | 426 | 5.2 | 410 | 428 | 1026 | 4.0 | 416 | 510 | 4431 | 4.1 |
| THF | 398 | 407 | 556 | 4.0 | 402 | 409 | 426 | 4.5 | 409 | 449 | 2178 | 3.9 | 416 | 524 | 4954 | 4.4 |
| DCM | 399 | 408 | 553 | 4.2 | 403 | 413 | 601 | 4.7 | 410 | 457 | 2508 | 4.2 | 419 | 516 | 4486 | 4.5 |
| MeOH | 394 | 404 | 628 | 3.6 | 398 | 406 | 495 | 4.5 | 404 | 470 | 3476 | 4.1 | 410 | 548 | 6142 | 4.0 |
| $CH_3CN$ | 395 | 407 | 746 | 3.9 | 399 | 408 | 553 | 4.4 | 404 | 472 | 3566 | 4.2 | 412 | 576 | 6911 | 4.2 |
| DMF | 400 | 413 | 787 | 3.7 | 404 | 413 | 539 | 3.7 | 409 | 476 | 3441 | 3.9 | 418 | 569 | 6349 | 3.9 |

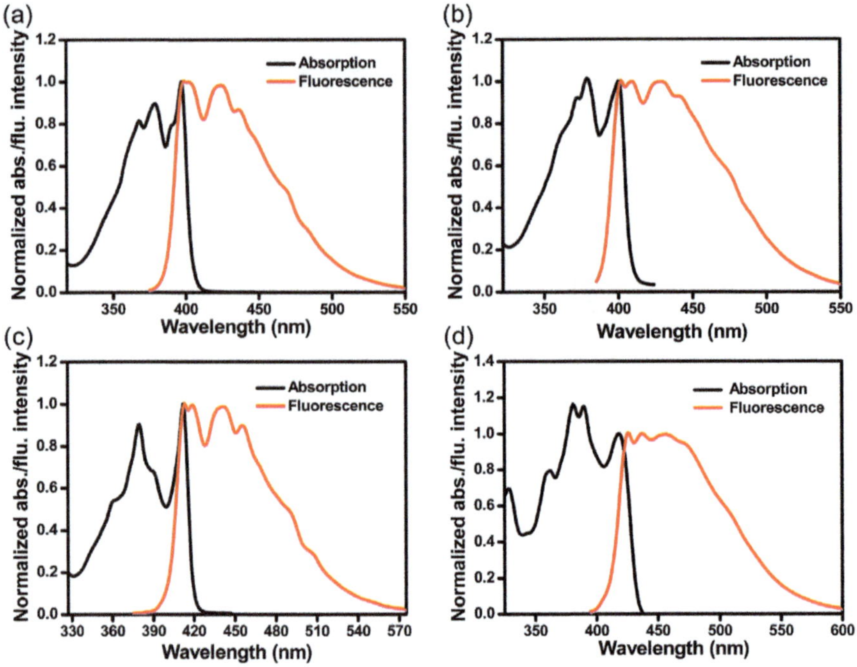


Fig. 5 Absorption and emission spectra of (a) **PyBP**, (b) **PyBPOMe**, (c) **PyBPCN**, and (d) **PyBPNMe$_2$** in cyclohexane.

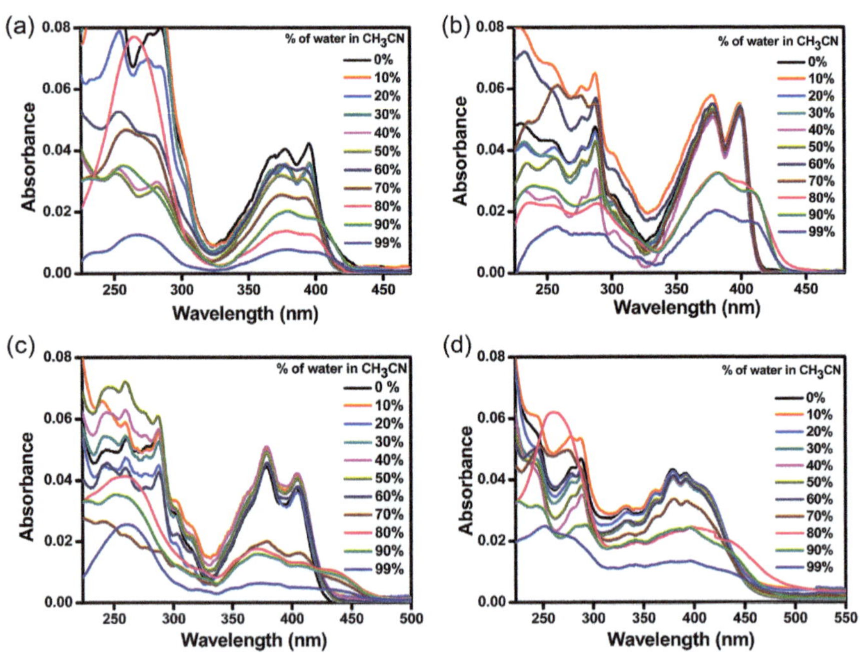


Fig. 6 Absorption spectra of (a) **PyBP**, (b) **PyBPOMe**, (c) **PyBPCN**, (d) **PyBPNMe$_2$** in water–$CH_3CN$ mixtures ($C = 1 \times 10^{-6}$ M, instrumental band width = 0.5 nm).

(~200 $M^{-1}$ $cm^{-1}$). Thus, the absorption and emission spectra of pyrene do not appear to show a mirror image relationship and the apparent Stokes' shift of pyrene is large. The intensities of vibronic emission bands of pyrene are also highly solvent dependent. Here, for the butadiyne bridged pyrene derivatives **PyBP**, **PyBPOMe**, **PyBPCN**, and **PyBPNMe$_2$**, the absorption and emission spectra showed a mirror image relationship in cyclohexane (Fig. 5). In particular, **PyBP** and **PyBPOMe** exhibited very close mirror image symmetry of the absorption and emission spectra. The 0–0 vibronic maxima of the absorption and emission spectra of the derivatives highly overlapped with each other. This suggested that the longest wavelength absorption band of the butadiynyl derivatives corresponds to the $S_0 \rightarrow S_1$ transition. The $S_0 \rightarrow S_1$ transition of the derivatives is allowed, resulting in considerably high molar extinction co-efficient values (Table 1).

### Absorption properties in mixed aqueous solvents

In order to understand the absorption spectral behaviors of the butadiynyl derivatives in mixed aqueous solvents, the absorption spectra of all the derivatives were recorded in water–$CH_3CN$ solvent compositions (Fig. 6). The longest wavelength absorption maxima ($\lambda_{max}$) of the derivatives remained almost unchanged in the 0–60% or 0–70% water–$CH_3CN$ mixtures. The $\lambda_{max}$ showed a red shift in 80–99% water–$CH_3CN$ compared to that observed in the 0–60% or 0–70% water–$CH_3CN$ mixtures. A slight red shift of the $\lambda_{max}$ was observed from the 80% to 99% water–$CH_3CN$ mixtures for all the derivatives. **PyBP** showed $\lambda_{max}$ at around 395 and 404 nm in 0–70% and 99% water–$CH_3CN$, respectively. For **PyBPOMe**, the $\lambda_{max}$ was at around 399 and 410 nm in the 0–70% and 99% water–$CH_3CN$ mixtures, respectively. A large red shift of the $\lambda_{max}$ was observed for **PyBPCN** and **PyBPNMe$_2$**. The derivative **PyBPCN** exhibited a $\lambda_{max}$ at around 404 and 440 nm in the 0–60% and 99% water–$CH_3CN$ mixtures, respectively. The $\lambda_{max}$ of the derivative **PyBPNMe$_2$** showed a 30 nm red shift in the 80–99% water–$CH_3CN$ mixtures compared to that in $CH_3CN$. The red shifted absorption band of all the butadiynyl derivatives was attributed to the aggregates. The vibrational fine structures in the absorption spectra of all the derivatives became almost blurred in the 80–99% water $CH_3CN$ mixtures. The absorption band at around 265 nm was found to be generally broad in the 80–99% water–$CH_3CN$ mixtures for the derivatives.

### Emission properties in mixed aqueous solvents

The interesting absorption spectral behaviour of the butadiynyl fluorophores in water–$CH_3CN$ solvents led us to investigate their emission spectral character in water–$CH_3CN$ mixtures (Fig. 7). The emission maxima of the derivatives **PyBP** and **PyBPOMe** remained almost unaltered in the 0–70% water–$CH_3CN$ mixtures. The fluorescence intensity was found to decrease on gradual addition of water. Notably, the emission maxima of **PyBP** and **PyBPOMe** got red shifted from around 400 nm in pure $CH_3CN$ to 513 nm in the 80–99% water–$CH_3CN$ mixtures (Fig. 7a and b). The emission maxima of the derivative **PyBPCN** exhibited a gradual red shift on addition of water from 0% to 70% into $CH_3CN$ with a concomitant decrease of intensity. This gradual red shift of emission maxima with the decrease of intensity was attributed to the ICT character of the derivative **PyBPCN**. The emission band of **PyBPCN** showed an abrupt red shift in the 80–99% water–$CH_3CN$ mixtures like the derivatives **PyBP** and **PyBPOMe**. The emission maximum

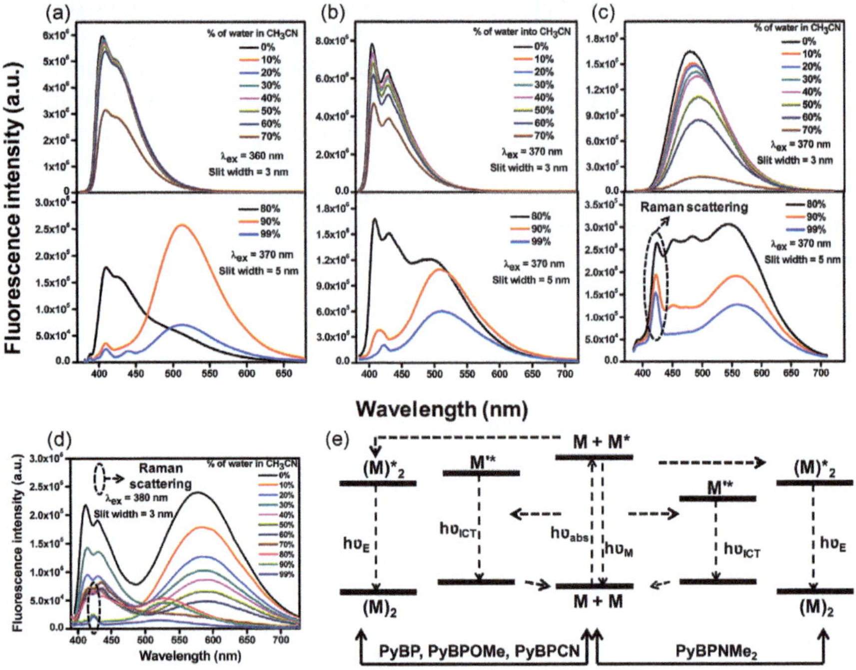

Fig. 7 Emission spectra of (a) PyBP, (b) PyBPOMe, (c) PyBPCN, and (d) $PyBPNMe_2$ in water–$CH_3CN$ mixtures ($C = 1 \times 10^{-6}$ M, dotted circle indicates the Raman scattering). (e) Schematic representation of various emissions of the derivatives. M*, M′*, $(M)_2$* representing the local, ICT and dimeric excited states, respectively.

of **PyBPCN** which was at 497 nm in 70% water–$CH_3CN$, got red shifted at 560 nm in the 99% water–$CH_3CN$ mixture (Fig. 7c). The red shifted emission band in the 80–99% water–$CH_3CN$ mixture suggested emission from the aggregate state for the derivatives **PyBP**, **PyBPOMe**, and **PyBPCN**. It was observed that the ICT character of **PyBPCN** having a strong acceptor group (–CN) was much stronger than that of the parent derivative **PyBP** and the derivative **PyBPOMe** having a weak donating group (–OMe). **$PyBPNMe_2$** having a strong donor group (–$NMe_2$) showed much stronger ICT character upon addition of water into the $CH_3CN$ solution. Upon addition of increased amounts of water, the ICT emission maxima of **$PyBPNMe_2$** were found to be red shifted with a concomitant decrease of intensity, which was persistent up to addition of 60% water into $CH_3CN$. Unlike the derivatives **PyBP**, **PyBPOMe**, and **PyBPCN**, a sudden blue shift of emission maxima of **$PyBPNMe_2$** was observed in the 70–99% water–$CH_3CN$ mixtures. The abrupt blue shift of the emission maxima upon addition of increased % of water into $CH_3CN$ was attributed to emission from the aggregate state of the fluorophore **$PyBPNMe_2$**. A maximum intensity for the blue shifted emission band of **$PyBPNMe_2$** was noted in 80% water–$CH_3CN$ whereas the intensity started decreasing in the 90% and 99% water–$CH_3CN$ mixtures (Fig. 7d). The maximum intensity of the blue emission band in the 80% water–$CH_3CN$ mixture could be attributed to the formation of micro aggregates whereas the decrease of intensity at 90% and 99% could be reasoned to be due to the formation of higher

 

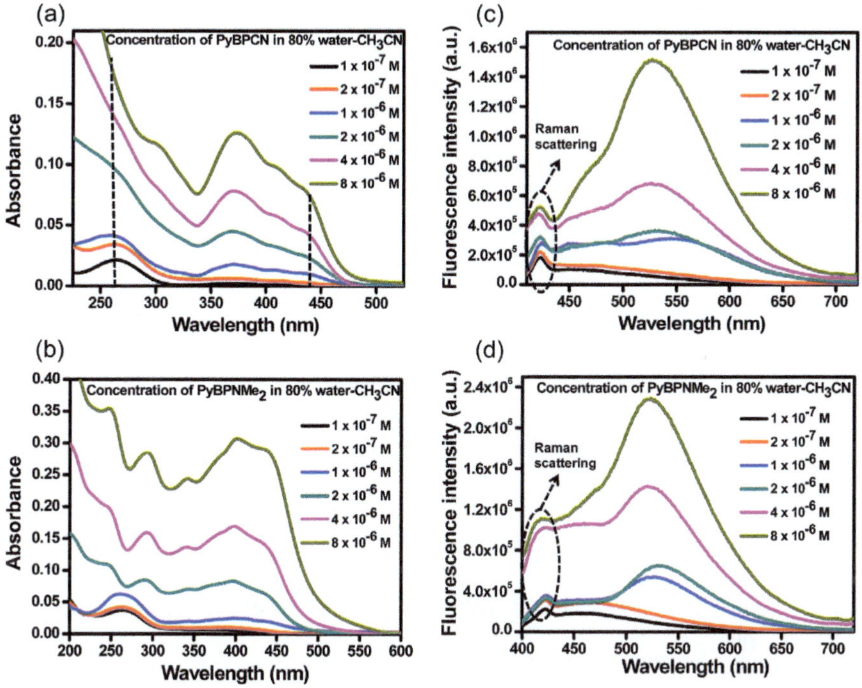


Fig. 8 Concentration dependent absorption spectra of (a) **PyBPCN** and (b) **PyBPNMe$_2$** (instrumental band width = 0.5 nm), and emission spectra of (c) **PyBPCN** ($\lambda_{ex}$ = 370 nm) and (d) **PyBPNMe$_2$** ($\lambda_{ex}$ = 380 nm) (instrumental slit width = 5 nm) in 80% water–$CH_3CN$ (dotted circle indicates the Raman scattering).

aggregates. Raman scattering was observed to appear in the solutions containing higher water percentages, probably due to the presence of aggregate species in the solutions. The Raman scattering which closely overlapped with the LE band, was identified through the standard photophysical technique of collecting fluorescence at different excitation wavelengths. It is important to note that the ICT emission band of the derivatives **PyBP**, **PyBPOMe**, and **PyBPCN** appeared at shorter wavelengths than their aggregate emission bands whereas the ICT

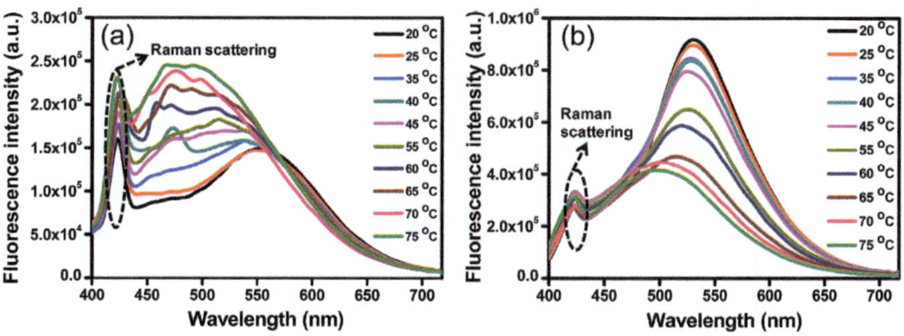


Fig. 9 Temperature dependent emission spectra of (a) **PyBPCN** ($\lambda_{ex}$ = 370 nm) and (b) **PyBPNMe$_2$** ($\lambda_{ex}$ = 380 nm) in 80% water–$CH_3CN$ ($C = 10^{-6}$ M, slit width = 5 nm, dotted circle indicates the Raman scattering).

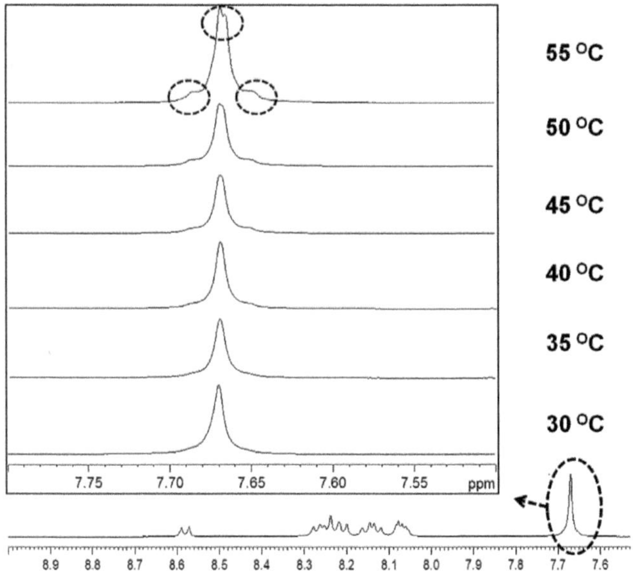


Fig. 10 Temperature dependent $^1$H NMR (500 MHz) spectra of **PyBPCN** in $CDCl_3$ ($C = 1.5 \times 10^{-2}$ M).

emission band of the derivative **PyBPNMe$_2$** appeared in a longer wavelength than its aggregate emission band. Thus, possibly, tuning of the highest occupied molecular orbital (HOMO) and lowest unoccupied molecular orbital (LUMO) through variation of different substitutions at the *para* position of the phenyl ring of the butadiynyl fluorophores was able to show a shift of the position of the ICT bands among the fluorophores. Fig. 7e represents a schematic diagram showing monomer emission (LE), ICT emission and emission from the dimeric excited state (excimer). On considering (M)$_2$ as a stable ground state dimer of the

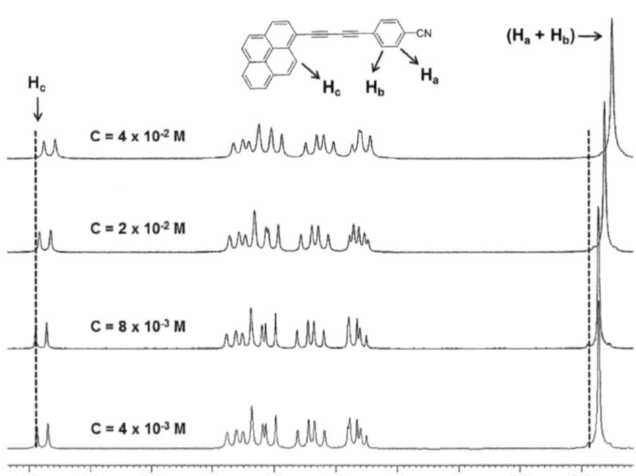


Fig. 11 Concentration dependent $^1$H NMR (500 MHz) spectra of **PyBPCN** in $CDCl_3$.

 

butadiynyl derivatives, its ground state energy level is expected to be lower than that of the corresponding monomer **M**.[45]

### Concentration dependent steady state absorption and emission spectral study

The derivatives **PyBPCN** and **PyBPNMe$_2$** were further investigated through concentration dependent steady state absorption and emission studies to understand the character of the red shifted absorption and emission bands in the 80–99% water–$CH_3CN$ mixtures (Fig. 8). The red shifted absorption band of **PyBPCN** and **PyBPNMe$_2$** at around $\lambda_{max}$ 440 nm which was attributed to the aggregate state, was not observed in very dilute solutions ($10^{-7}$ M). The absorption band ($\lambda_{max}$ 440 nm) started appearing on increasing concentration (Fig. 8a and b). Very dilute solutions ($10^{-7}$ M) of **PyBPCN** and **PyBPNMe$_2$** mainly exhibited the shorter wavelength emission band centred at around 450 nm, which was due to LE emission (Fig. 8c and d). At higher concentration, the intensity of the aggregate emission band centred at around 510–560 nm was high compared to the LE emission band. As the intensity of the aggregate emission band increased on increasing concentration of **PyBPCN** and **PyBPNMe$_2$**, the band was attributed to the excimer type of emission band. It is worth noting that the excimer emission bands of the butadiyne bridged pyrene derivatives **PyBPCN** and **PyBPNMe$_2$** showed more red shift than the excimer band of the alkyne containing pyrene derivatives reported in the literature.[24]

### Temperature dependent steady state fluorescence spectral study

To better understand the aggregate emission band, temperature dependent fluorescence experiments were carried out for the derivatives **PyBPCN** and **PyBPNMe$_2$** (Fig. 9). The emission bands centered at around 510–560 nm in **PyBPCN** and **PyBPNMe$_2$** were found to be blue shifted on increasing temperature.

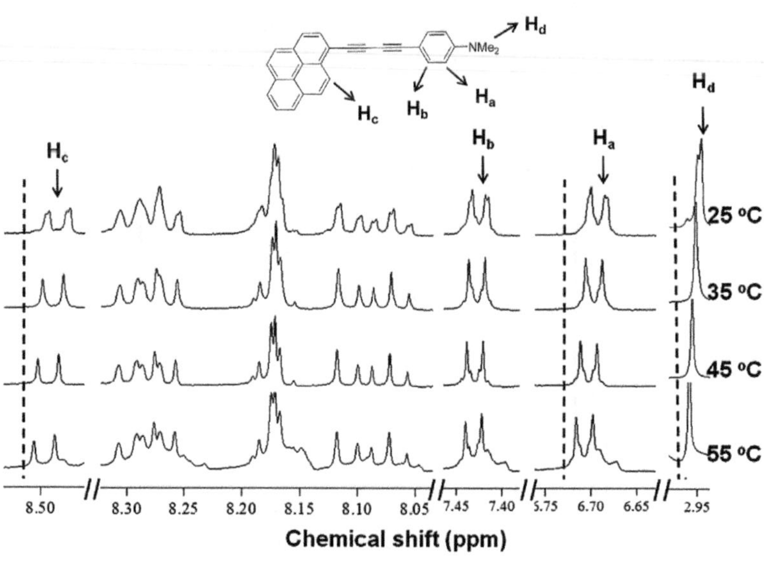


Fig. 12 Temperature dependent $^1H$ NMR (500 MHz) spectra of **PyBPNMe$_2$** in 10% $D_2O$–$CD_3CN$ solvent ($C = 4 \times 10^{-3}$ M).

At higher temperature, the LE band appeared at shorter wavelengths (430–500 nm). The disappearance of the longer wavelength band at higher temperatures appears to correspond to disaggregation. This indicates that this band relates to excimer emission. The intensity of the LE emission band for the derivative **PyBPCN** increased on increasing temperature whereas the intensity of the LE band for the derivative **PyBPNMe$_2$** decreased on increasing temperature.

### Temperature and concentration dependent $^1$H NMR study

An understanding of the nature of aggregation at molecular level is better achieved through $^1$H NMR experiments (Fig. 10, 11 and 12). The four phenyl protons of the derivative **PyBPCN** appeared as a singlet instead of two doublets at low temperature (Fig. 10). The increase of temperature showed the signature of two doublets at 55 °C, suggesting breakage of the aggregates at higher temperature. On the other hand, the concentration dependent $^1$H NMR spectra of **PyBPCN** showed an upfield shift of resonances of the phenyl protons $H_a$ and $H_b$ on increasing concentration (Fig. 11). The resonance of the $H_c$ proton of the pyrene ring[45] also got upfielded on increasing concentration. This indicated that the phenyl protons and the $H_c$ proton of the pyrene moiety took part in the aggregation process of the derivative **PyBPCN**. Fig. 12 shows that the resonances of the pyrene proton $H_c$ and phenyl proton $H_a$ of **PyBPNMe$_2$** slightly appeared in the upfield on decreasing temperature. This indicated the possible intermolecular interactions between the acceptor pyrene ring and the donor dimethyl amino substituted phenyl moiety. Notably, the pyrene proton $H_c$ and phenyl proton $H_a$ did not properly split into a doublet at 25 °C. The methyl proton $H_d$ was found to be a doublet instead of a singlet at the same temperature. On increasing temperature, $H_c$ and $H_a$ appeared as a doublet and $H_d$ exhibited a singlet signal individually. The doublet nature of the methyl proton $H_d$ at 25 °C indicated its involvement in the aggregation process of **PyBPNMe$_2$**. Although the $H_c$ proton belongs to the pyrene moiety, its exact assignment (Fig. 11 and 12) was tentatively done based on the literature.[46]

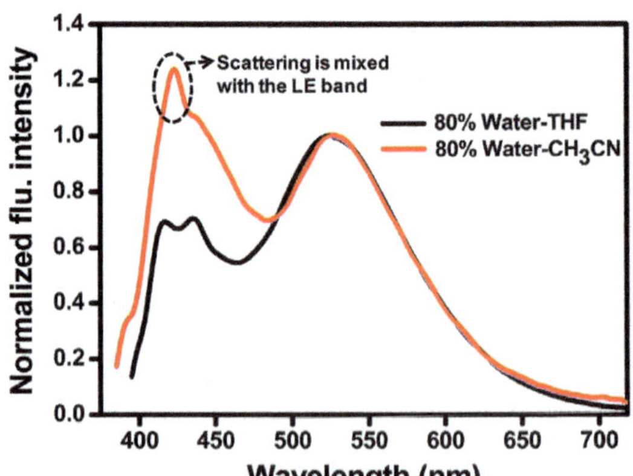


Fig. 13 Solvent dependent aggregate emission (steady state) of the derivative PyBPNMe$_2$ ($C = 1 \times 10^{-6}$ M, dotted circle indicates the Raman scattering).

 

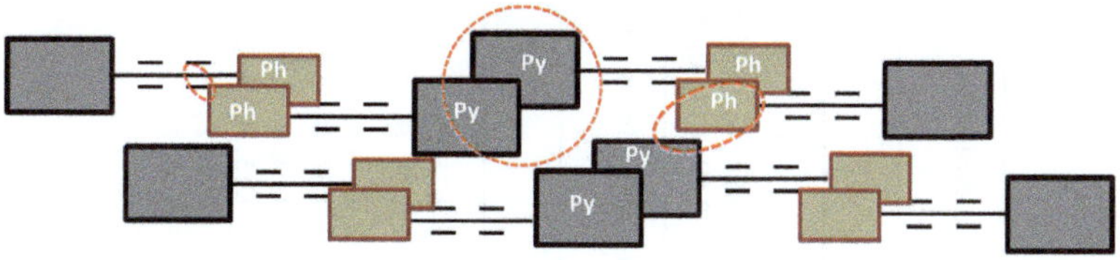


Fig. 14 Schematic representation of stacking diagram for the butadiynyl derivatives (the red circles indicate the different types of plausible interactions).

## Nature of emission from aggregation

In the 80–99% water–$CH_3CN$ mixtures, due to the presence of water as a poor solvent, intermolecular interactions in the derivatives are strong enough such that on excitation, the excimer emission predominates over the ICT emission. Although the nature and extent of aggregation at present is not completely clear, the aggregate emission resembles the pyrene excimer emission and hence it can be conjectured that there is a strong pyrene–pyrene stacking interaction in the aggregate structure. Thus, irrespective of the number of molecules involved in the aggregate, it is probably the dimeric aggregate unit that is controlling the emission spectral behaviour. The insensitivity of the aggregate emission band towards solvent polarity further suggests the pyrene excimer type of emission (Fig. 13).

NMR experiments suggested the involvement of both the phenyl moiety and the pyrene ring in the intermolecular interactions. The concentration used in the NMR experiments was about 1000 times or higher than the concentration used in the steady state absorption and fluorescence experiments. At this higher concentration, the interactions of the phenyl ring with the pyrene moiety in addition to the pyrene–pyrene interactions are not surprising. Even the involvement of the methyl proton of the $NMe_2$ group of the derivative **PyBPNMe$_2$** was observed in the aggregation process. A solid state structure of the derivatives would shed better light on such interactions. Rodriguez *et al.* have shown that the crystal structure of the butadiynyl derivative 1,4-bis[2-(*N*,*N*-dimethylamino)phenyl]buta-1,3-diyne contained a H–π interaction between the methyl proton of the $NMe_2$ group and the π system of the butadiyne moiety.[47] Thus, the crystal structure of the butadiynyl derivative reported by Rodriguez *et al.* supports our observation in the $^1H$ NMR experiments. Based on this observation (H–π interaction), coupled with the observations of the pyrene–phenyl interaction from $^1H$ NMR and pyrene–pyrene interaction from steady state fluorescence experiments, a tentative scheme of molecular packing has been proposed in Fig. 14.

## Time resolved fluorescence studies

Further, to understand the excited state characters of the fluorophores, time resolved fluorescence experiments were carried out at different emission wavelengths. The fluorescence decay profile of the parent derivative **PyBP** showed a mono-exponential decay in the non-polar solvent heptane at $\lambda_{em}$ 410, 460 and 510 nm with a lifetime of *ca.* 1 ns (Table 2). The fluorophore also exhibited an excited state lifetime of *ca.* 1 ns in the polar solvent $CH_3CN$ at $\lambda_{em}$ 410, 460 and 510 nm. Pyrene is known to have an excited state lifetime of *ca.* 450 ns in non-polar solvents. As discussed earlier, the long lifetime of pyrene indicates that the $S_0 \rightarrow S_1$

Table 2 TCSPSC experimental data of the parent derivative **PyBP**

| Solvents | Lifetime values ($\tau$) with relative excited state population ($\alpha$) and relative contribution to total emission ($B$) at different emission wavelengths using a 370 nm LED | | |
|---|---|---|---|
| | $\lambda_{em} = 410$ nm | $\lambda_{em} = 460$ nm | $\lambda_{em} = 510$ nm |
| Heptane | $\tau = 0.98$ ns, $B = 1$, $\alpha = 1$ | $\tau = 0.99$ ns, $B = 1$, $\alpha = 1$ | $\tau = 1.00$ ns, $B = 1$, $\alpha = 1$ |
| $CH_3CN$ | $\tau = 1.10$ ns, $B = 1$, $\alpha = 1$ | $\tau = 1.11$ ns, $B = 1$, $\alpha = 1$ | $\tau_1 = 1.12$ ns, $\alpha_1 = 0.97$, $B_1 = 0.99$<br>$\tau_2 = 0.50$ ns, $\alpha_2 = 0.03$, $B_2 = 0.01$ |

| Mixed aqueous solvents | Lifetime values ($\tau$) with relative excited state population ($\alpha$) and relative contribution to total emission ($B$) at different emission wavelengths using a 370 nm LED | |
|---|---|---|
| | $\lambda_{em} = 425$ nm | $\lambda_{em} = 525$ nm |
| 40% water–$CH_3CN$ | $\tau = 1.08$ ns, $B = 1$, $\alpha = 1$ | $\tau_1 = 1.00$ ns, $\alpha_1 = 0.75$, $B_1 = 0.93$<br>$\tau_2 = 0.45$ ns, $\alpha_2 = 0.25$, $B_2 = 0.07$ |
| 80% water–$CH_3CN$ | $\tau_1 = 0.92$ ns, $\alpha_1 = 0.99$, $B_1 = 0.85$<br>$\tau_2 = 5.26$ ns, $\alpha_2 = 0.01$, $B_2 = 0.15$ | $\tau_1 = 1.74$ ns, $\alpha_1 = 0.28$, $B_1 = 0.43$<br>$\tau_2 = 0.46$ ns, $\alpha_2 = 0.67$, $B_2 = 0.28$<br>$\tau_3 = 6.84$ ns, $\alpha_3 = 0.05$, $B_3 = 0.29$ |
| 90% water–$CH_3CN$ | $\tau_1 = 0.97$ ns, $\alpha_1 = 0.99$, $B_1 = 0.86$<br>$\tau_2 = 5.14$ ns, $\alpha_2 = 0.01$, $B_2 = 0.14$ | $\tau_1 = 2.09$ ns, $\alpha_1 = 0.25$, $B_1 = 0.41$<br>$\tau_2 = 0.67$ ns, $\alpha_2 = 0.71$, $B_2 = 0.37$<br>$\tau_3 = 8.02$ ns, $\alpha_3 = 0.04$, $B_3 = 0.22$ |

 

transition of pyrene is forbidden, resulting in a large apparent Stokes' shift and lack of mirror image symmetry between its absorption and emission spectra. Here, for the butadiyne bridged pyrene derivative, the lifetime (around 1 ns) is short in comparison with the lifetime of pyrene (around 450 ns). The presence of mirror image symmetry of absorption and emission spectra (Fig. 5), and a small Stokes' shift (Table 1) suggest that the derivatization of pyrene has enabled its $S_0 \rightarrow S_1$ transition to be allowed.[48] The 1 ns lifetime component was assigned to the LE state of the derivative. A short lifetime of *ca.* 1.2 ns was also reported for the local emissive state of a butadiyne bridged pyrene dimer by Benniston *et al.*[36] In addition to the 1 ns lifetime component, a small population (3%) of an excited state of the derivative **PyBP** with a lifetime of 0.50 ns was observed at $\lambda_{em}$ 510 nm in $CH_3CN$. The relatively shorter lifetime (0.50 ns) in comparison with the LE component (1 ns) was assigned to the ICT excited state of the derivative owing to the fact that intramolecular charge transfer is generally known to produce a shorter lifetime component than the LE one. On adding 40% water into the $CH_3CN$ solution of the derivative **PyBP**, the fluorescence decay showed the presence of both LE and ICT states at $\lambda_{em}$ 525 nm with significant population of the ICT excited state (25%), while the relative contribution to total emission from the ICT excited state was less (7%). In the 80% and 90% water–$CH_3CN$ mixtures, in addition to the LE state component with a lifetime of *ca.* 1 ns; a comparatively longer lifetime component of *ca.* 5 ns (5.14 and 5.26 ns) than the LE and ICT states was observed at $\lambda_{em}$ 425 nm. This *ca.* 5 ns lifetime component is believed to

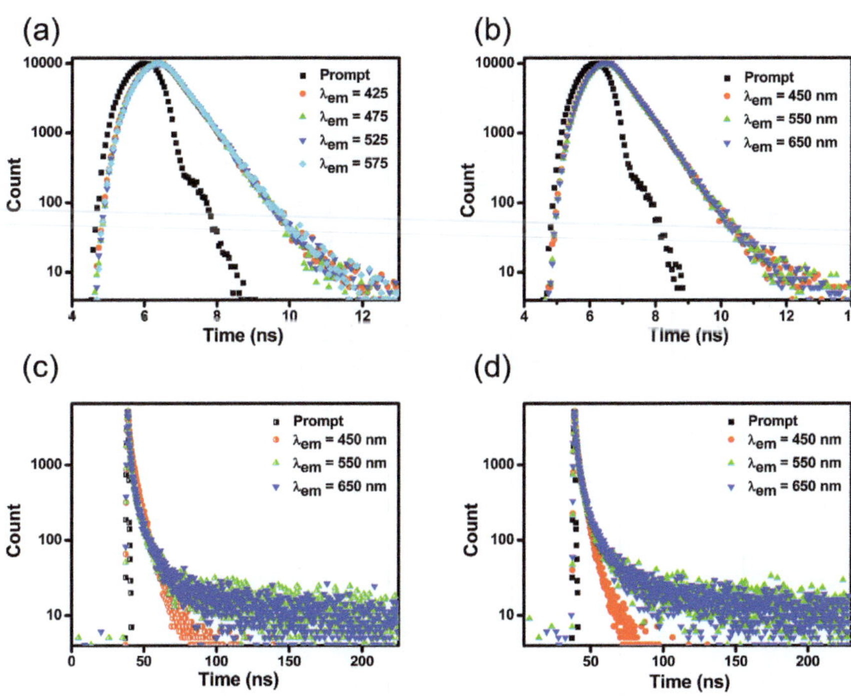


Fig. 15 Fluorescence lifetime decay profile of **PyBPCN** in (a) $CH_3CN$, (b) 40% water–$CH_3CN$, (c) 80% water-$CH_3CN$ (scattering corrected decay), and (d) 90% water–$CH_3CN$ (scattering corrected decay).

Table 3 TCSPC experimental data of the derivative **PyBPCN**

| Solvents | Lifetime values ($\tau$) with relative excited state population ($\alpha$) and relative contribution to total emission ($B$) at different emission wavelengths using a 370 nm LED | | |
|---|---|---|---|
| | $\lambda_{em} = 425$ nm | $\lambda_{em} = 525$ nm | $\lambda_{em} = 575$ nm |
| Heptane | $\tau = 0.94$ ns, $B = 1$, $\alpha = 1$ | $\tau = 0.98$ ns, $B = 1$, $\alpha = 1$ | $\tau = 0.99$ ns, $B = 1$, $\alpha = 1$ |
| $CH_3CN$ | $\tau_1 = 0.96$ ns, $\alpha_1 = 0.05$, $B_1 = 0.09$ | $\tau_1 = 0.93$ ns, $\alpha_1 = 0.06$, $B_1 = 0.10$ | $\tau_1 = 0.90$ ns, $\alpha_1 = 0.09$, $B_1 = 0.15$ |
| | $\tau_2 = 0.52$ ns, $\alpha_2 = 0.95$, $B_2 = 0.91$ | $\tau_2 = 0.52$ ns, $\alpha_2 = 0.94$, $B_2 = 0.90$ | $\tau_2 = 0.51$ ns, $\alpha_2 = 0.91$, $B_2 = 0.85$ |

| Mixed aqueous solvents | Lifetime values ($\tau$) with relative excited state population ($\alpha$) and relative contribution to total emission ($B$) at different emission wavelengths using a 370 nm LED | | |
|---|---|---|---|
| | $\lambda_{em} = 450$ nm | $\lambda_{em} = 550$ nm | $\lambda_{em} = 650$ nm |
| 40% water–$CH_3CN$ | $\tau_1 = 0.98$ ns, $\alpha_1 = 0.08$, $B_1 = 0.13$ | $\tau_1 = 1.40$ ns, $\alpha_1 = 0.01$, $B_1 = 0.03$ | $\tau_1 = 0.93$ ns, $\alpha_1 = 0.10$, $B_1 = 0.06$ |
| | $\tau_2 = 0.55$ ns, $\alpha_2 = 0.92$, $B_2 = 0.87$ | $\tau_2 = 0.58$ ns, $\alpha_2 = 0.99$, $B_2 = 0.97$ | $\tau_2 = 0.53$ ns, $\alpha_2 = 0.90$, $B_2 = 0.84$ |
| 80% water–$CH_3CN$ | $\tau_1 = 3.30$ ns, $\alpha_1 = 0.20$, $B_1 = 0.48$ | $\tau_1 = 4.40$ ns, $\alpha_1 = 0.11$, $B_1 = 0.40$ | $\tau_1 = 3.6$ ns, $\alpha_1 = 0.13$, $B_1 = 0.39$ |
| | $\tau_2 = 8.14$ ns, $\alpha_2 = 0.04$, $B_2 = 0.24$ | $\tau_2 = 35.0$ ns, $\alpha_2 = 0.01$, $B_2 = 0.12$ | $\tau_2 = 19.1$ ns, $\alpha_2 = 0.01$, $B_2 = 0.16$ |
| | $\tau_3 = 0.50$ ns, $\alpha_3 = 0.76$, $B_3 = 0.28$ | $\tau_3 = 0.60$ ns, $\alpha_3 = 0.88$, $B_3 = 0.48$ | $\tau_3 = 0.61$ ns, $\alpha_3 = 0.86$, $B_3 = 0.45$ |
| 90% water–$CH_3CN$ | $\tau_1 = 3.29$ ns, $\alpha_1 = 0.17$, $B_1 = 0.55$ | $\tau_1 = 4.30$ ns, $\alpha_1 = 0.17$, $B_1 = 0.44$ | $\tau_1 = 4.40$ ns, $\alpha_1 = 0.18$, $B_1 = 0.45$ |
| | $\tau_2 = 9.79$ ns, $\alpha_2 = 0.01$, $B_2 = 0.11$ | $\tau_2 = 27.8$ ns, $\alpha_2 = 0.01$, $B_2 = 0.15$ | $\tau_2 = 25.1$ ns, $\alpha_2 = 0.01$, $B_2 = 0.16$ |
| | $\tau_3 = 0.43$ ns, $\alpha_3 = 0.82$, $B_3 = 0.34$ | $\tau_3 = 0.81$ ns, $\alpha_3 = 0.82$, $B_3 = 0.41$ | $\tau_3 = 0.84$ ns, $\alpha_3 = 0.81$, $B_3 = 0.39$ |

 

originate from the aggregate state of the derivative. At $\lambda_{em}$ 525 nm in the 80–90% water–$CH_3CN$ mixtures of the derivative, in addition to the ICT component (lifetime 0.46, 0.67 ns); two other components with lifetimes of *ca.* 2 ns (1.74 and 2.09 ns) and *ca.* 7–8 ns (6.84 and 8.02 ns) were observed (Table 2). These two lifetimes (around 2 ns and 7–8 ns) could be arising due to the presence of mixed aggregates. The mixed aggregates coupled with the ICT state were responsible for the tri-exponential fluorescence decay of **PyBP**. Although the excited state populations of the aggregate state with a lifetime of *ca.* 5 ns at $\lambda_{em}$ 425 nm and the aggregate state with a lifetime of *ca.* 7–8 ns at $\lambda_{em}$ 525 nm in 80–90% water–$CH_3CN$ were less (1–5%), the population of the aggregate state having a lifetime of *ca.* 2 ns was significant (25–28%) at $\lambda_{em}$ 525 nm. The less populated aggregate states (1–5%) of **PyBP** were not omitted owing to the fact that the relative contribution to total emission from these states was not negligible (14–29%) in 80–90% water–$CH_3CN$. Further, it is important to note that although the excited state population of the ICT state (67–71%) was higher than the population of the aggregate state (25–28%) with a lifetime of *ca.* 2 ns at $\lambda_{em}$ 525 nm in 80–90% water–$CH_3CN$; the relative contribution to total emission from the aggregate state (41–43%) was higher than that of the ICT state (28–37%) of the derivative.

The fluorescence decay profile of **PyBPCN** showed bi-exponential to tri-exponential behaviour on moving from polar $CH_3CN$ and 40% water–$CH_3CN$ to 80% and 90% water–$CH_3CN$ mixtures, respectively (Fig. 15). The derivative showed mono-exponential decay with a lifetime of *ca.* 1 ns in non-polar heptane (Table 3). In polar $CH_3CN$ and the 40% water–$CH_3CN$ mixture, the presence of two components with lifetimes of *ca.* 1 ns and 0.5 ns were observed in all the emission wavelengths (Table 3). Similar to the parent derivative **PyBP**, the 0.5 ns lifetime was assigned to the ICT state, while the lifetime around 1 ns was set to the LE state of **PyBPCN**. In $CH_3CN$ and the 40% water–$CH_3CN$ mixture, the population of the ICT excited state (90–99%) as well as the relative contribution to total emission from the ICT excited state (84–97%) predominated over the LE state population (1–10%) and relative contribution to total emission (3–15%) from the LE state respectively. The addition of a higher percentage of water (80–90%) into the $CH_3CN$ solution of **PyBPCN** produced tri-exponential character in the decay profile at all emission wavelengths. Similar to the derivative **PyBP**, the tri-exponential decay of the derivative **PyBPCN** in the 80–90% mixtures was attributed to the ICT and mixed aggregates in the excited state. For example, in the 80% water–$CH_3CN$ mixture, three components with lifetimes of *ca.* 0.5, 3 and 8 ns were observed at $\lambda_{em}$ 450 nm and lifetimes of 0.6, 4, and 35 ns were found at $\lambda_{em}$ 550 nm. The lifetimes of 0.6, 3.6, and 19 ns were noted at $\lambda_{em}$ 650 nm. While the shorter lifetime components (0.5, 0.6 ns) were due to the ICT state, the relatively longer lifetime components could be due to the different sizes of aggregates present in the excited state. Although the excited state population of one of the aggregate states of the derivative **PyBPCN** was less (1–4%), it was not ignored since the relative contribution to total emission (11–24%) from this state was not negligible (Table 3). Although in the 90% water–$CH_3CN$ mixture of the derivative **PyBPCN**, the excited state population of the ICT states was higher (81–82%) than the population of the aggregate states (17–18%) having lifetimes around 3–4 ns; the relative contribution to total emission from the aggregate states (44–55%) was higher than the relative contribution to total emission from the ICT states

| M.O.s | Butadiynyl derivatives | | | |
|---|---|---|---|---|
| | **PyBP** | **PyBPOMe** | **PyBPCN** | **PyBPNMe$_2$** |
| LUMO |  |  | 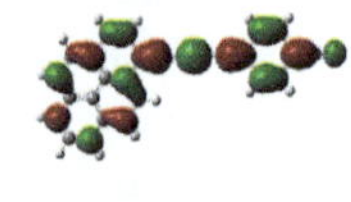 |  |
| HOMO |  |  | 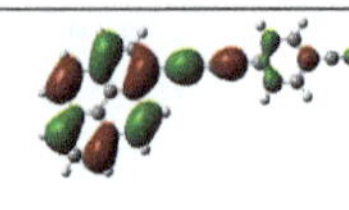 | 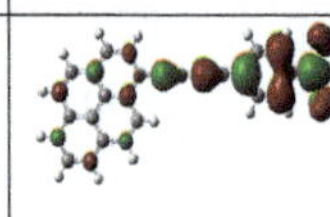 |

Fig. 16 HOMO and LUMO of the butadiynyl derivatives from B3LYP/6-311G(d,p) optimized ground state geometries in $CH_3CN$.

(34–41%). Such an observation was also found in the excited state of the parent derivative **PyBP**.

## Computational investigations on electronic transitions

Finally, to get insight into the electronic transitions of the derivatives **PyBP**, **PyBPOMe**, **PyBPCN**, and **PyBPNMe$_2$**, time dependent density functional theory (TDDFT) based vertical absorption energy calculations were carried out using the Gaussian 09 computational suit. The highest occupied molecular orbital (HOMO) and lowest unoccupied molecular orbital (LUMO) mainly contributed to the $S_0 \rightarrow S_1$ vertical transition of the butadiyne bridged pyrene derivatives. The HOMO and LUMO of the parent derivative **PyBP** were found to be completely delocalized over the entire molecule (Fig. 16). For **PyBPOMe**, although the HOMO spread over the whole molecule, the LUMO mainly remained concentrated on the pyrene ring and the butadiyne moiety. While the LUMO of the derivative **PyBPCN** got diffused throughout the whole molecule, its HOMO mainly resided over the pyrene ring and the butadiyne moiety. Thus, the molecular orbitals showed some ICT character of the derivatives **PyBPOMe** and **PyBPCN**. **PyBPNMe$_2$** showed much stronger ICT character as the HOMO and LUMO of the derivative remained localized on the butadiyne containing *N*,*N*-dimethylaniline and butadiyne containing pyrene rings, respectively. The molecular orbital studies show that the pyrene ring acts as an acceptor in the derivatives **PyBPOMe** and **PyBPNMe$_2$** whereas it behaves as a donor in the derivative **PyBPCN**. The butadiyne moiety was found to be the bridge between the donor and acceptor. The computed vertical absorption energies of

Table 4 Vertical absorption wavelength, oscillator strength, and orbital contributions of the derivatives using CAM-B3LYP/6-311+G(d,p) in $CH_3CN$

| Derivatives | $\lambda_{expt.}$ (nm) | $\lambda_{TD\text{-}FT}$ (nm) | Oscillator strength (f) | Orbital contributions |
|---|---|---|---|---|
| **PyBP** | 395 | 387.6 | 1.41 | H → L (91%), H−1 → L+1 (3%) |
| **PyBPOMe** | 399 | 390.6 | 1.55 | H → L (91%), H−1 → L+1 (3%) |
| **PyBPCN** | 404 | 400.5 | 1.60 | H → L (87%), H−1 → L (2%), H → L+1 (2%) |
| **PyBPNMe$_2$** | 412 | 405.6 | 1.80 | H → L (82%), H−1 → L (6%), H → L+1 (3%) |

the butadiynyl derivatives are tabulated in Table 4. The computed absorption energies are in close agreement with the experimental data.

## Conclusions

The butadiyne bridged pyrene–phenyl conjugates show interesting differences in their photophysical properties with respect to substituted diphenyl butadiynes studied recently.[26] Some of the salient features include:

(1) All the butadiynyl derivatives showed LE emissions originating from the pyrene moiety in solvents of different polarities, which is in contrast to the butadiyne moiety origin of the LE emission in diphenyl butadiynes. In addition to the LE emission, the derivatives with donor–acceptor character (**PyBPCN** and **PyBPNMe$_2$**) exhibited ICT emissions in polar solvents. The parent derivative **PyBP** and the derivative **PyBPOMe** containing a weak donor (–OMe), which did not show ICT emission in pure solvents, showed ICT emission in mixed aqueous solvents (water–$CH_3CN$). Further, the derivatization of pyrene has made the $S_0 \rightarrow S_1$ transition allowed.

(2) In the 80–99% water–$CH_3CN$ mixtures, all the fluorophores formed aggregates. The aggregate emissions were found to be excimer type; the excimer emission being present in addition to LE and ICT emissions. The excimer emission bands were found to be much red shifted compared to the pyrene excimer, probably due to the diacetylenic conjugation in the derivatives.

(3) While the excimer emission maxima of the different derivatives were in the range 510–560 nm, the ICT emission maxima varied over a wide range of wavelengths (475–600 nm). The ICT emission bands of the derivatives **PyBP**, **PyBPOMe**, and **PyBPCN** appeared in a shorter wavelength than their excimer emission bands whereas the ICT emission band of the derivative **PyBPNMe$_2$** appeared in a longer wavelength than its excimer emission band.

(4) The significant red shift of the longest wavelength absorption maxima in the mixed aqueous solvents and the absence of negative amplitude at the aggregate emission wavelength in the fluorescence decay studies suggested that the emissions originated from a static excimer.

(5) $^1$H NMR studies also suggested the formation of aggregates. A possible model has been proposed which involves pyrene–pyrene and pyrene–phenyl interactions.

(6) Molecular orbital calculations indicated that the pyrene moiety acts as an acceptor in the derivatives **PyBPOMe** and **PyBPNMe$_2$** whereas it behaves as a donor in the derivative **PyBPCN**.

## Acknowledgements

We thank DST, New Delhi for financial support. The high performance computing facility of IIT Madras is greatly acknowledged. A. K. P. thanks CSIR, New Delhi for a research fellowship.

## References

1 M. Goes, J. W. Verhoeven, H. Hofstraat and K. Brunner, *ChemPhysChem*, 2003, **4**, 349–358.

2 A. Kraft, A. C. Grimsdale and A. B. Holmes, *Angew. Chem., Int. Ed.*, 1998, **37**, 403–428.
3 A. R. Brown, A. Pomp, C. M. Hart and D. M. de Leeuw, *Science*, 1995, **270**, 972–974.
4 R. Badugu, J. R. Lakowicz and C. D. Geddes, *J. Am. Chem. Soc.*, 2005, **127**, 3635–3641.
5 D. R. Kanis, M. A. Ratner and T. J. Marks, *Chem. Rev.*, 1994, **94**, 195–242.
6 J.-L. Bredas, *Science*, 1994, **263**, 487–488.
7 B. Champagne, E. Deumens and Y. Ohrn, *J. Chem. Phys.*, 1997, **107**, 5433–5444.
8 A. M. Rao, P. C. Eklund, S. Bandow, A. Thess and R. E. Smalley, *Nature*, 1997, **388**, 257–259.
9 R. Kumai, Y. Okimoto and Y. Tokura, *Science*, 1999, **284**, 1645–1647.
10 K. S. Oh, C.-W. Lee, H. S. Choi, S. J. Lee and K. S. Kim, *Org. Lett.*, 2000, **2**, 2679–2681.
11 D. J. Cowley, *Nature*, 1986, **319**, 14–15.
12 W. Rettig, *Angew. Chem., Int. Ed.*, 1986, **25**, 971–988.
13 V. Bonačić-Koutecký, J. Koutecký and J. Michl, *Angew. Chem., Int. Ed.*, 1987, **26**, 170–189.
14 B. Carlotti, R. Flamini, I. Kikas, U. Mazzucato and A. Spalletti, *Chem. Phys.*, 2012, **407**, 9–19.
15 B.-S. Kim, H. Yu and Y.-A. Son, *Mol. Cryst. Liq. Cryst.*, 2012, **563**, 257–271.
16 Y. Zhu, S. Guang, X. Su, H. Xu and D. Xu, *Dyes Pigm.*, 2013, **97**, 175–183.
17 Y. Li, T. Ren and W.-J. Dong, *J. Photochem. Photobiol., A*, 2013, **251**, 1–9.
18 L. Zhu, C. Zhong, C. Liu, Z. Liu, J. Qin and C. Yang, *ChemPhysChem*, 2013, **14**, 982–989.
19 N. M. Dickson-Karn, C. M. Olson, W. C. W. Leu and C. S. Hartley, *J. Phys. Org. Chem.*, 2014, **27**, 661–669.
20 W. Z. Yuan, Y. Gong, S. Chen, X. Y. Shen, J. W. Y. Lam, P. Lu, Y. Lu, Z. Wang, R. Hu and N. Xie, *Chem. Mater.*, 2012, **24**, 1518–1528.
21 S. A. Jenekhe, L. Lu and M. M. Alam, *Macromolecules*, 2001, **34**, 7315–7324.
22 M. R. Bryce, *Adv. Mater.*, 1999, **11**, 11–23.
23 A. A. Vasilev, K. D. Mey, I. Asselberghs, K. Clays, B. Champagne, S. E. Angelova, M. I. Spassova, C. Li and K. Muellen, *J. Phys. Chem. C*, 2012, **116**, 22711–22719.
24 S.-W. Yang, A. Elangovan, K.-C. Hwang and T.-I. Ho, *J. Phys. Chem. B*, 2005, **109**, 16628–16635.
25 U. Subuddhi, S. Haldar, S. Sankararaman and A. K. Mishra, *Photochem. Photobiol. Sci.*, 2006, **5**, 459–466.
26 A. K. Pati, M. Mohapatra, P. Ghosh, S. J. Gharpure and A. K. Mishra, *J. Phys. Chem. A*, 2013, **117**, 6548–6560.
27 A. L. K. S. Shun and R. R. Tykwinski, *Angew. Chem., Int. Ed.*, 2006, **45**, 1034–1057.
28 G. S. Kottas, L. I. Clarke, D. Horinek and J. Michl, *Chem. Rev.*, 2005, **105**, 1281–1376.
29 S.-T. Wu, H.-H. B. Meng and L. R. Dalton, *J. Appl. Phys.*, 1991, **70**, 3013–3017.
30 S.-T. Wu, J. D. Margerum, H. B. Meng, L. R. Dalton, C.-S. Hsu and S.-H. Lung, *Appl. Phys. Lett.*, 1992, **61**, 630–632.
31 S.-T. Wu, M. E. Neubert, S. S. Keast, D. G. Abdallah, S. N. Lee, M. E. Walsh and T. A. Dorschner, *Appl. Phys. Lett.*, 2000, **77**, 957–959.

 

32 M. F. Beristain, T. Ogawa, G. Gomez-Sosa, E. Munoz, Y. Maekawa, F. Halim, F. Smith, A. Walser and R. Dorsinville, *Mol. Cryst. Liq. Cryst.*, 2010, **521**, 237–245.
33 J. S. Kim, M. G. Choi, K. C. Song, K. T. No, S. Ahn and S.-K. Chang, *Org. Lett.*, 2007, **9**, 1129–1132.
34 S. Werder, V. L. Malinovskii and R. Haener, *Org. Lett.*, 2008, **10**, 2011–2014.
35 L. Zeng, P. Wang, H. Zhang, X. Zhuang, Q. Dai and W. Liu, *Org. Lett.*, 2009, **11**, 4294–4297.
36 A. C. Benniston, A. Harriman, D. J. Lawrie and S. A. Rostron, *Eur. J. Org. Chem.*, 2004, 2272–2276.
37 A. K. Pati, S. J. Gharpure and A. K. Mishra, *Phys. Chem. Chem. Phys.*, 2014, **16**, 14015–14028.
38 M. J. Frisch, *et al.*, *Gaussian 09, Revision C.01*, Gaussian, Inc., Wallingford CT, 2010.
39 A. D. Becke, *J. Chem. Phys.*, 1993, **98**, 5648–5652.
40 T. Yanai, D. P. Tew and N. C. Handy, *Chem. Phys. Lett.*, 2004, **393**, 51–57.
41 J. Tomasi, B. Mennucci and R. Cammi, *Chem. Rev.*, 2005, **105**, 2999–3093.
42 K. Balaraman and V. Kesavan, *Synthesis*, 2010, 3461–3466.
43 S. Techert, S. Schmatz, A. Wiessner and H. Staerk, *J. Phys. Chem. A*, 2000, **104**, 5700–5710.
44 D. S. Karpovich and G. J. Blanchard, *J. Phys. Chem.*, 1995, **99**, 3951.
45 F. M. Winnik, *Chem. Rev.*, 1993, **93**, 587414.
46 A. R. Katritzky, G. J. Hitchings, R. W. King and D. W. Zhu, *Magn. Reson. Chem.*, 1991, **29**, 2–8.
47 J. G. Rodriguez, A. Lafuente, R. Martin-Villamil and M. P. Martinez-Alcazar, *J. Phys. Org. Chem.*, 2001, **14**, 859–868.
48 T. Shyamala, S. Sankararaman and A. K. Mishra, *Chem. Phys.*, 2006, **330**, 469–477.

Faraday Discussions

PAPER

# Time resolved *in situ* X-ray diffraction study of crystallisation processes of large pore nanoporous aluminophosphate materials

Kerry Simmance,[a] Wouter van Beek[b] and Gopinathan Sankar*[a]

Received 21st September 2014, Accepted 22nd October 2014
DOI: 10.1039/c4fd00179f

Time resolved high-resolution X-ray powder diffraction was utilized to obtain detailed changes in the crystal structure parameters during the hydrothermal crystallization process of the nanoporous aluminophosphate AlPO-5 (AFI) structure. This *in situ* study offered not only the influence of metal ions on the onset of crystallization and estimation of the activation energy of the process, but also allowed us to determine in detail the changes in lattice parameters during this process. More importantly the time-resolved study clearly showed the lattice expansion in the divalent metal ions substituted system right from the on-set of crystallization process, compared to the one without any dopant ions, which suggest that an amorphous or poorly crystalline network is formed prior to crystallization that contains the large divalent ions (compared to Al(III), the substituting element), which is in agreement with the combined XAS/XRD study reported earlier. A mechanism based on this and the earlier study is suggested.

## Introduction

Understanding the crystallisation mechanisms taking place during hydrothermal synthesis of open framework zeo-type materials plus the metal substitution mechanism is of considerable interest, since it could lead to a more rational approach towards the design and synthesis of new materials with specific crystalline structure, morphology and particle size for catalytic applications. A large number of studies have been devoted to studying the transformation of gel to nanoporous materials using a variety of *in situ* techniques. However, these studies have, generally, only been able to obtain information on the kinetics of crystallisation (using X-ray diffraction (XRD) methods),[1,2] particle size (through small-angle X-ray scattering (SAXS) techniques)[3,4] and local structural change (through X-ray absorption spectroscopy (XAS)).[4,5] Although a combined technique

[a]*Department of Chemistry, University College London, 20 Gordon Street, London WC1H 0AJ, United Kingdom. E-mail: g.sankar@ucl.ac.uk*
[b]*Swiss Norwegian Beamline ESRF, F-38000 Grenoble, France*

approach using all these methods has been used to derive mechanistic aspects of the crystallisation process,[4,6] recent developments, particularly in the area of angular-dispersive High Resolution Synchrotron Powder XRD (HRXRD) (which can also be combined with Raman and XAS)[7], make it possible to collect high-quality diffraction data in a very short time scale, which will not only determine the kinetics, but also the precise structural parameters, in particular the variation in lattice parameters of the reacting system. To utilise this approach, we investigated the formation of a metal substituted nanoporous aluminophosphate catalytic material under hydrothermal conditions.

This family of microporous materials plays a very important role in the field of shape selective catalysis and are employed in several industrially important processes, such as methanol to olefin (MTO) conversion.[8,9] In the last two decades the development of crystalline microporous materials has been extended to include a variety of other compositions such as aluminophosphates (AlPOs) and metal–organic frameworks (MOFs) as their surface areas and pore volumes are comparable with, if not greater than those of the aluminosilicate zeolites. Discovery of AlPOs dates back to 1982 (ref.10), since then a large range of open-framework aluminophosphate, $AlPO_4$-n (where n denotes a particular structure type), materials have been synthesised.

The structures of $AlPO_4$-n consist of strict alternation of corner sharing Al(III) and P(V) tetrahedra (with a strict framework composition of $Al/P = 1$) forming an electrically neutral open-framework which unlike aluminosilicate zeolites, requires no extra-framework cations. AlPOs are isoelectronic with $SiO_2$. Therefore, in theory, a neutral AlPO framework can be derived from a neutral pure silica zeolite by replacement of two Si(IV) cations with one Al(III) and one P(V) cation. This concept led to the synthesis of a variety of AlPOs by isomorphism substitution of P(V) with Si(IV) to form silicoaluminophosphates (SAPO),[11] or substitution of Al(III) with divalent metal cations, Me(II), which are able to adopt tetrahedral coordination (*e.g.* Co, Zn, Fe, Mn and Mg) to form metalaluminophosphates (MeAlPO), or with a combination of both to form metalsilicoaluminophosphates (MAPSO).[12] This isomorphous substitution renders the framework negative which leads to the AlPO materials having interesting acid and redox catalytic applications.

Furthermore, to be able to synthesise a porous material instead of a dense phase, AlPOs must be synthesised with an organic amine or quaternary ammonium ion, more frequently the former, which acts as a ‘structure directing’ agent (SDA) or template. The majority of AlPOs are usually stable upon removal of the organic template and have been found to exhibit thermal stability between 600 and 1000 °C.[13] The role which the SDA plays in the formation of the pore dimensions and connectivity is still widely discussed and it has not been possible to assign a distinct attribute of the SDA to its ability to form specific AlPO topologies. Several different purposes of the SDA have been proposed, ranging from it acting purely as a space-filler[14] to true structure directing.[15]

Among the various hetero-atom substituted AlPO's, AlPO-5 (AFI), AlPO-18 (AEI), AlPO-36 (ATS), AlPO-34 (CHA) and AlPO-11 (AEL) have attracted considerable interest due to their ability to induce shape selective catalytic reactions. Within this range of systems, AlPO-5 is one of the most extensively studied, since it has a large pore one-dimensional channel system and more importantly it is the only system which can be prepared in a relatively short time comprising of a range of metal ions (hetero-atoms such as Zn, Co, Mg, Cr, Fe or Mn)[14,16,17] and using a large number of structure directing agents of the type quaternary ammonium salts and amines. For example, this one-dimensional $AlPO_4$ structure, AlPO-5 can be synthesised using triethylamine (TEA),[18] tripropylamine (TPA),[19] ethyldicyclohexylamine (ECHA)[20] or tetraethylammonium hydroxide (TEAOH). However, most of these organic SDAs only form the AFI topology under very specific conditions. Methyldicyclohexylamine (MCHA), on the other hand, has been shown to only form phase pure AFI over a wide range of synthesis conditions, *i.e.* pH and temperature.[21] The AFI framework also shows a high thermal stability during catalysis and it does not collapse upon calcination, when removing the SDA, showing complete integrity.[22,23] Therefore, it is of general interest to understand the detailed processes involved in the synthesis of this system, in particular how the metal ions can affect the synthesis process. To do so, it is necessary to use *in situ* techniques that are sensitive to the crystal chemistry, to understand the crystallisation process.

Here, we demonstrate the power of time-resolved HRXRD techniques in understanding the structural changes that takes place during the crystallisation of metal ions substituted aluminophosphate with AFI structure using MCHA as the structure directing agent. The results clearly show that a network containing divalent metal ions along with Al(III) and P(V) are formed prior to the crystallisation process. The study also shows clearly that the lattice parameters are different when metal ions are incorporated into the framework. Based on this and previous study a mechanism for the formation is proposed.

# Experimental

### Hydrothermal synthesis

The AFI type materials were synthesized by dissolving aluminium hydroxide in an aqueous solution of phosphoric acid. In the case of MeAlPO-5, an aqueous solution of cobalt(II) acetate tetrahydrate (Aldrich) or zinc(II) acetate dihydrate (Aldrich) was also added to the mixture. After rigorous stirring for *ca.* 10 minutes the gel attains homogeneity and the organic SDA, methyldicyclohexylamine (MCHA) (Aldrich), was added to form the final gel which was stirred for *ca.* 1 h.

### *In situ* characterisation

A gel with composition $(1 - x)$ Al; 1.5 P : $x$Me; 0.8 MCHA; $25H_2O$, $x = 0$ or 0.05, was transferred to a specially designed hydrothermal synthesis cell with an approximate 60% fill volume, and heated to a temperature between 160–170 °C for an appropriate amount of time. Three different gels were studied using synchrotron radiation; unsubstituted AlPO, CoAlPO and ZnAlPO, with the HRXRD study being performed on the BM01B Swiss-Norwegian beamline of the ESRF at a fixed wavelength of 0.5 Å. The details of the beam line and the diffractometer are given elsewhere.[24] The duration of each data collection was tuned with the reaction speed in which the optimized collection period was found to be 2 minutes to cover a 2-theta range of 2–10° (2.89 to 11.89 d-spacing). The data were processed using TOPAS software[25] to extract the lattice parameter of the crystallising system as a function of time.

In a typical experiment, a prepared gel was introduced in to a custom made hydrothermal cell which has a path length of *ca.* 2 mm and a window material made of mica (50 micron thick) so that minimum loss of intensity due to the cell window and amount of sample in the beam is achieved. The heating bloc was heated to a specific temperature before introducing the cell containing gel and the data collection was started once the cell was introduced into the heating block.[3] Typically each data collection took approximately 4 minutes.

## Results and discussion

Crystallisation of three different precursor gels was monitored by HRXRD (wavelength = 0.5 Å) so that we could study the formation of AlPO-5, Zn–AlPO-5 and Co–AlPO-5 with MCHA as the organic template. In Fig. 1 we show stacked plots of the HRXRD data of pure, cobalt and zinc substituted AlPO-5. It is clear from the stacked plots that only phase pure AFI structure is formed, irrespective of the metal content; independent studies at different temperatures, pH of the starting gel and metal content also showed that only the AFI structure is formed when MCHA is used as the template implying the true structure-directing aspect of MCHA.[21]

The evolution of peak intensities as a function of time can be used to derive valuable information on the kinetics and crystallisation mechanism. The area under the Bragg peaks of each crystalline phase was determined by using a fundamental parameters profile fitting through the TOPAS-*academic* programme.[25] Comparison of the area under a number of reflections from a single *in situ* experiment revealed crystallisation curves which were generally superimposable over the duration of the experiment, indicating that the growth is isotropic on all crystallographic planes, and therefore all peak areas should be representative of the growth mechanism that takes place during the reaction. For subsequent analysis, the total summed area of all Bragg reflections of each crystalline AFI phase was used; the areas were then converted to the extent of the reaction ($\alpha$), scaled from 0 to 1, using the relationship $\alpha(t) = I_{\text{hkl}}(t)/I_{\text{hkl}}(\text{max})$, where $I_{\text{hkl}}(t)$ is the area of a given peak at a given time $t$ and $I_{\text{hkl}}(\text{max})$ is the maximum area of this peak. The resulting sigmoidal crystallisation curves are

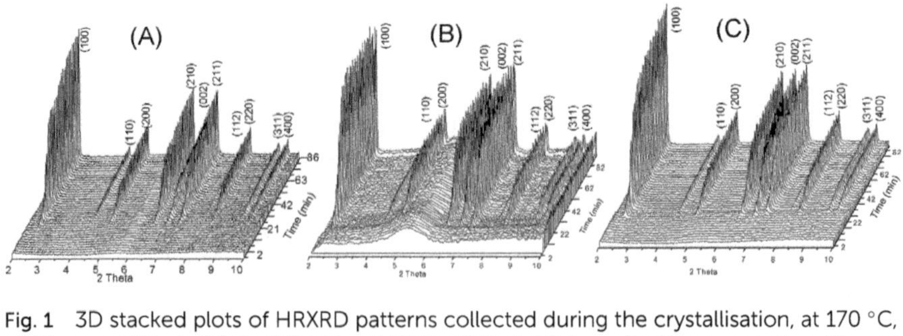


Fig. 1 3D stacked plots of HRXRD patterns collected during the crystallisation, at 170 °C, of (A) pure AlPO-5, (B) CoAlPO-5 and (C) ZnAlPO-5. All the reflections are indexed based on the AFI structure and no other phases are seen to form under the experimental conditions. It should be noted that the small amorphous hump observed throughout the synthesis of CoAlPO-5 (B) is due to a higher water amount in the synthesis cell.

 

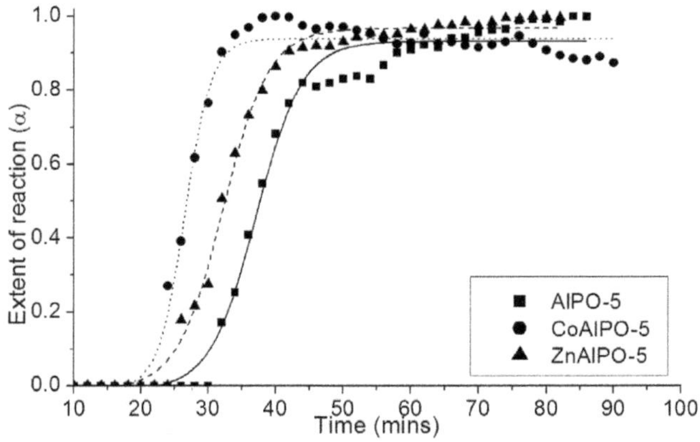


Fig. 2 A plot of the crystallisation curves (Boltzmann sigmoid function) of AlPO-5 (■), CoAlPO-5 (●) and ZnAlPO-5 (▲); each point represents the sum of the experimental integrated values of each reflection as a function of time.

shown in Fig. 2 and are typical for a crystallisation process consisting of an initial period of induction/nucleation followed by rapid crystallisation of the material and gradual growth of the crystallites until a constant value is reached implying completion of crystal growth.

These results show that in the presence of a divalent metal cation, in particular Cobalt(II), there is a reduction in the induction period compared to unsubstituted AlPO-5, indicating that metal ions (5%) promote the formation of AFI type materials. Similar observations have been made by others;[1] however, a previous investigation[4] on the kinetics of AFI growth contradict our observation, which found that cobalt significantly retards nucleation, growth and crystallisation of this system. Apart from various experimental factors, differences may occur due to the use of triethylamine as the organic template, which can produce more than one phase, unless a specific pH is used for a given metal ion containing system. Another factor, is the metal concentration, which has been found to reverse this trend when the cobalt concentration is above 6 percent.[1]

We analysed the HRXRD data in detail to extract the unit cell parameters and crystallite sizes of each crystalline AFI phase employing a fundamental parameters approach using the TOPAS-*academic* program[25] with the initial unit cell parameters taken from the single crystal X-ray diffraction study of Klap *et al.*[26] ($a\ (= b) = 13.718$ Å, $c = 8.4526$ Å, $\alpha\ (= \beta) = 90°$, $\gamma = 120°$ and space group $P6/mcc$). The results yield interesting additional information on the structural aspects during the growth process. Information on the crystallite size can be obtained from analysis of the full width at half maximum (FWHM) of the observed diffraction reflections. Fig. 3 shows the variation in the crystallite size (Lorentzian component) during the crystallization of AlPO-5, CoAlPO-5 and ZnAlPO-5 as a function of time. Results show that in all three AFI materials, the crystallite size was large at the onset of crystallization, ranging between 80 to 120 nm, and grew by approximately 30 percent during the hydrothermal crystallization.

The plots of the $a\ (= b)$ and $c$ parameters over crystallisation time including the cell volume are shown in Fig. 4. The $a\ (= b)$ parameter remained relatively constant during the crystallisation process and was found to be largest for

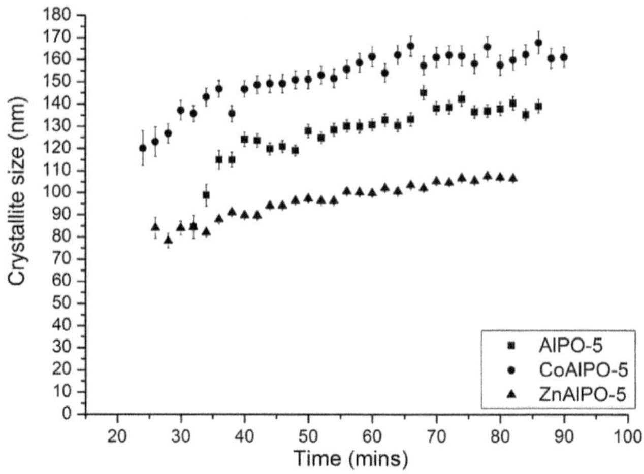


Fig. 3 The variation in the crystallite size of the AFI particles is plotted against crystallization time for AlPO-5 (■), CoAlPO-5 (●) and ZnAlPO-5 (▲).

MeAlPO-5, 13.781 Å (CoAlPO-5) and 13.780 Å (ZnAlPO-5), compared to unsubstituted AlPO-5, 13.735 Å, indicative of the metal cations substituting for aluminium. Tetrahedrally coordinated Co(II) and Zn(II) (direct evidence for the

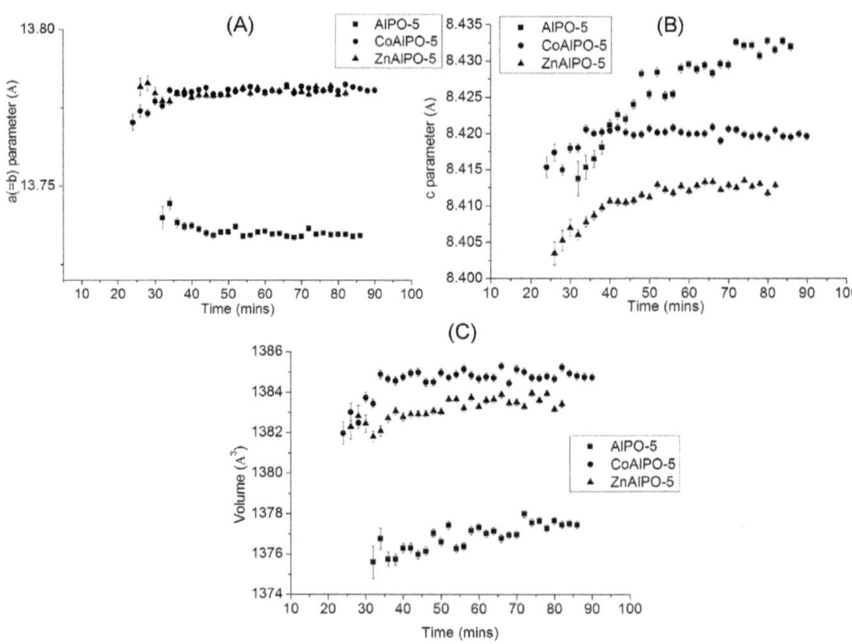


Fig. 4 Variation in the lattice parameters obtained from HRXRD data recorded at the BMO1B (SNBL) beam line, as a function of time during the crystallisation of AlPO-5 (■), CoAlPO-5 (●) and ZnAlPO-5 (▲); (A) shows the *a* (= *b*) parameter change, (B) the *c* parameter variation and (C) is the change in the cell volume over crystallisation time. The analysis was performed only on data above *ca.* 20 minutes, below which the reflections were absent or too few.

 

tetrahedral coordination is obtained from respective metal K-edge X-ray absorption spectroscopic data reported elsewhere)[21,22,27,28] cations have larger ionic radii, 0.58 Å and 0.6 Å respectively, compared to Al(III) which has an ionic radius of 0.39 Å;[29] therefore an expansion in the $a$ (= $b$) direction is observed when a metal ion is substituted into the framework. This trend in the $a$ (= $b$) parameter is in contrast to the $c$ parameter change. The $c$ parameter was found to increase over crystallisation time, more predominantly for unsubstituted AlPO-5 (by approximately 0.02 Å) compared to CoAlPO-5 (increase of 0.01 Å) and ZnAlPO-5 (increase of 0.006 Å).

The AFI framework is known to be have mild hydrophilicity, in which water molecules either bind to framework Al(III) or are physisorbed and reorient isotropically within the channels.[30,31] Recently, through computational methods it has been shown that water molecules reside in the 6-membered ring channels or, with the organic template, in the 12-membered ring channels pointing in the same direction as crystal growth, the $c$ direction.[32] Therefore, this increase in the $c$ parameter, over crystallization time, could be due to the uptake of water molecules which may coordinate to Al(III) ions accessible in the channel direction, increasing the Al–O bond distance from 1.74 Å for tetrahedral Al(III) up to 1.9 Å for an octahedral environment; the more hydrophilic pure AlPO-5 material would therefore show a larger increase in the $c$ parameter compared to MeAlPO-5. It should be noted that template orientation within the structure could also play a role in this change in $c$ parameter and cell volume. The uptake of water molecules would also explain the differences seen on comparison of the $c$ parameter at the end of crystallization, in which AlPO-5 was found to have a larger $c$ parameter compared to MeAlPO-5. XRD patterns collected of the dry as-synthesized forms of these AFI-type materials revealed a $c$ parameter decrease with the following trend: MeAlPO-5 > AlPO-5, which is similar to that observed in the $a$ (= $b$) parameter. From these findings we can conclude that divalent metal cations, such as Co(II) or Zn(II), are incorporated into the AlPO-5 framework in such a way as to cause an expansion in the $a$ (= $b$) direction, which is perpendicular to the channel direction.[33] A change in the $c$ parameter and cell volume illustrates the flexibility of the AFI framework.

These results provide interesting new information about the crystallisation mechanism of AFI in the presence of divalent metal cations. By combining the information obtained from this *in situ* HRXRD experiment with our previous simultaneous XRD/XAS investigation[33] it is possible to derive a mechanism and a crystallization model for the formation of MeAlPO-5, in particular CoAlPO-5.

Our previous simultaneous XRD/XAS results showed cobalt transforming from octahedral to framework tetrahedral *via* a pseudo-octahedral intermediate, importantly, the second stage occurred simultaneously with the appearance of Bragg peaks in the diffraction data. This did not agree with previous studies by Grandjean *et al.*[4] using *in situ* XAS combined with wide angle X-ray scattering (WAXS) or by Weckhuysen *et al.*[34] using *in situ* UV-Vis, although both studies showed a two-stage transformation of octahedral Co(II) species to tetrahedral Co(II), they reveal Co(II) as tetrahedral species just prior to the onset of crystallization. It is important to note that both these studies referred to above used triethylamine as the organic template, which is well-known to produce a CHA phase as a competing material, under certain conditions.[35] Although the CHA phase was not observed during the crystallisation process (with the given XRD detection limit), the use of this template may have a different effect on the Co(II) ions. For example, an *ex situ* XRD/Diffuse Reflectance Spectroscopy (DRS) study[36] on the formation of MeAlPOs observed a pseudo-octahedral Co(II) species during the formation of CoAlPO-5 (AFI). However, this species was not present during the formation of CoAlPO-34 (CHA).

Therefore by combining the findings from our previous simultaneous XRD/XAS investigation with our recent *in situ* HRXRD study; results suggest that the CoAlPO-5 material is formed *via* a solid hydrogel transformation mechanism. In this mechanism it is believed that the structure of the molecular sieve material is obtained by reorganization of the framework of solid phase aluminophosphate hydrogel formed from the condensation of phosphate and aluminate ions in the early stages of crystallization.[37] The crystallographic $a$ (= $b$) parameter is large even at the onset of phase formation (compared to un-substituted AlPO-5) and remained constant throughout the reaction, indicating that the cobalt(II) ions are part of the solid-phase network prior to crystallization and it is unlikely that the Co(II) ions are incorporated into the structure from solution phase. This is also supported by evaluation of the crystallite size over time, which revealed the formation of large particles at the onset of crystallization.

A schematic illustration of the mechanism of formation and substitution of cobalt(II) into the AFI structure is given in Fig. 5. It is generally accepted that the synthesis gel consists of $AlO_4$ and $PO_4$ tetrahedrons which form a 1D chain unit[38–41] along with a $[Co(H_2O)_6]^{2+}$ species. The octahedral Co(II) transforms to a $[CoO_4(H_2O)_2]^{2+}$ or $[CoO_5(H_2O)]^{2+}$ species, in which some of the coordinating oxygen atoms bridge between two cations (P(V)–O–Co(II)). At the onset of crystallization, Co(II) ions start to lose the bound water molecules to form a tetrahedral environment and, simultaneously, water molecules bind to Al(III) ions accessible in the channels. The majority of these water molecules bound to the Al(III) ions are then lost upon dehydration of the solid. A solid hydrogel transformation mechanism and similar phenomena have previously been observed in zeolite and aluminophosphate synthesis[42] In relation to other divalent metal substituted AFI

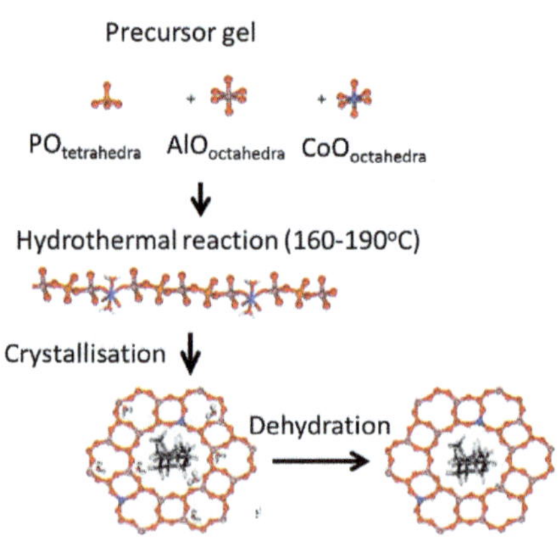


Fig. 5 A schematic illustration of the formation of CoAlPO-5 based on our combined X-ray absorption spectroscopy at the Co K-edge and X-ray diffraction studies reported here and elsewhere. Co(II) = blue, P(V) = orange, Al(III) = light pink, O = red, H = white. The organic template is shown in grey scale.

 

materials, such as ZnAlPO-5, the unit cell parameters were found to have the same trend over crystallization time, indicating a similar mechanism of formation. Although, XAS does not provide additional supporting information on the zinc ion incorporation, Zn(II) ions adopt a tetrahedral coordination environment in the synthesis gel. This study has shown that HRXRD can be a very powerful tool in elucidating information on the crystallization mechanism.

## Summary

In this study, we have determined not only the kinetics using time-resolved angular dispersive HRXRD, but also the structural parameters, in particular the variation in lattice constants, during the hydrothermal synthesis of AFI type materials. By combining results from our previous *in situ* simultaneous XRD/XAS (reported elsewhere) with this *in situ* high-resolution XRD study we propose that the crystallisation of CoAlPO-5 may occur *via* a solid hydrogel transformation mechanism. In this mechanism it is believed that the structure of the molecular sieve material is obtained by reorganisation of the framework of solid phase aluminophosphate hydrogel formed from the condensation of phosphate and aluminate ions in the early stages of crystallisation.[43] The crystallographic $a$ ($= b$) parameter is large even at the onset of phase formation (compared to unsubstituted AlPO-5) and remained constant throughout the reaction, indicating that the divalent metal ions are part of the solid-phase network (amorphous or poorly crystalline solid) formed prior to the crystallisation process and it is unlikely that the divalent ions are incorporated into the structure from solution phase.

## Acknowledgements

We thank the EPSRC and Royal Society (to GS) for their financial support and ESRF for the provision of beam time and other facilities.

## References

1 A. T. Davies, G. Sankar, C. R. A. Catlow and S. M. Clark, *J. Phys. Chem. B*, 1997, **101**, 10115–10120.
2 G. Sankar, T. Okubo, W. Fan and F. Meneau, *Faraday Discuss.*, 2007, **136**, 157–166.
3 G. Sankar and W. Bras, *Catal. Today*, 2009, **145**, 195–203.
4 D. Grandjean, A. M. Beale, A. V. Petukhov and B. M. Weckhuysen, *J. Am. Chem. Soc.*, 2005, **127**, 14454–14465.
5 M. Dong, G. F. Wang, Z. F. Qin, J. G. Wang, T. Liu, S. P. Yuan and H. J. Jiao, *J. Phys. Chem. A*, 2007, **111**, 1515–1522.
6 G. Sankar, J. M. Thomas, F. Rey and G. N. Greaves, *J. Chem. Soc., Chem. Commun.*, 1995, 2549–2550.
7 C. Kongmark, V. Martis, A. Rubbens, C. Pirovano, A. Loefberg, G. Sankar, E. Bordes-Richard, R.-N. Vannier and W. van Beek, *Chem. Commun.*, 2009, 4850–4852.
8 G. Bellussi and P. Pollesel, in *Molecular Sieves: from Basic Research to Industrial Applications, Pts A and B*, ed. J. Cejka, N. Zilkova and P. Nachtigall, 2005, vol. 158, pp. 1201–1212.

9 J. M. Thomas, R. Raja, G. Sankar and R. G. Bell, *Acc. Chem. Res.*, 2001, **34**, 191–200.
10 S. T. Wilson, B. M. Lok, C. A. Messina, T. R. Cannan and E. M. Flanigen, *J. Am. Chem. Soc.*, 1982, **104**, 1146–1147.
11 B. M. Lok, C. A. Messina, R. L. Patton, R. T. Gajek, T. R. Cannan and E. M. Flanigen, *J. Am. Chem. Soc.*, 1984, **106**, 6092–6093.
12 E. M. Flanigen, B. M. Lok, R. L. Patton and S. T. Wilson, *Pure Appl. Chem.*, 1986, **58**, 1351–1358.
13 J. Yu and R. Xu, *Chem. Soc. Rev.*, 2006, **35**, 593.
14 S. T. Wilson and E. M. Flanigen, *ACS Symp. Ser.*, 1989, **398**, 329–345.
15 M. G. O'Brien, M. Sanchez-Sanchez, A. M. Beale, D. W. Lewis, G. Sankar and C. R. A. Catlow, *J. Phys. Chem. C*, 2007, **111**, 16951–16961.
16 I. Arends, R. A. Sheldon, M. Wallau and U. Schuchardt, *Angew. Chem., Int. Ed. Engl.*, 1997, **36**, 1144.
17 M. Hartmann and L. Kevan, *Chem. Rev.*, 1999, **99**, 635–663.
18 Y. Liu, R. L. Withers and L. Noren, *Solid State Sci.*, 2003, **5**, 427–434.
19 O. Weiss, G. Ihlein and F. Schuth, *Microporous Mesoporous Mater.*, 2000, **35–6**, 617–620.
20 J. M. Bennett, J. P. Cohen, E. M. Flanigen, J. J. Pluth and J. V. Smith, *ACS Symp. Ser.*, 1983, **218**, 109–118.
21 M. Sanchez-Sanchez, G. Sankar, A. Simperler, R. G. Bell, C. R. A. Catlow and J. M. Thomas, *Catal. Lett.*, 2003, **88**, 163–167.
22 P. A. Barrett, G. Sankar, C. R. A. Catlow and J. M. Thomas, *J. Phys. Chem.*, 1996, **100**, 8977–8985.
23 G. Sankar, R. Raja and J. M. Thomas, *Catal. Lett.*, 1998, **55**, 15–23.
24 E. Boccaleri, F. Carniato, G. Croce, D. Viterbo, W. van Beek, H. Emerich and M. Milanesio, *J. Appl. Crystallogr.*, 2007, **40**, 684–693.
25 A. A. Coelho, TOPAS-Academic: General Profile and Structure Analysis Software for Powder Diffraction Data, 2007.
26 G. J. Klap, H. van Koningsveld, H. Graafsma and A. M. M. Schreurs, *Microporous Mesoporous Mater.*, 2000, **38**, 403–412.
27 P. A. Barrett, G. Sankar, R. H. Jones, C. R. A. Catlow and J. M. Thomas, *J. Phys. Chem. B*, 1997, **101**, 9555–9562.
28 G. Sankar, G. Muncaster, D. Gleeson, C. R. A. Catlow, J. M. Thomas, J. F. W. Mosselmans, I. Harvey and A. J. Dent, *J. Synchrotron Radiat.*, 2001, **8**, 622–624.
29 D. B. Akolekar, *Appl. Catal., A*, 1998, **171**, 261–272.
30 D. Goldfarb, H. X. Li and M. E. Davis, *J. Am. Chem. Soc.*, 1992, **114**, 3690–3697.
31 R. S. Pillai and R. V. Jasra, *Langmuir*, 2010, **26**, 1755–1764.
32 L. Gomez-Hortigueela, J. Perez-Pariente and F. Cora, *Chem.–Eur. J.*, 2009, **15**, 1478–1490.
33 K. Simmance, G. Sankar, R. G. Bell, C. Prestipino and W. van Beek, *Phys. Chem. Chem. Phys.*, 2010, **12**, 559–562.
34 B. M. Weckhuysen, D. Baetens and R. A. Schoonheydt, *Angew. Chem., Int. Ed.*, 2000, **39**, 3419–3422.
35 F. Rey, G. Sankar, J. M. Thomas, P. A. Barrett, D. W. Lewis, C. R. A. Catlow, S. M. Clark and G. N. Greaves, *Chem. Mater.*, 1995, **7**, 1435–1436.
36 B. M. Weckhuysen, R. R. Rao, J. A. Martens and R. A. Schoonheydt, *Eur. J. Inorg. Chem.*, 1999, 565–577.

 

37 R. R. Xu, P. Wenqin, J. H. Yu, H. Qisheng and J. Chen, *Chemistry of Zeolites and Related Porous Materials – Synthesis and Structure*, John Wiley and sons Ltd, Singapore, 2007.
38 B. A. Adair, S. Neeraj and A. K. Cheetham, *Chem. Mater.*, 2003, **15**, 1518–1529.
39 A. A. Ayi, S. Neeraj, A. Choudhury, S. Natarajan and C. N. R. Rao, *J. Phys. Chem. Solids*, 2001, **62**, 1481–1491.
40 A. M. Beale, A. M. J. van der Eerden, D. Grandjean, A. V. Petukhov, A. D. Smith and B. M. Weckhuysen, *Chem. Commun.*, 2006, 4410–4412.
41 J. H. Yu and R. R. Xu, *Acc. Chem. Res.*, 2003, **36**, 481–490.
42 W. Y. Xu, J. Q. Li, W. Y. Li, H. M. Zhang and B. C. Liang, *Zeolites*, 1989, **9**, 468–473.
43 P. Xu, W. Pang, Q. Huo and J. Chem, *Chemistry of Zeolites and Related Porous Materials – Synthesis and Structure*, John Wiley and sons Ltd, Singapore, 2007.

Faraday Discussions

PAPER

# *In situ* characterization of Cu–Co oxides for catalytic application†

Z. Y. Tian,*[ab] H. Vieker,[c] P. Mountapmbeme Kouotou[b] and A. Beyer[c]

Received 4th October 2014, Accepted 28th October 2014
DOI: 10.1039/c4fd00192c

*In situ* emission and absorption FTIR methods were employed to characterize the spatially resolved structure of binary Co–Cu oxides for low-temperature oxidation of CO and propene. Co–Cu oxide catalysts were controllably synthesized by pulsed-spray evaporation chemical vapor deposition. XRD, FTIR, XPS, UV-vis and helium ion microscopy (HIM) were employed to characterize the as-prepared thin films in terms of structure, composition, optical and thermal properties as well as morphology. *In situ* emission FTIR spectroscopy indicates that $Co_3O_4$, $CuCo_2O_4$ and CuO are thermally stable at 650, 655 and 450 °C, respectively. The catalytic tests with absorption FTIR display that the involvement of Co–Cu oxides can initiate CO and $C_3H_6$ oxidation at lower temperatures. The results indicate that *in situ* emission and absorption FTIR are useful techniques to explore the thermal properties and catalytic performance of functional materials, allowing many potential applications in tailoring their temporally and spatially resolved structure-property relationships.

## Introduction

From a catalytic point of view, a good catalyst for a real application should meet the criteria of low cost, attractive performance and good thermal stability. It is generally accepted that some binary transition metal oxides, in contrast to pure metal oxides, are potential candidates for the development of active catalysts. Binary oxides, especially cobalt-based ones, including Co–Mn,[1,2] Co–Zn[3] and Co–Fe oxides,[4] have been reported to be more active and have better performance than the pure oxide. Compared to these oxides, copper–cobalt oxides with the general formula $Cu_xCo_{3-x}O_4$ represent a family that are used for alcohol

[a]*Institute of Engineering Thermophysics, Chinese Academy of Sciences, 11 Beisihuanxi Road, Beijing 100190, China. E-mail: tianzhenyu@iet.cn; Fax: +86 10 8254 3184; Tel: +86 10 8254 3184*
[b]*Department of Chemistry, Bielefeld University, Universitätsstraβe. 25, D-33615 Bielefeld, Germany*
[c]*Department of Physics, Bielefeld University, Universitätsstraβe. 25, D-33615 Bielefeld, Germany*

† Electronic supplementary information (ESI) available. Tables ESI 1–3: the curve fittings results of Co $2p_{3/2}$, Cu $2p_{3/2}$ and O 1s and satellites of Co–Cu oxides; Table ESI 4: overview of the catalytic performance of CO and propene oxidation with Co–Cu oxides and some catalysts in the literature; Table ESI 5: experimental conditions for the preparation of Co–Cu oxides; Fig. ESI 1–3: additional HIM micorgraphs of $CuCo_2O_4$ films. See DOI: 10.1039/c4fd00192c

synthesis,[5] automobile pollution control[6] and oxygen evolution.[7] Copper–cobalt oxide spinel exhibits interesting performance for CO oxidation.[8] Although a series of works have been reported for the electrocatalytic application of copper cobaltite,[9–12] its application for the oxidation of common emissions such as CO and propene and the evaluation of the thermal stability of such materials has been scarcely investigated.

According to Gautier *et al.*,[11] the catalytic performances and thermal stability of oxides are correlated to their method of preparation. For copper–cobalt spinel, $CuCo_2O_4$ was reported to be difficult to synthesize, which is different from other cobalt-based spinels.[11] Several methods, including the sol–gel,[13] co-precipitation,[10,12] one-pot simple molten salt method,[14] thermal decomposition[15–17] and chemical spray pyrolysis,[11] have been reported in the literature as $CuCo_2O_4$ preparation methods. The stability of $CuCo_2O_4$ was measured to be 400 to 450 °C (ref. 16) and the increase of Cu/Co could lead to a decrease of the thermal stability of the Co–Cu spinel.[11]

To synthesize the copper–cobalt spinel and reveal its thermal stability, a recently developed pulsed-spray evaporation chemical vapor deposition (PSE-CVD) method and *in situ* emission Fourier transform infrared (FTIR) spectroscopy were employed in the present work. Compared to other preparation methods mentioned above, PSE-CVD exhibits various advantages such as relatively low cost, simplicity, high throughput, easy control of the thickness and morphology of the film and particularly, PSE-CVD can deliver single-phase spinel with tailored composition on different supports.[18] In the conventional methods to measure the thermal stability, tiny changes in the dynamically renewed gas were generally detected with low sensitivity. With emission FTIR, samples with surfaces as small as $1 \times 0.5$ cm$^2$ can be reliably analyzed and the sensitivity can be dramatically improved.

This work focuses on the preparation and *in situ* characterization of copper–cobalt oxides for catalytic applications. PSE-CVD was used to synthesize the copper–cobalt oxides and the phases, composition, morphology, thermal and optical properties were comprehensively characterized with X-ray diffraction (XRD), emission FTIR, X-ray photoelectron (XPS), UV-visible and helium ion microscopy (HIM). It should be mentioned that part of the information is provided in ref. 19 for pure CuO. The catalytic potential of the synthesized copper–cobalt oxides was studied for the total oxidation of CO and propene with online absorption FTIR.

# Results and discussion

## Phase identification

The structure and the purity of the prepared samples were revealed by both XRD and online emission FTIR spectroscopy. The XRD patterns of the thin films coated on silicon wafers at 400 °C are shown in Fig. 1. Besides the (400) peak of the silicon wafer (68.74°), the observed peaks at 31.14, 36.68, 38.35, 44.63, 55.40, 59.16 and 65.13° fit well with the planes (220), (311), (222), (400), (422), (511) and (440) of $CuCo_2O_4$, indicating the formation of a Cu–Co binary oxide phase. No peaks corresponding to impurity phases were indicated, revealing the existence of a pure spinel structure. By applying the Debye-Scherrer formula to the two most intense diffraction peaks (220) and (311), the average crystallite size was

 

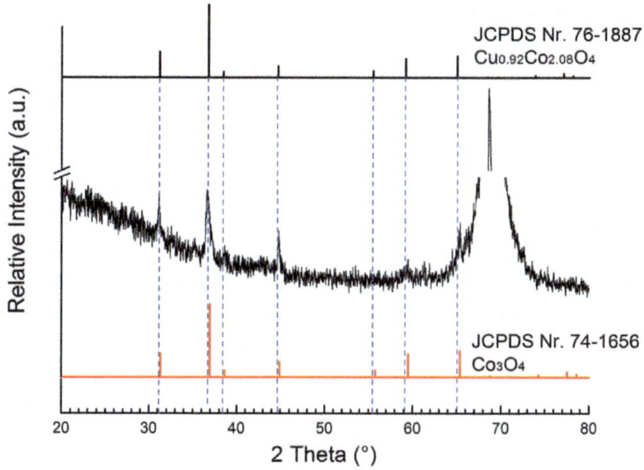


Fig. 1 XRD patterns of Cu–Co oxide thin films.

determined to be about 20 nm. This value is much smaller than that of the pure cobalt oxide prepared under the same conditions.[20] The lattice parameter was calculated to be 0.8108 nm, which is quite close to the reported values by Chi *et al.*[10] and the JCPDS 76-1887 file for spinel $Cu_{0.92}Co_{2.08}O_4$.

Fig. 2 displays the emission FTIR spectrum of Co–Cu binary oxides measured at 140 °C with $Co_3O_4$ (ref. 18) and CuO[19] as references. As reported in our previous work,[18] the two strong and sharp bands peaked at around 650 $cm^{-1}$ and 545 $cm^{-1}$ correspond to the $\nu$ (Co–O) modes. With the doping of Cu, the two bands shifted to higher wavenumbers at 668 and 567 $cm^{-1}$, respectively. These shifts result from the progressive incorporation of copper into the cubic structure. For the

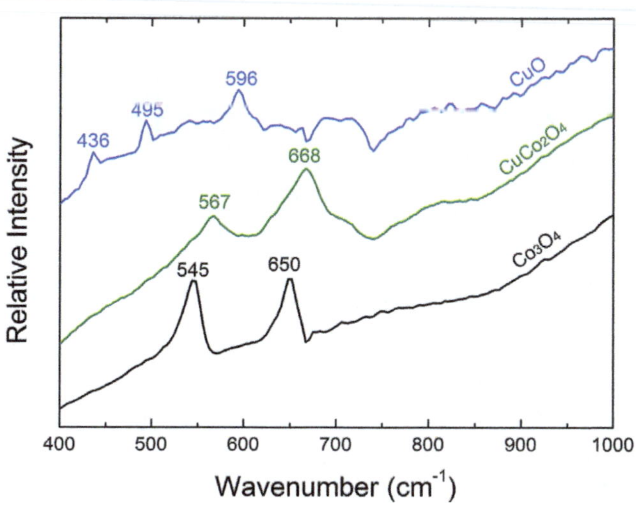


Fig. 2 Emission FTIR spectra of Co–Cu oxide thin films. The samples $Co_3O_4$ and CuO are reproduced from ref. 18 and 19.

pure CuO, three bands peaking at 436, 495 and 596 $cm^{-1}$ were observed and attributed to the $A_u$ (436 and 495 $cm^{-1}$, Cu–O stretching) and $B_u$ (596 $cm^{-1}$) modes of CuO.[19,21,22]

### Thermal properties

To obtain the thermal properties of the prepared Co–Cu oxides, *in situ* emission FTIR spectroscopy was performed at ambient atmosphere and using a temperature-programmed mode in the current work. In the measurement, the lattice stability was determined by integrated intensity of the feature peaks of the spinel as a function of the temperature which reflects the phase transformation. Fig. 3 shows the stability limit of the Co–Cu binary oxide. From 205 to 650 °C, the characteristic bands of Co–Cu oxides were very strong. When the temperature increases to 655 °C, the two bands get weaker. Further increase of the temperature leads to the disappearance of the two bands, revealing the stability limit of Co–Cu binary oxide is 655 °C. According to our pervious work,[20] cobalt spinel and copper oxide[19] are stable up to 650 and 450 °C, respectively. The substitution of Co with Cu favors the formation of Co–Cu binary oxides with slightly better thermal stability relative to cobalt spinel. In the study of cobalt-based binary oxides, slight insertion of manganese[20] and zinc[3] leads to higher stability, which is resulted by the formation of smaller solid solution particles. In the present work, the calculated crystallite size is slightly smaller than that of the pure cobalt spinel.[20] This could increase the content of active oxygen and introduce the unusual cation distribution, which is beneficial for better stability. The formation of the ion groups such as $OH^{-1}$ and carbonates which contribute to the cation distribution will be discussed in the XPS analysis results. This result will implement the pool of cobalt-based binary oxides with better stability by adding Zn, Cr and Mn established by Bahlawane *et al.*[23] and Tian *et al.*[18,20] with Cu.

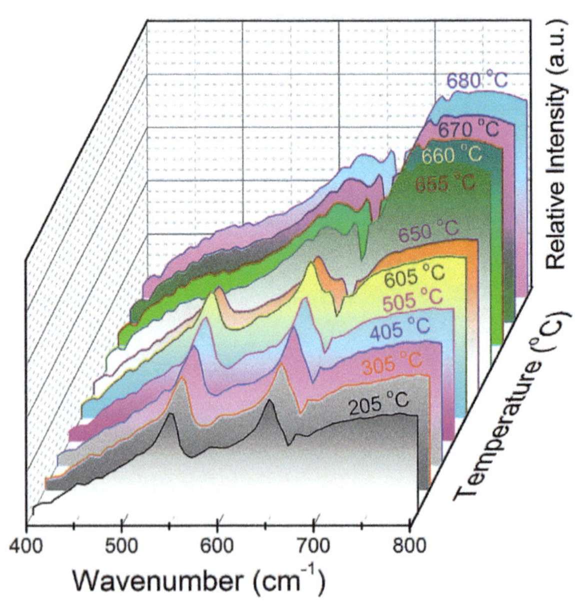


Fig. 3 Thermal stability of Cu–Co oxide thin films.

 

### Surface components

To correlate the surface components of the prepared Co–Cu oxides with their catalytic application, XPS was performed. Fig. 4 and 5 depict the spectra of Co 2p, Cu 2p and O 1s of the Co–Cu oxides. The details of the curve fitting results are summarized in Tables ESI 1–3 (ESI).† For $Co_3O_4$ and $CuCo_2O_4$ (Fig. 4a and b), the binding energies of Co $2p_{3/2}$ and Co $2p_{1/2}$ observed at 779.9 and 795.0 eV, respectively, corresponding to the characteristic spin–orbital splitting of 15.1 eV reported in the literature.[24–26] The fitting of the Co $2p_{3/2}$ was done according Biesinger *et al.*,[27] with very good agreement to the $Co_3O_4$ reference spectra. Therefore, we conclude that the ratio of $Co^{2+}/Co^{3+}$ is around 1/2. Fig. 4c and d show the XPS spectra of Cu 2p and deconvolution of Cu $2p_{3/2}$ peaks of Co–Cu oxide and CuO, respectively. The two major peaks of Cu $2p_{3/2}$ and Cu $2p_{1/2}$ are located at 934.2 and 954.0 eV, respectively. The splitting energy is 19.8 eV, which agrees well the literature reported value[28] and indicates the formation of $Cu^{2+}$. The satellite peaks on the high energy side, arise from the shake-up transition by a ligand–metal 3d charge transfer, and also confirm the presence of $Cu^{2+}$. The deconvolution is in agreement with reference data for CuO,[29] where there may be an additional small (<4%) fraction of $Cu^{1+}$.

Fig. 5 presents the XPS spectra of O 1s which are deconvoluted into two main components. Based on reference data for cobalt and copper oxides,[27,29] the component with binding energy located at about 530 eV is characteristic of lattice oxygen contribution and is attributed to the oxygen present in the form of oxide ($O^{2-}$). The component at a binding energy of ~532 eV is assigned to the surface oxygen from a wide variety of species such as chemisorbed or dissociated oxygen or hydroxyl species as well as carbonate groups.[27–29] The small shoulder at around

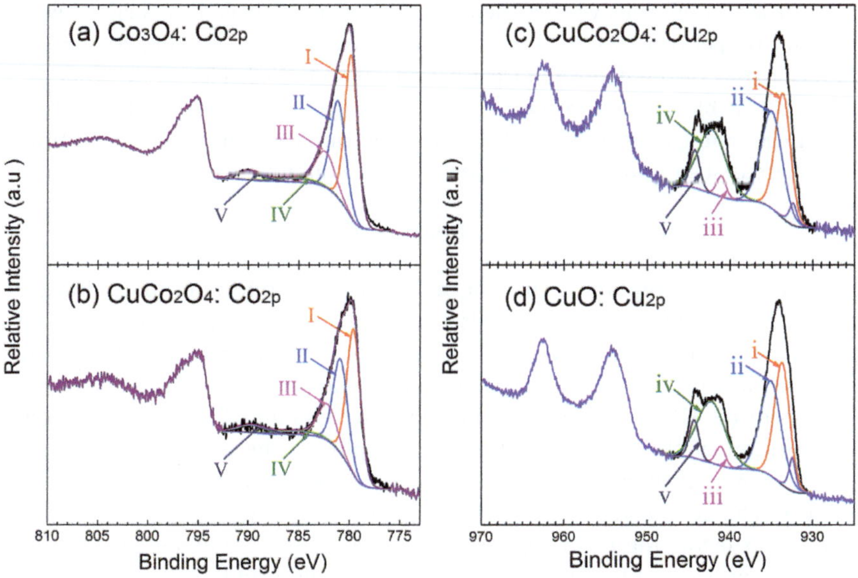


Fig. 4 XPS of Co 2p and Cu 2p of Co–Cu oxides. The samples $Co_3O_4$ and CuO are reproduced from ref. 18 and 19.

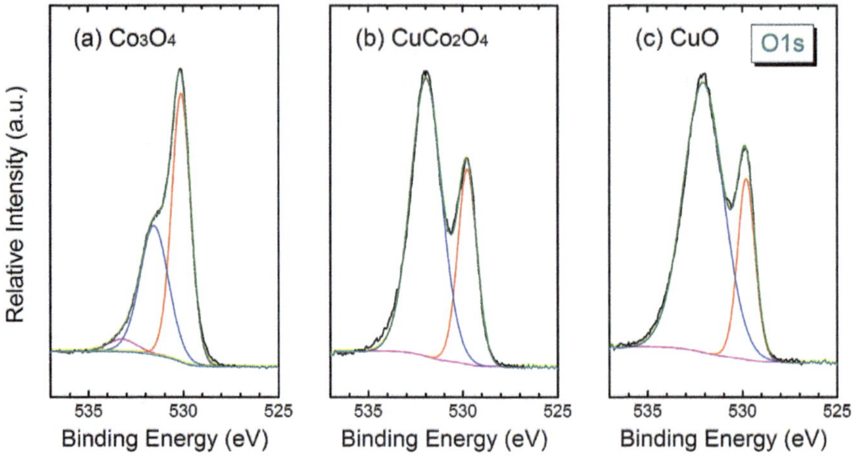


Fig. 5 XPS of O 1s of Co–Cu oxides. The samples $Co_3O_4$ and CuO are reproduced from ref. 18 and 19.

533 eV could be due to subsurface $O^-$ species.[30] Among the Co–Cu oxides, both CuO and $CuCo_2O_4$ exhibit abundant adsorbed and lattice oxygen, while a lot of lattice oxygen and certain amount of subsurface $O^-$ are observed in $Co_3O_4$. It has been widely recognized that the catalytic oxidation could happen on the lattice and/or the adsorbed oxygen. The different amount of lattice and adsorbed oxygen present in the Co–Cu oxides could lead to different behavior in their catalytic applications.

## Optical properties

As proposed in previous work,[18] the optical property is strongly associated with the catalytic performance of transition metal oxides. In this work, the optical characteristics of the deposited Co–Cu oxide was studied. Fig. 6a presents the typical UV-visible absorption spectra of Co–Cu oxide films grown on planar glass. The direct band gap energy ($E_g$) of the prepared samples can be calculated from the collected absorption spectra through Tauc's equation:

$$\alpha h\nu = A\ (h\nu - E_g)^n,$$

where $\alpha$ is the absorption coefficient, $A$ is a constant relying on the refractive index, $h\nu$ is the photon energy, $E_g$ is the band gap depending on the crystal structure and phase composition, and $n = 1/2$ for allowed direct transitions. The corresponding plot of $(\alpha h\nu)^2$ as a function of $h\nu$ is shown in Fig. 6b. The band gap energies of the Co–Cu oxide are estimated to be 1.43 and 2.05 ± 0.05 eV from extrapolation of the straight portion of this plot to the abscissa. These two energies can be attributed to the $O^{2-}$-octahedral ions (mainly $Co^{3+}$ and $Cu^{2+}$), and $O^{2-}$–$Co^{2+}$ charge transfer, respectively. Compared to the corresponding values of pure cobalt oxide, the band gaps obtained with Co–Cu oxide shifted to lower values. According to Jibril,[31] the lower band gap could lead to better mobility of lattice oxygen and enhance the catalyst reducibility. Thus, the Co–Cu oxides are expected to have higher catalytic activity.

 

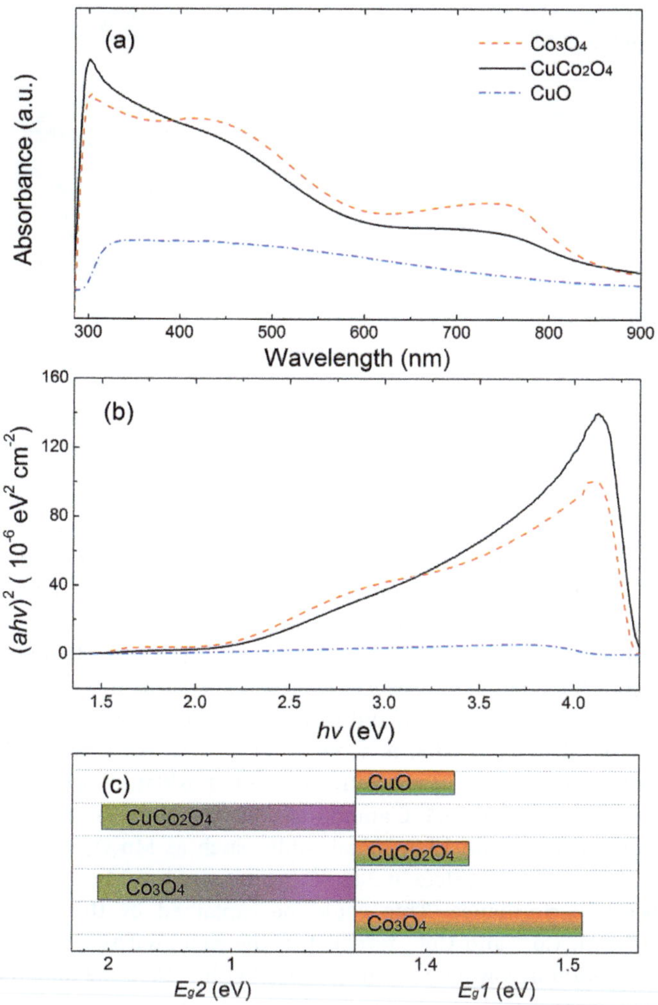


Fig. 6 Optical properties of Co–Cu oxides, with data of $Co_3O_4$ and CuO reproduced from ref. 19 and 20.

## Morphology

The surface morphology of the prepared samples is displayed in Fig. 7. As a newly developed imaging method similar to SEM, HIM was used in the present work. With the involvement of a focused beam of helium ions with a diameter down to 0.35 nm, secondary electrons (SE) generated by the $He^+$ impact are detected in HIM. In SE imaging, the topology of the sample produces contrast as more electrons are ejected when the $He^+$ beam hits the sample at glancing incidence. Since the SE-energy is very low, HIM exhibits superior resolution and outstanding performance on insulating samples.[32–34] As presented in Fig. 7, the Co–Cu binary oxides show uniformly dispersed particles. The particle size is estimated to be 20 nm, which agrees well with the value calculated by the XRD analysis. This highly dispersed morphology with small particle size could produce more active sites to

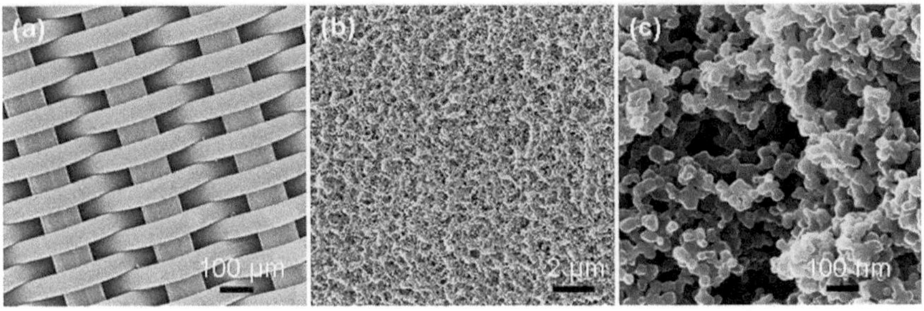


Fig. 7 HIM images of $CuCo_2O_4$ films on stainless steel mesh used for the catalytic test.

hold the adsorbed oxygen and be beneficial to the catalytic application of the Co–Cu binary oxides. Additional high resolution images are in the ESI (Fig. ESI 1–3).†

## Catalytic performance

To investigate the potential application of the grown Co–Cu oxides, the catalytic performance of the prepared samples was tested with respect to the total oxidation of CO and propene. The same tests were also carried out on bare meshes to exclude the background effect of the mesh structures on the oxidation processes. Fig. 8 compares the light-off curves of CO and propene by using $Co_3O_4$ and Co–Cu oxide referring to the bare mesh. With $Co_3O_4$, the conversion of CO occurs at 270 °C and complete conversion is achieved at 362 °C. For Co–Cu oxide, these values shift to 125 and 221 °C, respectively. The CO oxidation over bare mesh is observed to begin at around 395 °C and finish at 833 °C. Under similar conditions, CO oxidation on other transition metal oxides such as $Mn_3O_4$ was reported to happen at 190 °C and complete at 343 °C.[35] Compared to $Mn_3O_4$, Co–Cu oxide exhibits better performance. This could be explained by the abundance of adsorbed oxygen, $Co^{3+}$ and $Cu^{2+}$ revealed in the XPS analysis. By applying the Arrhenius equation at low conversion profiles (within 5%), the apparent activation energies were calculated to be 68.2, 77.3 and 82.9 kJ mol$^{-1}$ for the CO oxidation with $Co_3O_4$, Co–Cu oxide and bare mesh, respectively.

As shown in Fig. 8b, both Co–Cu oxide and $Co_3O_4$ improve the catalytic oxidation of propene compared to the bare mesh. In the presence of Co–Cu oxide, the propene conversion becomes detectable at 220 °C. This value is 30 and 120 °C lower than that with $Co_3O_4$ and bare mesh, respectively. With Co–Cu oxide and $Co_3O_4$, complete conversion of propene was reached within 412 and 454 °C, while this value is observed to be 772 °C for the reaction on bare mesh. It should be mentioned that $CO_2$ is the unique measurable product and no trace of CO was detected in the oxidation process with the prepared $CuCo_2O_4$ and $Co_3O_4$, as depicted in Fig. 9. This result agrees well that the conclusion drawn in our recent work that the active transition metal oxides prevent the formation of CO in the oxidation of low-rank hydrocarbons.[20,36] In the reaction on bare mesh, a large amount of CO peaking at 758 °C was produced, which comes from partial oxidation.

Compared to the reported $T_{50}$ values (corresponding to 50% propene conversion) in the temperature-programmed reaction, 330 °C obtained with $CuCo_2O_4$ is lower than that measured with $Au/Al_2O_3$ (349 °C),[37] $Co_3O_4$ (354 °C),[36] Au/MgO/

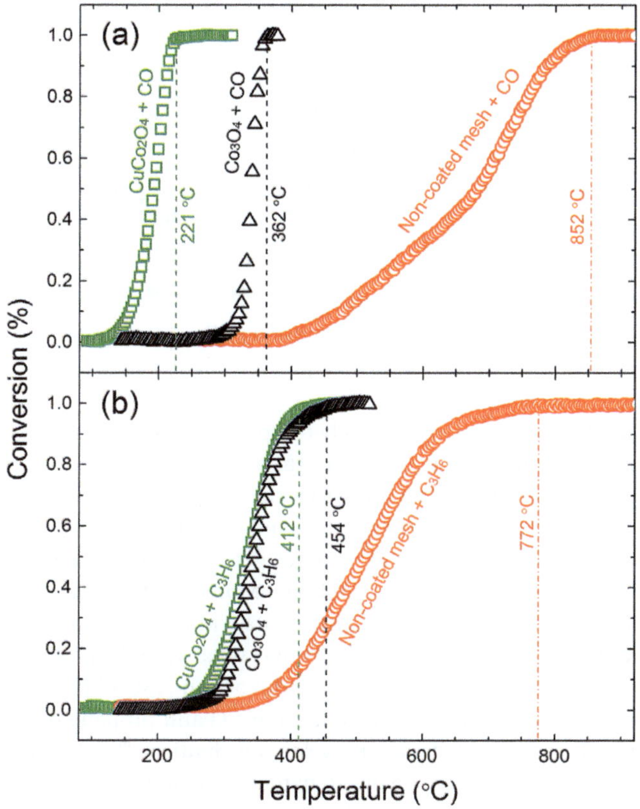


Fig. 8 Catalytic performance of Co–Cu oxides: oxidation of CO (a) and propene (b).

$Al_2O_3$ (359 °C),[38] $La_{1.7}Sr_{0.3}CuO_4S_{0.2}$ (419 °C)[39] and exhibits competitive performance with that of $Au/Li_2O/Al_2O_3$ (327 °C)[38] and $Co_{2.66}Mn_{0.34}O_4$ (321 °C).[20] The detailed information of experimental conditions and $T_{50}$ values is summarized in

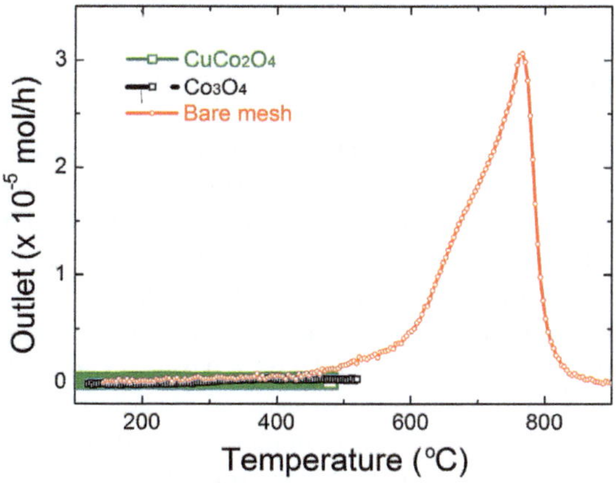


Fig. 9 Outlet profiles of CO in the propene oxidation over $CuCo_2O_4$, $Co_3O_4$ and bare mesh.

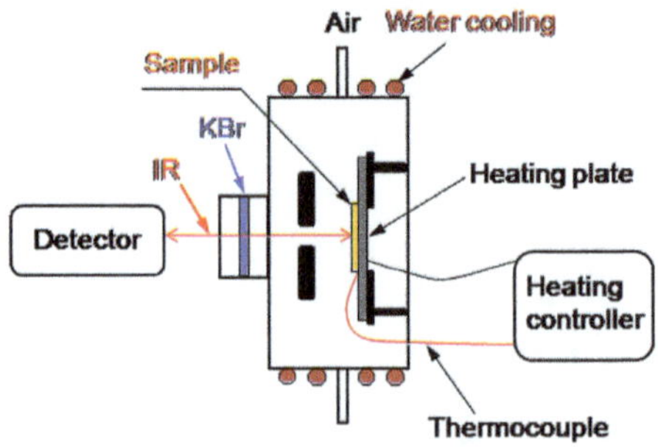


Fig. 10 Experimental scheme of the *in situ* emission FTIR spectroscopy.

Table ESI 4 (ESI).† Although CuO shows better performance with a $T_{50}$ of 272 °C,[19] $CuCo_2O_4$ has more application prospects than CuO by considering the higher thermal stability. It is worth mentioning that the magnitude order of $T_{50}$ for CuO, $CuCo_2O_4$ and $Co_3O_4$ follows the order of the band gaps. It can be explained that the migration of $O^{2-}$ from the bulk to the surface gets easier for samples with lower band gaps which have better electron mobility. Moreover, the apparent activation energy for propene oxidation with $CuCo_2O_4$ is 82.6 kJ mol$^{-1}$, which is lower than that obtained with $Co_3O_4$ (130 kJ mol$^{-1}$),[20] CuO (109.5 kJ mol$^{-1}$),[19] $Co_{2.66}Mn_{0.34}O_4$ (115.5 kJ mol$^{-1}$)[20] as well as the non-catalyzed reaction (138.3 kJ mol$^{-1}$).

It has been widely accepted that oxidation of CO and propene over transition metal oxides follows the Mars-van Krevelen mechanism.[40–42] In the oxidation process, two reversible steps are involved, namely interaction of trapped or lattice oxygen with CO/propene, leading to the reduction of $Cu^{2+}$ to $Cu^{1+}$ and $Co^{3+}$ to $Co^{2+}$, and then a subsequent re-oxidation of the reduced metal oxide site to higher ionic states. This mechanism has been established for CO and propene oxidation over CuO,[19] Co–Mn oxides,[18,20,35] and Co–Fe oxides.[43,44] The same redox behavior could exist in the reaction with Co–Cu oxides. Besides the role of lattice oxygen, the adsorbed oxygen, oxygen-vacant sites[45] and cooperative effect among metallic species in the deposited samples could also contribute to the attractive performance of Co–Cu mixed oxides.

## Conclusions

Co–Cu binary oxides were synthesized with a facile PSE-CVD approach to control the film composition and microstructure. The physicochemical properties, including structure, thermal stability, morphology, surface composition and band gap energy of the obtained thin film were comprehensively characterized by XRD, *in situ* emission FTIR, XPS, HIM and UV-vis spectroscopy. The structure study indicates the obtained Co–Cu oxide samples are thermally stable up to 655 °C and exhibit band gaps of 1.43 and 2.05 ± 0.05 eV. The catalytic tests show that the prepared Co–Cu binary oxides are very active catalysts for the oxidation of CO and propene, with a performance comparable to that of supported noble metals. The attractive activity of Co–Cu binary oxides by favoring $CO_2$ selectivity at low temperature is assigned to the redox properties and decrease of the activation energy.

 

# Experimental

### Synthetic procedures

The deposition of Co–Cu binary oxides was carried out in a cold-wall stagnation point flow CVD reactor employing a PSE unit for the delivery of the precursor liquid feedstock.[18,20] To obtain stoichiometric single-phase films and keep the precursors soluble in ethanol as the solvent, cobalt acetylacetonate ($Co(acac)_2$) and copper acetylacetonate ($Cu(acac)_2$) were kept at respective concentrations of 5 mM. During the deposition process, the feedstock was kept at room temperature under atmospheric pressure. The PSE delivery was achieved with a valve opening time of 2.5 ms and a frequency of 4 Hz. The feedstock was injected as a fine spray into a 30 cm-long evaporation chamber kept at 200 °C. The resulting vapor was transported to the deposition chamber kept at 220 °C, with $N_2/O_2$ flow rates of 0.5/1 standard liters per minute (SLM). Glass, silicon, stainless steel and mesh of stainless steel substrates were heated using a flat resistive heater to 400 °C during deposition and the total pressure in the reactor was kept at 30 mbar. The thickness of the deposited films was determined gravimetrically by measuring the weight change of the substrates with a microbalance (Mettler ME30, digital resolution of 1 µg). The experimental conditions for the synthesis are summarized in Table ESI 5 (ESI).†

### Characterizations and catalytic test

To characterize the crystalline phase of the deposited films, both XRD and *in situ* emission FTIR spectroscopy were used. The XRD (Phillips X'Pert Pro MDR diffractometer (Cu K$\alpha$ radiation, $\lambda = 0.154056$ nm) with PW 3830 X-ray generator) was carried out at 40 kV and 30 mA under ambient conditions. Data were recorded in a $2\theta$ range from 20° to 80°, with a scanning step of 0.05°. The crystalline phases were identified by referring to the powder XRD database (JCPDS-ICDD). The crystallite sizes were calculated using the value of full width at half maximum (FWHM) for the highest intensity diffraction peak, by means of the following well known Debye-Scherrer equation. An *in situ* emission FTIR spectrometer (Nicolet FTIR 5700) with resolutions of 2 cm$^{-1}$ was used to confirm the structure of the oxides.

To explore the thermal property of the Co–Cu oxides, emission FTIR spectroscopy was employed as an *in situ* analysis tool under temperature-programmed mode with a rate of 3 °C min$^{-1}$ (Fig. 10).[18] A thermal emission caused by gray body radiation can be generated by heating a stainless steel sample with a coated thin film. The thermal emissions are superposed with characteristic bands in the mid-infrared range of 500–700 cm$^{-1}$. By integrating the characteristic IR bands of the oxides, the materials' structures can be monitored online.

XPS was performed in a multi-technique UHV instrument (Multiprobe, Omicron Nanotechnology) using a monochromatic Al K$\alpha$ X-ray source (1486.7 eV, 250 W) and a hemispherical electron energy analyzer (Sphera) in constant analyzer energy mode. The sample was under an angle of 13° from the surface normal to the electron detector. The chamber base pressure was $2 \times 10^{-10}$ mbar. Casa XPS was used for the analysis of the XP spectra, and a Shirley background subtraction procedure was employed.

The recently developed imaging technique, HIM (Carl Zeiss Orion Plus®), was used to study the microstructure of the Co–Cu binary oxides. The helium ion beam was operated at 35 kV acceleration voltages at currents of 0.3–0.8 pA. The working distance was 10.7 mm at a sample tilt of 0°. Secondary electrons were collected by an Everhart-Thornley detector at 500 V grid voltage. The samples were plasma-cleaned in the HIM load-lock for 8 minutes before measurement. More details of the HIM measurement could be found in our previous work.[19]

The catalytic performance of $CuCo_2O_4$ for the oxidation of CO and propene was investigated in a fixed bed quartz reactor at atmospheric pressure. The catalytic tests were respectively performed on the fresh mesh samples with 12 mg catalyst grown as a thin film and pretreated ones. The samples were treated by a flow of 10% $O_2$ stream diluted in Ar at 300 °C for 1 hour. The experimental setup has been described in detail elsewhere.[36] The total flow rate was kept at 15 ml $min^{-1}$ with 1% of fuel gas and 10% $O_2$ diluted in Ar. The weight hourly space velocity (WHSV) was controlled at 75 000 ml ${g_{cat}}^{-1}$ $h^{-1}$. The gas flow rates were controlled by MKS mass-flow controllers, whereas the temperature inside the reactor was raised with a ramp of 3 °C $min^{-1}$ using a HT60 controller (Horst). The temperature of the mesh inside of the reactor was recorded using K-thermocouples and a digital thermometer (Greisinger GMH3250). The exhaust gases were analyzed using a home-made KBr-transmission cell and an FTIR spectrometer in the wavelength range of 400–4000 $cm^{-1}$. Details of the data treatment were described elsewhere.[18,20] Blank experiments were performed using a reactor filled with bare mesh of stainless steel.

## Acknowledgements

ZYT thanks the financial support from the Recruitment Program of Global Youth Experts (Y41Z024BA1 and Y31Z061BA1). ZYT and HV thank Prof. Katharina Kohse-Höinghaus and Prof. Armin Gölzhäuser for useful discussions.

## Notes and references

1 Q. Liang, K. D. Chen, W. H. Hou and Q. J. Yan, *Appl. Catal., A*, 1998, **166**, 191–199.
2 D. A. Aguilera, A. Perez, R. Molina and S. Moreno, *Appl. Catal., B*, 2011, **104**, 144–150.
3 N. B. J. Feldmann and K. Kohse-Höinghaus, Elecrochem. Proceedings of EUROCVD-15, *J. Electrochem. Soc.*, 2005, **2005–09**, 659–666.
4 P. M. Kouotou, H. Vieker, Z. Y. Tian, et al., *Catal. Sci. Technol.*, 2014, **4**, 3359–3367.
5 A. Segier, E. Freud, USA Patent 4,122,110, 1978.
6 G. N. Bliznakov and D. R. Mekhandzhiev, *Kinet. Catal.*, 1987, **28**, 100.
7 E. J. Sullivan and E. Calvo, *Reaction Kinetics*, ed.R. G. Compton, Elsevier, Amsterdam, 1989, Vol. 27.
8 R. P. Bonchev, T. Zheleva and S. C. Sevov, *Chem. Mater.*, 1990, **2**, 93–95.
9 I. Nikolov, R. Darkaoui, E. Zhecheva, R. Stoyanova, N. Dimitrov and T. Vitanov, *J. Electroanal. Chem.*, 1997, **429**, 157–168.
10 B. Chi, H. Lin and J. Li, *Int. J. Hydrogen Energy*, 2008, **33**, 4763–4768.

11 J. L. Gautier, E. Trollund, E. Rios, P. Nkeng and G. Poillerat, *J. Electroanal. Chem.*, 1997, **428**, 47–56.
12 B. Lal, N. K. Singh, S. Samuel and R. N. Singh, *J. New Mater. Electrochem. Syst.*, 1999, **2**, 59–64.
13 A. Amri, Z. T. Jiang, T. Pryor, C. Y. Yin, Z. H. Xie and N. Mondinos, *Surf. Coat. Technol.*, 2012, **207**, 367–374.
14 M. V. Reddy, C. Yu, J. H. Fan, K. P. Loh and B. V. R. Chowdari, *RSC Adv.*, 2012, **2**, 9619–9625.
15 B. Marzan, N. Fradette and G. Beaudoin, *J. Electrochem. Soc.*, 1992, **139**, 1889–1896.
16 K. Petrov, T. Karamaneva, S. Angelov and D. Mehandjiev, *Mater. Res. Bull.*, 1983, **18**, 637–642.
17 X. Wu and K. Scott, *J. Mater. Chem.*, 2011, **21**, 12344–12351.
18 Z. Y. Tian, N. Bahlawane, V. Vannier and K. Kohse-Höinghaus, *Proc. Combust. Inst.*, 2013, **34**, 2261–2268.
19 Z. Y. Tian, H. J. Herrenbrück, P. Mountapmbeme Kouotou, H. Vieker, A. Beyer, A. Gölzhäuser and K. Kohse-Höinghaus, *Surf. Coat. Technol.*, 2013, **230**, 33–38.
20 Z. Y. Tian, P. H. Tchoua Ngamou, V. Vannier, K. Kohse-Höinghaus and N. Bahlawane, *Appl. Catal., B*, 2012, **117–118**, 125–134.
21 A. S. Ethiraj and D. J. Kang, *Nanoscale Res. Lett.*, 2011, **7**, 70.
22 K. Kliche and Z. V. Popovic, *Phys. Rev. B: Condens. Matter Mater. Phys.*, 1990, **42**, 10060–10066.
23 N. Bahlawane, P. H. Tchoua Ngamou, V. Vannier, T. Kottke, J. Heberle and K. Kohse-Höinghaus, *Phys. Chem. Chem. Phys.*, 2009, **11**, 9224–9232.
24 W. F. Wei, W. X. Chen and D. G. Ivey, *Chem. Mater.*, 2008, **30**, 1941–1947.
25 M. Kim, X. M. Chen, E. Fradkin, P. Abbamonte and S. L. Cooper, *Phys. Rev. Lett.*, 2010, **105**, 026402.
26 D. Barreca, C. Massignan, S. Daolio, et al., *Chem. Mater.*, 2001, **13**, 588–593.
27 M. C. Biesinger, B. P. Payne, A. P. Grosvenor, L. W. M. Lau, A. R. Gerson and R. S. C. Smart, *Appl. Surf. Sci.*, 2011, **257**, 2717–2730.
28 A. Amri, Z. T. Jiang, X. L. Zhao, et al., *Surf. Coat. Technol.*, 2014, **239**, 212–221.
29 M. C. Biesinger, L. W. M. Lau, A. R. Gerson and R. S. C. Smart, *Appl. Surf. Sci.*, 2010, **257**, 887–898.
30 S. Royer, A. V. Neste, R. Davidson, S. Mcintyre and S. Kaliaguine, *Ind. Eng. Chem. Res.*, 2004, **43**, 5670–5680.
31 B. Y. Jibril, *React. Kinet. Catal. Lett.*, 2005, **86**, 171–177.
32 B. W. Ward, J. A. Notte and N. P. Economou, *J. Vac. Sci. Technol., B: Microelectron. Nanometer Struct.*, 2006, **24**, 2871–2874.
33 R. Hill, J. A. Notte and L. Scipioni, in: Chapter 2-Scanning Helium Ion Microscopy, *Advances in Imaging and Electron Physics*, Elsevier, 2012, pp. 65–148.
34 F. Marabelli, G. B. Parravicini and F. Salghetti-Drioli, *Phys. Rev. B: Condens. Matter Mater. Phys.*, 1995, **52**, 1433–1436.
35 Z. Y. Tian, P. Mountapmbeme Kouotou, N. Bahlawane and P. H. Tchoua Ngamou, *J. Phys. Chem. C*, 2013, **117**, 6218–6224.
36 Z. Y. Tian, N. Bahlawane, F. Qi and K. Kohse-Höinghaus, *Catal. Commun.*, 2009, **11**, 118–122.
37 S. Ivanova, C. Petit and V. Pitchon, *Gold Bull.*, 2006, **39**, 3–8.

38 A. C. Gluhoi, N. Bogdanchikova and B. E. Nieuwenhuys, *J. Catal.*, 2005, **232**, 96–101.
39 M. Machida, K. Ochiai, K. Ito and K. Ikeue, *J. Catal.*, 2006, **238**, 58–66.
40 G. Sedmak, S. Hočevar and J. Levec, *J. Catal.*, 2004, **222**, 87–99.
41 H. H. Voge and C. R. Adams, *Adv. Catal.*, 1967, **17**, 151–221.
42 C. Doornkamp and V. Ponec, *J. Mol. Catal. A: Chem.*, 2000, **162**, 19–32.
43 P. Mountapmbeme Kouotou, Z. Y. Tian, H. Vieker, A. Beyer, A. Gölzhäuser and K. Kohse-Höinghaus, *J. Mater. Chem. A*, 2013, **1**, 10495–10504.
44 Z. Y. Tian, P. Mountapmbeme Kouotou, A. El Kasmi, P. H. Tchoua Ngamou, K. Kohse-Höinghaus, H. Vieker, A. Beyer and A. Gölzhäuser, *Proc. Combust. Inst.*, 2015, **35**, 2207–2214.
45 H. Noller and H. Vinek, *J. Mol. Catal.*, 1989, **51**, 285–294.

 

Faraday Discussions

DISCUSSIONS

# Time and Space resolved Methods: general discussion

Martin Zanni, Jemmis E D, Sankarampadi Aravamudhan, Anuradha Pallipurath, Elangannan Arunan, Christoph Schnedermann, Ashok Kumar Mishra, Mark Warren, Jonathan D. Hirst, Franklin John, R. Pal, John R. Helliwell, Kiran Moirangthem, Shamik Chakraborty, Arend G. Dijkstra, Priyadarshi Roy Chowdhury, Kenneth Ghiggino, R J Dwayne Miller, Stephen Meech, Himani Medhi, Mahesh Hariharan, Freek Ariese, Alison Edwards, Ajith R. Mallia, Siva Umapathy, Martin Meedom Nielsen, Neil Hunt, Zhen-Yu Tian, Jonathan Skelton, Gopinathan Sankar and Debabrata Goswami

DOI: 10.1039/c5fd90017d

**John R. Helliwell** opened a general discussion of the paper by **Franklin John**: I think your paper didn't prove encapsulation, can you please summarise your poster results? *I.e.* as they truly prove the encapsulation of the drug.

**Franklin John** responded: The *in vitro* release profile indicated that nearly 80% of SCR7 was released within 4 days. In 24 h, which is the time scale of the cytotoxic studies, *in vitro* release showed 50% drug release.

The best fit values of the Dynamic Light Scattering (DLS) data indicate the formation of micelles with a hydrodynamic diameter of 22.8 nm with a poly-dispersity index (PDI) of 0.188. Further evidence for drug encapsulated micelles was obtained from Small Angle Neutron Scattering (SANS) measurements.

Analysis of the SANS data indicated the formation of micelles with a core radius of 35 Å and a shell thickness of 46 Å.

**John R. Helliwell** asked: What is the anticipated timescale of the release of the anticancer agent from its capsule?

**Franklin John** responded: Nearly 80% release occurred *in vitro* within 4 days. In 24 h, about 50% release was observed.

**Siva Umapathy** enquired: How do we know that the encapsulated drug goes in and out of the cell membrane and then further into the nuclear membrane. What is the mechanism that induces the drugs to enter the cell?

**Franklin John** answered: Further cytotoxic assays and a comet assay would possibly answer this question. *In vivo* cell imaging can also give some insight. The mechanism of the drug entry is not understood yet.

**Stephen Meech** commented: In Fig. 1(a) of your paper there is a large blue shift between DMSO and P123/DMSO. What is the origin of this blue shift? Can you relate it to the molecular structure of SCR7. Does it indicate solvatochromism? If so, have the solvent dependent spectra been reported?

**Franklin John** replied: We have to further investigate that. The molecular structure of SCR7 may not have an influence on this blue shift.

**Ashok Kumar Mishra** remarked: The blue shift of the fluorescence maximum in Fig. 1(a) could be due to the presence of SCR7 in a nonpolar environment in the P123 copolymer. A regular solvatochromism study of SCR7 using some solvents of different polarity could help understanding the blue shift.

**Franklin John** answered: Thank you so much for the valuable comment.

(305:[306]306) **Sankarampadi Aravamudhan** addressed **Franklin John**:

(i) The hydrophobicity of SCR7 decreases its bioavailability which is a major setback in the utilization of this compound as a therapeutic agent. See page 1 lines 13–15 of your paper.

(ii) Polyethylene glycol (PEG)19 based block copolymers have the distinct advantage as compared to other delivery systems due to their ability to encapsulate large amounts of a drug. See page 2 lines 16–18 of your paper.

(iii) The shift in the $\lambda_{max}$ of the fluorescence spectrum is an indication of the interaction between SCR7 and the copolymer P123. See page 2 lines 26 and 27 of your paper.

(iv) This indicates the ability of P123 in encapsulating the drug and its release upon treatment with hydrophobic solvents like chloroform. See page3 lines 5 and 6 of your paper.

The *in vitro* study reported in your paper establishes the aspects which are useful criteria for favouring the bioavailability of a drug for *in vivo* applications. If the drug : capsulating copolymer ratio is 1 : 40, (if this value is as it was in the presentation materials) then, is this in accordance with the "large amounts of" drug as stated in point 2 above?

The actual *in vivo* application of the drug would require bioavailability at the site of the cancerous cells at the specific organ tissues. Even though drug movement and transport in living systems is facilitated by encapsulation as the drug is not exposed to hydrophobic interactions, the drug being targeted to the specific affected region and released at the site does not seem to have been ensured by the encapsulation, inferring from the reported results in accordance with the current *in vitro* studies. Maybe there were certain *in vitro* studies for the conditions of several organs reported during the presentation. However, nothing much conclusive was said which is reassuring that those would ensure proper functioning in *in vivo* conditions. More clarifications/citation from earlier studies could be convincing for this aspect: *i.e.* efficacy of these encapsulating materials evidenced from *in vitro* studies.

 

**Franklin John** answered:

(i) Bioavailability: Pluronic copolymers are a well characterized system for the encapsulation and delivery of hydrophobic drugs.

(iii) Regarding the shift in the fluorescence spectrum, this shift may be due to the result of a change in the micro environment of SCR7 within the hydrophobic core of the micelles.

(iv) The ability of P123 to encapsulate the drug – page 3 lines 5 and 6 in "large amounts". This is in accordance with the statement in point 2 "*in vivo* applications". An extensive *in vivo* study has been published in the paper by Franklin John *et al.*[1]

1 Franklin John *et al.*, *Macromol. Biosci.*, 2014, DOI: 10.1002/mabi.201400480.

**Anuradha Pallipurath** asked: Apart from drug release studies, have you carried out any drug dissolution studies? Smaller crystallites are known to dissolve much faster and drugs are generally classified under the Biopharmaceutical classification system based on their solubility and permeability. Do you have an idea about what the crystallite size might be, of the drug in the carrier?

**Franklin John** responded: Drug dissolution studies? No, drug dissolution studies have not been done so far. Crystallite size? This has to be further investigated.

**Kenneth Ghiggino** remarked: The fluorescence emission spectrum of the inhibitor SCR7 shows a significant blue shift upon encapsulation in the P123 copolymer compared to DMSO solvent. Have you used this fluorescence shift to follow the efficiency and rate of controlled drug release? If not, could you speculate on the usefulness of fluorescence to monitor the drug release?

**Franklin John** replied: We haven't followed the fluorescence shift to follow drug release. We are planning to study the solvatochromic shift by different solvents. We hope that this will give more insight into drug encapsulation.

**John R. Helliwell** remarked that an alternative means for drug encapsulation would be a protein. An example can be found here.[1]

1 O. K. Gasymov, A. R. Abduragimov, E. O. Gasimov, T. N. Yusifov, A. N. Dooley, B. J. Glasgow, Tear lipocalin: potential for selective delivery of rifampin, *Biochim. Biophys. Acta*, 2004, 1688(2), 102–111.

**John R. Helliwell** opened a discussion of the paper by Neil Hunt: In your paper please explicitly indicate what are the differences between the peaks displayed in the frames in Fig. 3 and 4. Presumably it is the change in extension of the peaks from bottom left to top right of each peak in each frame?

**Neil Hunt** answered: The spectra differ in terms of the shape and diagonal elongation of the peaks in the spectra as a function of waiting time. The NO stretching absorption is inhomogeneously broadened by fluctuations of the molecular environment within the haem pocket. When the dynamics of these fluctuations take place on timescales that are slower than the experimental waiting time, this causes a correlation between pump and probe frequency at early waiting times. This is lost as the waiting time increases because spectral diffusion occurs on the timescale of the underlying dynamics. The effect is described in more detail in the references contained in the text and below.[1–4]

1 N. T. Hunt, *Dalton Trans.*, 2014, **43**, 17578–17589.
2 K. Adamczyk *et al.*, *Meas. Sci. Technol.*, 2012, **23**, 062001.
3 N. T. Hunt, *Chem. Soc. Rev.*, 2009, **38**, 1837–1848.
4 P. Hamm and M. Zanni, *Concepts and Method of 2D Infrared Spectroscopy*, Cambridge University Press, Cambridge, 2011.

**Freek Ariese** asked:

(i) You are using NO as a sentinel compound to provide information on other ligands binding to the central $Fe^{3+}$ ion. To what extent could the presence of NO itself be influencing the binding processes?

(ii) Would it be possible to study these systems also with resonance Raman? And, if someone has already done that, were the results in agreement with your IR studies?

**Neil Hunt** responded:

(i) Based on the crystal structures of HRP with BHA bound in the substrate binding pocket, the effect of a diatomic haem ligand is minimal. In the case that no haem ligand is present there is evidence for a single water molecule being loosely bound to the haem when BHA is present (Ref. 5,6,17 in the main text). No structural data is available for the nitrosylated protein but if a CN ligand is introduced at the haem then the substrate binding is unaffected except for a rotation of the head group of BHA to accommodate the optimal H-bonding interactions with ligand and protein (Ref. 5 in the main text).

(ii) Haem proteins have been widely studied using resonance Raman and this technique provides valuable data on the vibrational modes of the haem. In our study, we were interested in the presence and dynamics of water near the haem and its interactions with a ligand that mimic in some way the biological binding of hydrogen peroxide that is central to the enzyme mechanism. As a result, resonance Raman is not directly applicable to the problem studied but does provide useful complementary information.

**Christoph Schnedermann** commented: By using benzohydroxamic acid you have chosen a non physiological substrate for your investigation. Did you measure other substrates to verify if your conclusions are transferable, *i.e.* are also valid for physiological substrates?

**Neil Hunt** replied: The physiological substrate is unfortunately not known but HRP has been shown to accommodate a large number of diverse organic molecules in the substrate binding position. We did not measure any other substrate candidates in this instance. Our conclusion that the water is excluded from the region near the haem ligand by the occurrence of competitive H-bonding between BHA and nearby residue side chains of HRP is likely to be important in the transferability of the effect. Substrates that are able to replicate this H-bonding would be expected to show similar behaviour whereas structurally very different molecules may interact with the haem ligand *via* a different process. These would be interesting experiments.

**Jonathan D. Hirst** remarked: To investigate the solvent accessibility of the haem binding pocket further, are there complementary techniques, such as hydrogen–deuterium exchange NMR measurements[1] that would reveal differences for residues in the pocket?

1 Englander and Kallenbach, Hydrogen exchange and structural dynamics of proteins and nucleic acids, *Q. Rev. Biophys.*, 1984, **16**(4), 521–655, DOI:10.1017/S0033583500005217.

**Neil Hunt** answered: There are some well-established methods using isotope substitution and labelling such as those mentioned (and reported) by Zanni, which can be used to provide site-specific information in the 2D-IR spectroscopy of proteins. In the case of HRP, these might be challenging to employ from the point of view of preparing the protein due to its large size but could, in principle, shed more light on the issues raised by our studies. Realistically, single point mutations are more straightforward to achieve in the case of HRP and can be valuable, but provide less direct information than single residue labelling.

**Martin Zanni** continued the discussion: **Jonathan D. Hirst**'s suggestion was interesting. I don't know if H/D has been done with mass spec, but I'd expect that it would have by now if it were possible. It might be possible with 2D IR. We published a paper[1] in JPC for the Fayer special issue a little while ago in which we measured H/D exchange with 2D IR for a single residue in a protein. Might be useful here. How many histidines are in your protein?

1 E. B. Dunkelberger, A. M. Woys and M. T. Zanni, 2D IR Cross Peaks Reveal Hydrogen–Deuterium Exchange with Single Residue Specificity, *J Phys Chem B.*, 2013, **117**, 15297.

**Neil Hunt** responded: This is an interesting suggestion, the use of isotope labelling has certainly been shown to be a powerful tool for extracting residue-specific information from 2D-IR spectroscopy on proteins and related systems. The sequence of HRP shows that there are 3 His residues. All three are located within the region close to the haem however, which may limit the amount of specific insight that one could extract from isotope labelling unless the three could be labelled differentially.

**Jonathan D. Hirst** replied: The H/D exchange experiments I had in mind were ones done with NMR.[1]

1 Englander and Kallenbach, Hydrogen exchange and structural dynamics of proteins and nucleic acids, *Q. Rev. Biophys.*, 1984, **16**(4), 521–655, DOI:10.1017/S0033583500005217.

**John R. Helliwell** asked: Please explain why you have chosen horseradish as the source of your enzyme?

**Neil Hunt** answered: This enzyme is available from commercial sources and is widely taken to be an archetypal example of the peroxidase enzymes.

**Franklin John** enquired: Is it known that Fe–Horse Raddish peroxidase interacts with carbohydrate binding proteins, especially lectins?

**Neil Hunt** responded: We have focussed upon the binding of organic substrates to HRP and so are not in a position to comment on its interactions with other proteins.

**John R. Helliwell** asked: Can you please tell us about the dynamics of the horseradish active site and how the water is displaced to its new position?

**Neil Hunt** replied: The dynamics of the active site are described by the observed spectral diffusion, as shown in Fig. 5 and Table 1 and take place on 4–6 ps timescales alongside a slower component that is unresolved by the 2D-IR experiment because the NO probe has a lifetime of 16.5 ps. The mechanism of water exclusion is concluded to be replacement by competitive H-bonding of the hydrophilic head-group of BHA to the important distal residue side chains near the haem centre.

**R J Dwayne Miller** remarked: The fluctuations of the protein involve correlated motions over different length and time scales, with motions as small as 0.1 Å being highly relevant given the integrated effect over the large number of residues involved in collective mode fluctuations. These timescales and relevant motions coupling to water channels, and water accessibility, may be outside the dynamic range probed within the lifetime of the NO reporter mode. I realize the main focus is using 2D IR to determine the effects of spectral diffusion in comparing with substrate and without substrate binding as a probe of water within the binding site. I am wondering whether the NO vibrational mode is an orthogonal coordinate to the important protein fluctuations gating water accessibility.

**Neil Hunt** answered: In this case, the 2D-IR of the NO stretching vibration was used as a probe of the local chemical environment near the haem ligand and, specifically, to determine whether water was present. In a more general sense, the NO vibrational mode, or that of any other haem ligand, is not expected to be related to water accessibility. However, the NO stretching vibration is analogous to the O–O motion that will result in $H_2O_2$ cleavage during the formation of **Compound1** in the peroxidase and catalase mechanisms. Thus, a more interesting question pertaining to these systems is how water may influence that particular motion, which may be linked to the reaction coordinate. Here we cite recent work on the closely-related catalase enzyme[1] where coherent oscillations in the IR pump–probe signal of the haem-bound NO stretching vibration indicated coupling of a well-defined low frequency vibrational mode to the haem ligand motion. This was attributed to motions coordinated by the water network present in the catalase active site and suggests that water motion may indeed influence the reaction coordinate.

1 K. Adamczyk *et al.*, *Chem. Sci.*, 2015, **6**, 505–516.

**Martin Zanni** enquired: The question has been raised whether the lifetime of the chromophore is long enough to measure functionally important motions. But the question you are asking, "is there water?" is a straightforward question for the dynamics that you are measuring.

**Neil Hunt** answered: I would agree with this observation; the lifetime of the NO stretching vibration was observed to be ~16.5 ps. This is long enough to observe water-related dynamics, which occur on timescales of 1-2 ps in bulk solution and which have been shown to increase slightly under dynamically-restricted conditions such as in the active site of enzymes[1] or in reverse micelles.[2]

In the case of the submitted article, the lack of dependence of the vibrational lifetime upon H/D exchange of the solvent was sufficient to address the question of whether there is direct interaction between the haem ligand and water molecules.

1 Adamczyk *et al.*, *Chem. Sci.*, 2015, **6**, 505–506.
2 Fayer *et al.*, *J. Phys. Chem. B*, 2011, **115**, 11658.

**Arend G. Dijkstra** opened a general discussion of the paper by Jonathan Skelton: In your presentation, you compare several methods to simulate quantum dynamics. You show that there is no good method to simulate non-adiabatic dynamics in crystals. Could you comment on how to proceed in this direction?

**Jonathan Skelton** replied: Performing non-adiabatic dynamics simulations to model excited-state dynamics in periodic systems is challenging due to the fact that they need to be coupled to time-dependent density-functional theory (TD-DFT) calculations in order to model the excited states. TD-DFT calculations generally require a high level of theory to yield acceptable results (*i.e.* at least a hybrid functional such as PBE0[1] or B3LYP,[2] rather than a standard (*meta-*)GGA such as PBE[3] or TPSS[4] ), but the computational cost of using such functionals for dynamics simulations is currently prohibitive, at least for the large unit cells (~100+ atoms) common to molecular crystals.

In addition to this issue, at the time of writing the infrastructure required for non-adiabatic dynamics simulations is not widely implemented in solid-state computational chemistry codes. Those interested in the details and issues involved in implementing TD-DFT, including dynamics, in a periodic code may wish to look at the reports from a pair of projects led by the UK national HPC service to add this functionality to the CASTEP code.[5] At present, there are really only two viable options for non-adiabatic dynamics simulations. If the crystal environment can be represented well by a dielectric continuum, one could adopt the approach tested in this work, and perform molecular calculations using a continuum model. In other cases, an embedding approach, where the active part of the crystal is modelled at an appropriate level of theory and the rest treated more approximately, is probably the most viable way to proceed. A particularly interesting paper from Kochman *et al.* shows how a molecular TD-DFT calculation can be embedded within a periodic DFT calculation to model the dynamics of a single photoexcited molecule in the crystallographic unit cell.[6]

1 Adamo and Barone, *J. Chem. Phys.*, 1999, **110**, 6158, DOI: 10.1063/1.478522.
2 Becke, *J. Chem. Phys.*, 1993, **98**, 1372, DOI: 10.1063/1.464304.
3 Perdew *et al.*, *Phys. Rev. Lett.*, 1996, 77, 3865, DOI: 10.1103/PhysRevLett.77.3865.
4 Tao *et al.*, *Phys. Rev. Lett.*, 2003, **91**, 146401, DOI: 10.1103/PhysRevLett.91.146401.
5 http://www.hector.ac.uk/cse/distributedcse/reports/castep02/castep02.pdf, http://www.hector.ac.uk/cse/distributedcse/reports/castep03/castep03.pdf.
6 Kochman *et al.*, *J. Chem. Theor. Comput.*, 2013, **9**(2), 1182, DOI: 10.1021/ct3008149.

**Siva Umapathy** asked: In your periodic calculation, the orientation of the solvent in the crystal structure and the location would make a difference to the results of the calculation. How did you decide the orientation and distance?

**Jonathan Skelton** responded: I presume this question refers to how the orientation and distance of neighbouring molecules in the crystal lattice, which will play a role in defining the dielectric environment of the component species, are accounted for when using continuum models to approximate this in molecular calculations.

In a general continuum model, the shape and orientation of the solvent are parameters which play a role in defining the shape of the cavity in which the molecule sits, and hence the interaction of the screening charges on the cavity surface with its charge-density cloud. Different variants of the continuum model approach this in different ways.

In the COSMO model,[1] which we used in the majority of our work, the only solvent parameter is the dielectric constant. The details of how the cavity is constructed are implementation dependent, but in most cases it is done by assembling atom-centred spheres around the molecule, whose size is based on the van der Waals' radii. In the implementation of COSMO in the NWChem code,[2] the van der Waals' radii of H, C, N and O are the optimal values determined by Klamt *et al.*,[3] while the radius of Ni is taken to be 2.223 Å.

In our work, we also performed selected calculations with the Gaussian code, using its polarisable-continuum model. The model implemented in Gaussian contains more parameters than COSMO, and the shape and orientation of the solvent can play a role in defining the cavity, *e.g.* by defining the solvent-accessible surface. In the present calculations, we did not attempt to optimise these parameters, and instead selected a pre-defined solvent with a suitable dielectric constant. While this is obviously a considerable approximation, it seemed to work reasonably well in this case.

In more general terms, it might be possible to improve the approach presented in our paper by optimising these additional parameters for the crystalline environment; this might be thought of as an intermediate step between the COSMO model and an atomistic embedding scheme.

1 Klamt and Schüürmann, *J. Chem. Soc., Perkin Trans. 2*, 1993, 799, DOI: 10.1039/P29930000799.
2 http://www.nwchem-sw.org/index.php/COSMO_Solvation_Model.
3 Klamt *et al.*, *J. Phys. Chem. A*, 1998, **102**(26), 5074, DOI: 10.1021/jp980017s.

**Alison Edwards** commented: In your paper you note "missing peaks" – as predicted from your calculations in the 200 nm region. Display of the UV-vis spectrum in the dominant linear in wavelength mode may be limiting your ability to observe the peaks you expect. Have you considered presenting and analysing your spectra linear in frequency?

Your paper states that three of the peaks you have observed are symmetry forbidden for the molecule in question. Is it not the case that your molecule is not symmetric and thus the peaks observed are in fact precisely those which would be expected for this d8 octahedral case?

 

**Jonathan Skelton** answered: To clarify, the absent peak we refer to in the section on optical properties is a broad feature at 650 nm in the experimental spectrum.[1] In the molecular TD-DFT calculation, electronic transitions close to this wavelength are found, but are predicted to have negligible oscillator strength. It is possible that transitions at longer wavelengths than the main absorption tail are also found in the periodic calculation, but without explicit excited states, as are obtained from molecular calculations, this is difficult to assess.

The explanation given for this in the paper – that the transitions are forbidden by symmetry – is incorrect, as you note, since the complex is not centrosymmetric. However, the suggestion that the transitions are otherwise forbidden in the equilibrium geometry, but become weakly allowed when coupled to vibrations at finite temperature, may nonetheless be a possibility. An alternative explanation could be that reproducing the finite oscillator strength observed experimentally requires a more accurate description of many-body effects, which are missing from the adiabatic TD-DFT approach we took in our molecular calculations (see also the discussion in response to question 316).

Your suggestion of presenting spectra linear in frequency (*i.e.* energy) rather than wavelength is a good one; indeed, for many periodic codes, this is the default output. As you note, however, plotting the spectra linear in wavelength is the more common practice among experimentalists. For comparison, we reproduce our Fig. 8 from the paper here with the *x*-axis in energy units (Fig. 1); doing so in effect spreads out the peaks, allowing the fine structure, and also the subtle differences between the spectra of the two isomers, to be more easily discerned.

1 Hatcher *et al.*, *Chem.–Eur. J.*, 2014, **20**(11), 3218, DOI: 10.1002/chem.201304172.

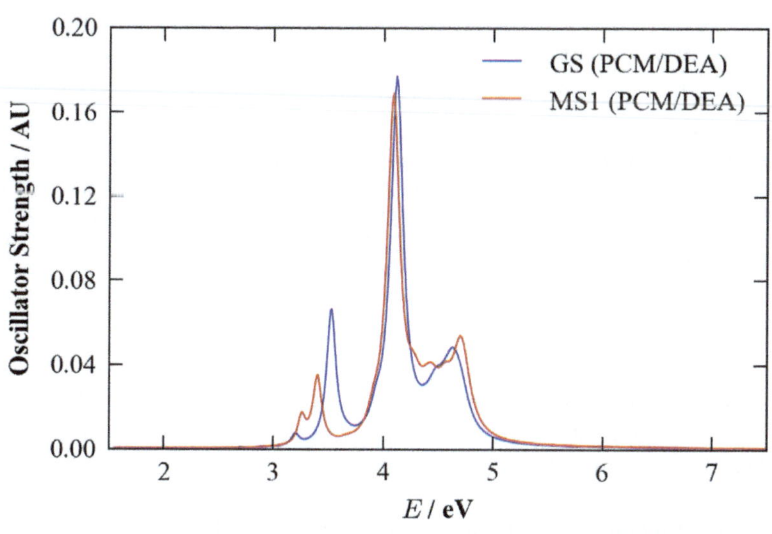


Fig. 1 Simulated UV-visible spectra of the ground-state (GS) and metastable-state (MS1) isomers of the $[Ni(Et_4dien)(\eta^2\text{-}O,ON)(\eta^1\text{-}NO_2)]$ system. The spectra were modelled using linear-response TD-DFT with the M06 functional and a continuum of diethylamine ($\varepsilon_{static} = 3.6$), and the peaks were broadened in wavelength space using a Lorentzian function with a full-width at half-maximum (FWHM) of 10 nm.

**John R. Helliwell** said: On page 19 of your paper Fig. 9 – regarding the dielectric constant of 30 behaviour, you have a lot resting on the toluene data point and no data point between that and the ethanol data point. Are there examples to improve the sampling of such dielectric data values?

**Jonathan Skelton** responded: We have computed additional points along this part of the curve, and attach a revised figure including these data.

With the extra points, there is still some apparent correlation between the energy differences and the dielectric constant, but the new data suggest that the energy difference is most sensitive over a considerably smaller range of values of $\varepsilon_{static}$ than the 4–24.5 range suggested by the more limited data in Fig. 9 of our paper (Fig. 2).

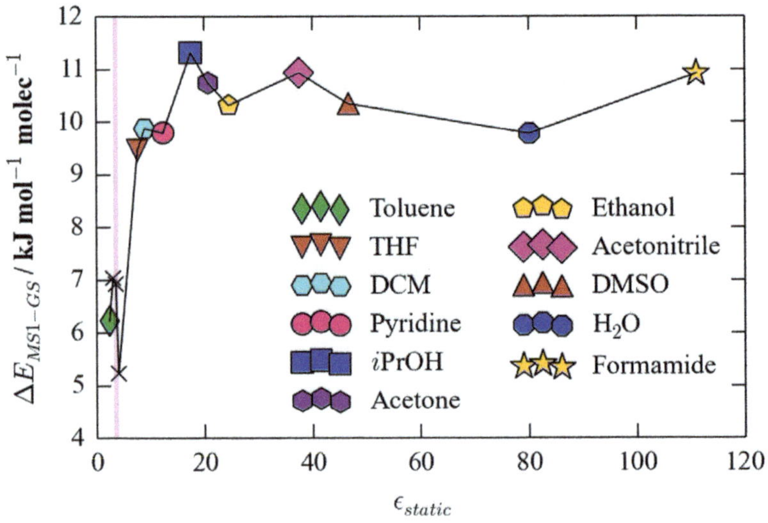


Fig. 2 Dependence of the energy difference between the ground-state (GS) and metastable-state (MS1) isomers of the [Ni(Et$_4$dien)($\eta^2$-O,ON)($\eta^1$-NO$_2$)] linkage–isomer system on the dielectric constant, $\varepsilon_{static}$, used in the continuum model. As in Fig. 9 of our paper, the black crosses in the shaded region correspond to values of $\varepsilon_{static}$ obtained from the periodic calculations, and points outside this region are labelled with solvents with comparable dielectric constants.

**John R. Helliwell** remarked: On page 12 of your article you use 3 and 2 decimal places then integers for the dielectric constant values you quote. What precision for dielectric constant values do you in fact have?

**Jonathan Skelton** replied: I make the distinction here between "precision" and "accuracy", where the former is set by the level of numerical noise in the simulations, and the latter refers to how quantitatively the calculations can reproduce experimental measurements. We applied a fairly tight convergence criterion for technical parameters such as the basis set, and in testing we found that changing less important parameters changed the calculated dielectric constants in the fifth decimal place. We therefore expect that the dielectric constants quoted in Tables 2 and 3 are precise to at least 3 decimal places (as quoted) for the given functionals and pseudopotential set. The accuracy of these values is more difficult to assess, as the dielectric constant of the molecular crystal has not yet been measured. Considering the optical absorption profile computed from the solid-state hybrid-functional calculations (Fig. 7), the agreement with the experimental spectrum[1] is not good, which could indicate issues in the underlying electronic structure, and hence perhaps significant error in the electronic-polarisation component of the dielectric constant. If we take the difference between the GGA and hybrid values as an indication of the likely order of magnitude of this, it would be more like $10^0$–$10^{-1}$ in this case. According to our calculations, ionic relaxation makes a more minor contribution to the dielectric constant in this material, but comparing the difference between the PBE and PBEsol values, and assuming the latter functional generally yields more accurate phonon frequencies compared to experiment, then again the accuracy would be on the order of $10^{-1}$. There has been a recent study in which the dielectric constant of the hybrid halide perovskite $(CH_3NH_3)PbI_3$ was calculated using a similar approach to that taken in this work.[2] Comparison with the experimental measurements[3] is complicated in this case by the fact that the cubic phase of the material on which the calculations were performed is only observed above ~300 K, whereas the calculations were done on the optimised 0 K structure; the smaller lattice volume is likely to lead to stiffer phonons,[4] and hence a reduced ionic-relaxation contribution to the dielectric constant. With this in mind, the calculations yielded a value of 25.7, compared to the experimental measurement of around 32 at 300 K. If used for a direct comparison, the accuracy here is approximately to within 20–30%. Based on these considerations, we suggest that the dielectric constants we calculated for the Ni–NO2 complex we studied are probably accurate to within ~0.5–1 unit, but again this is impossible to quantify in the absence of experimental measurements.

1 Datta *et al.*, *Chem.–Eur. J.*, 2014, **20**(11), 3218, DOI: 10.1002/chem.201304172.
2 Brivio *et al.*, *APL Mat.*, 2013, **1**, 042111, DOI: 10.1063/1.4824147.
3 Poglitsch and Weber, *J. Chem. Phys.*, 1987, **87**, 6373, DOI: 10.1063/1.453467.
4 Skelton *et al.*, *Phys. Rev. B*, 2014, **89**, 205203, DOI: 10.1103/PhysRevB.89.205203.

**R J Dwayne Miller** said: I think this is extremely important work. The recent advances in X-ray and electron sources open up the use of time resolved crystallography to give an atomic view of reaction dynamics. Theory is needed to understand the reaction forces involved and you have pointed out the challenges and highlighted important test beds for checking the accuracy of quantum-chemical techniques with various levels of approximation. Ultimately, we need to get the excited state potential energy surface. An important check is to calculate the absorption spectra. The spectral width indicates whether coupling of the photoactive chromophore to the surrounding lattice is properly captured. You mention that the calculated spectra do not get the broad features at 650 nm, using a dielectric continuum model for the surrounding lattice. Can you comment on what is needed to improve the calculations? I expect the continuum model may not give you the right spectral widths but it seems some electronic states are missing.

What do you think is currently the best theoretical method (with current state of the art computational facilities) to theoretically treat reaction dynamics in crystals? There is so much information from these studies. Experimentally, we now effectively have atomic level details for both the system and the system–bath interaction. We really need theory to extract the details related to the reaction forces.

**Jonathan Skelton** responded: We thank you for your comment, and we agree that combining state-of-the-art experimental work and theoretical modelling has the potential to be a powerful combination for characterising reaction dynamics.

As discussed in the paper, from the output of the TD-DFT calculation it is not that these electronic states are missing, but rather that they are predicted to have zero oscillator strength, and therefore do not appear as peaks in the simulated spectrum. There are two possible reasons for this. One is that these electronic transitions are formally forbidden at 0 K, but could become weakly allowed at finite temperature due to thermal motion leading to structural perturbations. The second is that an accurate description of these transitions requires many-body effects (*e.g.* exciton formation) which are missing from the adiabatic TD-DFT approach we took in our work. If this is the case, a higher level of theory would be required to reproduce the finite oscillator strengths.

Incidentally, I note that we did not attempt to calculate the spectral widths in any of our absorption spectra. In principle, this could be done, for example by performing a vibrational calculation and averaging the absorption profiles over realistic displacements along the modes, but this would be an expensive calculation.

In my opinion, the best method to treat reaction dynamics in crystals depends on the nature of the reaction.

Modern periodic codes provide a suite of tools suitable for characterising thermally-activated processes, for example molecular dynamics, lattice dynamics and transition-state modelling techniques such as the (nudged) elastic band method. An example of some of these being used for the system discussed in the paper can be seen in our recent work.[1] As discussed in the answer to question 311, modelling photochemical reactions is considerably more challenging, as it requires the exploration of the reaction energy surface to be coupled to a time-dependent DFT (TD-DFT) calculation to model the electronic excited state of the system. At present, the TD-DFT support in most periodic codes does not usually extend to the computation of forces, plus the level of theory needed to accurately model electronic excitations is often prohibitively expensive for dynamics calculations, at least for molecular crystals with large unit cells (*i.e.* 100+ atoms). For photochemical reactions, it is my feeling that the most feasible approach currently would be to use an embedding technique. Some development may be needed in this area to make the method more accessible, but I feel that QM/QM methods,[2] which retain the advantages of *ab initio* techniques over parameterised force fields while keeping the computational cost manageable, represent a promising future direction.

1 Walsh *et al.*, *CrystEngComm*, 2015, **17**, 383–394, DOI: 10.1039/C4CE01411A.
2 Kochman and Morrison, *J. Chem. Theory Comput.*, 2013, **9**(2), 1182, DOI: 10.1021/ct3008149, for example.

 

**Sankarampadi Aravamudhan** asked: I refer you to the "coarse graining" technique in Computational Chemistry by Warshel and co-authors[1] and the Nobel Lecture of Arieh Warshel on multiscale modelling in computational chemistry.[2] The techniques described by Jonathan Skelton are similar to the discrete lattice sum in the immediate vicinity and applying continuum description with effective parameters substituted. The Clausius–Mosetti equation is a well known context.[3–5] Also consider the contents of the links in which I have made reference to specific water interactions with biological sites and the general bulk water disposition.[6–8]

1 Shina C.L. Kamerlin, Spyridon Vicatos, Anatoly Dryga, and Arieh Warshel, Coarse-Grained (Multiscale) Simulations in Studies of Biophysical and Chemical Systems, *Annu. Rev. Phys. Chem.*, 2011, **62**, 41–64.
2 A. Warshel, Multiscale Modeling of Biological Functions: From Enzymes to Molecular Machines (Nobel Lecture), *Angew. Chem., Int. Ed. Engl.*, 2014, **53**, 10020–10031.
3 http://nehuacin.tripod.com/id5.html.
4 http://nehuacin.tripod.com/sitebuildercontent/sitebuilderpictures/abstract_45pc.jpg.
5 http://nehuacin.tripod.com/sitebuildercontent/sitebuilderpictures/358_18april2008.jpg.
6 http://www.ugc-inno-nehu.com/DBIBT.ppt.
7 http://www.ugc-inno-nehu.com/icetcs-cug2013/0-2-Specific-Water-interct-Amino/.
8 http://www.ugc-inno-nehu.com/icetcs-cug2013/0-3-HydroniumIon/.

**Jonathan Skelton** responded: Thankyou for your comment, and for these interesting references. I think the techniques presented therein are essential components in "multiscale" modelling, a research direction which is increasingly being explored in the theoretical literature. This puts a broader perspective on the work presented in our paper: whereas our focus here was really on describing the properties and behaviour of the component species in a molecular crystal, one can see how this information might ultimately be used to build a coarse-grained model to look at larger-scale behaviour, such as the nucleation and spread of the excited-state species within the crystal lattice.

**John R. Helliwell** commented: You presented a challenges matrix for computer simulations. One of these challenges is the length of time for a molecular dynamics (MD) simulation. A recent paper[1] containing the MD simulation of a protein lasting one microsecond, is a very impressive result and a demonstration of the power of computer hardware today compared with my own results of 15 years ago or so comprising 10 nanoseconds length of time simulations for similar sized proteins (see references below).[2,3] Indeed it shows the beneficial change that computer hardware can bring us over that number of years. If you assume a Moore's law further progression of computer calculation capability, in 20 years time are those extremely difficult challenges (your red boxes) still going to be red at that future time?

1 M. E. Wall, A. H. van Benschoten, N. K. Sauter, P. D. Adams, J. S. Fraser and T. C.Terwilliger, Conformational dynamics of a crystalline protein from microsecond-scale molecular dynamics simulations and diffuse X-ray scattering, *Proc. Natl Acad. Sci. USA*, 2014, **111**, 17887–17892.
2 G. M. Bradbrook, T. Gleichmann, S. J. Harrop, J. Habash, J. Raftery, A. J. Kalb (Gilboa), J. Yariv, I. H. Hillier and J. R. Helliwell, X-ray and molecular dynamics studies of concanavalin A glucoside and mannoside complexes: Relating structure to thermodynamics of binding, *J. Chem. Soc., Faraday Trans.*, 1998, **94**(11), 1603–1611.

3 J. V. Pratap, G. M. Bradbrook, G. B. Reddy, A. Surolia, J. Raftery, J. R. Helliwell and M. Vijayan, The combination of molecular dynamics with crystallography for elucidating protein–ligand interactions: a case study involving peanut lectin complexes with T-antigen and lactose, *Acta Cryst.*, 2001, **D57**, 1584–1594.

**Jonathan Skelton** replied: The "red boxes" to which this question refers (Fig. 3) are part of a figure presented at the meeting, which is not included in the paper, and so is reproduced here for reference.

For the purpose of this discussion, I divide the tasks I identified as being intractably difficult in solid-state chemistry at present into three classes: (a) running otherwise routine calculations at the hybrid level of theory, (b) running highly-accurate electronic-structure calculations using "post-DFT/HF" methods, and (c) performing non-adiabatic dynamics simulations.

Hybrid functionals[1,2] are widely implemented in periodic codes, and with advances in both computing power and algorithmic efficiency are becoming more routinely accessible for increasingly larger systems; as a result, if computing power continues to scale favourably, I would expect most tasks in class (a) to be possible in the fairly near future.

Post-DFT methods (*e.g.* MP2,[3] CCSD(T),[4] *etc.*) are widely available in molecular codes, but are less common in solid-state codes. This can be attributed to the higher complexity of implementing them in periodic codes, and to computational cost. Recently, there has been significant interest in this area, however, in particular in using the *GW* method[5] and solving the Bethe–Salpeter equation[6] to more accurately treat many-body effects in absorption spectra. I also mention a recent paper by Booth *et al.*[7] in which the accuracy of properties of some simple bulk materials predicted by "gold-standard" electronic-structure methods were compared. If the present progress in algorithmic development continues over the next decade, then, combined with increases in available computing power, I would expect post-DFT calculations to be widely used in 20 years' time, at least for the more difficult problems for which even hybrid functionals do not yield satisfactory results.

Finally, at present TD-DFT support in periodic codes could be described as "patchy", and where it is available it is usually set up for modelling optical properties rather than for studying excited-state dynamics. Readers interested in the practicalities and issues associated with the implementation of TD-DFT and excited-state dynamics are referred to the references given in the response to question 311. In my opinion, therefore, progress in this area is likely to be limited by development effort more than by advances in computing power, and for this reason I am not sure whether these calculations will become more widely possible within a ten-year timeframe.

With regard to standard ("ground-state") molecular dynamics, this is a very versatile technique which has benefitted both from advances in computing power and developments in algorithmic efficiency, as is nicely illustrated by the contrast in the references given in the question. Over the next 20 years, I would expect the length- and timescales accessible to MD to continue to grow with computing power, and, in particular, for *ab initio* molecular dynamics, *i.e.* simulations using first-principles quantum-mechanical methods such as DFT, to become more accessible, helping to offset the challenges which are sometimes associated with parameterising force fields to study complex systems.

1 Adamo and Barone, *J. Chem. Phys.*, 1999, **110**, 6158, DOI: 10.1063/1.478522, for example.
2 Krukau *et al.*, *J. Chem. Phys.*, 2006, **125**, 224106, DOI: 10.1063/1.2404663, for example.
3 Møller and Plesset, *Phys. Rev.*, 1934, **46**, 618, DOI: 10.1103/PhysRev.46.618.
4 Pople *et al.*, *J. Chem. Phys.*, 1987, **87**, 5968, DOI: 10.1063/1.453520.
5 Hedin, *Phys. Rev.*, 1965, **136**, A796, DOI: 10.1103/PhysRev.139.A796.
6 Salpeter and Bethe, *Phys. Rev.*, 1951, **84**, 1232, DOI: 10.1103/PhysRev.84.1232.
7 Booth *et al.*, *Nature*, 2013, **493**, 365, DOI: 10.1038/nature11770.

| Property | LDA | GGA | Meta-GGA | Hybrid | Post-DFT/HF |
|---|---|---|---|---|---|
| **Periodic Calculations** | | | | | |
| Equilibrium structure/geometry | ☺ | ☺ | ☺ | ☹ | - |
| Energetics | ☺ | ☺ | ☺ | 😐 | ☹ |
| Structural properties (e.g. moduli) | ☺ | ☺ | ☺ | ☹ | - |
| IR intensities | ☺ | ☺ | ☺ | ☹ | - |
| Raman polarisabilities | 😐 | 😐 | 😐 | ☹ | - |
| Dielectric properties | ☺ | ☺ | 😐 | 😐 | ☹ |
| Electronic excitations | 😐 | 😐 | 😐 | ☹ | ☹ |
| Non-adiabatic dynamics | ☹ | ☹ | ☹ | ☹ | - |
| **Molecular Calculations** | | | | | |
| Equilibrium structure/geometry | ☺ | ☺ | ☺ | ☺ | 😐 |
| Energetics | ☺ | ☺ | ☺ | ☺ | 😐 |
| Structural properties (e.g. moduli) | - | - | - | - | - |
| IR intensities | ☺ | ☺ | ☺ | ☺ | 😐 |
| Raman polarisabilities | ☺ | ☺ | ☺ | ☺ | 😐 |
| Dielectric properties | - | - | - | - | - |
| Electronic excitations | ☺ | ☺ | ☺ | ☺ | 😐 |
| Non-adiabatic dynamics | 😐 | 😐 | 😐 | 😐 | 😐 |

Fig. 3 Matrix illustrating the relative complexity of calculating some common properties at five different levels of theory in periodic (upper rows) and molecular (lower rows) calculations. The levels of theory run from the local-density approximation (LDA; most approximate/least expensive) to "post-DFT/HF" methods (least approximate/most expensive). Cells are colour-coded according to whether the property calculation at that level of theory is routine (green), difficult (yellow), or intractable (red). Combinations which are unlikely to be used together are shaded in grey (*e.g.* it is not possible to compute bulk structural properties such as elastic moduli from molecular calculations).

**John R. Helliwell** said: How do you see the future of molecular dynamics simulations in say 20 years time?

**Jonathan Skelton** answered: This question is also discussed in the responses to questions 326 and 423.

Molecular dynamics is a powerful tool for approaching a number of problems, the main restriction being the length- and timescales accessible to simulations. Some recent examples from the literature which push the current limits are simulations of microsecond protein dynamics,[1] large-scale atomistic modelling of radiation damage in pyrochlores,[2] and *ab initio* simulations of the crystallisation of phase-change materials.[3,4] Over the next 20 years, I would expect continued increases in computing power, together with improvements in software efficiency and parallelisation, to broaden the range of problems which can be feasibly tackled with MD, perhaps making this a routine tool for modelling *e.g.* structural dynamics and crystallisation.

1 Wall *et al.*, *Proc. Natl. Acad. Sci. USA*, 2014, **111**(50), 17887.
2 Todorov *et al.*, *J. Mater. Sci.* 2007, **42** (6), 1920.
3 Kalikka *et al.*, *Phys. Rev. B*, 2012, **86**, 144113, DOI: 10.1103/PhysRevB.86.144113.
4 Skelton *et al.*, *Adv. Func. Mater.*, 2014, **24**(46), 7291, DOI: 10.1002/adfm.201401202.

**Jemmis E D** addressed **Jonathan Skelton**: This is a question prompted by the general matrix of computational developments. Experimental observations are nearing their observational limits. What is the projection for computational capability for the next 20 years? When will computations catch up to predict the details of, say, the process of crystallization? What are major bottle necks?

**Jonathan Skelton** replied: The matrix referred to in this question (Fig. 3) was presented at the discussion and is not part of the paper itself – it is reproduced here as part of the response to question 326.

I think we can expect computing power to continue to increase over the next 20 years. This will likely come about through improvements in hardware efficiency and a move from the "traditional" x86-64 architecture towards "accelerators" such as graphics processing units (GPUs) and the Intel Phi. At present, fully harnessing the power of these coprocessors usually requires major changes to codes, *i.e.* optimising core algorithms for the hardware, and hence substantial development effort, but based on some of the success stories so far this could result in step changes in capability.

I also anticipate that software development, in addition to implementing new features, will focus on improving parallelism to enable scaling to larger resources (*e.g.* CPU core counts). This could take the form of optimising the implementation of existing algorithms, or developing new ones which are inherently better suited to contemporary hardware. Parallel scaling is currently a bottleneck in a lot of codes, and prevents large problems being tackled despite the required computer power, in principle, being available.

With regard to the problem of predicting crystallization, I consider here performing molecular-dynamics simulations to model nucleation and growth from a solution or amorphous phase. The main requirements for this are large models and long simulation times, to give a realistic statistical sampling of rare events.

The requirements for modelling crystallisation with molecular dynamics will be system dependent. For some systems, however, modelling crystallisation this way is already possible. A nice example is the work done by Gale and coworkers on the crystallisation of minerals from solution.[1] A second is that small-scale *ab initio* molecular-dynamics simulations have been shown to be capable of reproducing the sub-nanosecond crystallisation of the Ge–Sb–Te alloys used in phase-change memory devices (*e.g.* rewritable DVD/Blu-Ray discs),[2] and such simulations are now a routine tool in this field.[3,4]

In general, I anticipate that, as the length- and timescales accessible to MD improve over the next 20 years, this technique will be feasible as a means to tackle an increasingly broad range of problems.

 

1 Gebauer *et al.*, *Chem. Soc. Rev.*, 2014, **43**, 2348, DOI: 10.1039/C3CS60451A.
2 Hegedüs and Elliott, *Nat. Mater.*, 2008, **7**, 399, DOI: 10.1038/nmat2157.
3 Kalikka *et al.*, *Phys. Rev. B*, 2014, **90**, 184109, DOI: 10.1103/PhysRevB.90.184109.
4 Skelton and Elliott, *J. Phys. Condens. Matter*, 2013, **25**(20), 205801, DOI: 10.1088/0953-8984/25/20/205801.

**Martin Zanni** opened the discussion of the paper by **Debabrata Goswami:** Are you confusing molecular discrimination with spatial resolution enhancement? You are looking at fluorescent lifetimes to discriminate different dyes, which is very nice, but that doesn't change the *XY* spatial resolution. Why does the *z* axial resolution change? Why is there discrimination when you introduce delays?

**Debabrata Goswami** responded: We are, in fact, using molecular discrimination as an approach for getting more spatial information. Using the molecular discrimination between two competing fluorophores, it can be argued, in the same spirit as in case of PALM[1] technique (but at much lesser photon collection time), the fluorescence discrimination of one fluorophore versus the other allows "seeing" the otherwise "not possible to be seen" fluorophore in the same illuminated focal volume. The principle of depleting the excited fluorophore population by stimulated emission is akin to STED,[2] however, performing this in a timed fashion allows us to take advantage of collecting the fluorescence from the florophore, which is otherwise not possible to be observed. Thus it can effectively lead to a better spatial information. Overall, since the entire process that we are involved in requires at least two or more interacting fields, we will see an improvement along the Fourier plane. Thus, the *z*-resolution is enhanced due to the nonlinear interaction. Similarly, since it is due to multiple fields interacting (the delayed pulse being responsible for excited state depletion with stimulated emission), it is imperative that the discrimination remains dependent on the relative pulse delays between the pulses. The possibility of relative difference in the excited state behavior of the dyes exists, which is presently being investigated by us.

1 E. Betzig, G. H. Patterson, R. Sougrat, O. W. Lindwasser, S. Olenych, J. S. Bonifacino, M. W. Davidson, J. Lippincott-Schwartz, H. F. Hess, Imaging Intracellular Fluorescent Proteins at Nanometer Resolution, *Science*, 2006, **313**(5793), 1642–1645.
2 Thomas A. Klar, Stefan W. Hell, Subdiffraction resolution in far-field fluorescence microscopy, *Opt. Lett.*, 1999, **24**(14), 954–956.

**R J Dwayne Miller** commented: You are using a time delay between pulse pairs for improved axial contrast. Since you are both stimulating absorption and emission from the system of interest, you have increased the overall number of field interactions. If these were 1-photon processes, the improved spatial resolution would be the square root in the confocal parameter over that of a single interaction. Here you are using 2-photon transitions for an overall 4-field interaction or a factor of two improvement in the confocal resolution over that of the fundamental. Is this the correct picture?

**Debabrata Goswami** answered: This is indeed the correct picture and is responsible for the narrowing of the point spread function along the Fourier plane in the *z*-direction of the image.

**R Pal** remarked: In the paper the authors claim to have increased axial resolution achievable by two photon microscopy below the diffraction barrier limited resolution. It is clearly stated in the introduction that this work and previous work found that the only confocal factor determining improved axial resolution in this form of 2PE microscopy is the time delay between pulses (line10 page 2). Based on the control experiments presented in this paper this could be a conclusion of a misinterpreted basic cross excitation *via* a 2PE absorption saturation phenomenon. They have been employing cellular dyes DAPI and Texas Red (corresponding absorption and excitation spectra are displayed in Fig. 4(c)) and claims have been made to suggest that using an excitation laser at 730 nm both have been excited by a 2PE process which suggests that they both simultaneously absorb two 730 nm photons to promote a 365 nm 1PE like photon absorption. This is extremely unlikely as supported by Fig. 4(c), DAPI will indeed undergo this excitation process (based on its 2PE cross section which needs mentioning) but Texas Red does not possess sufficient (or in this case it is almost zero) absorption below 480nm to promote such a 2PE process. The process which is observed by the researchers in their cell experiments (these cell experiments should also be detailed more alongside with calculations or corresponding 1PE observable sample thickness using the confocal set up detailed on page 4 section 2.2) is a clear cut case of unwanted cross excitation due to an unfavourable experimental set up. So the most likely case is that regardless of 730 or 750 nm 2PE excitation DAPI will be excited and the overlapping DAPI emission will cross excite the close proximity Texas RED *via* a 1PE excitation process which therefore results in observable Texas Red emission. The experiments to prove the Texas Red 2PE claims is to do an only Texas Red loaded cell experiment with a similar 2PE setup (all experimental parameters kept the same as with the DAPI–Texas Red pair experiment) and see if similar or in fact any corresponding Texas Red emission can be detected above 550 nm upon 730 nm 2PE excitation. Without this control experiment the above claim is invalid.

Despite all pulse delay details some of the audience still find it hard to understand how a 1 ps time delay could help distinguish the two dyes when their lifetime is in the ns (needs stating which dye has what lifetimes) regime. Delayed pulses will deliver overall excitation energy to the applied fluorophore in a different quantitative manner resulting in lower level promotion of absorption and subsequent lower levels of promoted emission (ns life time, fs pulses and ps delays). So 2 identically overlapped identical excitation pulses will promote higher levels of dye fluorescence than 2 identical pulses applied in a delayed manner, especially with 2PE microscopy where the promoted emission and all important confocal voxel is determined by an intensity squared manner. This can be underpinned with a simple experiment using the suggested Rhodamine 6G dye keeping all experimental parameters identical (pulse overlapping *vs.* delayed pulse as a function of detected emission). Fig. 4(b) is also underpinning the cross excitation nature as with almost linearly decreasing DAPI excitation upon increasing pulse delay (resulting in less saturated absorption of the dye) leads to an exponential fluorescence decay for Texas Red.

I would suggest better explanation of Fig. 5 in respect to its displayed error bars and also with regard to the overall pulse intensity fluctuation and fluorescence experimental S/N justification. Due to the applied experimental parameters the "observed" axial resolution improvement could be explained by and associated to the decreasing contrast (preserving all experimental parameters constant as stated and inherently the overall crucial experimental S/N ratio too) as a result of the delayed pulse pairs upon delay increase which cannot promote identical levels of 2PE absorption in the same (I square *vs.* excited voxel nature of 2PE microscopy) diffraction limited voxel (as mentioned above). But all in all these experiments will still be diffraction limited, regardless, hence any observed resolution enhancement or improvement could be misinterpreted. I would also like to ask the authors to clarify that using the applied 30 mW total laser power what is the exact calculated power (nj/voxel/s) in their experiments as it is crucial to underpin claims regarding 2PE excitation and Stimulated emission depletion. The latter case when presenting findings in the paper regarding Texas Red STED depletion using the detailed 750 nm set up also needs clarification. The experiment which could underpin these claims is to use a Texas Red only stained cell slide excite it with a suitable laser (514, 596, or even 633) and record an image to observe 1PE emission then keeping all parameters identical record an image where the suggested depletion laser is applied and highlight the observed difference in detected Texas Red emission. A small note that on Fig. 6 the 1000 fs and 2000 fs Gaussian (e and f) graphs appear to be identical replicas of each other. Fig. 7 could benefit from the same detailed explanation as suggested for Fig. 5 above in this comment. Taking all the observations detailed above resolution enhancement as a direct result of 2PE excitation pulse pair relative delay has not been fully demonstrated and underpinned.

**Debabrata Goswami** answered: This work is done on a fixed slide that has the cell stained with the two dyes adhering differentially to the two different parts of the cell structure, *i.e.*, DAPI adheres selectively to the nucleus while Texas Red gets attached to the α-tubulin of the cell. Thus, the possibility of the proposed energy transfer mechanism (misrepresented as cross-excitation) is of minimum concern for such microscopic studies. The entire argument and discussion for different experiments and suggestions are therefore superfluous. Furthermore, it is important to point out that both DAPI and Texas Red undergo the two-photon excitation unlike what is being conjectured! There is an absorption shoulder at ~365 nm for the $S_0 \rightarrow S_n$ transition for Texas Red and there is a substantial two-photon cross-section at 365 nm for DAPI. Only the stimulated emission is due to the single-photon condition at 730 nm, which is applicable only to Texas Red when a 730 nm ultrafast laser used. When we use a 750 nm instead of 730 nm ultrafast laser, we specifically address the case of DAPI fluorescence only. We have also taken two-channel confocal single-photon images for the same slide with different cw excitation wavelengths, where we can get beautifully resolved high contrast images for each of the two dyes (once again no energy transfer issues are seen). I hope this addresses the confusions in the questioner's mind. There have been some confusions due to errors in the figure legends which have now been corrected in the final manuscript. Thanks for pointing these errors out. In our particular experiments, the pulse intensity fluctuations and the signal-to-noise ratio related to the fluorescence measurements are both less than 0.1% for our microscope. So, they are all included within the error-bars as shown in all our figures presented (Fig. 5 and Fig. 7). The amount of energy that is of relevance as mentioned is 12 nJ/voxel/sec, which is certainly enough for two-photon absorption as well as stimulated emission for the dyes that we are working with. Finally, as discussed before, since the entire process that we are involved in requires at least two or more interacting fields, we will see an improvement along the Fourier plane. Thus, the *z*-resolution is enhanced due to the nonlinear interaction. Similarly, since it is due to multiple fields interacting (the delayed pulse being responsible for excited state depletion with stimulated emission), it is imperative that the discrimination remains dependent on the relative pulse delays between the pulses. Relative difference in the excited state behavior of the dyes might also exist, which is presently being investigated by us.

**Sankarampadi Aravamudhan** addressed **Debabrata Goswami**: On page 7 Fig. 5(a) and 5(b), and on page 9 Fig. 7, the resolution for 400 fs pulse pair separation is better than that for 250 fs whereas Fig. 6(c) and 6(d) indicate that 400 fs resolution is poorer than 232 fs. This contrast between the fluorophores at a ~1 ps time delay between the pulse pair can be critical in distinguishing the overlapping fluorophores. Hence using a pulse pair separation of more than 1 ps is not any more advantageous than running a single pulse experiment. However the situation of poster 03 seems different – and a comment on this is requested. See page 5 lines 39 to 42. However, if a time-delayed probe pulse that is wavelength-tuned to the red edge of fluorescence sends the population back to the ground electronic state by stimulated emission in competition with the fluorescence process, fluorescence suppression is observed. This results seems the main advantage in the two pulse experiments.

**Debabrata Goswami** responded: We are thankful for the valuable discussions and the nice comments on our work. However, one needs to be careful while extracting resolution information from Fig. 6, since the different plots have different *x*-axis scales. Indeed, our 400 fs delay does have better resolution as compared to the 232 fs delay as we report in the paper.

**R J Dwayne Miller** said: It is interesting that you see an enhanced contrast between dyes (as well as in studies of improved axial resolution) for pulse pair delays on the order of 500–900 fs. You are using 2-photon absorption with a central excitation of 750 nm. This puts the initially prepared excited state significantly above the So to S1 origin, which generally has the maximum emission probability due to better Franck–Condon overlap that is dictating the radiative rate. Taking the DAPI and Texas Red study, the initially prepared state is approximately 1670 cm$^{-1}$ above the S1 origin whereas it is 10,000 cm$^{-1}$ above the S1 origin for Texas red. It would take much longer for the excess vibrational energy to be dissipated in the Texas Red case relative to DAPI. I am wondering if the additional contrast you are seeing reflects differences in the excited state relaxation dynamics. Can you comment on the reason for the observed timed dependence for the contrast?

**Debabrata Goswami** replied: The excited state relaxation dynamics as suggested may play a very important role in this study. This is highly likely from the fact that the delay between the pulses is important. Understanding this point, in fact, is an ongoing effort in our lab where we are trying to observe the competing excited state dynamics of these dyes. An appropriate answer to this will soon be available from our experiments that are presently in progress.

**Stephen Meech** remarked: Could you explain the origin of the intensity change seen for Texas Red compared with DAPI in Fig. 4a? Since the time between the two pulses (900 fs) is quite long then, in the simplest case, one might expect all phase information to be lost, and the pulses could be regarded as well separated. In that case one expects a balance between any stimulated emission induced by the second pulse quenching the emission and additional 2 photon excitation adding to it. Neither should be strongly dependent on interpulse decay on the sub picosecond time scale. However, when the time delay is increased by only 100 fs the measured emission intensity decreases by 45%. Clearly some other factor must operate, could you speculate as to what it is?

**Debabrata Goswami** replied: We have invoked the stimulated emission process as the origin of the intensity change seen for Texas Red compared with DAPI. The effect of interference between the pulses are too strong near the pulse widths of the pulses concerned, so we are unable to be definitive at relative time delays between pulses on the order of the first 100 fs. Thereafter, there could be important effects of excited state relaxation behaviour of the dyes, which would reflect strongly on how the pulse delays impact. Specifically, our conjecture relies on the fact that there are enough excited state Texas Red species that are ready to be depleted by single photon Stimulated Emission Process (SEP). Since our laser wavelengths would most likely create vibrationally excited states through two-photon excitation, they need to vibrationally relax within the excited electronic state such that they are just right and ready to be depleted through SEP. Thus, the excited state vibrational dynamics may play an important role in this process, which is our present research investigation. I hope an appropriate answer to this would soon be available from our experiments that are presently in progress.

**Priyadarshi Roy Chowdhury** asked: If we use different pulse for photon excitation, could you suggest whether the response to the dyes that you have used in this work will monotonically increase or decrease?

**Debabrata Goswami** answered: We have used a two-photon excitation condition, which is highly dependent on the pulse width as well as the center wavelength of the laser pulse excitation. Beyond the starting threshold of the two-photon induced fluorescence process, the fluorescence yield would increase as the square of the applied excitation laser intensity for any particular dye.

**Priyadarshi Roy Chowdhury** enquired: What will happen if the mixture of DAPI or Texas red is used at different concentrations as well as in equimolar concentrations? What will be the effect on the intensity in varying excitation ranges?

**Debabrata Goswami** answered: It is important to mention here that this work is done on a fixed slide that has the cell stained with the two dyes adhering differentially to the two different parts of the cell structure, *i.e.*, DAPI adheres selectively to the nucleus while the Texas Red gets attached to the α-tubulin of the cell. So, for all practical purposes these dyes behave as if they are independent of each other and do not influence each other in any particular way as is expected in liquid mixtures! The only influence therefore comes through the effect of the excitation light on the individual dyes. Since the excitation is through a two-photon process, there is a threshold to that and a very careful study that can measure the fluorescence yield to the laser intensity provided should show a quadratic behavior. However, within the possible range of "safe" intensities that we can work on such a microscopic arrangement, it is practically impossible to uncover the quadratic relationship of the fluorescence yield to the pulsed intensity applied. We definitely do see the threshold intensity effect.

**Priyadarshi Roy Chowdhury** queried: What will happen if the pulses are delayed? As you are using a pulse shaper – if the first pulse excites, is there any possibility that the second pulse will deactivate the molecule partially on mutual interaction between its atoms?

**Debabrata Goswami** responded: We have shown the effect of pulse delays to show that the best possible resolution is at some non-zero delay. We are using simple interferometry to delay the pulses. Indeed we use the principle of two-photon excitation and single photon stimulated emission pumping for depletion of the excited molecules.

**Priyadarshi Roy Chowdhury** remarked: What filters have you used in delaying the pulses? What will happen if the pulse is not delayed using a interferometer? Could you explain what will be the effect if the pulses are delayed in the femto-second range rather than a pico-second range delay?

**Debabrata Goswami** answered: We are using simple interferometry to delay the pulses. The effect of interference between the pulses is too strong near the pulse widths of the pulses concerned, so we are unable to be definitive of relative time delays between pulses on the order of the first 100 fs.

**Priyadarshi Roy Chowdhury** asked: What will be the effect on the resolution in laser microscopy of using the multifocal system rather than a confocal system?

**Debabrata Goswami** responded: It is expected that the resolution would increase as has been seen in Structured Illumination Microscopy (SIM) studies.

**Priyadarshi Roy Chowdhury** addressed Professor **Debabrata Goswami**: This paper by Debabrata Goswami discusses mainly resolution enhancement through confocal multi-photon fluorescence imaging microscopy. Very fruitful interpretation has been given with respect to confocal systems. However, I feel that if multifocal systems would have been used instead, better resolution could have been achieved. Moreover, a comparative study with respect to DAPI (blue fluorophore staining nuclei) and Texas Red-X phalloidin (red fluorophore staining mitochondria) with that of other staining dyes of previous studies available; would have established this work with DAPI and Texas Red-X phalloidin to a much greater extent than the previous studies.

**Debabrata Goswami** answered: That is a very useful suggestion indeed. Using a spatially modulated light could enable the multifocal condition. We have also used Mito-Tracker Red dye as another dye for our studies. Definitely more such studies with various kinds of dyes would establish the larger extent and applicability of the representative studies presented here.

**Kenneth Ghiggino** opened a discussion of the paper by Ashok Kumar Mishra: You refer to the presence of aggregate emission bands of your compounds with high water content in mixed solvent (water–acetonitrile) conditions. Absorption spectral shifts are also noted under the same conditions. Is the emission you see arising from photoexcited ground state dimers rather than true excimer fluorescence (excimers are dissociative in the ground state)? Under some conditions the aggregate emission band is blue shifted compared to the ICT emission while in other cases it is the other way around. Do you see any evidence for energy transfer between the excited species in the aggregates present in solutions of your compounds?

**Ashok Kumar Mishra** responded: According to Birks' definition, "excimers" are associative in the electronic excited state and dissociative in the ground state.[1] Often, the excimer emission originating from photo-excitation of pre-associated ground state molecules is termed as "static excimer" emission; and in contrast, excimers that follow Birks' definition are termed as "dynamic" excimers.[2] In our work, (i) aggregation induced steady state absorption spectral changes, (ii) the absence of rise time of the aggregate emission in lifetime studies, and (iii) the existence of Py···Ph interactions as observed in proton-NMR studies on aggregates, suggest that the emission originates from the "static excimer".

The aggregate emission bands do not shift much for the derivatives (510-560 nm), however, the ICT emission bands span over a wide range (475–600 nm) depending on the donor–acceptor character of the derivatives and the choice of solvent.

In the present work, we did not look for the possibility of energy transfer between the excited species in the aggregates. We thank you for this suggestion.

1 J. B. Birks, *Rep. Prog. Phys.*, 1975, **38**, 903–974.
2 F. M. Winnik, *Chem. Rev.*, 1993, **93**, 587–614.

**Elangannan Arunan** asked: The schematic shown in Fig. 14 has three circles pointing out Py···Py, Py···Ph and Ph···C≡C interactions. Why not the Ph···Ph interaction? Is this schematic reasonable in a dynamic solution?

**Ashok Kumar Mishra** replied: The schematic shown (Fig. 14) is not meant for dynamic intermolecular interactions of completely solvated monomeric molecules; it is for the aggregates that form in water–MeCN solvent mixtures. The concentration and temperature dependent $^1$H-NMR experiments of the derivatives suggest intermolecular interactions in the aggregates between pyrene and phenyl rings (Py···Ph), in addition to the Ph···C≡C interactions. The Py···Py interaction is proposed based on the observation of pyrene type excimer emission. The possibility of Ph···Ph interactions can neither be completely excluded nor be ascertained at the present stage. It may require further investigations; possibly single crystal X-ray diffraction studies would shed better light on it.

**Mahesh Hariharan** enquired: Your molecule looks simple and is likely to form lots of reactive intermediates and products like exciplets. Can you have pyrene interacting with the amine part to form exciplets? Can this aggregate lead to photodimerisation? In water do you see dicarbonyl formation in acetylenes? Are you able to see these products on TLC? There may be micromolar concentrations of the aggegrate.

**Ashok Kumar Mishra** responded: The possibility of photochemical reactions of the types you describe do exist for some of the molecules we studied. Under the low intensity light conditions of the fluorimeter and low (micromolar) concentrations that we used, we did not observe such reactions. We thank you for the suggestion, we will plan a regular photochemical reaction study using photoreactors.

**Ajith R. Mallia** remarked:

(i) The manuscript deals with the intramolecular electron transfer (ICT) occurring in butadiyne bridged pyrene–phenyl molecular conjugates. I would be glad to know whether the authors were successful in probing the charge transfer intermediates formed during the ICT processes using any spectroscopic techniques?

(ii) The authors presented the solvatochromism observed in the molecular conjugates using solvent dependent absorption and emission studies. I would like to know whether the authors have calculated the change in dipole moment using Lippert–Mataga analysis which can further justify the solvatochromism.

(iii) Fig. 6 shows absorbance changes of the conjugates with different water–ACN percentages but the graph is irregular without any trend. Can you comment?

**Ashok Kumar Mishra** answered:

(i) If there is any charge transfer intermediate for an ICT process, it would not be possible to probe it with nanosecond time scale resolution. We have not carried out any other spectroscopic studies in probing the existence of charge transfer intermediates.

(ii) For this set of pyrene–phenyl derivatives, the Lippert–Mataga plot was not carried out. Our main focus here was to understand different kinds of emissive behaviours and the origin of local transitions for such derivatives. The solvatochromism behaviour of a similar set of butadiynyl fluorophores was previously discussed through Lippert–Mataga analysis in our earlier article published elsewhere.[1]

(iii) From our experience of working on butadiynyl derivatives, we have observed that the absorption spectra (longest wavelength absorption band) of such derivatives generally do not change significantly with solvent polarity. Fig. 6 in the manuscript does not show any appreciable change in 0-60% or 0-70% water–$CH_3CN$ mixtures. However, Fig. 6 shows that there is a red shift of the longest wavelength absorption band in higher fraction of water (such as 80-99%), which is attributed to aggregates that lead to static excimers on photoexcitation.

1 Mishra *et al.*, *J. Phys. Chem. A*, 2013, **117**, 6548–6560.

(413:[417]417) **Sankarampadi Aravamudhan** asked:

This question pertains to the mention of the structure of first excited state of diphenyldiyne having two phenyl rings twisted to an out of plane configuration. The ground state of diphenyldiyne has a planar configuration for the two end-phenyl groups.

Such a planar disposition in the ground state would have the possibility of full-length extended conjugation with the corresponding delocalization energy contributing to the stability of the molecule. When the structure has the end-phenyl groups twisted to out-of-the-common-plane, then in the molecule the conjugation of the central diyne part does not extend to the end-phenyl groups. Then, there would be reduction in the delocalization energy resulting in a higher energy of the molecule. Does the excitation (transition frequency) energy for the two level system, equal the extent of delocalization caused by the twist-to-out-of-plane configuration of the two phenyl groups?

A break up of energy (total excitation energy) in terms of destabilization due to reduced delocalization and any other structural changes of the excited state configuration could provide better insight for interpretation of the experimental results.

**Ashok Kumar Mishra** responded: We believe that there is no structural change in the molecule that aggregates in water–MeCN solvent mixtures and shows static excimer emission.

**Shamik Chakraborty** asked: For the 80% water–$CH_3CN$ mixture there is a significant change in the emission spectrum. Is this a critical concentration for aggregate formation? Is it possible to control the size distribution of the aggregates?

**Ashok Kumar Mishra** replied: We believe that the aggregates, formed with increasing fraction of an unfavourable solvent like water, are non-specific aggregates. Thus, there may not be any "critical" concentration of aggregation. It may not be possible to control the size distribution of non-specific aggregates.

**Martin Meedom Nielsen** opened a general discussion of the paper by Gopinathan Sankar: Understanding the formation of order from disorder is a fundamental problem, and methods such as you describe are beginning to be able to address this. Can you comment on the underlying mechanisms responsible for the kinetics you observe?

**Gopinathan Sankar** replied: It is difficult to generalise the underlying mechanism responsible for the kinetics observed in our work. For example, some framework structures take several days to form and some can be obtained in a few hours. Several factors may control this, for example formation of tetrahedral units, subsequently the required ring structures (sub units) and then crystallisation. Both solution mediated crystallisation and solid–solid transformation are proposed as the underlying mechanism. These aspects have been reviewed in the literature.[1,2]

1 C. S. Cundy and P. A. Cox, *Chem. Rev.*, 2003, **103**, 663–701.
2 C. S. Cundy and P. A. Cox, *Microporous Mesoporous Mater.*, 2005, **82**, 1–78.

**Mark Warren** asked: In the article the authors account for the initial increase in cell dimensions observed in X-ray powder diffraction analysis (Fig. 4) by the "uptake of water molecules which may coordinate to Al(III) ions" (page 7, line 15). In Fig. 5, the authors suggest that a dehydration may occur after crystallisation, which one might expect would lead to a lattice contraction.

Is this consistent with the diffraction data? Also, is there any experimental evidence for this dehydration (*e.g.* TGA, vibrational spectroscopy)?

**Gopinathan Sankar** answered: During synthesis we find that the lattice parameter, in particular the "*c*" parameter increases during the crystallisation process. Based on other experiments, in particular, X-ray total scattering experiments (pair-distribution function analysis) of various similar systems, water molecules that are coordinated to Al(III) ions making them partly octahedral can be removed easily by dehydrating at low temperatures. What we suggest is that this change in "*c*" parameter is likely to be associated with water uptake during the hydrothermal synthesis.

**Kiran Moirangthem** enquired: It is claimed/proposed in this paper that the crystallization of the nanoporous aluminophosphate (AlPO-5) takes place *via* the formation of a "hydrogel intermediate".

In this era of technological advancement, I would like to know:

Is it possible to experimentally monitor/detect the "hydrogel intermediate" with real time measurement (like what is normally done in ultrafast spectroscopy)?

If not, what limits this kind of real time measurement in the crystallisation process and what is the best possible option so far available?

**Gopinathan Sankar** responded: What we claim in the paper is that the nanoporous aluminophosphate synthesis, irrespective of the structure type, takes place *via* "hydrothermal methods" and we do not say that hydrogel intermediates are formed. It forms from hydrogels or colloids or some other kinds of precipitates. The measurements we carry out in understanding the crystallisation process were to monitor the initial state of the starting and intermediate states and also we were equally interested in the final states. Whilst ultrafast spectroscopy will be very useful, for the measurements we employ, to my knowledge for these systems (X-ray diffraction, small-angle X-ray scattering, X-ray absorption spectroscopy) it is difficult go beyond a millisecond time scale to obtain useful information. Furthermore, some other metal ion species in these systems are in low concentrations.

**R J Dwayne Miller** remarked: There have been major advances in nanofluidics that have enabled *in situ* studies of structures within liquid environments using real space imaging with a TEM. Spatial resolution of less than 1 nm is possible. It seems to me that one could observe the crystal growth giving rise to these interesting nanoporous structures using TEM. One can envisage even using 10 nm spot sizes for diffraction analysis of growth. Would this be of any value? The current cell designs can handle the temperature (170 °C) but what are the pressures involved during hydrothermal growth? The nanocell design would likely need to incorporate a more rigid window design if the pressures are much above 10 atm.

**Gopinathan Sankar** answered: This is certainly a challenging area which I and the rest of the community working on solution processing of materials will be very interested in. The main challenge is temperature (*ca.* 100 to 200 °C) and pressures of *ca.* 1 to 10 atmospheres need to be addressed. These are typically used in the synthesis of zeolitic materials. Another issue we face is that some of the materials are not stable in the electron beam and may decompose. One should consider their stability in the electron beam as well. In summary, this is certainly an important area to consider and see whether we can advance further the understanding of the formation of these materials.

**John R. Helliwell** commented: On page 7 of your article you compare your results with those of ref. 4 and 34. You identify a transformation between octahedral and tetrahedral and that there is a pseudo-controversy due to the different template. This led you to make comments in your talk where you oppose the notion of a standard nucleation mechanism for crystal growth of your sample. But could the Co(II) octahedral species itself be regarded as the nucleus for crystal growth? Obviously my suggestion involves a deviation from the usual definition of crystal nucleation and crystal growth, *i.e.* in which weak intermolecular interactions stabilise a crystal lattice.

**Gopinathan Sankar** replied: In short summary, this is an interesting topic for which we certainly do consider the definition of nuclei in the system. Firstly, what we claim is that we start with an octahedrally coordinated Co(II) in the solution and that converts over a period of time through interaction with the remaining chain of Al(III) and P(V) slowly into tetrahedral, the required coordination geometry. Whether a template (organic amines) influences this process, is certainly an interesting topic to consider. But it is difficult to say whether small amounts of Co(II) octahedra act as the nuclei as one can form similar materials with identical structures with or without (pure aluminophosphate systems) Co(II) and other metal ions.

**Elangannan Arunan** enquired: Did you mention that the nucleation site may not be important in crystal formation? Your conclusions about nucleation may not be applicable for polymorphs, which seem prevalent. Any comments?

**Gopinathan Sankar** responded: I did not mention that the nucleation site may not be important in the crystal formation. Evidence suggests that we can form amorphous intermediates through mixing these chemicals and use these powders to convert under steam (in a closed vessel called “steam assisted synthesis”) to form the final crystalline product. This is believed to be through solid–solid transformations. We are not dealing with polymorphs at least in this case.

**Priyadarshi Roy Chowdhury** asked: What are the present day developments in structural analysis and what is the most significant characterisation for determination of the structure of compounds to date?

**Gopinathan Sankar** answered: This is very difficult to answer. There are several ways we can look into it. The development in structural analysis and what is most significant characterisation. Each participant or many other scientist will interpret this in different ways. Is it the local (first few coordination) structure? Is it the long or medium range structures? If we define this, it could be slightly easier. Furthermore, we have ignored several advanced computational methods which can assist us greatly in defining "determining the structure of compounds".

**Priyadarshi Roy Chowdhury** queried: Why is nucleation not possible in your compounds? Justify.

**Gopinathan Sankar** answered: I am not saying nucleation is not possible. How does one define a nuclei. Is it a molecular species? We observe a molecular species in our experiments and in some cases we have observed a system of more than nanometer size. If we consider this is the nuclei, then yes, we can define this.

**Priyadarshi Roy Chowdhury** remarked: Which analytical technique or combined techniques available today give the maximum possible resolution as well as structural information of a compound other than Atomic Force Microscopy?

**John R. Helliwell** replied: AFM has no data base, to my knowledge, of 3D structures resolved by Atomic Force Microscopy. Whereas crystal structure analysis has more than 750,000 crystal structures in the Cambridge Crystallographic Database, more than 100,000 in the Protein Data Bank, thousands in the Inorganic Crystal Structure Database and many in the metals structures data base and finally of course a huge number of diffraction profiles in the Powder Diffraction Data file.

 

**Gopinathan Sankar** commented: It depends on what "resolution" means – is this time or spatial or atomic/structure resolution. Diffraction methods can provide very detailed structural information and can be combined with time-resolution – again one needs to define what time-resolution is required which depends on the system and the chemistry we are trying to understand. In summary, one needs to know the specific aim and objective of the work and accordingly explore the techniques available to extract structural information.

**Priyadarshi Roy Chowdhury** questioned: How do you combine the different characterization techniques together?

**Gopinathan Sankar** answered: Different characterisation techniques can be combined in several ways. Firstly in the last 20 years it has been established that several X-ray techniques can be combined together in a measurement. In addition, if we can match the time-resolution with techniques based on optical spectroscopy, it is not too difficult to get information simultaneously. It has been established through various studies that Raman Spectroscopy with X-ray absorption spectroscopy and diffraction is feasible. Similarly, IR and XAS in various forms have been established to investigate in the same system under reaction conditions. Furthermore, it is now possible to combine data analysis of different systems, together, to analyse data with a single starting model so that they yield more unique structural information.

**John R. Helliwell** opened the discussion of the paper by **Zhen-Yu Tian**: This seems to be a temperature resolved study, as highlighted in Fig. 3, 8 and 9, but your abstract refers to a temporally resolved aspect. Please clarify.

**Zhen-Yu Tian** responded: As can be seen in our recent paper[1] regarding the catalytic performance as a function of time, the conversion was observed to be quite stable. It was reasonable in this work to correlate the thermal stability and catalytic performance with the structure property in a temperature-programmed mode. Since the temperature is directly related to the time and changes gradually during the tests, this correlation is also a temporally resolved behavior.

1 Assebban *et al.*, *Chem. Eng. J.*, 2015, **262**, 1252–1259.

**Himani Medhi** commented: As per Fig. 7 of your paper, the material is highly porous and porosity obviously plays an important role in selective catalysis. But in your paper, the authors haven't mentioned the pore size of the material. To be a good quality paper it should be mentioned.

**Zhen-Yu Tian** replied: Thanks for the comment. According to the HIM images, the pore size is estimated to be 5–20 nm.

**Priyadarshi Roy Chowdhury** addressed **Zhen-Yu Tian**: In this paper Zhen-Yu Tian has discussed the characterization of Cu–Co oxides. However, as I go through this work, I find that the XPS full scan survey is missing, only portions of the scans are depicted. A full scan XPS survey would have provided detailed analysis of the Cu–Co oxide catalyst, and whether impurities have crept in within the catalyst framework. I have doubts that without a full scan XPS survey, that was not shown, the compounds would be free from impurities. Secondly, the surface area measurements were not done, so we have no idea of the pore size and pore volume as well as the specific surface area. Thirdly, multi metal oxides must possess different phases, whose analysis was missing. A HR-TEM image would have been much more fruitful. Fourthly, the transition metal oxide catalysts of this type possess defects, so photoluminescence analysis would definitely give an idea of the defects. Fifthly, in the determination of the band gap by Tauc plots, the Tauc plots were not shown, the band gap of the synthesized catalyst should have been compared with respect to commercial oxide catalysts like ZnO, CuO, CoO, *etc.*, so that a conclusion could be drawn from it whether the synthesized material would be photocatalytically active in the UV or visible region or not. Lastly, the elemental composition of the composite Cu–Co oxide catalyst should have been clearly shown by EDX measurements. So, I would suggest that this amount of work should have been done/can be done in future.

**Zhen-Yu Tian** responded: Thanks for the insightful comments. We would like to reply to the questions one by one:

(i) The full scan XPS surveys of the Cu–Co oxides are shown below (Fig. 4). The purity of the prepared samples can be clearly seen.

(ii) It is generally accepted that the surface area measurement for thin films cannot be easily be done, especially for the samples with only 300 nm thickness. According to the HIM images, the pore size is estimated to be 5–20 nm.

(iii, iv, vi) We agree that it is a good idea to involve the accurate measurements with HR-TEM, photoluminescence analysis and EDX, but they are unavailable in our institute. However, we would like to perform such measurements in our future work.

(v) The Tauc's equation is shown in the context of the paper and Fig. 6b shows the plot of $(ah\nu)_2$ as a function of $h\nu$, which is a transformation of the Tauc's plot.

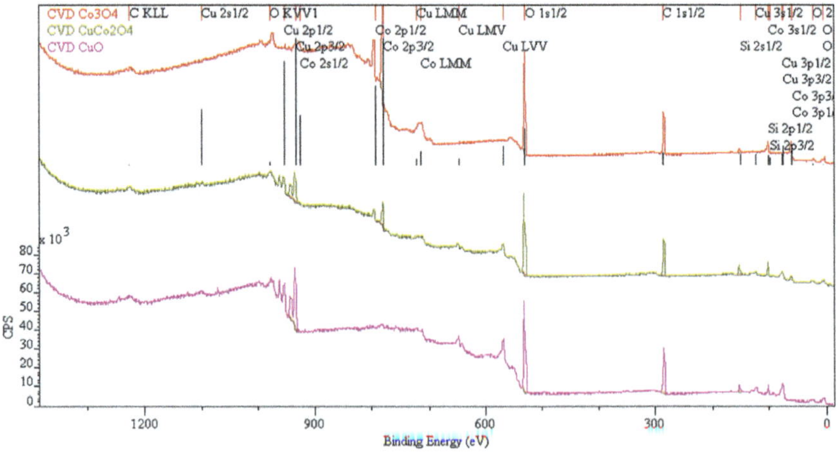


Fig. 4 Full scan XPS survey of Cu–Co oxides.

Faraday Discussions

PAPER

# Electron transfer quenching in light adapted and mutant forms of the AppA BLUF domain

Sergey P. Laptenok,[b] Andras Lukacs,[bd] Richard Brust,†[a] Allison Haigney,‡[a] Agnieszka Gil,[a] Michael Towrie,[c] Gregory M. Greetham,[c] Peter J. Tonge*[a] and Stephen R. Meech*[b]

Received 2nd October 2014, Accepted 14th October 2014
DOI: 10.1039/c4fd00189c

The Blue Light Using Flavin (BLUF) domain proteins are an important family of photoreceptors controlling a range of responses in a wide variety of organisms. The details of the primary photochemical mechanism, by which light absorption in the isoalloxazine ring of the flavin is converted into a structure change to form the signalling state of the protein, is unresolved. In this work we apply ultrafast time resolved infra-red (TRIR) spectroscopy to investigate the primary photophysics of the BLUF domain of the protein AppA ($AppA_{BLUF}$) a light activated antirepressor. Here a number of mutations at Y21 and W104 in $AppA_{BLUF}$ are investigated. The Y21 mutants are known to be photoinactive, while W104 mutants show the characteristic spectral red-shift associated with BLUF domain activity. Using TRIR we observed separately the decay of the excited state and the recovery of the ground state. In both cases the kinetics are found to be non-single exponential for all the proteins studied, suggesting a range of ground state structures. In the Y21 mutants an intermediate state was also observed, assigned to formation of the radical of the isoalloxazine (flavin) ring. The electron donor is the W104 residue. In contrast, no radical intermediates were detected in the studies of the photoactive dark adapted proteins, $dAppA_{BLUF}$ and the dW104 mutants, suggesting a structure change in the Y21 mutants which favours W104 to isoalloxazine electron transfer. In contrast, in the light adapted form of the proteins ($lAppA_{BLUF}$, lW104) a radical intermediate was detected and the kinetics were greatly accelerated. In this case the electron donor was Y21 and major structural changes are associated with the enhanced quenching. In $AppA_{BLUF}$ and the seven mutants studied

[a]Department of Chemistry, Stony Brook University, Stony Brook, New York 11794-3400, USA. E-mail: ptonge@notes.cc.sunysb.edu
[b]School of Chemistry, University of East Anglia, Norwich NR4 7TJ, UK. E-mail: s.meech@uea.ac.uk
[c]Central Laser Facility, Research Complex at Harwell, Harwell Science and Innovation Campus, Didcot, Oxon OX11 0QX, UK
[d]Department of Biophysics, Medical School Szigeti str. 12#, H-7624 Pécs, Hungary
† Present address: The Scripps Research Institute, 130 Scripps Way, Jupiter, FL 33458, USA.
‡ Present address: University of Pennsylvania, Philadelphia, PA 19104, USA.

radical intermediates are readily observed by TRIR spectroscopy, but there is no correlation with photoactivity. This suggests that if a charge separated state has a role in the BLUF photocycle it is only as a very short lived intermediate.

# Introduction

A wide range of living organisms have evolved the means to detect light, either to exploit it as a source of energy or to avoid its harmful effects. Consequently an array of photoreceptors have been discovered, with perhaps the most famous being the rhodopsins, the photoreceptor cells in the retina.[1,2] These have been studied intensely for a number of years, allowing the excited state reaction, an ultrafast isomerization, to be characterised in considerable detail.[2,3] More recently a wide array of new photoreceptors has been discovered. Some, such as the photoactive yellow protein, depend on an excited state structure change similar to the photoisomerization in rhodopsin.[4,5] However, in many others the blue light absorbing isoalloaxazine ring of either flavin adenine dinucleotide (FAD) or flavin mononucleotide (FMN) acts as the chromophore.[6–9] In these cases the mechanism by which this planar aromatic chromophore converts the energy of an absorbed photon into a structure change in the surrounding protein matrix is incompletely understood. Thus, there is considerable interest in determining the mechanism of operation of these flavoprotein based blue light sensors. This fundamental interest has been further stimulated by the demonstration that these proteins have applications in the emerging field of 'optogenetics' or, more generally, 'optobiology'.[10–12] In optobiology the light activated function of a protein is recruited by genetically encoding it with a specific target. For example, light activated ion channel rhodopsins were shown to mediate signalling in neuronal cells.[13] More recently blue light sensing flavoproteins have been used to introduce a range of light activated functionalities, for example allowing optical excitation to modify protein–protein interactions or DNA binding.[11] In principle, this methodology affords the experimenter both spatial and temporal control over a specific cellular response, and is thus a potentially powerful tool in cell biology.

Three families of flavoprotein based photosensors have been identified: blue light using flavin (BLUF) domains; light-oxygen-voltage (LOV) domains; cryptochromes and photolyases.[8,9] Collectively these are involved in controlling a wide variety of light activated processes including phototaxis, stomatal opening, and circadian rhythms. Each family has distinctive aspects to their light sensing mechanisms. For cryptochromes and photolyase, which are found in plants, animals and fungi, there is strong evidence for radical intermediates suggesting an electron transfer mechanism.[14] The modular LOV domain unit, found in a variety of proteins in organisms from plants to bacteria, exhibits a photochemical mechanism which is mediated by the triplet state of the isoalloxazine ring, populated *via* intersystem crossing.[15,16] This triplet state reacts with an adjacent cysteine residue to form a metastable adduct, which is sufficient to initiate a structure change in the protein, leading to signalling state formation. The BLUF domain is also a modular unit found in bacterial proteins, which operates by a third mechanism which is not yet fully resolved, and is the topic of this investigation.

 

One of the best characterised BLUF domains is the light activated anti-repressor protein AppA found in *Rhodobacter sphaeroides*, where it is responsible for controlling the expression of the photosynthetic apparatus in response to both light and oxygen levels.[17–21] In this work we focus on the isolated BLUF domain, $AppA_{BLUF}$. BLUF domain photoactivity is indicated by a small (10–15 nm) red-shift in the electronic absorption spectrum of the isoalloxazine ring on irradiation with light at around 450 nm.[20] Thus, in both the dark state ($dAppA_{BLUF}$) and the red-shifted light adapted signalling state ($lAppA_{BLUF}$) the flavin remains in its fully oxidized form (the reduced forms such as radicals and ions have very different absorption spectra[22]). The photoactivity is reversible, with the relaxation from $lAppA_{BLUF}$ to $dAppA_{BLUF}$ occurring with a half-life of about thirty minutes, after which the protein can be re-excited.[23,24]

Time resolved studies showed that the red-shift develops within 1 ns of electronic excitation, suggesting a role for ultrafast processes.[25] However, recent time resolved IR measurements on the femtosecond to millisecond time scale revealed that further evolution in the protein structure occurs on multiple time scales, suggesting that the red shift is the start of a more complex structure change.[26] In addition, stationary state IR difference spectra showed that the flavin carbonyl has a different H-bonding environment in $lAppA_{BLUF}$ compared to $dAppA_{BLUF}$, and that there was an associated change in the protein structure.[27,28] Thus, although the electronic structure of the flavin is unchanged between $dAppA_{BLUF}$ and $lAppA_{BLUF}$ the H-bond environment is restructured on multiple time scales, from sub-nanosecond to at least microseconds.

Crystal structures are an invaluable guide to the mechanism. A structure of $AppA_{BLUF}$ is shown in Fig. 1 along with the detail of the H-bonding environment of what is assumed to be $dAppA_{BLUF}$.[17] The published structures all suggest that residues Y21, Q63, N45, W104 and M106 are important in the flavin binding pocket. Mutations to the conserved residues Y21 and Q63 show that they are essential for observation of the red-shift, and N45 is crucial to flavin binding. Mutations in W104 are known to support the red-shift, but biological activity is greatly reduced.[29,30] A femtosecond to microsecond time resolved IR experiment showed that in W104A the microsecond timescale light induced structure change in $AppA_{BLUF}$ was short circuited, implicating this residue in communicating the local structure change to the protein.[26]

Unfortunately, there is some disagreement among the published crystal structures, particularly with regard to the important residues Q63 and W104.[17,31] The originally proposed mechanism for signalling state formation was that Q63 isomerized between dark and light adapted states giving rise to a change in H-bonding at the flavin carbonyl, in agreement with IR difference spectra.[17,32] However, other structures suggested a different conformation for Q63 and a subtly different H-bonding environment, with W104 and M106 occupying different positions;[31] these have been labelled the $W_{in}$ and $W_{out}$ models. Associated with this discussion has been the role of light induced structure changes in Q63, with different models favouring either the original isomerization mechanism[17,32] or an initial light induced tautomerization.[33,34] Two recent theoretical models favoured a $W_{out}$ structure with a tautomerization reaction in Q63 as the light activated trigger for structural change,[35,36] while another favoured Q63 rotation and a $W_{in}$ to $W_{out}$ structure change.[37] Interestingly Udvarhelyi and Domratcheva recently suggested that the orientation of Q63 is not well defined in

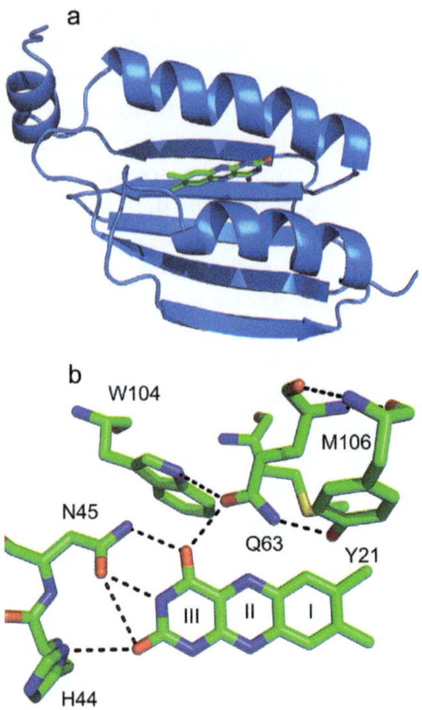


Fig. 1 (a) Structure of $AppA_{BLUF}$ determined by crystallography (PDB 1YRX). (b) Details of the H-bonding environment around the isoalloxazine ring of FAD. Note that alternative structures exist which differ on the orientation and location of Q63 and W104/M106 (PDB 2IYG – see text).

the ground state, in which case light induced isomerization alone is unlikely to be the sole structural trigger.[38]

A second point of debate concerns the driving force for the structure change. Gauden *et al.* proposed on the basis of transient absorption data that an excited state electron transfer reaction between the conserved Y21 residue and the isoalloxazine ring forms a radical pair state which undergoes proton transfer, giving rise to isomerization in Q63, and thus the altered H-bond network seen in IR difference measurements.[25,32,39] A transient radical state was indeed observed in the BLUF protein PixD. However, early time resolved IR (TRIR) measurements on $dAppA_{BLUF}$ observed only an instantaneous perturbation to the protein H-bond environment and no further kinetic intermediates on the nanosecond time-scale.[33] Such an instantaneous perturbation was also demonstrated in the photoinactive Q63E mutant.[40] As a result it was proposed that optical excitation itself may be sufficient to perturb the H-bond environment, and specifically that optical excitation might favour tautomerization in Q63.[33] These data suggest that the role of the radical intermediates in the BLUF photocycle, which is assumed to be important in a number of calculations, requires further clarification. Recently Lukacs *et al.* presented a detailed study of the TRIR of three BLUF domains and contrasted the results with model spectra of radical isoalloxazines and with flavoproteins which unambiguously show electron transfer reactions.[41] Although radical states are observed in some cases there was no correlation between BLUF photoactivity and electron transfer. Further, unnatural amino acid substitutions were used to artificially modulate the redox potential of the putative Y21 donor, but the results were not consistent with a simple electron transfer reaction.

In this work we extend our TRIR studies of the $AppA_{BLUF}$ domain to a number of mutants of Y21 and W104 and seek a correlation between the observability of radical intermediates, the rate of reaction and photoactivity. Radical intermediates are observed in a number of cases involving both W104 and Y21 as donors, but no correlation is observed between their rate of formation and photoactivity. The results are discussed in terms of the local structure and the environment of the isoalloxazine chromophore.

# Experimental

### Protein expression and purification

Site-directed mutagenesis on $AppA_{BLUF}$ in a pET-15b vector was performed using pfuTurbo (Agilent). Primer sequences for W104A were: forward: 5′-TTTGCGGGAGCGCACATGCAGC TCTCCTGCTCG-3′, reverse 5′ CGAGCAGGA-GAGCTGCATGTGCGCTCCCGCAAA-3′. Primer sequences for Y21C were 5′ CTG GTT TCC TGC TGC TGC CGC AGC CTG GCG GCC 3′ (forward) and 5′ GGC CGC CAG GCT GCG GCA GCA GCA GGA AAC CAG 3′ (backward); for Y21S were 5′ CTG GTT TCC TGC TGC AGC CGC AGC CTG GCG GCC 3′ (forward) and 5′ GGC CGC CAG GCT GCG GCT GCA GCA GGA AAC CAG 3′ (backward); for Y21I were 5′ CTG GTT TCC TGC TGC ATC CGC AGC CTG GCG GCC 3′ (forward) and 5′ GGC CGC CAG GCT GCG CAT GCA GCA GGA AAC CAG 3′ (backward). All constructs were verified by DNA sequencing. Plasmids were transformed into BL21 (DE3) *E. coli* cells. A single colony was used to inoculate 10 mL starter cultures containing LB media with 200 μg $mL^{-1}$ ampicillin at 37 °C which was grown overnight and used the following day to inoculate a 1 L LB per ampicillin culture. Cells were grown to an OD600 of about 1 at 30 °C for 5 hours. Induction was performed by adding 0.8 mM IPTG at 18 °C, followed by overnight incubation in the dark. The following day cells were harvested by centrifugation at 5000 rpm at 4 °C and stored at −20 °C until needed.

Purification of wild type and the various mutants were performed in the dark. Thawed cell pellets were resuspended in 40 mL of wash buffer (10 mM NaCl, 50 mM $Na_2H_2PO_4$, pH 8.0), to which 200 μL of 50 mM PMSF was added to inhibit endogenous protease activity. Cells were lysed by sonication and cell debris was removed by ultracentrifugation (33 000 rpm for 90 minutes). The cell pellet was discarded, and to the soluble fraction 1 mL of a 10 mg $mL^{-1}$ solution of FAD was added to the supernatant and incubated for 45 minutes on ice. The protein was loaded onto an equilibrated Ni-NTA affinity column. The bound protein was then washed using increasing amounts of imidazole and finally eluted with 250 mM imidazole. The fractions were then dialyzed overnight in the dark in the 3 L of wash buffer. Protein purity and yield were determined using UV-Vis spectroscopy and SDS-PAGE.

### Ultrafast measurements

All measurements reported here used the 10 kHz ULTRA facility developed at the Central Laser facility of the Rutherford Appleton Laboratories which are described in detail elsewhere.[42,43] ULTRA offers wide tunability in the visible region, a broad bandwidth in the IR probe, sub 100 fs time resolution and excellent stability. In the present experiments it was used in the visible pump – IR probe geometry. The excitation (pump) wavelength was 450 nm with a pulse energy of a few hundred nJ focussed to a 200 micron spot size. It was checked that the spectra and kinetics were independent of the pump wavelength and pulse energy. IR transmission was measured sequentially for pump-on and pump-off using a 5 kHz mechanical chopper, and the data processed to provide the TRIR difference spectra as described elsewhere.

Many dark adapted BLUF domains exhibit a light to dark recovery time of several minutes. Thus, to obtain transient data of the dark state the sample has to be both rastered in the beam, so that the sample is not subject to multiple excitations, and flowed so that fresh sample is present after each cycle of the raster. In the present case *ca.* 1 mL of 1 mM protein solution in phosphate buffer made up in $D_2O$ at pD 8 (pD = pH reading + 0.4) was circulated through a 50 micron path length $CaF_2$ cell. For the study of light adapted states the sample was not flowed and was irradiated for 3–4 min at 385 nm then held in the beam under continuous irradiation during the measurement.

### Data analysis

Data were assessed in two ways. First, we are interested in the formation (or otherwise) of intermediate states in the ultrafast kinetics. Using TRIR we have previously identified marker modes for the excited state decay and ground state recovery.[41] Thus these can be fit separately, and any difference between them suggests the formation of a metastable intermediate, which should also be visible in TRIR spectra (for example in the case of electron transfer, the radical state of the isoalloxazine ring). Thus transient data at the two characteristic wavenumbers were analysed separately. It was found that the kinetics were non-single exponential in all cases but could be fit with a bi-exponential function with fast and slow components, suggesting an inhomogeneous system; including further exponentials in the fit gave a statistically better description but with increasing errors. Thus all marker mode data were fit to the bi-exponential function and the separate decay times are reported, along with their weighted average for the mean decay time. Second, we have conducted a global analysis of the complete data set using established methods.[44] Initially simple sequential kinetic models were assumed, giving evolution associated difference spectra (EADS). However, these simple models do not reproduce the non-single exponential relaxation. Thus, in some cases specific target models were assumed in an effort to represent the underlying processes, this analysis results in species associated difference spectra (SADS).

## Results and discussion

Y21 is essential to photoactivity in AppA and PixD, and exchanging it for any other residue suppresses the light induced red-shift in the flavin absorption. The W104 residue appears to have a role in transmitting the primary events of optical excitation to the rest of the protein to form the signalling state. Mutations at W104 in $AppA_{BLUF}$ result in a protein which does exhibit a red-shift in the flavin spectrum, but recovery of the dark state is accelerated, and biological activity correspondingly attenuated. Our recent TRIR studies suggest that this arises because the primary events in the photocycle occur, but these are not communicated to the rest of the protein.[26] Here we report TRIR data for the single AppA mutants Y21S, Y21I, Y21C, W104Y, W104T and W104A and for the double mutant Y21I/W104A; Gauden *et al.* previously studied a similar series using transient absorption.[45] As expected, all W104 mutants showed a red-shift under blue light irradiation while Y21 mutants did not. However, the TRIR spectra reveal a variety of excited state chemistry.

## Photoinduced electron transfer in Y21 mutants

TRIR spectra for Y21S recorded one ps after excitation are compared with those for $dAppA_{BLUF}$ in Fig. 2. In TRIR spectra the negative ΔOD peaks appearing immediately upon excitation (bleaches) can be assigned to modes of the ground state whose populations have been reduced by electronic excitation. These modes may be localised either in the chromophore or the surrounding protein matrix; it was shown elsewhere that excitation of the isoalloxazine ring can instantaneously perturb the local protein H-bond environment.[40] The highest frequency bleach in $dAppA_{BLUF}$ (*ca.* 1701 $cm^{-1}$) was assigned to a mainly C4=O localised mode (although C4=O and C2=O modes are coupled, with the degree of coupling depending on the environment[46]). In Y21S (Fig. 2), and all the other Y21 mutants studied, the frequency of C4=O mode is unchanged on mutation, suggesting that although photoinactive, the H-bonding environment of the isoalloxazine ring remains similar to that in $dAppA_{BLUF}$. Note also that the transient absorption signal (positive ΔOD) seen at 1666 $cm^{-1}$ in $dAppA_{BLUF}$ is suppressed in Y21S (Fig. 2) and the other Y21 mutants. As previously noted[33] this instantaneously formed bleach acts as a marker mode for photoactivity, being absent or obscured in photoinactive mutants of $AppA_{BLUF}$, which is also the case here.

The temporal evolution of the TRIR for Y21S is shown in Fig. 3a, and is representative of other Y21 mutants. Most significantly a transient absorption 1531 $cm^{-1}$ is seen to grow in with time, reaching its maximum ΔOD in several

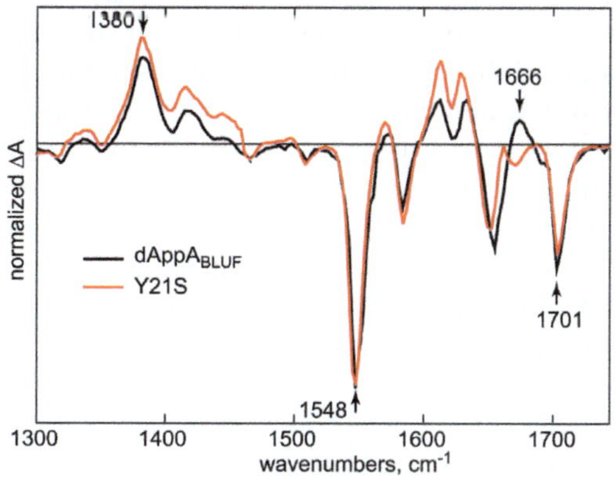


Fig. 2 TRIR spectra for $dAppA_{BLUF}$ and Y21S recorded straight after electronic excitation. Negative ΔOD signals correspond to bleaching of ground state vibrational modes, whereas positive ΔOD indicates promptly formed transients.

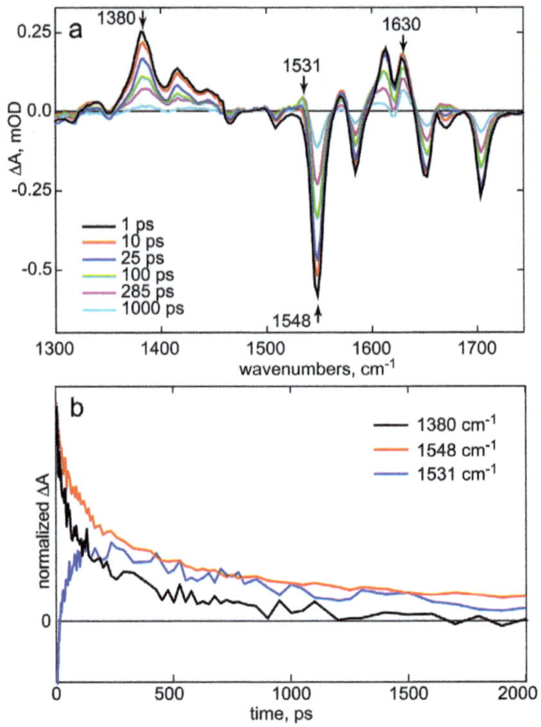


Fig. 3 (a) TRIR spectra of Y21S measured as a function of time after excitation. Note the appearance over tens of picoseconds of the transient absorption around 1530 cm$^{-1}$. (b) Transient data for Y21S at 1380 cm$^{-1}$ and (inverted for comparison) 1548 cm$^{-1}$; also shown are the intermediate kinetics at 1531 cm$^{-1}$.

tens of picoseconds; this wavenumber was shown in model studies to be associated with an IR active vibrational absorption of the flavin radical.[47] Thus, the appearance of this mode suggests that these photoinactive mutants support an electron transfer reaction. The same result was found earlier by Gauden *et al.* for Y21F, Y21I and Y21C in transient absorption.[45] However, the vibrational resolution of TRIR allows for a clearer separation of transient species, including in this case the identification of the isoalloxazine radical absorption. Further the decay of the singlet excited state and the recovery of the ground state can be monitored separately in TRIR, from the decay of the 1380 cm$^{-1}$ transient absorption and the recovery of the 1548 cm$^{-1}$ ground state bleach respectively. As expected for the formation of an intermediate the excited state decays more rapidly than the ground state recovers (Table 1). However, the difference between the two is quite small, suggesting that once the radical is formed it decays quite quickly to the ground state. To illustrate this behaviour time resolved data for the 1380 cm$^{-1}$ decay and the 1548 cm$^{-1}$ ground state recovery of Y21S are compared in Fig. 3b. The separation between the two is apparent, whereas for dAppA$_{BLUF}$ the traces overlapped,[41] suggesting the excited state decay and ground state recovery match, consistent with a short lived (or absent) intermediate in dAppA$_{BLUF}$.

The Y21 mutant data can be globally analysed assuming an FAD* → FAD$^{\bullet -}$ → FAD kinetics scheme. The data were fit to this simplest possible model, which

**Table 1** Excited state decay (1380 $cm^{-1}$) and ground state recovery (1548 $cm^{-1}$) kinetics for dAppA, dW104 and Y21 mutants. Tau 1 and 2 were recovered from a biexponential fit to the data and the weight is the amplitude of the fast component. A constant offset was included, and has a significant amplitude in the 1548 $cm^{-1}$ recovery (Fig. 3 and 5)

| BLUF | Frequency/$cm^{-1}$ | Weight tau 1 | Tau 1/ps | Tau 2/ps | ⟨Tau⟩/ps |
|---|---|---|---|---|---|
| dAppA | 1380 | 0.38 | 17 ± 8 | 401 ± 110 | 255 |
| | 1548 | 0.42 | 29 ± 5 | 512 ± 60 | 309 |
| Y21S | 1380 | 0.48 | 25 ± 7 | 350 ± 100 | 194 |
| | 1548 | 0.36 | 38 ± 10 | 413 ± 70 | 278 |
| Y21C | 1380 | 0.71 | 49 ± 20 | 420 ± inf | 156 |
| | 1548 | 0.41 | 50 ± 13 | 431 ± 90 | 274 |
| Y21I | 1380 | 0.56 | 26 ± 7 | 313 ± 105 | 152 |
| | 1548 | 0.48 | 48 ± 9 | 431 ± 98 | 247 |
| Y21IW104F | 1380 | 0.32 | 7 ± 4 | 152 ± 30 | 105 |
| | 1548 | 0.53 | 81 ± 30 | 294 ± 120 | 181 |
| W104A | 1380 | 0.37 | 4 ± 3 | 663 ± 290 | 419 |
| | 1548 | 0.42 | 21 ± 6 | 598 ± 140 | 355 |
| W104T | 1380 | 0.27 | 4 ± inf | 300 ± 85 | 219 |
| | 1548 | 0.28 | 20 ± 4 | 465 ± 40 | 340 |
| W104Y | 1380 | 0.27 | 9 ± 3 | 488 ± 70 | 358 |
| | 1548 | 0.27 | 11 ± 3 | 478 ± 45 | 351 |

assumes that the excited state decays can be represented approximately by a single exponential decay function, which is not the case; the non-single exponential kinetic introduced considerable ambiguity into the analysis. Nevertheless, the fit to the data by the simplest model is reasonable, with the singlet decaying to the intermediate in 31 ps which goes to a final state in 400 ps; the EADS are shown in Fig. 4a. In this case the 1531 $cm^{-1}$ transient absorption associated with the $FAD^{\bullet -}$ appears in the spectrum of the intermediate, consistent with the scheme. However, the 1380 $cm^{-1}$ transient associated with FAD* persists for longer times than the radical formation, which is unphysical, but is a consequence of imposing a simple single exponential decay scheme on non-single exponential data. Note also that the final state has finite amplitude suggesting some long lived modification to the spectrum in Y21 mutants. This could be a change in the H-bond environment of the FAD or in principle a long lived triplet state (although the triplet state of FMN does not have any contribution around 1630 $cm^{-1}$, which is an important part of the final spectrum, Fig. 4a (ref. 26)). In an effort to model these inherently non-single exponential data they were analysed assuming two initial states (A and A′), one of which decays to the radical ($A^{\bullet -}$) and one directly to the ground state; the condition that the 1380 $cm^{-1}$ peak associated with FAD* is absent in the $FAD^{\bullet -}$ spectrum was also imposed. The radical state then relaxes to the final state (B), which is again slightly perturbed compared to the ground state. The scheme is shown in Scheme 1 and the SADS and relaxation times in Fig. 4b.

Again the fit is good, with the radical mainly associated with the intermediate $A^{\bullet -}$, the decay of A and A′ reproducing the nonexponential decay at 1380 $cm^{-1}$ and the extended $A^{\bullet -}$ decay giving rise to the slower recovery at 1548 $cm^{-1}$. However, this scheme is not unique, many others could be tried, and even with the high signal-to-noise presently available (Fig. 3) the data are not sufficient to

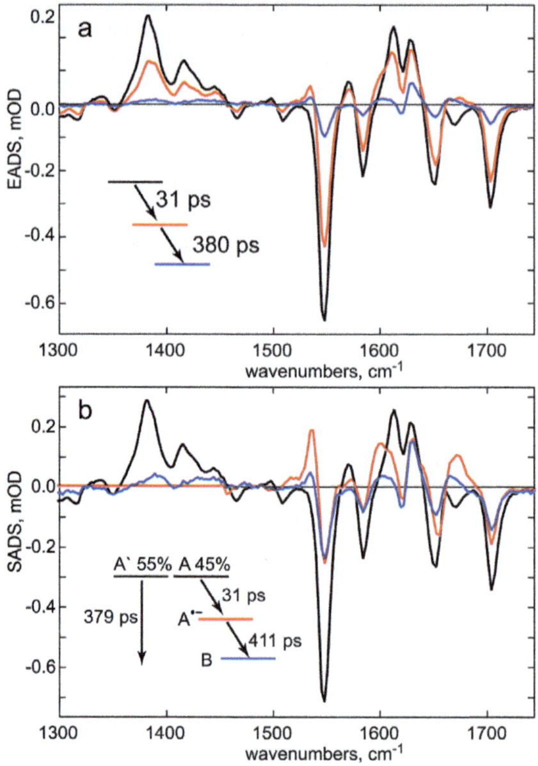


Fig. 4 (a) EADS from analysis of the Y21S data (Fig. 3a) using a sequential kinetic scheme with a single intermediate (b) SADS associated with Scheme 1 (see text). Timescales associated with each step are shown. Note the long lived final structure consistent with a metastable light induced change in the protein structure.

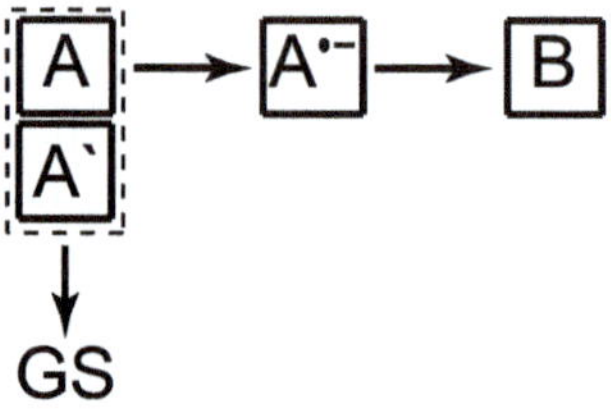


Scheme 1

distinguish between them. This simple Scheme 1 is interesting because it raises the possibility of an inhomogeneous ground state with one population able to undergo radical formation and one which cannot. This is consistent with the non-exponential decay of FAD* and also with recent calculations which found (for the $dAppA_{BLUF}$ structure) that Q63 could occupy different conformers.[38] It is plausible that each conformer has associated with it a re-organised H-bond structure, with each of the various orientations having distinct photochemistry.

 

Concerning the identity of the electron donor, in the absence of Y21 the most likely candidate is the adjacent tryptophan residue W104.[45] In line with this, the double mutant Y21I/W104F did not show any radical like absorption at 1531 $cm^{-1}$. The previously studied (photoinactive) mutant Y21W also showed rapid and efficient electron transfer and clear intermediate state kinetics; thus, in that case Trp is unambiguously the electron donor.[41] The rate of electron transfer in Y21W was much higher than for the present Y21 mutants, which is consistent with W21 being located close to the isoalloxazine ring. Interestingly along with the expected radical isoalloxazine feature at 1531 $cm^{-1}$, Y21W showed an additional intense sigmoidal transient/bleach feature about 1630 $cm^{-1}$ in both the experimental spectra and the EADS recovered from global analysis.[41] This 1630 $cm^{-1}$ feature is also evident in the EADS of Y21S (Fig. 4a) and other Y21 mutants, but is of lower amplitude than for Y21W. Thus, these data are consistent with Trp being the electron donor in both cases, and the difference in the intermediate spectra suggests that W104 and W21 occupy different H-bond environments, with different amplitude vibrational modes.

### Photodynamics in dW104 mutants

These mutants all showed a red-shift on irradiation at 450 nm, a necessary (but not sufficient[29]) condition for photoactivity. However, the signalling state (light) to dark relaxation was accelerated compared to $dAppA_{BLUF}$, by a factor which varies between three (for W104M) and greater than twenty (for W104A). The TRIR spectrum recorded shortly after excitation of dW104T is shown in Fig. 5a and compared with that for $dAppA_{BLUF}$. As was observed for the Y21 mutants (Fig. 3) there are only small effects of W104 mutation on the spectra when compared to $dAppA_{BLUF}$. The C4=O mode is shifted slightly to higher wavenumber (by 3 $cm^{-1}$, which is close to our instrumental resolution, but a similar shift was seen by Masuda and co-workers[29]) while the 1666 $cm^{-1}$ photoactive marker mode is retained. This is consistent with a $dAppA_{BLUF}$ rather than a $lAppA_{BLUF}$ environment (see below) for these mutants.

The temporal evolution for W104T is shown in Fig. 5b, and is representative of the other W104 mutants. In contrast to what was observed for the photoinactive Y21 mutants, the photoactive (as judged by red-shift under irradiation) W104 mutants do not show the appearance of a flavin radical peak in the 1531 $cm^{-1}$ region, a result which they share with the photoactive $dAppA_{BLUF}$[33] and BlsA[48] (though not PixD[32,39,41]). Further, as was also found for these photoactive BLUF domains,[41] the decay of the excited state (1380 $cm^{-1}$) matches the recovery of the ground state (1548 $cm^{-1}$), which is again inconsistent with a long lived charge separated intermediate (Table 1). Thus, there is no obvious correlation between the observation of an intermediate radical state in the BLUF photocycle and BLUF photoactivity; in fact the mutations studied here suggest an anti-correlation, with the flavin radical intermediate only being observed in photoinactive Y21 mutants. The absence of a radical feature in the photoactive W104 mutants thus requires that either the Y21-FAD → $Y21^{\bullet +}FAD^{\bullet -}$ electron transfer reaction does not occur, or that the decay of the radical state once formed is accelerated compared to Y21, such that the concentration is very low. This point is discussed further below.

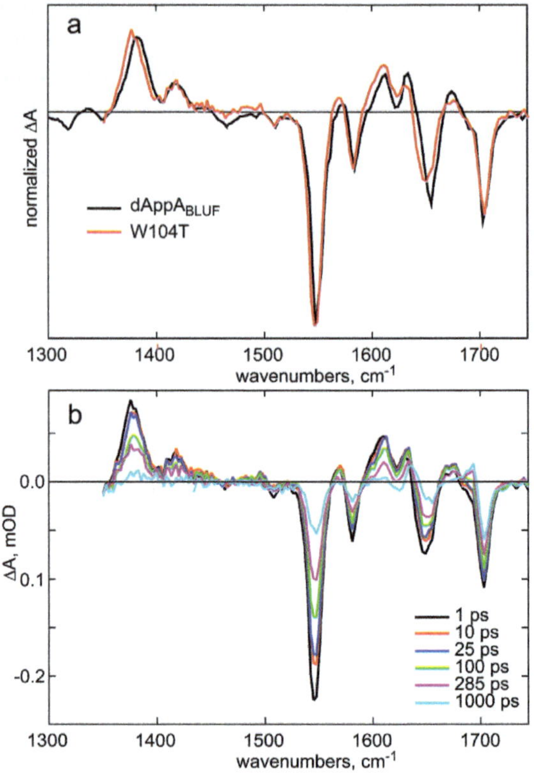


Fig. 5 (a) TRIR of W104T compared with dAppA$_{BLUF}$ immediately after excitation (b) Time dependence of TRIR for W104T.

## Radical intermediates in lAppA$_{BLUF}$ and lW104 mutants

The light adapted form lAppA$_{BLUF}$ is readily generated by irradiation and persists for some time because of its slow light to dark recovery. For lW104 mutants the recovery is faster, but they may also be studied if the sample is continuously irradiated. Thus, TRIR of the metastable light adapted states can be measured (Fig. 6a). Two features are immediately apparent. First, as noted previously,[25,33,49] the excited state decay and ground state recovery are >10 times faster than for dAppA$_{BLUF}$, and second that the electron transfer reaction proceeds quite efficiently in lAppA$_{BLUF}$, as indicated by the transient formation of the flavin radical intermediate (here at 1525 cm$^{-1}$) and the much faster decay at 1380 cm$^{-1}$ compared to the 1548 cm$^{-1}$ recovery (Table 2, Fig. 6b). The changes in the bleach spectrum between the TRIR of d and lAppA$_{BLUF}$ (comparing with the data in Fig. 2 and 5) were described elsewhere, and can be understood on the basis of the modified H-bonding around the isoalloxazine ring on formation of the light adapted state, with the red-shift in the C4=O ground state bleach aligning with the light minus dark steady state IR difference spectra reported by Masuda and co-workers.[29,50]

The complex lineshape of the TRIR spectra in the 1510–1530 cm$^{-1}$ region for lAppA$_{BLUF}$ (Fig. 6a) suggests a fast evolution of the spectral profile in the flavin

 

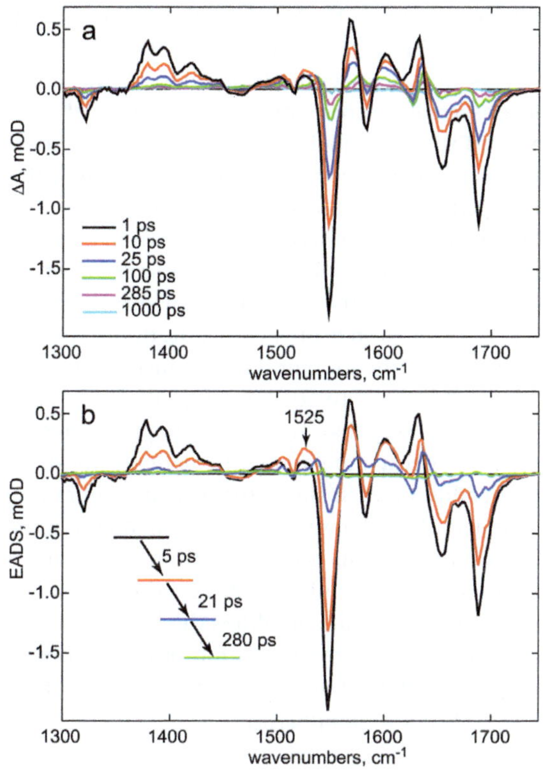


Fig. 6 (a) Evolution of TRIR for $lAppA_{BLUF}$ (b) EADS for $lAppA_{BLUF}$ recovered from a sequential analysis assuming two intermediate states. Note that the final spectrum has essentially zero amplitude suggesting repopulation of the ground state.

Table 2 Excited state decay (1380 $cm^{-1}$) and ground state recovery (1548 $cm^{-1}$) kinetics for $lAppA_{BLUF}$ and lW104 mutants

| | Frequency/$cm^{-1}$ | Weight $\tau_1$ | $\tau_1$/ps | $\tau_2$/ps | ⟨Tau⟩/ps |
|---|---|---|---|---|---|
| $lAppA_{BLUF}$ | 1380 | 0.26 | 1.4(+1) | 21(±3) | 15 |
| | 1548 | 0.77 | 16.3(±1) | 190(±40) | 56 |
| lW104A | 1380 | 0.89 | 1 ± inf | 16 ± inf | 2 |
| | 1548 | 0.8 | 17 ± 9 | 474 ± 1inf | 108 |
| lW104T | 1380 | 0.71 | 8 ± 4 | 108 ± 70 | 37 |
| | 1548 | 0.63 | 6 ± 91 | 102 ± 15 | 41 |
| lW104Y | 1380 | 0.65 | 7 ± 3 | 93 ± 40 | 37 |
| | 1548 | 0.67 | 9 ± 1 | 161 ± 25 | 59 |

radical region, which is different to the behaviour seen in the Y21 mutants. This is highlighted by the global analysis of $lAppA_{BLUF}$ (Fig. 6b), in which a simple sequential model was assumed (*i.e.* again imposing single exponential decay on FAD*) but with a second intermediate. In this model the excited state decays to generate a transient species around 1520 $cm^{-1}$ which subsequently evolves to yield a second transient species at 1530 $cm^{-1}$. This may be consistent with a fast ground state proton transfer reaction $FAD^{\bullet-} \rightarrow FADH^{\bullet}$ following excited state electron transfer. However, changes in H-bonding to the protein could also contribute to the observed spectral shift; unfortunately the available model spectra[47,51] are not sufficiently detailed to resolve the spectrum further. The second intermediate essentially repopulates the ground state without leaving any long lived structural perturbation (unlike Y21 mutants).

The enhanced quenching in $lAppA_{BLUF}$ must arise from a reorganisation of the protein in the light adapted state to dramatically enhance electron transfer. One possibility for $lAppA_{BLUF}$ is that W104 moves from an 'out' to an 'in' configuration, and this gives rise to fast excited state quenching, the driving force for a Trp donor being favourable.[52] However, the mutant data are inconsistent with this assignment. All the lW104A lW104T and lW104Y mutants also show faster excited state decay and ground state recovery than their dark adapted states, the effect being noticeable in the average decay time and particularly in the slow component of the decay/recovery. In addition each shows the formation of radical intermediate states. These mutant data show that the likely electron donor in the light adapted state is Y21. Further support for this is obtained from our studies with unnatural amino acid substitution, where modulation of the redox potential of Y21 through substitution with fluorinated tyrosine modifies the kinetics in the light adapted state (unpublished). Significantly, the formation of observable radical intermediates is only readily detected in a photo-inactivated forms of the protein, in this case $lAppA_{BLUF}$, but is absent in $dAppA_{BLUF}$.

## Relation of AppA structure to quenching

The classical Marcus expression for the rate constant of an electron transfer reaction, $k_{et}$, is:[53]

$$k_{et} = \frac{2\pi\langle V_{el}\rangle^2}{\hbar(4\pi\lambda kT)^{1/2}} \exp - \left[\frac{(\Delta G_0 + \lambda)^2}{4\lambda kT}\right],$$

in which $\Delta G_0$ is the free energy change associated with the reaction, which is related to the reduction and oxidation potentials of the reactants, $V_{el}$ is the electronic coupling parameter between donor and acceptor and $\lambda$ the reorganization energy. Both of the latter parameters depend on the details of the local structure and environment, while $\lambda$ is also a function of $\Delta G_0$. The other parameters have their usual meaning. This equation defines the Marcus curve, a parabola with maximum at $-\Delta G_0 = \lambda$ separating the normal and Marcus inverted regions. As discussed by Lukacs *et al.*, the failure to observe a radical intermediate following photoexcitation of the isoalloaxazine ring in $dAppA_{BLUF}$, an observation which is repeated here for the photoactive dW104 mutants, may be explained by charge recombination (CR) occurring faster than charge separation (CS) in the FAD*–Y21 pair.[41] With the available information on $\Delta G_0^{CR}$ and $\Delta G_0^{CS}$ this suggested a reaction in the normal region of the Marcus parabola. However, modulation of $\Delta G_0^{CS/CR}$, achieved by replacing Y21 with fluorinated tyrosine derivatives, did not result in the expected acceleration of the excited state decay at 1380 $cm^{-1}$.[41] Thus for $dAppA_{BLUF}$ (and, by extension, for the dW104 mutants) there is no conclusive experimental evidence that charge separation is actually the driving force for BLUF photoactivation. What is clear is that electron transfer accompanied by formation of a radical intermediate state is possible in a number of cases (lAppA, the lW104 mutants, the Y21 mutants) in which the BLUF photoactivation pathway is suppressed. In these cases both Y21 (in the lW104 mutants) and W104 (in the Y21 mutants) have been shown to act as the electron donor, where electron transfer was inferred from observations of a transient absorption at a frequency associated with the isoalloxazine radical. This is turn suggests that in these cases the charge separation reaction occurs in the Marcus inverted region, which could arise from a change in the reorganization energy or the driving force by a few hundred meV.

For the Y21 mutants, the electron donor is proposed to be W104 and the excited state decay/ground state recovery times are quite similar to that of $dAppA_{BLUF}$. That the radical intermediate is not seen in $dAppA_{BLUF}$ suggests either that the electron transfer does not occur or that the CS state is much shorter lived compared with the Y21 mutants. In either case, this implies a significant change in the protein environment (compared to $dAppA_{BLUF}$) around W104/FAD in the Y21 mutants. In contrast to the Y21 mutants the 'photoactive' (*i.e.* red-shifting) dW104 mutants show no intermediate, although they also have similar spectra and decay kinetics to $dAppA_{BLUF}$. The change in biological activity for these mutants probably arises from their inability to communicate the photoinduced change in the local structure to the remainder of the protein.[26] In $lAppA_{BLUF}$ and in the lW104 mutants the excited state kinetics are greatly accelerated and an intermediate assigned to a radical state is observed. This more efficient quenching (which has been noted before) indicates a substantial change in the excited state chemistry. In the most extreme case this could arise from a change in the charge state of the reactants (*e.g.* from tyrosine to tyrosinate), thus modifying $\Delta G_0$. Alternatively, the change in structure associated with formation of the light adapted state must significantly alter the electronic coupling as well as the reorganization energy.

The question which is not really addressed by these considerations is the origin of the nonexponential kinetics in all BLUF proteins so far studied. This observation suggests that a distribution of ground state structures exists in solution in thermal equilibrium, and that each structure has a different excited state chemistry, with perhaps only the minor populations being photoactive. Recent calculations indeed suggested a distribution of Q63 rotamers.[38] A similar distribution may also exist in the excited state. It is then possible to imagine a fluctuation in the structure in one of the conformers about FAD* leading to a reactive conformation. In this conformation the reaction (be it tautomerization, electron or proton transfer, or some combination thereof) must occur on an ultrafast timescale returning the protein directly to a ground electronic state which, on further relaxation, traps the protein in its light adapted state, characterized by a reorganized H-bond network around FAD. Such a distribution would further complicate the analysis of BLUF domain photoactivity, as the distribution itself might also be sensitive to interactions between the protein and its complex partners. Thus investigation of these complexes is an important goal.

## Summary

The photodynamics of a number of Y21 and W104 mutants of the $dAppA_{BLUF}$ domain have been studied by ultrafast TRIR spectroscopy. In all cases excited state decay and ground state recovery could be monitored by the study of distinct marker modes. In every case the kinetics departed from simple single exponential behavior, which is assigned to a distribution of ground states structures. In the cases where the light induced red-shift of the isoalloxazine absorption occurs, which indicates the onset of photoactivity in BLUF domains (*i.e.* in $dAppA_{BLUF}$ and the dW104 mutants) no intermediate states could be detected in TRIR spectra. In contrast, in the photoinactive Y21 mutants and in the light adapted $lAppA_{BLUF}$ and lW104 mutants an intermediate could be observed, with an absorption associated with the isoalloxazine radical anion. In the Y21 mutants the donor is likely to be W104 and in the light adapted forms it is Y21. There is no indication in these measurements for formation of a metastable charge separated state in the BLUF photocycle. Finally, the efficient quenching in the light adapted states suggests a very significant change in the local structure or in the reactants of the Y21/FAD pair.

## Acknowledgements

We are grateful to EPSRC and NSF for financial support of this project (EP/K000764/1 to SRM, CHE-1223819 to PJT) and to STFC for programme access to the Central Laser Facility. AL is a Bolyai Janos Research Fellow and was supported by OTKA NN113090.

## References

1 S. O. Smith, in *Annual Review of Biophysics*, ed. D. C. Rees, K. A. Dill and J. R. Williamson, 2010, vol. 39, pp. 309–328.
2 A. Wand, I. Gdor, J. Y. Zhu, M. Sheves and S. Ruhman, Shedding New Light on Retinal Protein Photochemistry, *Annu. Rev. Phys. Chem.*, 2013, **64**, 437–458, DOI: 10.1146/annurev-physchem-040412-110148.
3 A. Wand, N. Friedman, M. Sheves and S. Ruhman, Ultrafast Photochemistry of Light-Adapted and Dark-Adapted Bacteriorhodopsin: Effects of the Initial Retinal Configuration, *J. Phys. Chem. B*, 2012, **116**, 10444–10452, DOI: 10.1021/jp2125284.
4 F. Schotte, *et al.*, Watching a signaling protein function in real time via 100-ps time-resolved Laue crystallography, *Proc. Natl. Acad. Sci. U. S. A.*, 2012, **109**, 19256–19261, DOI: 10.1073/pnas.1210938109.
5 T. W. Kim, *et al.*, Protein Structural Dynamics of Photoactive Yellow Protein in Solution Revealed by Pump-Probe X-ray Solution Scattering, *J. Am. Chem. Soc.*, 2012, **134**, 3145–3153, DOI: 10.1021/ja210435n.
6 S. Masuda, Light Detection and Signal Transduction in the BLUF Photoreceptors, *Plant Cell Physiol.*, 2013, **54**, 171–179, DOI: 10.1093/pcp/pcs173.
7 A. Losi, Flavin-based blue-light photosensors: A photobiophysics update, *Photochem. Photobiol.*, 2007, **83**, 1283–1300, DOI: 10.1111/j.1751-1097.2007.00196.x.
8 A. Losi and W. Gartner, Old Chromophores, New Photoactivation Paradigms, Trendy Applications: Flavins in Blue Light-Sensing Photoreceptors, *Photochem. Photobiol.*, 2010, **87**, 491–510, DOI: 10.1111/j.1751-1097.2011.00913.x.

 

9 K. S. Conrad, C. C. Manahan and B. R. Crane, Photochemistry of flavoprotein light sensors, *Nat. Chem. Biol.*, 2014, **10**, 801–809, DOI: 10.1038/nchembio.1633.

10 K. Deisseroth, Optogenetics, *Nat. Methods*, 2011, **8**, 26–29, DOI: 10.1038/nmeth.f.324.

11 B. Kim and M. Z. Lin, Optobiology: optical control of biological processes via protein engineering, *Biochem. Soc. Trans.*, 2013, **41**, 1183–1188, DOI: 10.1042/bst20130150.

12 A. Moglich, X. J. Yang, R. A. Ayers and K. Moffat, in *Annual Review of Plant Biology*, ed. S. Merchant, W. R. Briggs and D. Ort, Annual Reviews, 2010, vol. 61, pp. 21–47.

13 P. Hegemann and A. Moglich, Channelrhodopsin engineering and exploration of new optogenetic tools, *Nat. Methods*, 2010, **8**, 39–42.

14 A. Lukacs, A. P. M. Eker, M. Byrdin, K. Brettel and M. H. Vos, in *Ultrafast Phenomena XVI*, ed. P. Corkum, S. DeSilvestri, K. A. Nelson and E. Riedle, Springer Series in Chemical Physics, 2009, vol. 92, pp. 604–606.

15 S. Raffelberg, M. Mansurova, W. Gärtner and A. Losi, Modulation of the Photocycle of a LOV Domain Photoreceptor by the Hydrogen-Bonding Network, *J. Am. Chem. Soc.*, 2011, **133**, 5346–5356, DOI: 10.1021/ja1097379.

16 T. Bednarz, A. Losi, W. Gartner, P. Hegemann and J. Heberle, Functional variations among LOV domains as revealed by FT-IR difference spectroscopy, *Photochem. Photobiol. Sci.*, 2004, **3**, 575–579, DOI: 10.1039/b400976b.

17 S. Anderson, *et al.*, Structure of a novel photoreceptor, the BLUF domain of AppA from Rhodobacter sphaeroides, *Biochemistry*, 2005, **44**, 7998–8005.

18 V. Dragnea, *et al.*, Crystallographic and functional studies of the AppA blue-light receptor BLUF domain from Rhodobacter sphaeroides, *Abstracts of Papers of the American Chemical Society*, 2005, **230**, U2809.

19 S. Masuda, Molecular mechanism of light-dependent signal transduction pathway in AppA, *Plant Cell Physiol.*, 2006, **47**, S7.

20 S. Masuda and C. E. Bauer, AppA is a blue light photoreceptor that antirepresses photosynthesis gene expression in Rhodobacter sphaeroides, *Cell*, 2002, **110**, 613–623.

21 S. Masuda, K. Hasegawa and T. Ono, Light-induced structural changes of apoprotein and chromophore in the BLUF domain of AppA for a signaling state, *Plant Cell Physiol.*, 2005, **46**, S103.

22 V. Massey, The chemicaland biological versatility of riboflavin, *Biochem. Soc. Trans.*, 2000, **28**, 283–296.

23 V. Dragnea, A. I. Arunkumar, C. W. Lee, D. P. Giedroc and C. E. Bauer, A Q63E Rhodobacter sphaeroides AppA BLUF Domain Mutant Is Locked in a Pseudo-Light-Excited Signaling State, *Biochemistry*, 2010, **49**, 10682–10690, DOI: 10.1021/bi1002162.

24 W. Laan, *et al.*, On the mechanism of activation of the BLUF domain of AppA, *Biochemistry*, 2006, **45**, 51–60.

25 M. Gauden, *et al.*, Photocycle of the flavin-binding photoreceptor AppA, a bacterial transcriptional antirepressor of photosynthesis genes, *Biochemistry*, 2005, **44**, 3653–3662.

26 R. Brust, *et al.*, Proteins in Action: Femtosecond to Millisecond Structural Dynamics of a Photoactive Flavoprotein, *J. Am. Chem. Soc.*, 2013, **135**, 16168–16174, DOI: 10.1021/ja407265p.
27 S. Masuda, K. Hasegawa, A. Ishii and T. Ono, Light-induced structural changes in a putative blue-light receptor with a novel FAD binding fold sensor of blue-light using FAD (BLUF); Slr1694 of Synechocystis sp PCC6803, *Biochemistry*, 2004, **43**, 5304–5313.
28 S. Masuda, K. Hasegawa and T. Ono, Light-induced structural changes of apoprotein and chromophore in the sensor of blue light using FAD (BLUF) domain of AppA for a signaling state, *Biochemistry*, 2005, **44**, 1215–1224.
29 S. Masuda, K. Hasegawa and T. A. Ono, Tryptophan at position 104 is involved in transforming light signal into changes of beta-sheet structure for the signaling state in the BLUF domain of AppA, *Plant Cell Physiol.*, 2005, **46**, 1894–1901.
30 S. Masuda, Y. Tomida, H. Ohta and K. I. Takamiya, The critical role of a hydrogen bond between Gln63 and Trp104 in the blue-light sensing BLUF domain that controls AppA activity, *J. Mol. Biol.*, 2007, **368**, 1223–1230, DOI: 10.1016/j.jmb.2007.02.087.
31 A. Jung, J. Reinstein, T. Domratcheva, R. L. Shoeman and I. Schlichting, Crystal structures of the AppA BLUF domain photoreceptor provide insights into blue light-mediated signal transduction, *J. Mol. Biol.*, 2006, **362**, 717–732, DOI: 10.1016/j.jmb.2006.07.024.
32 M. Gauden, *et al.*, Hydrogen-bond switching through a radical pair mechanism in a flavin-binding photoreceptor, *Proc. Natl. Acad. Sci. U. S. A.*, 2006, **103**, 10895–10900, DOI: 10.1073/pnas.0600720103.
33 A. L. Stelling, K. L. Ronayne, J. Nappa, P. J. Tonge and S. R. Meech, Ultrafast structural dynamics in BLUF domains: Transient infrared spectroscopy of AppA and its mutants, *J. Am. Chem. Soc.*, 2007, **129**, 15556–15564.
34 T. Domratcheva, B. L. Grigorenko, I. Schlichting and A. V. Nemukhin, Molecular models predict light-induced glutamine tautomerization in BLUF photoreceptors, *Biophys. J.*, 2008, **94**, 3872–3879.
35 M. G. Khrenova, A. V. Nemukhin and T. Domratcheva, Photoinduced Electron Transfer Facilitates Tautomerization of the Conserved Signaling Glutamine Side Chain in BLUF Protein Light Sensors, *J. Phys. Chem. B*, 2013, **117**, 2369–2377, DOI: 10.1021/jp312775x.
36 F. Collette, T. Renger and M. Schmidt am Busch, Revealing the Functional States in the Active Site of BLUF Photoreceptors from Electrochromic Shift Calculations, *J. Phys. Chem. B*, 2014, **118**, 11109–11119, DOI: 10.1021/jp506400y.
37 Y.-W. Hsiao, J. P. Goetze and W. Thiel, The Central Role of Gln63 for the Hydrogen Bonding Network and UV-Visible Spectrum of the AppA BLUF Domain, *J. Phys. Chem. B*, 2012, **116**, 8064–8073, DOI: 10.1021/jp3028758.
38 A. Udvarhelyi and T. Domratcheva, Glutamine Rotamers in BLUF Photoreceptors: A Mechanistic Reappraisal, *J. Phys. Chem. B*, 2013, **117**, 2888–2897, DOI: 10.1021/jp400437x.
39 C. Bonetti, *et al.*, Hydrogen Bond Switching among Flavin and Amino Acid Side Chains in the BLUF Photoreceptor Observed by Ultrafast Infrared Spectroscopy, *Biophys. J.*, 2008, **95**, 4790–4802, DOI: 10.1529/biophysj.108.139246.

 

40 A. Lukacs, *et al.*, Photoexcitation of the Blue Light Using FAD Photoreceptor AppA Results in Ultrafast Changes to the Protein Matrix, *J. Am. Chem. Soc.*, 2011, **133**, 16893–16900, DOI: 10.1021/ja2060098.

41 A. Lukacs, *et al.*, BLUF Domain Function Does Not Require a Metastable Radical Intermediate State, *J. Am. Chem. Soc.*, 2014, **136**, 4605–4615, DOI: 10.1021/ja4121082.

42 G. M. Greetham, *et al.*, ULTRA: A Unique Instrument for Time-Resolved Spectroscopy, *Appl. Spectrosc.*, 2011, **64**, 1311–1319.

43 M. Towrie, *et al.*, Development of a broadband picosecond infrared spectrometer and its incorporation into an existing ultrafast time-resolved resonance Raman, UV/visible, and fluorescence spectroscopic apparatus, *Appl. Spectrosc.*, 2003, **57**, 367–380.

44 J. J. Snellenburg, S. P. Laptenok, R. Seger, K. M. Mullen and I. H. M. van Stokkum, Glotaran: A Java-Based Graphical User Interface for the R Package TIMP, *J. Stat. Software*, 2012, **49**, 1–22.

45 M. Gauden, *et al.*, On the role of aromatic side chains in the photoactivation of BLUF domains, *Biochemistry*, 2007, **46**, 7405–7415, DOI: 10.1021/bi7006433.

46 A. Haigney, *et al.*, Ultrafast Infrared Spectroscopy of an Isotope-Labeled Photoactivatable Flavoprotein, *Biochemistry*, 2011, **50**, 1321–1328, DOI: 10.1021/bi101589a.

47 A. Lukacs, *et al.*, Excited State Structure and Dynamics of the Neutral and Anionic Flavin Radical Revealed by Ultrafast Transient Mid-IR to Visible Spectroscopy, *J. Phys. Chem. B*, 2012, **116**, 5810–5818, DOI: 10.1021/jp2116559.

48 R. Brust, *et al.*, Ultrafast Structural Dynamics of BlsA, a Photoreceptor from the Pathogenic Bacterium Acinetobacter baumannii, *J. Phys. Chem. Lett.*, 2014, **5**, 220–224, DOI: 10.1021/jz4023738.

49 V. Dragnea, M. Waegele, S. Balascuta, C. Bauer and B. Dragnea, Time-resolved spectroscopic studies of the AppA blue-light receptor BLUF domain from Rhodobacter sphaeroides, *Biochemistry*, 2005, **44**, 15978–15985.

50 K. Hasegawa, S. Masuda and T. Ono, Structural intermediate in the photocycle of a BLUF (sensor of blue light using FAD) protein Slr1694 in a cyanobacterium Synechocystis sp PCC6803, *Biochemistry*, 2004, **43**, 14979–14986.

51 R.-K. Zhao, *et al.*, Ultrafast transient mid IR to visible spectroscopy of fully reduced flavins, *Phys. Chem. Chem. Phys.*, 2011, **13**, 17642–17648.

52 H. Ishikita, Light-induced hydrogen bonding pattern and driving force of electron transfer in AppA BLUF domain photoreceptor, *J. Biol. Chem.*, 2008, **283**, 30618–30623, DOI: 10.1074/jbc.m803864200.

53 R. A. Marcus and N. Sutin, Electron transfers in chemistry and biology, *Biochim. Biophys. Acta*, 1985, **811**, 265–322, DOI: 10.1016/0304-4173(85)90014-x.

Faraday Discussions

PAPER

# Spatio-temporal correlations in aqueous systems: computational studies of static and dynamic heterogeneity by 2D-IR spectroscopy†

Rikhia Ghosh, Tuhin Samanta, Saikat Banaerjee, Rajib Biswas and Biman Bagchi*

Received 10th October 2014, Accepted 5th November 2014
DOI: 10.1039/c4fd00201f

Local heterogeneity is ubiquitous in natural aqueous systems. It can be caused locally by external biomolecular subsystems like proteins, DNA, micelles and reverse micelles, nanoscopic materials *etc.*, but can also be intrinsic to the thermodynamic nature of the aqueous solution itself (like binary mixtures or at the gas–liquid interface). The altered dynamics of water in the presence of such diverse surfaces has attracted considerable attention in recent years. As these interfaces are quite narrow, only a few molecular layers thick, they are hard to study by conventional methods. The recent development of two dimensional infra-red (2D-IR) spectroscopy allows us to estimate length and time scales of such dynamics fairly accurately. In this work, we present a series of interesting studies employing two dimensional infra-red spectroscopy (2D-IR) to investigate (i) the heterogeneous dynamics of water inside reverse micelles of varying sizes, (ii) supercritical water near the Widom line that is known to exhibit pronounced density fluctuations and also study (iii) the collective and local polarization fluctuation of water molecules in the presence of several different proteins. The spatio-temporal correlation of confined water molecules inside reverse micelles of varying sizes is well captured through the spectral diffusion of corresponding 2D-IR spectra. In the case of supercritical water also, we observe a strong signature of dynamic heterogeneity from the elongated nature of the 2D-IR spectra. In this case the relaxation is ultrafast. We find remarkable agreement between the different tools employed to study the relaxation of density heterogeneity. For aqueous protein solutions, we find that the calculated dielectric constant of the respective systems unanimously shows a noticeable increment compared to that of neat water. However, the 'effective' dielectric constant for successive layers shows significant variation, with the layer adjacent to the protein having a much lower value. Relaxation is also slowest at the surface. We find

*SSCU, Indian Institute of Science, Bangalore 560012, India. E-mail: bbagchi@sscu.iisc.ernet.in*

† Electronic supplementary information (ESI) available: Figures showing layerwise calculation of frequency fluctuation correlation functions with increasing reverse micelle size, and comparison with that of bulk. See DOI: 10.1039/c4fd00201f

that the dielectric constant achieves the bulk value at distances more than 3 nm from the surface of the protein.

# I. Introduction

Understanding the structure and dynamics of water molecules, in bulk as well as in aqueous solutions of biomolecules (proteins, DNA, lipids) has been the subject of continuing interest for many decades.[1–10] A common characteristic of these systems has been found to be the appearance of a local heterogeneity in the liquid near the interface that may propagate only up to 5–10 molecular layers (that is, 2–4 nm).[11] Existence of heterogeneity at such a small length scale makes interfacial water a difficult system to study using conventional techniques, such as light and neutron scattering, nuclear magnetic resonance (NMR) or linear spectroscopic methods like Raman line width measurements. Since the heterogeneity is found to be dynamic in nature, associated with ultrafast time scales in many cases, even X-ray scattering or electron microscopic techniques are not very useful.

Fortunately, recently developed two dimensional infra-red spectroscopy (2D-IR) can serve as a powerful technique for measuring the water dynamics with its femtosecond time resolution.[12,13] In 2D-IR spectroscopy,[14] one can use the –O–H stretching frequency of water, or some other molecule, as a reliable marker of the position of the molecule. The –O–H stretching frequency in water is known to undergo substantial variation.[15] For example, for those water molecules which are hydrogen bonded to a charged group at an inner surface or reverse micelle, the –O–H stretching frequency can be as low as 2800 $cm^{-1}$ while the corresponding value in the bulk is $\sim$3600 $cm^{-1}$. 2D-IR spectroscopy thus provides a valuable method to obtain spatio-temporal resolution that is not achievable with other conventional methods. Water dynamics at the interface of biomolecules can also be explored through dielectric relaxation measurements, albeit with much poorer spatial resolution.

Water at the surface of biological macromolecules, such as proteins, DNA and at extended biological surfaces such as lipid bilayers, is now known as biological water.[4,9,16–18] Many of the characteristics of water at such surfaces are markedly different from those of bulk water, marked by lower values of effective dielectric constant, slower dynamical responses and higher density. In fact, proper hydration of proteins and enzymes is predominantly important for the stability of the structure as well as functionality at the specified site. This explains the largely growing efforts to understand the underlying characteristics of biological water. Dielectric spectra of aqueous protein solutions have been thoroughly studied for more than 60 years. However, the effect of large dipole moments of proteins on polarization fluctuation of the aqueous solution has still not been adequately addressed.

In this work, we employ 2D-IR spectroscopy to quantitatively measure the nature of water dynamics (i) inside reverse micelles and (ii) in supercritical water. (iii) We also explore the heterogeneity of water at the biomolecular interface, selectively measuring the polarization fluctuation of water for aqueous protein solutions. To understand the altered dynamics of water we perform a layerwise decomposition of water inside the reverse micelle, with an aim to enumerate the relative contributions of different layers of water molecules to the calculated 2D-IR spectrum. We further quantify the variation of static and dynamic heterogeneity in supercritical water across the Widom line by constructing a 2D-IR spectrum of the '–O–H' stretch as well as from the non-linear density response function, $\chi_4(t)$ and non-Gaussian parameter $\alpha_2(t)$. All the variables beautifully capture the ultrafast time scale of heterogeneity prevalent near the Widom line.

Next we carry out a systematic study of the polarization fluctuation by considering two proteins, immunoglobin binding domain protein G (GB1) and lysozyme (Lys), immersed in a large number of water molecules. Analysis of the total and local polarization fluctuation shows surprisingly different water dynamics for the two proteins.

In this report, we have attempted to bring together three different types of aqueous systems, by using mostly the 2D-IR spectroscopic technique, described in section II. For two of them, discussed in sections III and IV, we could construct 2D-IR spectra, and study the heterogeneous dynamics through anisotropy. In the third case, discussed in section V, we study the spatio-temporal correlation in polarization fluctuations.

# II. Spectroscopic modeling of 2D-IR spectra

The experimentally observed 2D-IR response functions $R_{l,m,n}$ are successfully obtained theoretically from the line shape function $g(t)$ as follows[14]

$$R_{1,2,3} = 2i\mu_{01}^4\left(\mathrm{e}^{-i\omega_{01}(t_3-t_1)} - \mathrm{e}^{-i((\omega_{01}-\Delta)t_3-\omega_{01}t_1)}\right)\\ \mathrm{e}^{-g(t_1)+g(t_2)-g(t_3)-g(t_1+t_2)-g(t_2+t_3)+g(t_1+t_2+t_3)} \tag{1}$$

$$R_{4,5,6} = 2i\mu_{01}^4\left(\mathrm{e}^{-i\omega_{01}(t_3+t_1)} - \mathrm{e}^{-i((\omega_{01}-\Delta)t_3+\omega_{01}t_1)}\right)\\ \mathrm{e}^{-g(t_1)-g(t_2)-g(t_3)+g(t_1+t_2)+g(t_2+t_3)-g(t_1+t_2+t_3)} \tag{2}$$

where $\mu_{10}$ is the 1–0 matrix element of the dipole operator and $\omega_{10}(t)$ is the classical time dependent transition frequency. $g(t)$ is the corresponding lineshape function that is obtained analytically from Kubo's lineshape theorem using the frequency fluctuation correlation function (FFCF)$c(t) = \langle\delta\omega(t)\delta\omega(0)\rangle$. The exact form of $g(t)$ is given by the following expression,

$$g(t) = \int_0^t \mathrm{d}\tau' \int^{\tau'} \langle\delta\omega(\tau'')\delta\omega(0)\rangle \mathrm{d}\tau'' \tag{3}$$

All the response functions are Fourier transformed to reveal purely absorptive 2D IR spectra.

Eqn (2) employs the Condon approximation. It has been shown that non-Condon effects are significant for the –O–H vibrational mode of water. We are currently investigating such non-Condon effects. Results presented here, however, use eqn (2), and therefore, are only semi-quantitatively reliable. However, some of the results are quite striking and we expect them to capture much of the true behavior.

# III. Reverse micelles: from spectral resolution to spatial decomposition

As mentioned earlier, reverse micelles (RM) provide us with a unique opportunity to study the ability of 2D-IR spectroscopy to explore the length and time scales of local heterogeneity. Here we adopt a two fold approach. First, we carry out detailed simulations to obtain microscopic details. Second, we use the simulation trajectories to construct the 2D-IR spectrum where water molecules are selected from different layers from the surface. Comparison between the two allows us to use the spectral resolution available from 2D-IR spectroscopy to do a spatial decomposition of the structure and dynamics.

### A. Simulation details

We use the united atom description for Isooctane and all atom description for the surfactant and water molecules. We use TIP4P-2005 water-like molecules.[19] The united atom description of isooctane is obtained from a GROMOS53A6 force field. The initial spherical RM structures are prepared using packmol and then solvated with isooctane like molecules in a cubic box. We use GROMACS, version 4.5.5 (ref. 20) for all the MD simulations in this study. Steepest descent method is used for energy minimization of the initial configuration; followed by position restrained simulation in the isothermal isobaric ensemble (constant pressure, NPT) for 2 ns, by imposing position restrain condition on oxygen of water and sulfur of AOT. We use fourth order Particle Mesh Ewald (PME) summation for long range interaction with grid spacing of 0.16. We equilibrate the system initially in NPT ensemble for 10 ns followed by 10 ns in the canonical ensemble (constant volume, NVT). Finally data acquisitions are performed in the NVT ensemble. The system is maintained at constant temperature of 300 K and 1 atm pressure by using the Nosé–Hoover thermostat[21,22] and Parrinello–Rahman barostat.[23] Periodic boundary conditions are applied in all three directions. The details of the different RM constituents are summarized in Table 1 (size index is given as $w_0$, which is the number of water molecules per surfactant molecule).

We calculate bulk properties by taking 1000 TIP4P-2005 water molecules in a cubic box.

### B. Evolution of water dynamics with increasing size of reverse micelle: 2D-IR spectroscopic view of heterogeneity

Water dynamics inside reverse micelles of varying sizes is a striking example of heterogeneous spatio-temporal correlations. We explore these aspects by considering reverse micelles of different sizes with different numbers of layers based on the radial density distribution of water. The layerwise classification of water in the RMs based on the radial density distribution of the water is summarized in Table 2.

From the radial density distribution of water from the surface to the center of a reverse micelle, we conclude the existence of distinct layering of water molecules (schematic diagram shown in Fig. 1). In the case of the lowest RM size *i.e.* $w_0 = 2.0$, it is observed that almost all water molecules are restricted in the vicinity of the

Table 1 Compositions of simulated RMs

| $w_0$ | $n_{H_2O}$ | $n_{AOT}/n_{Na}$ | $n_{ISO}$ |
|---|---|---|---|
| 2.0 | 52 | 26 | 504 |
| 4.0 | 140 | 35 | 930 |
| 7.5 | 525 | 70 | 1138 |
| 10.0 | 980 | 98 | 1377 |

 

Table 2 Layerwise classification of water in reverse micelles based on the radial density distribution of the water. Distance from sulfur atom of AOT is denoted as *r*, and $\bar{n}_{H_2O}$ is the average number of water molecules present in the different layers. For $w_0 = 2.0$ all AOT head groups are not solvated by water, which led us to assume a single water layer in this RM

| | Surface layer (SL) | | Intermediate layer (IL) | | Central layer (CL) | |
|---|---|---|---|---|---|---|
| Reverse micelle | $r$ (Å) | $\bar{n}_{H_2O}$ | $r$ (Å) | $\bar{n}_{H_2O}$ | $r$ (Å) | $\bar{n}_{H_2O}$ |
| $w_0 = 4.0$ | $r \leq 4.96$ Å | 111 | $r > 4.96$ Å | 29 | — | — |
| $w_0 = 7.5$ | $r \leq 5.0$ Å | 332 | $5.0\ \text{Å} \geq r > 5.9$ Å | 72 | $r > 5.9$ Å | 121 |
| $w_0 = 10.0$ | $r \leq 5.0$ Å | 474 | $5.0\ \text{Å} \geq r > 6.9$ Å | 269 | $r > 6.9$ Å | 237 |

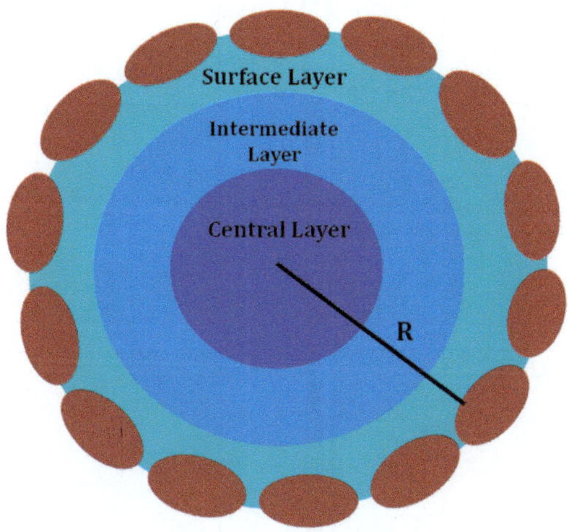


Fig. 1 Schematic diagram showing layering of water inside a representative reverse micelle of effective radius *R*.

surface head groups of the AOT molecules, forming a single water layer. However, with increasing RM size the presence of three distinct layers of water is observed, based on structural and dynamical properties. The water molecules bound to the surface head groups of AOT are named as the *surface layer* (SL); subsequent to the SL there is an intermediate *water* (IL) layer followed by the *central water pool* (CL).

Next we calculate the layerwise hydrogen bond dynamics of water for reverse micelles of different sizes by evaluating the corresponding frequency fluctuation correlation function (FFCF) of all the –O–H bonds. The FFCF efficiently evaluates the spectral diffusion in terms of time scales of the dynamics. Layerwise decomposition of the FFCF gives rise to the following equation

$$\langle \delta\omega(t)\delta\omega(0)\rangle = \frac{1}{N_L}\sum_{L=1}^{N_L}\frac{1}{{N_W}^{LL}}\sum_{i=1}^{{N_W}^{LL}}\left[\left\langle \delta{\omega_i}^{L}(t)\delta{\omega_i}^{L}(0)\right\rangle\right] + \frac{1}{N_L(N_L-1)}\sum_{L_1=1}^{N_L} \times \sum_{L_2=1}^{N_L}\frac{\delta_{L_1L_2}}{{N_W}^{L_1L_2}}\sum_{i=1}^{{N_W}^{L_1L_2}}\left[\left\langle \delta{\omega_i}^{L_1}(t)\delta{\omega_i}^{L_2}(0)\right\rangle\right] \quad (4)$$

where $\delta{\omega_i}^L$ is the frequency fluctuation of the O–H bond of the *i*-th water molecule staying in layer *L*, $N_L$ is the total number of layers, ${N_W}^{LL}$ is the number of water

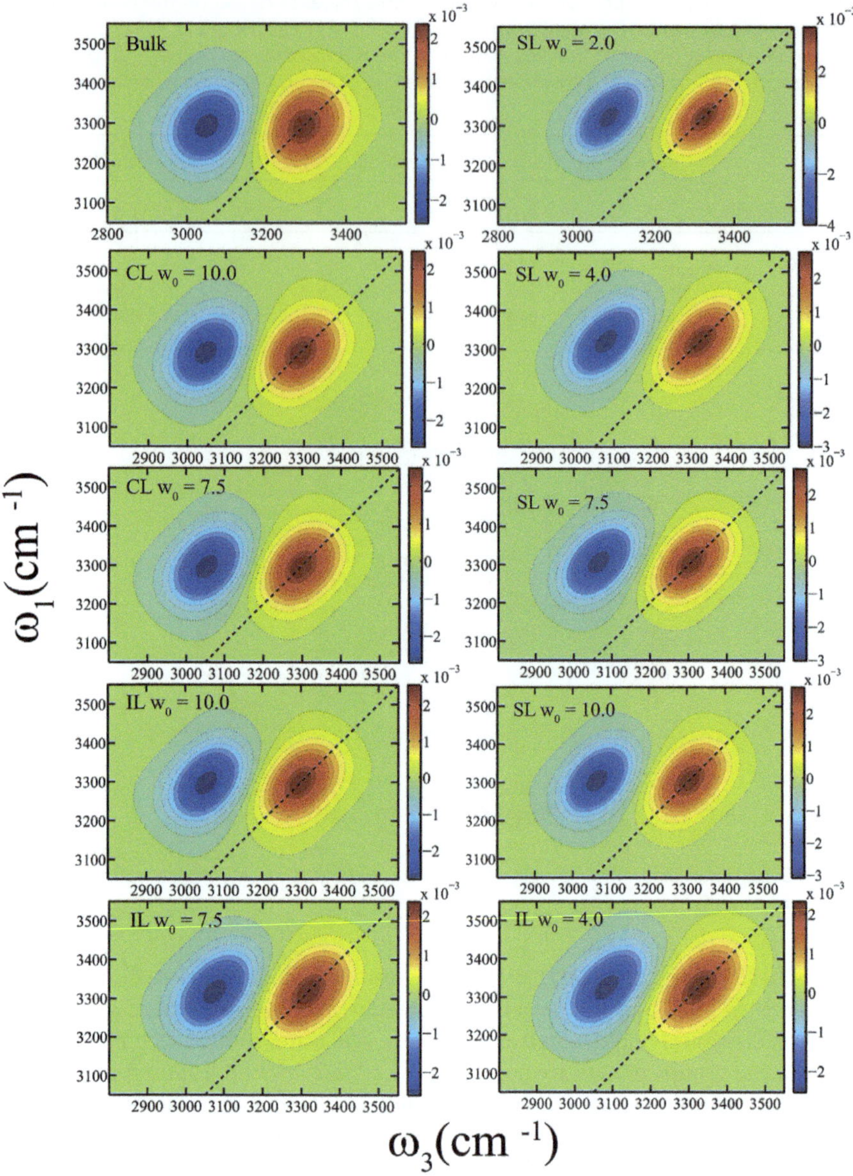


Fig. 2 The 2D IR spectra of the O–H stretch in bulk water and in different layers of reverse micelles (presented clockwise starting from the one for bulk water). Here the layerwise 2D IR spectrum is calculated by tagging the water molecules which stay continuously in that layer during the data acquisition. All the spectra are obtained at a waiting time of 500 fs. The shape of the spectrum depends on two factors: the size of the reverse micelles and the position of the layer we are looking at. It is clear from the figures that as the reverse micelle size increases the spectrum becomes more bulk-like in the central layer.

molecules present in layer $L$ throughout the detection time and gives the self-term $\langle\delta\omega_i^{L}(t)\delta\omega_i^{L}(0)\rangle$, similarly $N_W^{L_1L_2}$ is the total number of water molecules which gives the cross term $\langle\delta\omega_i^{L_1}(t)\delta\omega_i^{L_2}(0)\rangle$, $\delta_{L_1L_2}$ has a value 0 if $L_1 = L_2$ and otherwise is

 

1.0. We compute the self-term of the FFCF using the layer definition. The layer-wise calculation of the FFCFs (shown in ESI Fig. S1†) reveals a clear distinction of hydrogen bond dynamics in different reverse micelles.

To extract the corresponding 2D-IR spectra, we use the following fitting function to fit the FFCFs,

$$c(t) = a_1 \cos(\omega_{OO}t)e^{-t/\tau_1} + a_2\,e^{-t/\tau_2} + a_3\,e^{-t/\tau_3} \tag{5}$$

Using the fitting parameters we calculate the lineshape function $g(t)$, and finally we obtain the 2D IR spectra from the corresponding response functions as described in eqn (1) and (2). The 2D IR spectra of different layers of reverse micelles as well as that of the bulk are shown in Fig. 2, obtained using a waiting time of 500 fs.

The red bands arise from the $0 \rightarrow 1$ vibrational transition whereas the blue bands arise from the vibrational echo emission at $1 \rightarrow 2$ vibrational frequency. The spectra of the $0 \rightarrow 1$ transition clearly show substantial elongation along the diagonal for the surface water layer (SL) with increasing reverse micelle size which is a measure of increasing inhomogeneity. In the case of the subsequent intermediate layers (IL) also, considerable heterogeneity is prevalent, as can be seen

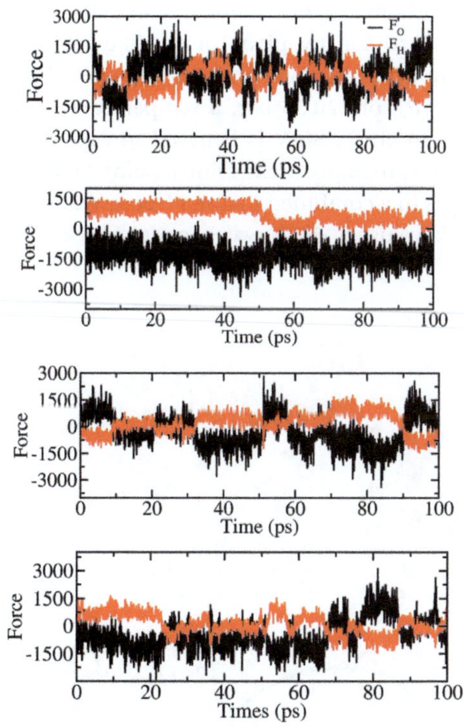


Fig. 3 Force trajectories (projected force in units of kJ mol$^{-1}$ nm$^{-1}$) acting separately on the oxygen atom (in black) and hydrogen atom (in red) of water in the cases of (a) bulk water, (b) tagged surface water of a reverse micelle of size $w_0 = 2.0$, (c) tagged water in the intermediate layer of a reverse micelle of size $w_0 = 4.0$ and (d) tagged central water of a reverse micelle of size $w_0 = 10.0$.

from the elongated $0 \rightarrow 1$ transition spectra. However the central layers (CL) for the corresponding reverse micelles are found to show bulk-like behavior. The altered dynamics of water with increasing reverse micelle size serves as a powerful example of the spatio-temporal correlation of heterogeneous water dynamics.

It is broadly acknowledged that anti-correlation existing between forces on the oxygen and hydrogen atoms of –O–H bonds plays an important role in the decay of FFCFs.[24] Bulk water is found to exhibit ultrafast decay in its FFCF, which essentially arises due to large amplitude angular jumps of water molecules and the appearance of a large cross correlation of projected force on the oxygen and hydrogen atoms of the –O–H bonds. We find such evidence from the corresponding force trajectories (Fig. 3). It can be seen that the surface effect significantly reduces anti-correlation between the forces on oxygen and hydrogen. However, substantial anti-correlation is prevalent at the center of the larger reverse micelles (Fig. 3(c)) which eventually reflects faster decay of the FFCF in this case (as can be seen in Fig. S1†)

## IV. Enhanced heterogeneity of supercritical water across the Widom line

The upward line that originates at the critical temperature ($T_c$) and separates (in the temperature–pressure–density plane) a gas-like region from a liquid-like region above $T_c$, is known as the Widom line. The critical point of real water is located at a pressure ($P_c$) of 220.55 bar, a temperature ($T_c$) of 647.13 K and a density ($\rho_c$) of 0.322 gm cm$^{-3}$. SCW is believed to exist at temperatures above 647.13 K. Because of the proximity to the critical point, SCW exhibits properties of both a liquid and a gas in a combined manner (representative snapshot provided in Fig. 4). In our work we use a SPC/E water model whose critical parameters are

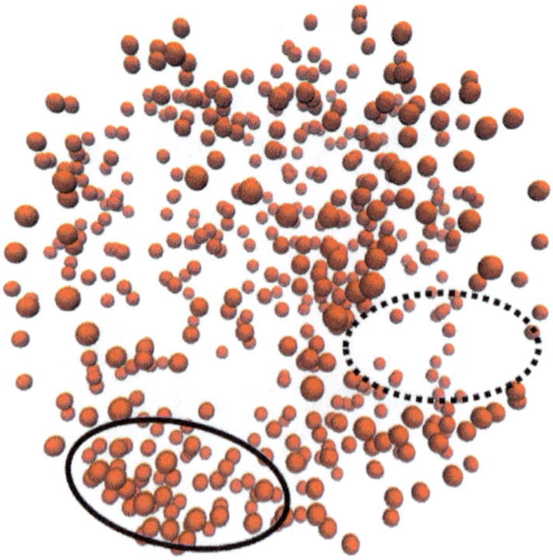

Fig. 4 Representative snapshot of supercritical water. Prominent signature of co-existing high density region (shown by solid circle) and low density region (marked by dashed circle) is found here.

$T_c = 651.7$ K, $\rho_c = 0.326$ g cm$^{-3}$, and $P_c = 189$ bar. These values are very close to the experimental gas–liquid coexistence point.

Although the fluid here is macroscopically homogeneous, supercritical water (SCW) near the Widom line is known to exhibit pronounced density fluctuations, remnant of critical phenomena. Variation of static and dynamic heterogeneity in SCW across the Widom line is studied (i) by constructing a 2D-IR spectrum of the '–O–H' stretch, (ii) from the non-linear density response function $\chi_4(t)$ and (iii) non-Gaussian parameter $\alpha_2(t)$. All of them show a sudden and sharp change close to the Widom line.

### A. Simulation details

All the simulations for supercritical water are carried out using 2048 SPC/E water in a cubic box. Similar steps of equilibration, as that for water inside the reverse micelle, are followed in this case also. The temperature is kept constant at 670 K using a Nosé–Hoover thermostat[21,22] and the pressure is kept at 217.3 atm. Density is varied consistently along the Widom line keeping the temperature constant ($\rho \sim 0.32$, 0.33 and 0.34 gm cm$^{-3}$ respectively).

### B. 2D-IR spectroscopic study of supercritical water: evidence of static and dynamic heterogeneity

Using the same technique as described above, we plot the 2D-IR spectra of the –O–H stretch for supercritical water measured at different waiting times (Fig. 5). Due to vibrational anharmonicity the red and blue bands are shifted toward

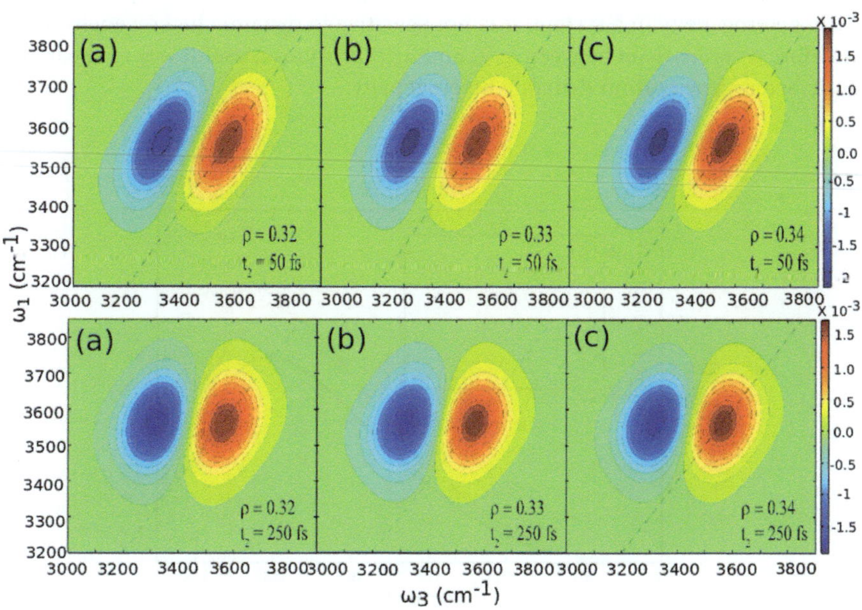


Fig. 5 2D-IR spectra for different densities: (a) $\rho = 0.32$ gm cm$^{-3}$, (b) $\rho = 0.33$ gm cm$^{-3}$, (c) $\rho = 0.34$ gm cm$^{-3}$, with delay time $t_2 = 50$ fs (upper panel) and 250 fs (lower panel). Note that at a short time for all densities the inhomogeneous environments of the system are well captured by the diagonal elongation of the 2D-IR spectra, at temperature 670 K.

lower values of the $\omega_3$ axis. The spectra with a short delay time (50 fs) show a pronounced diagonal elongation, arising due to incomplete spectral diffusion and inhomogeneous broadening. However, an increase in the waiting time causes the –O–H bond to start sampling an average environment that is homogenous in nature and thereby leads to complete spectral diffusion. For a larger waiting time we expect to get a symmetric shape of the 2D-IR spectra. The observed 2D-IR spectra reflect the enhanced inhomogeneity in the vicinity of the critical point ($\rho \sim 0.33$ gm cm$^{-3}$). This essentially gives a slower spectral diffusion. However, such spectral diffusion is relatively faster for other densities.

We establish an idea about the time-scale of the heterogeneity by the plotting dynamical four point susceptibility, $\chi_4(t)$. $\chi_4(t)$ essentially measures the spatio-temporal correlation between the local density at two points in space at two different times. Thus, as the critical point is approached and the system becomes intrinsically more heterogeneous, $\chi_4(t)$ is generally found to be increasingly more pronounced.

In Fig. 6 we plot $\chi_4(t)$ for the density range of 0.32–0.34 gm cm$^{-3}$ (around critical density) at $T = 670$ K. For all densities, $\chi_4(t)$ is nearly zero at shorter times and peaks at around 100 fs. This essentially suggests that the dynamic heterogeneity present in SCW has a lifetime of the order of 100 fs. The variation of peak height of $\chi_4(t)$ with density is shown in the inset of Fig. 6. The peak height shows a maximum at the density ($\rho$) 0.33 gm cm$^{-3}$, as expected. As we go beyond the density ($\rho$) 0.33 gm cm$^{-3}$ the peak height value of $\chi_4(t)$ again decreases. This signifies that the system becomes more homogeneous as we move away from the critical point.

The well-known non-Gaussian parameter $\alpha_2(t)$ demonstrates the presence of static heterogeneity in a system. It is not sensitive to dynamic heterogeneity and therefore a comparison between $\chi_4(t)$ and $\alpha_2(t)$ provides a useful way to separate static heterogeneity from dynamic heterogeneity.

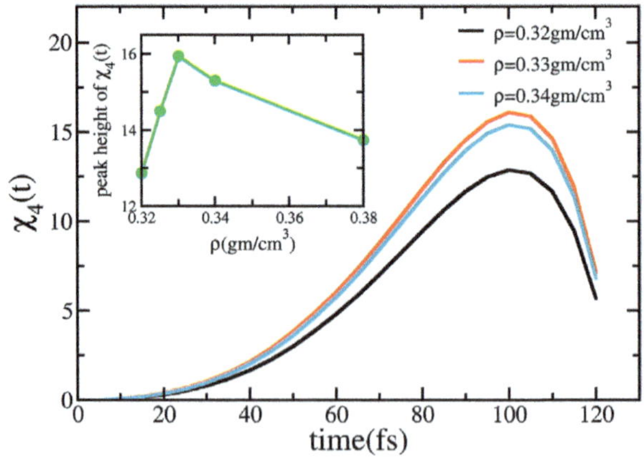


Fig. 6 Four point density susceptibility *versus* time for different average densities at 670 K. In the inset, the respective peak maxima of four point susceptibility are plotted with density, all at $T = 670$ K. Note that the inset plot, density dependent peak height of four point susceptibility shows a maximum at $\rho = 0.33$ gm cm$^{-3}$.

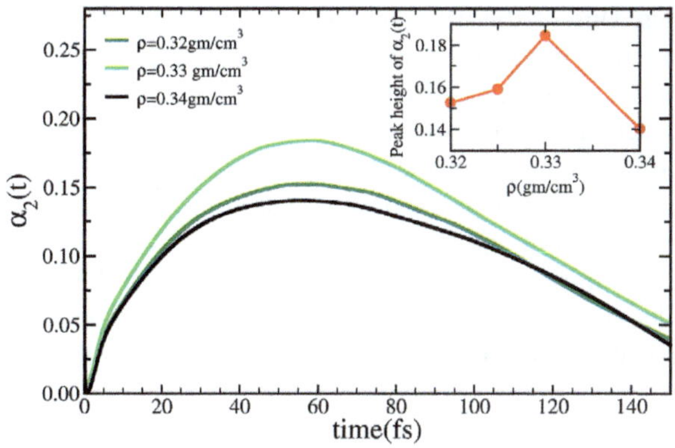


Fig. 7 Temporal evolution of the non-Gaussian parameter for different densities (inset: variation of peak maxima of non-Gaussian parameter along with varying density). Note that the inset plot shows a transition of peak maximum again at $\rho = 0.33$ gm cm$^{-3}$ where non-Gaussian behavior and heterogeneity are maximum.

In Fig. 7, we observe a large value of $\alpha_2(t)$ due to the presence of long lived heterogeneity near the critical density ($\rho_c = 0.33$ gm cm$^{-3}$). Deviation from Gaussian behavior becomes less prominent in all densities except that of the critical density. The time scale of static heterogeneity is found to prevail at ~70 fs (as observed from the peak position).

## V. Spatio-temporal resolution of polarization fluctuations in aqueous protein solutions

Aqueous protein solutions serve as an important subject of study, as the large dipole moment of the protein molecules can significantly perturb the dynamics of

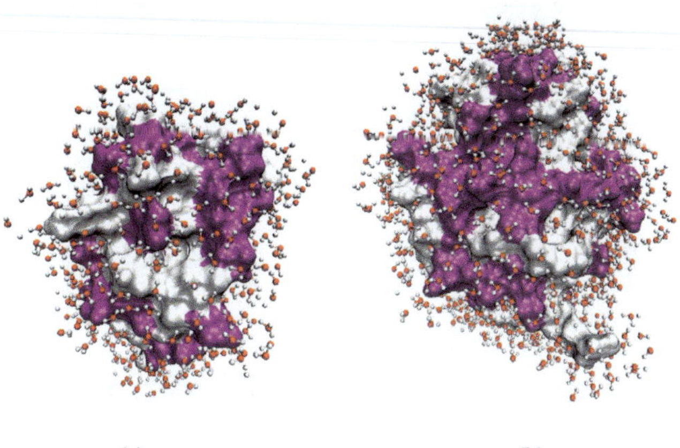


Fig. 8 Representative conformational surface of the two proteins investigated in this work: (a) protein G (56 residues) and (b) lysozyme (129 residues). The purple surface represents the hydrophilic patch whereas the white surface represents the hydrophobic patch. The water layer around the protein up to 5 Å is shown. Note that the two proteins are quite different.

water molecules. One of the most relevant properties that is anticipated to be affected due to the presence of such large dipoles, is collective polarization fluctuation, related to the dielectric constant of the solution.

For this particular problem we have not yet been able to generate the 2D-IR spectra of the different layers. The difficulty is to account for the effects of all the atoms of the protein on the spectral diffusion of the –O–H vibrational mode. Even the calculation of the average frequency within the Condon approximation is a non-trivial (but not impossible) task. We leave this work for the future. But we can obtain some degree of information from the study of various time correlation functions, particularly from the comparison of water dynamics for two different proteins.

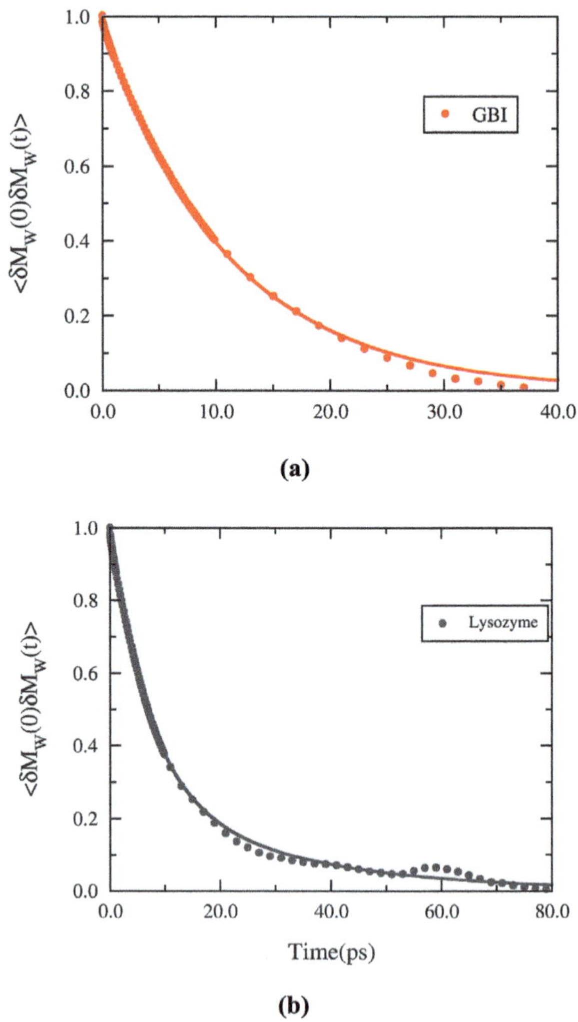


Fig. 9 Dipole–dipole auto-correlation functions of water $\langle \delta M_W(0) \delta M_W(t) \rangle$ for (a) protein G (GB1) and (b) lysozyme (Lys). In both cases the correlation is found to decay to zero within 100 ps.

 

We explore these aspects by studying the aqueous solutions of two different proteins, immunoglobulin binding domain protein G (GB1) and hen-egg white lysozyme (Lys) with different characteristics (shown in Fig. 8). We find that the calculated dielectric constant of the system shows a noticeable increment in all cases compared to that of neat water (for SPC/E model water the estimated value is ~68). The total dipole moment auto time correlation function of water $\langle \delta M_W(0) \delta M_W(t) \rangle$ is found to be sensitive to the nature of the protein. A shell-wise decomposition of water molecules around the protein reveals a higher density of the first layer compared to the succeeding ones.

### A. Simulation details

We performed an atomistic molecular dynamics simulation of the two proteins with a sufficiently large number of water molecules (number varied according to size of the proteins) and at 300 K temperature and 1 bar pressure. For protein G (PDB ID:GB1) we used an ~8.5 nm box and 20 893 water molecules and for lysozyme (PDB ID:1AKI) we used an ~9.5 nm box and 29 204 water molecules. We used the OPLS-AA[25] force field and extended simple point charge model (SPC/E)[26] for water. After performing a steepest descent energy minimization, the solvent was further equilibrated for 10 ns by restraining the position of the protein. Finally a production run was performed for 20 ns in the NPT ensemble. Temperature was kept constant using a Nosé–Hoover[21,22] Thermostat and a Parrinello–Rahman Barostat[23] was used for pressure coupling. After final equilibration of the system, a further 5 ns run was performed with 10 fs resolution to generate the trajectory for calculating the correlation functions.

### B. Evaluation of total and local polarization fluctuation of water in aqueous protein solutions

In Fig. 9 we plot the correlation in the fluctuation of the total dipole moment of water $\langle \delta M_W(0) \delta M_W(t) \rangle$ for the two proteins. The detailed nature of the correlation functions is expected to be sensitive to protein size, amino acid sequence as well as orientation of the corresponding dipoles. Whereas the correlation functions decay within 80 ps for lysozyme, in the case of GB1 the decay is found to be comparatively faster and reaches 0 within 40 ps. The correlation functions are fitted to exponential forms and the corresponding amplitudes and time constants are given in Table 3. We find that the corresponding correlation function for GB1 fits to a single exponential form with a time constant of 10.9 ps which is a signature of fast water. However lysozyme fits to a bi-exponential with two time constants; the first time constant signifies the fast dynamics of free water and the second time constant is a signature of slow biological water. Interestingly, in both cases we find that the estimated value of the dielectric constant of water is higher than the normal value (for SPC/E water the dielectric constant is ~68).

In Fig. 10 we show the shell-wise decomposition of the partial dipole moment–dipole moment time correlation function $\langle \delta M_i(0) \delta M_i(t) \rangle (i$ – shell index) for the first two shells (shells are taken to be approximately ~1 nm thick). For lysozyme a striking observation is the considerable slowing down of polarization fluctuation for the first shell. By the second shell, relaxation achieves almost bulk-like behavior. However the GB1 protein shows bulk-like decay even for the first shell. The slowing down of relaxation for lysozyme in the first shell emerges due to the

Table 3 Estimation of fitting parameters obtained by fitting the $\langle \delta M_W(0)\delta M_W(t)\rangle$ function to exponential forms

| Protein | $a_1$ | $\tau_1$ (ps) | $a_2$ | $\tau_2$ (ps) | $\varepsilon$ |
|---|---|---|---|---|---|
| GB1 | 0.97 | 10.9 | — | — | 75.4 |
| LYS | 0.75 | 6.9 | 0.25 | 31.7 | 70.9 |

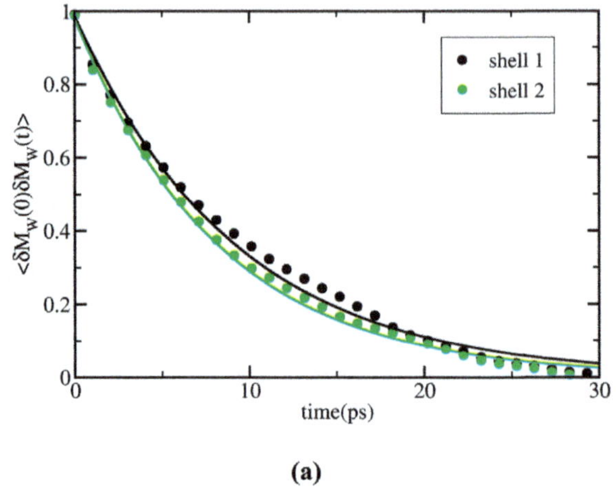


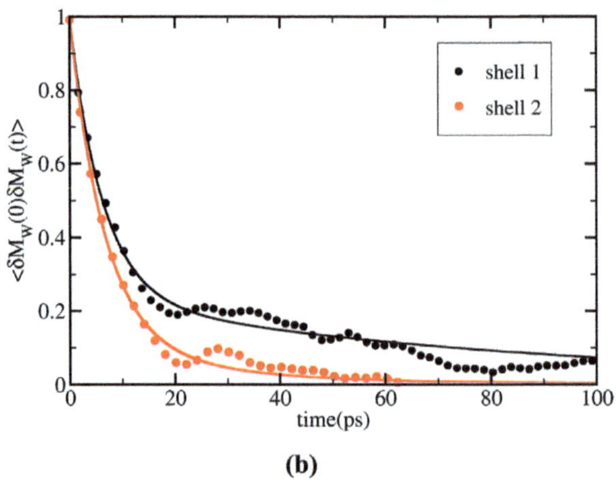


Fig. 10 Demonstration of the shell-wise polarization fluctuation of water for (a) GB1 and (b) lysozyme.

appearance of an additional relaxation channel in the relaxation of biological water that is found to be missing in the case of GB1. The associated slow relaxation time is of the order of 40 ps. Surprisingly, the shell-wise calculation of the dielectric constant gives a much lower value for both the shells (first shell: 49.2 for GB1 and 48.31 for lysozyme; second shell: 59.3 for GB1 and 53.3 for lysozyme) compared to that of the bulk.

 

# VI. Conclusion

The omnipresence of heterogeneity in natural aqueous systems has always been an intriguing subject of study. The heterogeneity is mostly local, a few molecular layers thick and leads to unusual dynamics. In this work we explore the consequences (both spatial and temporal) of heterogeneity in different aqueous systems. In all cases, we find that the heterogeneity is reflected through significant alteration of water dynamics. In the case of the reverse micelle, we find that with increasing system size, the dynamical behavior of water in different layers varies substantially. Thermodynamics induced heterogeneity in supercritical water is also well reflected in the corresponding spectral diffusion of the 2D-IR spectra. Here the time scales of density relaxation are ultrafast and captured faithfully by 2D-IR spectra. Finally, we evaluate local and collective polarization fluctuation in aqueous protein solutions, where we also find the prominent signature of heterogeneity. The present work is thus successful in identifying and generalizing the existence of heterogeneity in various aqueous systems.

In future we plan to calculate the 2D-IR spectra of the aqueous –O–H mode in different hydration layers of protein solutions.

# Acknowledgements

We thank Ms. Susmita Roy and Milan Hazra for helpful discussions. This work was supported in parts by DST (India) and BRNS (India). BB thanks DST for a Sir JC Bose Fellowship.

# References

1 B. Bagchi, *Water in Biological and Chemical Processes: From Structure and Dynamics to Function*, Cambridge University Press, 2014.

2 M. D. Fayer and N. E. Levinger, Analysis of Water in Confined Geometries and at Interfaces, *Annu. Rev. Anal. Chem.*, 2010, **3**, 89–107.

3 S. K. Pal and A. H. Zewail, Dynamics of Water in Biological Recognition, *Chem. Rev.*, 2004, **104**, 2099–2124.

4 N. Nandi and B. Bagchi, Dielectric Relaxation of Biological Water, *J. Phys. Chem. B*, 1997, **101**, 10954–10961.

5 T. Telgmann and U. Kaatze, On the Kinetics of the Formation of Small Micelles. 1. Broadband Ultrasonic Absorption Spectrometry, *J. Phys. Chem. B*, 1997, **101**, 7758–7765.

6 B. M. Ladanyi, Computer simulation studies of counterion effects on the properties of surfactant systems, *Curr. Opin. Colloid Interface Sci.*, 2013, **18**, 15–25.

7 S. Vajda, R. Jimenez, S. J. Rosenthal, V. Fidler, G. R. Fleming and E. W. Castner, Femtosecond to nanosecond solvation dynamics in pure water and inside the [gamma]-cyclodextrin cavity, *J. Chem. Soc., Faraday Trans.*, 1995, **91**, 867–873.

8 S. C. Tucker, Solvent Density Inhomogeneities in Supercritical Fluids, *Chem. Rev.*, 1999, **99**, 391–418.

9 B. Bagchi, Water Dynamics in the Hydration Layer around Proteins and Micelles, *Chem. Rev.*, 2005, **105**, 3197–3219.

10 R. Biswas, J. Furtado and B. Bagchi, Layerwise decomposition of water dynamics in reverse micelles: A simulation study of two-dimensional infrared spectrum, *J. Chem. Phys.*, 2013, **139**, 144906.

11 S. Pronk, E. Lindahl and P. M. Kasson, Dynamic heterogeneity controls diffusion and viscosity near biological interfaces, *Nat. Commun.*, 2014, **5**, 1–7.

12 M. T. Zanni and R. M. Hochstrasser, Two-dimensional infrared spectroscopy: a promising new method for the time resolution of structures, *Curr. Opin. Struct. Biol.*, 2001, **11**, 516–522.

13 J. Zheng, K. Kwak and M. D. Fayer, Ultrafast 2D IR Vibrational Echo Spectroscopy, *Acc. Chem. Res.*, 2006, **40**, 75–83.

14 P. Hamm and M. Zanni, *Concepts and Methods of 2D Infrared Spectroscopy*, Cambridge University Press, 2011.

15 S. T. Roberts, K. Ramasesha and A. Tokmakoff, Structural Rearrangements in Water Viewed Through Two-Dimensional Infrared Spectroscopy, *Acc. Chem. Res.*, 2009, **42**, 1239–1249.

16 N. Nandi and B. Bagchi, Anomalous Dielectric Relaxation of Aqueous Protein Solutions, *J. Phys. Chem. A*, 1998, **102**, 8217–8221.

17 R. Pethig, Protein-Water Interactions Determined by Dielectric Methods, *Annu. Rev. Phys. Chem.*, 1992, **43**, 177–205.

18 N. Nandi, K. Bhattacharyya and B. Bagchi, Dielectric Relaxation and Solvation Dynamics of Water in Complex Chemical and Biological Systems, *Chem. Rev.*, 2000, **100**, 2013–2046.

19 J. L. F. Abascal and C. Vega, A general purpose model for the condensed phases of water: TIP4P/2005, *J. Chem. Phys.*, 2005, **123**, 234505.

20 B. Hess, C. Kutzner, D. van der Spoel and E. Lindahl, GROMACS 4: Algorithms for Highly Efficient, Load-Balanced, and Scalable Molecular Simulation, *J. Chem. Theory Comput.*, 2008, **4**, 435–447.

21 S. Nosé, A unified formulation of the constant temperature molecular dynamics methods, *J. Chem. Phys.*, 1984, **81**, 511–519.

22 W. G. Hoover, Canonical dynamics: Equilibrium phase-space distributions, *Phys. Rev. A*, 1985, **31**, 1695–1697.

23 M. Parrinello and A. Rahman, Polymorphic transitions in single crystals: A new molecular dynamics method, *J. Appl. Phys.*, 1981, **52**, 7182–7190.

24 S. Roychowdhury and B. Bagchi, Vibrational phase relaxation of O–H stretch in bulk water: Role of large amplitude angular jumps and negative cross-correlations among the forces on the O–H bond, *Chem. Phys.*, 2008, **343**, 76–82.

25 W. L. Jorgensen, D. S. Maxwell and J. Tirado-Rives, Development and Testing of the OPLS All-Atom Force Field on Conformational Energetics and Properties of Organic Liquids, *J. Am. Chem. Soc.*, 1996, **118**, 11225–11236.

26 H. J. C. Berendsen, J. R. Grigera and T. P. Straatsma, The missing term in effective pair potentials, *J. Phys. Chem.*, 1987, **91**, 6269–6271.

 

Faraday Discussions

PAPER

# Vibronic structure in the far-UV electronic circular dichroism spectra of proteins

Zhuo Li, David Robinson and Jonathan D. Hirst*

Received 29th August 2014, Accepted 5th November 2014
DOI: 10.1039/c4fd00163j

The Franck–Condon effect is considered and the vibrational structure of the $\pi_{nb}\pi^*$ transition of the peptide backbone is incorporated into matrix method calculations of the electronic circular dichroism (CD) spectra of proteins in the far-ultraviolet. We employ the state-averaged CASPT2 method to calculate the ground and $\pi_{nb}\pi^*$ excited state geometries and frequencies of *N*-methylacetamide (NMA), which represents the peptide chromophore. The results of these calculations are used to incorporate vibronic levels of the excited states into the matrix method calculation. The CD spectra of a set of 49 proteins, comprising a range of structural types, are calculated to assess the influence of the vibrational structure. The calculated spectra of α-helical proteins are better resolved using the vibronic parameters and correlation between the experimental and the calculated intensity of less regular β structure proteins improves over most wavelengths in the far-UV. No obvious improvement is observed in the calculated spectra of regular β-sheet proteins. Our high-level *ab initio* calculations of the vibronic structure of the $\pi_{nb}\pi^*$ transition in NMA have provided some further insight into the physical origins of the nature of protein CD spectra in the far-UV.

## 1 Introduction

Electronic circular dichroism (CD) spectroscopy in the far-UV is a widely used technique for studying chiral molecules. Since almost all biological systems and many non-living systems are chiral, CD spectroscopy has become an important tool for structure determination in several research areas. Applications include establishing that a chiral molecule has indeed been synthesized,[1] determining the absolute configuration of a molecule,[2] studying molecular interactions by comparing the calculated spectra of different models with the experimental spectrum,[3] and probing the structures of biological macromolecules, such as proteins[4] or DNA,[5] as well as their folding[6,7] and binding[8] processes. These studies can benefit from an accurate calculation from first principles of the CD spectrum.

Much effort has been made to improve the accuracy of calculations of CD spectra, in part to understand better the relationship between molecular structure

*School of Chemistry, University of Nottingham, Nottingham NG7 2RD, UK. E-mail: jonathan.hirst@nottingham.ac.uk; Fax: +44 115 9513562; Tel: +44 115 9513478*

and the associated spectra. For small molecules such calculations have become fairly routine and can be performed with density functional theory (DFT),[2,9] Hartree–Fock[10] or coupled cluster theory.[11] High-level calculations have been performed on simple amides and dipeptides.[12–14] However, calculation of the CD spectra of proteins remains a challenge, due to their size and flexibility. There are several approximate methods for polymeric systems like proteins, such as the dipole interaction model[15–19] and the matrix method.[20–23] The calculations in this study are based on the matrix method, an exciton approach,[24] which has proven to be quite successful.[25,26] The calculations are almost, but not fully, quantitative for proteins with a high amount of α-helical structure; the spectra for proteins with other secondary types are less well reproduced. Additional aspects may need to be considered to obtain more accurate calculated spectra.

The parameterization in the matrix method is from *ab initio* calculation of *trans N*-methylacetamide (NMA), which represents the chromophoric group of the protein backbone.[27] Fig. 1 shows the two electronic transitions of NMA located in the far-UV region, which are from the amide non-bonding π orbital ($\pi_{nb}$) to the anti-bonding π orbital ($\pi^*$) and from the oxygen lone pair orbital n to the $\pi^*$ orbital. The parameters used in the CD calculation describe the charge distributions of the different electronic states. Deviation from planarity of the peptide bond will have some influence on the electronic structure,[28] but this is not the focus of the current study. Hydrogen bonding may also be important, but previous work on the explicit consideration of this did not find any significant improvement for protein calculations.[29,30]

Vibrational fine structure is common in high-resolution gas phase spectra. These fine features may be not obvious in solution, but can contribute to the broadening of bands. The application of a (uniform) bandwidth to calculated line spectra to model the broadening is an approximation. To explore this aspect further, we have examined explicit incorporation of vibronic effects of the $\pi_{nb}\pi^*$ transition of the peptide chromophore.

The Franck–Condon effect has been considered in previous calculations of the CD spectra of various molecules. The CD spectrum of dimethyloxirane in the gas phase was calculated using time-dependent DFT (TD-DFT) including the envelopes of several states.[31] Some narrower bands arise from excitations to the 1B, 2A

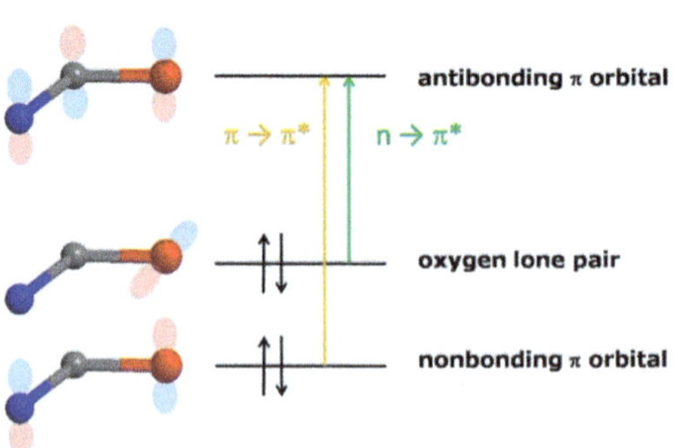


Fig. 1 Molecular orbitals relevant to electronic transitions in the far-UV region. The ball and stick model represents *N*-methylacetamide (blue-nitrogen, grey-carbon, red-oxygen).

 

and 3B states, while other broader bands were assigned to transitions to the 4A and 5A states. The CD spectrum was reproduced well once the vibrational structure for the various transitions was considered. The disappearance of a negative band in the experimental spectrum was rationalised by the calculation: the band was cancelled by other positive bands because of a strong vibrational progression. Vibronic contributions were taken into account in calculations of the CD spectra of naphthalenediimide (NDI) dimers[32] and oligomers.[33] The CD spectra of NDI helical nanotubes were calculated with the matrix method considering several vibronic transitions associated with two electronic transitions. Inclusion of the Franck–Condon effect was essential in reproducing the key features of the CD spectra.

An early study considering vibronic coupling in proteins investigated absorption in the far-UV of an idealized α-helix[34] and poly-L-lysine as β-sheet[35] and calculated the spectrum, focusing on the C–N stretch mode in the $\pi_{nb}\pi^*$ band. The predominant role of the C–N stretch in this transition has been confirmed by resonance Raman experiments.[36] No combination transition along the amide II′ overtones was observed in $D_2O$. Although the role of protic solvent is not clear yet, the band of the carbonyl stretch, which is the primary mode in the gas phase and in acetonitrile, was not apparent in aqueous solution.

In this study, we calculate the vibronic structure of the $\pi_{nb}\pi^*$ transition of the backbone chromophore and incorporate this into the calculations of protein CD spectra. There are various approaches to exciton-vibrational coupling.[37] Yang[38] described a reduced density-matrix theory for the case where a single vibration mode is predominant in the coupling, with a spectral density used to represent a bath of nuclear degrees of freedom. We take advantage of this model in constructing the system Hamiltonian, which will be described in detail in the next section.

## 2 Methods

We have considered the *trans* form of NMA, which is much more common in proteins than the *cis* form[39] (0.34% peptides are *cis* in our dataset). The electronic ground and $\pi_{nb}\pi^*$ excited state geometries and vibrational frequencies of NMA were computed using multireference perturbation theory[40] (CASPT2) based on a complete active space self-consistent field (CASSCF) reference and the 6-31G* basis set with the MOLPRO package (Version 2010.1).[41] A multi-configurational method is necessary for a proper description of the electronic structure of amides, because of a significant interaction between the ground and $\pi_{nb}\pi^*$ states. The active space we used is four electrons in five active orbitals (two occupied and three virtual), which, although minimal, suffices for our purposes. The excited state optimization used state-averaged CASPT2, averaged over the ground, $n\pi^*$ and $\pi_{nb}\pi^*$ states, which should give molecular orbitals that are not biased toward any particular electronic state. Although the ground state of NMA is planar,[42] we calculated both the ground and excited states with no symmetry imposed.

The Franck–Condon overlap integral, $f_v$, characterizes the vibronic transitions, which govern the intensity distribution along the progression in the absorption spectrum. The intensity is proportional to the Franck–Condon factor, which is the square of the overlap integral of the vibrational wave function between the ground and the excited states. ezSpectrum[43] was used to calculate the Franck–Condon overlap integrals within the double-harmonic approximation. The calculations require the equilibrium geometries of the ground and the excited states as well as the frequencies and vibrational mode vectors, which are obtained from the *ab initio* calculations. The Duschinsky rotation is not included in the calculation because there is no severe distortion of the C–N stretch mode between the ground and excited vibrational states. It means the structure of the excited state can align well with the ground state geometry after rotation and translation. The temperature is set to 0 K so that the system populates the lowest vibrational level of the ground electronic state. No combinations of normal modes were considered, since the C–N stretch is the predominant vibration in the absorption spectrum of NMA in aqueous solution.

In the matrix method, the protein is considered as being composed of $M$ chromophores. The wave function $\psi^{\mathrm{k}}$ is expressed as a linear combination of the wave function $\Phi_{ia}$, which is a product of $M$ chromophoric wave functions $\phi$. Each chromophoric group $i$ is considered separately and, in the simplest form of the approach, can only be excited to higher electronic states, $\phi_{ia}$, of itself, while others stay in the ground state, $\phi_{j0}$.

$$\Phi_{ia} = \phi_{10}\cdots\phi_{ia}\cdots\phi_{j0}\cdots\phi_{M0} \tag{1}$$

$$\psi^{\mathrm{k}} = \sum_{i}^{M}\sum_{a}^{ni} c_{ia}^{\mathrm{k}}\Phi_{ia} \tag{2}$$

where $c_{ia}^{\mathrm{k}}$ are expansion coefficients, which reflect the interactions of the states, to be determined.

The ground state of the protein is

$$\psi^{0} = \phi_{10}\cdots\phi_{i0}\cdots\phi_{j0}\cdots\phi_{M0} \tag{3}$$

With the above approximation, the wave function $\psi^{\mathrm{k}}$ of the electronic excited state k of the protein associated with the energy can be calculated by solving the Schrödinger equation

$$\hat{H}\psi^{\mathrm{k}} = E^{\mathrm{k}}\psi^{\mathrm{k}} \tag{4}$$

The Hamiltonian of the system $\hat{H}$ is constructed as the sum of the local Hamiltonian $\hat{H}_i$ of the independent group $i$ and the interactions between two groups which is denoted as intergroup potential $\hat{V}_{ij}$.

$$\hat{H} = \sum_{i=1}^{M}\hat{H}_i + \sum_{i=1}^{M-1}\sum_{j=i+1}^{M}\hat{V}_{ij} \tag{5}$$

The interaction energy between transitions a ← 0 on group $i$ and transition b ← 0 on group $j$ is expressed as:

$$V_{i0a;j0b} = \iint \phi_{i0}\phi_{ia}\hat{V}_{ij}\phi_{j0}\phi_{jb}\mathrm{d}\tau_i\mathrm{d}\tau_j \tag{6}$$

The interactions between chromophores are assumed to be purely electrostatic in nature, involving charge densities $\rho$ of separation $r$.

 

$$V_{i0a;j0b} = \iint \frac{\rho_{i0a}(r_i)\rho_{j0b}(r_j)}{4\pi\varepsilon_0 r_{ij}} \mathrm{d}\tau_i \mathrm{d}\tau_j \tag{7}$$

where $\rho_{i0a}(r_i)$ and $\rho_{j0b}(r_j)$ are the transition electron densities of chromophores $i$ and $j$, $\varepsilon_0$ is the vacuum permittivity and $r_{ij}$ is the separation of the chromophores.

The monopole–monopole approximation is used to make the calculation of $V_{i0a;j0b}$ tractable, whereby monopole charges are fitted to the atoms of the group to reproduce the permanent and transition densities. The Coulomb interactions between these monopoles account for the interaction between groups.

$$V_{i0a;j0b} = \sum_{s=1}^{N_s} \sum_{t=1}^{N_t} \frac{q_s q_t}{r_{st}} \tag{8}$$

where $q_s$ and $q_t$ are the monopole charges on chromophores, and $N_s$ and $N_t$ are the number of charges fitted on the chromophore. The inter-chromophore interaction depends only on the magnitudes and locations of these point charges and their distance and orientation with respect to each other will reflect the secondary and tertiary structure of the protein.

The incorporation of vibrational structure is within the harmonic approximation. The corresponding illustrative Hamiltonian matrix for a dipeptide in eqn (9) considers the n$\pi^*$ transition and two vibrational states, $\nu1$ and $\nu2$, in the $\pi_{\mathrm{nb}}\pi^*$ transition. In our protein calculations, we consider five vibrational states. The energy spacing between these vibrational states is derived from the vibrational frequencies of the relevant mode obtained from the *ab initio* calculation.

$$\hat{H} = \begin{pmatrix} E_{1n\pi^*} & V_{1n\pi^*;1\pi\pi^*\_\nu1} & V_{1n\pi^*;1\pi\pi^*\_\nu2} & V_{1n\pi^*;2n\pi^*} & V_{1n\pi^*;2\pi\pi^*\_\nu1} & V_{1n\pi^*;2\pi\pi^*\_\nu2} \\ V_{1n\pi^*;1\pi\pi^*\_\nu1} & E_{1\pi\pi^*\_\nu1} & 0 & V_{1\pi\pi^*\_\nu1;2n\pi^*} & V_{1\pi\pi^*\_\nu1;2\pi\pi^*\_\nu1} & V_{1\pi\pi^*\_\nu1;2\pi\pi^*\_\nu2} \\ V_{1n\pi^*;1\pi\pi^*\_\nu2} & 0 & E_{1\pi\pi^*\_\nu2} & V_{1\pi\pi^*\_\nu1;2n\pi^*} & V_{1\pi\pi^*\_\nu2;2\pi\pi^*\_\nu1} & V_{1\pi\pi^*\_\nu2;2\pi\pi^*\_\nu2} \\ V_{1n\pi^*;2n\pi^*} & V_{1\pi\pi^*\_\nu1;2n\pi^*} & V_{1\pi\pi^*\_\nu2;2n\pi^*} & E_{2n\pi^*} & V_{2n\pi^*;2\pi\pi^*\_\nu1} & V_{2n\pi^*;2\pi\pi^*\_\nu2} \\ V_{1n\pi^*;2\pi\pi^*\_\nu1} & V_{1\pi\pi^*\_\nu1;2\pi\pi^*\_\nu1} & V_{1\pi\pi^*\_\nu2;2\pi\pi^*\_\nu1} & V_{2n\pi^*;2\pi\pi^*\_\nu1} & E_{2\pi\pi^*\_\nu1} & 0 \\ V_{1n\pi^*;2\pi\pi^*\_\nu2} & V_{1\pi\pi^*\_\nu1;2\pi\pi^*\_\nu2} & V_{1\pi\pi^*\_\nu2;2\pi\pi^*\_\nu2} & V_{2n\pi^*;2\pi\pi^*\_\nu2} & 0 & E_{2\pi\pi^*\_\nu2} \end{pmatrix} \tag{9}$$

where the n$\pi^*$ transition energy is $E_{\mathrm{n}\pi^*}$; the $\pi_{\mathrm{nb}}\pi^*$ transition has two associated elements, $E_{\pi\pi^*\_\nu1}$ and $E_{\pi\pi^*\_\nu2}$, representing the different vibrational levels in the excited state.

The interactions between transitions are modified by multiplying the monopole charges by the Franck–Condon overlap integral, $f_{\mathrm{v}}$. It is assumed that there is no interaction between the vibrational levels of the same chromophore. The Franck–Condon overlap integral of the n$\pi^*$ transition was taken as unity.

$$V_{i0a;j0b} = \sum_{s=1}^{N_s} \sum_{t=1}^{N_t} \frac{q_s q_t f_a f_b}{r_{st}} \tag{10}$$

Diagonalization of the Hamiltonian matrix by a unitary matrix U can be used to calculate the energies and the coefficients $c_{ia}^{\mathrm{k}}$ . The eigenvalues are the transition energies of the system because the ground state energy is set to be zero and the eigenvectors form the unitary matrix.

$$U^{-1} \cdot \hat{H} \cdot U = H_{\mathrm{diag}} \tag{11}$$

The electric and magnetic transition dipole moments from the ground state to $k^{th}$ excited state of the interacting system are calculated from the transition dipole moments of individual chromophore using the unitary matrix:

$$\vec{\mu}_i = \sum_a U_{ai}\vec{\mu}_a^0 \tag{12}$$

$$\vec{m}_i = \sum_a U_{ai}\vec{m}_a^0 \tag{13}$$

With the electric and magnetic transition dipole moments and wave functions of the excited state of the protein, the CD spectrum (rotational strengths in the interacting system) is readily calculated using Rosenfeld equation. The rotational strength ($R^{0k}$) of an excitation from the ground state 0 to the excited state k is the product of the electric transition dipole moment, $\vec{\mu}$, and the magnetic transition dipole moment, $\vec{m}$ , given by Rosenfeld equation:

$$R^{0k} = \text{Im}\left(\left\langle \psi^0 \middle| \vec{\mu} \middle| \psi^k \right\rangle \left\langle \psi^k \middle| \vec{m} \middle| \psi^0 \right\rangle\right) \tag{14}$$

where $\psi^0$ and $\psi^k$ represent the wave functions of different states. Im represents the imaginary part of the product.

Calculations using the matrix method generate line spectra, in which the lines represent the calculated rotational strength at the specific wavelength. A Gaussian function is applied to the line spectra, broadening bands, to mimic the spectra in solution. This permits a direct comparison between the calculations and the experimental spectra. The calculated CD spectra are sensitive to the bandwidth,[45] which is effectively an empirical parameter. A single bandwidth is used for all peaks in all proteins in this study. A bandwidth of 10 nm was used for spectra

Table 1 Proteins considered in this study

| Class | Protein (PDB code) |
|---|---|
| α-Helix | Lysozyme (193l), myoglobin whale (1a6m), leptin (1ax8), c-phycocyanin (1ha7)*[a], hemoglobin (1hda), cytochrome C (1hrc), calmodulin (1lin), insulin (1trz), phospholipase A2 (1une), myoglobin horse (1ymb) |
| Mixed-αβ | Aldolase (1ado)*, haloalkane dehydrogenase (1bn6), carbonic anhydrase I (1hcb), PNMT (1hnn)*, NmrA (1k6j)*, monellin (1mol), papain (1ppn), dehydroquinase type 1 (1qfe), rhodanese (1rhs), ubiquitin (1ubi), dehydroquinase type 2 (2dhq), deoxyribonuclease-1 (3dni), ribonuclease A (3rn3), carboxypeptidase A (5cpa), triosephosphate isomerase (7tim) |
| β-I | Pectate lyase C (1air), β-bactoglobulin (1b8e), γ-D-crystallin (1elp), γ-S-crystallin (C-term) (1ha4), γ-D-crystallin (human) (1hk0), jacalin (1ku8)*, lentil lectin (1les), γ-E-crystallin (1m8u), concanavalin A (1nls), pea lectin (1ofs), avidin (1rav), thaumatin (1thw), β-B2-crystallin (2bb2), pepsinogen (2psg), γ-B crystallin (4gcr) |
| β-II | Cu,Zn superoxide dismutase (1cbj), α-bungarotoxin (1hc9), rubredoxin (1r0i), streptavidin (1stp), α-chymotrypsinogen (2cga), ferredoxin (2fdn), elastase (3est), α-chymotrypsin (5cha), aprotinin (5pti) |

[a] For computational tractability, only the unique chains of multi-meric proteins (denoted by asterisk) containing more than 500 residues were considered. The validity of this approximation has been established elsewhere.[44]

 

considering vibronic coupling, while 12.5 nm was used for the calculations without vibrational structure. Our study, thus, in part, investigates one of the contributing factors to line broadening.

CD spectra were calculated for a set of 49 proteins (Table 1) with the above approach. The crystal structures of these proteins are in the Protein Data Bank (PDB). These proteins are taken from the database SP175 of the Protein Circular Dichroism Data Bank[46] (PCDDB) in which the experimental spectra are available. To assess our calculations, the CD spectra of these proteins were calculated and compared with the experimental data in the far-UV region.

# 3 Results and discussion

The active space employed in the CASPT2 calculation comprises two occupied and three virtual orbitals. The first transition is the $n\pi^*$ transition, followed by the $\pi_{nb}\pi^*$ transition. The lowest energy conformers of ground and $\pi_{nb}\pi^*$ excited states of NMA are shown in Fig. 2 and the optimized geometric parameters are summarized in Table 2. The structure of NMA has been previously studied experimentally[47–50] and theoretically.[53–55] Our calculation suggests a planar symmetry for the ground state which is consistent with X-ray crystallography[51] and gas electron diffraction experiments.[56] The $\pi_{nb}\pi^*$ excited state has been less well studied, due to the challenges of characterising the geometries of electronically excited states. The geometry distortion and the displacements involved in the excited states are known mainly from Raman experiments.[36,55] Chen *et al.*[55] concluded that C–N elongation is the largest displacement of NMA in the aqueous phase, which agrees with Mayne and Hudson's analysis[36] of the resonance Raman spectrum that the C–N stretch is the predominant mode in water. They estimated the vibrational frequency of the C–N stretch to be 1400 cm$^{-1}$, with which they obtained a good fit to the absorption spectrum of NMA. Li *et al.*[53] calculated, using the configuration interaction between all singly excited configurations (CIS) approach, that the excited state of NMA possesses a pyramidal carbonyl and a planar amine conformation. Our CASPT2 calculation suggests a similar amide geometry. However, our calculation indicated a shorter C–N bond in the excited state structure compared to the ground state geometry, while this bond appeared to be elongated in the Raman experiments. This discrepancy may have arisen from the neglect of solvent effects in our calculation. Nevertheless, the absorption

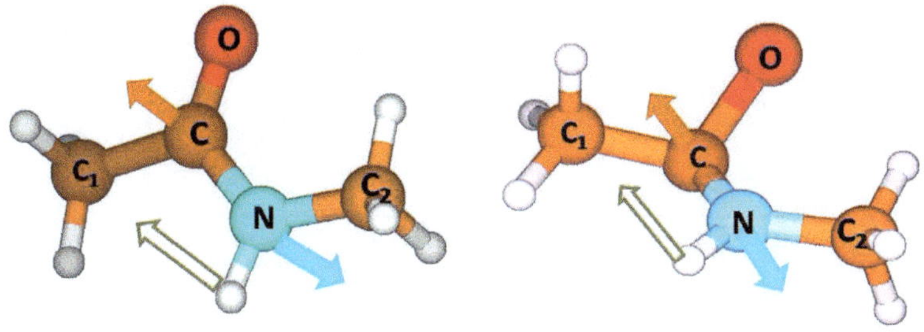


Fig. 2 Calculated geometries and C–N stretching mode of NMA in the ground state (left) and the $\pi_{nb}\pi^*$ excited state (right).

Table 2 Ground and excited state geometries of NMA, optimized at the CASPT2 level, along with the frequencies of C–N stretch in both states. Experimental results[51,52] are listed as a reference

| | Exp. G.s.[51] | CASPT2 | |
|---|---|---|---|
| | | G.s. | $\pi_{nb}\pi^*$ e.s. |
| C–N/Å | 1.386 | 1.368 | 1.326 |
| C=O/Å | 1.224 | 1.231 | 1.587 |
| N–H/Å | 1.106 | 1.010 | 1.021 |
| C–$C_1$/Å | 1.520 | 1.517 | 1.485 |
| N–$C_2$/Å | 1.469 | 1.450 | 1.457 |
| O=C–N/° | 121.8 | 123.2 | 108.4 |
| C–N–H/° | 110.0 | 118.9 | 120.2 |
| N–C–$C_1$/° | 114.1 | 115.2 | 117.1 |
| C–N–$C_2$/° | 119.6 | 122.2 | 121.3 |
| H–N–C=O/° | | −179.9 | 138.6 |
| C–N stretch[52]/$cm^{-1}$ | 1581 | 1610 | 1696 |

spectrum of NMA in water (Fig. 3) can be reproduced quite well by the Franck–Condon factors generated with these *ab initio* data. We incorporate the vibronic coupling into the CD calculations with the Franck–Condon overlap integrals calculated from the geometries and frequencies of the current results. The calculated vibrational frequencies of the C–N stretch in the ground and excited states are 1610 $cm^{-1}$ and 1696 $cm^{-1}$, respectively. Compared to the experimental

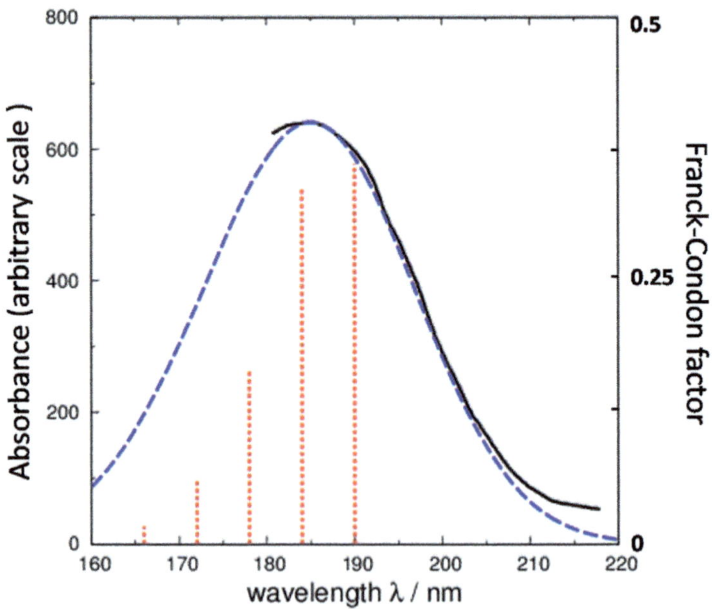


Fig. 3 Absorption spectra. The black solid line is the experimental spectrum, the dashed line is the calculated spectrum and the dotted vertical lines are proportional to the Franck–Condon factors.

Table 3 Franck–Condon overlap integrals of the $\pi\pi^*$ transition in NMA, considering only the C–N stretching mode

| Transition | $0 \leftarrow 0$ | $1 \leftarrow 0$ | $2 \leftarrow 0$ | $3 \leftarrow 0$ | $4 \leftarrow 0$ |
|---|---|---|---|---|---|
| FC integral | 0.62 | 0.60 | 0.42 | 0.25 | 0.13 |

frequency of the ground state, which is 1581 $cm^{-1}$, the calculated ground state frequency is within the known systematic errors.[57]

The parallel approximation in ezSpectrum was used to calculate the Franck–Condon overlap integral of NMA. Only the C–N displacement was considered in this study and the Franck–Condon overlap integrals of C–N stretch mode are summarized in Table 3. The transitions are denoted as [excited vibronic state $\leftarrow$ ground state]. Fig. 3 compares the experimental absorption spectrum[58] with the simulated absorption intensity from our *ab initio* calculation. The origin of the transition ($0 \leftarrow 0$) was placed at 190 nm (52 632 $cm^{-1}$) and the energy gap between different vibrational levels is set to 1696 $cm^{-1}$, taken from the excited state frequency calculation. The intensities, which are represented by the vertical lines, are proportional to the Franck–Condon factors. A Gaussian function with a bandwidth of 15 nm was applied to simulate other broadening effects.

In order to assess the CD calculations, we used the Spearman rank correlation coefficient and mean absolute error (MAE). The Spearman rank correlation coefficient measures how the calculated intensities correlate with the experimental ranks at a specific wavelength (1 corresponds to a perfect correlation, zero to no correlation, and $-1$ to a perfect inverse correlation). The CD calculations are discussed below based on various secondary structure types. The proteins are grouped (Table 1) based on assignments by DSSP.[59] Proteins containing more than 40% $\alpha$-helical or $\beta$ structures are classified as the $\alpha$-helix and $\beta$-sheet groups, respectively. The $\beta$ type is further divided into $\beta$-I and $\beta$-II subtype based on the nature of the $\beta$ structure. The $\beta$-I type proteins have regular sheets while the $\beta$-II type contains more $\beta$-bulges and irregular strands. The rest of the proteins in the dataset are categorized as mixed-$\alpha\beta$.

Fig. 4(a) compares the experimental and calculated spectra of myoglobin (PDB code: 1YMB), which with 73% helix is a representative of the $\alpha$-helix subset. The main features of the CD spectra of $\alpha$-helical proteins are an intense positive band at around 190 nm and two negative bands at 208 nm and 222 nm. The bands at 190 nm and 208 nm arise from the $\pi_{nb}\pi^*$ transition, while the band at 222 nm is due to the $n\pi^*$ transition. For myoglobin, the intensity of peaks at 190 nm and 208 nm are reproduced equally well by the calculations with and without the vibronic structure considered with bandwidths of 10 nm and 12.5 nm, respectively. The double negative bands are better reproduced when the vibronic structure is included. CD spectra calculated with a 10 nm bandwidth for the non-vibronic-parameter are also shown in Fig. 4 to examine the influence of the bandwidth in the calculations. The 208 and 222 nm peaks are distinct with a narrower bandwidth without considering vibrational structure. However, both peaks arising from the $\pi_{nb}\pi^*$ transition are overestimated. This is because a narrower bandwidth is balanced by the vibronic coupling, which broadens the peak. Calculating a spectrum simply using a narrower bandwidth without incorporating any other broadening effect in the absorption process is, thus, less

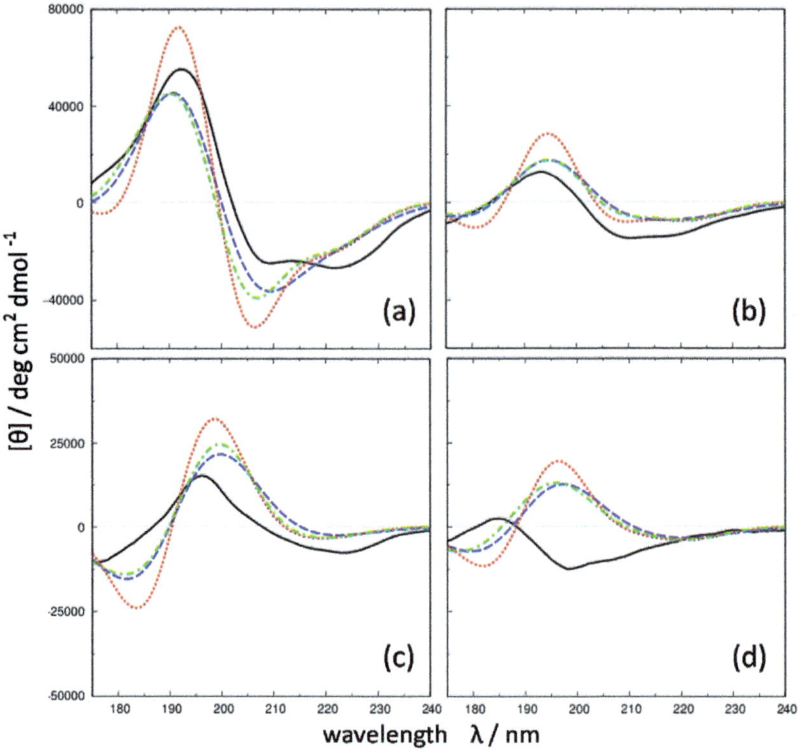


Fig. 4 Experimental and calculated CD spectra for (a) myoglobin, α-helical protein, (b) deoxyribonuclease-1, mixed-α,β protein, (c) concanavalin A, β-I protein, (d) elastase, β-II protein: solid line, experiment; dashed line, non-vibronic calculation with a bandwidth, $\Delta = 12.5$ nm; dotted line, non-vibronic calculation with $\Delta = 10$ nm; dash-dotted line, vibronic calculation with $\Delta = 10$ nm.

satisfactory. However, the MAE between 210 nm and 240 nm is a little higher for the vibronic calculation and the correlation is a little lower between the calculated and experimental intensity (Fig. 5(a)) around 215 nm and 235 nm.

Although the spectra are reproduced well for the intense positive peak, there is poor correlation around 200 nm, which is the descending edge of the most intense peak in the spectrum of α-helical proteins. This is the point that the intensity changes from positive to negative with a large gradient, which means a small error in the wavelength calculation will lead to a large difference in intensity. In the α-helical proteins subset, six out of ten calculated spectra are blue-shifted for both non-vibronic and vibronic calculations. This zero-crossing issue may also explain the difference in the MAEs (Fig. 5(a)). Compared to the non-vibronic calculations, those with vibronic structure are slightly shifted to higher energy due to the position of the origin transition. This deviation causes a larger difference when the CD intensity changes sign.

The features of the mixed-αβ type (Fig. 4(b)) are quite similar to the spectra of the α-helical proteins in most cases. However, the positive peak positions and the intensities, the zero crossing points and the ratio of the double negative bands at 208 and 220 nm all fluctuate, due to the variation in secondary structure content.

 

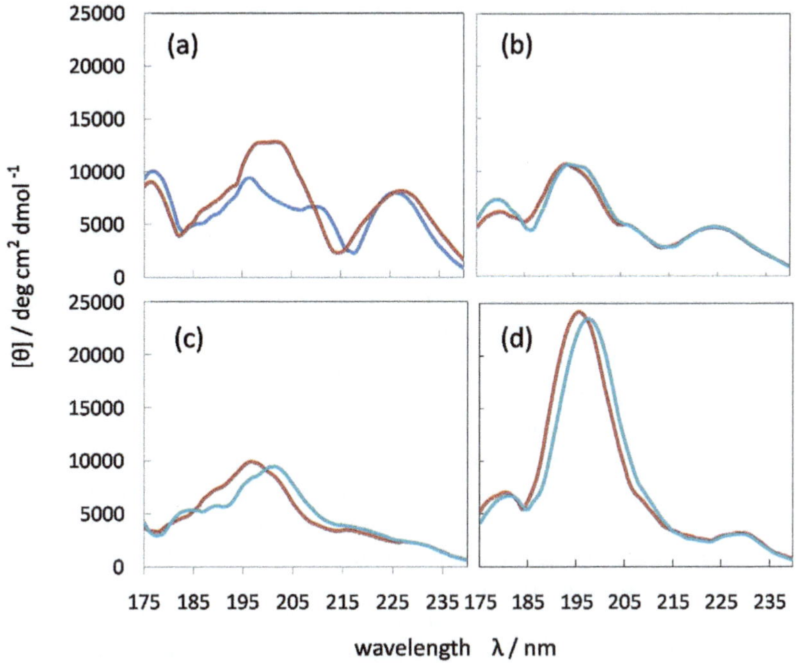


Fig. 5 Mean absolute errors of calculations without (blue line) and with vibronic structure (red line). (a) α-helix; (b) mixed-αβ; (c) β-I; (d) β-II.

The Spearman rank correlation coefficient for the mixed-αβ proteins is similar to that of the α-helical group between 175 and 240 nm. The MAE also seems comparable. However, for 10 out of 15 mixed-αβ proteins the peaks at 190 nm are overestimated, while these peaks in the α-helical proteins are either reproduced well or underestimated. The differences between calculated and experimental spectra for mixed-αβ proteins with greater β content are much larger than the other proteins in this subset. There are also double negative bands in this subtype between 205 nm and 225 nm. The vibronic calculation tends to capture this feature in the mixed-αβ group, although the ratio of these negative peaks is not well-reproduced, as is the case with the α-helical subset.

Fig. 4(c) and (d) compare the spectra for β-I and β-II type proteins, respectively. The general features of β-I spectra are a negative band at about 216 nm and a positive band with comparable magnitude near 195 nm. When there are irregular β structures, the main features are a minimum at about 200 nm and a positive band in the range of 185 to 190 nm. However, the band positions vary dramatically for different β-sheet proteins. For concanavalin A (Fig. 4(c), 1NLS), the intensity of the negative peak was reproduced better with the vibronic structure considered, though both calculations (with and without vibronic structure) show a red shift. The vibronic parameters tend to give more intense bands around 195 nm for β-I subtype proteins. This causes a larger error (Fig. 5(c)) between 185 nm and 200 nm; both calculations overestimate the positive bands at 195 nm in most cases in this protein group. For 10 out of 15 β-I proteins, the calculated intensity of the negative peak at 216 nm is too low.

The β-II type proteins have less regular structure and possibly greater conformational dynamics[60] than β-I proteins and their CD spectra are usually poorly reproduced. The MAE is up to 15 000 degree $cm^2$ $dmol^{-1}$ greater in the region 185 to 210 nm, compared to other secondary structure types (Fig. 5). There is an improvement in the Spearman rank correlation (Fig. 6(d)) at most wavelengths for the β-II proteins using the vibronic parameters; the correlations at the two peak positions, 187 and 201 nm, are summarized in Table 4. The coefficients are averaged over a range of ±2 nm to cover the appropriate peak locations. Both peaks listed in Table 4 arise from the $\pi_{nb}\pi^*$ transitions. Calculations considering vibronic coupling of the $\pi_{nb}\pi^*$ transition have a positive effect for β-II proteins. Although the correlation coefficient is over 0.75 between 200 and 205 nm, six out of nine proteins in this type, have calculated spectra with a band of the opposite sign to experiment in this region. Both parameter sets overestimate the positive peaks and underestimate the negative ones.

A comparison of the Spearman rank correlation coefficient and MAEs of the entire protein data set for the two sets of calculations (Fig. 7) reveals no obvious improvement for most regions in the calculated CD spectra with vibronic parameters. However, there is a smaller error around 208 nm for α-helix, β-I and β-II groups. Including the vibronic structure of the $\pi_{nb}\pi^*$ transition may enhance the coupling between the $\pi_{nb}\pi^*$ transition and the $n\pi^*$ transition. The MAE difference between 185 and 200 nm is mainly from the wavelength where the origin was set. The large error between 190 and 210 nm comes predominantly from the β-II group.

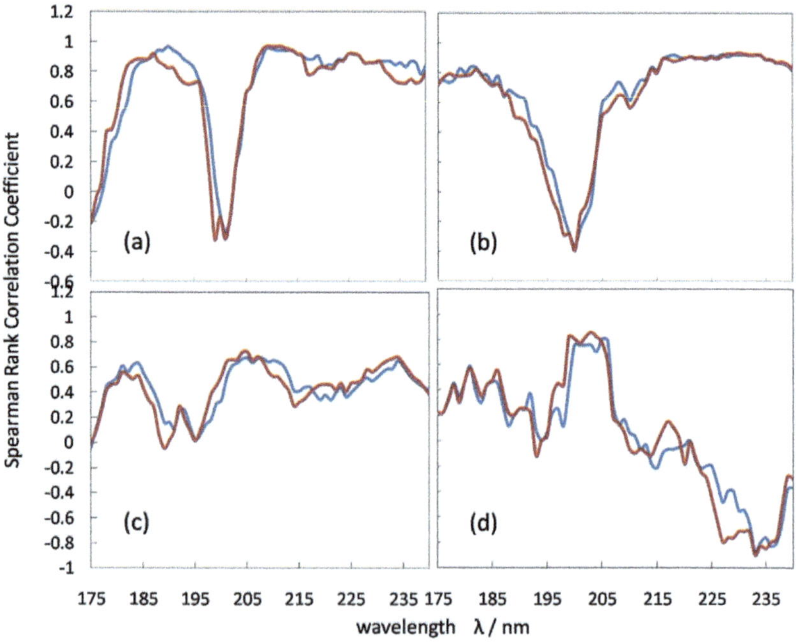


Fig. 6 The Spearman rank correlation coefficient between the experimental and calculated intensities for the non-vibronic (blue line) and vibronic parameters (red line) of different structural types of proteins: (a) α-helix; (b) mix-α,β; (c) β-I; (d) β-II.

 

Table 4 Spearman rank correlation coefficients between the calculated and experimental intensity at the peaks close to 187 and 201 nm, for β-II proteins calculated by with and without vibronic structure

| | Correlation coefficient | |
|---|---|---|
| Parameter | 187 ± 2 nm | 201 ± 2 nm |
| Non-vibronic | 0.30 | 0.69 |
| Vibronic | 0.36 | 0.81 |

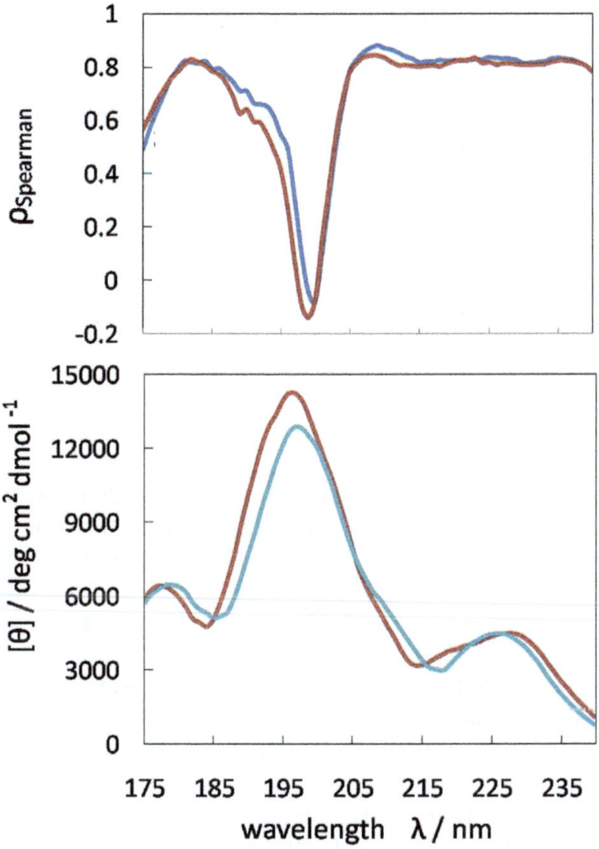


Fig. 7 The Spearman rank correlation coefficient (upper) and MAE (lower) of all 49 proteins in the dataset. The blue lines correspond to the non-vibronic-parameters and the red lines to the vibronic-parameters.

## 4 Conclusions

To incorporate the vibronic coupling in the matrix method, Franck–Condon overlap integrals have been calculated using the ground and excited state geometries and frequencies of NMA. CASPT2 calculations gave a planar geometry of the ground state; in the $\pi_{nb}\pi^*$ excited state the carbonyl is pyramidal and the amine is planar. The conformations are consistent with the previous experimental and theoretical results. Five vibronic transitions in the $\pi_{nb}\pi^*$ excited state were taken into account in the matrix method calculation.

A set of 49 proteins, whose experimental spectra are available, allows us to assess the influence of incorporation of vibrational structure in the CD calculations. The negative peaks between 205 and 225 nm are reproduced with better resolution in α-helical proteins using vibronic parameters. No obvious improvement is observed for mixed-αβ and β-I proteins. For β-II proteins, whose structures are less regular and whose spectra are difficult to reproduce, the Spearman correlation coefficients between the experimental and the calculated intensities show improvement over most wavelengths in the far UV region when taking the vibrational structure into account. It is possible that the calculation may be further improved by considering the conformational dynamics and flexibility of β-II proteins. Incorporation of the vibrational structure of NMA in the matrix method calculations provides us with a template to consider Franck–Condon effects in the near-UV,[23] and will be the subject of future work.

## Acknowledgements

We are grateful for access to the University of Nottingham High Performance Computing facility. ZL is partially supported by a University of Nottingham Vice-Chancellor's scholarship.

## References

1 N. Berova, L. Di Bari and G. Pescitelli, *Chem. Soc. Rev.*, 2007, **36**, 914.
2 D. M. McCann and P. J. Stephens, *J. Org. Chem.*, 2006, **71**, 6074.
3 G. Pescitelli, S. Gabriel, Y. K. Wang, J. Fleischhauer, R. W. Woody and N. Berova, *J. Am. Chem. Soc.*, 2003, **125**, 7613.
4 K. Matsuo, H. Hiramatsu, K. Gekko, H. Namatame, M. Taniguchi and R. W. Woody, *J. Phys. Chem. B*, 2014, **118**, 2785.
5 A. M. O'Mahony, M. F. Cronin, A. McMahon, J. C. Evans, K. Daly, R. Darcy and C. M. O'Driscoll, *J. Pharm. Sci.*, 2014, **103**, 1346.
6 J. L. S. Lopes, D. Orcia, A. P. U. Araujo, R. DeMarco and B. A. Wallace, *Biophys. J.*, 2013, **104**, 2512.
7 M. Vanhove, A. Lejeune and R. H. Pain, *Cell. Mol. Life Sci.*, 1998, **54**, 372.
8 A. J. Miles, N. U. Fedosova, S. V. Hoffmann, B. A. Wallace and M. Esmann, *Biochem. Biophys. Res. Commun.*, 2013, **435**, 300.
9 T. Mori, Y. Inoue and S. Grimme, *J. Phys. Chem. A*, 2007, **111**, 7995.
10 K. L. Bak, A. E. Hansen, K. Ruud, T. Helgaker, J. Olsen and P. Jorgensen, *Theor. Chim. Acta*, 1995, **90**, 441.
11 T. B. Pedersen, H. Koch and K. Ruud, *J. Chem. Phys.*, 1999, **110**, 2883.
12 J. D. Hirst and B. J. Persson, *J. Phys. Chem. A*, 1998, **102**, 7519.
13 N. A. Besley, M. J. Brienne and J. D. Hirst, *J. Phys. Chem. B*, 2000, **104**, 12371.
14 M.-P. Gaigeot, N. A. Besley and J. D. Hirst, *J. Phys. Chem. B*, 2011, **115**, 5526.
15 H. Devoc, *J. Chem. Phys.*, 1964, **41**, 393.
16 H. Devoe, *J. Chem. Phys.*, 1965, **43**, 3199.

 

17 J. Applequist, K. R. Sundberg, M. L. Olson and L. C. Weiss, *J. Chem. Phys.*, 1979, **70**, 1240.
18 J. Applequist, *J. Chem. Phys.*, 1979, **71**, 4332.
19 K. A. Bode and J. Applequist, *J. Phys. Chem.*, 1996, **100**, 17825.
20 P. M. Bayley, E. B. Nielsen and J. A. Schellma, *J. Phys. Chem.*, 1969, **73**, 228.
21 R. W. Woody and I. Tinoco, *J. Chem. Phys.*, 1967, **46**, 4927.
22 R. W. Woody, *J. Chem. Phys.*, 1968, **49**, 4797.
23 R. W. Woody and N. Sreerama, *J. Chem. Phys.*, 1999, **111**, 2844.
24 A. S. Davydov, *Theory of molecular excitations*, Plenum Press, New York, 1971.
25 N. A. Besley and J. D. Hirst, *J. Am. Chem. Soc.*, 1999, **121**, 9636.
26 D. M. Rogers and J. D. Hirst, *Biochemistry*, 2004, **43**, 11092.
27 J. D. Hirst, D. M. Hirst and C. L. Brooks, *J. Phys. Chem. A*, 1997, **101**, 4821.
28 N. A. Besley, M. T. Oakley, A. J. Cowan and J. D. Hirst, *J. Am. Chem. Soc.*, 2004, **126**, 13502.
29 N. A. Besley and J. D. Hirst, *J. Mol. Struct.: THEOCHEM*, 2000, **506**, 161.
30 Z. J. Dang and J. D. Hirst, *Angew. Chem., Int. Ed.*, 2001, **40**, 3619.
31 J. Neugebauer, E. J. Baerends, M. Nooijen and J. Autschbach, *J. Chem. Phys.*, 2005, **122**, 234305.
32 T. Seidler, M. Andrzejak and M. T. Pawlikowski, *Chem. Phys. Lett.*, 2010, **496**, 74.
33 B. M. Bulheller, G. D. Pantos, J. K. M. Sanders and J. D. Hirst, *Phys. Chem. Chem. Phys.*, 2009, **11**, 6060.
34 T. Sanematu and Y. Mizuno, *J. Phys. Soc. Jpn.*, 1976, **40**, 1733.
35 T. Sanematu, *J. Phys. Soc. Jpn.*, 1977, **43**, 600.
36 L. C. Mayne and B. Hudson, *J. Phys. Chem.*, 1991, **95**, 2962.
37 M. Volkhard and K. Oliver, *Charge and Energy Transfer Dynamics in Molecular Systems*, Wiley-VCH, Weinheim, 2011.
38 M. Yang, *J. Mol. Spectrosc.*, 2006, **239**, 108.
39 M. S. Weiss, H. J. Metzner and R. Hilgenfeld, *FEBS Lett.*, 1998, **423**, 291.
40 B. O. Roos and K. Andersson, *Chem. Phys. Lett.*, 1995, **245**, 215.
41 H.-J. Werner, P. J. Knowles, G. Knizia, F. R. Manby and M. Schütz, *Wiley Interdiscip. Rev.: Comput. Mol. Sci.*, 2012, **2**, 242.
42 N. G. Mirkin and S. Krimm, *J. Am. Chem. Soc.*, 1991, **113**, 9742.
43 V. A. Mozhayskiy and A. I. Krylov, ezSpectrum, http://iopenshell.usc.edu/downloads.
44 B. M. Bulheller, A. J. Miles, B. A. Wallace and J. D. Hirst, *J. Phys. Chem. B*, 2008, **112**, 1866.
45 J. D. Hirst, K. Colella and A. T. B. Gilbert, *J. Phys. Chem. B*, 2003, **107**, 11813.
46 L. Whitmore, B. Woollett, A. J. Miles, D. P. Klose, R. W. Janes and B. A. Wallace, *Nucleic Acids Res.*, 2011, **39**, D480.
47 *Higher Excited States of Polyatomic Molecules*, ed. M. B. Robin, New York, 1985.
48 A. T. Hagler, L. Leiserowitz and M. Tuval, *J. Am. Chem. Soc.*, 1976, **98**, 4600.
49 K. Itoh and T. Shimanou, *Biopolymers*, 1967, **5**, 921.
50 F. Fillaux and C. Deloze, *Chem. Phys. Lett.*, 1976, **39**, 547.
51 J. L. Katz and B. Post, *Acta Crystallogr.*, 1960, **13**, 624.
52 L. C. Mayne, L. D. Ziegler and B. Hudson, *J. Phys. Chem.*, 1985, **89**, 3395.
53 Y. Li, R. L. Garrell and K. N. Houk, *J. Am. Chem. Soc.*, 1991, **113**, 5895.
54 X. G. Chen, R. Schweitzerstenner, S. Krimm, N. G. Mirkin and S. A. Asher, *J. Am. Chem. Soc.*, 1994, **116**, 11141.

55 X. G. Chen, S. A. Asher, R. Schweitzerstenner, N. G. Mirkin and S. Krimm, *J. Am. Chem. Soc.*, 1995, **117**, 2884.
56 M. Kitano, T. Fukuyama and K. Kuchitsu, *Bull. Chem. Soc. Jpn.*, 1973, **46**, 384.
57 J. A. Pople, A. P. Scott, M. W. Wong and L. Radom, *Isr. J. Chem.*, 1993, **33**, 345.
58 K. Kaya and S. Nagakura, *Theor. Chim. Acta*, 1967, **7**, 124.
59 W. Kabsch and C. Sander, *Biopolymers*, 1983, **22**, 2577.
60 J. D. Hirst, S. Bhattacharjee and A. V. Onufriev, *Faraday Discuss.*, 2003, **122**, 253.

 

Faraday Discussions

PAPER

# Probing deactivation pathways of DNA nucleobases by two-dimensional electronic spectroscopy: first principles simulations†

Artur Nenov,*[a] Javier Segarra-Martí,‡[b] Angelo Giussani,‡[b] Irene Conti,[b] Ivan Rivalta,[c] Elise Dumont,[c] Vishal K. Jaiswal,[b] Salvatore Flavio Altavilla,[b] Shaul Mukamel[d] and Marco Garavelli*[bc]

Received 13th September 2014, Accepted 11th November 2014
DOI: 10.1039/c4fd00175c

The SOS//QM/MM [Rivalta *et al.*, *Int. J. Quant. Chem.*, 2014, **114**, 85] method consists of an arsenal of computational tools allowing accurate simulation of one-dimensional (1D) and bi-dimensional (2D) electronic spectra of monomeric and dimeric systems with unprecedented details and accuracy. Prominent features like doubly excited local and excimer states, accessible in multi-photon processes, as well as charge-transfer states arise naturally through the fully quantum-mechanical description of the aggregates. In this contribution the SOS//QM/MM approach is extended to simulate time-resolved 2D spectra that can be used to characterize ultrafast excited state relaxation dynamics with atomistic details. We demonstrate how critical structures on the excited state potential energy surface, obtained through state-of-the-art quantum chemical computations, can be used as snapshots of the excited state relaxation dynamics to generate spectral fingerprints for different de-excitation channels. The approach is based on high-level multi-configurational wavefunction methods combined with non-linear response theory and incorporates the effects of the solvent/environment through hybrid quantum mechanics/molecular mechanics (QM/MM) techniques. Specifically, the protocol makes use of the second-order Perturbation Theory (CASPT2) on top of Complete Active Space Self Consistent Field (CASSCF) strategy to compute the high-lying excited states that can be accessed in different 2D experimental setups. As an

[a]*Dipartimento di Chimica G. Ciamician, Università di Bologna, Via F. Selmi 2, 40126 Bologna, Italy. E-mail: Artur.Nenov@unibo.it; Tel: +39 051 2099495*
[b]*Dipartimento di Chimica G. Ciamician, Università di Bologna, Via F. Selmi 2, 40126 Bologna, Italy*
[c]*Laboratoire de Chimie, Ecole Normale Superieure de Lyon, 46 allée d'Italie, 69364 Lyon, France*
[d]*Department of Chemistry, University of California, Irvine, California 92697-2025, USA*

† Electronic supplementary information (ESI) available: Reference gas-phase calculation of the adenine monomer in gas-phase, adenine $\pi$-orbitals, and Cartesian coordinates of all geometries used in this work. See DOI: 10.1039/c4fd00175c

‡ These authors contributed equally to this work.

example, the photophysics of the stacked adenine–adenine dimer in a double-stranded DNA is modeled through 2D near-ultraviolet (NUV) spectroscopy.

# 1 Introduction

Ultrafast pump–probe spectroscopy experiments backed up by theoretical models at the quantum-mechanical level have achieved a detailed understanding of the photophysics of the DNA building blocks adenine, guanine, thymine and cytosine in gas-phase and solution. A thorough overview of the current status of research is given in several review articles.[1–4] Multimeric nucleobase systems formed through stacking (single-stranded DNA) and base-pairing (double-stranded DNA) add many new facades to the photophysics of single bases: formation and dynamics of exciton states,[5,6] population and decay of intra- and inter-strand charge transfer (CT) states,[7,8] charge separation and recombination,[9–11] ultrafast decay through photoinduced proton transfer,[7,12–15] and formation of lesions.[16–18] Interpretation of the electronic spectra of DNA multimers, which is the entirety of all these processes, is a challenge for the conventional one-dimensional (1D) pump–probe spectroscopy. When multiple electronic transitions come into resonance with the excitation wavelength the spectroscopic signatures of individual de-excitation pathways overlap making their disentanglement virtually impossible. Population transfer between different channels cannot be tracked because the correlation between pump and probe wavelengths is not resolved. This complexity calls for a spectroscopic method with an enhanced spectral resolution. Ultrafast bi-dimensional (2D) spectroscopy holds the premise to disentangle multiple deactivation pathways. First applications in the visible (Vis) and in the near-ultraviolet (NUV) have been documented, resolving energy transfer mechanisms in light harvesting proteins[19,20] and solvation dynamics in nucleotides.[21,22]

The evolution of experimental techniques raises the expectations to theoretical modelling of nonlinear spectra. The theoretical framework of nonlinear electronic spectroscopy has been worked out[23–25] and nowadays methods are sought to feed the resulting equations with system-specific data.[26–28] Static and dynamic studies at the quantum-mechanics level have been successfully applied to study the excited state potential energy surfaces of DNA dimers in gas-phase,[8] in aqueous solution[29] and in the native environment of the solvated single- and double-stranded DNA,[30–32] thus providing a tool for obtaining information about excited state energetics and geometrical deformations. Time-resolved Vis-pump IR/Vis-probe[33] and UV-pump stimulated-Raman-probe[28] spectra simulations demonstrate how quantum-mechanical data can be translated into experimental observables. Simulating ultrafast 2D electronic spectroscopy from first principles is challenging as it requires the incorporation of knowledge about the relaxation dynamics of singly and doubly excited states accessible along the de-excitation pathways into the non-linear spectroscopy simulations. Then, it is the excited state absorption (ESA) and the stimulated emission (SE) whose temporal evolution reflects the geometrical and electronic changes along the potential energy surface. Novel computational and conceptual strategies are required.

The SOS//QM/MM protocol is the first method which couples quantum-mechanics/molecular mechanics (QM/MM) calculations, multi-configurational excited state calculations with non-linear spectroscopy to generate time-resolved

 

2D electronic spectra.[34] It allows for the characterization of the vertical 1D and 2D electronic spectra of monomeric and dimeric systems with unprecedented details and accuracy.[27,35,36] Prominent features like doubly excited local and excimer states, accessible in multi-photon processes, as well as CT states are revealed. In this paper we present an extension of this protocol to simulate coherent excited state dynamics based on static and dynamic computational data.[37] Under the assumption that the system's time evolution is slow compared to signal generation, we are able to probe single snapshots along the excited state relaxation pathways, which can be achieved experimentally in a pump–probe pulse arrangement through scanning the delay time between the pump and the probe pulse sequences. The protocol is used to generate the spectroscopic signatures at selected geometries along the major deactivation channels of the stacked adenine–adenine (A–A) dimer in an explicitly solvated double-stranded DNA model.

## 2 Bidimensional time-resolved electronic spectroscopy from first principles

We treat the molecular system as a three-level model, whose eigenstates can be formally divided into: $g$ the ground state; $e$ a set of singly excited states, accessible *via* the pump pulse sequence; and $f$ a set of higher excited states, which are probed after a time delay $t_2$. The nonlinear response functions for ground state bleaching (GSB), SE and ESA, whose Feynman pathways are shown in Fig. 1 for the rephasing pulse orientation technique, read:[23,24,38]

$$
\begin{aligned}
R^{(SOS)}_{\mathbf{k}_1,GSB}(t_1,t_2,t_3) &= -i\hbar^{-3}\Theta(t_1)\Theta(t_2)\Theta(t_3)\sum \boldsymbol{\mu}_{ge}\boldsymbol{\mu}_{eg}\boldsymbol{\mu}_{eg}\boldsymbol{\mu}_{ge} \\
&\quad \times \exp\left[-i(\omega_{eg}-i\gamma)t_3 - i(\omega_{ge}-i\gamma)t_1\right] \\
R^{(SOS)}_{\mathbf{k}_1,SE}(t_1,t_2,t_3) &= -i\hbar^{-3}\Theta(t_1)\Theta(t_2)\Theta(t_3)\sum \boldsymbol{\mu}_{g'e'}\boldsymbol{\mu}_{e'g'}\boldsymbol{\mu}_{eg}\boldsymbol{\mu}_{ge} \\
&\quad \times \exp\left[-i(\omega_{e'g'}-i\gamma)t_3 - i(\omega_{ge}-i\gamma)t_1\right] \times \mathbb{G}_{e\to e'}(t_2) \\
R^{(SOS)}_{\mathbf{k}_1,ESA}(t_1,t_2,t_3) &= i\hbar^{-3}\Theta(t_1)\Theta(t_2)\Theta(t_3)\sum \boldsymbol{\mu}_{e'f'}\boldsymbol{\mu}_{f'e'}\boldsymbol{\mu}_{eg}\boldsymbol{\mu}_{ge} \\
&\quad \times \exp\left[-i(\omega_{fe'}-i\gamma)t_3 - i(\omega_{ge}-i\gamma)t_1\right] \times \mathbb{G}_{e\to e'}(t_2)
\end{aligned}
\tag{1}
$$

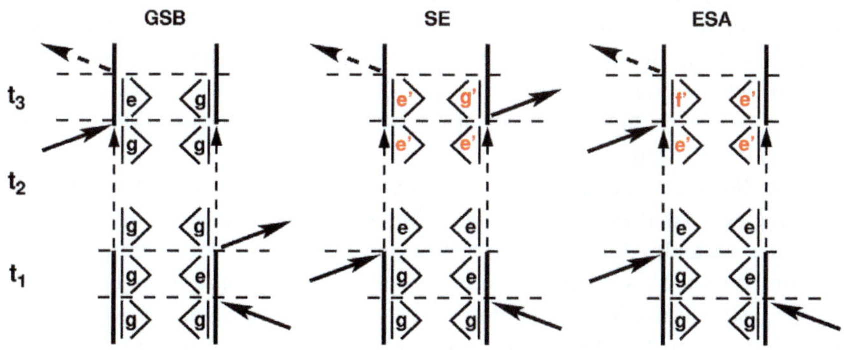


Fig. 1 Feynman diagrams for the rephasing signal including evolution during $t_2$; GSB = ground state bleaching, SE = stimulated emission, ESA = excited state absorption.

with $\Theta(t)$ the Heavyside step-function, $\boldsymbol{\mu}_{ab}$ the vector of the transition dipole moment between states $a$ and $b$, $\omega_{ab}$ the frequency of the electronic transition between states $a$ and $b$, $\gamma$ a phenomenological dephasing rate constant and $\mathbb{G}_{e\rightarrow e'}(t_2)$ a function describing the evolution of the system during the time interval $t_2$. The transition $e \rightarrow e'$ could describe both evolution in the same electronic state and population transfer to another electronic state. It determines the electronic spectrum of the system after the delay time $t_2$ which enters eqn (1) through transition frequencies and dipole moments (shown in red). Ideally, quantum or mixed quantum-classical dynamics simulations are conducted to obtain these parameters. Parametrized models which solve the secular Redfield equations have been used to account for population relaxation and Stokes shifts at early times $t_2$ without explicit dynamics simulations.[39]

The temporal evolution of the nonlinear response is a function of the electronic and geometrical changes in the system modulated by the surroundings. Due to their nature GSB signals do not exhibit time-dependent spectral shifts and their recovery is determined by the decay rates of all excitations resonant with the pump pulse. These features make bleach signals excellent probes for extracting decay rates by means of time-resolved 1D ultrafast spectroscopy. However, the correlation between decay rates and de-excitation pathways cannot be resolved. On the other hand, SE and ESA are specific to each excited state, which makes them eligible probes for pathway-specific extraction of decay rates. Tracking SE and ESA signals by means of 1D ultrafast spectroscopy is challenging as signals from competing channels overlap spectrally and temporally. Furthermore, the temporal evolution of the SE and ESA often span over a large spectral range. 2D time-resolved spectroscopy utilizing broadband probe pulses disentangles spectrally and resolves temporally the fingerprints of the individual de-excitation channels.[40–42]

In this work we employ geometry optimizations in the excited state and probe the higher excited manifold $f'$ and the ground state $g'$ at the obtained stationary points, while keeping the pump pulse pair in resonance with the singly excited state manifold $e$ at the Franck–Condon (FC) point. The local minima are believed to trap excited state population and are, thus, responsible for decay times in the picosecond (ps) time range. With this approach the characteristic spectral signatures of different de-activation channels can be obtained. To compute the fingerprints of each de-excitation channel we regard the channels as independent, *i.e.* we assume that no population is transferred between them or to the ground state (GS) during relaxation from the FC region. With this simplification the 2D spectrum of the FC point can be formally divided along $\omega_1$ into traces resonant with either transition. The time-dependent variation of the traces is analyzed by comparing the signals at the FC point and at the stationary points obtained in this study.

The following approximations are adopted: the system evolves coherently along the populated excited state, *i.e.* inhomogeneous broadening is neglected; coherence dynamics during $t_1$ and $t_3$ are neglected, *i.e.* the system's time evolution is assumed to be slow compared to the signal generation; inter-state coherences (*i.e.* the system is not in a population state during $t_2$ in the SE and ESA diagrams, Fig. 1) are assumed to decay prior to probing. FC active vibrational modes were considered out of the scope of the present work, although several studies on DNA/RNA nucleobases[43–45] report their influence on the photochemical decay path, as well as on the inherent spectroscopic fingerprints in techniques like electronic circular dichroism.[46,47] To the best of our knowledge simulations considering FC active modes refrain to the use of only a few degrees of freedom considered of importance in the molecule, a strategy that is not always accurate enough to embody the essentials of the molecular system under study.[48] Multidimensional problems could become affordable with recent techniques based in multilayer approaches,[49] making the FC active vibrational modes and its role on the excited state deactivation pathways a whole different avenue of research.

Note that geometry optimizations do not contain temporal information and, therefore, we cannot provide time delays. For simulating quasi-absorptive (rephasing and non-rephasing) 2D electronic spectra we have combined the QM/MM methodology, which provides the transition dipole moments at the Complete Active Space Self Consistent Field (CASSCF) level[50] and transition energies corrected at the second-order Perturbation Theory (CASPT2),[51] with the sum-over-states (SOS) approach.[52] SOS calculations were performed with Spectron 2.7 (ref. 24) readapting the energy levels calculated at each stationary point in order to include the GS bleaching at the FC, *i.e.* by maintaining the FC $g$–$e$ energy gap and rigidly shifting all the $f'$ energies calculated at the stationary point in order to align the $e'$ energy with the FC $e$ energy.

Using the fingerprint traces and adopting some assumptions supported by experimental data and the mechanistic picture obtained at the *ab initio* computational level ideal 2D spectra for different scenarios can be simulated. These spectra show how static data from excited state optimizations and minimum energy path calculations can be translated into spectroscopic signatures.

# 3 Application to the adenine–adenine dimer in solvated DNA

## 3.1 Experimental and theoretical observations for poly-deoxyadenosine systems

Femtosecond (fs) transient absorption experiments by Kohler and co-workers support a model in which DNA multimer excitations decay to long lived exciplex states with a decay time constant ranging from ten to hundreds of ps.[53] Adenine homopolymers and homoduplexes made of adenine–thymine base pairs show much slower decay times compared to those of isolated bases.[1,2,54,55] More recent findings support the fact that long lived excited states decay more rapidly in double-stranded poly-deoxyadenosine–deoxythymine (*i.e.* poly(dA–dT)) than in single stranded poly(dA) sequences,[56,57] addressing the role of inter-strand proton transfer in the deactivation mechanism.[7,12,13] Long-lived signals have also been probed by means of fluorescence up-conversion techniques,[5,58,59] and more recently through femtosecond infrared spectroscopy by probing the distinct cationic and anionic species formed in the process.[10] The experimental findings suggest that excited states localized on just two stacked bases are the common trap states independently of the number of stacked nucleotides.[6,60] On the other hand, Markovitsi and co-workers have postulated the possible delocalization of the exciton over several nucleobases[5,58,59] (up to six according to some theoretical models[61]) by means of fluorescence up-conversion experiments. They suggest that the long-lived component is caused by charge relocalization on the excited monomers. The de-excitation channels present in systems such as the adenine–thymine base pair have been recently studied with 2D photon echo IR spectroscopy, yielding novel fingerprints that embody the different motions undergone by the molecules during the relaxation process.[62,63]

Several computational works have been undertaken on DNA multimer models to elucidate how light interacts with these electronically and structurally complex systems. Merchán and co-workers computed energies and structure of adenine homodimers in vacuum and in water at CASPT2 level.[8] They support the idea that the long-lived excited states seen in DNA are of CT type, formed by stacking of two bases. Two stacked conformations were characterized, a perfectly stacked "face-to-face" orientation and a poorly stacked orientation where both nucleobases are twisted. Going from the poorly stacked towards the perfectly stacked conformation the CT states decrease in energy, therefore concluding that the population of the CT states upon photoexcitation strongly depends on the conformational properties of DNA. They associate the sub-ps life-time with two parallel routes: one leads to the formation of the excited CT complex, the other to the decay to the ground state directly from $L_a$ state. The formed CT state would display much slower decay times (*i.e.* in the ps regime), as the system has to overcome an energy barrier to access a conical intersection (CI) with the GS. A time-dependent density functional theory (TD-DFT) study by Barone and co-workers concerning different single- and double-stranded poly(dA) multimers embedded in water (treated as a polarizable continuum) also indicate that the long living component of the excited state population correspond to a dark exciplex produced by a charge transfer between stacked adenines.[29] In their study the exciplex constitutes the absolute excited state minimum. A QM/MM study by Plasser *et al.* utilizing the multi-reference configuration interaction (MRCI) *ab initio* method on the adenine dinucleotide in water predicts the formation of several stable low lying exciplexes of $n\pi^*$ and $\pi\pi^*$ character with short inter-molecular separation but without notable charge transfer. The authors suggest that the long life-time results from trapping the system in these minima.[31] Based on QM/MM computations of the adenine monomer and homodimer embedded in a double-stranded DNA, instead, Conti *et al.* propose that an intra-monomer mechanism associated with relaxation of the $L_a$ channel is compatible with the observed multi-exponential decay, including the longer (>100 ps) life-time component.[64,65] Summing up, there is still no agreement on the de-excitation mechanisms in nucleobase multimers, particularly regarding the participation of CT states.

### 3.2 Mechanistic picture and de-excitation pathways

In the following we outline the recent results of Conti *et al.* for the adenine dimer embedded in a double-stranded DNA model.[65] Both nucleobases were treated quantum-mechanically at CASPT2//CASSCF level with the surrounding DNA chain and solvent represented through point charges. The bright excited states with $L_a(3')$ and $L_a(5')$ character (3′ and 5′ denote the upper and lower nucleobase in the stacked dimer in Fig. 2) couple and form an exciton pair $L_a(3') - L_a(5')/L_a(3') + L_a(5')$ which absorbs around 5.00 eV with a small splitting of 0.05 eV (500 $cm^{-1}$). After excitation the exciton states evolve toward monomer localized excitations with a deactivation mechanism similar to the one reported in gas-phase, occurring through a $C_2$ puckering distortion[66–70] (see Fig. 2 ~65° at the CI). After the

 

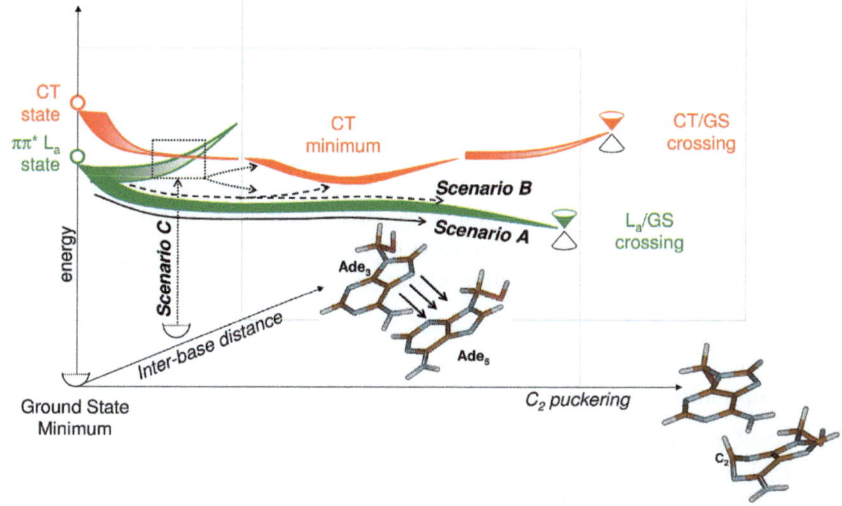


Fig. 2 Schematic representation of the photophysics of the stacked adenine–adenine dimer embedded in a double-stranded DNA in the space of $C_2$ puckering and inter-base distance following the findings of ref. 65; three hypothetical de-excitation mechanisms are shown: (A) exclusively through locally excited states of $L_a$ character (solid lines); (B) *via* populating the CT states out of the $L_a$ plateau through intra-molecular vibrational energy redistribution (dashed lines); (C) *via* populating both La and CT states directly in the FC region (dotted lines); GS = ground state, CT = charge transfer.

initial energy minimization, associated with bond rearrangements, a broad plateau is reached (Fig. 2). Only at large twisting angles (*i.e.* above 50°) the excited state energy decreases again towards the CI. Conti *et al.* suggest that the lack of a steep gradient towards the CI may be the cause for the long life-time encountered in poly(dA) multimers.

At the relaxed GS geometry the lowest pair of CT states (highest occupied molecular orbital HOMO(3′) → lowest unoccupied molecular orbital LUMO(5′) and HOMO(5′) → LUMO(3′)) are located above 6.00 eV, *i.e.* well above the $L_a$ and $L_b$ bands (Fig. 2).[32] They are found to mix with the energetically close lying bright $B_a$ and $B_b$ bands borrowing partially their oscillator strength. The energetic position of the CT states makes them inaccessible *via* NUV pumping. However, the energetic position of CT states is a dynamical feature modulated by thermal fluctuations which might allow low-lying CT states with energies even in the range of the $L_a$ absorptions.[8,71] Upon optimization the CT character can be traced down to the first excited state where local CT minima are encountered. These minima are characterized through shortening of the inter-base distance of *ca.* 0.6–0.9 Å, accompanied by significant bond rearrangements in both chromophores aiming at stabilizing the charges. Two CI mediated deactivation pathways out of the minima are encountered for each CT state: a proton-transfer and an intramolecular deformation. Both CIs are accessible only *via* barriers which offer another potential source of the long life-time encountered in poly(dA) multimers.[5,53,58,59,72] We note, however, that the barriers evaluated by Conti *et al.* are too large (~1 eV) to correlate with ps life-times. Furthermore, despite a thorough search conformations with a CT state below 5.50 eV could not be located.

In the course of geometry optimization Conti *et al.* observe that the $C_2$ puckering distortion is a common feature of both the $L_a$ and CT relaxation pathways, even though the $C_2$ puckering along the CT pathway is less pronounced. Along the $L_a$ deactivation pathway the puckering is a result of the ethylene-like twist around the $C_2$–$N_3$ double bond, necessary to reach the CI. Along the CT deactivation pathway the puckering reduces the electron repulsion of the excess charge density in the negatively charged monomer. A 2D map in the space of the puckering and inter-base distance coordinates indicates that population transfer through vibrational energy redistribution can occur in both directions *via* small barriers.[65] The oscillation between deactivation channels is likely to slow down the de-excitation process.

### 3.3 Computational details

The critical geometries obtained in the mechanistic study of the bright exciton states $L_a(3') - L_a(5')$ and $L_a(3') + L_a(5')$ and of the lowest CT states $CT(5' \rightarrow 3')$ and $CT(3' \rightarrow 5')$ are used to generate time domain 2D electronic spectra. For this purpose excited state calculations of the refined GS snapshot and of the excited state minima were performed with Molcas 7.7 (ref. 73) at the state averaged (SA)-CASSCF level including two occupied (HOMO − 1, HOMO, see Fig. S1 in the ESI†) and two virtual (LUMO, LUMO + 1, see Fig. S1 in the ESI†) π-orbitals on each chromophore in the active space (*i.e.* CASSCF(8,8)). Calibration against full active space calculations of the adenine monomer in gas-phase shows that this highly reduced AS does not permit quantitative predictions (details in the ESI†). In fact, a number of transitions accessible by visible light out of the $L_a$ band, which involve lower occupied and higher virtual π-orbitals, are missed. Despite this, the reference calculations demonstrate that doubly excited states which arise *via* excitations among HOMO − 1, HOMO, LUMO and LUMO + 1 acquire the highest oscillator strengths and, thus, dominate the NUV spectral window. We expect these excited states to be affected the most along the de-excitation pathways, considering that they involve the same orbitals as the optimized $L_a$ and CT states. Thus, our efforts focus in resolving the qualitative spectral dynamics of the ESA into the doubly excited states in the NUV spectral window between 20 000 $cm^{-1}$ and 40 000 $cm^{-1}$ from the $L_a$ band (*i.e.* between 60 000 $cm^{-1}$ and 80 000 $cm^{-1}$ from the GS). A shift of 3000 $cm^{-1}$ was applied to all states (implies a 6000 $cm^{-1}$ shift of all mixed doubly excited states) in order to better reproduce the reference calculations. This shift has no effect on the position of the SE and ESA in the spectra, but blue-shifts the GSB by 3000 $cm^{-1}$. The generally contracted ANO-L basis set was utilized and the following contraction scheme was adopted: C, O/[3s2p1d] and H/[2s].[74] Subsequent energy refinement was done perturbationally with the multi-configurational counterpart of the Møller–Plesset method, denoted as CASPT2, in its single state (SS) version. An imaginary shift of 0.2 (ref. 75) and an ionization potential electron affinity (IPEA) shift of 0.0 (ref. 76) were used. 60 states were included in the state-averaging procedure. The number of roots was chosen to include excitations which upon CASPT2 correction lie in the energy ranges reported in the electronic spectra. Transition dipole moments were calculated at the CASSCF level. The MM part of each snapshot was treated as a set of external point charges in both CASSCF and CASPT2 calculations. The Cholesky decomposition approximation was used to speed up the calculation of two-electron integrals.[73] A constant dephasing of 250 $cm^{-1}$ was employed with finite transform-limited Gaussian pulse envelopes corresponding to a bandwidth of 10 000 $cm^{-1}$. While this broadening is unrealistic it serves the purpose of showing the fingerprints of each channel in a broad spectral window, which would provide valuable information for designing experiments with tailored probe pulses. Spectra are plotted on a linear scale.

### 3.4 Traces for $L_a$ and CT

First, we discuss the traces of the individual excited state relaxation pathways out of the FC region towards the local minima.

We begin with the trace of the $L_a$ de-excitation channel from the FC region to the broad plateau in the relaxed potential energy surface. Due to the flat energy profile the position of the local minimum depends crucially on the computational level used to describe the dynamic correlation. To deal with this issue we selected two representative geometries with $C_2$ puckering 11° and 40° for generating the spectra. The $L_a$ band gives rise to the most intense transitions in the NUV, absorbing around 39 000 $cm^{-1}$. They are delocalized over both nucleobases due to exciton coupling forming the exciton states $L_a(3') - L_a(5')$ and $L_a(3') + L_a(5')$. Fig. 3 shows the positive linear combination which collects most of the oscillator strength in the NUV. The region around the GSB of the $L_a(3') + L_a(5')$ (37 000–41 000 $cm^{-1}$) is congested with off-diagonal bleach signals (*e.g.* peak 2) and ESA to mixed doubly excited states (*e.g.* $L_a(3' + 5')$ peak 5) and is hard to disentangle. Therefore, we encourage the performance of experiments in the low energy window below the bleach region. Pairs of ESA peaks appear (peaks 3|4, and 6|7), which arise *via* probing the local doubly excited states $D_1^*(3'/5')$ and $D_2^*(3'/5')$ out of the $L_a$ band. The ESA doublets 3|4, and 6|7 split as the system leaves the FC region and the exciton localizes on one of the bases. Thereby, the ESA to doubly excited states on the puckered adenine exhibit no blue-shift (peak 3) or red-shifts (peaks 6 and 8) while ESA to doubly excited states on the unmodified adenine exhibit a blue-shift (peaks 4 and 7) equal to the red-shift of the SE (peak 1′). The pronounced red-shift of peaks 6 and 8 indicates that the geometrical

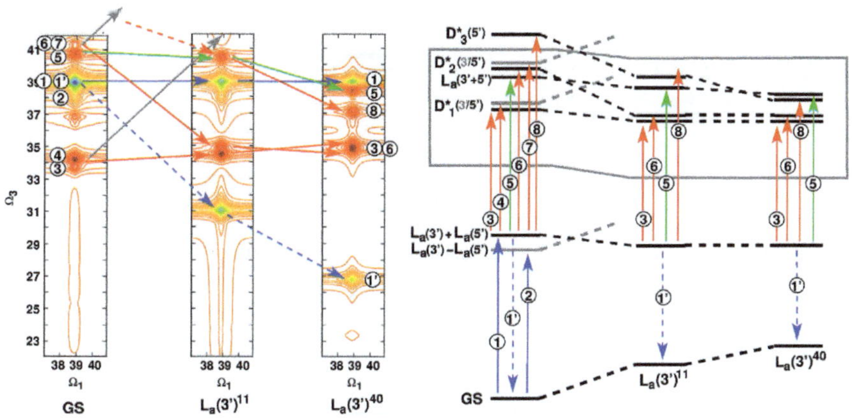


Fig. 3 ESA and SE evolution along the $L_a$ relaxation pathway: (left) spectral traces at critical points; (right) level scheme.

deformations along the $L_a$ relaxation pathway stabilize local doubly excited states even stronger which is a remarkable finding. The SE red-shifts continuously even as the excited state reaches a plateau since the GS exhibits constant destabilization with $C_2$ puckering deformation.

We note that in a real spectrum the low energy window (*i.e.* below 33 000 $cm^{-1}$) will exhibit peaks coming from ESA to singly excited states which remain unresolved with the small active space used in this study. Nevertheless, since these states involve orbitals other than HOMO − 1, HOMO, LUMO and LUMO + 1, we expect that their signals will blue-shift along the relaxation pathway. We also note that the absolute positions of the fingerprint ESA are likely to change following more accurate computations. Exemplarily, comparison with reference gas-phase calculations indicates that peak 3 appears around 30 000 $cm^{-1}$. Nevertheless, our findings make peaks 3, 6 and 8 unique for the $L_a$ relaxation pathway.

In contrast to local excited states, the spectroscopic trace of the CT states is dynamic, depending on the energetic fluctuations of the CT energy levels given by the GS dynamics and on the mixing with degenerate transitions. Our calculations indicate that stabilization down to the NUV region alone is not enough to promote a notable fraction of molecules into the CT states upon NUV pumping, as the oscillator strength of CT states remains generally weak, albeit increasing with respect to unstacked conformations. A noteworthy gain is only possible through mixing with the $L_a$ band. As a consequence, the CT state would inherit oscillator strength and ESA features of the $L_a$ band, whereas the positions of these features would fluctuate with the energy of the CT. In Fig. 4, we focus solely on the spectral characteristics of a pure CT state in order to disentangle in such ideal spectrum its characteristic components that are hard to monitor experimentally, due to the weak oscillator strength of pure CT states. In the relaxed GS geometry the lowest CT state (*i.e.* CT(3′ → 5′)) lies at 6.10 eV with respect to the GS, therefore, pumping occurs with a far-UV (FUV) pulse centered at 49 000 $cm^{-1}$. Since we are interested in the NUV spectral window, probing is done in the 20 000–40 000 $cm^{-1}$ range, where the diagonal GSB of the CT (peak 3) cannot be resolved. An off-diagonal bleach signal at 38 900 $cm^{-1}$ attributed to the $L_a$ bleaching (peak 1) is visible. In the FC region the spectral window below the $L_a$ bleach is lacking intense ESA,

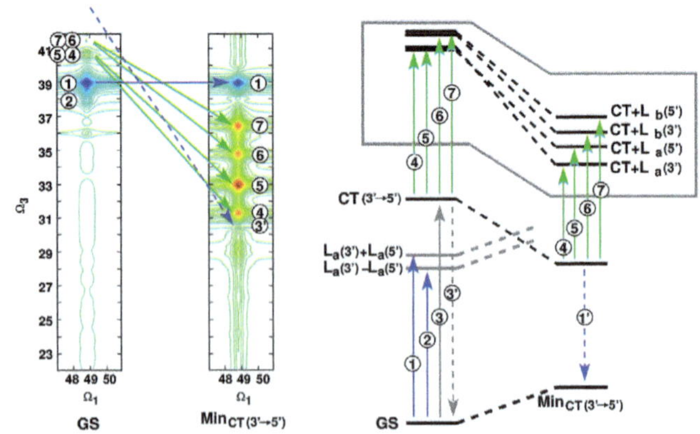


Fig. 4 ESA and SE evolution along the CT relaxation pathway: (left) spectral traces at critical points; (right) level scheme.

 

whereas four intense peaks appear at the CT minimum. They originate from strongly red-shifted transitions to mixed doubly excited states which, based on wavefunction analysis, we label CT + $L_a(3')$ (peak 4), CT + $L_a(5')$ (peak 5), CT + $L_b(3')$ (peak 6) and CT + $L_b(5')$ (peak 7). Exemplarily, CT + $L_a(3')$ denotes a local HOMO → LUMO excitation in the positively charged moiety of the exciplex. The red-shifted broad ESA feature is the fingerprint of the lowest CT states in the aggregate. The strong red-shift indicates that the $L_a/L_b$ transitions are stabilized along the CT de-excitation pathway in support of the finding that geometrical deformations along these pathways present similarities.

### 3.5 2DNUV spectra for hypothetical deactivation scenarios

At this point we have enough information to generate ideal 2DNUV ultrafast spectra for various scenarios. Thereby, we make use of the experimental observation that in a poly-adenine DNA single strand the decay times ($\tau_1 = 0.39$ ps, $\tau_2 = 4.3$ ps, $\tau_3 = 182$ ps)[77] differ by an order of magnitude. We assume that leaving the FC region, fast and slow decay do not overlap temporally. Thus, in the ps regime only ESA and SE originating from the deactivation channel responsible for the long life-time survive.

Fig. 5 shows ultrafast 2D spectra for three hypothetical de-excitation mechanisms. Pumping is done with 733 cm$^{-1}$ broad pulses (corresponds to a 20 fs Fourier limited pulse) centered at 38 000 cm$^{-1}$, while 2932 cm$^{-1}$ broad pulses centered at 34 000 cm$^{-1}$ are used for probing. Plotting is done on a linear scale.

**3.5.1 Mechanism A: deactivation exclusively through local excited states.** The first hypothetical mechanism (A, Fig. 5 upper panel) that we consider involves the following features:

• Both exciton states $L_a(3') + L_a(5')$ and $L_a(3') - L_a(5')$ are populated upon excitation;

• The CT states are not populated in the FC and do not play a role in the deactivation process;

• the $L_a$ plateau can be reached from each exciton state populated in the FC point through bifurcation of the wavepacket in the femtosecond regime;

• Due to the plateau on the potential energy surface of $L_a$ the wavepacket undergoes spreading in the ps regime.§

According to mechanism A the exciton state $L_a(3') + L_a(5')$, which absorbs at 38 900 cm$^{-1}$ dominates the spectrum, the weaker $L_a(3') - L_a(5')$ state, which absorbs at 38 500 cm$^{-1}$, introduces asymmetry in the signals. Two ESA bands, one around 33 000–35 000 cm$^{-1}$ and another one around 36 000–37 000 cm$^{-1}$ are observed. On a femtosecond timescale the SE red-shifts to 31 000–32 000 cm$^{-1}$, the ESA around 33 000–35 000 cm$^{-1}$ blue-shifts by 1000 cm$^{-1}$ and becomes more intense, while the ESA around 36 000–37 000 cm$^{-1}$ disappears. On a ps timescale the SE spreads in the Vis part of the spectrum between 26 000–32 000 cm$^{-1}$ and consequently its intensity decreases. The ESA around 34 000–36 000 cm$^{-1}$ stays compact and retains its intensity. Intuitively, this ESA behavior could be attributed to a compactness of the wavepacket and coherent dynamics in the excited

§ The wavepacket spreading along the $L_a$ is accounted for by averaging the spectra over 10 snapshots generated by interpolating energies and transition dipole moments between $L_a(5')$ (ref. 7) and $L_a(5')$,[40] as well as between $L_a(3')$[11] and $L_a(3')$.[40]

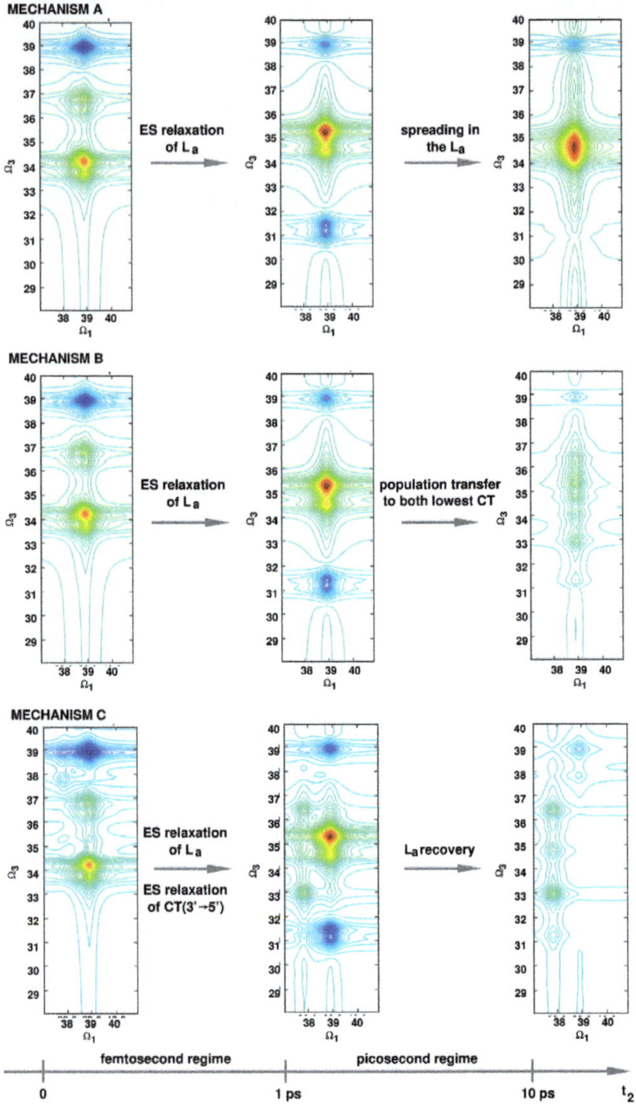


Fig. 5 2DNUV spectra for three hypothetical deactivation mechanisms. The details are given in Sec. 3.5.

state. This behavior reflects the presence of a flat excited state profile along the $L_a$ de-excitation pathway.

**3.5.2 Mechanism B: deactivation involving population of CT states through intra-molecular vibrational energy redistribution.** The second hypothetical deactivation mechanism (B, Fig. 5 middle panel) involves the following features:

• Both exciton states $L_a(3') + L_a(5')$ and $L_a(3') - L_a(5')$ are populated upon excitation;

• The CT states are not populated in the FC;

• The $L_a$ plateau can be reached from each exciton state populated in the FC point through bifurcation of the wavepacket in the femtosecond regime;

• When reaching the $L_a$ plateau of each base the wavepacket bifurcates again; most of the $L_a$ population decays rapidly (*i.e.* in the few-ps regime) to the GS, while a non negligible fraction is transferred to a CT channel through intra-molecular vibrational energy redistribution where it gets trapped in the corresponding minimum;

• As a result of the observations of Conti *et al.* CT(3′ → 5′) is accessible only from the $L_a(5')$ plateau, while CT(5′ → 3′) is accessible only from the $L_a(3')$ plateau.

The femtosecond photophysics of the first and second mechanisms are identical. With the majority of $L_a$ population decaying back to the GS in the ps regime the overall spectral intensity decreases. The decay of the $L_a$ population is coded in the partial recovery of the GSB. The population transfer to the CT shows in the broad ESA band between 31 000 $cm^{-1}$ and 37 000 $cm^{-1}$ which emerges along the $L_a(3') - L_a(5')$ and $L_a(3') + L_a(5')$ traces. Computations by Conti *et al.* suggest that mechanisms A and B are correlated, whereby slowing down the relaxation along the flat $L_a$ plateau (mechanism A) facilitates the back-and-forward population transfer between the $L_a$ and CT channels. Provided this mechanism is confirmed experimentally, it is intriguing whether it is a coherent process showing through oscillations between the $L_a$ and CT traces.

**3.5.3 Mechanism C: deactivation involving population of CT states in the FC region.** The third hypothetical deactivation mechanism (C, Fig. 5 lower panel) involves the following features:

• In 50% of the snapshots the CT(3′ → 5′) is stabilized dynamically below the $L_a$ region to ~38 000 $cm^{-1}$ and can be directly populated;¶ note that the percentage and the energetic position of the CT(3′ → 5′) state are arbitrary despite some theoretical support;[71]

• The CT(3′ → 5′) state mixes with the $L_a$ band in the FC region, borrows oscillator strength and adopts its ESA features;||

• Local transitions ($L_a$, $D_1^*(3'/5')$, $D_2^*(3'/5')$ and $D_3^*(5')$) are not affected by the deformations (shortening of the inter-base distance) which stabilize the CT state;

• The $L_a$ plateau can be reached from each exciton state populated in the FC point through bifurcation of the wavepacket in the femtosecond regime;

• The $L_a$ population decays rapidly (*i.e.* in the few-ps regime) to the GS, the CT population is trapped in the local minimum on the excited state.

Although this mechanism is not supported by QM/MM computations,[31,33,65] other groups have considered the possibility to populate low lying CT states directly in the FC region.[8,71] In this regard we would like to draw attention to a recent paper by Rohlfing and co-workers which argues that aqueous solvation may lower CT states by more than 1 eV based on many-body Green's function computations.[71] We believe that only a detailed analysis of the conformational dynamics considering thermal fluctuations instead of analyzing a single representative relaxed geometry will give a definite answer to this pending question.

¶ This is accomplished by red-shifting the energy of the CT(3′ → 5′) state, as well as the energies of the mixed doubly excited states CT + $L_a(3')$, CT + $L_a(5')$, CT + $L_b(3')$ and CT + $L_b(5')$ (giving rise to peaks 4–7 in Fig. 4) by 8000 $cm^{-1}$.

|| This is accomplished by assigning finite transition dipole moments from the CT(3′ → 5′) state to the local doubly excited states accessible from the $L_a$ states ($D_1^*(3'/5')$, $D_2^*(3'/5')$ and $D_3^*$ (5′) in Fig. 3).

At the FC point the CT trace appears in the long-wavelength region of the NUV absorption band. Due to mixing with the $L_a$ band $L_a$ features appear as weak ESA along the CT trace at ~35 500 $cm^{-1}$ and ~38 000 $cm^{-1}$, blue-shifted by 1000 $cm^{-1}$ with respect to the intense ESA along the $L_a(3') + L_a(5')$ trace. In the femtosecond timescale the $L_a$ channel relaxes to its plateau, while the CT channel relaxes to its local minimum $Min_{CT(3' \rightarrow 5')}$ and the CT fingerprints are clearly recognized. In the ps regime only the CT fingerprints survive. The fs population recovery is again coded in the GSB of both the $L_a$ and CT bands. Note, that neither GSB recovers completely as there is still population in the excited state. Comparison between the second and third mechanisms demonstrates the sensitivity of 2D spectroscopy in resolving population transfer (B). Furthermore, 1D pump–probe spectroscopy, in which the 2D spectra collapse along $\omega_1$, will hardly discriminate the ps signatures of these two mechanisms.

# 4 Conclusion

In this contribution we provided a framework for simulating time-resolved 2D electronic spectroscopy experiments relying on accurate *ab initio* methods. The SOS//QM/MM protocol requires transition energies of singly and doubly excited states together with the transition dipole moments among them. While dynamics simulations are necessary to cover all feasible de-excitation pathways and to collect enough information for simulating non-homogeneous broadenings, already static computations (*i.e.* geometry optimizations, minimum energy path calculations) may provide information on the temporal evolution of individual deactivation channels and may be used to generate ideal 2D spectra in support of hypothetical scenarios, derived from the mechanistic picture. To assess the validity of these ideal spectra a thorough modelling on top of mixed quantum-classical dynamics simulations is required.

The SOS//QM/MM protocol is a hybrid approach and as such it can profit from the progress in three different fields of research: QM/MM dynamics, excited state computations and non-linear spectroscopy. Potential improvements range from employing polarizable force fields[78] or orbital-free embedding[79] to treat the environment in the QM/MM dynamics *via* using more flexible wavefunction-based techniques for computing excited states (*e.g.* generalized active space SCF,[80] the density matrix renormalization group approach,[81] the *n*-electron valence state perturbation theory[82]) to using real pulse shapes and accounting for coherence evolution when computing the nonlinear response of the system.[28]

The adenine–adenine stacked dimer embedded in an explicitly solvated double-stranded DNA model was chosen to demonstrate the application of the protocol in practice. Characteristic points along the first excited state potential energy surface were used to generate time-dependent traces of the local $L_a$ and CT exciplex channels. Subsequently, the 2DNUV spectra were simulated for three de-excitation mechanisms: a deactivation exclusively along the $L_a$ surface (mechanism A), a deactivation involving population of CT states through intra-molecular vibrational energy redistribution (mechanism B) and a deactivation associated with populating both the $L_a$ and the CT states in the FC region. A more detailed study will follow using more characteristic points along the potential energy surface, applying more accurate computational methods (*e.g.* increasing the active space), considering the role of the $L_b$ band, including coherence evolution and parametrized inhomogeneous broadenings, as well as applying a kinetic model to simulate population transfer along and between deactivation channels.

## Acknowledgements

M.G. acknowledges support by the European Research Council Advanced Grant STRATUS (ERC-2011-AdG no. 291198). S.M. gratefully acknowledges the support of the National Institute of Health Grant No. GM-59230 and the National Science Foundation through grant no. CHE-1361516.

## References

1 C. T. Middleton, K. D. L. Harpe, C. Su, Y. K. Law, C. E. Crespo-Hernández and B. Kohler, *Annu. Rev. Phys. Chem.*, 2009, **60**, 217–239.
2 C. E. Crespo-Hernández, B. Cohen, P. M. Hare and B. Kohler, *Chem. Rev.*, 2004, **104**, 1977–2020.
3 M. S. de Vries and P. Hobza, *Annu. Rev. Phys. Chem.*, 2007, **58**, 585.
4 K. Kleinermanns, D. Nachtigallova and M. S. de Vries, *Int. Rev. Phys. Chem.*, 2013, **32**, 308.
5 I. Vayá, T. Gustavsson, T. Douki, Y. Berlin and D. Markovitsi, *J. Am. Chem. Soc.*, 2012, **134**, 11366–11368.
6 J. Chen and B. Kohler, *J. Am. Chem. Soc.*, 2014, **136**, 6362–6372.
7 C. E. Crespo-Hernández, B. Cohen and B. Kohler, *Nature*, 2005, **436**, 1141–1144.
8 G. Olaso-González, M. Merchán and L. Serrano-Andrés, *J. Am. Chem. Soc.*, 2009, **131**, 4368–4377.
9 F. D. Lewis, H. Zhu, P. Daublain, K. Sigmund, T. Fiebig, M. Raytchev, Q. Wang and V. Shafirovich, *Photochem. Photobiol. Sci.*, 2008, **7**, 534.
10 D. B. Bucher, B. M. Pilles, T. Carell and W. Zinth, *Proc. Natl. Acad. Sci. U. S. A.*, 2014, **111**, 4369–4374.
11 Y. Zhang, J. Dood, A. A. Beckstead, X.-B. Li, K. V. Nguyen, C. J. Burrows, R. Improta and B. Kohler, *Proc. Natl. Acad. Sci. U. S. A.*, 2014, **111**, 11612.
12 N. K. Schwalb and F. Temps, *J. Am. Chem. Soc.*, 2007, **129**, 9272–9273.
13 N. K. Schwalb and F. Temps, *Science*, 2008, **322**, 243–245.
14 V. Saurí, J. P. Gobbo, J. J. Serrano-Pérez, M. Lundberg, P. B. Coto, L. Serrano-Andrés, A. C. Borin, R. Lindh, M. Merchán and D. Roca-Sanjuán, *J. Chem. Theory Comput.*, 2013, **9**, 481–496.
15 J. P. Gobbo, V. Saurí, D. Roca-Sanjuán, L. Serrano-Andrés, M. Merchán and A. C. Borin, *J. Phys. Chem. B*, 2012, **116**, 4089–4097.
16 R. P. Sinha and D.-P. Häder, *Photochem. Photobiol. Sci.*, 2002, **1**, 225.
17 M. Hada and A. G. Georgakilas, *J. Radiat. Res.*, 2008, **49**, 203.
18 A. C. Kneuttinger, G. Kashiwazaki, S. Prill, K. Heil, M. Müller and T. Carell, *Photochem. Photobiol.*, 2014, **90**, 1.
19 E. Collini, C. Y. Wong, K. E. Wilk, P. M. G. Curmi, P. Brumer and G. D. Scholes, *Nature*, 2010, **463**, 644.
20 T. Brixner, J. Stenger, H. M. Vaswani, M. Cho, R. E. Blankenship and G. R. Fleming, *Nature*, 2005, **434**, 625.
21 C. H. Tseng, S. Matsika and T. C. Weinacht, *Opt. Express*, 2009, **17**, 18788–18793.

22 B. A. West and A. M. Moran, *J. Phys. Chem. Lett.*, 2012, **3**, 2575–2581.
23 S. Mukamel, *Principles of Nonlinear Optics and Spectroscopy*, Oxford Unversity Press, Oxford, UK, 1995.
24 D. Abramavicius, B. Palmieri, D. V. Voronine, F. Sanda and S. Mukamel, *Chem. Rev.*, 2009, **109**, 2350–2408.
25 M. Cho, *Two-Dimensional Optical Spectroscopy*, CRC Press, Boca Raton, Florida, USA, 2009.
26 J. Jiang and S. Mukamel, *Phys. Chem. Chem. Phys.*, 2011, **13**, 2394.
27 A. Nenov, I. Rivalta, G. Cerullo, S. Mukamel and M. Garavelli, *J. Phys. Chem. Lett.*, 2014, **5**, 767.
28 B. P. Fingerhut, K. E. Dorfman and S. Mukamel, *J. Chem. Theory Comput.*, 2014, **10**, 1172.
29 F. Santoro, V. Barone and R. Improta, *Proc. Natl. Acad. Sci. U. S. A.*, 2007, **104**, 9931–9936.
30 R. Improta and V. Barone, *Angew. Chem., Int. Ed.*, 2011, **50**, 12016.
31 F. Plasser and H. Lischka, *Photochem. Photobiol. Sci.*, 2013, **12**, 1440.
32 V. A. Spata and S. Matsika, *J. Phys. Chem. A*, 2014, **118**, 12021.
33 D. Polli, P. Altoe, O. Weingart, K. M. Spillane, C. Manzoni, D. Brida, G. Tomasello, G. Orlandi, P. Kukura, R. A. Mathies, M. Garavelli and G. Cerullo, *Nature*, 2010, **467**, 441.
34 I. Rivalta, A. Nenov, G. Cerullo, S. Mukamel and M. Garavelli, *Int. J. Quantum Chem.*, 2014, **114**, 85–93.
35 A. Nenov, I. Rivalta, G. S. Mukamel and M. Garavelli, *Comput. Theor. Chem.*, 2014, **1040**, 295.
36 A. Nenov, S. a. Beccara, I. Rivalta, G. Cerullo, S. Mukamel and M. Garavelli, *ChemPhysChem*, 2014, **15**, 3282.
37 I. Rivalta, A. Nenov, O. Weingart, G. Cerullo, G. S. Mukamel and M. Garavelli, *J. Phys. Chem. B*, 2014, **118**, 8396.
38 P. Hamm and M. Zanni, *Concepts and Methods of 2D Infrared Spectroscopy*, Cambridge Unversity Press, Cambridge, UK, 2011.
39 D. Abramavicius, L. Valkunas and S. Mukamel, *Europhys. Lett.*, 2007, **80**, 17005.
40 G. Aubock, C. Consani, F. van Mourik and M. Chergui, *Opt. Lett.*, 2012, **37**, 2337.
41 N. Krebs, I. Pugliesi, J. Hauer and E. Riedle, *New J. Phys.*, 2014, **15**, 085016.
42 R. B. Varillas, A. Candeo, D. Viola, M. Garavelli, S. D. Silvestri, G. Cerullo and C. Manzoni, *Opt. Lett.*, 2014, **39**, 3849.
43 D. Picconi, V. Barone, A. Lami, F. Santoro and R. Improta, *ChemPhysChem*, 2011, **12**, 1957.
44 D. Picconi, A. Lami and F. Santoro, *J. Chem. Phys.*, 2012, **136**, 244104.
45 D. Picconi, F. J. A. Ferrer, R. Improta, A. Lami and F. Santoro, *Faraday Discuss.*, 2013, **163**, 223.
46 G. Pescitelli, V. Barone, L. D. Bari, A. Rizzo and F. Santoro, *J. Org. Chem.*, 2013, **78**, 7398.
47 D. Padula, D. Picconi, A. Lami, G. Pescitelli and F. Santoro, *J. Phys. Chem. A*, 2013, **117**, 3355.
48 R. Borrelli, M. Thoss, H. Wang and W. Domcke, *Mol. Phys.*, 2012, **110**, 751.
49 H. Wang and M. Thoss, *J. Chem. Phys.*, 2003, **119**, 1289.
50 B. O. Roos, *Ab Initio Methods in Quantum Chemistry: Part II*, Wiley, Chicester, UK, 1987.

 

51 K. Andersson, P. A. Malmqvist, B. O. Roos, A. J. Sadlej and K. Wolinski, *J. Phys. Chem.*, 1990, **94**, 5483–5488.
52 G. H. Chen, S. Mukamel, D. Beljonne and J. L. Bredas, *J. Chem. Phys.*, 1996, **104**, 5406–5414.
53 T. Takaya, C. Su, K. D. L. Harpe, C. E. Crespo-Hernández and B. Kohler, *Proc. Natl. Acad. Sci. U. S. A.*, 2008, **105**, 10285–10290.
54 C. E. Crespo-Hernández, K. D. L. Harpe and B. Kohler, *J. Am. Chem. Soc.*, 2008, **130**, 10844–10845.
55 K. D. L. Harpe, C. E. Crespo-Hernández and B. Kohler, *J. Am. Chem. Soc.*, 2009, **131**, 17557–17559.
56 K. D. L. Harpe and B. Kohler, *J. Phys. Chem. Lett.*, 2011, **2**, 133–138.
57 B. Kohler, *J. Phys. Chem. Lett.*, 2010, **1**, 2047–2053.
58 I. Vayá, T. Gustavsson, F.-A. Miannay, T. Douki and D. Markovitsi, *J. Am. Chem. Soc.*, 2010, **132**, 11834–11835.
59 I. Vayá, F.-A. Miannay, T. Gustavsson and D. Markovitsi, *ChemPhysChem*, 2010, **11**, 987–989.
60 J. Chen, A. K. Thazhathveetil, F. D. Lewis and B. Kohler, *J. Am. Chem. Soc.*, 2013, **135**, 10290–10293.
61 I. Bucharov, Q. Wang, M. Raytchev, A. Trifonov and T. Fiebig, *Proc. Natl. Acad. Sci. U. S. A.*, 2007, **104**, 4794–4797.
62 C. Greve, N. K. Preketes, R. Costard, B. Koeppe, H. Fidder, E. T. J. Nibbering, F. Temps, I. Andricioaei, S. Mukamel and T. Elsaesser, *J. Phys. Chem. A*, 2012, **116**, 7636–7644.
63 C. Greve, N. K. Preketes, H. Fidder, R. Costard, B. Koeppe, I. A. Heisler, S. Mukamel, F. Temps, E. T. J. Nibbering and T. Elsaesser, *J. Phys. Chem. A*, 2013, **117**, 594–606.
64 I. Conti, P. Altoè, M. Stenta, M. Garavelli and G. Orlandi, *Phys. Chem. Chem. Phys.*, 2010, **12**, 5016–5023.
65 I. Conti, A. Nenov, S. Höfinger, S. F. Altavilla, I. Rivalta, E. Dumont, G. Orlandi and M. Garavelli, *Phys. Chem. Chem. Phys.*, 2015, DOI: 10.1039/C4CP05546B.
66 L. Serrano-Andrés, M. Merchán and A. C. Borin, *Proc. Natl. Acad. Sci. U. S. A.*, 2006, **103**, 8691–8696.
67 L. Serrano-Andrés, M. Merchán and A. C. Borin, *Chem.–Eur. J.*, 2006, **12**, 6559–6571.
68 M. Merchán, R. González-Luque, T. Climent, L. Serrano-Andrés, E. Rodríguez, M. Reguero and D. Peláez, *J. Phys. Chem. B*, 2006, **110**, 26471–26476.
69 A. Giussani, J. Segarra-Mart, D. Roca-Sanjun and M. Merchn, *Excitation of Nucleobases from a Computational Perspective I: Reaction Paths*, Springer, Berlin Heidelberg, 2013, pp. 1–41.
70 I. Conti, M. Garavelli and G. Orlandi, *J. Am. Chem. Soc.*, 2009, **131**, 16108–16118.
71 H. Yin, Y. Ma, J. Mu, C. Liu and M. Rohlfing, *Phys. Rev. Lett.*, 2014, **112**, 228301.
72 C. Su, C. T. Middleton and B. Kohler, *J. Phys. Chem. B*, 2012, **116**, 10266–10274.
73 F. Aquilante, L. De Vico, N. Ferre, G. Ghigo, P. A. Malmqvist, P. Neogrady, T. B. Pedersen, M. Pitonak, M. Reiher, B. O. Roos, L. Serrano-Andres, M. Urban, V. Veryazov and R. Lindh, *J. Comput. Chem.*, 2010, **31**, 224–247.
74 P. O. Widmark, P. A. Malmqvist and B. O. Roos, *Theor. Chim. Acta*, 1990, **77**, 291–306.
75 N. Forsberg and P. A. Malmqvist, *Chem. Phys. Lett.*, 1997, **274**, 196–204.

76 G. Ghigo, B. O. Roos and P. A. Malmqvist, *Chem. Phys. Lett.*, 2004, **396**, 142–149.
77 W.-M. Kwok, C. Ma and D. L. Phillips, *J. Am. Chem. Soc.*, 2006, **128**, 11894.
78 Y. Shi, Z. Xia, J. Zhang, R. Best, C. Wu, J. W. Ponder and P. Ren, *J. Chem. Theory Comput.*, 2013, **9**, 4046.
79 T. A. Wesolowski, *Computational Chemistry: Reviews of Current Trends*, World Scientific, 10th edn, 2006.
80 D. Ma, G. L. Manni and L. Gagliardi, *J. Chem. Phys.*, 2011, **135**, 044128.
81 Y. Kurashige and T. Yanai, *J. Chem. Phys.*, 2011, **135**, 094104.
82 C. Angeli, M. Pastore and R. Cimiraglia, *Theor. Chem. Acc.*, 2007, **117**, 743.

 

Faraday Discussions

PAPER

# Out-of-equilibrium dynamics of photoexcited spin-state concentration waves

A. Marino,[a] M. Buron-Le Cointe,*[a] M. Lorenc,[a] L. Toupet,[a] R. Henning,[b] A. D. DiChiara,[c] K. Moffat,[b] N. Bréfuel[d] and E. Collet*[a]

Received 29th August 2014, Accepted 9th October 2014
DOI: 10.1039/c4fd00164h

The spin crossover compound $[Fe^{II}H_2L^{2\text{-}Me}][PF_6]_2$ presents a two-step phase transition. In the intermediate phase, a spin state concentration wave (SSCW) appears resulting from a symmetry breaking (cell doubling) associated with a long-range order of alternating high and low spin molecular states. By combining time-resolved optical and X-ray diffraction measurements on a single crystal, we study how such a system responds to femtosecond laser excitation and we follow in real time the erasing and rewriting of the SSCW.

## Introduction

$Fe^{II}$ spin-crossover (SCO) complexes have been widely studied over the last decades, because of their ability to show reversible switching between low-spin (LS, $S = 0$) and high-spin (HS, $S = 2$) states, controlled by a change in temperature or pressure and more interestingly under light irradiation. SCO complexes are now considered as prototype photo-switchable molecular materials.[1,2] One of the important successes of ligand field theory was to establish the origin of SCO in the coupling between electronic and structural degrees of freedom. As the electronic configuration changes from $t_{2g}^6e_g^0$ in the LS state to the less bonding $t_{2g}^4e_g^2$ in the HS state, an important structural reorganization occurs. It is associated with an expansion of the average distance between the Fe and the 6 N atoms bonding it to the ligand.[1] Typical values are $\langle Fe–N\rangle_{LS} \approx 2.0$ Å and $\langle Fe–N\rangle_{HS} \approx 2.2$ Å.[3–11] In the solid state, the coexistence of short- and long-range interactions between molecules yields cooperative effects leading to continuous conversions or first order

[a]Institut de physique de Rennes, UMR 6251 Université Rennes 1-CNRS, 35042 Rennes, FRANCE. E-mail: marylise.buron@univ-rennes1.fr; eric.collet@univ-rennes1.fr
[b]Center for Advanced Radiation Sources, The University of Chicago, Chicago, IL 60637, USA
[c]Argonne National Laboratory, 9700 S. Cass Av. Argonne, IL 60439-4857, USA
[d]OMNT, CEA-CNRS, Campus Minatec, 38053 Grenoble, FRANCE

phase transitions with hysteresis.[1,2,12] The SCO phenomenon can be followed through the evolution of the fraction of HS molecules:

$$X_{HS} = N_{HS}/N$$

where $N_{HS}$ is the number of molecules in the HS state among the $N$ molecules of the crystal. This is a totally symmetric order parameter. In many systems, the LS state is the ground state. When temperature is increased, the HS state of higher entropy is favoured, and $X_{HS}$ increases from 0 to 1. It is convenient to monitor the evolution of $X_{HS}$ by using complementary probes sensitive to the electronic state, such as optical spectroscopy,[13] or structural changes such as X-ray diffraction.[6] These probes can also be used in the time domain to follow the HS state conversion, when induced by light excitation. Femtosecond optical and X-ray spectroscopy studies of isolated molecules in solution evidenced that the molecular photo-switching occurs on the sub-picosecond timescale.[14–19] More recently we combined time-resolved X-ray diffraction and optical spectroscopy on weakly cooperative systems to investigate how SCO solids respond to femtosecond (fs) light excitation.[20–25] We have shown that the dynamics induces a multi-step response. First, the sub-picosecond non-thermal molecular photo-switching occurs, resulting from a local trapping similar to the one observed in solution studies. Then, the molecular swelling and lattice heating induced by light drive lattice expansion within tens of ns. Finally, a thermal switching occurs on the microsecond timescale as the HS state is favoured once the temperature of the macroscopic crystal is higher.

Some SCO compounds, instead of showing a single step conversion, undergo two-step transitions. In addition to the HS and LS phases, an intermediate (INT) phase appears, for which $X_{HS}$ is close to 1/2,[9,15] 1/3,[26] 1/4[27] or is irrational.[5] For some compounds, the appearance of the INT phase results from symmetry breaking and a long-range ordering of molecules in HS and LS states. This is the case for the material investigated here: $[Fe^{II}H_2L^{2\text{-}Me}][PF_6]_2$ ($H_2L^{2\text{-}Me}$ denotes the organic ligand bis[((2-methylimidazol-4-yl)methylidene)-3-aminopropyl] ethylenediamine).[4]

### Spin state concentration wave (SSCW)

Magnetic susceptibility measurements on $[Fe^{II}H_2L^{2\text{-}Me}][PF_6]_2$,[4] give the temperature evolution of the HS fraction $X_{HS}$, indicating a two-step SCO conversion (Fig. 1). Above 250 K, all molecules are in the HS state, and the two molecules of the unit-cell are symmetrically equivalent (one independent crystallographic site). Below 250 K, $X_{HS}$ continuously decreases to reach a pseudo-plateau in the [97–142 K] domain. In this INT phase, approximately 50% of the molecules are in LS and HS states. X-ray diffraction data revealed the occurrence of symmetry breaking in this INT phase. It corresponds to a doubling of the crystalline ***c*** axis, characterized by the appearance of additional Bragg reflections. In the doubled cell, there are now two crystallographically independent molecular sites (Fig. 1 and 2). X-ray diffraction data at 130 K (ref. 4) show that site 1 is mainly HS, as characterized by $\langle Fe_1–N\rangle = 2.145(5)$ Å, and site 2 is mainly LS with $\langle Fe_2–N\rangle = 2.055(5)$ Å. The fact that the sites are not completely HS or LS, *i.e.* $\langle Fe_1–N\rangle$ is different from $\langle Fe–N\rangle_{HS} = 2.190$ Å and $\langle Fe_2–N\rangle$ is different from $\langle Fe–N\rangle_{LS} = 2.012$ Å, results from the partial spin-state ordering.[4,5,9]

 

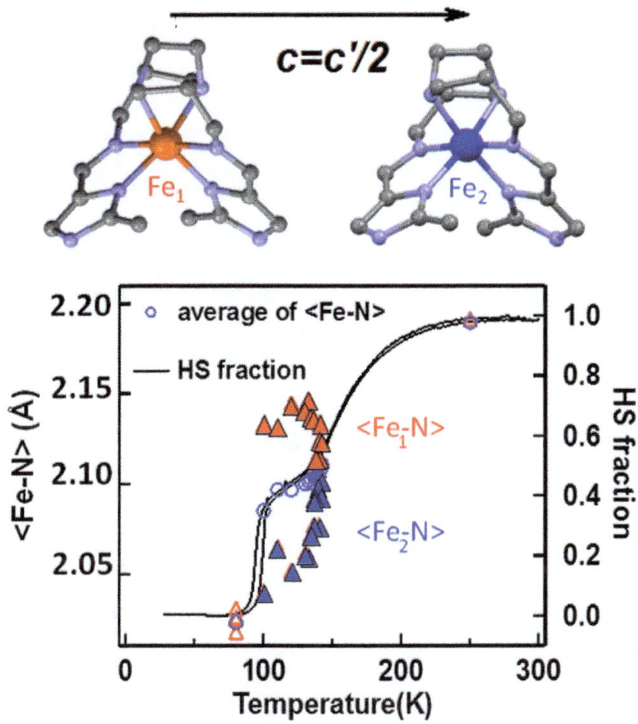


Fig. 1 (top) Structure of the INT phase at 110 K made of two independent sites around $Fe_1$ and $Fe_2$. Only the two symmetry independent sites are shown for clarity. Above 142 K $Fe_1$ and $Fe_2$ are equivalent by translation $\mathbf{c}$, but not in the INT phase because of the cell doubling with $\mathbf{c}' = 2\mathbf{c}$. (bottom) Evolution of the HS fraction (right axis) deduced from magnetic measurements (solid line) and X-ray diffraction (blue circles) through the ⟨Fe–N⟩ bond length (left axis) on site $Fe_1$ (red triangle) and $Fe_2$ (blue triangle) given in ref. 4.

The X-ray diffraction data and this partial order can be well described in the framework of the Landau theory of phase transitions.[28] This theory postulates that any density $\rho$ in the low symmetry phase can be described as made of two components $\rho_0$ and $\delta\rho$. In the HS phase, any density $\rho_0$ belongs to the totally symmetric transformations of the high symmetry crystal. In particular, the spin state concentration (or density) $X_{HS}(\mathbf{r})$ is the same for every molecular site in position $\mathbf{r}$, which are all equivalent by applying the symmetry operators of the HS space group and especially the translation symmetry $\boldsymbol{c}$:

$$X_{HS}(\mathbf{r}) = X_{HS}(\mathbf{r} + n\boldsymbol{c}) = X_{HS},\ n \in |N \text{ (integer)}$$

In the high-temperature/high-symmetry phase the different molecular sites along $\mathbf{r}//\boldsymbol{c}$ have the same probability to be in the HS state. The scheme in Fig. 2c shows that $X_{HS}(\mathbf{r}) \approx 0.5$ at 142 K, whereas $X_{HS}(\mathbf{r}) \approx 1$ above 250 K. In the INT phase of lower symmetry, the density starts to change and becomes $\rho = \rho_0 + \eta\delta\rho$. $\delta\rho$ has the symmetry of the INT phase, which is lower than that of $\rho_0$ because of the cell doubling along $\boldsymbol{c}$. Therefore $\rho = \rho_0 + \eta\delta\rho$ describes any density with the

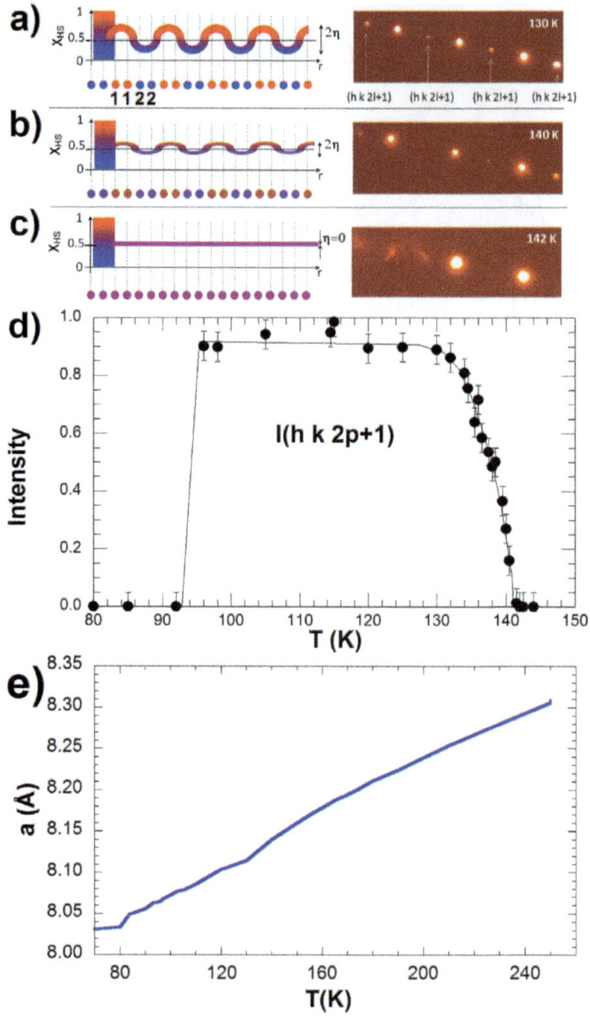


Fig. 2 (a) Spin-state concentration wave among molecular sites 1 and 2 schematically represented by $X_{HS}(\mathbf{r}) = X_{HS} + \eta \times \cos(\mathbf{q}.\mathbf{r})$ on the plateau where $X_{HS} \approx 0.5$. The periodicity of the modulation is 2**c**, where **c** is the distance between sites 1 and 2 which are symmetry equivalent at high temperature in the high symmetry phase. The spin-state concentration wave gives rise to diffracted intensity in the Bragg peaks indexed ($h\ k\ 2p + 1$) in the doubled cell (right). Below 135 K, the order is maximum with a strong diffracted intensity $I(h\ k\ 2p + 1)$ on superstructure peaks. (b) At 140 K the diffracted intensity $I(h\ k\ 2p + 1) \propto \eta^2$ is $\approx$25% of the saturated value, where $\eta = 0.25$, indicating $\eta \approx 0.125$ at 140 K. Above 142 K $I(h\ k\ 2p + 1) = 0$ indicating $\eta = 0$. (d) Temperature dependence of the Bragg peaks characterizing the spin-state concentration wave. (e) Thermal expansion of the lattice parameter *a*.

symmetry of the INT phase (and the new periodicity $\boldsymbol{c}' = 2\boldsymbol{c}$), and so does $X_{HS}(\mathbf{r})$ in the INT phase. $\eta$ is the symmetry breaking order parameter measuring the deviation of the INT phase from the high symmetry phase. In the doubled cell, there are two symmetry-independent molecular sites alternating along $\boldsymbol{c}'$ and now

 

$$X_{HS}(\mathbf{r}) = X_{HS}(\mathbf{r} + c') = X_{HS}(\mathbf{r} + 2nc) \neq X_{HS}(\mathbf{r} + (2n + 1)c)$$

The INT phase can then be simply described as a spatial modulation of the probability for a site in position **r** to be in the HS state:

$$X_{HS}(\mathbf{r}) = X_{HS} + \eta \times \cos(\mathbf{q}.\mathbf{r}),$$

where $X_{HS}$ is the average HS fraction in the crystal ($X_{HS} \approx 0.5$ on the plateau, Fig. 1), **q** is the wave vector corresponding to a cell doubling along the $c$ axis ($\mathbf{q} = \mathbf{c}^*/2 = \mathbf{c}'^*$) and the order parameter $\eta$ measures the amplitude of the wave. As schematically represented in Fig. 2a, the spin-state ordering corresponds to a spin-state concentration wave (SSCW), which forms on the plateau around an average value. Fig. 2a depicts $X_{HS}(\mathbf{r})$ on the lattice sites, with a colour code indicating that sites are mainly HS (red) when $X_{HS}(\mathbf{r})$ is above 0.5, or mainly LS (blue) below. There are 4 molecules per unit cell in the INT phase but only 2 symmetry independent sites (two sites 1 and two sites 2). These are equivalent in the high symmetry phase by the translation $\mathbf{c}$. That is the reason why the sequence of sites 1122 along $\mathbf{c}$ corresponds to HS-HS-LS-LS order (Fig. 2a).

It is possible to describe the evolution of $X_{HS}$ and $\eta$ from the X-ray diffraction data. The ⟨Fe–N⟩ bond is a well-known marker of the HS fraction[29] as $X_{HS}$ weights the contribution of HS and LS molecules in the observed value of ⟨Fe–N⟩:

$$\langle \text{Fe–N} \rangle = X_{HS}\langle \text{Fe–N} \rangle_{HS} + (1 - X_{HS})\langle \text{Fe–N} \rangle_{LS}$$

with $\langle \text{Fe–N} \rangle_{HS} = 2.190$ Å and $\langle \text{Fe–N} \rangle_{LS} = 2.012$ Å (Fig. 1). This is also true for the independent molecular sites 1 and 2 in the INT phase with local HS fractions $X_{HS1}$ and $X_{HS2}$:

$$\langle \text{Fe}_1\text{–N} \rangle = X_{HS1}\langle \text{Fe–N} \rangle_{HS} + (1 - X_{HS1})\langle \text{Fe–N} \rangle_{LS}$$

$$\langle \text{Fe}_2\text{–N} \rangle = X_{HS2}\langle \text{Fe–N} \rangle_{HS} + (1 - X_{HS2})\langle \text{Fe–N} \rangle_{LS}$$

Since at 130 K $\langle \text{Fe}_1\text{–N} \rangle = 2.145(5)$ Å and $\langle \text{Fe}_2\text{–N} \rangle = 2.055(5)$ Å, we can estimate $X_{HS1}$ (130 K) = 0.75 and $X_{HS2}$ (130 K) = 0.25 (Fig. 1). This gives the two order parameters describing the wave at 110 K:

$$X_{HS} = (X_{HS1} + X_{HS2})/2 = 0.5 \text{ and } \eta = (X_{HS1} - X_{HS2})/2 = 0.25$$

The spin-state concentration wave at 130 K is therefore defined by:

$$X_{HS}(\mathbf{r}) = 0.5 + 0.25 \times \cos(\mathbf{q}.\mathbf{r}).$$

It is shown in Fig. 1 that $\langle \text{Fe}_1\text{–N} \rangle$ and $\langle \text{Fe}_2\text{–N} \rangle$ evolve with temperature and become equal at 142 K, where the symmetry breaking disappears, *i.e.* where the two sites become symmetry equivalent. Above 142 K the symmetry imposes $X_{HS1} = X_{HS2}$. The structure of the unit in the INT phase (with a lattice parameter $c' = 2c$)

contains two independent molecular sites shifted by $\mathbf{c} = \mathbf{c}'/2$. The structure factor is then the sum of the contribution of each site:

$$F(h\,k\,l) = F_1(h\,k\,l) + F_2(h\,k\,l).$$

For site 1, the molecule can be in the HS state, with a probability $X_{HS1}$ and corresponding to a HS molecular structure factor $F_{HS}$, or in the LS state with a probability $(1 - X_{HS1})$ and corresponding to a structure factor $F_{LS}$. The structure factor on site 1 is then:

$$F_1(h\,k\,l) = X_{HS1} \times F_{HS} + (1 - X_{HS1}) \times F_{LS}.$$

For site 2, the molecule can also be in the HS state with a probability $X_{HS2}$ and corresponding to a structure factor $F_{HS2}$ or to a molecule in the LS state with a probability $(1 - X_{HS2})$ and corresponding to a structure factor $F_{LS2}$. But since the atoms in site 2 are shifted in space in the unit cell by $\mathbf{c} = \mathbf{c}'/2$ with respect to site 1, a phase term appears in $F_{HS2}$ and $F_{LS2}$ for all atoms, with respect to $F_{HS}$ and $F_{LS}$ on site 1

$$F_{HS2} = F_{HS}\mathrm{e}^{-i\pi l} \text{ and } F_{LS2} = F_{LS}\mathrm{e}^{-i\pi l}$$

For $l = 2p + 1$ (odd), $\mathrm{e}^{-i\pi l} = -1$, one obtains

$$\begin{aligned} F_2(h\,k\,2p+1) &= X_{HS2} \times F_{HS2} + (1 - X_{HS2}) \times F_{LS2} \\ &= -X_{HS2} \times F_{HS} - (1 - X_{HS2}) \times F_{LS} \end{aligned}$$

and

$$\begin{aligned} F(h\,k\,2p+1) &= F_1(h\,k\,2p+1) + F_2(h\,k\,2p+1) \\ &= X_{HS1} \times F_{HS} + (1 - X_{HS1}) \times F_{LS} - X_{HS2} \times F_{HS} - (1 - X_{HS2}) \times F_{LS} \end{aligned}$$

Since the concentration wave has an amplitude $\eta$ around an average value $X_{HS}$:

$$\begin{aligned} X_{HS1} &= X_{HS} + \eta,\ (1 - X_{HS1}) = 1 - X_{HS} - \eta, \\ X_{HS2} &= X_{HS} - \eta,\ (1 - X_{HS2}) = 1 - X_{HS} + \eta \end{aligned}$$

Then for Bragg peaks $(h\,k\,l)$ with $l$ odd:

$$F(h\,k\,2\mathrm{p}+1) = (X_{HS} + \eta) \times F_{HS} + (1 - X_{HS} - \eta) \times F_{LS} - (X_{HS} - \eta) \times F_{HS} - (1 - X_{HS} + \eta) \times F_{LS}$$

$$F(h\,k\,2p+1) = (2\eta) \times F_{HS} + (-2\eta) \times F_{LS} = (2\eta) \times (\mathrm{F}_{HS} - F_{LS}), \text{ and}$$

$$I(h\,k\,2p+1) \propto \eta^2 \times |F_{HS} - F_{LS}|^2$$

 

In other words, the intensity of the Bragg reflections indexed as $(h\ k\ 2p + 1)$ in the $\mathbf{c}' = 2\mathbf{c}$ cell, which characterize the spin-state ordering between molecular sites 1 and 2, is proportional to the square of the order parameter $\eta$ measuring the amplitude of the wave. The intensity of these peaks $I(h\ k\ 2p + 1) = 0$ when $\eta = 0$. Fig. 2d, shows the evolution of the intensity of such Bragg peaks, which vanishes at 142 K and characterizes the temperature above which the SSCW is erased and the high symmetry phase is reached. These data also show that the order rapidly saturates below 135 K as $I(h\ k\ 2p + 1)$ remains constant. This is also in agreement with Fig. 1, showing $a \approx 0.1$ Å difference between $\langle Fe_1–N\rangle$ and $\langle Fe_2–N\rangle$ below 135 K, whereas the difference also vanishes at 142 K. When the temperature is decreased below 90 K, the compound completely switches to the LS state and the spin concentration wave disappears. Finally, the lattice parameters are also sensitive to the thermal spin conversion, especially the parameter $a$ (Fig. 2e). The expansion arises from continuous conversion from INT to HS phases. In the 100–250 K range, the thermal expansion is nearly linear, with a slope of 0.0015 Å $K^{-1}$.

These results differ from those obtained on the derivative compound with a $SbF_6$ counter anion, for which a SSCW forms with a periodicity incommensurate with the one of the lattice.[5] It remains at low temperature and can be erased by light below 40 K, as the photoinduced HS state is stable. The erasing of the SSCW was due to the complete conversion to the HS state, indicated by the disappearance of the characteristic satellites reflections and the change of $\langle Fe–N\rangle$.

The spin-state concentration wave being so characterized, we now discuss how it responds in the time domain to femtosecond light excitation.

## Results

### Optical pump-probe studies

First, we performed ultrafast optical pump-probe measurements to study the photo-switching dynamics of $[Fe^{II}H_2L^{2\text{-}Me}][PF_6]_2$ in the INT phase where the SSCW forms. We used the femtosecond pump-probe set-up at IPR in Rennes. The effects of temperature and excitation density on the time course of $X_{HS}$ were investigated. Details are given in the experimental section.

Fig. 3 shows how optical density (OD), measured on a single crystal, changes in the visible range as the SCO occurs from INT to HS phases. The characteristic Metal-to-Ligand Charge-Transfer (MLCT) band of the LS state appears around 500 nm in the INT phase. We photoexcited the system in the tail of this band at 530 nm to ensure a larger penetration depth and an efficient photo-switching from LS to HS states as observed under continuous light irradiation at 10 K.[4,5] A global decrease of OD in the visible range is observed when temperature increases and LS to HS conversion occurs. We used OD change (probe at 610 nm) to monitor the evolution of $X_{HS}$ after fs laser excitation, as successfully performed in other SCO solids.[20–24] Typical results are presented in Fig. 4a and b. Experiments were performed at different excitation energies at 135 K, where the SSCW is ordered, and 142 K, where it disappears.

We observe a 3-step process, similar to the one reported for the out-of-equilibrium spin-state switching induced by fs laser excitation in other single-step spin-crossover crystals.[20–24] First, the absorption of light at the molecular level locally photo-switches a small fraction $\Delta X_{HS}{}^{h\nu}$ of molecules from LS to HS state. The inset of Fig. 4b shows the evolution of optical density, ΔOD, following fs laser excitation. The

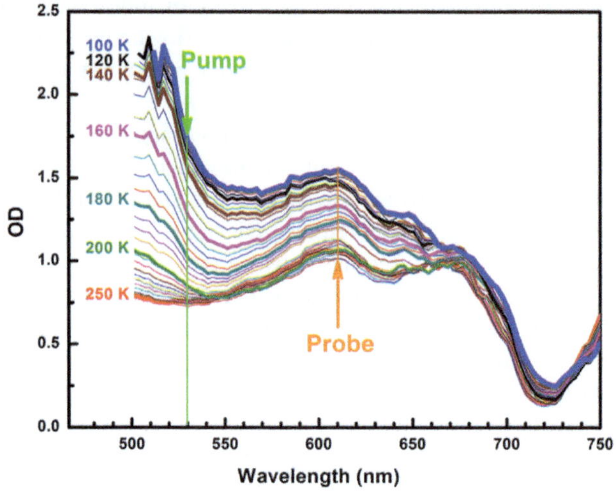


Fig. 3 Temperature dependence of optical density measured on a single crystal, revealing the MLCT band of the LS state at around 500 nm. For time-resolved studies, the pump was set to 530 nm and the probe to 610 nm.

transient peak, which appears immediately after excitation is associated with the MLCT state, whereas the decrease of OD onto a plateau within hundreds of fs is the optical fingerprint of the formation of HS state. This step is referred to as the photo-switching step. Since a low excitation density was used, a small fraction of LS molecules was photo-switched to the HS state. From the OD change shown in the inset, it is estimated that 0.2% of the molecules switch from LS to HS state at the ps timescale. Higher excitation density resulted in sample damage.

The responses at 135 and 142 K show similar features and in particular, a linear dependence of the fraction of photo-switched molecules $\Delta X_{HS}{}^{h\nu}$ with the excitation energy per pump pulse (Fig. 4c). This linear response to excitation density was already reported for cooperative $Fe^{III}$ SCO material,[14] and underlined the local nature of this step. The fraction of HS molecules increases anew after tens of ns as lattice expansion occurs (discussed below under time-resolved diffraction) and reaches $\Delta X_{HS}{}^{El}$. Previous studies indicated that an internal pressure drives the lattice expansion.[20–24] Finally, the thermal step occurs resulting from macroscopic crystal heating by the laser flash, which gives rise to a thermal population of the HS state $\Delta X_{HS}{}^{Th}$ on the μs timescale. We observe in Fig. 4c at 135 K, where the SSCW forms, a change of thermal regime: below 0.8 μJ the thermal population $\Delta X_{HS}{}^{Th}$ is moderate, whereas it increases above that energy. At 142 K the change of regime occurs around 0.4 μJ. In order to clarify the origin of this effect, we studied the response to fs laser excitation at different temperatures with a fixed pump energy of 0.7 μJ per pulse. The data shown in Fig. 5 clearly indicate that the thermal population is maximal around 150 K and moderate around 135 K.

As explained in detail in ref. 21, this thermal population can be described as resulting from a temperature jump $\Delta T$ of the crystal due to laser excitation. It depends on the absorbed optical energy and therefore of $X_{LS}$ the initial fraction of molecules in the LS state at a given temperature $T$, which is the absorbing species. The optical energy absorbed by the crystal is redistributed over phonons and the

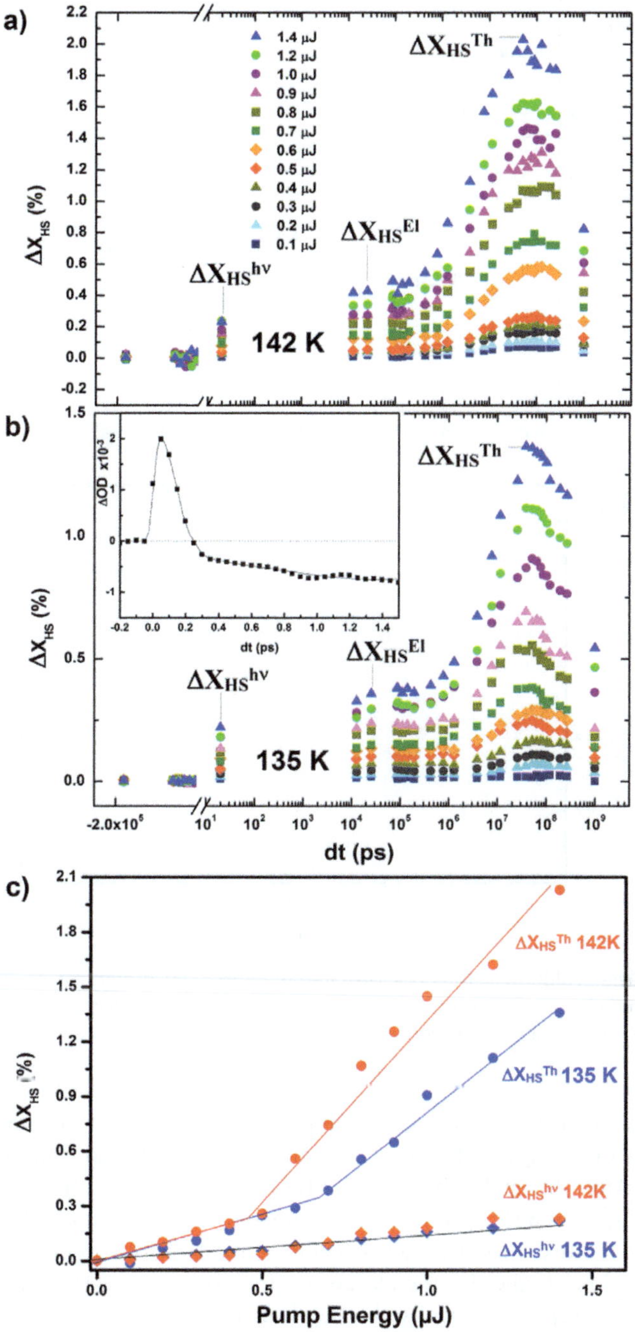


Fig. 4 Time evolution of the photo-excited fraction of HS molecules at 142 K (a) and 135 K (b) for different pump energy, revealing the 3 steps of photoinduced ($\Delta X_{HS}^{h\nu}$), elastic ($\Delta X_{HS}^{El}$) and thermal ($\Delta X_{HS}^{Th}$) nature. The inset in (b) shows the OD change during the photo-switching: the transient peak is due to MLCT state and the decrease to HS state. (c) Shows the evolution of the photoinduced and thermal steps with the pump energy.

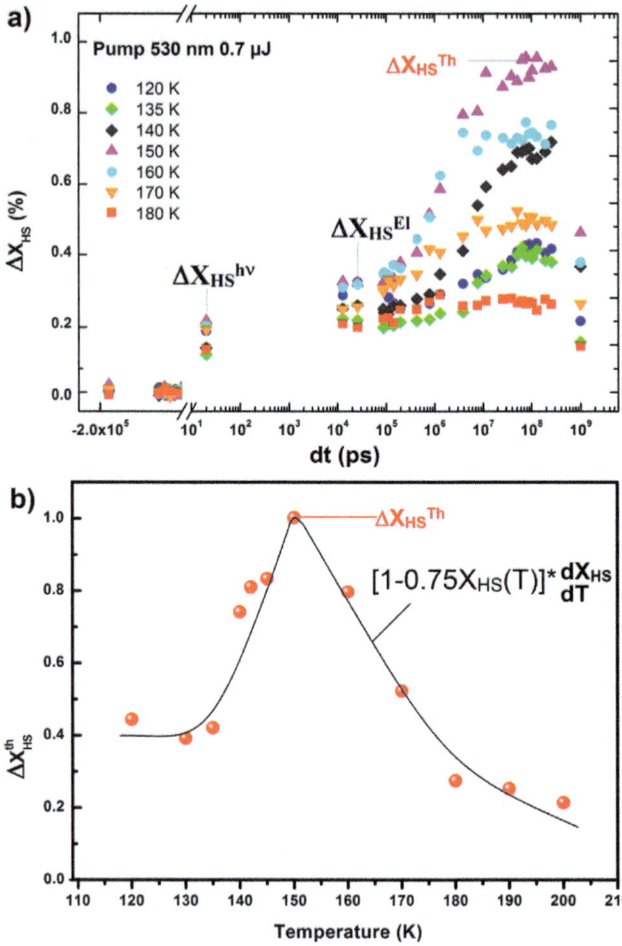


Fig. 5 (a) Time evolution of the photo-excited fraction of HS molecules for 0.7 μJ pump energy for different temperature. (b) Evolution of the thermal step with temperature and model used for describing the thermal response (solid line).

corresponding heat causes a temperature increase $\Delta T$ over the macroscopic length scale of the crystal. This occurs on the μs time scale and is limited by heat diffusivity. After this global heating, the HS fraction reaches a transient equilibrium value $X_{HS}(T + \Delta T)$. Then when the temperature jump is small, $\Delta X_{HS}{}^{Th} = \Delta T \times dX_{HS}(T)/dT$. Since the optical density at 530 nm, describing optical absorption, is around 4 times higher for the LS species (fraction $1 - X_{HS}(T)$) than for the HS (fraction $X_{HS}(T)$) species, the absorbed optical energy is:

$$E_a(T) \propto \Delta T(T) \propto X_{LS} + 1/4X_{HS} = 1 - 0.75X_{HS}(T)$$

Then we can calculate how the thermal population of the HS state $\Delta X_{HS}{}^{Th}$ changes with temperature from the temperature dependence of $X_{HS}$ provided in Fig. 1:

 

$$\Delta X_{HS}{}^{Th} = [1 - 0.75X_{HS}(T)] \times dX_{HS}(T)/dT$$

This curve shown in Fig. 5 reproduces well the maximum of thermal population at around 150 K, where the slope of the thermal evolution of $X_{HS}(T)$ is maximum (Fig. 1). It also indicates that the thermal effect is minimum at around 135 K for an excitation pump energy of 0.7 μJ, where the thermal population is negligible and smaller than 1%. Both the data at 135 K for different pump energies and the data at 0.7 μJ for different temperatures indicate that thermal effects are small for a pump energy of 0.7 μJ at 135 K, where the SSCW is formed. We use these experimental conditions to study the response of the SSCW to fs laser excitation by using time-resolved X-ray diffraction.

## Time-resolved diffraction studies

Time-resolved diffraction studies were performed at the BioCARS beamline at the Advanced Photon Source, Argonne National Laboratory. The optical pump laser beam was similar to the one used for optical measurements in terms of pump wavelength (530 nm), laser spot size (300 μm) and pump energy (0.7 μJ). Fig. 6a shows how $X_{HS}$ obtained by optical spectroscopy changes in time, with a negligible thermal population of HS state (<1%). X-ray diffraction experiments reveal a multi-step expansion of the lattice parameter *a* (quite sensitive to the spin-state conversion). Just after laser excitation, there is no change of the lattice parameter

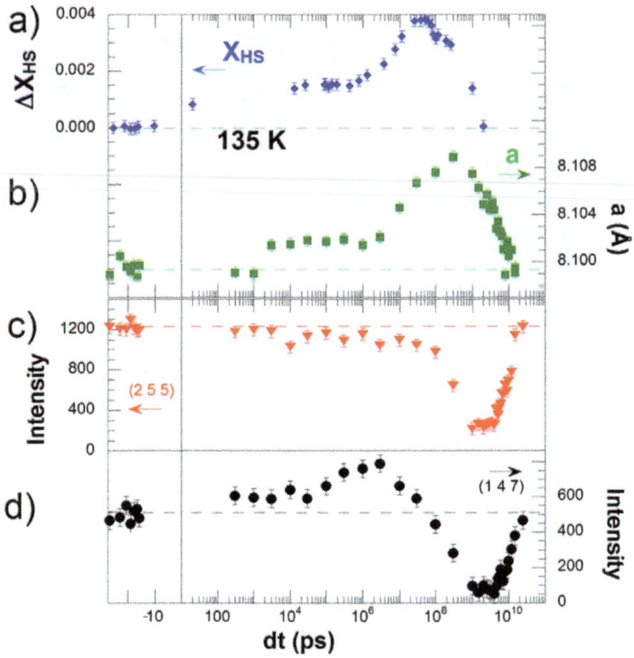


Fig. 6 Time dependence of $X_{HS}$ obtained by optical studies (a), of the lattice parameters *a* (b), and of the intensity of Bragg peaks characterizing the presence of the SSCW (c and d). Data are shown on a log scale for positive time delays.

(Fig. 6b) and a first expansion is only observed after a few ns, corresponding to the so-called elastic step.[21–24] It results from an internal pressure induced by the larger volume of the photoswitched molecules and by the lattice heating. A second and larger expansion is observed with a maximum 200 μs after laser excitation, when other molecules thermally populate the HS state.

These data correlate quite well with the optical data presented in Fig. 4 and with previous reports on other single-step spin-crossover materials.[20–24] What is new in the present compound is the symmetry breaking aspect associated with the formation of the SSCW, which can be probed by following the time course of the characteristic Bragg reflections indexed ($h\ k\ 2p + 1$) in the doubled cell. Fig. 6c shows the typical response of such Bragg peaks. Just after laser excitation, the intensity of some peaks decreases *e.g.* for the (2 5 5) peak or increases *e.g.* for the (1 4 7) peak.

On longer timescales, the intensity of all the Bragg peaks indexed as ($h\ k\ 2p + 1$) monotonically decreases and approaches 0 within 1.5 ms. Fig. 7 shows the evolution

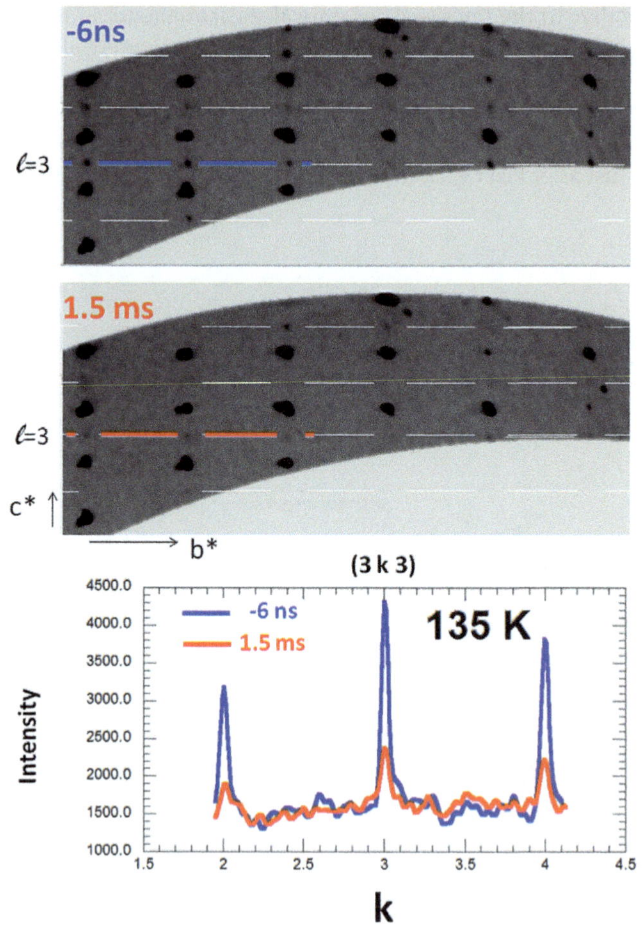


Fig. 7 (top) Diffracted intensity in the reciprocal (3 $k$ $l$) plane indexed in the doubled cell, showing the decrease of (3 $k$ 2$p$ + 1) Bragg peaks before (−6 ns) and 1.5 ms after laser excitation. (bottom) Slice of the diffracted intensity along (3 $k$ 3).

 

of the diffracted intensity in the (3 $k$ $l$) reciprocal plane 6 ns before and 1.5 ms after laser excitation. The decrease of intensity of the Bragg reflections down to ≈20% is similar for all the Bragg peaks indexed ($h$ $k$ $2p+1$), as observed in Fig. 7 for the (3 $k$ 3) Bragg peaks. The intensity then recovers its initial value within 15 ms as shown in Fig. 8 on a linear timescale; so does the lattice parameter $a$.

From the crystal structures obtained by time-resolved X-ray diffraction, we could also extract the evolution of the order parameters $X_{HS}$ from the time evolution of the $\langle Fe_1\text{–}N\rangle$ and $\langle Fe_2\text{–}N\rangle$ bond lengths and $\eta$ from their difference, in a similar way to that explained above for characterizing the SSCW at thermal equilibrium. Data shown in Fig. 9 indicate that $X_{HS}$ does not change significantly during the out-of-equilibrium dynamics (in agreement with the optical data), whereas $\eta$ decreases by a factor ≈2 around 1 ms in agreement with the decrease down to ≈20% of $I(h\ k\ 2p+1)$ Bragg peaks.

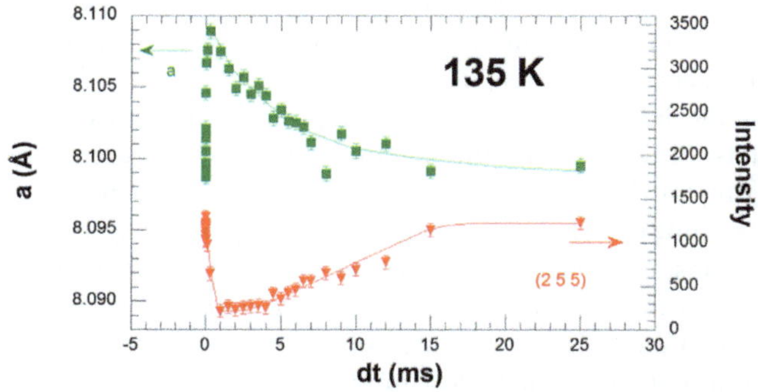


Fig. 8 Recovery of the lattice parameters $a$, and of the intensity of the (2 5 5) Bragg peaks characterizing the presence of the SSCW.

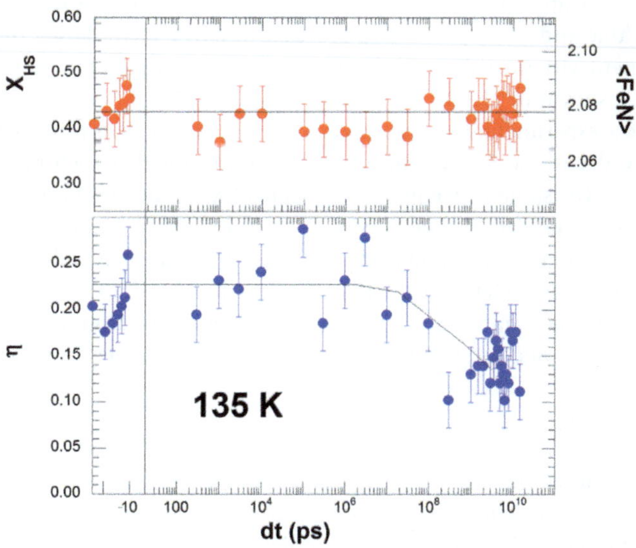


Fig. 9 Evolution of the order parameters $X_{HS}$ and $\eta$ obtained by time-resolved X-ray diffraction.

# Discussion

Combined time-resolved optical spectroscopy and X-ray diffraction studies provide an overview of the out-of-equilibrium dynamics of SSCW. During the three steps associated with the out-of-equilibrium dynamics in the solid state, there are a few differences compared to the response of single-step materials.

On the photoswitching step, we could only photoswitch a small fraction $\Delta X_{HS}{}^{h\nu} \approx 0.2\%$ of the molecules. This is too small to erase the wave: as the initial fraction of HS molecules on site 2 is 25%, light cannot bring this value up-to 0.75 to make it equal to the fraction on site 1, which would erase the wave. The number of photons required in a fs pulse would simply destroy the sample. This is however possible in some materials[5] with cw light at low temperature, since the photo-induced HS state is long-lived.

However, some molecules absorb optical energy and switch from LS to HS states and this is enough to drive lattice expansion as an internal pressure appears after photo-excitation. As molecules have to rearrange in this step, the structure factor is slightly modified: the intensity of some $(h\ k\ 2p + 1)$ peaks increases, whereas for others it decreases (Fig. 5).

Then, as the lattice warms up, the temperature increases and the HS/LS equilibrium is shifted and an additional fraction of molecules reach the HS state. But again, this fraction of thermally populated HS state is small with $\Delta X_{HS}{}^{Th} < 0.005$. It is possible to estimate the temperature jump of the crystal by comparing the 0.01 Å expansion of the lattice parameter $a$ (Fig. 5b) to the thermal expansion coefficient of 0.0015 Å $K^{-1}$ observed in Fig. 2e. We estimate then the temperature jump $\Delta T$ to be in the 5–7 K range, which should bring the crystal from the initial 135 K to a transient temperature close to 142 K within hundreds of µs. Indeed, in this temperature interval the average HS fraction $X_{HS}$ remains almost constant at thermal equilibrium. But at 142 K, the SSCW is erased. We indeed observe in Fig. 6 and 7 through the vanishing of the $I(h\ k\ 2p + 1)$ peak intensities that the SSCW almost disappears within 1.5 ms. The intensity decreases down to 15% of the initial value and corresponds to a decrease of $\eta$ by a factor larger than 2. This is in agreement with the value estimated from the bond length difference between sites 1 and 2 shown in Fig. 9.

From these experimental results, we can conclude that it takes time to destroy the long-range order of the SSCW. Erasing the SSCW means that each molecular site needs to reach the same new equilibrium value of $X_{HS}$. Therefore we can conclude that the typical timescale for the system to explore different HS/LS configurations and to reach thermal equilibrium is in the 100 µs–1 ms range. Finally, as the crystal cools down by heat exchange with the cryostat, the initial 135 K temperature is regained. The lattice parameter relaxes to its initial value, as observed on single step SCO compounds,[24] and the SSCW reappears on the 15 ms timescale.

# Conclusion

In addition to providing molecular movies, time-resolved diffraction is a powerful tool for investigating symmetry breaking aspects of photoinduced phase transitions and out-of-equilibrium thermodynamics.

Our study reveals that the photo-switching step alone cannot erase the SSCW by promoting 50% of the molecules from LS to HS: this requires an excitation density well above the sample damage. It is only in the transient thermal equilibrium, reached with a moderate temperature jump only slightly increasing the average HS fraction, that the spin state concentration wave can be erased. It requires the system to explore different HS/LS configurations, so that $X_{HS}$ reaches the same value on the different crystalline sites.

## Experimental

Experiments were performed on single crystals with typical dimensions of (250 ± 50) × (200 ± 50) × (30 ± 5) $\mu m^3$. Crystals were cooled with a liquid nitrogen cryostream. The optical pump-probe experiments were configured in transmission geometry with a quasi-collinear configuration. We used the set-up described in ref. 21, combining a mechanical translation stage to adjust the optical path difference for sub-ns measurements and the synchronization of two femtosecond amplifiers for 10 ns to ms delays. The pump wavelength was set to 532 nm on the MLCT band where it efficiently induces LS-to-HS conversion of the $[Fe^{II}H_2L^{2\text{-}Me}]$ cation.[4,5] To excite the crystal, we used energies per pump pulse in the 0.1–1.4 μJ range focused in a 300 μm spot, which corresponds to excitation fluences in the 1–16 μJ $mm^{-2}$ range. The probe was set to 610 nm to monitor the LS-to-HS photoswitching dynamics through the variation of OD, and from which $X_{HS}$ was extracted (see also ref. 14 and 21). Time-resolved X-ray diffraction experiments were performed at the BioCARS beamline at the APS synchrotron,[30] where individual X-ray pulses were selected by a fast chopper. X-ray diffraction data were collected at 15 keV with a MAR-CCD detector and varying delays dt between the laser pump and the X-ray probe. For measuring the time dependence of the lattice parameters and diffracted intensity, we performed partial data collection. At each time delay, we collected 60 frames each with a 10 s exposure, at 1° steps of the diffractometer $\varphi$ axis. We used a typical pump excitation density of 10 μJ $mm^{-2}$. The unit cell parameters and data reduction were obtained with CrysAlis software.[31]

## Acknowledgements

This work was supported by the CNRS and Région Bretagne (PhD support of A.M.), the Institut Universitaire de France, Rennes Métropole, the ANR (ANR-13-BS04-0002) and the European Regional Development Fund (FEDER). BioCARS is funded by NIH grant R24 GM111072 to K.M. K.M. and E.C. thank the University of Chicago for funding through the "France and Chicago Collaborating in the Sciences" (FACCTS) program. This research used resources of the Advanced Photon Source, a U.S. Department of Energy (DOE) Office of Science User Facility operated for the DOE Office of Science by Argonne National Laboratory under Contract No. DE-AC02-06CH11357.

## Notes and references

1 A. Bousseksou, G. Molnar, L. Salmon and W. Nicolazzi, *Chem. Soc. Rev.*, 2011, **40**, 3313.

2 *Spin-Crossover Materials: Properties and Applications*, ed. M. A. Halcrow, John Wiley & Sons, 2013.

3 P. Guionneau, M. Marchivie, G. Bravic, J.-F. Létard and D. Chasseau, *Top. Curr. Chem.*, 2004, **234**, 97–28.

4 N. Bréfuel, H. Watanabe, L. Toupet, J. Come, M. Kojima, N. Matsumoto, E. Collet, K. Tanaka and J. P. Tuchagues, *Angew. Chem., Int. Ed.*, 2009, **48**, 9304.

5 E. Collet, H. Watanabe, N. Bréfuel, L. Palatinus, L. Roudaut, L. Toupet, K. Tanaka, J.-P. Tuchagues, P. Fertey, S. Ravy, B. Toudic and H. Cailleau, *Phys. Rev. Lett.*, 2012, **109**, 257206.

6 E. Collet, M. L. Boillot, J. Hebert, N. Moisan, M. Servol, M. Lorenc, L. Toupet, M. Buron-Le Cointe, A. Tissot and J. Sainton, *Acta Crystallogr., Sect. B: Struct. Sci.*, 2009, **65**, 474.

7 J. Kusz, M. Zubko, A. Fitch and P. Gütlich, *Z. Kristallogr.*, 2011, **226**, 576–584.

8 S. Pillet, V. Legrand, M. Souhassou and C. Lecomte, *Phys. Rev. B: Condens. Matter Mater. Phys.*, 2006, **74**, 140101.

9 D. Chernyshov, M. Hostettler, K. W. Törnroos and H.-B. Bürgi, *Angew. Chem., Int. Ed.*, 2003, **42**, 3825–3830.

10 M. Yamada, M. Ooidemizu, Y. Ikuta, S. Osa, N. Matsumoto, S. Iijima, M. Kojima, F. Dahan and J. P. Tuchagues, *Inorg. Chem.*, 2003, **42**, 8406–8416.

11 M. Marchivie, P. Guionneau, J. A. K. Howard, G. Chastanet, J.-F. Létard, A. E. Goeta and D. Chasseau, *J. Am. Chem. Soc.*, 2002, **124**, 194–195.

12 M. Buron-Le Cointe, J. Hébert, C. Baldé, N. Moisan, L. Toupet, P. Guionneau, J.-F. Létard, E. Freysz, H. Cailleau and E. Collet, *Phys. Rev. B: Condens. Matter Mater. Phys.*, 2012, **85**, 064114.

13 A. Goujon, F. Varret, K. Boukheddaden, C. Chong, J. Jeftic, Y. Garcia, A. D. Naik, J. C. Ameline and E. Collet, *Inorg. Chim. Acta*, 2008, **361**, 4055–4064.

14 R. Bertoni, M. Lorenc, A. Tissot, M.-L. Boillot and E. Collet, *Coord. Chem. Rev.*, 2015, **282–283**, 66–76.

15 R. Bertoni, M. Lorenc, A. Tissot, M. Servol, M.-L. Boillot and E. Collet, *Angew. Chem., Int. Ed.*, 2012, **51**, 7485–7489.

16 A. Marino, P. Chakraborty, M. Servol, M. Lorenc, E. Collet and A. Hauser, *Angew. Chem., Int. Ed.*, 2014, **53**, 3863–3867.

17 A. Marino, M. Servol, R. Bertoni, M. Lorenc, C. Mauriac, J.-F. Létard and E. Collet, *Polyhedron*, 2013, **66**, 123–128.

18 H. T. Lemke, C. Bressler, L. X. Chen, D. M. Fritz, K. J. Gaffney, A. Galler, W. Gawelda, K. Haldrup, R. W. Hartsock, H. Ihee, J. Kim, K. H. Kim, J. H. Lee, M. M. Nielsen, A. B. Stickrath, W. Zhang, D. Zhu and M. Cammarata, *J. Phys. Chem. A*, 2013, **117**, 735.

19 C. Bressler, C. Milne, V. T. Pham, A. El Nahhas, R. M. van der Veen, W. Gawelda, S. Johnson, P. Beaud, D. Grolimund, M. Kaiser, C. Borca, G. Ingold, R. Abela and M. Chergui, *Science*, 2009, **323**, 489.

20 M. Cammarata, R. Bertoni, M. Lorenc, H. Cailleau, S. Di Matteo, C. Mauriac, S. F. Matar, H. Lemke, M. Chollet, S. Ravy, C. Lauhlé, J.-F. Létard and E. Collet, *Phys. Rev. Lett.*, 2014, **113**, 227402.

21 M. Lorenc, C. Balde, W. Kaszub, A. Tissot, N. Moisan, M. Servol, M. Buron-Le Cointe, H. Cailleau, P. Chasle, P. Czarnecki, M. L. Boillot and E. Collet, *Phys. Rev. B: Condens. Matter Mater. Phys.*, 2012, **85**, 054302.

22 H. Cailleau, M. Lorenc, L. Guérin, M. Servol, E. Collet and M. Buron-Le Cointe, *Acta Crystallogr., Sect. A: Found. Crystallogr.*, 2010, **66**, 189.

 

23 E. Collet, M. Lorenc, M. Cammarata, L. Guérin, M. Servol, A. Tissot, M. L. Boillot, H. Cailleau and M. Buron-Le Cointe, *Chem.–Eur. J.*, 2012, **18**, 2051.

24 E. Collet, N. Moisan, C. Baldé, R. Bertoni, E. Trzop, C. Laulhé, M. Lorenc, M. Servol, H. Cailleau, A. Tissot, M. L. Boillot, T. Graber, R. Henning, P. Coppens and M. Buron-Le Cointe, *Phys. Chem. Chem. Phys.*, 2012, **14**, 6192.

25 S. M. Neville, B. A. Leita, G. J. Halder, C. J. Kepert, B. Moubaraki, J.-F. Létard and K. S. Murray, *Chem.–Eur. J.*, 2008, **14**, 10123–10133.

26 S. Pillet, E.-E. Bendeif, S. Bonnet, H. J. Shepherd and P. Guionneau, *Phys. Rev. B: Condens. Matter Mater. Phys.*, 2012, **86**, 064106.

27 K. D. Murnaghan, C. Carbonera, L. Toupet, M. Griffin, M. M. Ditru, C. Desplanches, Y. Garcia, E. Collet, J. F. Létard and G. Morgan, *Chem.–Eur. J.*, 2014, **20**, 5613–5618.

28 L. Landau, *Ukr. J. Phys.*, 2008, **53**, 25–35; L. Landau, *Zh. Eksp. Teor. Fiz.*, 1937, **7**, 19–32.

29 P. Guionneau and E. Collet, Piezo- and Photo-Crystallography Applied to Spin-Crossover Materials, in *Spin-Crossover Materials: Properties and Applications*, ed. M. A. Halcrow, John Wiley & Sons Ltd, Oxford, UK, 2013, ch. 20, p. 508.

30 T. Graber, P. Anfinrud, H. Brewer, Y.-S. Chen, H.-S. Cho, N. Dashdorj, R. W. Henning, I. Kosheleva, G. Macha, M. Meron, R. Pahl, Z. Ren, S. Ruan, F. Schotte, V. Šrajer, P. J. Viccaro, F. Westferro and K. Moffat, *J. Synchrotron Radiat.*, 2011, **18**, 658–670.

31 CrysAlis RED, *Version 1.171.26*, Oxford Diffraction Ltd, Abingdon, Oxfordshire, England.

Faraday Discussions

DISCUSSIONS

# Local and Global Dynamics: general discussion

Martin Zanni, E. D. Jemmis, Sankarampadi Aravamudhan, Elangannan Arunan, Neil Hunt, Sanghamitra Mukhopadhyay, Helena J. Shepherd, Srihari Keshavamurthy, Arend G. Dijkstra, Mike Ashfold, Himangshu Prabal Goswami, Artur Nenov, Himani Medhi, Kenneth Ghiggino, Kiran Moirangthem, R. J. Dwayne Miller, Debabrata Goswami, Siva Umapathy, John R. Helliwell, Jonathan D. Hirst, Stephen Meech, Shaul Mukamel, Biman Bagchi and Marylise Buron-Le Cointe

DOI: 10.1039/C5FD90018B

**Kenneth Ghiggino** opened a general discussion of the paper by **Stephen Meech**: You refer to the origin of the non-exponential kinetics you observe in the BLUF proteins as arising from a distribution of ground state structures (or rotamers). Are you able to expand on what other evidence is available, or could be obtained, to confirm this hypothesis?

**Stephen Meech** responded: The evidence for multiple conformations of amino acids around the isoalloxazine ring was summarized by Udvarhelyi and Domratcheva.[1] There are experimental differences between NMR and X-ray structures, especially regarding Y21 and W104, and calculations show multiple conformers of Q63, separated by low barriers. It will be interesting to investigate the correlation of non-exponentiality and photoactivity as a function of temperature.

1 A. Udvarhelyi and T. Domratcheva, *J. Phys. Chem. B*, 2013, **117**, 2888–2897.

**Neil Hunt** asked: What was the rationale for the specific choice of the Y21 mutants used? What is the effect on the H-bonding structure observed in the wild-type protein upon mutation; is this conserved or broken?

**Stephen Meech** answered: The key point of these Y21 mutations is to remove the H-bond between Y21 and Q63. This was already known to be critical in photoactivity, and the effect is to trap the BLUF domain in its dark adapted state (as judged by the S0 to S1 absorption spectrum). All the Y21 mutants we have studied achieve this. The specific mutants were chosen to see the effect of polarity

(Y21I compared to Y21S) and electron donating ability (Y21C and Y21F (data not shown) compared to Y21I). The only Y21 mutant which behaved differently was Y21W, where the Trp to FAD electron transfer was clearly observed in TRIR and electronic spectroscopy, although the protein remained photoinactive.[1]

1 A. Lukacs *et al.*, *J. Am. Chem. Soc.*, 2014, **136**, 4605–4615.

**John R. Helliwell** remarked: On page 3 of your article you refer to a "disagreement among the published (protein) crystal structures with regard to the important (amino acid) residues Q63 and W104". This might be accounted for from possible differences between diffraction (and spectroscopic) measurements at 100 K and room temperature, *i.e.* for which structural differences in conformation and in structural dynamics do occur in proteins. See for example the references below (although unfortunately not for your studied proteins):

1 A. Deacon, T. Gleichmann, A. J. Kalb (Gilboa), H. Price, J. Raftery, G. Bradbrook, J. Yariv and J. R. Helliwell, The structure of concanavalin A and its bound solvent determined with small-molecule accuracy at 0.94 Å resolution, *Faraday Transactions*, 1997, **93**(24), 4305–4312.
2 Simon W. M. Tanley and John R. Helliwell, Structural dynamics of cisplatin binding to histidine in a protein, *Struct. Dyn.*, 2014, **1**, 034701.

**E. D. Jemmis** queried: Intramolecular ene-reactions with high specificity are known in organic chemistry, but achieved with difficulty. The specific template provided by the enzyme in your example makes this very easy. Any comments on this specific reaction to organic chemists, other than the evolutionary perfection that the enzyme has achieved?

**Stephen Meech** replied: Your question already hints at my answer. Proteins are typically exquisitely well adapted for their precise function, but their 'skills' are not always very transferable. I think that is the case here, where the mechanism we propose requires a difference in H-bonding ability between FAD ground and excited states, and a rather specific H-bonding arrangement among the key amino acids. It would be very challenging to reproduce this synthetically. Having said that, a synthetic system involving the isoalloxazine ring substituted with a moiety exhibiting intramolecular H-bonding might be an interesting one to study.

**Himani Medhi** asked: What do you mean by the Marker modes for the excited state decay and ground state recovery?

**Stephen Meech** responded: Through a study of model compounds we can identify signals at particular frequencies in the transient difference spectra where the dominant contribution is from one species. For example 1380 wavenumbers are dominated by the singlet excited state of the isoalloxazine ring, so this signal is the 'marker' for that state. If we follow its time dependence it tells us about the decay (or formation) of that state. We find similar markers for the ground state recovery (1548 $cm^{-1}$) and the radical intermediate (around 1530 $cm^{-1}$).

**Srihari Keshavamurthy** opened the discussion of the paper by **Biman Bagchi**: In case of the reverse micelles, it seems like the frequency–frequency correlation functions already give a rather detailed insight regarding the heterogeneity inherent to the system. What additional/different insights does one obtain from the 2D-IR studies and results?

**Biman Bagchi** answered: I like the question and this was also my own initial response to 2D-IR and higher order non-linear techniques. I am surprised that more questions like this have not been asked. Yes, the core quantity remains the frequency–frequency time correlation function (FF-TCF). The strength of 2D-IR lies in its ability to use spectral differences among different "chromophores" in different locations to achieve the desired spatial resolution. Then one can associate a time constant to a given location. This is not possible easily by other methods. We can of course calculate FF-TCF but we need to measure it.

**Arend G. Dijkstra** remarked: Professor Bagchi, you are discussing the calculation of two-dimensional infrared spectra of water. Now, in water, all molecules will have similar vibrational frequencies. When two water molecules are close together, they will be coupled. This leads to the delocalization of the vibrational excitation, which should be observable in the 2D-IR spectrum. In you theoretical analysis, you use the line shape function, which assumes that the system Hamiltonian commutes with the system bath interaction. Therefore, you don't include the delocalization effect. Could you comment on this?

**Biman Bagchi** replied: Of course. We are effectively considering a dilute solution of H–O–D and –O–D stretch. We neglect the difference between –O–H and –O–D which could be significant. If I have –O–D stretch in a dilute solution, that will weakly couple with the surrounding –O–H stretch. So, by neglecting the coupling and consequent delocalization among the –O–H stretch, we are essentially mimicking the –O–D stretch. One can of course do –O–D stretch.

**R. J. Dwayne Miller** commented: Your calculations of the 2D-IR spectra of liquid water were for pure $H_2O$; yet the comparison is being made to H–O–D experimental studies in terms of probing the hydrogen bond dynamics. However, for pure $H_2O$, the O–H stretch is resonantly coupled to adjacent waters and this resonant coupling greatly enhances the overall coupling to all the bath modes, collective modes of the liquid water state. We developed nanofluidic cells to enable the study of pure $H_2O$ and the spectral diffusion is much faster than what you are calculating (Cowan *et al.*, *Nature*, 2005).[1] Generally, it was thought that the energy transfer for the purely resonant case for neat $H_2O$ would dominate the spectral diffusion dynamics, *i.e.* the spatial propagation of the O–H vibrational excitation would average out structural variations in the water structure in relation to hydrogen bond dynamics. We did a temperature dependence in which the energy transfer rate is unaffected and found that the spectral diffusion dramatically slowed down at lower temperatures within changes as small as 10 degrees (Kraemer *et al.*, *PNAS*, 2008).[2] This study demonstrated that the spectral diffusion and relation to the hydrogen bond dynamics, under the full resonant conditions of neat water, is extremely sensitive to the details of the degree of hydrogen bonding. We ascribed the fast spectral diffusion in $H_2O$ to the higher frequency collective librational modes of liquid water, consistent with the observed times scales, and the degree of hydrogen bonding with respect to the rms motion of such modes. The bottom line is that the vibrational modes of pure water, under fully resonant conditions, are extremely sensitive to relatively small changes in the correlation lengths within the hydrogen bond network. It would be great if just the temperature dependence for the 2D spectra of liquid $H_2O$ could be recovered. The problem is in treating the excitonic coupling and the accuracy of the potentials used. We developed a split operator approach (Paarmann *et al.*, *J. Chem. Phys.*, 2009)[3] that handles this aspect of the problem. This aspect of the problem can be treated fairly well with this approach. My understanding (personal discussions with James Skinner, University of Wisconsin) is that it is very difficult to treat the temperature dependence of neat water as the potentials are highly tweaked for the room temperature properties of liquid water. For example, the melting point found by MD for ice is well below the thermodynamic value. It is difficult to give a corrected temperature for the MD to correlate to the experiment.

The above discussion is to put in context how sensitive the O–H vibrational stretch is to the details of the hydrogen bond network and that the fully resonant terms need to be included to properly treat pure $H_2O$. I realize that the main intent of these calculations is to theoretically treat the time dependent frequency correlations, *i.e.* the observable in 2D-IR, to infer structural correlations of water within biologically relevant environments. Given the relatively high sensitivity of the water potential and the importance of resonant interaction terms within the hydrogen bond network, these factors may be important in how water interacts with membranes and biomolecule surfaces. The O–D stretch of H–O–D is generally used to probe these issues, given the much longer lifetime and much slower spectral diffusion, which gives a larger dynamics range for separating different contributions to the water dynamics. However, the resonant terms in neat $H_2O$ give rise to enhanced coupling to intermolecular modes, as in the case for librations, that may even affect structural correlations. The intermolecular potential is different from H–O–D and these would not be taken into account. Can you comment on what needs to be done theoretically to treat water in its full splendour? As a first starting point, would it be possible to treat the extremely sensitive temperature dependence for spectral diffusion? This high sensitivity may help both refine the potentials used for water and for probing the effects of biological interfaces on the dynamic structure of water.

1 M. L. Cowan, B. D. Bruner, N. Huse, J. R. Dwyer, B. Chugh, E. T. J. Nibbering, T. Elsaesser and R. J. D. Miller, *Nature*, 2005, **434**, 199–202.
2 D. Kraemer, M. L. Cowan, A. Paarmann, N. Huse, E. T. J. Nibbering, T. Elsaesser and R. J. Dwayne Miller, *PNAS*, 2008, **105**(2), 437–442.
3 A. Paarmann, T. Hayashi, S. Mukamel and R. J. D. Miller, *J. Chem. Phys.*, 2009, **130**, 204110.

 

**Biman Bagchi** answered: As I stated before, we are effectively considering a dilute solution of H–O–D and –O–D stretch. We neglect the difference between –O–H and –O–D which could be significant. If I have –O–D stretch in a dilute solution, that will weakly couple with the surrounding –O–H stretch. So, by neglecting the coupling and consequent delocalization among –O–H stretch, we are essentially mimicking –O–D stretch. One can of course do –O–D stretch. I agree that one should directly treat H–O–D or $D_2O$ in dilute solutions. That should not be difficult – maybe the statistics will be an issue. As to the sensitivity of water potential to charged surfaces, there is a real problem. Our force-fields are not good enough (as far I know) to pick-up small differences between various surfaces and molecules at various locations.

**Himangshu Prabal Goswami** asked:

(i) Do 2D-IR response functions need to always decay to an equilibrium value?

(ii) If answer to question 1 is yes, then for the exponential in eqn 2 of your paper to be negative, such that the 2D-IR response function decays to an equilibrium value, the following mathematical inequality must hold good among the line-shape functions $g(t)$:

$$g(t_1) + g(t_2) + g(t_3) + g(t_1 + t_2 + t_3) > g(t_1 + t_2) + g\,(t_2 + t_3). \qquad (1)$$

What underlying physics takes care of this inequality?

(iii) If the answer to question number 1 is no, the above inequality may sometimes hold and sometimes not. Then what physical processes are involved in having control over the value of the line shape functions so that the response increases or decreases? How do we control these parameters or processes so that response function reaches a saturation value or indefinitely increases?

**Biman Bagchi** responded: The force–force time correlation function (FF-RCF) always decays to zero. At long times, the 2D-IR spectra become symmetric and spherical. I do not see any issue here.

**Sanghamitra Mukhopadhyay** addressed **Biman Bagchi** and **Martin Zanni**: Using 2D-IR spectroscopy to find spatio-temporal correlations in aqueous systems is very promising. My question is whether neutron spectroscopy can be used in the same manner to find the same correlations? In this context I would like to mention that there is the VESUVIO instrument at ISIS (http://www.isis.stfc.ac.uk/instruments/vesuvio/) which is used for neutron Compton scattering and femtosecond dynamics. Since the neutron is very much sensitive to hydrogen, do you think this instrument has potential for this type of spectroscopy?

**Martin Zanni** replied: I believe that the neutron source would need to be coherent, although there may be incoherent 2D analogs.

**Biman Bagchi** commented: The uniqueness of the 2D-IR method lies in its ability to probe the response of a chemical bond (in our case –O–H stretch), and also our ability to interpret the results. I am not too aware of the abilities of neutron Compton scattering. But it seems it will provide different kinds of information, like spatial structure and translational dynamics. It will certainly be worthwhile to do neutron Compton experiments near an interface to learn more. This can nicely complement 2D-IR studies.

**Debabrata Goswami** asked: The basic premise of the presented work on 2D-IR calculation is the dephasing across the entire system. This knowledge could give you better insight into the different structures, the theory might be very complex and it might be hard to use all that information. Please comment.

**Stephen Meech** responded: This is an interesting question. My first thought is that the large number of protein modes contributing to the spectral region of interest here (which encompasses the amide backbone) is too large to be addressed by 2D-IR. It is however true that NMR methods routinely tackle small proteins where all the very numerous nuclei in question contribute. 2D-NMR had a considerable head start on 2D-IR, so it is possible that there is further technical, experimental and theoretical progress to be made which might clean up or simplify the 2D-IR spectra. Meanwhile, one approach is to incorporate unnatural amino acids at key locations which have intense IR transitions in regions of the spectrum where the protein does not absorb. A second way forward, specific to photoactive proteins, is to perform pump–2D-IR probe measurements. We have shown for flavoproteins that the fact of electronic excitation perturbing the protein structure[1] and the subsequent relaxation could make them probes for 2D-IR.

1 A. Lukacs *et al.*, *J. Am. Chem. Soc.*, 2011, **133**, 16893–16900.

**Biman Bagchi** also commented: I should clarify that we tried to correlate our results with experiments that treat dilute H–O–D or $D_2O$ in water. That was done in our paper in a crude way by turning off the interactions among –O–H vibrations that can lead to the delocalization of the vibration. In experiments such a delocalization is not usually probed for H–O–D or $D_2O$. The question of coupling between –O–H modes in neat water is certainly relevant. In the old days this was called "exchange dephasing" and has been considered. In experiments one always uses a "chromophore" (like H–O–D) to avoid such problems.

**Martin Zanni** remarked: It seems that a lot of these questions hinge around what 2D-IR is good for. Experimental methodology of 2D-IR is very advanced now. Theory is also very good. What can it do and what are the future directions? As Stephen Meech just said, for biology we have coupled 2D-IR with sophisticated labeling methods, cell free protein expression and isotope labels. There is a whole community that can put non-natural amino acids into proteins. The future is very bright for interrogating important questions in biology, as well as other topics.

**Stephen Meech** answered: I agree that progress in 2D-IR has been spectacular, and that its combination with unnatural amino acid (UAA) substitution presents a means to escape from the spectral congestion we typically encounter in proteins. As you suggest, in the longer term UAA substitution offers far more than the simple placement of convenient IR probes. For example, the possibility of placing photoactive molecules at critical points in the protein structure suggests a large number of exciting experiments.

**R. J. Dwayne Miller** opened the discussion of **Jonathan D. Hirst**'s paper: Your group has worked on fully accounting for all the interactions of a system with the surrounding bath for theoretically modelling 2D spectra. The first input is to get the 1D absorption spectra right and to do that it is essential to use an accurate density of states for the bath–solvent interactions. This approach works very well for absorption spectra and one can get extremely good agreements. However, CD spectra are much more sensitive to anisotropic interactions locally to the chromophore, which gives rise to the unique ability to probe chiral structures or excitonic features involving coupling between chromophores. I would think that fitting CD spectra is significantly more involved in properly accounting for these interactions. Can anyone comment on the challenges to properly treat CD spectra theoretically to put this present work in the proper context?

**Shaul Mukamel** replied: The numerical effort involved in calculating non-chiral or chiral tensor components of the linear response is comparable. Circular dichroism (CD) signals are related to the chiral response and usually carry more information since they involve an interference between contributions of different parts of the molecule due to the field phase variation across the molecule. This gives the CD signals an extra sensitivity to the local structure compared to non-chiral absorption spectra. This is why the technique is widely used for determining the secondary structure of proteins. Averaging over random orientations can be performed analytically using tensor calculus, yielding universal compact expressions. My group has modeled CD and chiral third order signals in chromophore aggregates and proteins in the visible and infrared respectively.[1,2]

1 D. Abramavicius, B. Palmieri, D. Voronine, F. Sanda and S. Mukamel, Coherent Multidimensional Optical Spectroscopy Excitons in Molecular Aggregates; Quasiparticle *vs.* Supermolecule Perspectives, *Chem. Rev.*, 2009, **109**, 2350–2408.
2 W. Zhuang, T. Hayashi and S. Mukamel, Coherent Multidimensional Vibrational Spectroscopy of Biomolecules; Concepts, Simulations and Challenges, *Angew. Chem., Int. Ed.*, 2009, **48**, 3750–3781.

**John R. Helliwell** said: The understanding of the CD solution spectra of proteins is an important object of study as we cannot crystallize every protein. You remark that "No obvious improvement is observed in the calculated spectra of beta sheet proteins" when considering the "influence of vibrational structure". Beta sheets are relatively rigid objects in protein fold examples and so this is presumably to be expected. Could you please clarify the choice of protein classes, *i.e.* where I would assume that those that are intrinsically dynamic would be optimal?

**Jonathan D. Hirst** responded: We have studied four broad classes of protein: α-helical, mixed α–β, and two classes of β-sheet protein: β-I and β-II proteins. The β-I and β-II proteins show different circular dichroism spectra (Manavalan and Johnson, 1983).[1] This may in part be explained by the greater conformational flexibility that β-II proteins exhibit compared to β-I proteins (Hirst *et al.*, 2003).[2] The rigidity or conformational dynamics referred to here is in the context of roughly nanosecond timescales. Such motion is of a much lower frequency than the frequency of the C–N bond stretch. Its influence in our calculations would be directly manifested in the off-diagonal elements of the exciton Hamiltonian matrix. In contrast, the dynamics associated with the vibronic structure of the ππ* transition is on a sub-picosecond scale. Our consideration of the vibronic structure is directly manifested in both the diagonal and the off-diagonal terms of the Hamiltonian matrix. Thus, *a priori*, one would not particularly expect the influence of the vibronic structure to be more important in the circular dichroism spectroscopy of more intrinsically (conformationally) dynamic proteins.

1 P. Manavalan and W. C. Johnson, Jr., *Nature*, 1983, **305**, 831.
2 J. D. Hirst, S. Bhattacharjee and A. V. Onufriev, *Faraday Discuss.*, 2003, **122**, 253.

**Mike Ashfold** remarked: At what energies do we need to worry about contributions from higher excited states to the electronic absorption spectra of proteins? What are the main contributors to the 'other broadening effects' that you accommodate by modelling each vibronic transition within the ππ* absorption band with a 15 nm Gaussian function?

**Jonathan D. Hirst** answered: There are, of course, higher excited states of interest. In other work, we have characterised charge transfer states (Gilbert and Hirst, 2004; Oakley and Hirst, 2006)[1,2] in dipeptides and in proteins, which have helped assign peaks observed between 160 nm and 180 nm in synchrotron radiation circular dichroism spectra of proteins (Bulheller *et al.*, 2008).[3] At yet shorter wavelengths, we have calculated the $\pi_b\pi^*$ and $n'\pi^*$ excitations to occur at around 130 nm (Besley and Hirst, 1998).[4] The coupling of these higher energy transitions to those at 190 nm and 220 nm has been explored (Bulheller *et al.*, 2008),[3] but does not lead to a quantitative improvement in the accuracy of the computed protein circular dichroism spectra. An important contributor to broadening will come from the thermal effects, which lead to a conformational ensemble of ground state geometries. We have established this in an earlier study (Besley *et al.*, 2004)[6] for small amides using molecular dynamics (MD) simulations at various temperatures to sample the conformational ensemble, followed by time-dependent density functional theory calculations on a series of snapshots from the MD trajectory.

1 A. T. B. Gilbert and J. D. Hirst, Charge-Transfer Transitions in Protein Circular Dichroism Spectra, *J. Mol. Struct.: THEOCHEM*, 2004, **675**, 53–60.
2 M. T. Oakley and J. D. Hirst, Charge-Transfer Transitions in Polypeptides, *J. Am. Chem. Soc.*, 2006, **128**, 12414–12415.
3 B. M. Bulheller, A. J. Miles, B. A. Wallace and J. D. Hirst, Charge-Transfer Transitions in the Vacuum-Ultraviolet of Protein Circular Dichroism Spectra, *J. Phys. Chem. B*, 2008, **112**, 1866–1874.
4 N. A. Besley and J. D. Hirst, *Ab Initio* Study of the Effect of Solvation on the Electronic Spectra of Formamide and *N*-Methylacetamide, *J. Phys. Chem. A*, 1998, **102**, 10791–10797.
5 N. A Besley, M. T. Oakley, A. J. Cowan and J. D. Hirst, A Sequential Molecular Mechanics/Quantum Mechanics Study of the Electronic Spectra of Amides, *J. Am. Chem. Soc.*, 2004, **126**, 13502–13511.

**Siva Umapathy** enquired: how would you consider the electric transition moments of transitions which are close to each other where one transition is less chiral and the other is more chiral? Would this not affect the line shape? The absorption coefficient of one transition could be very large, so if there is overlap of the two transitions the line shape will be influenced by the non chiral residue. Is it possible to use the same approach for drug protein interactions? And if the drug itself is highly chiral? Can they be measured with a similar technique?

**Jonathan D. Hirst** replied: The circular dichroism of a protein in the far-ultraviolet is dominated by the local, regular repeating secondary structure. Helices and beta-strands are chiral, *e.g.* helices are typically right-handed and strands also have a distinctive twist. This "macromolecular" chirality is much more important than the chirality of individual amino acids. If a ligand binding to a protein has chromophoric groups with electronic transitions in the far- or near-ultraviolet, then these, in principle, regardless of whether the ligand itself is chiral or achiral, could be affected by binding and differences in circular dichroism could (and have been) used to study protein–ligand interactions.

**Elangannan Arunan** asked: You mentioned the Rydberg state being pushed away due to intermolecular interactions and hence the two bands observed are clearly from $n\pi^*$ and $\pi\pi^*$. In general, when one orbital is perturbed by another, the interaction could lead to two new orbitals one getting stabilized and another de-stabilized. Would this not affect the nature of the two states observed?

**Jonathan D. Hirst** responded: The Rydberg states present in the gas phase are destabilised by Pauli repulsion interactions with the solvent (Besley and Hirst, 1998).[1] This destabilisation is sufficiently large that the interaction between the Rydberg states and the low-lying valence states is small. For the (planar) *N*-methylacetamide molecule in isolation, the $n\pi^*$ and $\pi\pi^*$ states have different symmetries and do not interact. In the asymmetric environment of the (rest of the) protein, the two states will (as you suggest) mix, and this interaction is explicitly present in the corresponding off-diagonal term of the exciton Hamiltonian matrix.

1 N. A. Besley and J. D. Hirst, *Ab Initio* Study of the Effect of Solvation on the Electronic Spectra of Formamide and *N*-Methylacetamide, *J. Phys. Chem. A*, 1998, **102**, 10791–10797.

**Artur Nenov** remarked: In the construction of the Hamiltonian matrix (eqn 9 of your paper) the off-diagonal elements (chromophore–chromophore couplings) are assumed to be purely electrostatic. Could you comment on the role of other interactions like induction, dispersion and quantum-mechanical exchange?

**Jonathan D. Hirst** answered: The magnitude of non-electrostatic interactions in the calculation of the off-diagonal elements of the exciton Hamiltonian is a potentially important question. For various conformations of a model dipeptide, the *ab initio* transition energies at the CASSCF level (Oakley and Hirst, 2006)[1] cannot be reproduced by a simple splitting of states due to coupling *via* electrostatic interactions. In several other molecular systems, including DNA (Nachtigallova *et al.*, 2008)[2] and naphthalene (Scholes and Ghiggino, 1994),[3] non-classical short range exchange interactions have been considered in the illustrative cases of dimers. Thus, the question does warrant further investigation.

1 M. T. Oakley and J. D. Hirst, *J. Am. Chem. Soc.*, 2006, **128**, 12414–12415.
2 D. Nachtigallova *et al.*, *Phys. Chem. Chem. Phys.*, 2008, **10**, 5689–5697.
3 G. D. Scholes and K. P. Ghiggino, *J. Phys. Chem.*, 1994, **98**, 4580–4590.

**Artur Nenov** asked: Was the origin of the $0 \leftarrow 0$ transition (190 nm) computed (adiabatic S0-S1 energy gap) or taken from experimental data?

**Jonathan D. Hirst** replied: The vertical transition energy computed at the CASPT2 level with a large active space, a large basis set and a continuum model of solvent, is 193 nm (Besley and Hirst, 1998),[1] which is close to the value of 190 nm that we have taken from the experiment.

1 N. A. Besley and J. D. Hirst, *Ab Initio* Study of the Effect of Solvation on the Electronic Spectra of Formamide and *N*-Methylacetamide, *J. Phys. Chem. A*, 1998, **102**, 10791–10797.

**Kiran Moirangthem** remarked: The paper discusses the computational study of interpreting the CD spectra of proteins, with the incorporation of vibronic coupling where the present approach well explains even the negative bands in the CD spectra of alpha-helical proteins (which occur around 208 nm and 222nm). But I am curious to know why we don't see this absorption at 208 nm (ignoring the forbidden $n\pi$ absorption at 222 nm) in the absorption spectrum shown in Fig. 3 of your paper? Why do we not see even the slightest hump corresponding to the absorption at 208 nm in the spectrum shown in Fig. 3?

 

**Jonathan D. Hirst** responded: The figure in our paper shows the absorption spectrum of *N*-methylacetamide, which only has one peptide chromophore. Thus, there is only a single peak at ~190 nm. For polypeptides, where there are several peptide chromophores, then the band or shoulder at ~208 nm, arising from the coupling of ππ* transitions on different peptide groups, is evident, *e.g.* early experiments showed this for poly-L-lysine (Rosenheck and Doty, 1961).[1]

1 K. Rosenheck and P. Doty, The far ultraviolet absorption spectra of polypeptide and protein solutions and their dependence on conformation, *Proc. Natl. Acad. Sci. U. S. A.*, 1961, **47**, 1775–1778.

**R. J. Dwayne Miller** opened discussion of the paper by **Artur Nenov**: To help the discussion and to illustrate the intellectual challenges of this topic, it is important to realize that DNA needs to have some degree of what is called "radiation hardening" to minimize UV induced damage. DNA is the "hard drive backup" for the cell. The genetic code gives the instruction set for replication that would be lost with UV induced damage to DNA, *e.g.* in the formation of T-T dimers. The cell has developed an elaborate correction mechanism based on DNA photolyase but there has also been early evidence that DNA is intrinsically immune to UV induced damage or is inherently radiation hardened. Early studies based simply on fluorescence quantum yields on individual bases showed there was essentially no fluorescence ($10^{-4}$ or less)[1] and by extension even smaller fluorescence quantum yields would be expected for DNA. The typical radiative rate for molecules representative of nucleotides would put the excited state lifetime on the picosecond timescale. At this time, the energy gap law for nonradiative relaxation was formulated. The extremely fast relaxation of the excited states of DNA was thought to proceed through the very high density of vibrational states within the low frequency modes of the DNA strand that would enhance the Franck–Condon factors (lead to resonant terms) connecting the S0 and S1 ground and excited states. The additional low frequency modes effectively create new vibrational relaxation pathways from that of the isolated nucleotide bases and this would reduce the excited state lifetimes further from estimates based on just the nucleotide bases. The excited state lifetimes of DNA strands would be expected to be subpicosecond based on the fluorescent quantum yields of the individual bases, with accurate fluorescence quantum yields under ambient conditions beyond the detection capabilities at the time. This was the picture for many years. There was always the doubt (RJDM) that there were dark states that had gone undetected in the fluorescence studies. Then came the work of Kohler's group and others that did the first femtosecond studies of the isolated DNA bases to fully characterize the excited state dynamics (see reference 1 in Artur Nenov's paper). The individual nucleotide bases showed extremely fast excited state relaxation processes on the picosecond to sub-picosecond time scale, corroborating the early fluorescence quantum yield studies. For a rigid molecular system, this observation is in violation of the classical energy gap law as formulated based on measurements of cyclic aromatic systems. (It is interesting to speculate how the theoretical developments would have advanced with this glaring exception to the breakdown in the formation of the Franck–Condon overlap or nuclear wavefunction overlap between coupled electronic states.) These observations have led to a much improved understanding of the electronic structure of the DNA bases in which conical intersections provide the very fast nonradiative relaxation.

At this point, the state of affairs in understanding DNA photophysics seemed to be secure. The DNA bases themselves exhibited extremely fast nonradiative relaxation to provide the inherent radiation hardening against UV induced damage. The full DNA single and/or double strand structure was not essential to provide new nonradiative relaxation channels to quickly dissipate the UV generated excited states. This present paper represents yet another new twist in the DNA story. The high level time dependent *ab initio* calculations find evidence for excimer and charge transfer (CT) states that necessarily involve very reactive photointermediates. Experimentally there is recent evidence for long lived excited states on the 10–100 ps timescale based on ground state recovery studies of model oligomers of DNA (see for example cited references 1, 6, 60 and 77 of Artur Nenov's paper). These excited state dynamics are 1–2 orders of magnitude slower than observed for individual bases. On a preliminary basis, my group has recently completely a series of 2D-UV studies of DNA with sufficient bandwidth in the UV range. The advantage of 2D spectroscopy is that the spectra remove inhomogeneous broadening and it is possible to directly observe coupled states that would be hidden in these previous studies. We find evidence for the very CT states predicted by this work. Such states would be much more reactive than other possible excited state intermediates. One can understand this based on the close proximity of the nucleotides within the DNA chain that open up electronic coupling between the excited state of adjacent bases both for excimer formation and for electron hopping and the ground valence states for hole hopping. This interaction will lead to a very rapid charge separation and the creation of very reactive intermediates – the very thing DNA needs to avoid. It appears DNA still holds some surprises. With this above narrative on the key issues in understanding DNA photophysics, what do you think are the key questions regarding the excited state dynamics of DNA that may be probed by 2D methods or others? Are the initial ultrafast processes the relevant photophysics to understand the presumably inherent UV protection of DNA (now in question)?

1 M. Daniels and W. Hauswirth, *Science*, 1971, **171**(3972), 675–677.

**Sankarampadi Aravamudhan** answered: In the variety of pulse experiments in the UV, Vis, and IR region of the electromagnetic spectrum, what was obviously a prime concern was the time scales (slow or fast) of the relaxation processes. The spontaneous emission is a radiative relaxation and with the internal conversion and inter system crossings preceding the spontaneous emissions, the terms fluorescence, delayed fluorescence and phosphorescence are applicable. All these are explainable when the excited state (population) occupation number has been changed due to the stimulated absorption. All these processes thus characterize the change in population differences between the lower ground state and the higher excited state. If the population differences have to be altered, then the stimulated absorption must result in a transition to account for the complete transition of the electron. The induced (stimulated) transition probabilities depend on the radiation power experimentally applied. The reciprocal of the transition probabilities are the time of transit of the electron under perturbation.

 

If the transition times corresponding to the induced transitions are larger than the pulse widths, then the perturbation lasts less than the time required for the electron to make a transition to the other level and thus non-stationary states are created which radiate with characteristic decoherent times. This decoherent process is not appropriately described by population difference changes. Maybe these decoherence times characterize the homogeneous line broadening or what is referred to as the natural line width spectroscopically. And this lifetime and the decoherence times are related by the uncertainty relation to the line widths. Whereas the relaxation time which governs changes in population differences are to be reckoned among the processes which couple the molecule to the medium degrees of freedom and the fluctuations induce these relaxations. These two kinds of relaxation times are well distinguished in the time scales of magnetic resonance experiments. This difference between the magnetic resonance descriptions under pulsed excitations and the optical transients were highlighted in the early papers by A. H. Zewail[1] and R. G. Brewer.[2] When ultra fast events under high power laser pulses are concerned, it is yet to be reconciled as to the distinction between the decoherence and consequent radiation transients and the transients of the type which arise by spontaneous emissions.

As an additional comment, the paper by Artur Nenov deals with "deactivation pathways". In this context it is my current understanding that unless the processes responsible for the occurrence of transients in the ultra fast regimes is well specified and the respective characteristic times (time scales) are known, interpretations on the basis of such transients at the inference for the chemical dynamics and biological functions would entail many ambiguities. I quote the statement on page 3 of this paper: "But this was exactly what Ahmed Zewail set out to do. He had realised, from an experiment in the 1970s on anthracene molecules at low temperature, that molecules could be brought to vibrate in pace. 'Coherent preparation' of a sample system is thus a key point in all his experiments. We shall return to this concept". For most of the contexts it should be quite clear whether the relevant transient time constant is due to the decoherence phenomenon or the spontaneous emission. Probably the two processes are distinctly different.

1 http://www.nobelprize.org/nobel_prizes/chemistry/laureates/1999/press.html
2 R. G. Brewer, *Coherence in Spectroscopy and Modern Physics*, *NATO Advanced Study Institutes Series*, Springer, 1978, vol. 37, pp. 41–84.

**Artur Nenov** answered: The mechanism explaining the existence of the long-lived signal in DNA polynucleotides and the nature of the excited state associated with it has been eagerly debated in the recent years. Both theoretical and experimental studies favor different hypotheses. A summary of the latest results can be found in the paper submitted by us for this Faraday meeting. We believe that 2D electronic spectroscopy (2D-ES) has the best prospects to link the long lifetime to a particular mechanism. The main goal of our paper is to demonstrate that every deactivation channel in DNA has its characteristic spectroscopic fingerprints in a 2D spectrum as a direct consequence of the different electronic nature of the excited states and the ability of 2D-ES to disentangle the population transfer between excited states populated upon UV radiation. These fingerprints relate to the excited state absorption (ESA) and stimulated emission (SE) bands. *Ab initio* theoretical simulations, as the ones performed in our group,[1,2] will help to identify these fingerprints, which has a twofold benefit: firstly, they can serve as a guide for setting up 2D-ES experiments through identifying spectral windows of interest; and secondly, they will aid the interpretation of the 2D spectra, with the ultimate goal of deciphering the nature of the electronic states involved in the formation and recovery of the long lived state. I would like to note that a sufficient bandwidth in the visible and UV is required in order to resolve the above-mentioned fingerprint bands. As demonstrated in our paper there exist bands which exhibit geometry dependent spectral shifts of several thousand $cm^{-1}$, which qualifies them as excellent fingerprints, but poses a challenge for their experimental detection. It is, therefore, all the more pleasing to hear that broad bandwidths are already feasible in the UV region as well. Related to this I would also like to draw your attention to our computational studies on aromatic compounds (adenine, indole, phenol, *etc.*),[3,4] which reveal a number of characteristic ESA bands that can be detected by probing in the visible region. Therefore, I would encourage the realization of "two-color" experiments using narrowband UV-pump pulses and supercontinuum Vis-probes, as recently reported for example by the group of Riedle.[5]

1 I. Rivalta, A. Nenov, O. Weingart, G. Cerullo, M. Garavelli and S. Mukamel, *J. Phys. Chem. B*, 2014, **118**(28), 8396–8405.
2 A. Nenov, S. a Beccara, I. Rivalta, G. Cerullo, S. Mukamel and M. Garavelli, *ChemPhysChem*, 2014, **15**(15), 3282–3290.
3 A. Nenov, I. Rivalta, S. Mukamel and M. GAravelli, *Comput. Theor. Chem.*, 2014, **1040–1041**, 295–303.
4 A. Nenov, I. Rivalta, G. Cerullo, S. Mukamel and M. Garavelli, *J. Phys. Chem. Lett.*, 2014, **5**(4), 767–771.
5 N. Krebs, I. Pugliesi, J. Hauer and E. Riedle, *New J. Phys.*, 2013, **15**, 085016.

**Sankarampadi Aravamudhan** commented: Those experiments which are carried out in such a way that the spectrum is obtained by slow scanning through the two level resonance are the continuous radiation or continuous wave regime with low power electromagnetic radiation and when only incoherent light source is used. Coherent sources are capable of delivering high powers. At high power radiation the induced transition rates of the stimulated transitions would be high and a continuous application of high power radiation would not result in a steady state population difference to be maintained and the observed absorption and emissions would be time dependent observable parameters, and hence not conducive for further inferences. Hence, when high powers are available the electromagnetic radiation (the light irradiation) is in the form of a pulse, the pulse widths are significantly less than the relaxation times and the induced transition times. If the induced transition rates are fast at high power levels, the reciprocal of the rate, which is the transition time, would be small for one molecule (particle) to undergo the transition from a lower energy level to a higher level (presuming the lower energy level is more populated under thermal equilibrium). If the pulse widths are small compared to the induced transition times ($1/W_{01} = 1/W_{10}$), then at the end of the pulse the particle would not have transited fully from say, the lower level to the higher level. Thus the particle would be left in a non stationary state since only the lower energy level or the higher energy level is an Eigen state.

 

Such a non stationary state would be a linear combination of the lower level $\psi_l$ and the upper level $\psi_u$. Since such linear combinations of $\psi_u$ and $\psi_l$ are generated by a coherent radiation which corresponds to the description as time dependent perturbation (corresponding to the electromagnetic radiation pulse), it is a time dependent non-stationary state at the end of the pulse, and matter in such a state would radiate energy. This radiation with an initial amplitude would soon tend towards the thermal equilibrium distributing the molecules (particles) with certain time independent population differences characteristic of a thermal equilibrium population distribution among the two levels. Thus these experiments should always be reckoned with the appropriate transition times for the induced transitions and relaxation transitions. Only then it would be convincingly set out for a specified coherent superposition of the two levels, the superposition arising due to the time dependent perturbation. Under time independent

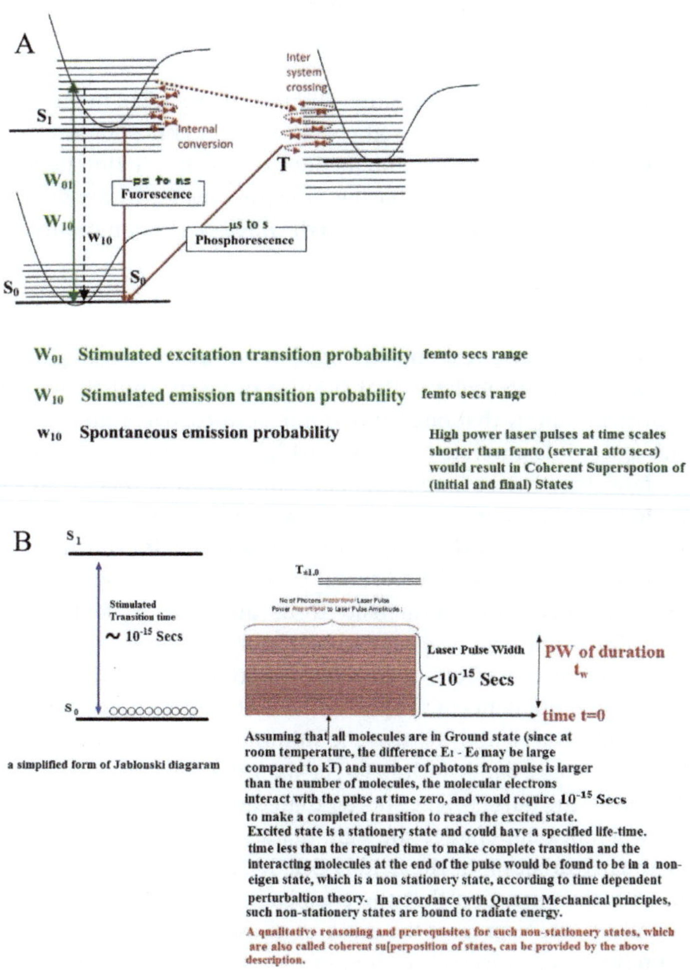


Fig. 1 (A) Jablonsky Diagram: description of related processes with typical time scales. (B) A simplified form of Jablonsky Diagram: a consideration of laser pulse characteristics with reference to transition time scales.

perturbations such states are time independent mixed states, but not non stationary coherent superpositions (Fig. 1).

These $W$ values are expressed as the number of molecules per second that make the transitions. The reciprocal of the $W$ values would then correspond to the time in secs required for one molecule to undergo a transition. Corresponding to every pair of two levels, there could be radiationless relaxation transitions. The reciprocal of the relaxation probabilities (different upward and downward rates) would correspond to the relaxation time which is defined conveniently as the time in terms of equilibrium population difference between the relevant 2-level system.

As an additional remark to the previous question by Dwayne Miller, I have mentioned the natural line width and indicated the inhomogeneous width due to overlap of lines. However, having to content myself with the qualitative assertion of the instantaneity of the event without proving that this "instantaneity" means a time scale several orders of magnitude less than the ultra fast pico/femto/atto seconds regime, may not be intellectually scientific, even if it sounds as unknown as origin of the universe or of life on planet earth.

**Shaul Mukamel** replied: This question should only be asked in the context of a specific experimental observable. Otherwise it is a philosophical issue. The absorption line shape involves two interactions with the radiation field and their separation is controlled by the dephasing time (the inverse line width). This can be interpreted as the time it takes to absorb the photon. Other measurements can be controlled by different processes.

**R. J. Dwayne Miller** also responded: The above discussion is all perfectly fine. The only comment I can make is that the use of lasers, and therein coherent radiation, does not imply that one is not in the perturbative limit as you seem to imply. With lasers it is possible to go to very high peak powers and couple many electronic states together in a nonstationary superposition that creates interferences that do not reflect the intrinsic dynamics of the individual states (in terms of nonradiative transitions). This feature of laser radiation is explicitly used in the Coherent Control (see the work of Brumer and Shapiro).[1] However, it is rather straight forward to control the intensity to be in the weak field limit. In this perturbative limit, one can use very short laser pulses with an appropriate bandwidth to probe the states of interest. For transform limited pulses, the different Fourier components of the pulse all have the same phase relationship at the carrier central wavelength to give the minimum pulse duration for a given bandwidth. If the laser bandwidth is larger than the spectral width of the transitions of interest, the pulse will be shorter than the intrinsic molecular dynamics of interest. One does not change the light-matter interaction. The Hamiltonian describing the perturbation by the field is exactly the same for the weak CW light condition, with the exception that the frequencies now have a common phase factor. This allows one to prepare the initial state as you describe and watch in the time domain the very processes you show in your figure. The information content is the same in both cases. Please note that the dynamics you depict in the figure contribute to the spectral linewidth in the frequency domain spectrum collected using CW excitation, which can be thought of as uncertainty broadening or more microscopically as relaxation induced Fourier components to the induced polarization in the medium. In this regard, the equivalence of frequency domain spectroscopy and time domain spectroscopy has been well established. The two methods of probing the system response are Fourier transforms of one another. (See the work by Loring and Mukamel on a similar issue raised in comparing time domain and frequency domain Coherent Anti-Stokes Raman.[2] Here the issue was whether the nonlinear aspects of the excitation could yield more information on homogeneous line widths. The equivalence of time domain and frequency domain for the same excitation processes was shown in this paper.) The attraction of using femtosecond pulses is that it lets you separate different possible pathways in the time domain, which becomes important if there are intermediate states and the time ordering of the relaxation pathways is key to understanding the molecular photophysics or photochemistry.

1 M. Shapiro and P. Brumer, *Rep. Prog. Phys.*, 2003, **66**(6), 859–942.
2 S. Mukamel and R. F. Loring, *JOSA B*, 1986, **3**(4), 595–606.

**Mike Ashfold** asked: Can someone briefly summarise the recent evidence for a long lived component to the excited state decay of DNA multimers? If correct, this finding surely necessitates some reappraisal of the photostability of the DNA bases – traditionally attributed to the efficiency of the radiationless processes by which their excited state population transfers back to the ground state.

**Artur Nenov** responded: From the Kerr-gated time-resolved pump–probe transient absorption and fluorescence experiments in (dA)20, a single-stranded adenine polynucleotide, Phillips and co-workers obtained three different time-constants, 0.39, 4.3 and 182 ps, the latter being the one responsible for the long-lived spectroscopic signal, which in this case was measured with time-resolved fluorescence detecting at ~390 nm.[1] Kohler and co-workers performed pump–probe transient absorption measurements (pump at ~266 nm/probe at ~252 nm) in adenine-based polynucleotides, obtaining different results depending on the length of the strand: 105 ps for ApA[2] and 207 ps for d(A)4.[3] Newer works from Kohler and co-workers[4,5] dwell on the relationship between the stacking motifs and the long-lived signal by constraining the adenosine chains with large groups to create steric hindrance and enhance the stacking, thereby relying on circular dichroism experiments to relate ground state conformations to the long-lived excited state. Markovitsi and co-workers have also studied adenine based polynucleotides by means of time-resolved fluorescence upconversion, obtaining similar results as those found by Kohler for natural DNA samples.[6,7]

Regarding double-stranded chains, the effect of the Watson–Crick base pairs (*e.g.* poly(dA-dT) *vs.* poly(dA))[8,9] quickly accelerates the decay lifetimes through proton–hydrogen transfer mechanisms that are ultrafast in nature.[10] Specific sequences, such as those alternating G/A in a double-stranded chain[11] have been reported to accelerate the decay times, especially those linked with G–C Watson–Crick base pairs.[10] Some recent works by Zinth and co-workers do point towards the quenching effect of the double-strand on the long-lived lifetime.[12]

1 W.-M. Kwok, C. Ma and D. L. Phillips, *J. Am. Chem. Soc.*, 2006, **128**, 11984.
2 T. Takaya, C. Su, K. de La Harpe, C. E. Crespo-Hernández and B. Kohler, *Proc. Natl. Acad. Sci. U. S. A.*, 2008, **105**, 10285.
3 C. Su, C. T. Middleton and B. Kohler, *J. Phys. Chem. B*, 2012, **116**, 10266.
4 J. Chen, A. K. Thazhathveetil, F. D. Lewis and B. Kohler, *J. Am. Chem. Soc.*, 2013, **135**, 10290.
5 J. Chen and B. Kohler, *J. Am. Chem. Soc.*, 2014, **136**, 6362.
6 D. Markovitsi, T. Gustavsson and I. Vayá, *J. Phys. Chem. Lett.* , 2010, **1**, 3271.
7 I. Vayá, T. Gustavsson, T. Douki, Y. Berlin and D. Markovitsi, *J. Am. Chem. Soc.* , 2012, **134**, 11366.
8 C. Greve, N. K. Preketes, R. Costard, B. Koeppe, H. Fidder, E. T. J. Nibbering, F. Temps, S. Mukamel and T. Elsaesser, *J. Phys. Chem. A*, 2012, **116**, 7636.
9 C. Greve, N. K. Preketes, H. Fidder, R. Costard, B. Koeppe, I. A. Heisler, S. Mukamel, F. Temps, E. T. J. Nibbering and T. Elsaesser, *J. Phys. Chem. A*, 2013, **117**, 594.
10 N. K. Schwalb and F. Temps, *J. Am. Chem. Soc.*, 2007, **129**, 9272.
11 N. K. Schwalb and F. Temps, *Science*, 2008, **322**, 243.
12 D. B. Bucher, A. Schlueter, T. Carell and W. Zinth, *Angew. Chem. Int. Ed.*, 2014, **53**, 11366.

**Kenneth Ghiggino** said: I would like to raise an issue about nomenclature. In your paper you have referred to CT excimer states. How do these states differ from the usual definition for an excimer or a CT state? Is it necessary to introduce the term CT excimer state?

**Artur Nenov** answered: Thanks to Prof. Ghiggino's comment we realized that the "CT excimer state" notation was inadequate. In fact, we refer to the existence of a low lying CT state formed through an electron hopping from an adenine to an adjacent one. As a consequence an excimer is formed, characterized by the shortening of the inter-base distance by approx. 0.5 Angstrom,[1] which is accompanied by geometrical deformations in both bases aiming at stabilizing the opposite charges. Thanks to the comment raised by Prof. Ghiggino we corrected the nomenclature in our paper.

1 I. Conti, A. Nenov, S. Hoefinger, S. F. Altavilla, I. Rivalta, E. Dumont, G. Orlandi and M. Garavelli, *Phys. Chem. Chem. Phys.*, 2015, **17**, 7291–7302.

**Helena J. Shepherd** opened the general discussion of the paper by **Marylise Buron-Le Cointe**: (1) You state that "...in the intermediate phase, a spin state concentration wave (SSCW) appears resulting from a symmetry breaking (cell doubling) associated with a long-range order of alternating high spin and low spin molecular states". Do you have any comments regarding the relationship (causal/non-causal, *etc.*) between the different symmetry breaking processes and the appearance/disappearance of the SSCW? This is a potentially interesting discussion in light of the fact that when the SSCW disappears (in the LS state), a further symmetry breaking transition occurs to a phase that is different from either the high spin or intermediate phases (*i.e.* the phase transition is not re-entrant). (2) You carried out time-resolved studies at temperatures corresponding to the plateau associated with the intermediate phase. Is it possible to perform similar experiments at *e.g.* ~90 K (in the LS state but above T(LIESST)), and what do you/would you expect to observe in this case (*e.g.* fraction of molecules excited, symmetry breaking, SSCW, relaxation times, *etc.*)?

 

**Marylise Buron-Le Cointe** replied: (1) There are different degrees of freedom in crystals which can break the symmetry. The SSCW is associated with a long-range order of molecules in the HS and LS states, which are symmetry-equivalent in the high temperature HS phase. Indeed, there are some SCO materials in the literature for which the HS and LS phases have the same symmetry, *i.e.* a re-entrant phase transition occurs (see ref. 26 and 27 of our paper for example). In the present case, the symmetry breaking in the LS phase results from a molecular tilting. This is a purely structural ordering as all the molecules have the same LS electronic state. This is also true for the photoinduced state where an ordering of tilted molecules appears: the photoinduced HS state is of lower symmetry than the HS state stable at high temperatures (see our ref. 4). Therefore the SSCW is a symmetry breaking process but in these materials other types of more conventional structural symmetry breaking can also appear, even though the SCO is usually an isostructural process.[1] (2) In the investigated material it is difficult to perform such a study (optical or X-ray diffraction) from the LS phase because the first-order phase transition from the SSCW plateau to the LS phase decreases the crystalline quality. In addition, the recovery time to the LS state increases at low temperatures and may be too long at 90 K for performing pump–probe studies with a 40 Hz repetition rate. This may not be a problem for other stepped SCO compounds for which the SSCW forms at higher temperatures and it might be interesting to investigate this aspect. We expect then to observe a similar multistep out-of-equilibrium dynamics. From the LS phase, a fs light excitation should promote locally some molecules to the HS state within 160 fs and the elastic and thermal steps will promote more molecules to the HS state. Then, it is possible that the SSCW forms on the few ms time-scale, when a transient temperature is reached and it should disappear in 10s ms when the sample relaxes to the initial temperature.

1 N. Bréfuel, E. Collet, H. Watanabe, M. Kojima, N. Matsumoto, L. Toupet, K. Tanaka and J. P. Tuchagues, Nanoscale self-hosting of molecular spin states in the intermediate phase of spin-crossover materials, *Chem. Eur. J.*, 2010, **16**, 14060–14068.

**R. J. Dwayne Miller** asked: The time scale you observe and assign to the relaxation within the spin manifold to reach the fully relaxed spin state (recovery to the low spin state) is comparable to the thermal diffusion. How do you separate the lattice relaxation from the thermal diffusion effects? One would think that an extremely fast forward transition, fast low spin to high spin transition, would have similarly fast high spin to low spin transitions upon relaxing back to the ground electronic state and relaxation of the Fe–N bond (the key coordinate) back to its original position, based on microscopic reversibility.

**Marylise Buron-Le Cointe** answered: Indeed, the recovery to the LS state is driven by the time it takes to reach thermal equilibrium with the sample environment and this is governed by the heat exchange with the cryostat and is therefore sample shape and size dependent. The recovery 10s ms time scale observed here for this stochastic dynamics is much longer than that of more elementary physical processes (molecular switching, unit-cell deformations, ...).

These are indeed hidden in a statistical average. We can only reveal the intrinsic time-scale of these elementary processes during the transformation, as their dynamics is clocked by the ultrashort laser pulse. The recent study of reverse LIESST has also shown that the photoinduced LS state is reached within 40 ps from the HS state. This photoinduced LS state recovers the HS state within less than 1 ms (see our ref. 16).

**Debabrata Goswami** queried: Since the concerned experiments involve Spin based processes – you would start seeing differences of linear *versus* circular polarization that could help with what you are looking at. The other approach is magnetic, can you isolate the temperature aspect? I feel the use of different polarizations would help.

**R. J. Dwayne Miller** replied: We use polarized excitation in some cases to maximally excite the sample. Since the sample orientations with respect to the transition dipole moments are not often known, at least not without detailed parallel studies, we also used circularly or depolarized excitations to provide constant excitation conditions independent of crystal orientation. The spin to which you refer is related to the magnetic spin transitions, say in the intersystem crossing or spin cross over phase transitions, which is not covered in this work. It is certainly of interest. In this case, we would have to use spin polarized electron sources to provide a contrast sufficient to definitively determine the spin coupled lattice dynamics. It is possible to generate spin polarized electron sources using negative affinity GaAs photocathodes. We would love to do such studies and I hope we can pursue such studies in the near future. There are important outstanding issues as discussed in the paper presented by M. Buron-Le Cointe *et al.* on spin crossover systems in these proceedings.

**Marylise Buron-Le Cointe** commented: The LIESST effect is driven by light absorption of the diamagnetic LS species. The use of polarization in such low symmetry crystals is therefore of interest only for optimizing the penetration depth of light (this was used in our ref. 14 and 15 for example). In the present sample, the light penetration depth dependence with polarization is weak and we observed similar responses for random orientations of the crystals with regard to light polarization. In addition, the HS state is paramagnetic (there is no magnetic order, at least in the investigated temperature range). There are other systems, such as the [Fe(1-propyltetrazole)6](BF4)2 SCO compound, where a single-molecule magnet behavior is observed in the photoexcited HS state.[1] However, this process only occurs at very low temperatures (around 20 K), where the lifetime of the photoinduced HS state is too long to perform kHz stroboscopic pump–probe studies.

1 X. Feng *et al.*, *J. Am. Chem. Soc.*, 2013, **135**, 15880.

**John R. Helliwell** asked: On page 8 of your article and in your talk you emphasized that (only) “0.2% of the molecules in your sample switch from the LS to HS state at the ps timescale”. You could of course use a smaller sample, to try and enhance this, and in any case preserve the X-ray diffraction intensities by using Laue diffraction rather than monochromatic diffraction.[1] However that is not the focus of your study, rather, it is seeing the doubled cell nicely evident in Figure 7 of your article and highlighted in your abstract. Am I correct?

1 Z. Ren, D. Bourgeois, J. R. Helliwell, K. Moffat, V. Srajer and B. L. Stoddard, Laue crystallography: coming of age, *J. Synchrotron Rad.*, 1999, **6**, 891–917.

**Marylise Buron-Le Cointe** responded: Yes it is correct. When one is interested in the structure of the photo-excited state, a matching of the sample size and of the laser penetration depth is necessary for photoswitching a large fraction of molecules. When this penetration depth is small, one has to use thin crystals and it is then necessary to increase the X-ray flux by using a broad band or "pink" X-ray spectrum. This is the Laue diffraction technique and for example P. Coppens *et al.* applied this technique to study photo-switching in different molecular crystals by using the ratio method.[1] This ratio method is well-adapted for small changes on Bragg peak intensities without considering changes of lattice parameters. In other words, it is not well adapted for observing the second elastic step and the third macroscopic heating step (see our Fig. 6b). In the out-of equilibrium dynamics of SCO systems investigated here, several steps are involved. During the ps non-thermal photo-switching step, we photo-excite typically 0.2% of the molecules, because in the μs delays the crystals warms up and the HS state is then thermally populated. On other SCO compounds, we could increase the fraction of photo-excited molecules up to a fraction closer to 1% and more (when the deposited laser energy increases). The crystal warming can then reach 20–30 K or more.[2,3] In the present study, we wanted to limit such heating effects to investigate the symmetry breaking aspects of the process, not the thermally-induced one. The process investigated here is not limited by the penetration of light since the symmetry changes globally in the crystal, as indicated by the almost complete disappearance of the peaks characteristic of the symmetry breaking. In addition, the use of Laue is not appropriate in the present study as we focus on the cell doubling aspects. We need therefore to separate the diffracted intensity by the Bragg peaks related to the symmetry breaking indexed (h k 2p + 1) from the diffracted intensity by the other Bragg peaks. With Laue techniques, some peaks indexed (h k 2p + 1) and (h′ k′ 2p′) can superpose on the same spot of the detector (if the X-ray spectrum is broad enough). It is therefore more difficult to reconstruct the diffracted intensity in the reciprocal space (see our Fig. 7) and especially for the (h k 2p + 1), directly related to the amplitude of the symmetry breaking order parameter.

1 P. Coppens, M. Pitak, M. Gembicky, M. Messerschmidt, S. Scheins, J. Benedict, S. Adachi, T. Sato, S. Nozawa, K. Ichiyanagi, M. Chollet and S. Koshihara, The RATIO method for time-resolved Laue crystallography, *J Synchrotron Rad.*, 2009, **16**, 226–230.
2 M. Lorenc, C. Baldé, W. Kaszub, A. Tissot, N. Moisan, M. Servol, M. Buron-Le Cointe, H. Cailleau, P. Chasle, P. Czarnecki, M. L. Boillot and E. Collet, *Phys. Rev. B*, 2012, **85**, 054302.
3 W. Kaszub, M. Buron-Le Cointe, M. Lorenc, M.-L. Boillot, M. Servol, A. Tissot, L. Guérin, H. Cailleau. and E. Collet, *Eur. J. Inorg. Chem.*, 2013, **5–6**, 992–1000.

**John R. Helliwell** commented in reply: I fully accept your clarification of the aim of your experiment, thank you. Regarding deconvoluting the overlapping of Laue diffraction spots, either exactly (namely 'multiplets') or spatial overlaps due to the finite diffraction spot size, the accurate estimation of the individual spot intensities was indeed a challenge but circumvented by several techniques, and indeed the measured accuracy of the deconvoluted multiples or the spatial overlaps is now as good as 'singlet' Laue spots,[1] which of course predominate.[2]

1 Y. P. Nieh, J. Raftery, S. Weisgerber, J. Habash, F. Schotte, T. Ursby, M. Wulff, A. Haedener, J. W. Campbell, Q. Hao and J. R. Helliwell, Accurate and highly complete synchrotron protein crystal Laue diffraction data using the ESRF CCD and the Daresbury Laue software, *J. Synchrotron Radiat.*, 1999, **6**, 995–1006.
2 D. W. J. Cruickshank, J. R. Helliwell and K. Moffat, Multiplicity Distribution of Reflections in Laue Diffraction, *Acta Cryst. A*, 1987, **43**, 656–674.

**R. J. Dwayne Miller** said: This question is aimed at trying to gain some physical insight into the mechanism for this extremely fast spin transition, giving rise to the formation of a spin concentration wave. For spin cross over materials, the transition from low to high spin occurs within a few 10s of a femtosecond. It seems clear that the spin transition is related to the bond displacement along the Fe–N coordinate as you discuss in your paper. What I find unusual is that the mechanism must involve a break down in the Born–Openheimer approximation in treating this electronic transition. The physics for the coupling of the bond displacement to the spin is not evident. For classic treatments of intersystem crossing of photoexcited molecular systems, the chemistry community has rationalize the spin transition in terms of <L.S><FC>, *i.e.* the coupling between the orbital angular momentum (L) and spin (S) weighted by the Franck–Condon (FC) overlap between the low spin and high spin transition electronic manifolds. This relationship gives the perturbative term to the Hamiltonian in which the force acting on the spin system is derived from the fluctuating magnetic moments associated with the photoexcited changes in electron distribution and the associated orbital angular momentum. The transition probability associated with this mechanism is generally many orders of magnitude smaller than what is observed for the photoexcited spin cross over materials. Can you give us some insight into the physical mechanism coupling the spin to the bond displacement coordinate? There needs to be some induced change in the local magnetic moments to couple to the spin, *i.e.* there needs to be a force to flip the spin.

**Marylise Buron-Le Cointe** replied: It is true that the mechanism behind the extremely fast LIESST, between states with different spin and structure, defies conventional descriptions. All the recent studies of the ultrafast LIESST effect indicate a breakdown of the Born–Oppenheimer approximation (see our ref. 16 and 20). The electronic and nuclear wave functions are strongly coupled during the process which occurs on a time-scale corresponding to elementary molecular deformations or vibrational periods (160 fs). LIESST in SCO systems is thus mediated by the spin–orbit coupling. There is also a vibronic coupling between the excited states involved during LIESST. The LS to HS LIESST can be induced from the relaxation of the initially excited $^1T_1$ ligand field state to the HS state or from the relaxation of $^1$MLCT states, as is the case here. van Veenendaal discussed theoretically the role of the significant changes of the metal–ligand (Fe–N) distance between LS and HS states: the coupling between the electronic state and the structure (the HS state is less bonding and the Fe-N bond expands) is similar to a constant force that displaces the equilibrium position of the ligand.[1] The ultrafast inter-system crossing was proposed to result from the dephasing of the photoexcited electronic state into the HS phonon states. This mechanism was recently evidenced by femtosecond pump–probe techniques, which revealed coherent breathing vibrations accompanying the LIESST and the change of molecular structure occurring within 160 fs (see our ref. 20). We should underline that the reverse LIESST process is different as it occurs on the 40 ps time-scale (see our ref. 16) and in this case the Born–Oppenheimer approximation seems valid.

1 M. van Veenendaal, J. Chang and A.J. Fedro, *Phys. Rev. Lett*, 2010, **104**, 067401.

Faraday Discussions

PAPER

# Probing electronic and vibrational dynamics in molecules by time-resolved photoelectron, Auger-electron, and X-ray photon scattering spectroscopy

Kochise Bennett,* Markus Kowalewski and Shaul Mukamel*

Received 17th September 2014, Accepted 4th November 2014
DOI: 10.1039/c4fd00178h

We present a unified description for time-resolved electron and photon scattering spectroscopies from molecules prepared in nonstationary states. Signals are expressed in terms of superoperator Green's functions and a systematic procedure for treating various degrees of freedom consistently at different levels of theory is developed. The standard Fermi Golden Rule expressions for photoelectron spectra, which are limited to broad, slowly-varying signals, are obtained as a limiting case of our more general theory that applies to broader parameter regimes.

## 1. Introduction

Molecular photophysical processes are widely monitored by detecting ejected electrons or scattered X-ray photons.[1-4] We present a computational framework for time-resolved photoelectron spectroscopy (TRPES), Auger-electron spectroscopy (AES), and off-resonant X-ray scattering (OXS). In these experiments, the common objective is to monitor the electronic and nuclear dynamics of nonstationary states (often prepared by excitation with a pump pulse). Electron spectroscopies in the X-ray regime will depend on valence, core, and continuum electronic states, and nuclear degrees of freedom as well as fluctuations from coupling to the environment (this could be a thermal bath or represent solvent degrees of freedom). Exactly including the nuclei quantum mechanically is often numerically too expensive and the Born–Oppenheimer approximation is made separate of electronic and nuclear degrees of freedom. This separation fails when electronic modes are vibronically coupled or at degeneracies (conical intersections).[5,6] Thus, the level of theory employed for the nuclei can vary greatly (*e.g.*, harmonic oscillators, Brownian oscillators, averaging over semiclassical trajectories, surface hopping, spawning[7]). In particular, if some nuclear degrees of freedom can be identified as the origin of the strongest intramolecular

*Chemistry Department, University of California, Irvine, California 92697-2025, USA. E-mail: kcbennet@uci.edu; smukamel@uci.edu; mkowalew@uci.edu*

interactions and fastest dynamics, it would be preferable to treat these explicitly. The remaining nuclear modes can then be treated implicitly, such an approach was advocated by Seel and Domcke for TRPES.[8] Except for model systems in which such effects can be explicitly included in the Hamiltonian and solved analytically, they are normally treated by numerical propagation or modification to the Fermi Golden Rule (FGR). The large number of choices of what level of theory to employ for various degrees of freedom calls for a consistent and comprehensive approach to these signals. Liouville space is a natural setting for handling bath degrees of freedom and making semiclassical approximations and can therefore provide a unified treatment of these experiments.

We start with time-resolved photoelectron spectroscopy (TRPES) which is a pump–probe technique in which a pump pulse prepares the system and, after some controlled delay time, a probe pulse ionizes it.[1] The photoelectron current is generally plotted as a function of the pulse parameters (principally the interpulse delay) and the photoelectrons' kinetic energy. This is given by the difference in energy between the ionized and unionized system. Comparison with the ground state spectra can thus be used to analyze the electronic and vibrational excited state dynamics and to monitor non-radiative decay channels. Photoelectron spectroscopy has been utilized to study a wide variety of material systems (*e.g.* atoms, molecules, films and surfaces, and metals[9–12]) and has recently been used to track the non-adiabatic radiationless decay of various organic molecules (*e.g.* uracil, thymine, benzene, and *cis*-dienes[13–15]) complementing optical methods (*e.g.* ref. 16). TRPES has less restrictive selection rules compared to optical detection schemes (any orbital may be ionized and the transition dipole to the continuum states does not depend much on the continuum state and may be well-approximated as flat in certain regions). The probabilities for excitation to various continuum states can still depend sensitively on the final molecular electronic state and the continuum states serve as a probe.[1,8]

In the X-ray regime, the TRPES process involves valence, core, and continuum electronic states. Photoionization prepares a core hole state, whose dynamics can be monitored. This state ordinarily rapidly decays as a valence electron fills the core hole while a second valence electron (the Auger electron, which is the object of study in AES) is ejected with kinetic energy equal to the core–valence energy gap. The Auger process is of interest in studying core-hole dynamics (the Auger decay process adds a linewidth to the TRPES spectra), and is governed by a Coulomb matrix element rather than a transition dipole. This difference in selection rules often leads to broader, less distinct spectra but can be useful for viewing dipole-forbidden transitions. Like TRPES, AES has been used to study the dynamics for diatomics, biomolecules, metals, oxide films, and electrolytes.[8,9,11,17–19] Because of the complexity of these processes, a variety of approaches have been adopted for their simulation including methods based on direct wave function propagation, many-body non-equilibrium Green's functions and the core-hole spectral function as well as a more straightforward quadratic-response function formalism.[20–24] These are frequently approximated in a "two-step" fashion in which the photoionization event is separated from the Auger decay (the ionization is also typically taken to be to the relaxed core-hole state), removing these restrictions is necessary to describe gain–loss peaks in the Auger spectra due to excited states induced by the ionization process.[25,26] Such considerations are obviously important when tracking the decay of electronic excited states *via*

 

TRPES/AES and, in this paper, we employ a one-step description so that the entire history of the system is properly included.

A third commonly employed technique for tracking the structure and dynamics of materials is the detection of scattered, off-resonant X-rays, whose elastic portion, diffraction, has been widely used to study the electronic charge density of materials.[3,4,27–29] In a recent paper, we examined time-, frequency-, and wavevector-resolved off-resonant X-ray scattering (OXS) from non-stationary states and found that the signal separates into one- and two-particle terms which carry different form factors.[30] For OXS from the ground state, the two-particle form factor was found to contain only elastic contributions while all inelastic processes were contained in the one-particle form factor. The inelastic contributions are thus dominated by the elastic for a many-particle sample. The one-particle form factor contains inelastic terms and is clearly the only contribution in a single molecule experiment. These inelasticities are therefore expected to be especially relevant for ultrafast diffraction from single-molecules, a regime that has drawn considerable interest recently.[31–34]

In this paper, we recast the theory of TRPES and AES in Liouville space; a previous treatment incorporated multidimensional processes and quantum fields but did not explicitly consider vibronic couplings between electronic states.[35] From this, we derive the FGR expression ordinarily associated with these signals revealing the underlying assumptions. The FGR should not hold when the nuclear dynamics is fast compared to the photoionization time. We thus expect it to break down near conical intersections where a more general formula will be needed. Following exposition of the Liouville space formalism, we provide a semiclassical simulation of the TRPES signal from acroleine following preparation of a valence excitation and discuss how this simulation procedure may be generalized to allow for nuclear motion during electronic coherences. We then discuss OXS as recently recast so as to explicitly include the time, frequency, and wavevector resolution of the photon detection event and analogies to TRPES/AES are discussed. In particular, all of these processes, in addition to being useful probes of nonstationary states at both the electronic and nuclear level, prepare interesting states in their own right. Following these processes, the system is left in a different nonstationary excited state. This state may then be probed by other nonlinear spectroscopic techniques that can also be described by the Liouville space theory presented here.

## 2. Photoelectron spectroscopy; Hamiltonian and the Fermi Golden Rule

We describe TRPES with the following Hamiltonian:

$$H = H_{\mathrm{M}} + H_{\mathbf{p}} + H_{\mathrm{x}}(t) + H_{\mathrm{v}}(t) \tag{1}$$

where

$$H_{\mathrm{M}} = \sum_{II'\mathrm{q}} |I_{\mathrm{q}}\rangle \mathscr{E}_{I_{\mathrm{q}}} \langle I_{\mathrm{q}}| + |I'_{q}\rangle H_{I'I} \langle I_{\mathrm{q}}| \tag{2}$$

represents the molecule with $\mathscr{E}_{I_q}$ the energy of the $I$-th electronic state with $q$ holes and $H_{I'I}$ the coupling between electronic states *via* nuclear coordinates. Both are operators in the nuclear subspace. In a fully microscopic theory, the states and parameters would be given in terms of single-particle occupation numbers (in a second-quantized representation). This is discussed further in the appendix.

$$H_{\mathbf{p}} = \sum_{\mathbf{p}} \varepsilon_{\mathbf{p}} c_{\mathbf{p}}^{\dagger} c_{\mathbf{p}} \tag{3}$$

is the Hamiltonian of the photoelectron states ($c_{\mathbf{p}}^{\dagger}$, $c_{\mathbf{p}}$ are creation/annihilation operators for the orbital with energy $\varepsilon_{\mathbf{p}}$) and $H_{\mathrm{x}}(t)$ and $H_{\mathrm{v}}(t)$ represent interactions with photoionizing X-ray and valence pump fields, respectively. We assume that $H_{\mathrm{M}}$ is block diagonal in the molecular charge number and label electronic states of $H_{\mathrm{M}}$ as $|I_q\rangle$ where the subscript indicates the molecular charge. The molecular Hamiltonian remains an operator in the nuclear subspace while we assume that the Hamiltonian of the continuum states is independent of the nuclei. In the simplest model, the nuclear coupling gives a vibrational splitting to each electronic state. More generally, the nuclear degrees of freedom can couple the electronic states (as in surface crossing and non-Born–Oppenheimer effects) leading to population transport, *etc.* In the Condon approximation, the dipolar interaction with the valence pump pulse is

$$H_{\mathrm{v}}(t) = -E_{\mathrm{v}}(t) \sum_{II'} \mu_{I'I}^{(q)} \left| I_q' \right\rangle \langle I_q | + \mu_{I'I}^{(q)*} | I_q \rangle \left\langle I_q' \right| \tag{4}$$

where $E_{\mathrm{v}}(t)$ is the temporal envelope of the valence pump pulse. We work in the product space of bound and continuum states and the interaction with the photoionizing X-ray pulse is then written as

$$H_{\mathrm{x}}(t) = -E_{\mathrm{x}}(t) \sum_{II'\mathbf{p}} \mu_{\mathbf{p}I'I}^{(q)} c_{\mathbf{p}}^{\dagger} \left| I_{q+1}' \right\rangle \langle I_q | + \mu_{\mathbf{p}I'I}^{(q)*} c_{\mathbf{p}} | I_q \rangle \left\langle I_{q+1}' \right| \tag{5}$$

where $E_{\mathrm{x}}(t)$ is the temporal envelope of the ionizing X-ray pulse and we have explicitly indicated the dependence of the transition dipoles on the charge number. In the following, we will only consider pumping of the neutral molecule and omit the $q = 0$ superscript on the transition dipoles. Throughout, we work in the interaction picture with respect to $H_{\mathrm{int}}(t) = H_{\mathrm{v}}(t) + H_{\mathrm{x}}(t)$ and use atomic units so that $\hbar = m_{\mathrm{e}} = e = 1$. A straightforward application of the Fermi Golden Rule gives the probability of generating a photoelectron in state $|\mathbf{p}\rangle$ as second order in the interaction with the ionizing pulse yielding

$$P(\mathbf{p}, T) = \sum_{II'} \rho_{I_0}(T) \left| E_{\mathrm{x}} \mu_{\mathbf{p}I'I} \right|^2 \delta\left(\omega_x - \varepsilon_{\mathbf{p}} - \mathscr{E}_{I_1'}(T) + \varepsilon_{I_0}(T)\right), \tag{6}$$

where

$$\rho_{I_0}(T) \propto |E_{\mathrm{v}} \mu_{Ig}|^2 \delta(\omega_{\mathrm{v}} - \mathscr{E}_{I_0}(0)) \tag{7}$$

is the population of state $|I_0\rangle$ and we have taken the snapshot limit of the pump and probe pulses (whereby the spectral and temporal pulse envelopes are both $\delta$-functions with $\omega_{\mathrm{x}}$ and $\omega_{\mathrm{v}}$ the frequencies of the X-ray and valence pump pulses, respectively) and the signal only depends on the relative delay time $T$. The bound state energies (denoted $\mathscr{E}$) and populations are operators in the nuclear subspace and this expression must still be averaged over the nuclear degrees of freedom. This can be done formally with a model for the nuclear–electronic coupling or semiclassically by averaging over stochastic trajectories. This approach is very intuitive and is readily amenable to semiclassical treatments. Similar formulas have been used to study the nuclear dynamics of the excited singlet state potential energy surfaces of a variety of molecules.[14,15]

The FGR makes several assumptions that may not apply for ultrafast experiments. One goal of this paper will be to elucidate these assumptions and how they may be relaxed for a better description of ultrafast dynamics.

## 3. Liouville space description of TRPES

The photoelectron signal is defined as the integrated, energy-resolved current:

$$S_{\mathrm{PES}} = \int \frac{\mathrm{d}}{\mathrm{d}t}\langle n_{\mathbf{p}}(t)\rangle. \tag{8}$$

Assuming that $[n_{\mathbf{p}}, H_{\mathrm{M}}] = 0$ (where $n_{\mathbf{p}} = c_{\mathbf{p}}^{\dagger}c_{\mathbf{p}}$ is the occupation number of the scattering state $|\mathbf{p}\rangle\rangle$), we have, from the Heisenberg equation of motion

$$\dot{n}_{\mathbf{p}} = -i[n_{\mathbf{p}}, H] = -2E_{\mathrm{x}}(t)\mathscr{I}\{\mu_{\mathbf{p}}\} \tag{9}$$

where

$$\mu_{\mathbf{p}} \equiv \sum_{I_0 I_1} \mu^{*}_{\mathbf{p}I_1 I_0} c_{\mathbf{p}} |I_0\rangle\langle I_1|. \tag{10}$$

The Liouville-space treatment can rigorously incorporate "bath" degrees of freedom without explicitly treating them quantum mechanically. We describe the dynamics with the stochastic Liouville equation (SLE). This assumes that the quantum system of interest is affected by a classical bath, whose stochastic dynamics follow a Markovian master equation. The SLE is an equation of motion for the field-free evolution of the joint system density matrix:

$$|\dot{\rho}(t)\rangle\rangle = \mathscr{L}|\rho(t)\rangle\rangle = -iH_{-}|\rho(t)\rangle\rangle + L_{\mathrm{b}}|\rho(t)\rangle\rangle \tag{11}$$

where $H_{-}$ is the superoperator corresponding to commutation with the electronic Hamiltonian (see Appendix) and $L_{\mathrm{b}}$ represents the stochastic Markovian dynamics of the bath. The versatility of this approach lies both in the choice of which degrees of freedom are explicitly incorporated into the Hamiltonian dynamics *versus* which are relegated to the bath, and at what level the bath dynamics is treated (commonly, some of the nuclear degrees of freedom are described as a semiclassical bath and solvent effects can be treated in this manner as well).

We consider a process in which a pump pulse populates the valence excited states, which then evolve for a delay period $T$ before interaction with the X-ray probe. Since the photoelectron state is initially vacant, the lowest order non-vanishing response has two interactions with the ionizing pulse. Since we are considering photoionization of an electronic excited state, we also expand the signal to the second order in the pump pulse. The space signal can be read from the diagram in Fig. 1:

$$S_{PES} = -2\int dt_4 dt_3 dt_2 dt_1 E_x(t_4)\mathscr{I}[(i)^3 \langle\langle 1|\mu_{\mathbf{p}L}(t_4)G(t_4,t_3)\mathscr{L}_x(t_3)G(t_3,t_2)\mathscr{L}_v(t_2)G(t_2,t_1)\mathscr{L}_v(t_1)|gg\rangle\rangle] \quad (12)$$

where

$$G(t,t') \equiv \theta(t-t')\exp[-i\mathscr{L}(t-t')] \quad (13)$$

is the Liouville space propagator and the $L$ subscript indicates action on the ket (see Appendix). The initial density matrix is given by the pure state $|gg\rangle\rangle$ which corresponds to the many body ground state in which the core is full, the continuum is empty, and the valence states are occupied up to some level. In principle all time propagations are with respect to $H$. However, assuming that $H$ conserves the charge, we may write $H_M = H_M(q)$ (as above, we assume $[n_{\mathbf{p}}, H] = 0$). We then use projection operators to simplify the Green's functions in eqn (12). We work with the product states $|I_q\rangle\cdot|\mathbf{p}\rangle \equiv |I_q\mathbf{p}\rangle$ and the associated Liouville-space projection operators

$$\mathscr{P}_1 = \sum_{I_0}|I_0, g\rangle\rangle\langle\langle I_0, g| + |g, I_0\rangle\rangle\langle\langle g, I_0| \quad (14)$$

$$\mathscr{P}_2 = \sum_{I_0}|I_0, I_0\rangle\rangle\langle\langle I_0, I_0|$$

$$\mathscr{P}_3 = \sum_{I_0 I_1 \mathbf{p}}|I_1\mathbf{p}, I_0\rangle\rangle\langle\langle I_1\mathbf{p}, I_0|$$

which allow to restrict attention to particular manifolds of states. In particular, $\mathscr{P}_1$ projects into the space of coherences between the ground state and the $|0\rangle$ manifold (the set of states with 0 holes), $\mathscr{P}_2$ projects into the space of populations of the $|0\rangle$ manifold and $\mathscr{P}_3$ projects into the space of coherences between the $|1\rangle$ and $|0\rangle$ manifolds (analogously to $\mathscr{P}_1$, $\mathscr{P}_3$ actually has two terms but we have kept only the one that contributes). Aside from the

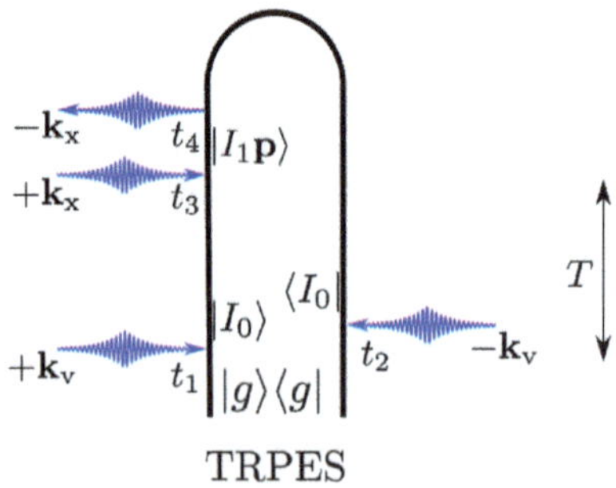


Fig. 1 Loop diagram for the TRPES process. The valence of both ket and bra are initially excited *via* an external pulse ($\mathbf{k}_v$). Following a delay $T$, the system interacts with an X-ray pulse ($\mathbf{k}_x$) which promotes an electron to the photoelectron continuum states ($\mathbf{p}$). This one loop diagram stands for two ladder diagrams since the valence pumping can occur first on the ket or first on the bra.

 

restriction to populations in $\mathscr{P}_2$ (this is justified for pulses separated by times larger than coherence decay times and is easily relaxed to account for overlapping pulses), inserting these projection operators entails no loss of generality since the Hamiltonian conserves the number of **p**-electrons and holes. This is equivalent to simply taking matrix elements by inserting resolutions of the identity operator. However, the use of projection operators makes clear how to include more complicated processes such as Auger decay which gives width to TRPES lines. We thus re-write the signal as

$$S_{\mathrm{PES}} = -2\int \mathrm{d}t_4\mathrm{d}t_3\mathrm{d}t_2\mathrm{d}t_1 E_{\mathrm{x}}(t_4)\mathscr{I}[(i)^3 \langle\langle 1|\mu_{\mathbf{p}L}(t_4)G^{(3)}(t_4,t_3)\mathscr{L}_{\mathrm{x}}(t_3)G^{(2)}(t_3,t_2)\mathscr{L}_{\mathrm{v}}(t_2)G^{(1)}(t_2,t_1)\mathscr{L}_{\mathrm{v}}(t_1)|gg\rangle\rangle] \tag{15}$$

where

$$G^{(i)}(t,t') \equiv \mathscr{P}_i G(t,t')\mathscr{P}_i. \tag{16}$$

Inserting the definitions and expanding in eigenstates, we obtain

$$S_{\mathrm{PES}} = -2\int \mathrm{d}t_4\mathrm{d}t_3\mathrm{d}t_2\mathrm{d}t_1 \mathscr{I}\left[(i)^3 E_{\mathrm{x}}^*(t_4)E_{\mathrm{x}}(t_3)\sum_{II'}|\mu_{Ig}\mu_{\mathbf{p}I'I}|^2 G^{(3)}_{I_1'\mathbf{p},I_0}(t_4,t_3)G^{(2)}_{I_0,I_0}(t_3,t_2) \times\left\{E_{\mathrm{UV}}^*(t_2)E_{\mathrm{UV}}(t_1)G^{(1)}_{I_0,g}(t_2,t_1)+E_{\mathrm{UV}}(t_2)E_{\mathrm{UV}}^*(t_1)G^{(1)}_{g,I_0}(t_2,t_1)\right\}\right] \tag{17}$$

where we have made the rotating wave approximation (requiring that field excitation is accompanied by material de-excitation and *vice versa*), and defined:

$$G^{(i)}_{n,m}(t,t') \equiv G^{(i)}_{n'm',nm}(t,t')\delta_{n'n}\delta_{m'm} = \langle\langle n,m|G^{(i)}(t,t')|n,m\rangle\rangle \tag{18}$$

valid when $H_{I'I} = 0$. Non-adiabatic effects (which are frequently the objects of study in TRPES experiments) can cause electronic transitions and their inclusion therefore requires the full tetradic structure of the Green's functions. Since, in obtaining the semiclassical FGR result from the Liouville-space treatment, these considerations are only relevant for the middle propagation period $G^{(2)}(t_3,t_2)$, we ignore this complication on this first approach. To discuss the snapshot limit of pump–probe signals, it is convenient to utilize the Wigner spectrogram for the incoming pulses

$$E_i^*(t)E_i(t') = \int\frac{\mathrm{d}\omega}{2\pi}\mathscr{W}_i\left(\frac{t+t'}{2},\omega\right)\mathrm{e}^{i\omega(t-t')} \quad i\in\{\mathrm{x},\mathrm{v}\} \tag{19}$$

and make the approximation of ideal time and frequency resolution

$$\mathscr{W}_i(t,\omega) = |E_i|^2\delta(t-t_i)\delta(\omega-\omega_i). \tag{20}$$

Since the field-free Hamiltonian is time-independent, we Fourier transform with respect to the difference times $\tau_i \equiv t_{i+1} - t_i$ giving

$$S_{\mathrm{PES}}(T) = \frac{-2}{(2\pi)^2}|E_{\mathrm{x}}|^2|E_{\mathrm{v}}|^2\mathscr{I}\left[i\sum_{II'}|\mu_{Ig}\mu_{\mathbf{p}I'I}|^2 G^{(3)}_{I_1'\mathbf{p},I_0}(\omega_{\mathrm{x}})G^{(2)}_{I_0,I_0}(T)\right.$$
$$\left.\times\left\{G^{(1)}_{I_0,g}(\omega_{\mathrm{v}}) + G^{(1)}_{g,I_0}(-\omega_{\mathrm{v}})\right\}\right] \tag{21}$$

where we have defined $T = t_{\mathrm{x}} - t_{\mathrm{v}}$ and assumed that $G^{(2)}_{I_0,I_0}\left(T - \frac{\tau_1 + \tau_3}{2}\right) \approx G^{(2)}_{I_0,I_0}(T)$ since $T \gg \tau_1, \tau_3$. Following ref. 7, the coherence Green's function is

$$G_{n,m}(t) = \theta(t)\mathrm{e}^{-iH_{nL}t}\,\mathrm{e}^{iH_{mR}t}. \tag{22}$$

Using the relation

$$G_{n,m}(t) = G_{m,m}(t)\,\exp_+^{-i\int_0^t \mathrm{d}\tau U_L(\tau)} \tag{23}$$

where

$$U(t) \equiv \mathrm{e}^{iH_m(t)}(H_n - H_m)\mathrm{e}^{-iH_m(t)} \tag{24}$$

and $\exp_+$ is the positive time-ordered exponential. Eqn (22) includes all degrees of freedom at the Hamiltonian level (and thus does not include bath degrees of freedom in the Liouvillian sense). This facilitates the semiclassical simulation of the nuclear dynamics and avoids the complication of solvent degrees of freedom (which are readily included in the $L_{\mathrm{b}}$ operator). Recall that all parameters are functions of the time-dependent nuclear coordinates $\{\mathbf{q}\}(t)$ and the signal given in eqn (21) must be averaged over the nuclear space. In the static (classical) limit, we neglect nuclear motions when the system is in an electronic coherence and $U(\tau)$ becomes time-independent

$$G_{n,m}(t) \approx \mathrm{e}^{-iU_Lt}. \tag{25}$$

Note that this is an operator only in the nuclear subspace and under the semiclassical approximation, is an ordinary exponential. In this limit, we have for the TRPES signal

$$S_{\mathrm{PES}}(\varepsilon_{\mathbf{p}}, T) = \sum_{I_1I_0}\rho_{I_0}(T)\left|E_{\mathrm{x}}\mu_{\mathbf{p}I_1I_0}\right|^2\delta(\omega_{\mathrm{x}} - \mathscr{E}_{I_1}(\{\mathbf{q}\}(T)) + \mathscr{E}_{I_0}(\{\mathbf{q}\}(T)) - \varepsilon_{\mathbf{p}}) \tag{26}$$

where we have explicitly notated that the time dependence is through the set of nuclear coordinates $\{\mathbf{q}\}$ and

$$\rho_{I_0}(T) \equiv \sum_{I_0'}\left|E_{\mathrm{v}}\mu_{I_0'g}\right|^2\delta\left(\omega_{\mathrm{v}} - \mathscr{E}_{I_0'}(\{\mathbf{q}\}(0))\right)G^{(2)}_{I_0I_0,I_0'I_0'}(T) \tag{27}$$

is the population of the $I$th state of $H(0)$ at time $T$ including the effects of vibronic coupling. Neglecting these effects amounts to the substitution $G^{(2)}_{I_0I_0,I_0'I_0'}(T) = G^{(2)}_{I_0,I_0}(T)\delta_{I_0,I_0'}$. This represents the classical limit of the TRPES current. For practical simulations, the quantity $G^{(2)}_{I_0I_0,I_0'I_0'}(T)$ is taken to be the population of the state initially excited to $I_0'$ and propagated along a trajectory for a time $T$ ($I_0$ is the final state to which the system is propagated) and is averaged over trajectories.

 

The finite widths of the pump and probe pulses can be included after the fact by convoluting with the field envelopes. This amounts to smearing the interaction time but keeping the bra and ket interaction times the same and neglecting electronic coherences. In cases where there is appreciable ultrafast nuclear motion (when the system is still in an electronic coherence), it will be necessary to go beyond this approximation and treat the coherence Green's function more exactly. Clearly, eqn (17) provides the most direct way to go beyond the static (classical) limit.

We represent the field amplitudes as the product of an envelope and a carrier frequency

$$E_i(t) = \tilde{E}_i(t - t_i)\mathrm{e}^{-i\Omega_i(t - t_i)} + \mathrm{c.c.},\ i \in \{\mathrm{x,v}\} \tag{28}$$

where $\Omega_i$ is the carrier frequency for the pulse and $\tilde{E}i(t)$ is the envelope (centered at $t = 0$). The time between the centers of the pulses is then $T = t_\mathrm{x} - t_\mathrm{v}$. To go beyond the static approximation, we allow for coupling between electronic levels, leaving the Green's function no longer diagonal.

$$\begin{aligned} S_\mathrm{PES}(\varepsilon_\mathbf{p}, T) = \mathscr{I}\Bigg[ & \sum_{\{I_0,I_1\}} \int_0^\infty \mathrm{d}\tau_3 \int_0^\infty \mathrm{d}\tau_2 \int_0^\infty \mathrm{d}\tau_1 \mu^*_{\mathbf{p}I_1' I_0'''} \mu_{\mathbf{p}I_1 I_0''} G^{(3)}_{I_1'\mathbf{p}I_0''',I_1\mathbf{p}I_0''}(\tau_3) G^{(2)}_{I_0''' I_0'', I_0' I_0'}(\tau_2) \\ & \times \left\{ \mu_{gI_0'}\mu_{I_0 g} G^{(1)}_{I_0' g, I_0 g}(\tau_1) + \mu_{gI_0}\mu_{I_0' g} G^{(1)}_{gI_0', gI_0}(\tau_1) \right\} \\ & \times \int \mathrm{d}t_4 \tilde{E}_\mathrm{x}(t_4)\tilde{E}_\mathrm{x}(t_4 - \tau_3)\tilde{E}_\mathrm{v}(t_4 - \tau_3 - \tau_2)\tilde{E}_\mathrm{v}(t_4 - \tau_3 - \tau_2 - \tau_1) \Bigg] \end{aligned} \tag{29}$$

where $\sum_{\{I_0,I_1\}}$ indicates summation over the set of state indices. As a first approximation to the extensions to the FGR expression, we treat the electronic gap coordinate as a function rather than an operator (treating the nuclear coordinates semiclassically) so that the time-ordered exponential reduces to a standard exponential. We then utilize eqn (23) to express the coherence Green's functions in terms of population Green's functions which have direct classical analogs. This results in

$$\begin{aligned} S_\mathrm{PES}(\varepsilon_\mathbf{p}, T) = \mathscr{I}\Bigg[ & \sum_{\{I_0,I_1\}} \int_0^\infty \mathrm{d}\tau_3 \int_0^\infty \mathrm{d}\tau_2 \int_0^\infty \mathrm{d}\tau_1 \mu^*_{\mathbf{p}I_1' I_0'''} \mu_{\mathbf{p}I_1 I_0''} \\ & \times \left\{ \mu_{gI_0'}\mu_{I_0 g}\,\mathrm{e}^{i(-\overline{U}_1 + \Omega_\mathrm{v})\tau_1} + \mu_{gI_0}\mu_{I_0' g}\,\mathrm{e}^{i(\overline{U}_1 - \Omega_\mathrm{v})\tau_1} \right\} \mathrm{e}^{i(\overline{U}_3 + \Omega_\mathrm{x} - \varepsilon_\mathbf{p})\tau_3} \\ & \times G^{(2)}_{I_0''' I_0''', I_0'' I_0''}(\tau_3) G^{(2)}_{I_0'' I_0'', I_0' I_0'}(\tau_2) G^{(2)}_{I_0' I_0', I_0 I_0}(\tau_1) \\ & \times \int \mathrm{d}t_4 \tilde{E}_\mathrm{x}(t_4)\tilde{E}_\mathrm{x}(t_4 - \tau_3)\tilde{E}_\mathrm{v}(t_4 - \tau_3 - \tau_2)\tilde{E}_\mathrm{v}(t_4 - \tau_3 - \tau_2 - \tau_1) \Bigg] \end{aligned} \tag{30}$$

where the auxilliary quantities

$$\overline{U}_1 \equiv \frac{1}{\tau_1}\int_0^{\tau_1} \mathrm{d}\tau' \mathscr{E}_{I_0 \to I_0'}(\tau') - \mathscr{E}_g(\tau') \quad \overline{U}_3 \equiv \frac{1}{\tau_3}\int_0^{\tau_3} \mathrm{d}\tau' \mathscr{E}_{I_1 \to I_1'}(\tau') - \mathscr{E}_{I_0'' \to I_0'''}(\tau') \tag{31}$$

are the time-averaged electronic gap coordinates for time periods $\tau_1$ and $\tau_3$ where the time averaging is done for each trajectory (symbolized by $\mathscr{E}_{i\to f}$). As a reminder, we have assumed the Franck–Condon approximation without which the transition dipoles would depend parametrically on time through their dependence on the nuclear coordinates. All Green's functions remain operators in the nuclear subspace and so eqn (26) and (30) must be averaged over these degrees of freedom (although this is not explicitly notated in the above).

For cases in which environmental and vibronic couplings between excited states are particularly simple, it is frequently more convenient to treat the problem in Hilbert space. We therefore give the TRPES signal in this form as well for completeness.

$$S_{\rm PES}(\varepsilon_{\bf p}, T) = \mathscr{I}\Bigg[\sum_{\{I_0,I_1\}}\int {\rm d}t_4{\rm d}t_3{\rm d}t_3{\rm d}t_1 \tilde{E}_{\rm x}(t_4)\tilde{E}_{\rm x}(t_3)\tilde{E}_{\rm v}(t_2)\tilde{E}_{\rm v}(t_1)$$

$$\times\left\langle \mu_{gI_0'}(t_2)\mathscr{G}^{-}_{I_0',I_0'''}(t_2,t_4)\mu^{*}_{{\bf p}I_1'I_0'''}(t_4)\mathscr{G}^{+}_{I_1'{\bf p},I_1{\bf p}}(t_4,t_3)\mu_{{\bf p}I_1I_0''}(t_3)\mathscr{G}^{+}_{I_0'',I_0}(t_3,t_1)\mu_{I_0g}(t_1)\right\rangle\Bigg] \quad (32)$$

where

$$\mathscr{G}^{+}_{ij}(t,t') = \theta(t-t')\langle i|\exp_{+}^{-i\int_t^{t'}H(\tau)d\tau}|j\rangle, \quad \mathscr{G}^{-}_{ij}(t,t') = \theta(t'-t)\langle i|\exp_{-}^{i\int_t^{t'}H(\tau)d\tau}|j\rangle \quad (33)$$

are the Hilbert space forward and backward propagators ($\exp_{-}$ is the negative time-ordered exponential) and we have explicitly notated the time dependence of the transition dipole moments (which is only relevant outside the Franck–Condon approximation) and the averaging over the nuclear subspace (represented by $\langle\ldots\rangle$).

## 4. Conical intersections in *trans*-acrolein

We present the simulation for a valence TRPES of *trans*-acrolein (propenal). The molecule has a conjugated π-system composed from the C=C double bond and the carbonyl group (C=O) making it an ideal system to investigate non-adiabatic excited state dynamics. Quantum chemistry calculations were carried out with MOLPRO[36] at the complete active space self-consistent field (CASSCF) level of theory (sa4-CAS(6/5)/6-31+G*) with 6 electrons in 5 active orbitals. The molecular orbitals (MOs) occupied in the electronic ground state are two π-orbitals for the CC and the CO bond as well as the oxygen lone pair. The lowest unoccupied molecular orbitals (LUMOs) correspond to the first two π*-orbitals. At this level of theory, 4 states have to be taken into account. The $S_2$ and $S_3$ states are separated by ≈0.65 eV at the Franck–Condon point. Moreover, the $S_2$ and $S_3$ states are coupled by a conical intersection which is in reach within the first 25 fs of the time

Table 1 Calculated excitation energies at the Franck–Condon point in eV. The experimental value is taken from ref. 37

| Transition | Calc. | Exp. |
|---|---|---|
| $S_0 \to S_1$ (n → π*) | 3.6279 | |
| $S_0 \to S_2$ (n → π*) | 6.8740 | |
| $S_0 \to S_3$ (π → π*) | 7.5165 | 6.41(193 nm) |

 

evolution. The excitation energies calculated at CAS level are given in Table 1. A vertical, resonant UV excitation is most favorable for the $S_0 \rightarrow S_3$ transition with a transition dipole moment of 0.18 au as opposed to the $S_0 \rightarrow S_2$ transition, which is much weaker ($\mu_{S_2,S_0} = 0.044$ au).

The nuclear dynamics triggered by the UV excitation was calculated by the Tully surface hopping (SH) algorithm[38] (NEWTON-X[39] with a home made interface to MOLPRO). The nuclear dynamics is described by the classical equations of motion, while the electronic degrees of freedom are treated quantum mechanically (sa4-CAS(5/6)/6-31+G*). The forces acting on the nuclei are derived from the gradient of a particular electronic state the trajectory is in at time $T$. We interpret the population $\rho_{I_0,I_0}$ of an electronic state $I_0$ as the average over all $N$ trajectories:

$$\rho_{I_0,I_0}(t) = \frac{1}{N}\sum_{i=1}^{N}\delta_{s_i(t),I_0}, \tag{34}$$

where $s_i(t)$ is the electronic state of trajectory $i$. The initial state populated by the UV pulse is the electronic $S_3$ state. We assume sudden excitation, thereby neglecting possible Franck–Condon factors. All states are coupled to each other by CoIn and consecutive state jumps back to the ground state are observed in our trajectory simulations as well as under experimental conditions.[37] The time evolution of $\rho_{I_0,I_0}$ is shown in Fig. 2 for the first 500 fs. The $S_3$ state undergoes a rapid decay *via* a close by non-adiabatic coupling to the $S_2$ state transferring half of its population into the $S_2$ state. The $S_1$ state is populated in a next step and reaches about 40% within 250 fs.

We use the semiclassical (FGR) expression for the TRPES spectra obtained in eqn (26). The population Green's function from eqn (27) is realized by the SH algorithm on the excited neutral states. In the SH method the electronic state

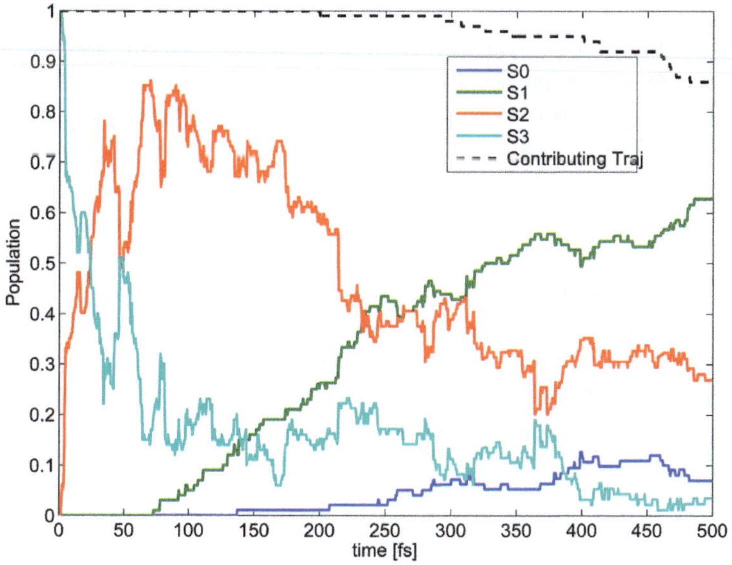


Fig. 2 Population of electronic states *vs.* time. The total number of contributing trajectories is 100 generated from a Wigner distribution at 0 K. The initial state is the $S_3$ state.

populations are obtained by averaging over a large number of trajectories. The TRPES signal $S_{\text{PES}}$ is then given by the sum over all $N$ trajectories:

$$S_{\text{PES,SH}} = \sum_{n=1}^{N} \sum_{I_1} \left|E_{\text{x}}\mu_{\mathbf{p}I_1I_0}\right|^2 \delta\left(\omega_{\text{x}} - \mathscr{E}_{I_1}(T) + \mathscr{E}_{I_0}(T) - \varepsilon_{\mathbf{p}}\right)\rho_{I_0,n}(T). \quad (35)$$

Note that the final state of the propagation ($I_0$) is not summed over since this averaging process is accounted for in the trajectory summation. Every trajectory is in a specific state $I_0$ at time $T$. Moreover it is assumed that the UV excitation pulse is resonant with a specific electronic state ($I_0'$) at $T = 0$ to yield the initial condition (state) for all trajectories.

$$\rho_{I_0,n}(T) \equiv \left|E_{\text{v}}\mu_{I_0g}\right|^2 G^{(2)}_{I_0I_0,I_0'I_0'}(T) \quad (36)$$

The X-ray ionization pulse is given a finite width in frequency by replacing the $\delta$ in function in eqn (35) by a Gaussian giving the final expression:

$$S_{\text{PES,SH}} \approx \sum_{n=1}^{N} \sum_{I_1} \rho_{I_0,n}(T)\left|E_{\text{x}}\mu_{\mathbf{p}I_1I_0}\right|^2 \text{e}^{-\left(\omega_{\text{x}} - \mathscr{E}_{I_1}(T) + \mathscr{E}_{I_0}(T) - \varepsilon_{\vec{p}}\right)^2 / 2\sigma_\omega^2}. \quad (37)$$

To incorporate the ionized states into the simulations the corresponding ionic doublet states $D_0$–$D_4$ are calculated at a similar level of theory (sa5-CAS(5/5)/6-31+G*) as the neutral states. The different size of the active space and the fact that CASSCF is not a consistent size introduces a small but negligible systematic error. The transition dipole moments $\mu_{\mathbf{p}I_1I_0}$ from the neutral to ionic states are approximated by utilizing Dyson orbitals:[14,40,41]

$$\Phi^{\text{D}}_{I_0,I_1}(\vec{r}) = \sqrt{N}\int \text{d}\vec{r}_1 \ldots \text{d}\vec{r}_{N-1} \Psi_{I_0}\left(\vec{r}_1 \ldots \vec{r}_N\right)\Psi_{I_1}\left(\vec{r}_1 \ldots \vec{r}_{N-1}\right) \quad (38)$$

where $\Psi_{I_0}$ is the electronic wave function of the neutral molecule and $\Psi_{I_1}$ the ion wave function for a particular snapshot. The norm of the Dyson orbital $\langle \Phi^{\text{D}}_{I_0,I_1} | \Phi^{\text{D}}_{I_0,I_1} \rangle$ is proportional to the ionization probability.[14,42] In a molecular orbital (MO) basis $\Phi^{\text{D}}_{I_0,I_1}$ can be written as:

$$\Phi^{\text{D}}_{I_0,I_1}(\vec{r}) = \sum_{\alpha} \phi_\alpha(\vec{r}) \langle \Psi_{I_1} | a_\alpha | \Psi_{I_0} \rangle. \quad (39)$$

Here $a_\alpha$ is an annihilation operator which removes an electron from the MO $\phi_\alpha$. In the case of a CASSCF wave function the index $\alpha$ runs over all orbitals in the active space and $\Phi^{\text{D}}$ can be expanded in the Slater determinants $\psi_{I_0,i}$ and $\psi_{I_1,j}$ and their corresponding configuration interaction (CI) coefficients $c_{I_0,i}$ and $c_{I_1,j}$:[14]

$$\Phi^{\text{D}}_{I_0,I_1}(\vec{r}) = \sum_{\alpha} \phi_\alpha(\vec{r}) \sum_{i,j} c_{I_0,i} c_{I_1,j} \langle \psi_{I_1,j} | a_\alpha | \psi_{I_0,i} \rangle. \quad (40)$$

The overlap factor between the different configurations in eqn (40) is determined by calculating the overlap between the Slater determinants of the ionic and neutral states:

$$\langle \psi_{I_1,j} | a_\alpha | \psi_{I_0,i} \rangle = \langle \phi'_1 \ldots \phi'_{\alpha-1} \phi'_{\alpha+1} \ldots \phi'_N | \phi_1 \ldots \phi_{\alpha-1} \phi_{\alpha+1} \ldots \phi_N \rangle = \det(\langle \phi_n' | \phi_m \rangle) \quad (41)$$

where $\phi_n'$ and $\phi_m$ are the MOs of the ionic and neutral wave function, respectively. The MOs of the neutral states form an orthonormal set and the norm of the Dyson orbitals, giving the weight factor for the ionization probability, can be written as:

$$\left\langle \Phi^{\mathrm{D}}_{I_0,I_1} | \Phi^{\mathrm{D}}_{I_0,I_1} \right\rangle = \sum_\alpha \left| \left\langle \Psi_{I_1} | a_\alpha | \Psi_{I_0} \right\rangle \right|^2. \quad (42)$$

By using the norm of the Dyson orbitals a possible dependence of the transition dipole moment on the photo electron kinetic energy is neglected as well as any directional dependence. Combining eqn (37) and (42) the approximated photo electron signal can then be written as:

$$S_{\mathrm{PES,SH}} \approx \sum_{n=1}^{N} \sum_{I_1} \rho_{I_0,n}(T) |E_{\mathrm{x}}|^2 \left\langle \Phi^{\mathrm{D}}_{I_0,I_1} | \Phi^{\mathrm{D}}_{I_0,I_1} \right\rangle \mathrm{e}^{-\left(\omega_{\mathrm{x}} - \mathscr{E}_{I_1}(T) + \mathscr{E}_{I_0}(T) - \varepsilon_{\vec{p}}\right)^2 / 2\sigma_\omega^2}. \quad (43)$$

The photoelectron spectrum calculated with eqn (43) is shown in Fig. 3. The photon energy of the pulse is chosen to be 12 eV (103 nm) with a FWHM of 0.71 eV which corresponds to a 5 fs temporal width. The shape of the spectrum reproduces the main features of the trajectory calculations as shown in Fig. 2. This is the fast decay to the $S_2$ state within the first 25 fs after the UV excitation and the increase in population in $S_1$ between 150 and 250 fs. The population in $S_1$ can attributed to the band around 2 eV, while the $S_2$ state population is visible through the band around 4.5 eV. Over the whole time evolution the variation due to nuclear motion is clearly visible in the TRPES. From these fast features the improvements to the TRPES model introduced in the previous sections can be directly motivated. If several femtosecond excitation and ionization pulses are used then corrections to the semiclassical limit can be substantial. A direct propagation approach for the fast molecular vibrations

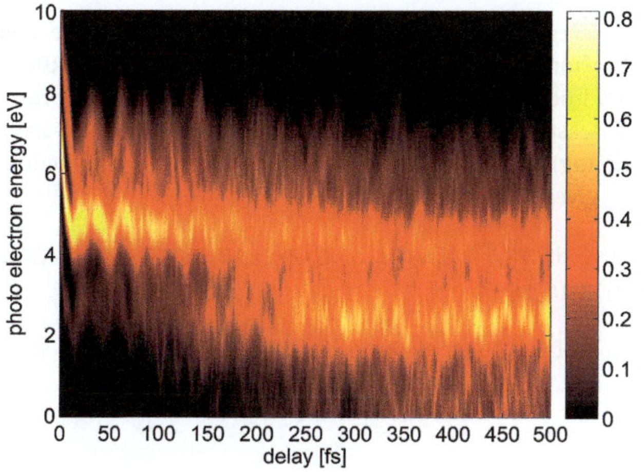


Fig. 3 TRPES for an ionization pulse with a photon energy of 12 eV and a FWHM of 5 fs. The intensity is given in arbitrary units.

(*e.g.* C–H stretch vibrations) which treats them by a full quantum model even opens up the route for coupling TRPES with vibrational spectroscopy methods like FSRS.

## 5. Time-resolved Auger signals

The above description of TRPES is suitable for the photoionization of both valence and core electrons. However, a treatment of core TRPES should distinguish between core and valence holes. Obviously, this is necessary for a description of Auger decay and Auger-electron spectroscopy. For this reason, in this section, the $q$ subscript on the molecular states will be reserved for core holes. States with two fewer valence electrons than the neutral molecule (dicationic valence states) will then be denoted by $\tilde{I}$.

In order to describe the Auger decay of core holes generated *via* TRPES processes, we add the following terms to the Hamiltonian:

$$H_{\mathbf{k}} = \sum_{\mathbf{k}} \varepsilon_{\mathbf{k}} c_{\mathbf{k}}^{\dagger} c_{\mathbf{k}} \quad A = \sum_{\mathbf{k}\tilde{I}I} A_{\mathbf{k}\tilde{I}I} c_{\mathbf{k}}^{\dagger} |\tilde{I}_0\rangle\langle I_1| + \text{h.c.} \tag{44}$$

where $H_{\mathbf{k}}$ is the Hamiltonian of the Auger electron continuum states (which, following ref. 23, we take to be energetically well-separated from any photoelectron states so that their Hamiltonians commute) and the $A$-operator induces Auger transitions. For clarity, this is represented in a fully microscopic manner with a fermionic second-quantized description of the molecular states (core and valence levels) in the Appendix. For the moment, we leave the expression in this more model-independent form. As was done for TRPES, the signal is defined as the integrated electron current

$$S_A = \int \frac{\mathrm{d}}{\mathrm{d}t} \langle n_{\mathbf{k}}(t) \rangle = -i[n_{\mathbf{k}}, H_T] = -2\mathscr{I}[A_{\mathbf{k}}] \tag{45}$$

where

$$A_{\mathbf{k}} \equiv \sum_{\tilde{I}I} A^{*}_{\mathbf{k}\tilde{I}I} c_{\mathbf{k}} |I_1\rangle\langle\tilde{I}_0|. \tag{46}$$

The Auger electrons originate from decay of the TRPES core hole and the description therefore contains more propagators compared to TRPES. Since the Auger states are initially unoccupied, the signal vanishes to zero order and must be expanded at least once in $A_{\mathbf{k}}$. Keeping only this term (which is second order in the Auger coupling) and expanding to second order in the interactions with the external pulses gives

$$S_A = -2\int \mathrm{d}t_6 \ldots \mathrm{d}t_1 \mathscr{I}[(i)^5 \langle\langle 1|A_{\mathbf{k}} G(t_6,t_5) A_{\mathbf{k}}^{\dagger} G(t_5,t_4) \mathscr{L}_{\mathrm{x}}(t_4) G(t_4,t_3) \mathscr{L}_{\mathrm{x}}(t_3) G(t_3,t_2) \mathscr{L}_{\mathrm{v}}(t_2) G(t_2,t_1) \mathscr{L}_{\mathrm{v}}(t_1)|gg\rangle\rangle]. \tag{47}$$

The inclusion of the Auger electron necessitates an additional state label

$$|\tilde{I}\rangle|\mathbf{p}\rangle|\mathbf{k}\rangle \equiv |\tilde{I}\mathbf{pk}\rangle. \tag{48}$$

As described above, when discussing core TRPES and Auger-electron spectroscopy, we use $I_q$ to denote states of the molecule that have $q$ core holes and the same number of valence electrons as the neutral molecule. In Auger processes (Fig. 4), we must also consider states of dicationic valence (valence states with two fewer electrons than valence states of the neutral molecule). Such states will be labeled with a "tilde" $\tilde{I}_q$. In addition to the projection operators defined previously, we introduce:

$$\mathscr{P}_3 = \sum_{I_0 I_1 \mathbf{p}} |I_1\mathbf{p}, I_0\rangle\rangle\langle\langle I_1\mathbf{p}, I_0| + |I_0 I_1\mathbf{p}\rangle\rangle\langle\langle I_0, I_1\mathbf{p}| \tag{49}$$

$$\mathscr{P}_4 = \sum_{I_1 I_1'\mathbf{p}} |I_1\mathbf{p}, I_1'\mathbf{p}\rangle\rangle\langle\langle I_1\mathbf{p}, I_1'\mathbf{p}|$$

$$\mathscr{P}_5 = \sum \sum_{I_1 \tilde{I}_0 \mathbf{pk}} |\tilde{I}_0\mathbf{pk}, I_1\mathbf{p}\rangle\rangle\langle\langle \tilde{I}_0\mathbf{pk}, I_1\mathbf{p}|$$

With which the signal is then expressed as

$$S_A = -2\int dt_6 dt_5 dt_4 dt_3 dt_2 dt_1 \mathscr{I}[(i)^5 \langle\langle 1|A_\mathbf{k} G^{(5)}(t_6,t_5) A_\mathbf{k}^\dagger G^{(4)}(t_5,t_4)\mathscr{L}_\mathrm{x}(t_4) G^{(3)}(t_4,t_3)\mathscr{L}_\mathrm{x}(t_3) G^{(2)}(t_3,t_2)\mathscr{L}_\mathrm{v}(t_2) G^{(1)}(t_2,t_1) \mathscr{L}_\mathrm{v}(t_1)|gg\rangle\rangle] \tag{50}$$

with $G^{(i)}$ defined as before. Substituting the interaction Liouvillians and formally evaluating matrix elements gives

$$\begin{aligned} S_A = -2\int & dt_6 dt_5 dt_4 dt_3 dt_2 dt_1 \mathscr{I}\left[(i)^5 \int d\varepsilon_\mathbf{p} \sum_{I_0 I_1 I_1' \tilde{I}_0} |\mu_{I_0 g}|^2 \right. \\ & \times A_{\mathbf{k}\tilde{I}_0 I_1} A^*_{\mathbf{k}\tilde{I}_0 I_1'} \mu_{\mathbf{p}I_1 I_0} \mu^*_{\mathbf{p}I_1' I_0} G^{(5)}_{\tilde{I}_0\mathbf{pk}, I_1'\mathbf{p}}(t_6, t_5) G^{(4)}_{I_1\mathbf{p}, I_1'\mathbf{p}}(t_5, t_4) \\ & \times \left\{ E^*_\mathrm{x}(t_4) E_\mathrm{x}(t_3) G^{(3)}_{I_1\mathbf{p}, I_0}(t_4, t_3) + E^*_\mathrm{x}(t_4) E_\mathrm{x}(t_3) G^{(3)}_{I_0, I_1\mathbf{p}}(t_4, t_3) \right\} G^{(2)}_{I_0, I_0}(t_3, t_2) \\ & \left. \times \left\{ E^*_\mathrm{v}(t_2) E_\mathrm{v}(t_1) G^{(1)}_{I_0, g}(t_2, t_1) + E_\mathrm{v}(t_2) E^*_\mathrm{v}(t_1) G^{(1)}_{g, I_0}(t_2, t_1) \right\}\right] \end{aligned} \tag{51}$$

where we have taken the Green's functions to be diagonal for now. Changing to delay times and employing the snapshot limit of pulses as before, this becomes

$$\begin{aligned} S_A = -2\int & dt_6 d\tau_5 |E_\mathrm{x} E_\mathrm{v}|^2 \mathscr{I}\left[(i)^3 \sum_{I_0 I_1 I_1' \tilde{I}_0 \mathbf{p}} |\mu_{I_0 g}|^2 A_{\mathbf{k}\tilde{I}_0 I_1} A^*_{\mathbf{k}\tilde{I}_0 I_1'} \mu_{\mathbf{p}I_1 I_0} \mu^*_{\mathbf{p}I_1' I_0} G^{(5)}_{\tilde{I}_0\mathbf{pk}, I_1'\mathbf{p}}(\tau_5) \right. \\ & \left. G^{(4)}_{I_1\mathbf{p}, I_1'\mathbf{p}}(t_6 - t_x - \tau_5) \left\{ G^{(3)}_{I_1\mathbf{p}, I_0}(\omega_\mathrm{x}) + G^{(3)}_{I_0, I_1'\mathbf{p}}(-\omega_\mathrm{x}) \right\} G^{2}_{I_0, I_0}(T) \left\{ G^{(1)}_{I_0, g}(\omega_\mathrm{v}) + G^{(1)}_{g, I_0}(-\omega_\mathrm{v}) \right\}\right] \end{aligned} \tag{52}$$

where we have Fourier transformed with respect to $\tau_1$ and $\tau_3$ and we have used the approximations $G^{(4)}_{I_1\mathbf{p}, I_1'\mathbf{p}}(t_6 - t_x - \tau_5 - \tau_3/2) \approx G^{(4)}_{I_1\mathbf{p}, I_1'\mathbf{p}}(t_6 - t_x - \tau_5)$ and

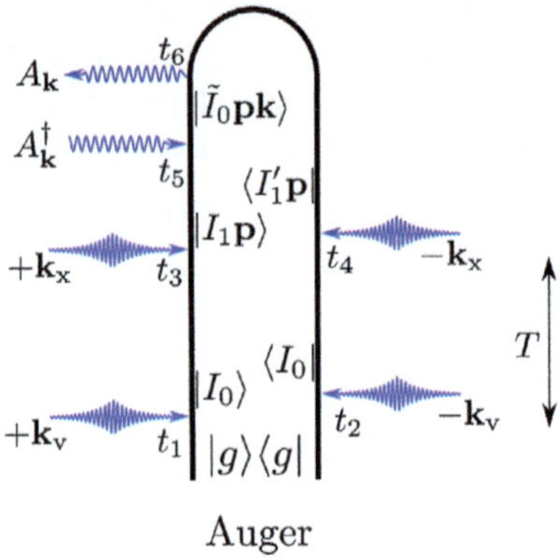


Fig. 4 Loop diagram for the Auger decay process. As before, the valence of both ket and bra are initially excited *via* an external pulse ($\mathbf{k}_{\mathrm{v}}$) and then photoionized *via* an X-ray pulse ($\mathbf{k}_{\mathrm{x}}$) following a delay time $T$. The system then evolves under the influence of the Hamiltonian including the Auger decay term and emits an Auger electron. This one loop diagram stands for 4 ladder diagram in Liouville space as each interaction with the external fields can be in either order (ket first and then bra or *vice versa*).

$G^{(2)}_{I_0,I_0}\left(T-\frac{\tau_1+\tau_3}{2}\right)\approx G^{(2)}_{I_0,I_0}(T)$ justified in the limit of ultrashort, well-separated excitation and ionization pulses. Substituting $G^{(4)}$ with its Fourier transform and carrying out the $\mathrm{d}t_6$ integration gives

$$S_A(\mathbf{k},T) = -2|E_{\mathrm{x}}E_{\mathrm{v}}|^2\mathscr{I}\left[i\sum_{I_0I_1I_1'\tilde{I}_0\mathbf{p}}|\mu_{I_0g}|^2A_{\mathbf{k}\tilde{I}_0I_1}A^*_{\mathbf{k}\tilde{I}_0I_1'}\mu_{\mathbf{p}I_1I_0}\mu^*_{\mathbf{p}I_1'I_0}G^{(5)}_{\tilde{I}_0\mathbf{pk},I_1'\mathbf{p}}(\omega=0)\right.$$
$$\left.G^{(4)}_{I_1\mathbf{p},I_1'\mathbf{p}}(\omega=0)\left\{G^{(3)}_{I_1\mathbf{p},I_0}(\omega_{\mathrm{x}})+G^{(3)}_{I_0,I_1'\mathbf{p}}(-\omega_{\mathrm{x}})\right\}G^{(2)}_{I_0,I_0}(T)\left\{G^{(1)}_{I_0,g}(\omega_{\mathrm{v}})+G^{(1)}_{g,I_0}(-\omega_{\mathrm{v}})\right\}\right] \tag{53}$$

where we have written the arguments of $G^{(4)}$ and $G^{(5)}$ so as to make clear that they are frequency-domain Green's functions. Since the ket and bra of $G^{(4)}_{I_1\mathbf{p},I_1'\mathbf{p}}$ are summed over the same manifold of states, their principal value vanishes and we make the substitution

$$G^{(4)}_{I_1\mathbf{p},I_1'\mathbf{p}}(\omega=0) = -i\pi\delta\big(\mathscr{E}_{I_1}(T)-\mathscr{E}_{I_1'}(T)\big) \tag{54}$$

which restricts the valence to being in a population during the $\tau_4$ time period. Combining the Green's functions for the periods $\tau_1$ and $\tau_3$ yields

$$S_A(\mathbf{k},T) = 2|E_{\mathrm{x}}E_{\mathrm{v}}|^2\int\mathrm{d}\varepsilon_{\mathbf{p}}\sum_{I_0I_1I_1'\tilde{I}_0}\left|\mu_{I_0'g}\mu_{\mathbf{p}I'I}A_{\mathbf{k}\tilde{I}I}\right|^2\delta\big(\varepsilon_{\mathbf{k}}+\mathscr{E}_{\tilde{I}_0}(T)-\mathscr{E}_{I_1}(T)\big)$$
$$\delta\big(\omega_{\mathrm{x}}+\mathscr{E}_{I_0}(T)-\mathscr{E}_{I_1}(T)-\varepsilon_{\mathbf{p}}\big)G^{(2)}_{I_0I_0,I_0'I_0'}(T)\delta\big(\omega_{\mathrm{v}}-\mathscr{E}_{I_0}(0)\big). \tag{55}$$

This is the semiclassical FGR result and is easily convoluted with spectral pulse envelopes, detector sensitivity functions, *etc.* to simulate the Auger spectra. As was done for TRPES, we have made several assumptions in arriving at this simplified expression. Principally, this takes the nuclei as frozen during electronic coherences. Although this is well-justified in the limit of ultrafast, well-separated pulses, it is certainly open to question in the case of excitation by overlapping or broader pulses during which processes the system may spend non-trivial amounts of time in electronic coherence. Eqn (51) provides an ideal starting point for going beyond this semiclassical limit.

$$
\begin{aligned}
S_A(\varepsilon_{\mathbf{k}}, T) = 2\mathscr{I}\Bigg[ i \int \mathrm{d}\varepsilon_{\mathbf{p}} \sum_{\{I_0, I_1 \bar{I}_0\}} \int \mathrm{d}t_6 \ldots \mathrm{d}t_1 \tilde{E}_{\mathrm{x}}(t_4)\tilde{E}_{\mathrm{x}}(t_3)\tilde{E}_{\mathrm{v}}(t_2)\tilde{E}_{\mathrm{v}}(t_1) \\
\times G^{(5)}_{\bar{I}_0\mathbf{kp}I_1''\mathbf{p},\bar{I}_0\mathbf{kp}I_1'\mathbf{p}}(t_6, t_5) G^{(4)}_{I_1'''\mathbf{p}I_1''\mathbf{p},I_1'\mathbf{p}I_1'\mathbf{p}}(t_5, t_4) \\
\times \Big\{ \mu^*_{\mathbf{p}I_1'I_0'''} \mu_{\mathbf{p}I_1 I_0''} G^{(3)}_{I_1'\mathbf{p}I_0''',I_1\mathbf{p}I_0''}(t_4, t_3) + \mu^*_{\mathbf{p}I_1 I_0''} \mu_{\mathbf{p}I_1' I_0''} G^{(3)}_{I_0'''I_1'\mathbf{p},I_0''I_1\mathbf{p}}(t_4, t_3) \Big\} \\
\times G^{(2)}_{I_0'''I_0'',I_0'I_0'}(t_3, t_2) \left\{ \mu_{gI_0'}\mu_{I_0 g} G^{(1)}_{I_0'g,I_0 g}(t_2, t_1) + \mu_{gI_0}\mu_{I_0' g} G^{(1)}_{gI_0',gI_0}(t_2, t_1) \right\} \Bigg]
\end{aligned}
\tag{56}
$$

where, for brevity, we have retained the interaction times ($t_i$) rather than the delay periods ($\tau_i$). As before, we may rewrite this in terms of population Green's functions and propagated phase factors representing the electronic energy gaps

$$
\begin{aligned}
S_A(\varepsilon_{\mathbf{k}}, T) = 2\mathscr{I}\Bigg[ i \int \mathrm{d}\varepsilon_{\mathbf{p}} \sum_{\{I_0, I_1 \bar{I}_0\}} \int \mathrm{d}t_6 \ldots \mathrm{d}t_1 \tilde{E}_{\mathrm{x}}(t_4)\tilde{E}_{\mathrm{x}}(t_3)\tilde{E}_{\mathrm{v}}(t_2)\tilde{E}_{\mathrm{v}}(t_1)\, \mathrm{e}^{i(\overline{U}_5 + \varepsilon_{\mathbf{k}})(t_6 - t_5)} \\
\times G^{(4)}_{I_1'''I_1''',I_1''I_1''}(t_6, t_5) G^{(4)}_{I_1''I_1'',I_1'I_1'}(t_5, t_4) \\
\times \Big\{ \mu^*_{\mathbf{p}I_1'I_0'''} \mu_{\mathbf{p}I_1 I_0''}\, \mathrm{e}^{i(\overline{U}_3 + \Omega_{\mathrm{x}} - \varepsilon_{\mathbf{p}})(t_4 - t_3)} + \mu^*_{\mathbf{p}I_1 I_0''} \mu_{\mathbf{p}I_1' I_0''}\, \mathrm{e}^{-i(\overline{U}_3 + \Omega_{\mathrm{x}} - \varepsilon_{\mathbf{p}})(t_4 - t_3)} \Big\} \\
\times G^{(2)}_{I_0'''I_0''',I_0''I_0''}(t_4, t_3) G^{(2)}_{I_0''I_0'',I_0'I_0'}(t_3, t_2) G^{(2)}_{I_0'I_0',I_0 I_0}(t_2, t_1) \\
\times \Big\{ \mu_{gI_0'}\mu_{I_0 g}\, \mathrm{e}^{i(-\overline{U}_1 + \Omega_{\mathrm{v}})(t_2 - t_1)} + \mu_{gI_0}\mu_{I_0' g}\, \mathrm{e}^{i(\overline{U}_1 - \Omega_{\mathrm{v}})(t_2 - t_1)} \Big\} \Bigg]
\end{aligned}
\tag{57}
$$

where we have dropped the continuum state subscripts on the Green's functions since they are assumed to be independent of the nuclear coordinates. The $\bar{U}_{1(3)}$ are defined as before and

$$
\overline{U}_5 \equiv \frac{1}{t_6 - t_5} \int_{t_5}^{t_6} \mathrm{d}t' \mathscr{E}_{\bar{I}_0 \to I_0'}(t') - \mathscr{E}_{I_1' \to I_1''}(\tau')
\tag{58}
$$

is the time-averaged electronic energy gap between the core-hole states and the dicationic-valence states. Eqn (55) and (57) must still be averaged over nuclear degrees of freedom. This may be performed, as in TRPES, by propagating the system along trajectories and then averaging. The expressions are written so as to facilitate this approach.

# 6. Off-resonant X-ray scattering

X-ray scattering is commonly employed for monitoring the electronic and nuclear structure of materials.[43–45] In the resonant case, this leads to X-ray absorption spectroscopy as well as higher-order multidimensional techniques which are commonly employed to obtain information on the electronic structure and track the dynamics of wavepackets of nuclear excitations. In the off-resonant case, this leads to diffraction which is used to determine the static structure of complex systems as well as to monitor the motion of nuclei in non-equilibrium processes (*e.g.* crystal melting[46,47]).

In a recent work, we analyzed the scattering of off-resonant X-rays coupled to matter *via* the $A^2$ term in the minimal-coupling Hamiltonian and gave a unified treatment of time, frequency, and wavevector resolution through a rigorous incorporation of the photon-detection event.[30] The signal is taken to be the integrated field intensity at the detector

$$S_{\mathrm{OXS}}(\bar{\omega},\bar{t},\bar{\mathbf{r}},\bar{\mathbf{k}}) = \int dt \int d\mathbf{r} \langle \mathbf{E}^{(t\mathbf{r}f\mathbf{k})\dagger}(\mathbf{r},t)\mathbf{E}^{(t\mathbf{r}f\mathbf{k})}(\mathbf{r},t)\rangle \tag{59}$$

where the field superscripts indicate multiplication (in appropriate domains) by detector gating functions and the arguments of the signal are the central parameters of these functions (we leave the dependence on the detector widths implicit). Since the detected field mode was initially unoccupied, we model the interaction as spontaneous emission into vacuum field modes (labeled with s and s′) and reabsorption at the detector. The probe field (labeled with p and p′) enters through the magnetic vector potential **A**. Since the vector potential is linearly coupled to the system's charge density (labeled $\hat{\sigma}$), the bare OXS signal (before consideration of the detector effects) will come as

$$S_{\mathrm{OXS,B}}(t',\omega',\mathbf{r}',\mathbf{k}') \propto \sum_{ss'} \int d\mathbf{r}_1 d\mathbf{r}_1' \int dt_1 dt_1' \left\langle \left\langle \mathbb{1} \middle| \mathbf{E}_R^{(s')\dagger}(\mathbf{r}',t')\mathbf{E}_L^{(s)}(\mathbf{r}',t')\mathbf{A}_R^{(s')}(\mathbf{r}_1',t_1') \right.\right.$$
$$\left.\left. \times \mathbf{A}_R^{(p)\dagger}(\mathbf{r}_1',t_1')\sigma_R(\mathbf{r}_1',t_1')\mathbf{A}_L^{(s)\dagger}(\mathbf{r}_1,t_1)\cdot\mathbf{A}_L^{(p)}(\mathbf{r}_1,t_1)\sigma_L(\mathbf{r}_1,t_1) \middle| \rho_T(0) \right\rangle \right\rangle \tag{60}$$

where $|\rho_T(0)\rangle$ is the total (field and matter) density matrix at time 0 (taken to be the terminus of a process that leaves the system in a non-equilibrium state). This is illustrated diagramatically in Fig. 5. The signal is not generally related to the time-dependent, single-particle momentum-space charge density but rather to its correlation function.[27,48] A compact expression for the signal in the limit of ideal spatial and spectral detector resolution is

$$S_{\mathrm{OXS}}(\bar{\omega},\Lambda) \propto \int d\omega_{\mathrm{p}} d\omega_{\mathrm{p}'} E_{\mathrm{p}}(\omega_{\mathrm{p}}) E_{\mathrm{p}}^*(\omega_{\mathrm{p}'}) \langle \sigma_T(-\mathbf{q}',\omega_{\mathrm{p}'} - \bar{\omega})\sigma_T(\mathbf{q},\bar{\omega} - \omega_{\mathrm{p}})\rangle \tag{61}$$

where $\mathbf{q}^{(\prime)} \equiv \frac{\overline{\omega}}{c}\hat{\bar{\mathbf{r}}} - \mathbf{k}_{\mathrm{p}^{(\prime)}}$ is the momentum transfer (not to be confused with the nuclear coordinates of previous sections), $\hat{\bar{\mathbf{r}}}$ is the normalized position vector of the detector pixel relative to the sample center, and $\Lambda$ represents the pulse parameters. Assuming that the sample is composed of $N$ identical, non-interacting molecules (indexed by $\alpha$, $\beta$), the correlation function of the total charge density operator ($\sigma_T$) may be expanded as

 

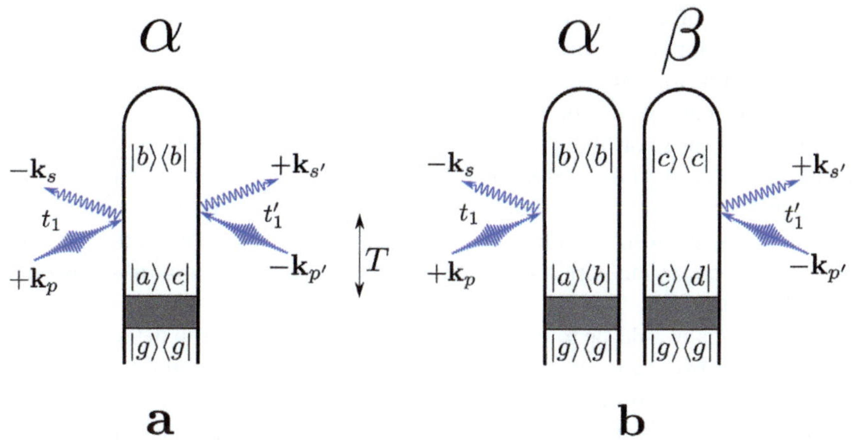


Fig. 5 Loop diagrams for incoherent (a) and coherent (b) X-ray scattering processes. The shaded area represents an unspecified process that prepares the system in an arbitrary state ($|g\rangle$ is the electronic ground state). We denote modes of the pump with p and p′ whereas s and s′ represent relevant scattering modes ($\mathbf{k}_{p^{(\prime)}}$ has frequency $\omega_{p^{(\prime)}}$ and $\mathbf{k}_{s^{(\prime)}}$ has frequency $\omega_{s^{(\prime)}}$). The time $T$ between the termination of this preparation process and the central time of the scattered pulse is shown *via* the arrow in the center of the figure. Elastic scattering corresponds to $\omega_{ab} = \omega_{bc} = 0$ (*i.e.* $\omega_{ac} = 0$) for the incoherent and $\omega_{bc} = \omega_{ed} = 0$ for the coherent contribution. Elastic scattering therefore originates from scattering off populations. For diagram rules, see ref. 52 and 53.

$$\langle\sigma_T(-\mathbf{q}',\omega-\omega_{p'})\sigma_T(\mathbf{q},\omega_p-\omega)\rangle = \sum_\alpha \mathrm{e}^{-i(\mathbf{q}-\mathbf{q}')\cdot\mathbf{r}_\alpha}\langle\sigma(-\mathbf{q}',\omega-\omega_{\mathrm{p}'})\sigma(\mathbf{q},\omega_p-\omega)\rangle$$
$$+\sum_\alpha\sum_{\beta\neq\alpha}\mathrm{e}^{-i(\mathbf{q}\cdot\mathbf{r}_\alpha-\mathbf{q}'\cdot\mathbf{r}_\beta)}\langle\sigma(-\mathbf{q}',\omega-\omega_{\mathrm{p}'})\rangle\langle\sigma(\mathbf{q},\omega_{\mathrm{p}}-\omega)\rangle \quad (62)$$

where $\mathbf{r}_\alpha$ is the position vector of the center of particle $\alpha$ and we have taken $\hat{\sigma}_T(\mathbf{r}) = \sum_\alpha \hat{\sigma}(\mathbf{r}-\mathbf{r}_\alpha)$ so that the total charge density is composed of a superposition of the (identical) charge densities associated with each molecule (assumed to be non-overlapping). We label the former (one-particle) terms in eqn (62) as incoherent and the latter (two-particle) terms as coherent, reflecting the way in which contributions from different molecules in these terms add together. Of particular note is that these two contributions come with different form factors. The incoherent form factor permits excitation (inelastic scattering) from the ground state while the coherent form factor only permits transitions between previously occupied states and is therefore purely elastic when considering scattering from the ground state. This is clearly seen from Fig. 5 since the coherent terms only have interactions on one side of the density matrix and therefore cannot cause transitions to unoccupied states.

X-ray diffraction is elastic and comes as the Fourier transform of the real-space charge density. Since the number of two-particle terms scales quadratically with the molecule number (*versus* the linear scaling of the one-particle terms), off-resonant scattering from the ground state is dominated by the elastic contribution. In the case of single-molecule samples or nonstationary initial states, the contribution from inelastic terms increases and can become substantial for certain values of **q**. The inelasticities are therefore overwhelmed by the elastic contributions for many-particle samples but are important in analyzing single-molecule scattering patterns. In particular, the off-resonant X-ray scattering from a single molecule cannot be represented as the ground state momentum-space charge density as in ordinary diffraction experiments. It is for this reason that we avoid the use of the term diffraction (which could be taken to imply this form for the signal) in favor of the more generic "scattering". This gives rise to the possibility of creating a coherent excitation *via* off-resonant scattering and probing this excitation with a second pulse, a process termed multidimensional X-ray diffraction.[49]

A future goal would be to introduce this level of theory to the detection event for photoelectron and Auger-electron experiments. This would require a quantum-field theoretical treatment of the continuum electrons but would facilitate the incorporation of propagation effects between the sample and detector. Such effects are necessary for a description of attosecond-resolution techniques such as the attosecond streak camera and the RABBIT technique.[2,50,51]

# 7. Conclusions

The electronic and nuclear structure and dynamics of molecules and materials are routinely probed by TRPES, AES, and OXS techniques. We have reported a Liouville-space description of these signals. The abundance of different degrees of freedom (core, valence, and continuum electronic states, vibrations, and solvent/environmental effects) is most easily handled in a Liouville-space framework since one may start with the full Liouvillian and approximate different degrees of freedom at different levels (retaining Hamiltonian evolution only for the fastest dynamics). For example, one can treat the electronic states and any nuclear coordinates that cause conical intersections at the Hamiltonian level while treating the remaining vibrational modes and the environment as a bath (which can be approximated in a variety of ways[7]). Moreover, the explicit incorporation of the detection event given in our account of OXS can be extended to TRPES and AES using a quantum field description of the electrons. This would then allow the formalism to handle propagation effects between the material and the detector such as multiple scatterings and external field effects as employed in the streak camera and RABBIT techniques to gain time resolution. Finally, all of these techniques leave the material system in a nontrivial excited state that can be the object of further study *via* traditional nonlinear spectroscopic techniques. Since, in Liouville space, the system (described by the $|\rho\rangle\rangle$ vector) travels only forward in time, the processes can be regarded in a modular fashion and further interactions are easily added to describe the probing of ionic states resulting from TRPES/AES processes.

# Appendix

### A: Liouville space

In Liouville space, we promote the density matrix $\rho$ to a vector denoted by a double ket $|\rho\rangle\rangle$. For every Hilbert space operator $O$, we associate Liouville space operators $O_L$ and $O_R$ that indicate action on the ket or bra of the density matrix respectively:

 

$$\begin{aligned} O_{\mathrm{L}}|\rho\rangle\rangle &\leftrightarrow O\rho \\ O_{\mathrm{R}}|\rho\rangle\rangle &\leftrightarrow \rho O \end{aligned} \tag{A1}$$

where the symbol "↔" indicates equivalence between Liouville and Hilbert space expressions. Additionally, we define an associated Liouvillian for every Hamiltonian defined above:

$$\mathscr{L} \equiv H_{-} \equiv H_{\mathrm{L}} - H_{\mathrm{R}}. \tag{A2}$$

This definition allows compact expressions for the commutations with the interaction Hamiltonians. In Hilbert space, operator expectation values are given by the trace of the operator product with the density matrix. The Liouville space analogy is

$$\langle\langle 1|O_{\mathrm{L}}|\rho\rangle\rangle \leftrightarrow \mathrm{Tr}[O\rho] \tag{A3}$$

where, as shorthand for the trace operator, we have used

$$\langle\langle 1| \equiv \sum_i \langle\langle ii| \tag{A4}$$

with $i$ indexing any complete basis.

## B: Second quantized fermionic description

In the main text, we left the description of the bound molecular states as arbitrary. In this section, we take some preliminary steps toward describing the states and operators used in the TRPES and Auger sections in a 2nd-quantized language in terms of the occupation numbers of the single-particle orbitals. Although this has already been done for the continuum states, we now examine the bound states in this manner as well. We will not specify a model for the Hamiltonian but will simply use this to clarify the form of the transition dipoles and the Auger decay operator. We index valence orbitals with l, m, n and the core orbitals with $\alpha$, $\beta$. In this basis, the Hamiltonians for the interaction with the valence pump and ionizing X-ray pulse are

$$H_{\mathrm{v}}(t) = -E_{\mathrm{v}}(t)\sum_{\mathrm{mn}} \mu_{\mathrm{mn}} c_{\mathrm{m}}^{\dagger} c_{\mathrm{n}} + \mathrm{c.c.} \quad H_{\mathrm{x}}(t) = -E_{\mathrm{x}}(t)\sum_{\mathbf{p}\alpha} \mu_{\mathbf{p}\alpha} c_{\mathbf{p}}^{\dagger} c_{\alpha} + \mathrm{c.c.} \tag{B1}$$

The $A$-operator (which controls the Auger decay of the core hole) is a generalized Coulomb matrix element and comes as

$$A = \sum_{\mathbf{k}\alpha\mathrm{mn}} A_{\mathbf{k}\alpha\mathrm{mn}} c_{\mathbf{k}}^{\dagger} c_{\alpha}^{\dagger} c_{\mathrm{m}} c_{\mathrm{n}} + \mathrm{c.c.} \tag{B2}$$

where

$$A_{\mathbf{k}\alpha\mathrm{mn}} = \int \mathrm{d}\mathbf{r}\mathrm{d}\mathbf{r}' \phi_{\mathbf{k}}(\mathbf{r})\phi_{\alpha}(\mathbf{r}') \frac{1}{|\mathbf{r}-\mathbf{r}'|} \phi_{\mathrm{m}}(\mathbf{r})\phi_{\mathrm{n}}(\mathbf{r}'). \tag{B3}$$

We may explicitly connect these to the forms given in the main text by supposing that we can construct the excited states from the ground state. We thus write

$$|I_0\rangle = \sum_{\mathrm{mn}} O_{I_0\mathrm{mn}} c^{\dagger}_{\mathrm{m}} c_{\mathrm{n}} |g\rangle \tag{B4}$$

$$|I_1\mathbf{p}\rangle = \sum_{\alpha I_0} O_{I_1\alpha I_0} c^{\dagger}_{\mathbf{p}} c_{\alpha} |I_0\rangle$$

$$|\tilde{I}_0\mathbf{pk}\rangle = \sum_{\alpha} O_{\tilde{I}_0\alpha\mathrm{mn}I_1} c^{\dagger}_{\mathbf{k}} c^{\dagger}_{\alpha} c_{\mathrm{m}} c_{\mathrm{n}} |I_1\mathbf{p}\rangle$$

where the $O$ coefficients are chosen to diagonalize the respective subspaces. With this we obtain the relationship between the transition dipoles and Auger elements in the two bases:

$$\mu_{I_0 g} = \sum_{\mathrm{mn}} O_{I_0\mathrm{mn}} \mu_{\mathrm{mn}} \tag{B5}$$

$$\mu_{\mathbf{p}I_1 I_0} = \sum_{\alpha} O_{I_1\alpha I_0} \mu_{\mathbf{p}\alpha}$$

$$A_{\mathbf{k}\tilde{I}_0 I_1} = \sum_{\alpha\mathrm{mn}} O_{\tilde{I}_0\alpha\mathrm{mn}I_1} A_{\mathbf{k}\alpha\mathrm{mn}}.$$

## Acknowledgements

The support of the Chemical Sciences, Geosciences, and Biosciences division, Office of Basic Energy Sciences, Office of Science, U.S. Department of Energy as well as from the National Science Foundation (grant CHE-1361516), and the National Institutes of Health (Grant GM-59230) is gratefully acknowledged. Kochise Bennett was supported by DOE.

## References

1 A. Stolow, A. E. Bragg and D. M. Neumark, *Chem. Rev.*, 2004, **104**, 1719.
2 M. Hentschel, R. Kienberger, C. Spielmann, G. A. Reider, N. Milosevic, T. Brabec, P. Corkum, U. Heinzmann, M. Drescher and F. Krausz, *Nature*, 2001, **414**, 509.
3 L. J. P. Ament, M. van Veenendaal, T. P. Devereaux, J. P. Hill and J. van den Brink, *Rev. Mod. Phys.*, 2011, **83**, 705.
4 P. Thibault and V. Elser, *Annu. Rev. Condens. Matter Phys.*, 2010, **1**, 237.
5 H. Meyer, *Annu. Rev. Phys. Chem.*, 2002, **53**, 141.
6 G. A. Worth and L. S. Cederbaum, *Annu. Rev. Phys. Chem.*, 2004, **55**, 127.
7 S. Mukamel, *Principles of Nonlinear Optical Spectroscopy*, Oxford University Press, USA, 1st edn, 1995.
8 M. Seel and W. Domcke, *J. Chem. Phys.*, 1991, **95**, 7806.
9 M. Ohno and G. Wendin, *J. Phys. B: At. Mol. Phys.*, 1979, **12**, 1305.
10 D. M. Neumark, *Annu. Rev. Phys. Chem.*, 2001, **52**, 255.

 

11 M. Chen, X. Wang, Y. Yu, Z. Pei, X. Bai, C. Sun, R. Huang and L. Wen, *Appl. Surf. Sci.*, 2000, **158**, 134.
12 H. Petek and S. Ogawa, *Prog. Surf. Sci.*, 1997, **56**, 239.
13 H. R. Hudock, B. G. Levine, A. L. Thompson, H. Satzger, D. Townsend, N. Gador, S. Ullrich, A. Stolow and T. J. Martínez, *J. Phys. Chem. A*, 2007, **111**, 8500.
14 A. L. Thompson and T. J. Martínez, *Faraday Discuss.*, 2011, **150**, 293.
15 T. S. Kuhlman, W. J. Glover, T. Mori, K. B. Møller and T. J. Martínez, *Faraday Discuss.*, 2012, **157**, 193.
16 B. P. Fingerhut, K. E. Dorfman and S. Mukamel, *J. Phys. Chem. Lett.*, 2013, **4**, 1933.
17 B. McFarland, J. Farrell, S. Miyabe, F. Tarantelli, A. Aguilar, N. Berrah, C. Bostedt, J. Bozek, P. Bucksbaum, J. Castagna, *et al.*, *Nat. Commun.*, 2014, **5**, 4235.
18 C.-M. Liegener, *J. Chem. Phys.*, 1983, **79**, 2924.
19 N. Ottosson, G. Öhrwall and O. Björneholm, *Chem. Phys. Lett.*, 2012, **543**, 1.
20 R. Manne and H. Ågren, *Chem. Phys.*, 1985, **93**, 201.
21 C.-M. Liegener, *Chem. Phys. Lett.*, 1982, **90**, 188.
22 M. Ohno, *Phys. Rev. B: Condens. Matter Mater. Phys.*, 1998, **58**, 12795.
23 O. Gunnarsson and K. Schönhammer, *Phys. Rev. B: Condens. Matter Mater. Phys.*, 1980, **22**, 3710.
24 L. Cederbaum and W. Domcke, *Adv. Chem. Phys.*, 1977, **36**, 205.
25 G. Wendin, *Breakdown of one-electron pictures in photoelectron spectra*, Springer, 1981.
26 L. Hedin, J. Michiels and J. Inglesfield, *Phys. Rev. B: Condens. Matter Mater. Phys.*, 1998, **58**, 15565.
27 W. Schulke, *Electron dynamics by inelastic X-ray scattering*, Oxford University Press, Oxford, New York, 2007.
28 J. Als-Nielsen and D. McMorrow, *Elements of modern X-ray physics*, Wiley, Hoboken, 2011.
29 C. Bressler and M. Chergui, *Chem. Rev.*, 2004, **104**, 1781.
30 K. Bennett, J. D. Biggs, Y. Zhang, K. E. Dorfman and S. Mukamel, *J. Chem. Phys.*, 2014, **140**, 204311.
31 J. Hajdu, *Curr. Opin. Struct. Biol.*, 2000, **10**, 569.
32 D. Starodub, A. Aquila, S. Bajt, M. Barthelmess, A. Barty, C. Bostedt, J. Bozek, N. Coppola, R. Doak, S. Epp, *et al.*, *Nat. Commun.*, 2012, **3**, 1276.
33 H. N. Chapman, *Nat. Mater.*, 2009, **8**, 299.
34 H. N. Chapman, A. Barty, M. J. Bogan, S. Boutet, M. Frank, S. P. Hau-Riege, S. Marchesini, B. W. Woods, S. Bajt and W. H. Benner, *Nat. Phys.*, 2006, **2**, 839–843.
35 S. Rahav and S. Mukamel, *Phys. Rev. A: At., Mol., Opt. Phys.*, 2010, **81**, 063810.
36 H.-J. Werner, P. J. Knowles, G. Knizia, F. R. Manby and M. Schütz, *Wiley Interdiscip. Rev.: Comput. Mol. Sci.*, 2012, **2**, 242.
37 A. M. D. Lee, J. D. Coe, S. Ullrich, M. L. Ho, S. J. Lee, B. M. Cheng, M. Z. Zgierski, I. C. Chen, T. J. Martinez and A. Stolow, *J. Phys. Chem. A*, 2007, **111**, 11948.
38 S. Hammes-Schiffer and J. C. Tully, *J. Chem. Phys.*, 1994, **101**, 4657.
39 M. Barbatti, M. Ruckenbauer, F. Plasser, J. Pittner, G. Granucci, M. Persico and H. Lischka, *Wiley Interdiscip. Rev.: Comput. Mol. Sci.*, 2014, **4**, 26.

40 T. Mori, W. J. Glover, M. S. Schuurman and T. J. Martinez, *J. Phys. Chem. A*, 2011, **116**, 2808.
41 C. M. Oana and A. I. Krylov, *J. Chem. Phys.*, 2007, **127**, 234106.
42 A. Perveaux, D. Lauvergnat, B. Lasorne, F. Gatti, M. A. Robb, G. J. Halász and A. Vibók, *J. Phys. B: At. Mol. Phys.*, 2014, **47**, 124010.
43 R. Neutze, R. Wouts, D. van der Spoel, E. Weckert and J. Hajdu, *Nature*, 2000, **406**, 752.
44 H. N. Chapman, P. Fromme, A. Barty, T. A. White, R. A. Kirian, A. Aquila, M. S. Hunter, J. Schulz, D. P. DePonte and U. Weierstall, *Nature*, 2011, **470**, 73.
45 R. Koopmann, K. Cupelli, L. Redecke, K. Nass, D. P. DePonte, T. A. White, F. Stellato, D. Rehders, M. Liang, J. Andreasson, *et al.*, *Nat. Methods*, 2012, **9**, 259.
46 C. W. Siders, A. Cavalleri, K. Sokolowski-Tinten, C. Tth, T. Guo, M. Kammler, M. H. v. Hoegen, K. R. Wilson, D. v. d. Linde and C. P. J. Barty, *Science*, 1999, **286**, 1340.
47 J. Larsson, R. W. Falcone, *et al.*, *Appl. Phys. A: Mater. Sci. Process.*, 1998, **66**, 587.
48 M. J. J. Vrakking and T. Elsaesser, *Nat. Photonics*, 2012, **6**, 645.
49 J. D. Biggs, K. Bennett, Y. Zhang and S. Mukamel, *J. Phys. B: At. Mol. Phys.*, 2014, **47**, 124037.
50 J. M. Dahlström, D. Guénot, K. Klünder, M. Gisselbrecht, J. Mauritsson, A. L'Huillier, A. Maquet and R. Taïeb, *Chem. Phys.*, 2013, **414**, 53.
51 P. M. Paul, E. Toma, P. Breger, G. Mullot, F. Augé, P. Balcou, H. Muller and P. Agostini, *Science*, 2001, **292**, 1689.
52 K. E. Dorfman, K. Bennett, Y. Zhang and S. Mukamel, *Phys. Rev. A: At., Mol., Opt. Phys.*, 2013, **87**, 053826.
53 J. D. Biggs, J. A. Voll and S. Mukamel, *Philos. Trans. R. Soc., A*, 2012, **370**, 3709.

 

Faraday Discussions

PAPER

# X-ray diffraction in temporally and spatially resolved biomolecular science†

John R. Helliwell,*[a] Alice Brink,[ab] Surasak Kaenket,[a] Victoria Laurina Starkey[a] and Simon W. M. Tanley[a]

Received 31st August 2014, Accepted 9th October 2014

DOI: 10.1039/c4fd00166d

Time-resolved Laue protein crystallography at the European Synchrotron Radiation Facility (ESRF) opened up the field of sub-nanosecond protein crystal structure analyses. There are a limited number of such time-resolved studies in the literature. Why is this? The X-ray laser now gives us femtosecond (fs) duration pulses, typically 10 fs up to ~50 fs. Their use is attractive for the fastest time-resolved protein crystallography studies. It has been proposed that single molecules could even be studied with the advantage of being able to measure X-ray diffraction from a 'crystal lattice free' single molecule, with or without temporal resolved structural changes. This is altogether very

[a]*School of Chemistry, University of Manchester M13 9PL, UK. E-mail: john.helliwell@manchester.ac.uk; Tel: +44 (0)1612754970*

[b]*Department of Chemistry, University of the Free State, Bloemfontein 9300, South Africa. E-mail: alice.brink@gmail.com*

† Electronic supplementary information (ESI) available: Fig. 2–4 were calculated using Refmac5 (main text ref. 17) using hen egg white lysozyme, coordinates PDB and code 2W1Y (main text ref. 18). The CCP4 mtz format processed structure factor amplitudes 'X-ray radiation damaged' diffraction data files used to make Fig. 2–4, displayed with COOT (main text ref. 19), are provided as ESI files to this paper. The cumulative absorbed X-ray dose by the time of the final data sets in the cryo cases was approximately 10 MGy and in the room temperature case approximately 2 MGy; uncertainties in these numbers arise from the X-ray generator flux at the sample being an estimate and the chemical composition of the protein crystal irradiated in each case is also somewhat approximate due to the inherent uncertainty in the bound atom occupancies and the final chemical composition of the protein crystal solvent channels. A detailed examination of such X-ray dose estimates can be found in main text ref. 32. Nevertheless the cumulative X-ray dose given is a good guide to the dose absorbed by the crystal samples. An absorbed X-ray dose of 30 MGy was recommended by Owen *et al.* 2006 (main text ref. 33) as an appropriate dose limit for typical macromolecular crystallography experiments conducted with the sample held at cryo temperatures *i.e.* typically 100 K; the cryo data collection experiments reported here are well within that absorbed X-ray dose limit as is also evident by the good quality of the protein electron density even at the final repeat X-ray diffraction data set with the exception of some disulphide bridge splitting. Fig. 2–4: HEWL crystal space group in each case was $P4_32_12$; the electron density maps displayed in Fig. 2–4 were respectively calculated at 1.7 Å, 1.6 Å and 2.2 Å. The PDB codes for the previously published 'HEWL $PtBr_6$' and 'HEWL $PtI_6$' X-ray crystal structures determined at 100 K (ref. 9) are respectively 4OWH and 4OWC. The PDB code for the 'HEWL $Ta_6Br_{12}$ $PtBr_6$' crystal structure determined at 100 K described for the first time here is 4R6C. Full details of the XRD data sets and molecular model parameters for these metal atom complexes bound to the HEWL protein can be found in these PDB files. See DOI: 10.1039/c4fd00166d

challenging R&D. So as to assist this effort we have undertaken studies of metal clusters that bind to proteins, both 'fresh' and after repeated X-ray irradiation to assess their X-ray-photo-dynamics, namely $Ta_6Br_{12}$, $K_2PtI_6$ and $K_2PtBr_6$ bound to a test protein, hen egg white lysozyme. These metal complexes have the major advantage of being very recognisable shapes (pseudo spherical or octahedral) and thereby offer a start to (probably very difficult) single molecule electron density map interpretations, both static and dynamic. A further approach is to investigate the X-ray laser beam diffraction strength of a well scattering nano-cluster; an example from nature being the iron containing ferritin. Electron crystallography and single particle electron microscopy imaging offers alternatives to X-ray structural studies; our structural studies of crustacyanin, a 320 kDa protein carotenoid complex, can be extended either by electron based techniques or with the X-ray laser representing a fascinating range of options. General outlook remarks concerning X-ray, electron and neutron macromolecular crystallography as well as 'NMR crystallography' conclude the article.

# 1 Introduction

The advent of intense synchrotron X-ray sources has allowed time-resolved diffraction in general, and macromolecular crystallography in particular, to have wide applicability to study time dependent phenomena at the molecular level; as spearhead time-resolved Laue protein crystallography at the ESRF (http://www.esrf.eu/UsersAndScience/Experiments/SoftMatter/ID09B) opened up the field of sub-nanosecond crystal structure analyses.[1] Biomolecular science 'fast time-resolved' examples include carbon monoxy myoglobin and phospho yellow protein (PYP).[2] More complex cases include following the enzyme reaction of hydroxymethylbilane synthase in the crystal which is slower by nature.[3] There are a limited number of such time-resolved studies in the literature. Why is this? Firstly, crystal lattice interactions can block the necessary structural changes for a given biochemical reaction to proceed. Secondly, crystal size determines a sample's scattering strength and thereby the required exposure time, clearly increasing as a sample gets smaller. The latter obviously can be at odds with the intrinsic time resolution required to monitor a given molecular structural change; different measuring protocols try and ease past this challenge such as the Hadamard measuring sequence[4] or the simpler approach of crystal to crystal averaging at equivalent time-slices.[3] Meanwhile the X-ray laser now gives us femtosecond (fs) duration pulses, typically 10 fs up to ~50 fs.[5] Their use is attractive for the fastest time-resolved protein crystallography studies. It has been proposed that single molecules could even be studied,[6] which would free us of the crystal lattice restrictions referred to above. Again though, the sample scattering strength along with the current X-ray laser maximum photons per pulse achievable from the source currently restrict the samples to crystalline arrays albeit micron to submicron ('nano') *i.e.* small crystal sizes.[7] In considering these challenges and the clear advantage of being able to measure X-ray diffraction from a single molecule, with or without temporal resolved structural changes, we have studied in detail a class of metal complexes, the platinum hexahalides, as well as the tantalum bromide cluster when bound to hen egg-white lysozyme (HEWL) as a test protein crystal including their 'X-ray-photo-dynamics.[8,9] These metal clusters have the major advantage of being very recognisable shapes (octahedral and pseudo spherical). So, where we can expect difficult to interpret electron density maps, and their time-resolved changes in biochemical reactions, these 'marker well-shaped complexes' will offer a start to electron density map interpretations.[9] A further approach is to investigate the X-ray laser beam diffraction strength of a well scattering nano-cluster; an example from nature being the iron containing ferritin (which has 2000 iron atoms in ferrihydrite form within a protein multi-subunit shell).[10] Another example would be, in a bionanotechnology approach, an encapsulated metal core within a virus protein shell.

## 2 Understanding the X-ray-photo-dynamics properties of the platinum hexahalides and of the tantalum bromide cluster ($Ta_6Br_{12}$) using X-ray crystallography

We have studied in detail a class of metal complexes, the platinum hexahalides, as well as the tantalum bromide cluster when bound to hen egg-white lysozyme (HEWL) as a test protein crystal.[8,9] Our interest first arose from the use of protein powder X-ray diffraction to exploit the wavelength dependent dispersive differences for phasing of extracted powder diffraction Bragg reflection amplitudes. The larger delta f ' signal arising from hexabromide was attractive.[8] The bromine K edge at 0.92 Å X-ray wavelength was very conveniently placed for two wavelength measurements involving 'on edge' and a remote X-ray wavelength. For most electron dense clusters of this type however, platinum hexaiodide is more attractive. In ref. 9 we reported the binding behaviour to histidine in HEWL as a test protein. In this paper we report the stability of the platinum hexahalides to prolonged X-ray irradiation. The platinum hexabromide and its photochemical response to UV light has been studied[11,12] but not the platinum hexaiodide to our knowledge. These metal clusters have the distinct advantage of being particularly recognizable shapes (octahedral and pseudo spherical). So, where we can expect difficult to interpret electron density maps, and their time-resolved changes in biochemical reactions, these 'marker well-shaped complexes' will offer a start to electron density map interpretations.[9] The break-up of platinum hexabromide, even when cryocooled, upon prolonged X-ray irradiation, akin to the UV study,[11] was reported previously.[13] Here we report the HEWL with bound $PtBr_6$ at room temperature also upon prolonged X-ray irradiation.

### 2.1 $PtBr_6$, $PtI_6$ and $Ta_6Br_{12}$ heavy atom clusters bound to a protein: their X-ray-photo-dynamics

**2.1.1 Crystallisation conditions.** HEWL crystals were prepared using the batch method as outlined by Blundell and Johnson (1976),[14] namely: HEWL (60 mg) was dissolved in a 0.04 M acetate buffer, pH 4.7 (1 mL), and 10% NaCl (1 mL) was added to the solution. HEWL crystals were then soaked for 24 h in a solution of either $K_2PtBr_6$ (10 mM) or $K_2PtI_6$ (10 mM). A solution of each heavy-atom compound was from a pre-made stock solution of 50 mM in acetate buffer.

In the case of 'HEWL $Ta_6Br_{12}$ $PtBr_6$' the protein was co-crystallised with 2 mM $Ta_6Br_{12}$, forming beautiful green coloured crystals (Fig. 1a), of the usual tetragonal HEWL crystal morphology. After being soaked in $PtBr_6$, the crystals turned

(a)

(b)

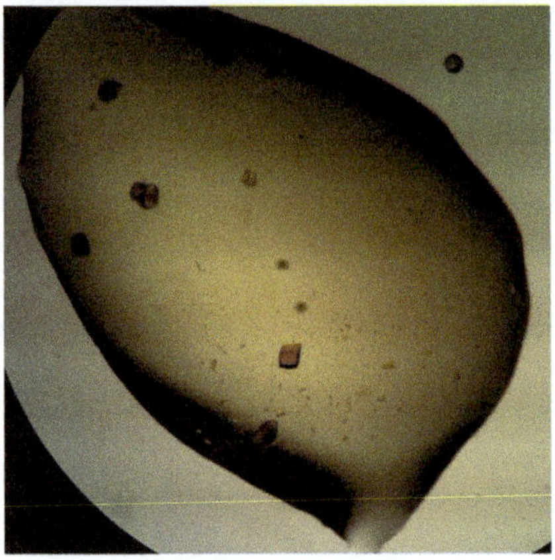

(c)

Fig. 1 Lysozyme crystals cocrystallized with $Ta_6Br_{12}$ (a. top) and soaked with $K_2PtBr_6$ (b. middle); the orange of $PtBr_6$ beats the green of $Ta_6Br_{12}$ and finally (c. bottom) HEWL crystals soaked in $K_2PtI_6$.

 

from green to the orange colour of the $PtBr_6$ (Fig. 1b). Fig. 1c shows the distinctive colour of the 'HEWL soaked in $K_2PtI_6$' crystals.

**2.1.2 X-ray diffraction data collection, protein crystal structure solution and molecular model refinement.** For the cryo diffraction data collections, a crystal from each soaking condition was scooped into a loop with paratone used as the cryoprotectant. These X-ray diffraction (XRD) data were measured on a Bruker CCD $Pt^{135}$ home source diffractometer at an X-ray wavelength of 1.5418 Å, carried out at a fixed temperature of 100 K. The XRD data collection strategy used led to a high completeness of unique data, high anomalous differences completeness and data redundancy at a reasonable level. All XRD data were processed using the Bruker software package, PROTEUM 2.[15] We collected repeated X-ray data sets on the same crystal to study the X-ray irradiation on the complexes.

The crystal structures were solved using PHASER_MR[16] and then rigid body and restrained refinement with CCP4i REFMAC5,[17] using the previously reported lysozyme structure 2W1Y as the molecular search model.[18] The use of PHASER_MR was probably not required as 2W1Y is relatively isomorphous with these crystals. Model building, adjustment and refinement were carried out respectively using the COOT molecular graphics programme[19] and REFMAC5 in CCP4i. The crystallographic and molecular model refinement parameters are summarized in the ESI† footnotes. All figures were produced using COOT.[19]

The experimental conditions for preparing the 'HEWL PtBr6 room temperature' crystal sample were basically the same as described above namely a native hen egg white lysozyme protein crystal was soaked in 10 mM $K_2PtBr_6$ for 24 hours. A crystal was selected to be smaller than the X-ray beam size (approx. 0.3 mm) but not too small to give too weak a diffraction on an APEXII microfocus sealed tube X-ray source. The crystal was mounted 'wet' in a glass capillary and thereafter remained held at room temperature, approximately 20° C. X-ray diffraction data were recorded over an angular range of rotation of the crystal of 210° by which time, *i.e.* X-ray dose (approximately 2 MGy), the diffraction pattern had noticeably deteriorated. Data sets were grouped according to 30° sweeps making seven data sets in all. The high symmetry of the tetragonal lysozyme space group ($P4_32_12$) yielded a reasonably good data set completeness for each 30° sweep of approximately 70%.

**2.1.3 Comparison of the X-ray and UV photochemical behaviour of $K_2PtBr_6$ and conclusions on the X-ray photochemical behaviour of $K_2PtI_6$ bound to HEWL**

*2.1.3.1 Cryo experiments.* $K_2PtBr_6$ and $Ta_6Br_{12}$ bound in the same crystal allow a direct comparison, side by side as it were, of the two metal complexes. As reported in ref. 13 a break up of $PtBr_6$ (Fig. 2), occurs *versus* a stability of $Ta_6Br_{12}$ throughout. The other sensitive marker of X-ray irradiation damage is the splitting of the protein disulphides[20] and such an effect was shown for two of the disulphides by a 7$^{th}$ repeat X-ray diffraction data set.

The extension of the cryocrystallographic studies to include the multiple X-ray data set irradiation stability of the iodo complex (*i.e.* $K_2PtI_6$) bound to lysozyme is a natural extension and complements the aim of using these platinum octahedral complexes to facilitate interpreting difficult electron density maps. The first X-ray diffraction data set electron density (green in Fig. 3) and the somewhat diminished (lilac in Fig. 3) electron density for $PtI_6$ at the 6$^{th}$ XRD data set are not as severe as the $PtBr_6$ break up upon X-ray irradiation. There is a $PtI_3$ moiety, directly

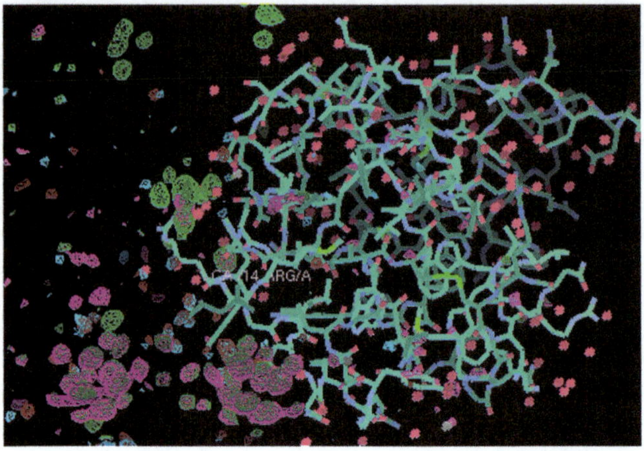


Fig. 2 Comparison of the XRD stability behaviour of $Ta_6Br_{12}$, which is stable, *versus* $PtBr_6$ and its clear break up as evidenced by X-ray data set 0 (electron density difference map contoured at 3 sigma green) and data set 6 (electron density difference map contoured at 3 sigma lilac). The nearest amino acid residue to the $PtBr_6$ complex, which is Arg 14, is labelled; note the nice octahedral electron density for the first X-ray diffraction data set (green) and the essentially absent lilac density for it at the 6th data set. At the lower portion of the figure the equivalent green and lilac electron densities are for the $Ta_6Br_{12}$ moieties, there are two of them, and their stabilities to the X-rays, which is very good, are essentially identical. The protein model is 2W1Y and the maps are each rigid body 2W1Y model refinement Fo-Fc 'omit' (*i.e.* neither $Ta_6Br_{12}$ or $PtBr_6$ are included in the model) electron density difference maps.

bound to one of the His15 imidazole nitrogens, which is stable *versus* X-ray dose (*i.e.* the green and lilac electron densities in Fig. 3 are identical).

*2.1.3.2 Room temperature X-ray crystallographic study of hen egg white lysozyme bound $PtBr_6$ and relation to its chemical kinetics.* The X-ray-photo-dynamics behaviour of $PtBr_6$ at room temperature when bound to hen egg white lysozyme and the stages of its stepwise removal of bromine atoms are shown in Fig. 4 which compares the metal complex at successive stages of X-ray irradiation, namely data sets 1, 3, 5 and 7 (data sets 2, 4 and 6 are not shown), *i.e.* with cumulatively increasing X-ray dose to the same crystal. The occupancies of Pt and Br are approximately 0.3; the bound water that would be expected (see Fig. 5) to complete the octahedral arrangement around the platinum centre at 0.3 occupancy is not visible in Fig. 4b–d; this is due to the limited diffraction resolution of 2.2 Å and a diffraction data completeness of ~70% in each case.

The chemical kinetics is basically that of a unimolecular reaction (*i.e.* break up) of the platinum hexabromide upon X-ray illumination. Since the X-ray illumination is at a continuous level, simply chosen to allow reasonable strength X-ray diffraction data to be recorded, we cannot determine an absolute rate constant for the break up. We note though that at room temperature the platinum hexabromide break up was more prompt than at cryo temperature in the lysozyme + $PtBr_6$ + $Ta_6Br_{12}$ study, since the room temperature X-ray diffraction data had to be subdivided into smaller angular portions than that for the cryo data. A future experiment could be undertaken whereby a larger X-ray intensity would be used to illuminate a crystal sample *e.g.* from an intense synchrotron undulator X-ray

 

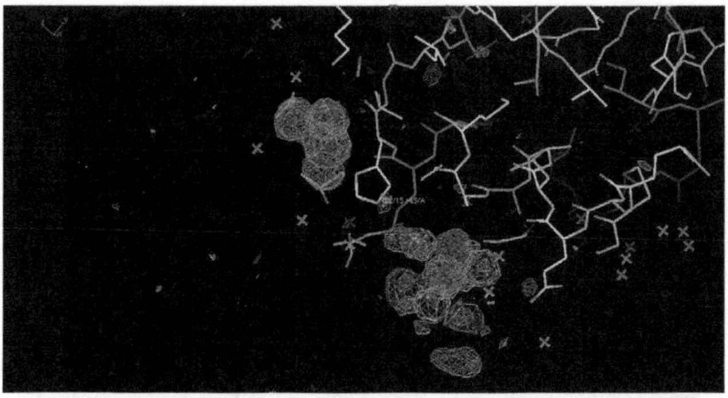

Fig. 3 Comparison of the XRD stability behaviour of $PtI_6$ and $PtI_3X$; comparison of data set 0 (electron density difference map contoured at 3 sigma green) and data set 6 (electron density difference map contoured at 3 sigma lilac). The nearest amino acid residue to both moieties, His15, is labelled; note the nice octahedral electron density for $PtI_6$ (right, green; viewed down the three fold axis of symmetry of the octahedron). The first X-ray diffraction data set electron density (green) and the somewhat diminished lilac electron density for $PtI_6$ at the 6th XRD data set are not as severe as the $PtBr_6$ break up upon X-ray irradiation. The $PtI_3X$ moiety, directly bound to one of the His15 imidazole nitrogens, is stable *versus* X-ray dose (*i.e.* the green and lilac electron densities are identical). The refined structure of the HEWL $PtI_6$ data set 0 PDB code is 4owc (ref. 9). The protein model shown here is 2W1Y and the maps are each rigid body 2W1Y model refinement Fo-Fc 'omit' (*i.e.* neither $PtI_6$ or the $PtI_3X$ are included in the model) electron density difference maps.

beam, which would form an X-ray flash photolysis approach, and then the irradiated sample be plunged frozen. This would be repeated with several samples each at different time lapse intervals after X-ray irradiation. Neutron macromolecular crystallography diffraction data sets would then be recorded for each time lapse plunge freeze sample; the neutron beam would then allow the molecular state of $PtBr_6$ to be probed without further radiation damage. Of course it would be entirely possible that the break up state would be identical in each neutron derived crystal structure, *i.e.* if the damage in an intense X-ray beam was 'instantaneous'. Glebov *et al.*[11] in their UV-vis studies indeed undertook flash photolysis and 'steady state illumination' experiments and kinetics analyses and from which we quote:-"The spectral curves obtained by steady state irradiation and pulse photolysis coincide, indicating that the first photoaquation step occurred within the time <50 ns".

# 3 A short summary of the UV photochemistry of $PtBr_6^{2-}$

The inorganic photophysics of $K_2PtBr_6$ in aqueous solution under steady state and laser flash photolysis (308 nm) has also been reported previously[11] including a reaction mechanism proposal taking place on a timescale of nanoseconds or shorter involving charge transfer to an excited d–d state for lengthening of the Pt–Br bond distance and subsequent departure of a bromine atom. It was also described that successive steps of bromine removal and take up of a water molecule is possible.

Glebov *et al.*[11] reported their experimental results as follows:-

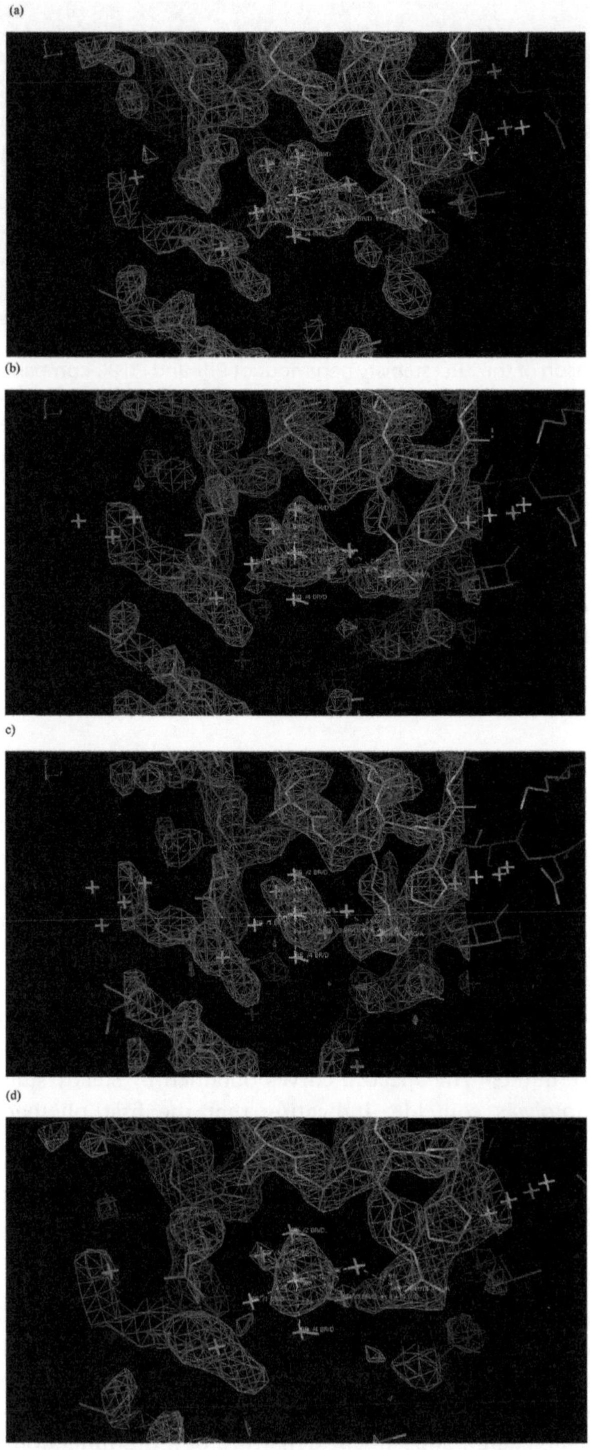


Fig. 4 The X-ray-photo-dynamics behaviour of $PtBr_6$ at room temperature when bound to hen egg white lysozyme and the stages of its stepwise removal of bromine atoms evidenced by a comparison of (a) data set 1, (b) data set 3, (c) data set 5 and (d) data set 7.

 

$$PtBr_6^{2-} \rightleftharpoons PtBr_5(H_2O)^- \rightleftharpoons$$
$$\rightleftharpoons PtBr_4(H_2O)_2 \rightleftharpoons PtBr_3(H_2O)_3^+$$

Fig. 5 Glebov *et al.*[11] reported results as follows:- *"Photochemistry of the $PtBr_6^{2-}$ complex in aqueous solution was studied by steady state and laser flash photolysis (308 nm). The multistep photo-aquation of the complex occurs in the nano- and micro-second time intervals without the formation of intermediate platinum(III) complexes."*

*"Photochemistry of the $PtBr_6^{2-}$ complex in aqueous solution was studied by steady state and laser flash photolysis (308 nm). The multistep photo-aquation of the complex occurs in the nano- and micro-second time intervals without the formation of intermediate platinum(III) complexes."* See their figure reproduced as Fig. 5 here. They conclude that *"Thus, experiments with pico- or femto-second time resolution are necessary to perform for detailed investigation of the processes that occur upon the excitation of the $PtBr_6^{2-}$ complex"*.

Our protein single crystal results are consistent with these models showing the successive steps in the removal of bromine atoms in the room temperature $PtBr_6$ experiment and crystal structure analyses (Fig. 4a–d).

# 4 Specific conclusions

$Ta_6Br_{12}$ is relatively the most stable under repeated X-ray irradiation (Fig. 2) with $PtI_6$ only showing some small changes (Fig. 3). Relatively less stable as a heavy atom complex is $PtBr_6$ (Fig. 2 and 4).

The study by Glebov *et al.*[11,12] of $PtBr_6$ shows that 'with and without UV-vis light flash' leads to photo dissociation of the platinum hexabromide (Fig. 5). The cryo 'HEWL $Ta_6Br_{12}$ $PtBr_6$' study (Fig. 2) and the room temperature XRD study of 'HEWL with $PtBr_6$' (Fig. 4a–d) concurs with the UV study, all showing a light induced break-up of $PtBr_6$.[11] Whilst the UV study of platinum hexaiodide seems not to have been undertaken, the HEWL $PtI_6$ behaviour shows that it is relatively more stable under X-ray irradiation than $PtBr_6$ and suggests therefore that the UV stability of $PtI_6$ would likewise be more stable.

The respective behaviours of the temporal and structural stabilities of these heavy atom clusters are clearly of interest for a number of reasons. With the femtosecond XRD 'diffract before destroy' method[5] the X-ray sensitivity of $PtBr_6$ is not relevant. However its irradiation by UV light before X-ray laser diffraction would prepare an altered oxidation state[11] which may be useful. The stability of $Ta_6Br_{12}$ and of $PtI_6$ upon prolonged X-ray irradiation is a good property when

Data sets 1 through to 7 are with cumulatively increasing X-ray dose to the same crystal. The nearest amino acid residue to the $PtBr_6$ complex is Arg 14, which is on the right hand side of the platinum complex and labelled. The protein model is the restrained refined 'HEWL_PtBr6 RT data set 1' and the electron density difference maps in blue are each rigid body 2W1Y model refinement 2Fo-Fc 'omit' maps, *i.e.* $PtBr_6$ is not included in the model (1.4 rms contour level). The occupancies of Pt and Br are approximately 0.3; the bound water that would be expected (see Fig. 5) to complete the octahedral arrangement around the platinum centre at 0.3 occupancy is not visible in Fig. 4b–d; this is due to the limited diffraction resolution of 2.2 Å and diffraction data completeness of ~70% in each case.

synchrotron or home lab X-ray sources are used for biological crystal structure determination.

## 5 General outlook

Whilst our article title is specifically on X-ray diffraction, a more general question is whether there will be a surge of electron based diffraction techniques for temporally and spatially resolved molecular science. Electrons interact much more strongly with atoms than do X-rays. Electron crystallography and single particle electron microscopy (EM) imaging offer alternatives to the trend with X-ray sources such as the X-ray laser for the study of ever smaller samples. Our X-ray crystal structural studies of the lobster carapace coloration protein beta-crustacyanin[21] have been extended to determine the layout at around 30 Å resolution of the alpha-crustacyanin complex, a 320 kDa complex of eight beta-crustacyanins,[22] using negative stain single particle EM and small angle X-ray scattering along with 'rigid body fitting' of the beta-crustacyanin crystal structure. This structural study of the alpha-crustacyanin complex can be extended either by electron based techniques or with the X-ray laser representing a fascinating range of options.[23] The technique of electron single crystal crystallography has very recently gained a major impetus with successful 3D structural analysis at 2 Å resolution, using molecular replacement, of hen egg white lysozyme as a test protein and using two individual microcrystals.[24] It may not be possible to make *de novo*, high resolution protein structure determinations unless accurate isomorphous derivative intensity differences can be measured.[25] However molecular replacement is now the predominant technique in protein crystallography not least due to the fact that the protein data bank contains more and more of the protein folds seen in nature (the PDB now contains over 100 000 entries). Thus alpha-crustacyanin might yet be solvable with electron (microcrystal) crystallography and using the beta-crustacyanin X-ray crystal structure determined with a xenon heavy atom derivative and softer X-rays from the Daresbury SRS. Clearly there are important general implications for biological crystallography from the electron technique results of ref. 24 using microcrystals of proteins and complexes of biological macromolecules. Besides, static structure determination dynamic studies based on electron crystallography from a microcrystal sample might be easier than from a larger crystal, *i.e.* preserving crystallinity and so on. Again though, like the comment regarding the challenge for electron based techniques of measuring accurate isomorphous heavy atom derivative intensity differences, Bragg reflection intensity changes from temporal structural sequences can be smaller than those induced by a heavy atom and thereby even more difficult to measure precisely. Nevertheless further development of electron crystallography may well open up new options in this fascinating field; Zewail and Thomas have championed electron based techniques for a general realisation of 4D, space and time, structural studies[26a] and the mapping of atomic motions with ultrabright electrons and the fundamental limits in space-time resolution are discussed by Dwayne Miller and coauthors in ref. 26*b*.

A regular challenge for X-ray crystal structure analysis of biological macromolecules is the determination of the protonation states of ionisable amino acid side chains of aspartic acid, glutamic acid and histidine. Likewise the orientation of bound water molecules or their state as hydroxyl or hydronium ions is even more challenging. Thus a growing development at neutron sources is the provision of instruments for neutron macromolecular crystallography notably in Japan at JAERI and JPARC, at the Institut Laue Langevin in Grenoble and at Oak Ridge National Lab, USA for both the research reactor and the SNS. Likewise a variety of macromolecular crystallographic software analysis packages now provide macromolecular refinement with neutron and X-ray scattering factors. Given the low flux of neutron beams at first sight it would seem that studies of structural and temporal dynamics using neutron macromolecular crystallography would not be possible. Is this the case? In fact 'no', it is possible with sample freezing to study structural intermediates albeit on a timescale set by the freezing time of the crystal sample.[27] This area of neutron macromolecular crystallography is expected to grow; a very recent example is ref. 28. One such 'next experiment', as mentioned above, is suggested from consideration of the chemical kinetics of the X-ray induced break up of $PtBr_6$.

Diffraction based techniques have limits. These have been highlighted in the recent book on NMR crystallography,[29] which I reviewed.[30] Quoting from Chapter 1 of this book:- *"A key advantage is the case of host–guest complexes where the guest is mobile and obviously thereby disordered and its visibility absent or limited in the diffraction-derived structural analyses…(thus)…diffraction-based information forms an important background to investigations and where NMR is essential to delineate motional effects'…NMR measurements are 'in addition to the relaxation times information which extends down to the microseconds region'. …There are two major areas in which complementary information (to diffraction) is desirable: Molecular-level mobility and purely spatial, i.e. static, disorder'."* Chapter 27 of the book[31] is devoted to structural biology and includes NMR measurements by Martin and Zilm of a protein, ubiquitin (consisting of 76 amino acids and which has a molecular mass of about 8.5 kDa), in three different sample states "of micro-crystals, nanocrystals and lyophilized (*i.e.* amorphous)". The solid state NMR spectra for the nanocrystals of ubiquitin were equivalent to the spectra from the larger crystals; the difference was in the amorphous solid state ubiquitin spectra. As macromolecular crystallographers we tend not to worry too much about the sample state details and yet the properties of metallic nanoparticles, for example, are known to be different from bulk solids. This is an interesting further question for debate as diffraction methods take macromolecular crystallography into the nanocrystal sample size domain.

## Acknowledgements

We are grateful to the University of Manchester for general support, to the EPSRC for a PhD studentship to SWMT, to the Thailand Government for studentship support to SK, to the SASOL Young Academic Collaboration Initiative for Dr Alice Brink, to the School of Chemistry for crystallization and computing facilities and to the Faculty of Life Sciences for X-ray diffraction facilities. We are very grateful to Dr Colin Levy and Dr Pat Bryant for their excellent stewardship of the Manchester Institute of Biotechnology ('MIB') X-ray Facility.

## Notes and references

1 *Time-Resolved Diffraction*, ed. J. R. Helliwell and P. M. Rentzepis, Oxford University Press, 1997; *Time-resolved Macromolecular Crystallography*, ed. D.

W. J. Cruickshank, J. R. Helliwell and L.N. Johnson, Oxford Science Publications,1992.

2 Z. Ren, D. Bourgeois, J. R. Helliwell, K. Moffat, V. Srajer and B. L. Stoddard, *J. Synchrotron Radiat.*, 1999, **6**, 891–917.

3 J. R. Helliwell, Y. P. Nieh, J. Raftery, A. Cassetta, J. Habash, P. D. Carr, T. Ursby, M. Wulff, A. W. Thompson, A. C. Niemann and A. Hädener, *J. Chem. Soc., Faraday Trans.*, 1998, **94**, 2615–2622.

4 B. A. Yorke, G. S. Beddard, R. L. Owen and A. R. Pearson, *Nat. Methods*, 2014, **11**(11), 1131–1134.

5 J. C. H. Spence and H. N. Chapman, Biology with free-electron X-ray lasers, *Philos. Trans. R. Soc., B*, 2014, **369**, 1647.

6 R. Neutze, R. Wouts, D. van der Spoel, E. Weckert and J. Hajdu, *Nature*, 2000, **406**, 752–757.

7 J. R. Helliwell, *Science*, 2013, **339**, 146–147.

8 J. R. Helliwell, A. M. T. Bell, P. Bryant, S. J. Fisher, J. Habash, M. Helliwell, I. Margiolaki, S. Kaenket, Y. Watier, J. Wright and S. Yalamanchilli, *Z. Krist.*, 2010, **225**, 570–575.

9 S. W. M. Tanley, L.-V. Starkey, L. Lamplough, S. Kaenket and J. R. Helliwell, *Acta Crystallogr., Sect. F: Struct. Biol. Cryst. Commun.*, 2014, 1132–1134.

10 J. M. A. Smith, J. R. Helliwell and M. Z. Papiz, *Inorg. Chim. Acta*, 1985, **106**, 193–196.

11 E. M. Glebov, V. F. Plyusnin, V. P. Grivin, A. B. Venediktov and S. V. Korenev, *Russ. Chem. Bull., Int. Ed.*, 2007, **56**, 2357–2363.

12 V. Balzani and V. Carassiti, *J. Phys. Chem.*, 1968, **72**, 383–388.

13 J. R. Helliwell, S. Kaenket, C. Levy, P. Bryant, L. Lamplough, J. Raftery, L.-V. Starkey, E. Kotsiliti, I. Margiolaki, A. Fitch and J. P. Wright, *Reported at 7th Radiation Damage 'RD7' Conference*, held Diamond Light Source, Harwell Science and Innovation Campus, Oxfordshire, UK, 2012.

14 T. L. Blundell, L. N. Johnson, *Protein Crystallography*, Academic Press, New York, 1976, pp. 66–77.

15 Bruker, *PROTEUM2. Version 2*, 2006, Bruker AXS Inc., Madison, Wisconsin, USA.

16 A. J. McCoy, R. W. Grosse-Kunstleve, P. D. Adams, M. D. Winn, L. C. Storoni and R. J. Read, *J. Appl. Crystallogr.*, 2007, **40**, 658–674.

17 G. N. Murshudov, P. Skubák, A. A. Lebedev, N. S. Pannu, R. A. Steiner, R. A. Nicholls, M. D. Winn, F. Long and A. A. Vagin, *Acta Crystallogr.*, 2011, **D67**, 355–367.

18 M. Cianci, J. R. Helliwell and A. Suzuki, *Acta Crystallogr.*, 2008, **D64**, 1196–1209.

19 P. Emsley and K. Cowtan, *Acta Crystallogr.*, 2004, **D60**, 2126–2132.

20 J. R. Helliwell, *J. Cryst. Growth*, 1988, **90**, 259–272.

21 M. Cianci, P. J. Rizkallah, A. Olczak, J. Raftery, N. E. Chayen, P. F. Zagalsky and J. R. Helliwell, *Proc. Natl. Acad. Sci. U. S. A.*, 2002, **99**, 9795–9800.

22 N. H. Rhys, M.-C. Wang, T. A. Jowitt, J. R. Helliwell, J. G. Grossmann and C. Baldock, *J. Synchrotron Radiat.*, 2011, **18**, 79–83.

23 J. R. Helliwell, *Crystallogr. Rev.*, 2010, **16**, 231–242.

24 B. L. Nannenga, D. Shi, A. G. W. Leslie and T. Gonen, *Nat. Methods*, 2014, **11**(9), 927–930.

 

25 M. E. Dumont, J. W. Wiggins and S. B. Hayward, *Proc. Natl. Acad. Sci. U. S. A.*, 1981, **78**, 2947–2951.

26 (*a*) A. Zewail and J. M. Thomas, *4D Electron Microscopy: Imaging in Space and Time*, Imperial College Press, London, 2010; (*b*) S. Manz, A. Casandruc, D. Zhang, Y. Zhong, R. A. Loch, A. Marx, T. Hasegawa, L. C. Liu, S. Bayesteh, H. Delsim-Hashemi, M. Hoffmann, M. Felber, M. Hachmann, F. Mayet, J. Hirscht, S. Keskin, M. Hada, S. W. Epp, K. Flöttmann and R. J. Dwayne Miller, *Faraday Discuss.*, DOI: 10.1039/c4fd00204k.

27 M. P. Blakeley, A. J. Kalb (Gilboa), J. R. Helliwell and D. A. A. Myles, *Proc. Natl. Acad. Sci. U. S. A.*, 2004, **101**, 16405–16410.

28 C. M. Casadei, A. Gumiero, C. L. Metcalfe, E. J. Murphy, J. Basran, M. G. Concilio, S. C. M. Teixeira, T. E. Schrader, A. J. Fielding, A. Ostermann, M. P. Blakeley, E. L. Raven and P. C. E. Moody, *Science*, 2014, **345**(6193), 193–197.

29 *NMR Crystallography*, ed. R. K. Harris, R. E. Wasylishen and M. J. Duer, John Wiley & Sons, Ltd., Chichester, 2009.

30 J. R. Helliwell, *Crystallogr. Rev.*, 2014, **18**, 241–245.

31 D. Middleton, in *NMR Crystallography*, ed. R. K. Harris, R. E. Wasylishen and M. J. Duer, John Wiley & Sons, Ltd., Chichester, 2009, ch. 27, and see specifically; R. W. Martin and K. W. Zilm, *J. Magn. Reson.*, 2003, **165**, 162–174.

32 J. R. Helliwell and S. W. M. Tanley, *Acta Crystallogr.*, 2013, **D69**, 121–125.

33 R. L. Owen, E. Rudino-Pinera and E. F. Garman, *Proc. Natl. Acad. Sci. U. S. A.*, 2006, **103**, 4912–4917.

Faraday Discussions

PAPER

# Disentangling detector data in XFEL studies of temporally resolved solution state chemistry

Tim Brandt van Driel,[a] Kasper Skov Kjær,[ab] Elisa Biasin,[a] Kristoffer Haldrup,[a] Henrik Till Lemke[c] and Martin Meedom Nielsen*[a]

Received 14th October 2014, Accepted 5th November 2014
DOI: 10.1039/c4fd00203b

With the arrival of X-ray Free Electron Lasers (XFELs), 2D area detectors with a large dynamic range for detection of hard X-rays with fast readout rates are required for many types of experiments. Extracting the desired information from these detectors has been challenging due to unpredicted fluctuations in the measured images. For techniques such as time-resolved X-ray Diffuse Scattering (XDS), small differences in signal intensity are the starting point for analysis. Fluctuations in the total detected signal remain in the differences under investigation, obfuscating the signal. To correct such artefacts, Singular Value Decomposition (SVD) can be used to identify and characterize the observed detector fluctuations and assist in assigning some of them to variations in physical parameters such as X-ray energy and X-ray intensity. This paper presents a methodology for robustly identifying, separating and correcting fluctuations on area detectors based on XFEL beam characteristics, to enable the study of temporally resolved solution state chemistry on the femtosecond timescale.

## 1. Introduction

The introduction of X-ray Free Electron Lasers (XFELs) has brought many X-ray techniques into a new regime. Unprecedented X-ray peak brilliance and exceptionally short pulse lengths make it possible to use X-rays to study many new phenomena, such as those occurring on time-scales shorter than atomic movement,[1] measure processes before heat effects and radiation-damage set in,[2] as well as using exceptional transverse coherence to probe nanoparticles.[3]

Along with new X-ray source and beam characteristics come a number of new challenges that have to be addressed. Among these challenges is the design and use of area detectors capable of operating at high readout rates as well as accurately

[a]*Molecular Movies, Department of Physics, Technical University of Denmark, 2800 Kgs. Lyngby, Denmark. E-mail: mmee@fysik.dtu.dk; Tel: +45 25 32 26*

[b]*Chemical Physics Department, Lund University, PO Box 124, S-22100 Lund, Sweden*

[c]*LCLS, SLAC National Laboratory, Menlo Park, California 94025, USA*

measuring X-rays in a large dynamic range for each individual X-ray pulse. The stochastic nature of the Self-Amplified Spontaneous Emission (SASE) process in XFELs results in significant pulse-to-pulse variation in both intensity, energy and time. Here, we present time-resolved X-ray diffuse scattering (XDS) from one of the first XDS experiments at the Linac Coherent Light Source (LCLS), measured with the first version of the Cornell-SLAC Pixel Array Detector (CSPAD), at the X-ray pump–probe beamline (XPP) at the LCLS. Upon analyzing XDS of solution samples measured with the CSPAD it was realized that the quality of the measurements was not limited by counting statistics, but instead limited by a number of fluctuating artefacts present in the measured scattering signals. Several different approaches to sort and filter the data have been applied with some success[4,5] but have not made it possible to model or explain the observed artefacts until now.

Here, we present a method to identify and remove some of these fluctuations. The method relies on using singular value decomposition (SVD) on a large number of reference images measured during typical XDS experiments. From such a subset of the measured data it is possible to use SVD to identify particular, characteristic, fluctuations in the detector signal. Further, these fluctuations can be directly correlated with variations in beam parameters such as X-ray energy and pulse intensity, which are measured simultaneously with the scattering data. Based on this analysis it is shown how the identified correlation between the observed fluctuations and the beam-parameter variations can be utilized to efficiently correct scattering images on an image-to-image basis prior to analysis.

# 2. XDS experiments

This paper is based on a typical pump–probe XDS experiment at the XPP beamline at LCLS. A liquid water sample is pumped through a sapphire nozzle producing a 100 μm liquid sheet that is continuously renewed and returned to a sample reservoir. The liquid jet ensures fresh sample for every pump–probe event in the interaction volume where the pump laser and probe X-rays are overlapped in time and space. The sample was repeatedly pumped with 1950 nm laser pulses, and probed with the XFEL X-ray pulses at 120 Hz. A series of time-delays were measured to study the temporal response of water-heating on ultra-fast timescales by collecting the diffuse scattering on an area detector (seen in Fig. 1).

Subtracting the un-pumped scattering images from the pumped results in difference-scattering images that can be azimuthally integrated to create 1D difference scattering curves $\Delta S(Q)$, where $Q = \dfrac{4\pi \sin(\theta)}{\lambda}$, $2\theta$ is the scattering angle and $\lambda$ is the wavelength. These curves can then be used to analyse the photo-induced structural changes in the sample. The procedure for conducting a full structural analysis of the XDS data is covered elsewhere.[6,7] Here we present a methodology to identify undesired signal fluctuations and create corrections, that, when applied to the individual images, yield a corrected dataset suitable for analysis.

# 3. The 2D CSPAD

Many X-ray experiments require 2D area detectors in order to measure scattering or dispersed spectral signals.[8] These 2D detectors need a large dynamic range to

 

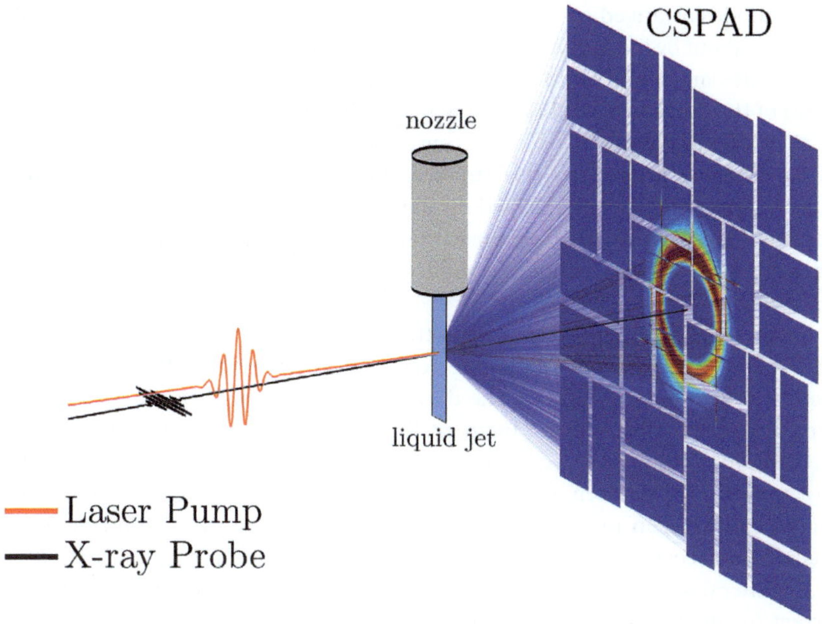


Fig. 1 A schematic of the experimental time-resolved XDS setup at XPP. The CSPAD is mounted downstream from the liquid jet, covering scattering angles up to 45°. Additionally the beam path and sample environment is mounted in a He filled chamber reducing the unwanted air scattering background as well as protecting the sample.

meet the requirements for many different techniques; from single photon detection[9–11] to many integrated photons/pixel,[12] in an energy range extending from soft to hard X-rays. At LCLS, the CSPAD[13] was constructed to satisfy these requirements for the Coherent X-ray Imaging (CXI) and XPP beamlines. Table 1 shows some of the technical specifications.

This 14 bit, 16 384 Analogue to Digital Units (ADU) detector has two different gain modes changing the sensitivity by a factor of 7. For XDS experiments using the full polychromatic X-ray beam, the low gain mode is necessary in order to avoid saturation. The large area detector shown in Fig. 1 consists of ~2.3 million

Table 1 Specifications of the CSPAD. Adapted from (ref. 12). The ADU/photon and readout noise are extracted from histograms of single photon counting on the CSPAD

| | | |
|---|---|---|
| Single ASIC | 185 × 194 pixels | |
| Single tile | 185 × 388 pixels | |
| Large 32 tile CSPAD | 1516 × 1516 pixels | 2.3 megapixel |
| Pixel size | 110 × 110 μm | |
| Detector area | 18.5 × 18.5 cm | |
| Frame rate | 120 Hz | |
| | **High Gain** | **Low Gain** |
| Pixel saturation | 350 photons | 2700 photons |
| ADU/Photon@8 keV | ~40 ADU | ~5 ADU |
| Readout noise | ~20 ADU | ~5 ADU |

silicon diodes arranged on 32 individual detector tiles. The diodes on each half-tile are read out by individual Application Specific Integrated Circuits (ASICs). The CSPAD tiles are arranged in a square, divided into 4 different quadrants. The centre of the detector has a hole, allowing the direct beam to pass through. Fig. 2 (a) shows a typical raw X-ray scattering image from water as recorded on the CSPAD. This raw image consists of the scattered intensity on top of a background. This background, sometimes called the dark image or dark current, is given by the intensity measured on the detector without incident X-rays. It can be measured independently and subsequently subtracted from the raw images prior to analysis.

In Fig. 2 the pixel intensities are plotted in the lab space coordinates and the gaps between the tiles as well as the central hole are evident from the lack of pixels. From the raw image (a), the dark image (b) containing the background intensities measured with no incident X-rays is subtracted. This contribution accounts for ~1800 ADU per pixel, reducing the dynamic range of the 14 bit detector by ~10% when subtracted. The standard deviation of 1000 dark images can be seen in (c) showing the tile-wise variation in readout noise. A mask (d) is necessary to remove unreliable pixels. In the early measurements, a grid of metal wires was placed on the CSPAD in an attempt to measure and correct for fluctuations. Pixels shadowed by this grid are also masked out, as are pixels that show significant deviations in total intensity, fluctuations or signal-to-noise between consecutive, nominally identical, measurements. After masking the image and subtracting the dark image, the XDS image shows a characteristic liquid ring on a weak background as seen in (e). Azimuthal integration of such isotropic XDS images reduces the data to 1D scattering curves $S(Q)$ without loss of information (f).

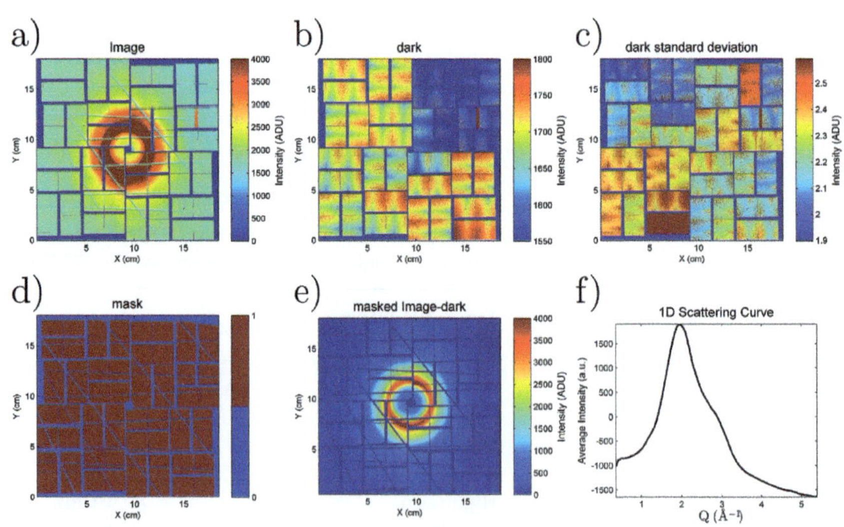


Fig. 2 CSPAD images showing: (a) Raw XDS image, (b) dark image, (c) standard deviation of 1000 dark images, (d) mask, (e) dark subtracted and masked XDS image of water and (f) azimuthally integrated scattering curve.

 

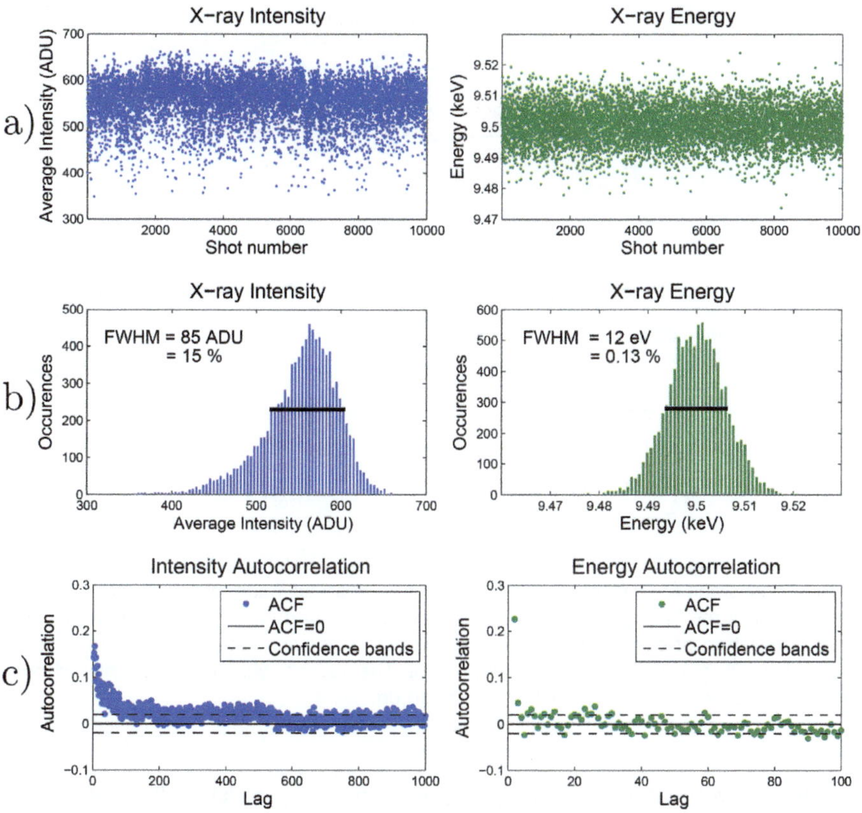


Fig. 3 The X-ray intensity and energy are measured for each X-ray shot (a). Histograms of the readout values allow for determining the FWHM of the intensity and energy distribution (b). The autocorrelation function of the intensity and energy show the time correlation of consecutive shots, confidence bands are added to evaluate when a significant autocorrelation is no longer present.

# 4. X-ray free electron lasers

The LCLS utilizes Self Amplified Spontaneous Emission (SASE) to produce short bursts of intense X-rays at 120 Hz. The typical pulse has a duration of tens of fs with up to $2 \times 10^{12}$ photons/pulse, which is many orders of magnitude higher in peak-brilliance than synchrotron X-ray pulses.[14] Due to the stochastic nature of the SASE process[15] the resulting XFEL pulses fluctuate in pulse intensity and X-ray energy. Furthermore, significant fluctuations have been identified in the pump–probe arrival time. The X-ray energy, intensity and timing fluctuations are discussed in more detail in the following and are shown in Fig. 3 and 4.

## 4.1. X-ray intensity

The X-ray intensity is measured as the integrated intensity on the CSPAD. Intensity Position Monitors (IPMs) are also available along the beam path, but require non-linear diode corrections.[16] The intensity fluctuations of the full polychromatic beam can be seen in Fig. 3 (a–c) where the average intensity is observed to fluctuate with 78 ADU (FWHM) corresponding to 14%.

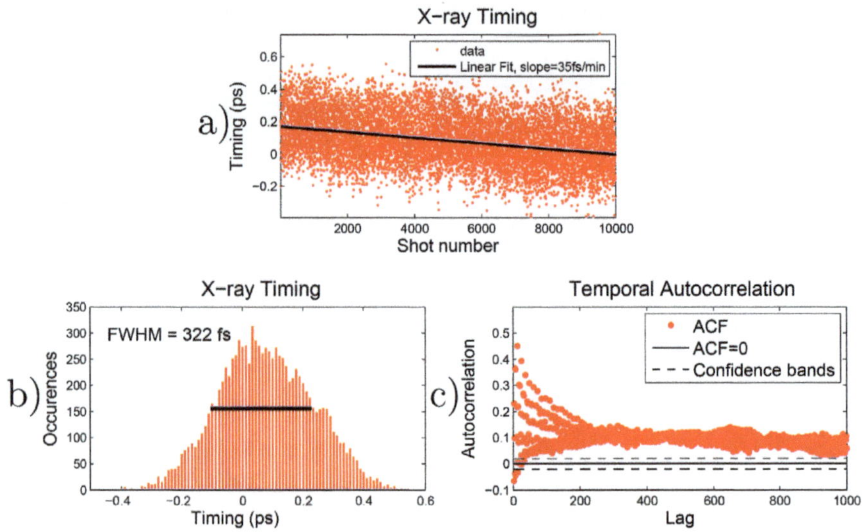


Fig. 4 The timing jitter is read out for each X-ray/laser shot (a) and the long term drift is fit with a linear slope of 35 fs $min^{-1}$. Histograms of the readout values allows for determining the FWHM of the timing jitter (b). The autocorrelation function of the timing shows the time correlation of consecutive shots, confidence bands are added to evaluate when a significant autocorrelation is no longer present. The autocorrelation of the timing shows significant autocorrelation as well as long term drifts.

The Autocorrelation Function (ACF) measures the correlation between observations at different times as a function of separation distance, or lag.[17]

As seen in Fig. 3 panel (c) the ACF of the X-ray intensity decays rapidly, approaching zero within 100 lags. However, the autocorrelation still remains larger than the confidence bands at late lags, thereby showing some correlation between intensity measurements with significant separation (several seconds) in time.

### 4.2. X-ray energy

As opposed to the X-ray undulator spectrum at a synchrotron, the XFEL SASE spectrum is narrow-band and exhibits spiky features with significant shot-to-shot variation. Detection schemes are currently being developed for determining the X-ray spectrum on a pulse to pulse basis.[18] Until such is available it is possible to estimate the pulse-to-pulse centre of mass X-ray energy from the electron beam energy. The energy histogram in Fig. 3 has a 12 eV FWHM corresponding to a 0.13% bandwidth at the nominal energy of 9.5 keV. The ACF of the X-ray energy shows no significant autocorrelation after a few lags, in contrast to the long-lived autocorrelation of the X-ray intensity.

### 4.3. X-ray timing

Both the X-ray and optical laser produce pulses at a 120 Hz frequency, in the following we describe the time jitter, which is the pulse-to-pulse deviation in actual pump–probe time difference from the nominal time difference $\Delta t$. The effective time resolution of the ~50 fs X-ray duration and ~70 fs laser duration is decreased by a large temporal jitter. The X-ray/laser time jitter is the combined jitter of both X-ray and laser pulse arrival time and was a significant challenge for the first time-resolved XFEL experiments. This has been solved at LCLS with the implementation of a timing tool.[1,19,20] The timing tool determines the arrival time of the X-ray with respect to the optical laser. Fig. 4 shows the time jitter for 10.000 shots and quantifies the variation similarly to the X-ray intensity and energy in Fig. 3. The time-jitter is 322 fs at FWHM and the timing tool allows us to measure the delay with ~15 fs accuracy.[1] The autocorrelation of the timing displays complicated behaviour, but the general trends can be separated into two different time-scales. The autocorrelation reflects the laser timing feedback loop giving a systematic correlation/anti-correlation between close shots, seen for the first 200 lags. Additionally the long term time drift results in a positive autocorrelation throughout the plot. A linear fit to the long time drift results in a drift of 35 fs $\text{min}^{-1}$ corresponding to a timing drift of 2.1 ps $\text{hour}^{-1}$.

The data presented here are measured with the implemented timing tool[1] which will not be covered further in this paper.

### 4.4. Other parameters

In addition to the three parameters described above, others may also vary. From experience it has been established that the X-ray position fluctuates up to 20 μm. For the presented XDS experiment this is not a concern as the sample is a liquid sheet, homogeneous on distances much larger than the position fluctuations, as well as 20 μm being significantly smaller than the dimensions of the pixels of the detector.

## 5. Detector fluctuations

As described below, the CSPAD displays a number of different fluctuations. Among these, tiles and quadrants exhibit so called common-mode fluctuations, where a common offset in intensity is measured in a particular well-defined part of the detector such as a tile or quadrant. Additionally each tile on the detector shows some spatial dependence in the noise behaviour, as shown above in the standard deviation of the dark measurements, see Fig. 2.

A linear area detector can in most cases be calibrated by subtracting a pixel-resolved dark background and measuring the individual pixel gain. The dark background provides a constant intensity offset when no X-rays are incident, and the individual pixel gain must be corrected, ensuring that each pixel has the same absolute scaling. This is described by eqn (1), where $I$ is the pixel intensity.

$$I_{\text{corrected}} = (I_{\text{measured}} - I_{\text{dark}}) \times \text{gain} \tag{1}$$

This description does not account for the observed fluctuations, but the observed fluctuations can be quantified as a deviation from this equation.

In the following sections we address fluctuations due to variations in X-ray energy and intensity. Plotting the X-ray energy against the X-ray intensity (see Fig. 5) shows that these two beam parameters are uncorrelated. Any fluctuations caused by either of these individual parameters can therefore be considered independently, making them easier to identify and separate.

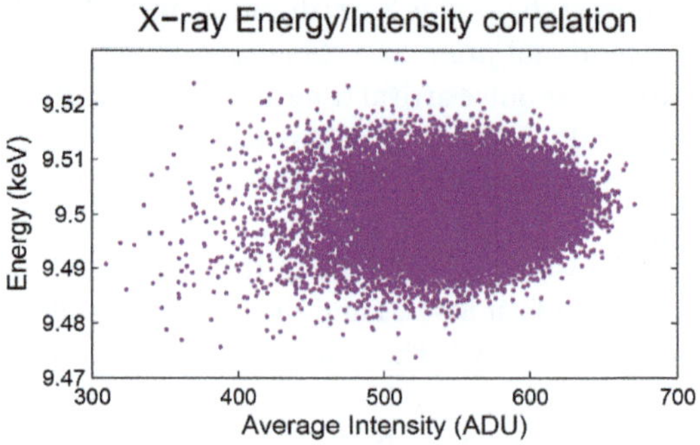


Fig. 5 X-ray beam energy plotted as a function of average intensity measured on the CSPAD for 23 000 shots. No correlation between these two beam parameters is evident.

In the following we investigate the relationships between variations in beam parameters and fluctuations in the signal observed on the CSPAD. Towards this goal we use Singular Value Decomposition (SVD)[21] as a tool to separate and identify characteristic detector fluctuations. In the case of non-linear/non-proportional detector behaviour, energy and intensity dependent signal fluctuations can conceivably be removed if a parametric model for the detector response as a function of X-ray intensity or energy can be established. Eqn (2) describes a detector correction scheme based on identified relationships between beam parameters and detector response.

$$I_{\text{corrected}} = I_{\text{measured}} - I_{\text{dark}} - \sum_i f_i(I) I_i^{\text{SVD}} - \sum_e f_e(E) I_e^{\text{SVD}} \tag{2}$$

By using SVD to identify fluctuating components ($I^{\text{SVD}}$) in combination with analysis of the relationship between each such component and key beam parameters, it may be possible to assign some of these to X-ray intensity variation ($I_i^{\text{SVD}}$) and some to variation in X-ray energy ($I_e^{\text{SVD}}$). Further, by establishing the parametric models $f_i(I)$ and $f_e(E)$ for the relationship between beam variations ($E$, $I$) and detector response ($I_i^{\text{SVD}}$, $I_e^{\text{SVD}}$), it is possible to correct for the undesired intensity and energy dependent behaviour solely based on the beam parameters, resulting in the corrected scattering $I_{\text{corrected}}$.

## 6. Singular value decomposition

SVD is a powerful tool to decompose a 2D matrix into orthogonal components. It is often used to extract the principal components as well as their kinetics from a time-series of spectra.[22] The large set of images acquired during an XDS experiment can be reshaped into such a 2D matrix. Fig. 2 shows how a set of acquired detector images can be recast as a 2D matrix amenable to SVD analysis. This is done by reorganizing each image into a vector of pixels and placing each image vector in a new column of the resulting [Data] matrix as shown in Fig. 6. Using SVD, the [Data] matrix is then decomposed into the product of $U$, an $m \times m$ set of orthonormal basis vectors, $S$, an $m \times n$ rectangular diagonal matrix and $V^{\text{T}}$, an $n \times n$ unitary matrix.

 

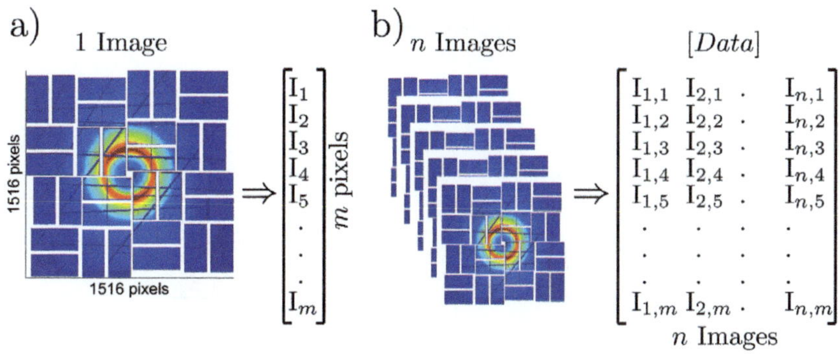


Fig. 6 (a) The individual images containing 2 296 960 pixels are reorganized into long vectors and (b) multiple images are reorganized into a large [Data] matrix.

$$[\mathrm{Data}] = USV^{\mathrm{T}} \tag{3}$$

The images contained in [Data] are decomposed into a number of significant image components, contained in the columns of $U$. The corresponding amplitude of each component in the different images, is given by $SV^{\mathrm{T}}$. For typical XFEL XDS experiments $n$ and $m$ are of the order of $10^4$ and $10^6$, respectively, making the construction of [Data] and the SVD analysis computationally quite intensive. However, as the columns of $U$ contain the orthonormal components in decreasing order of $S(i, i)$, it is possible to select a cutoff $i_r$ after which the contribution of the following components is judged to be sufficiently small. This reduced-rank approximation of the data limits the calculation to a "thin" SVD decomposition, where only the first $r$ components are computed. The thin SVD limits the size of the resulting $U$, $S$ and $V^{\mathrm{T}}$ to $m \times r$, $r \times r$ and $r \times n$, greatly reducing the computational cost and size of the decomposed matrix.

Still, the SVD computation requires the entire [Data]-matrix to be present in computer memory. Even though the thin SVD reduces the size of the resulting $U$, $S$, $V$, the size of the [Data]-matrix makes the thin SVD infeasible on a normal PC. However, it is possible to calculate a thin low-rank SVD from a subset of [Data] and update it incrementally[23] using MATLAB® code implemented by D. Wingate,[24] as briefly described here. Given the SVD of $[X]$, a subset of the columns in the [Data]-matrix,

$$[X] = USV^{\mathrm{T}} \tag{4}$$

$[X]$ can be updated to include a further partition of the columns in the [Data]-matrix, $[A]$, to yield updated $U$, $S$ and $V$, denoted $U_{\mathrm{p}}$, $S_{\mathrm{p}}$ and $V_{\mathrm{p}}^{\mathrm{T}}$ such that

$$[XA] = U_{\mathrm{p}}S_{\mathrm{p}}V_{\mathrm{p.}}^{\mathrm{T}} \tag{5}$$

This makes it possible to iteratively process the entire dataset in smaller partitions.

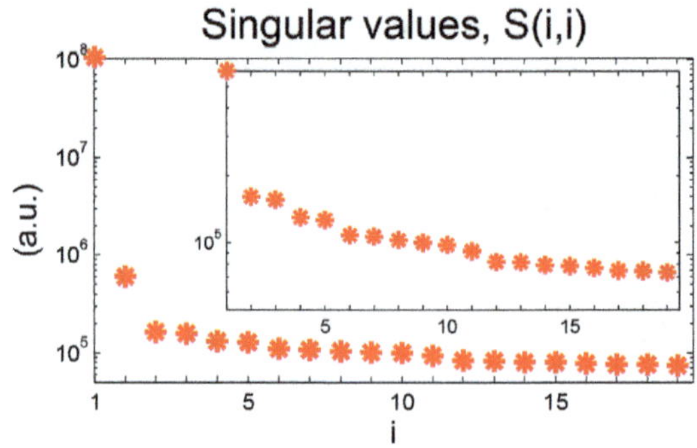


Fig. 7 The singular values of the diagonal $S(i, i)$ from the thin SVD of rank 20, for $i$ 1–20. The inset shows a zoom of $i$ 2–20.

# 7. Singular value decomposition of CSPAD data

In order to characterise the detector, a set of steady-state (un-pumped), dark-subtracted and masked images is used. These steady-state images are nominally identical measurements regularly inter-spaced in the data acquisition sequence, and thus contain the same detector fluctuations present in the time-resolved data. Usually, every 2nd–10th acquired image is such a reference, yielding a total of $\sim 10^4$ such reference images for a typical data acquisition. This set of reference data acquired with no laser excitation of the sample can be utilized as a separate dataset probing the detector response under the exact same experimental conditions as the time-resolved part of the acquisition sequence.

For the dataset described in this article, every fifth image was acquired with no laser pulse arriving at the sample. Following dark-subtraction and masking, these images were utilized in the detector characterization described in detail in the following. Here, we present the result of the thin SVD of those $\sim$5000 reference images of neat water. The rank cutoff was set to 20, and as Fig. 7 shows, the relative contribution of the components of rank greater than 10 are more than three orders of magnitude less than the primary component. Based on this observation, we consider a cutoff of 20 to be quite conservative and observe that the total contribution of excluded components account for less than $1 \times 10^{-5}$ percent of the full dataset. We thus regard this reduced representation of the acquired data as fully adequate to describe all salient parts of the detector response.

Fig. 8–11 characterize the first 8 components of $U$ as obtained from the SVD decomposition. Turning first to panel (a) of Fig. 8, this shows the components in $U$ re-plotted onto the detector coordinates. The left column contains the first component, and shows a reassuringly close likeness to the dark-subtracted image in Fig. 2. Turning next to the right-hand column of Fig. 8 as a representative example of these rather complex datasets, panel (a) shows the second SVD component, column two of $U$. Referring to Fig. 7, this second component accounts for about 1% of the signal in the series of reference images. Panel (b) shows the azimuthal integration of this component compared to the azimuthal integration of the averaged images. Panels (c) and (d) show plots of the amplitude

 

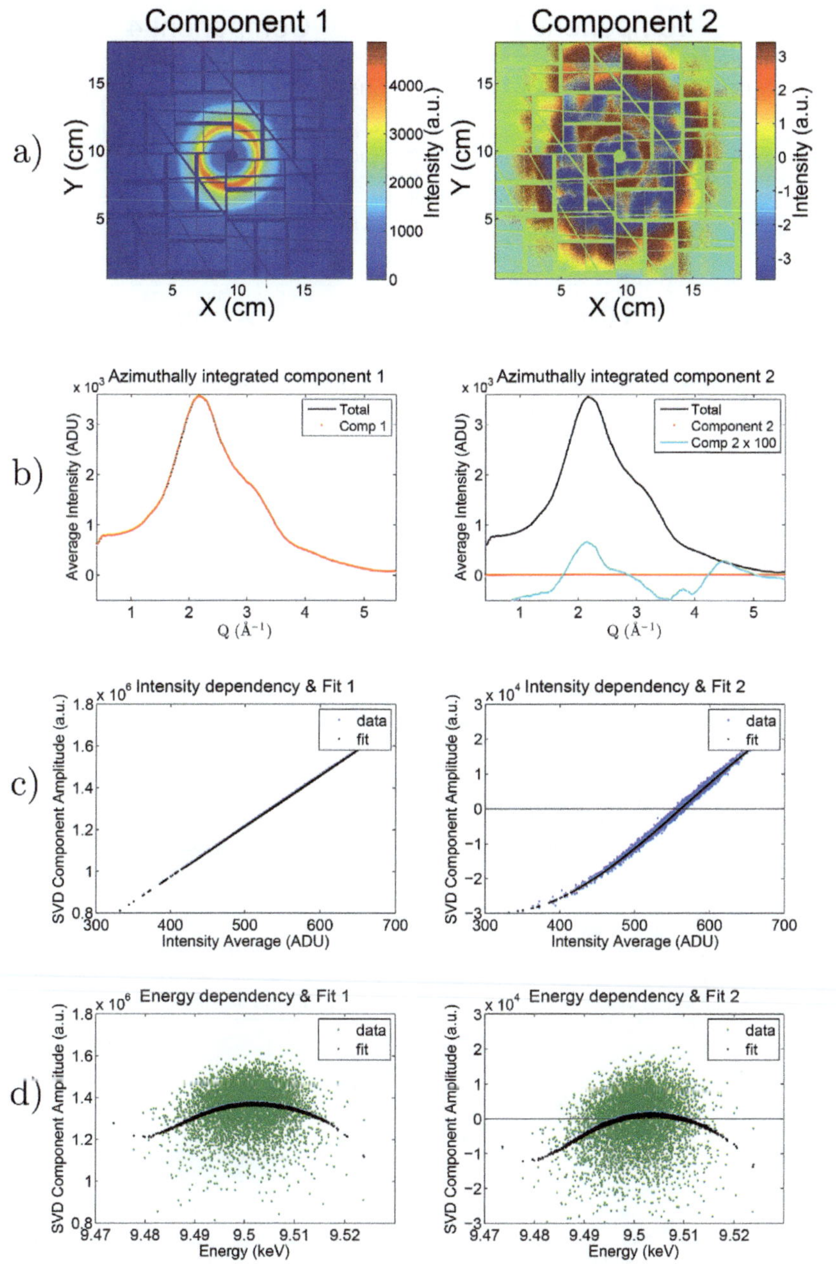


Fig. 8 Results of the SVD analysis of the data matrix, showing the first two components and their amplitude *versus* X-ray pulse intensity and X-ray pulse energy. The *U* components multiplied by their mean absolute amplitude have been mapped to the detector geometry in (a). Panel (b) shows the azimuthal integration of the image in (a) (red trace), compared with the azimuthal integration of the average image (black trace). Cyan trace shows the red trace scaled up for visibility. Panel (c) shows the amplitude of the *U* component plotted against X-ray pulse intensity, and panel (d) shows the amplitude of the *U* component plotted against X-ray pulse energy. Component 1 represents the linear response of the CSPAD seen from the linear intensity dependency. Component 2 describes a non-linear intensity dependency, as seen from panel (c).

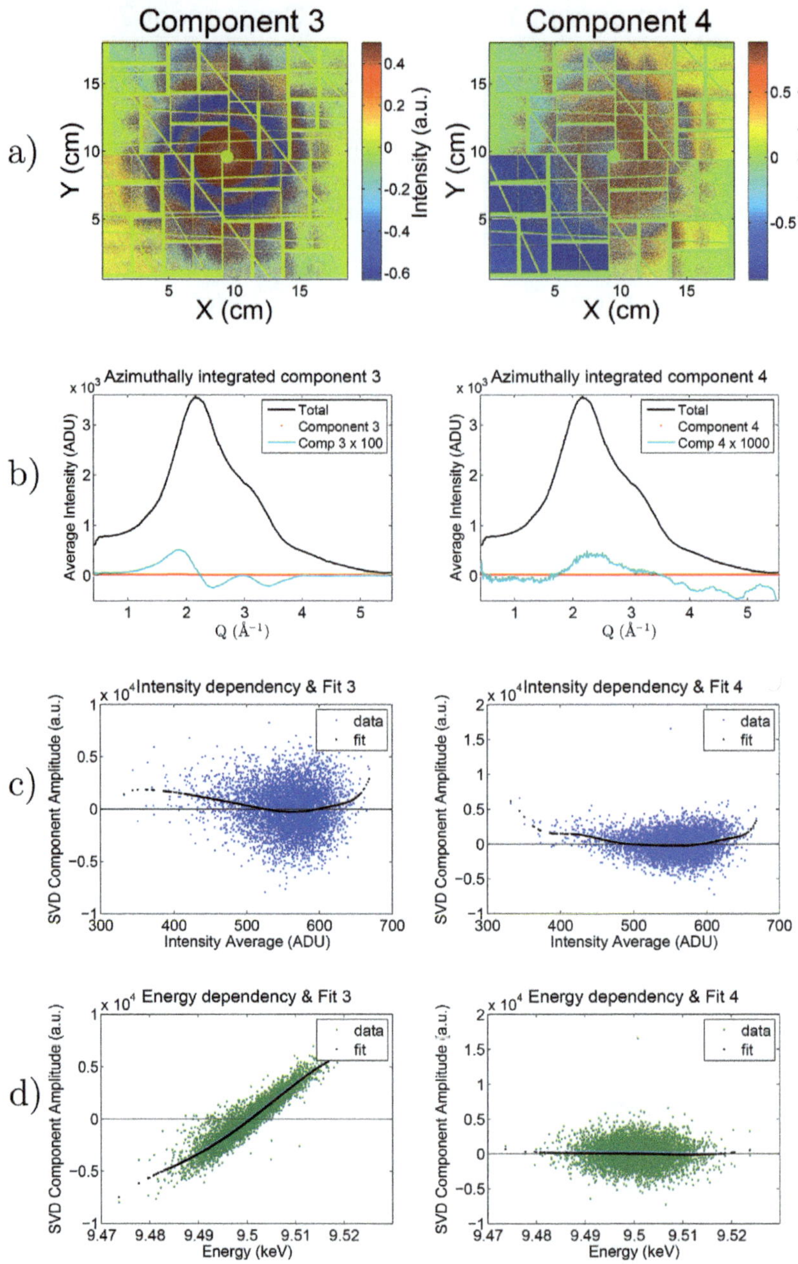


Fig. 9 Results of the SVD analysis of the data matrix, showing components 3 and 4 and their amplitude *versus* X-ray pulse intensity and X-ray pulse energy. The *U* components multiplied by their mean absolute amplitude have been mapped to the detector geometry in (a). Panel (b) shows the azimuthal integration of the image in (a) (red trace), compared with the azimuthal integration of the average image (black trace). Cyan trace shows the red trace scaled up for visibility. Panel (c) shows the amplitude of the *U* component plotted against X-ray pulse intensity, and panel (d) shows the amplitude of the *U* component plotted against X-ray pulse energy. Component 3 describes the X-ray energy dependency of the detector, seen from panel (d). Component 4 describes a random quadrant fluctuation and correlates poorly with X-ray intensity and energy.

 

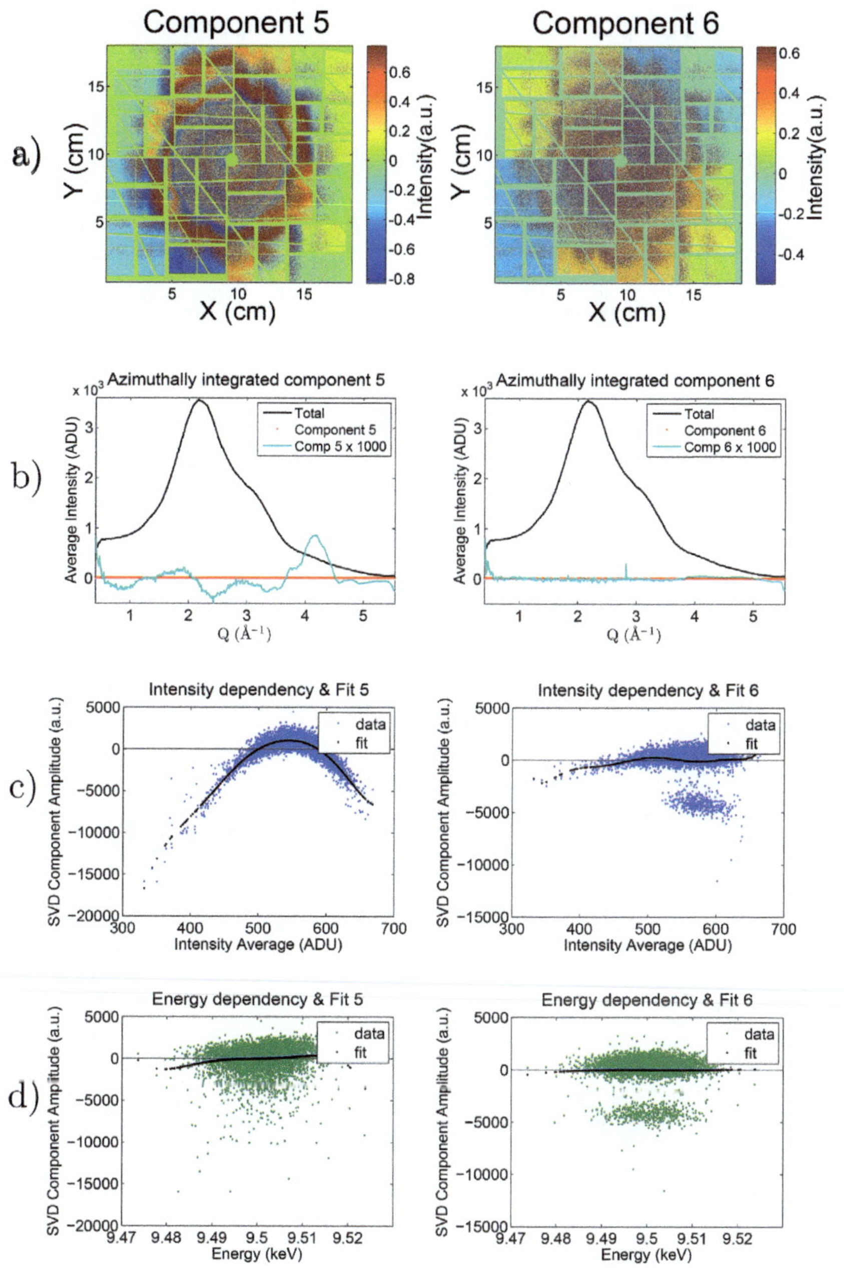


Fig. 10 Results of the SVD analysis of the data matrix, showing components 5 and 6 and their amplitude *versus* X-ray pulse intensity and X-ray pulse energy. The *U* components multiplied by their mean absolute amplitude have been mapped to the detector geometry in (a). Panel (b) shows the azimuthal integration of the image in (a) (red trace), compared with the azimuthal integration of the average image (black trace). Cyan trace shows the red trace scaled up for visibility. Panel (c) shows the amplitude of the *U* component plotted against X-ray pulse intensity, and panel (d) shows the amplitude of the *U* component plotted against X-ray pulse energy. Component 5 correlates with the pulse intensity and together with component 2 it describes the non-linear intensity response. Component 6 shows no clear intensity or energy dependency and resembles common-mode tile fluctuations, as seen in panel (a).

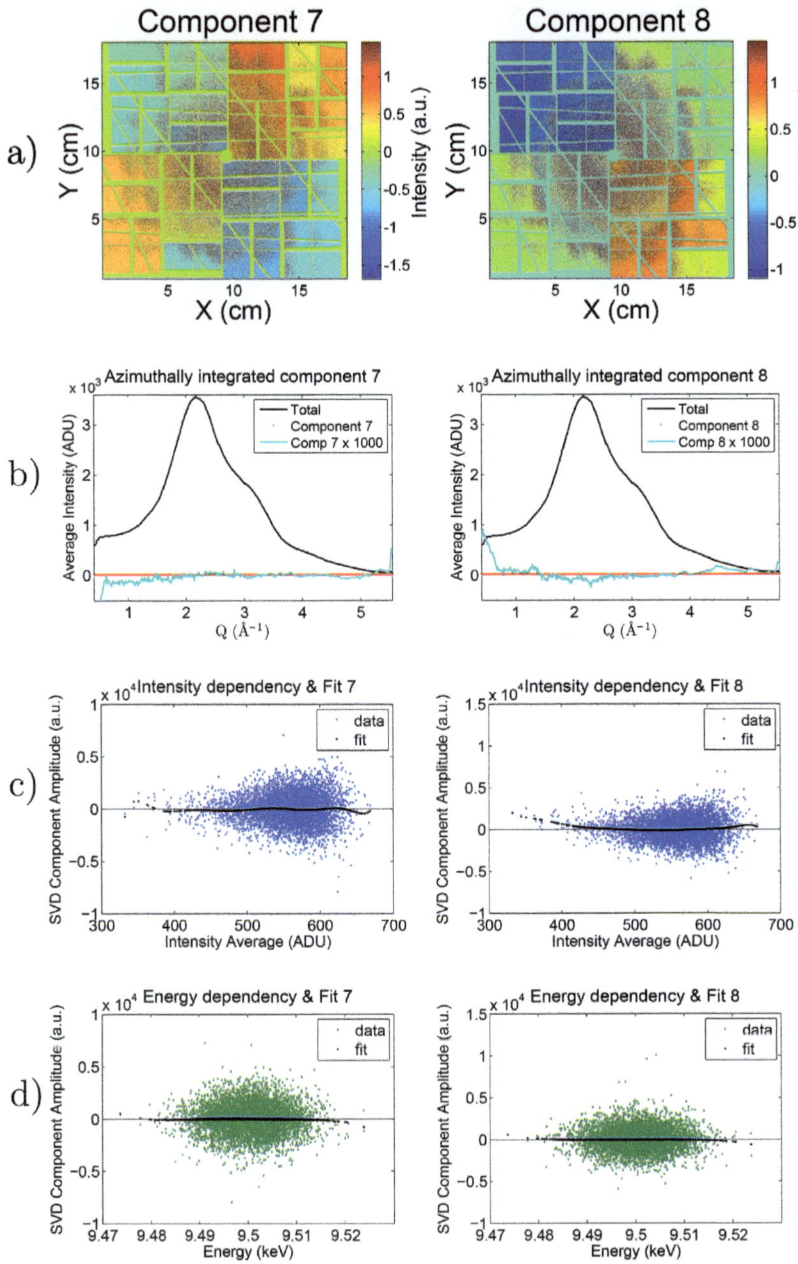


Fig. 11 Results of the SVD analysis of the data matrix, showing components 7 and 8 and their amplitude *versus* X-ray pulse intensity and X-ray pulse energy. The *U* components multiplied by their mean absolute amplitude have been mapped to the detector geometry in (a). Panel (b) shows the azimuthal integration of the image in (a) (red trace), compared with the azimuthal integration of the average image (black trace). Cyan trace shows the red trace scaled up for visibility. Panel (c) shows the amplitude of the *U* component plotted against X-ray pulse intensity, and panel (d) shows the amplitude of the *U* component plotted against X-ray pulse energy. Components 7 and 8 show no clear intensity or energy dependency and resemble common-mode fluctuations of a similar nature to components 4 and 6, as seen in panel (a).

 

of this component against the X-ray intensity and the X-ray energy for every image in the series of reference images. Panel (d) shows no correlation between the X-ray energy and the amplitude of component 2. Panel (c), on the other hand, shows a clear intensity dependency. The black points in this plot represent a polynomial fit to the data points, thus providing a parametric description of the magnitude of this component, $I_2^{\mathrm{SVD}}$ as a function of X-ray beam intensity. The polynomium describes $f_2(I)$ in eqn (2), identifying and characterizing a non-linear intensity dependency in the CSPAD response.

Continuing to component 3, shown in the left hand column of Fig. 9, panel (a) again maps the component onto the detector coordinates. Panel (b) shows the same component, azimuthally integrated. Component 3 does not correlate with the intensity (panel (c)) but has a clear correlation with the X-ray beam energy (panel (d)) and again the black points represent a polynomium describing the dependency of this component on the X-ray energy. We note that the shape of the azimuthally integrated curve corresponds to a shift in $Q$ of the scattering signal, further supporting the connection between this characteristic variation in the set of reference images and the X-ray beam energy.

Referring next to component 4 shown in Fig. 9, panel (b) shows a component predominantly localised on the lower left quadrant. This component does not correlate with the incoming intensity or energy and therefore cannot be corrected based on the X-ray intensity or energy.

### 7.1. Intensity and energy dependent components

In general, a component $I_k^{\mathrm{SVD}}$ is considered to be significantly correlated with either X-ray beam energy or X-ray beam intensity if either $f_k(I)$ or $f_k(E)$ can be described (fitted) by a $p$-order polynomial. In the analysis of the present dataset, $p = 9$ and the goodness of fit is determined by an $R^2$-measure. Which is based on inspection of typical fits and corresponding $R^2$ values for correlated components.

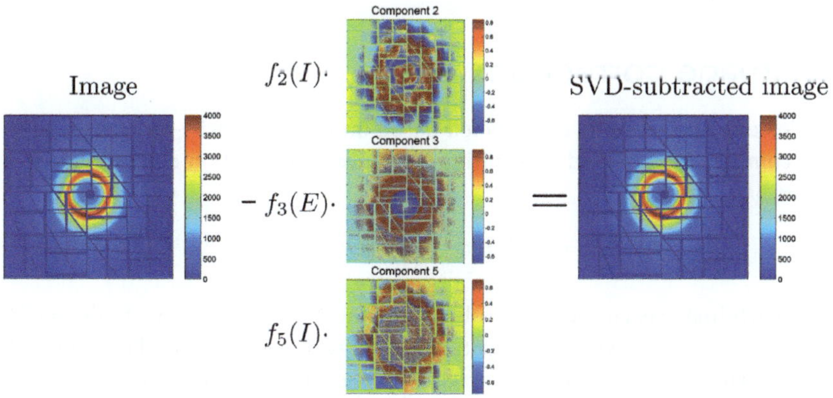


Fig. 12 Graphical representation of the correction of a single image. As discussed in the main text, three fluctuating components can be removed based on their individual intensity ($I$) and energy ($E$) dependencies, $f_2(I)$, $f_3(E)$ and $f_5(E)$. The effect of the ~1% corrections can be hard to see on the resulting image, but becomes highly evident when difference signals are calculated, as seen in Fig. 14.

Applying these criteria, we find that components 1, 2, 3 and 5 are significantly correlated with the beam parameters as follows.

The most significant SVD component, which is seen in Fig. 8, is observed to be directly proportional to the X-ray intensity, as expected if the incident X-ray intensity is fluctuating and the detector has a fully linear response. This component comprises 99% of the signal.

Component 2 accounts for 0.5% of the average signal, as seen from the ratio of the singular values for component 1 and 2 in Fig. 7. It can be seen from panel (c) that the amplitude of component 2 has a clear intensity dependency. Component 5 is likewise observed to have an intensity dependency. Thus components 2 and 5 describe a non-linear intensity response of the CSPAD. In order to correct for this behaviour, eqn (2) can be used where $I_2^{\mathrm{SVD}}$ and $I_5^{\mathrm{SVD}}$ are the second and fifth components in the columns of $U$ and $f_2(I)$ and $f_5(I)$ are the polynomial fits to the intensity dependency of the two components, respectively.

The amplitude of component 3 is observed to be strongly correlated with the X-ray energy. Thus, the energy dependency described by component 3 can be corrected using $I_3^{\mathrm{SVD}}$ and $f_3(E)$ in eqn (2).

### 7.2. Other components

The remaining fluctuations described by components 4, 6, 7 and 8 contain clear contributions from common-mode behaviour in either quadrants or tiles. These components appear random in time and do not correlate with the incoming intensity or energy and therefore cannot be corrected based on the measured beam parameters. The remaining components 9–20 (not shown) describe similar apparently random common-mode fluctuations.

The lower left quadrant and especially one of the tiles in this quadrant often contribute significantly to these common-mode fluctuations. This matches the observation that the tiles in this quadrant also exhibit the largest fluctuations in the standard deviation of the dark images in Fig. 2. To reduce some of the observed fluctuations this specific tile and quadrant could be masked out, at the cost of counting statistics.

## 8. Image corrections

The detector correction is constructed based on the 3 intensity- or energy-correlated components 2, 3 and 5, resulting in a specific correction equation for this dataset based on eqn (2).

$$I_{\mathrm{corrected}} = I_{\mathrm{measured}} - I_{\mathrm{dark}} - \sum_{i=\{2,5\}} f_i(I) I_i^{\mathrm{SVD}} - \sum_{e=\{3\}} f_e(E) I_e^{\mathrm{SVD}} \tag{6}$$

where each image is corrected by subtracting the 3 individual contributions $I^{\mathrm{SVD}}$ weighted by their individual intensity and energy dependent correlation fits $f_i(I)$ and $f_e(E)$. This of course relies on knowing the X-ray beam energy and intensity for each image acquired in the measurement, but as discussed above this information is provided as part of the standard data stream at LCLS.

Fig. 12 shows the example correction of an image, where the three corrections are subtracted. The corrections are small compared to the total scattering, however these fluctuations become clear when studying small differences. In the

 

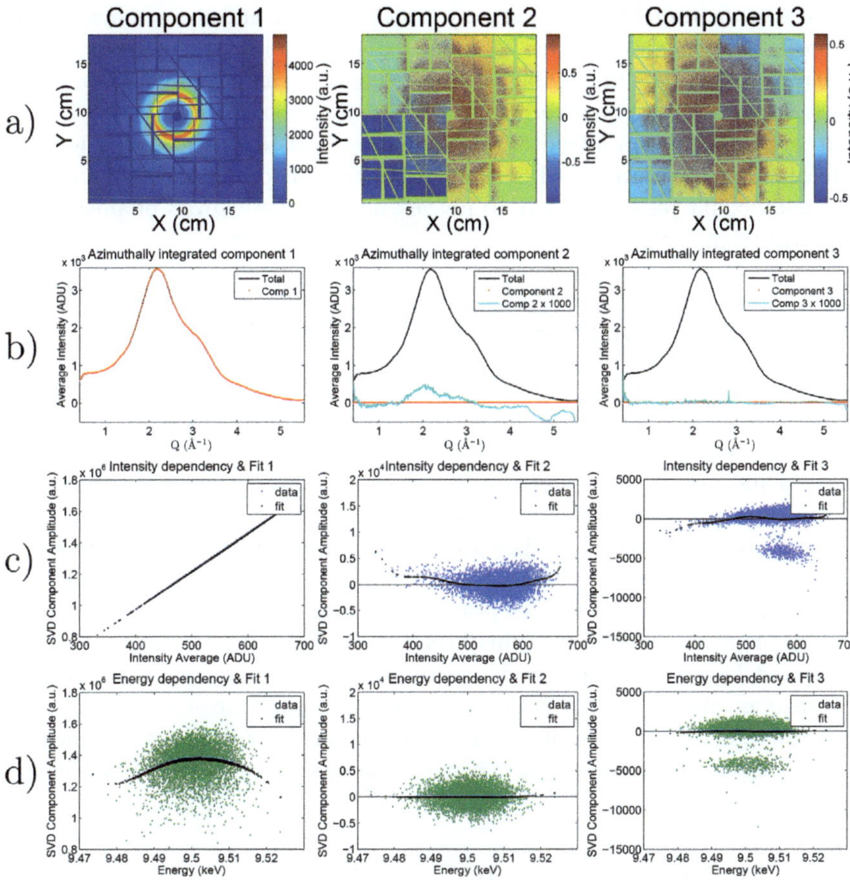


Fig. 13 Results of a second SVD analysis of the data matrix after subtracting artefacts based on the initial SVD analysis. The figure shows the first three components and their amplitude *versus* X-ray pulse intensity and X-ray pulse energy. The *U* components multiplied by their mean absolute amplitude have been mapped to the detector geometry in (a). Panel (b) shows the azimuthal integration of the image in (a) (red trace), compared with the azimuthal integration of the average image (black trace). Cyan trace shows the red trace scaled up for visibility. Panel (c) shows the amplitude of the *U* component plotted against X-ray pulse intensity, and panel (d) shows the amplitude of the *U* component plotted against X-ray pulse energy. The linear intensity dependency is described by component 1, where components 2 and 3 are the most significant remaining detector fluctuations. These two components are similar to components 4 and 6 seen in Fig. 9 and 10, respectively.

exemplified corrections in Fig. 12 the image has an average intensity of 554 ADU/pixel and the corrections account for an average of 0.6, 0.9 and 0.4 ADU/pixel for components 2, 3 and 5 respectively.

The successful correction of a given dataset can be verified by applying the corrections to the set of reference images and subsequently recalculating the singular value decomposition.

In Fig. 13 the resulting components from a second SVD are shown. The first component remains, and the second and third components resemble component

4 and 6 in the initial SVD, respectively. This shows the effective removal of the intensity- and energy-correlated components. After applying the corrections the observed fluctuations in the set of reference images is ten-fold smaller, with the most significant component accounting for only 0.1% of the variation. We note that in contrast to an effective, but ad-hoc method reported previously,[4] the artefact subtraction presented here allows for analysis of the underlying causes and additionally retains the full 2D images, allowing for analysis of anisotropic scattering signals.[25]

# 9. Artefact-subtracted difference signals

Typical time-resolved XDS investigations rely on analyzing difference scattering signals, obtained by subtracting unpumped scattering images from laser-pumped scattering images. As a starting point for such analysis, the acquired images are masked and corrected for setup specific effects such as X-ray polarization and X-ray absorption through the liquid sheet, as well as solid angle covered by each detector pixel. The images are then azimuthally integrated and scaled to produce 1D scattering curves. The difference scattering curves are then generated by subtracting an average of the two nearest references from each scattering curve.

Fig. 14 shows the output of this XDS data reduction scheme applied to the presented dataset, (a) without and (b) with the artefact subtraction introduced in the sections above. The full set of difference scattering curves (in total 20.000 individual pump–probe events) have had outliers removed[26] and are binned in ~10 fs steps to produce averaged difference scattering curves for each time delay,

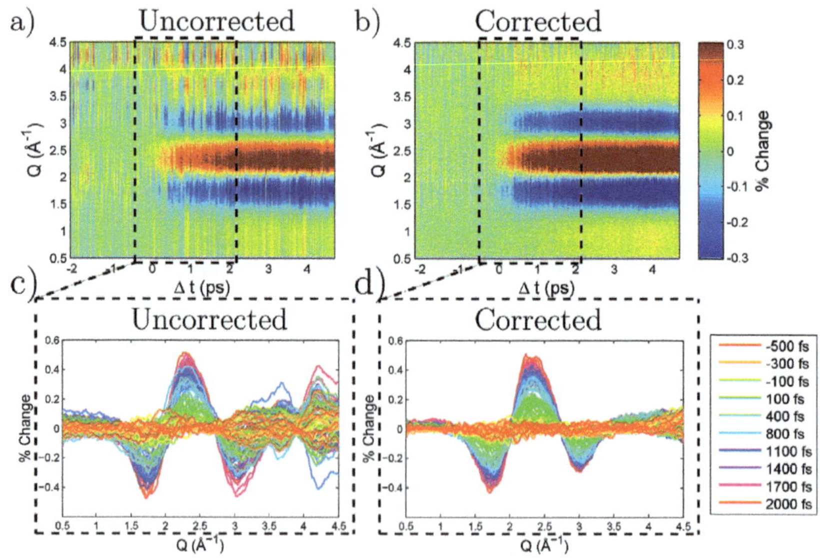


Fig. 14 Azimuthally integrated difference scattering curves as a function of time (a) following the established procedure for XDS data reduction.[27] (b) As the previous panel with the applied 2D corrections presented herein. Panels (c) and (d) show a zoom of −0.5 ps to 2 ps showing the 1D difference scattering curves of the uncorrected and corrected data. The 1D curves in (c) and (d) are smoothed with a 1%-wide moving-average filter.

 

$\Delta S(Q, \Delta t)$. Each bin contains 50 difference scattering curves. Apart from the artefact subtraction, the data in the two figures have been treated identically. Panels (c) and (d) show the individual difference scattering curves from −0.5 ps to 2 ps delays.

In comparison with Fig. 12, the effect of applying the SVD-based corrections is now highly evident. The artefact subtraction removes a large part of the fluctuations clearly evident in panel (a) and effectively makes it possible to analyse the signal in individual time-delay bins. This is prerequisite for robustly analyzing the sub ps signals from typical XDS experiments.

# 10. Gain correction

The corrections presented so far aim to correct the area detector down to a fully linear response. However, it is not a given that each of the millions of individual pixels making up the CSPAD all have the same slope of the response function. This slope represents the gain of each pixel, and they may be individually calibrated by extending eqn (2) with a pixel-to-pixel gain:

$$I_{\text{corrected}} = \left[I_{\text{measured}} - I_{\text{dark}} - \sum_i f_i(I) I_i^{\text{SVD}} - \sum_e f_e(E) I_e^{\text{SVD}}\right] \times \text{gain} \tag{7}$$

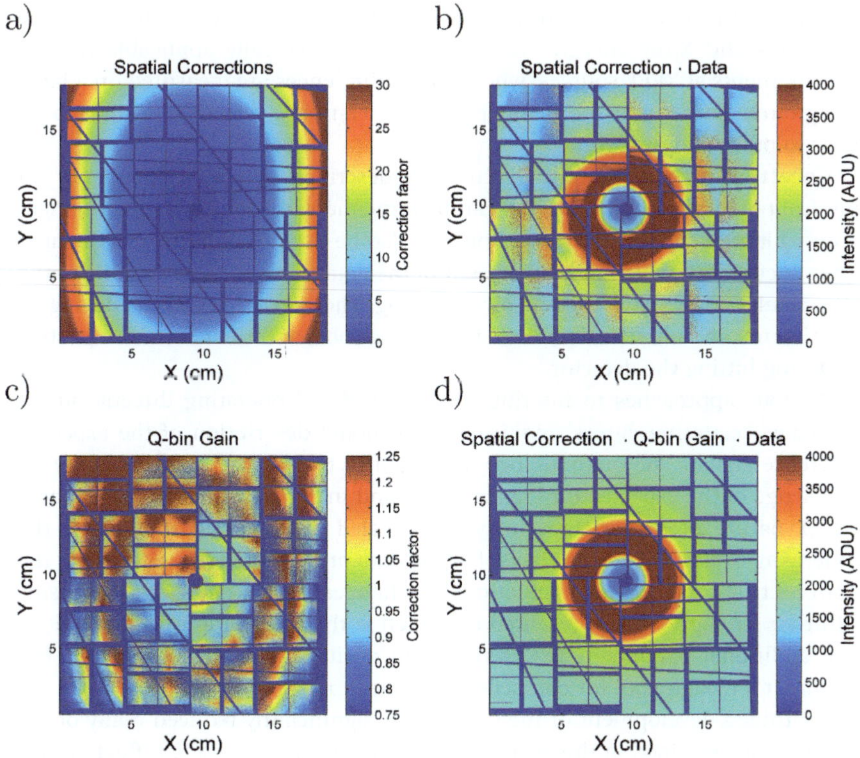


Fig. 15 (a) Calculated spatial corrections for polarisation, absorption through the liquid sheet and solid angle coverage. (b) The corrected data. (c) The $Q$ correction resulting from enforcing azimuthal isotropy. (d) The resulting gain corrected isotropic scattering image.

In order to identify the gain for each pixel we suppose that the steady-state scattering is azimuthally isotropic after spatial corrections have been applied. Assuming this, each pixel value in a given $Q$-bin can be individually scaled to yield the mean of all the pixels in that $Q$-bin, and the scaling value determined in this way represents the pixel-resolved gain function for the detector. This will not change the 1D $\Delta S(Q)$-curves resulting from the azimuthal integration, but the 2D images will show smaller pixel-to-pixel variation and correct for spatial, tile dependent effects.

Fig. 15 shows the procedure for such a $Q$-bin gain correction. Panel (a) shows the spatial/geometry corrections applicable for this particular experimental setup and panel (b) shows the data after applying this correction. Panel (c) shows the resulting $Q$-bin gain correction following the assumption that the scattering signal is isotropic and panel (d) shows the pixel-resolved gain-function applied to the data. The $Q$-bin gain correction is robust and yields the correct gain for each pixel, if both the average gain correction of the detector and the geometry corrections are correct. As an alternative, for an absolute gain correction, the pixels in each $Q$-bin can instead be scaled to the signal from reference measurements of the studied sample.[28]

## 11. Discussion and outlook

In this paper, we have presented a method for robustly removing detector fluctuations that can be correlated with changes in beam parameters such as pulse intensities and X-ray energy. This approach is specifically applicable to XDS measurements at XFEL sources, where small differences are measured on a large background and where the beam properties can vary very significantly, both pulse-to-pulse and over time.

Since the method removes fluctuations described by their dependency on physical parameters, the approach can be implemented robustly without danger of over-filtering the data. The method can also be used to identify and quantify other fluctuations which may be present in the data, such as those relating to the detector electronics or changes in air scattering. The use of this method during an experiment can help identify unwanted fluctuations, and their sources *e.g.* stray scattering hitting the detector.

Previous approaches to filtering XFEL XDS data,[4] operating directly on 1D difference scattering curves, require a good model description of the expected difference signal. The methodology presented in this work requires no *a priori* knowledge of the expected difference signal and in contrast to the 1D-approach, the 2D-nature of the artefacts may be revealed and utilized to identify the underlying causes of such artefacts. For the data presented here, we find that the main contributors to signal fluctuations are changes in X-ray energy and intensity. A further strength of the presented approach is that correlation of fluctuations with additional parameters (*e.g.* X-ray beam position) can easily be tested for, making it a flexible tool for characterizing detectors.

Continued development of the CSPAD has significantly reduced many of the fluctuations described in this work. However, the energy dependent fluctuations are not a detector artefact but instead a result of the shot-to-shot energy variation of the XFEL pulses. The energy variation can be significantly reduced by using a monochromator[29] or self seeding scheme[30] at the cost of X-ray flux. In pink-beam mode, the energy variation can be addressed by assigning new $Q$ vectors to each pixel, based on the incident energy. Implementing this method of correcting for the energy variation, yields qualitatively similar results to the energy-dependent "artefact" subtraction.

The method presented in this paper was developed to address problems arising from the transition of XDS experiments from synchrotrons to XFELs, in particular detector problems associated with fast readout speeds and shot-to-shot variations of the XFEL pulses. Current construction of the European XFEL as well as the LCLS-II upgrade includes an additional increase in repetition rate by 2 orders of magnitude, and thus requires even faster detector readout speeds. These upcoming facilities are currently developing new detectors to meet these requirements. Such new detectors may present similar challenges as encountered with the synchrotron to XFEL transition and we believe the methodology presented here can provide valuable feedback on detector performance.

## Acknowledgements

The authors acknowledge support from DANSCATT and wish to thank the detector group at SLAC National Accelerator Laboratory for discussions and feedback. Authors TBvD and MMN acknowledge support from the ReLiable project (Project no. 11-116792) funded by the Danish Council for Strategic Research Programme Commission on Sustainable Energy and Environment. Portions of this research were carried out at the Linac Coherent Light Source (LCLS) at the SLAC National Accelerator Laboratory. LCLS is an Office of Science User Facility operated for the U.S. Department of Energy Office of Science by Stanford University.

## References

1 M. Harmand, R. Coffee, M. R. Bionta, M. Chollet, D. French, D. Zhu, D. M. Fritz, H. T. Lemke, N. Medvedev, B. Ziaja, S. Toleikis and M. Cammarata, *Nat. Photonics*, 2013, 7, 215–218.

2 H. N. Chapman, P. Fromme, A. Barty, T. a. White, R. a. Kirian, A. Aquila, M. S. Hunter, J. Schulz, D. P. DePonte, U. Weierstall, R. B. Doak, F. R. N. C. Maia, A. V. Martin, I. Schlichting, L. Lomb, N. Coppola, R. L. Shoeman, S. W. Epp, R. Hartmann, D. Rolles, A. Rudenko, L. Foucar, N. Kimmel, G. Weidenspointner, P. Holl, M. Liang, M. Barthelmess, C. Caleman, S. Boutet, M. J. Bogan, J. Krzywinski, C. Bostedt, S. Bajt, L. Gumprecht, B. Rudek, B. Erk, C. Schmidt, A. Hömke, C. Reich, D. Pietschner, L. Strüder, G. Hauser, H. Gorke, J. Ullrich, S. Herrmann, G. Schaller, F. Schopper, H. Soltau, K.-U. Kühnel, M. Messerschmidt, J. D. Bozek, S. P. Hau-Riege, M. Frank, C. Y. Hampton, R. G. Sierra, D. Starodub, G. J. Williams, J. Hajdu, N. Timneanu, M. M. Seibert, J. Andreasson, A. Rocker, O. Jönsson, M. Svenda, S. Stern, K. Nass, R. Andritschke, C.-D. Schröter, F. Krasniqi, M. Bott, K. E. Schmidt, X. Wang, I. Grotjohann, J. M. Holton, T. R. M. Barends, R. Neutze and S. Marchesini, *Nature*, 2011, **470**, 73–77.

3 S. Lee, W. Roseker, C. Gutt and B. Fischer, *Opt. Express*, 2013, **21**, 2037–2041.

4 K. Haldrup, *Philos. Trans. R. Soc., B*, 2014, **369**, 20130336.

5 S. Canton, K. S. Kjær, G. Vanko, T. van Driel, S.-i. Adachi, A. Bordage, C. Bressler, P. Chabera, M. Christensen, A. Dohn, A. Galler, W. Gawelda, D. Gosztola, K. Haldrup, T. Harlang, Y. Liu, K. Møller, Z. Nemeth, S. Nozawa, M. P`apai, T. Sato, K. Suarez-Alcantara, T. Togashi, K. Tono, J. Uhlig, D. Amarasinghe Vithanage, K. Wärnmark, M. Yabashi, J. Zhang, V. Sundström and M. Nielsen, *Nat. Commun.,* (accepted).

6 K. Haldrup, M. Christensen, M. Cammarata, Q. Y. Kong, M. Wulff, S. O. Mariager, K. Bechgaard, R. Feidenhans'l, N. Harrit and M. M. Nielsen, *Angew. Chem., Int. Ed.*, 2009, **48**, 4180–4184.

7 K. Haldrup, M. Christensen and M. M. Nielsen, *Acta Crystallogr., Sect. A: Found. Crystallogr.*, 2010, **66**, 261–269.

8 R. Alonso-Mori, J. Kern, D. Sokaras, T.-C. Weng, D. Nordlund, R. Tran, P. Montanez, J. Delor, V. K. Yachandra, J. Yano and U. Bergmann, *Rev. Sci. Instrum.*, 2012, **83**, 073114.

9 R. Alonso-Mori, J. Kern, R. J. Gildea, D. Sokaras, T.-C. Weng, B. Lassalle-Kaiser, R. Tran, J. Hattne, H. Laksmono, J. Hellmich, C. Glöckner, N. Echols, R. G. Sierra, D. W. Schafer, J. Sellberg, C. Kenney, R. Herbst, J. Pines, P. Hart, S. Herrmann, R. W. Grosse-Kunstleve, M. J. Latimer, A. R. Fry, M. M. Messerschmidt, A. Miahnahri, M. M. Seibert, P. H. Zwart, W. E. White, P. D. Adams, M. J. Bogan, S. Boutet, G. J. Williams, A. Zouni, J. Messinger, P. Glatzel, N. K. Sauter, V. K. Yachandra, J. Yano and U. Bergmann, *Proc. Natl. Acad. Sci. U. S. A.*, 2012, **109**, 19103–19107.

10 S. Herrmann, S. Boutet, B. Duda, D. Fritz, G. Haller, P. Hart, R. Herbst, C. Kenney, H. Lemke, M. Messerschmidt, J. Pines, A. Robert, M. Sikorski and G. Williams, *Nucl. Instrum. Methods Phys. Res., Sect. A*, 2013, **718**, 550–553.

11 W. Zhang, R. Alonso-Mori, U. Bergmann, C. Bressler, M. Chollet, A. Galler, W. Gawelda, R. G. Hadt, R. W. Hartsock, T. Kroll, K. S. Kjær, K. Kubiček, H. T. Lemke, H. W. Liang, D. a. Meyer, M. M. Nielsen, C. Purser, J. S. Robinson, E. I. Solomon, Z. Sun, D. Sokaras, T. B. van Driel, G. Vankó, T.-C. Weng, D. Zhu and K. J. Gaffney, *Nature*, 2014, **509**, 345–348.

12 D. Arnlund, L. C. Johansson, C. Wickstrand, A. Barty, G. J. Williams, E. Malmerberg, J. Davidsson, D. Milathianaki, D. P. DePonte, R. L. Shoeman, D. Wang, D. James, G. Katona, S. Westenhoff, T. a. White, A. Aquila, S. Bari, P. Berntsen, M. Bogan, T. B. van Driel, R. B. Doak, K. S. Kjær, M. Frank, R. Fromme, I. Grotjohann, R. Henning, M. S. Hunter, R. a. Kirian, I. Kosheleva, C. Kupitz, M. Liang, A. V. Martin, M. M. Nielsen, M. Messerschmidt, M. M. Seibert, J. Sjöhamn, F. Stellato, U. Weierstall, N. a. Zatsepin, J. C. H. Spence, P. Fromme, I. Schlichting, S. Boutet, G. Groenhof, H. N. Chapman and R. Neutze, *Nat. Methods*, 2014, **11**, 923–926.

13 H. T. Philipp, M. W. Tate and S. M. Gruner, *J. Instrum.*, 2011, **6**, C11006.

14 S. Boutet and G. J. Williams, *New J. Phys.*, 2010, **12**, 035024.

15 P. R. Ribic and G. Margaritondo, *J. Phys. D: Appl. Phys.*, 2012, **45**, 213001.

16 H. T. Lemke, C. Bressler, L. X. Chen, D. M. Fritz, K. J. Gaffney, A. Galler, W. Gawelda, K. Haldrup, R. W. Hartsock, H. Ihee, J. Kim, K. H. Kim, J. H. Lee, M. M. Nielsen, A. B. Stickrath, W. Zhang, D. Zhu and M. Cammarata, *J. Phys. Chem. A*, 2013, **117**, 735–740.

17 G. E. Box, G. M. Jenkins and G. C. Reinsel, *Time series analysis: Forecasting and control*, Wiley, 2008, p. 746 s.


 

18 D. Zhu, M. Cammarata, J. Feldkamp, D. M. Fritz, J. Hastings, S. Lee, H. T. Lemke, A. Robert, J. Turner and Y. Feng, *J. Phys.: Conf. Ser.*, 2013, **425**, 052033.

19 M. R. Bionta, H. T. Lemke, J. P. Cryan, J. M. Glownia, C. Bostedt, M. Cammarata, J.-C. Castagna, Y. Ding, D. M. Fritz, a. R. Fry, J. Krzywinski, M. Messerschmidt, S. Schorb, M. L. Swiggers and R. N. Coffee, *Opt. Express*, 2011, **19**, 21855–21865.

20 H. T. Lemke, M. Weaver, M. Chollet, J. Robinson, J. M. Glownia, D. Zhu, M. R. Bionta, M. Cammarata, M. Harmand, R. N. Coffee and D. M. Fritz, *Proc. SPIE*, 2013, **8778**, 87780S.

21 W. Press, B. Flannery, S. Teukolsky and W. Vetterling, *Numerical Recipes - The Art of Scientific Computing.*, Cambridge University Press, 1986.

22 R. W. Hendler and R. I. Shrager, *J. Biochem. Biophys. Methods*, 1994, **28**, 1–33.

23 M. Brand, *Linear Algebra Appl.*, 2006, **415**, 20–30.

24 D. Wingate, http://www.mit.edu/wingated/resources.html, 2007.

25 U. Lorenz, K. B. Mller and N. E. Henriksen, *New J. Phys.*, 2010, **12**, 113022.

26 M. Christensen, K. Haldrup, K. Bechgaard, R. Feidenhans'l, Q. Y. Kong, M. Cammarata, M. Lo Russo, M. Wulff, N. Harrit and M. M. Nielsen, *J. Am. Chem. Soc.*, 2009, **131**, 502–508.

27 K. Haldrup, T. Harlang, M. Christensen, A. Dohn, T. B. van Driel, K. S. Kjær, N. Harrit, J. Vibenholt, L. Guerin, M. Wulff and M. M. Nielsen, *Inorg. Chem.*, 2011, **50**, 9329–9336.

28 K. Haldrup, *et al.*, in preparation.

29 D. Zhu, Y. Feng, S. Stoupin, S. a. Terentyev, H. T. Lemke, D. M. Fritz, M. Chollet, J. M. Glownia, R. Alonso-Mori, M. Sikorski, S. Song, T. B. van Driel, G. J. Williams, M. Messerschmidt, S. Boutet, V. D. Blank, Y. V. Shvyd'ko and A. Robert, *Rev. Sci. Instrum.*, 2014, **85**, 063106.

30 J. Amann, W. Berg, V. Blank, F. J. Decker, Y. Ding, P. Emma, Y. Feng, J. Frisch, D. Fritz, J. Hastings, Z. Huang, J. Krzywinski, R. Lindberg, H. Loos, A. Lutman, H. D. Nuhn, D. Ratner, J. Rzepiela, D. Shu, Y. Shvyd'ko, S. Spampinati, S. Stoupin, S. Terentyev, E. Trakhtenberg, D. Walz, J. Welch, J. Wu, A. Zholents and D. Zhu, *Nat. Photonics*, 2012, **6**, 693–698.

Faraday Discussions

PAPER

# Mapping atomic motions with ultrabright electrons: towards fundamental limits in space-time resolution

Stephanie Manz,[a] Albert Casandruc,[a] Dongfang Zhang,[a] Yinpeng Zhong,[a] Rolf A. Loch,[a] Alexander Marx,[a] Taisuke Hasegawa,[ab] Lai Chung Liu,[c] Shima Bayesteh,[d] Hossein Delsim-Hashemi,[e] Matthias Hoffmann,[e] Matthias Felber,[e] Max Hachmann,[e] Frank Mayet,[e] Julian Hirscht,[a] Sercan Keskin,[a] Masaki Hada,[af] Sascha W. Epp,[a] Klaus Flöttmann[e] and R. J. Dwayne Miller*[acg]

Received 14th October 2014, Accepted 12th November 2014
DOI: 10.1039/c4fd00204k

The long held objective of directly observing atomic motions during the defining moments of chemistry has been achieved based on ultrabright electron sources that have given rise to a new field of atomically resolved structural dynamics. This class of experiments requires not only simultaneous sub-atomic spatial resolution with temporal resolution on the 100 femtosecond time scale but also has brightness requirements approaching single shot atomic resolution conditions. The brightness condition is in recognition that chemistry leads generally to irreversible changes in structure during the experimental conditions and that the nanoscale thin samples needed for electron structural probes pose upper limits to the available sample or "film" for atomic movies. Even in the case of reversible systems, the degree of excitation and thermal effects require the brightest sources possible for a given space-time resolution to observe the structural changes above background. Further progress in the field, particularly to the study of biological systems and solution reaction chemistry, requires increased brightness and spatial coherence, as well as an ability to tune the electron scattering cross-section to meet sample constraints. The electron bunch density or intensity depends directly on the magnitude of the extraction field for photoemitted electron sources and electron

[a]The Max Planck Institute for the Structure and Dynamics of Matter, Center for Free Electron Laser Science, Luruper Chaussee 149, Hamburg 22761, Germany. E-mail: dwayne.miller@mpsd.mpg.de
[b]Department of Chemical Engineering, Faculty of Engineering Kyoto University Katsura, Nishikyo-ku, Kyoto, 615-8510, Japan
[c]Departments of Chemistry and Physics, University of Toronto, Toronto, Ontario M5S 3H6, Canada
[d]Institut für Experimentalphysik, Universität Hamburg, Luruper Chaussee 149, Hamburg 22761, Germany
[e]DESY, Notkestrasse 85, 22607 Hamburg, Germany
[f]Materials & Structures Laboratory, Tokyo Institute of Technology, Japan & JST-PRESTO, Yokohama 226-8503, Kawaguchi 332-0012, Japan
[g]The Hamburg Centre for Ultrafast Imaging CUI, Universität Hamburg, Luruper Chaussee 149, Hamburg 22761, Germany

 

energy distribution in the transverse and longitudinal planes of electron propagation. This work examines the fundamental limits to optimizing these parameters based on relativistic electron sources using re-bunching cavity concepts that are now capable of achieving 10 femtosecond time scale resolution to capture the fastest nuclear motions. This analysis is given for both diffraction and real space imaging of structural dynamics in which there are several orders of magnitude higher space-time resolution with diffraction methods. The first experimental results from the Relativistic Electron Gun for Atomic Exploration (REGAE) are given that show the significantly reduced multiple electron scattering problem in this regime, which opens up micron scale systems, notably solution phase chemistry, to atomically resolved structural dynamics.

# 1 Introduction

One of the long standing objectives in science has been to have sufficient spatial and temporal resolution to directly watch atomic motions during the primary motions governing structural transitions.[1,2] This quest is relevant to helping understand at the atomic level of detail, effectively, all classes of structural transitions from issues of condensed matter physics of strongly correlated electron-lattice materials, extreme states of matter, and biological functions. In chemistry, this objective is central to the discipline. The very notion of a chemical process involves the passage of a system from one stable structure to another in which the chemist tries to control conditions to direct the desired process over the myriad of other possible outcomes. The ability to control chemistry largely resides in arranging conditions to lower the barrier and entropic factors to favour a particular chemical pathway. The intellectual pursuit of chemistry is to understand the factors that control barrier heights, within the context of a complex many body potential, and to connect molecular structure to macroscopic properties. This statement is intended to include advances in *ab initio* theory, synthetic strategies, and experimental methods to probe different aspects of the system and system-bath interactions driving chemistry. The unifying concept in this pursuit is to try to identify the key atomic motions that lead the system over the barrier region, or stated differently to picture the transition state region, the critical "point of no return" that connects the reactant to the product free energy surfaces. A direct observation of atomic motions during the transition from one structure to another would provide the most rigorous test possible for various concepts that have evolved for inferring chemical pathways. We could see this directly. This objective has now been met.[3–5] The recent development of ultrabright electron and X-ray sources to provide the necessary structural probes have opened up the femtosecond time domain to atomically resolved dynamics.[2–8] The source technology is rapidly advancing to enable atomic imaging of structural dynamics of ever larger and more complex systems. There are, however, fundamental limits to source brightness and sample issues that need to be overcome to apply this new imaging modality to problems of general interest.

For the purpose of this discussion, we will focus on source requirements with respect to spatial and temporal resolving power to study molecular systems undergoing chemical reactions. In regard to time resolution, the relevant time scales to this problem have been well appreciated since the first connection of microscopic processes to reaction rates in the quest for an absolute rate theory.[9]

 

This work ultimately led to the development of transition state theory. The defining moment of chemistry is captured in the discussion of barrier crossing processes. This time scale varies for different systems but is on the order of few 10 s of femtoseconds (fs; 1 fs = $10^{-15}$ s) to 100 femtosecond time scales. There are faster motions in nature. For example, the OH stretch of water has a period of 10 fs but the motions involved are less than 0.05 Å, hardly chemically relevant. The time scale is defined by the relevant motions over length scales corresponding to 0.1 to 1 Å, leading to distinct changes in structure. More specifically, these motions must in turn be coupled to the nuclear continuum of states to relax the system to a displaced minimum in the many body potential corresponding to a long lived state or stable product state. There must be damping or energy dissipation in the process to relax on the product surface. It is the approximately 100 fs time scale of relaxation along a reaction coordinate that dictates the required time resolution. It is important to realize that not all motions are equally coupled to the reaction coordinate and the degree of coupling rapidly evolves in the barrier crossing region.[4] The highly anharmonic nature of the many body interactions at this far from equilibrium point in nuclear configuration space leads to strong mixing of the otherwise suitable normal mode basis to describe equilibrium fluctuations. Normal modes are accurate descriptions of atomic motions for small excursions over which the potential is well approximated to be harmonic, *i.e.* the nuclear fluctuations are within linear response limits. The far from equilibrium fluctuations that sample the barrier crossing region are highly nonlinear and the most anharmonic modes tend to be the low frequency modes that undergo the largest relative motions. These modes are also the most highly damped as they occur within the highest spectral density of modes for conserving energy and momentum in the relaxation process.[10,11] It is the net excursion along the modes most strongly coupled to the reaction coordinate that define the relevant time scales of chemistry. These modes, whether involving intramolecular processes, solvent controlled chemistry, or the chemistry controlled by protein environmental fluctuations, tend to be in approximately the 100 $cm^{-1}$ frequency range, which gives rise to the generalized requirement of 100 fs time resolution to capture the chemistry.[4]

The spatial resolution requirements are equally challenging in that one needs sub-Å spatial resolution to pull out the important relative motions directing atomic displacements from one stable structure to another. To a first approximation, the spatial resolution requirement exceeds that needed for static structure determination. For example, in the photo-isomerization of retinal, the primary event involved in vision and energy transduction within the rhodopsin family of proteins, the key motions are on the order of 0.1 Å for the carbon atoms at the central bond axis of the isomerization.[12] Save in the act of bond dissociation, chemically relevant motions are between 0.1 and 1 Å. To fully resolve the primary motions requires either hard X-rays or high energy electron probes with sufficiently short carrier wavelengths to resolve these motions. Even so, this level of spatial resolution would be out of range for most sources, however, time-resolved measurements involve differential measurements. The initial structure is known. It is only the relative changes from this known starting point that need to be resolved, not the entire structure for every atomic position to this level of accuracy. The key to being able to resolve chemically relevant motions is that the source serving as the structural probe must be sufficiently stable and bright enough to render signal to noise ratios (SNR) sufficient to pull out these small relative changes in intensity that report the atomic positions. The most sensitive method for determining atomic structure is the use of diffraction methods or imaging in reciprocal space. This observable takes advantage of the $N^2$ amplification of the diffracted signal intensity by scattering off N identically arranged molecules to amplify the scattered signal. The criteria of high brightness and high stability for the source reduce to achieving sufficient signal to noise ratios in the diffracted orders to pull out the structural changes of interest. The quality of sample often plays a deciding factor in the achievable resolution; however, it is only recently that the sources serving as structural probes have reached the level of brightness where sample issues are the main bottleneck.

Differential detection of the structural changes is achieved by comparing the changes in structure following an excitation pulse, which triggers the structural change of interest. Here, it needs to be fully appreciated that, in the absence of a "trigger", the act of barrier crossing driven by background thermal noise is a rare event. For even small barriers (*e.g.* 1 eV), there are less than $1:10^8$ molecules undergoing a barrier crossing event at any instant in time.[4] In principle, one could track the motions of an individual molecule undergoing first order reactions, or molecular collision partners for second order processes, to observe such crossings but the ability to observe such motions and the enormity of collecting sufficient atomic images at the required framing rate to get above background noise, never mind increased demands on source brightness, make this prospect intractable. The processes of interest must be optically triggered to observe the key relative atomic motions connecting two stable structures within the complex, highly dimensional, potential energy surfaces of interest to chemistry. This simple realization has a number of important consequences for the conduct of the experiment. First, the perturbation used to trigger the chemistry must be faster than the ensuing motions of interest. Only femtosecond laser excitation meets this requirement. Furthermore the optical excitation must prepare the system on excited state surfaces that intersect reactive crossings under barrierless conditions. If there is a barrier in the excited state, the time scale for the buildup of the product state is much slower than that of the primary motions involved. In the presence of a barrier, the system will reach a thermal equilibrium within the vibrational modes corresponding to the excited state surface. In this event, there are uncorrelated, thermally, sampled crossings and the details of the relevant motions will not be resolved. The problem reverts to the same statistics as trying to capture barrier crossing events along the ground state. This requirement for an effectively barrierless excited state process not only limits the number of potentially tractable systems for study but has additional consequences. To observe the relevant motions above the background of unexcited molecules requires that on the order of 10% of the molecules or lattice sites are undergoing the photoinduced structural change. The quantum yield for the desired photo-process must be at least this large and there is an upper limit to the degree of excitation. The <100 femtosecond requirements for the excitation pulse necessarily leads to high peak powers, however, the peak power must be maintained at excitation levels on the order of 100 GW $cm^{-2}$ or lower to avoid multi photon ionization artifacts.[13] Basically, above this peak power multi photon processes begin to dominate and even the degree of state preparation becomes ill defined as multi photon excitation to higher lying excited states begin to also significantly contribute. For typical molecular densities, this latter condition means that samples must be on the order of a few microns or thinner or else the required excitation levels (>10% photoproduct formation) lead to excessively high incident peak powers. These are nontrivial sample constraints as one must have sufficient surface area within this thickness restriction to attain adequate signal to noise to stitch together a movie of the atomic motions involved in the process of interest.

Of all the above discussed requirements, the most limiting is the requirement for approximately 10% of the system to be undergoing the phototriggered structural transition. Apart from gas phase systems, which can be rapidly exchanged between laser shots, this level of excitation leads to sample damage in a single shot. This statement is especially true for single crystals that give the highest structural resolution. Normally in diffraction experiments, the upper limit for source brightness is determined by X-ray or electron induced damage. In this class of experiments, the main limitation with respect to sample damage is not the structural probe but the excitation process to trigger the chemistry that damages the sample. It is not enough to have simultaneous atomic resolution with 100 fs time resolution or less but one must attain this experimental parameter space within single shot conditions. Herein lies the real challenge. The source brightness for the structural probe must not only be capable of high space-time resolution but of sufficient intensity to achieve atomic resolution at or approaching single shot conditions.[4,14]

Enormous gains in source brightness have been achieved for both electron and X-ray sources to meet this condition. The major advance in X-ray source brightness was achieved through technical advances made in undulator technology that enabled scaling the free-electron laser concept to the X-ray range. These sources are referred to as 4th generation light sources or X-Ray Free Electron Lasers (XFELs). In comparison to 3rd generation synchrotron sources, XFELs represent an increase in source brightness of over almost ten orders of magnitude.[7] XFELs are not true laser resonators but are based on self amplification of spontaneous emission or SASE sources. As with SASE sources in the visible range, there are huge stochastic fluctuations in intensity and spectrum that make shot to shot normalizing essential to improve the SNR. In addition, there are time jitter issues, with respect to synchronizing the laser excitation used to trigger structural changes and the RF phase of the electron acceleration, that gives rise to 200 fs timing jitter. In principle, XFELs are capable of <50 fs time resolution using time stamping methods as another normalization step to extract the time dependent changes in diffraction efficiency.[15] To date, time-resolved structural dynamics on the prerequisite time scales (100 fs to picosecond) have not been realized with sufficient numbers of diffraction orders to follow structural changes. There has been a recent report of resolved structural changes, albeit not the actual transient structures, on the microsecond time scale[16] and studies of a single diffraction order have been used to infer the role of the lattice in directing material properties for strongly correlated electron lattice systems.[17] There remain a number of technical challenges in the normalization procedures required to attain sufficient SNR and also in the large number of crystal projections needed for X-ray structure determination, prior to the onset of X-ray induced damage, that have hindered efforts in this direction. Further, these sources are not dedicated facilities to this line of inquiry. The most important application for XFELs appears to be in the use of the high spatial coherence and brightness to enable nano- to micro-protein crystallography, prior to the onset of X-ray induced damage.[8] Recent developments in self-assembly of up to M-pixel crystal arrays or photochips[18,19] (solid target solution to sample delivery) and aerosol injectors to give random orientations[20] may solve the last technical obstacle in providing a general solution for collecting sufficient reciprocal space to stitch together atomic movies on the femtosecond time scale.

With respect to electron sources, there are inherent electron–electron repulsion or space charge effects that limit source brightness. This problem was readily appreciated and it was felt that it would not be possible to achieve the needed brightness with electron sources. In this regard, there have been proposals to achieve the necessary space-time resolution with low intensity sources, with single electron pulses being the ultimate limit to completely avoid space charge limits in time resolution.[21,22] The basic problem is that one needs approximately $10^5$ to $10^6$ electrons to have sufficient SNR to invert diffraction patterns to structures and approximately 2–3 orders of magnitude more for real space imaging. In the single electron limit, this requirement translates to over $10^8$ photoexcitation events in the data collection process. It is not possible to have sufficient sample area for such a large number of excitation events, save for gas phase samples where other issues have limited the time resolution. One needs fully reversible systems that can withstand over $10^6$ photocycles between reactive excited and ground electronic states. The prospect of using single electrons remains an elusive prospect with respect to achieving simultaneous femtosecond time resolution and sub-Å spatial resolution to structural transitions. To date, there has only been one report on time resolved dynamics in the single electron limit in which simple thermally excited acoustic phonons were followed on the 100 ps timescale. This process is fully reversible by its very nature as it does not involve a structural change. This work rather reinforces the importance of brightness. In all cases, the brighter the source the better the space-time resolution is, as long as the time resolution is sufficient to follow the dynamics of interest. As will be seen below, ultrabright electron pulses on the order of 10 fs are now possible, which provides sufficient time resolution to follow even the fastest nuclear motions. The first atomic movie with sufficient space-time resolution, *i.e.*, sufficient number of diffraction orders, to resolve the relative atomic motions involved in a structural change was in fact captured with a high brightness electron source.[3] The major advance that made this possible was the realization that high bunch charge electron pulses do not lose space-time correlation at sufficient intensities to achieve single shot structure determination. This realization was made possible through an effectively exact solution to the electron propagation dynamics by solving the coupled equations of motions of some $10^4$ electrons,[23] sufficient for the structure determination of systems with simple unit cells (<3 nm). It was discovered that the transverse velocity spread, related to the transverse spatial coherence, was not significantly affected. The main issue was the longitudinal space charge effect. Two solutions were apparent from these calculations. One solution involved the use of extremely short propagation paths to the sample target to limit pulse broadening to retain 100 fs time resolution. Here, the key realization was that the transverse coherence, even at short propagation distances with typical transverse velocity spreads, was sufficient for atomic resolution for systems with unit cells of a few nm. The other solution to emerge from this work was to explicitly exploit the conserved space-time correlation and development of an extremely linear chirp that naturally develops for nonrelativistic electrons to compress the pulse at the sample position. In this respect, the electrons at the front of the electron pulse experience electron–electron repulsion effects that lead to acceleration or energy exchange between the electrons at the back of the pulse that experience deceleration. For nonrelativistic electrons, the higher energy electrons travel faster than the slow energy electrons. The electrons at the front stay at the front and the electrons at the back stay at the back to conserve the original space-time relationship and develop an extremely linear chirp with propagation. This observation led to the proposal for the use of electric field compression methods or other dispersive elements to compress the electron pulse.[23] The use of a longitudinal half cycle RF cavity to recompress the pulse conserves the transverse coherence and appears to be the best solution to pulse compression for high brightness applications,[24] although there are other applications for improved time resolution.[25] The overall gains in electron source brightness over previous low electron density pulsed sources approaches similar gains as XFELs over prior technology. The compact electron gun design is now capable of $10^5$ to $10^6$ electrons per pulse in a 200 fs pulse that is focusable down to 100 micron radius spot sizes typically used in femtosecond laser experiments.[2] New designs at higher energies will enable <100 fs pulses. The DC-RF pulse compression concept has achieved approximately a factor of 10 increase in brightness with attendant technical issues in RF timing jitter that currently limit the achieved time resolution to approximately 200 fs.[2,26–28] The time resolution can be improved to 30 fs with the use of time stamping methods.[29] In comparison, the compact electron gun is jitter free. These table top electron sources are exceptionally bright. For calibration, taking into account the factor of $10^5$ to $10^6$ higher scattering cross-section for electrons over X-rays for the same energy, this source technology is comparable to $10^{12}$ X-ray photons per pulse for practical laser excitation parameters, *i.e.* these electron sources are on par with XFELs in terms of observed signal levels. Here it is important to keep in mind that peak power limitations require the use of samples on the micron scale or smaller such that there is little distinguishing differences in sample requirements. The diffraction efficiency for X-rays is very small for sample thicknesses on the micron scale such that the observed signal levels are expected to be similar for these two different source technologies. The big differences are that the electron sources are dedicated table top facilities for this class of experiment and are orders of magnitude more stable than the X-ray counterpart. It is these decided advantages that has enabled atomic movies of the primary motions involved in structural transitions to be first captured using electron sources. Advances in laser generated X-ray plasma sources have likewise opened up atomically resolved structural dynamics on the prerequisite sub-ps timescale.[30–32] The most important difference in brightness levels of XFELs sources in relation to all other sources is the near perfect transverse coherence. This difference makes XFELs the ideal source for the study of complex large unit cell systems such as proteins; whereas electron sources are currently limited to transverse coherences suitable for the study of relatively small systems (<10 nm unit cells). This difference may not hold for long, as will be discussed in this report. Electron sources have at least 2 orders of magnitude possible increase in brightness that may close this gap in imaging resolution. In this respect, the recent introduction of relativistic electron sources for this application[33–38] promises to provide the highest spatial-temporal resolution due to the reduction of pulse broadening effects in the relativistic regime. The full potential of relativistic electron sources to achieve the highest possible time resolution relies on removing the initial velocity and temporal spread in the RF acceleration phase. This discussion paper focuses on the first results from the Relativistic Electron Gun for Atomic Exploration (REGAE) that introduces the use of a rebuncher cavity to achieve the ultimate limit in high brightness for relativistic electron sources.

## 2 Ultrafast diffraction and real space imaging with relativistic electrons

### 2.1 Coherence issues: diffraction

The focus of this discussion paper is on electron source technology for atomic imaging dynamics on the primary time scales of chemistry. The image resolution as with any source is related to both the transverse and spatial coherence of the source for diffractive imaging or aberrations in the lens system for real space reconstruction. For X-rays or other light sources, the degree of coherence is defined by the beam divergence, which is a constant after the modality of light generation is fixed. For electrons, the degrees of transverse and longitudinal coherence are coupled parameters that depend on the specific space-time focusing. As an electron pulse is made shorter in duration for a given bunch charge at some point the space charge effects and spatial inhomogeneity will lead to increased beam divergence or transverse velocity spread and corresponding loss of in-plane resolution. The transverse spatial coherence is defined by:[39–41]

$$L_x \approx \lambda/2\pi\sigma_\theta \approx \hbar/\sigma_{p_x}, \tag{1}$$

where $\sigma_{p_z}$ is the angular spread, and $\sigma_{p_x}$ is given by the transverse momentum spread. There is a similar relation for the longitudinal coherence, which is defined as:

$$L_z \approx \hbar/\sigma_{p_z}. \tag{2}$$

The longitudinal coherence is only an issue if one wants to coherently reconstruct an image as in holography. For diffractive imaging, the longitudinal coherence is not the limiting factor. The coherence length, even for strongly space charge broadened pulses, is generally much larger than the unit cells of even large unit cell crystals (>10 nm). Thus, the two most important, coupled, parameters to consider in attaining the required space-time resolution for a particular problem of interest is to match the transverse coherence to the unit cell of interest for the spatial resolution and to adjust the electron bunch density accordingly to give the required electron pulse duration for the desired time resolution. In the former case, the required spatial resolution is achieved by adjusting the source size at the photocathode or introducing an aperture to produce an effective source size that gives a transverse coherence length at the sample position that is a few times the lateral dimension of the unit cell.[41] In terms of time resolution, the required spatial resolution limits the bunch density for a given transverse energy spread and the minimum pulse duration on target is achieved by reducing the number of electrons for a given probe beam size at the sample position, until the longitudinal space charge effects are negligible relative to the time resolution needed. Again, these parameters are coupled.

For most problems of chemical interest, the pulse duration with sufficient electrons for single shot structure determination is the primary concern. The shortest electron pulses are achieved with the highest extraction fields as this minimizes the pulse propagation time to the sample position and thereby minimizes space charge broadening of the time resolution. The highest extraction fields are achieved with RF acceleration methods that avoid charge accumulation and breakdown. Extraction fields an order of magnitude larger than DC electron guns are possible. In addition, as one goes to higher electron energies, there is a reduction in space charge broadening due to the relativistic correction to the electron velocities. To first order, in the relativistic regime, all the electrons would travel at essentially the same velocity, near the speed of light. In this limit, the time broadening problem would reduce solely to issues related to the transverse coherence or transverse velocity spread that would increase with temporal compression in the longitudinal direction. What are the fundamental limits to space-time resolution with electrons? There are new developments both in the generation of ultrabright electrons sources for the shortest possible electron pulse durations/temporal resolution and photocathode concepts that promise to offer orders of magnitude increases in transverse coherence. These higher energy electron sources also have greater penetration depths to greatly reduce sample constraints. We are entering into a new regime for electron sources. The order of magnitude higher field gradients possible with relativistic electron guns and enormous reduction of longitudinal space charge broadening in the relativistic regime hold promise to provide the brightest sources possible for the highest possible spatial-temporal resolution. We highlight the promise to go beyond present limits in diffraction or reciprocal space imaging of chemical and biological problems – for which the dynamics are essential to understanding mechanism and functionality.

## 2.2 Coherence issues: real space imaging – dynamic high energy electron microscopy

The incoherent nature of the spatial phase of the electron pulses across the beam profile presents more difficulties for real space imaging than for diffraction, that are hard to compensate. To acquire an image with reasonable signal to noise ratio at the detector position, one aims for $10^6$ to $10^7$ electrons per acquisition, which needs to be matched to a typical sample area of a few $\mu m^2$. The electron beam needs to be tightly focused to achieve this density. A tight focus on the other hand means a reduction of the local coherence length, which calls for an improved total coherence of the electron pulse from its source on. The photoemitted electron image at the cathode has to be as close to point-like as possible. Laser spot sizes below 10 μm are difficult to achieve at the cathode position inside an RF cavity. Typically, the laser is coupled into the cavity from a window several tens of cm away from the cathode surface, which is insufficient working distance to focus tighter than 10 microns. It is possible to use structured cathodes to give the desired source size. However, for current designs, single shot imaging will only be possible with a partially coherent mode and thus relies on $Z$-contrast ($Z =$ atomic number) rather than phase contrast. By spatial filtering with an aperture, the scattered electrons lead to an intensity drop in the imaged electron beam relative to the position of the scattered object for bright-field imaging. In dark field imaging, the unscattered electron beam is stopped by a negative aperture (beam stop) and the scattered electrons are imaged. In both cases, the favorable situation is for each electron to scatter less than once on average. Too little scattering gives low imaging contrast and calls for high electron doses, whereas multiple scattering broadens the minimal spot size of point scatterers in the sample. The number of scattering events depends on the elastic cross-section $\sigma_{\mathrm{el}}$, which depends on the electrons' energy and the atomic number $Z$ of the scattering material. An approximate expression for $\sigma_{\mathrm{el}}$ is given by:[43]

$$\sigma_{\mathrm{el}} = \frac{c^2 h^2 Z^{4/3}}{\pi E_0 \beta}, \tag{3}$$

where $E_0$ is the electron rest energy, and $\beta = v/c$ with the electron's velocity $v$. The mean free path $\Lambda_{\mathrm{el}}$ of an electron in a material with density $\rho$ is related to the cross-section by $\Lambda_{\mathrm{el}} = A/N_{\mathrm{a}}\sigma_{\mathrm{el}}\rho$. While the expression for $\sigma_{\mathrm{el}}$ gives a helpful approximation, it can differ from the actual cross-section by a factor of 2. We use the approach presented in ref. 42 to include relativistic corrections in the calculation. In Fig. 1, we compare calculations for the elastic cross-section using the Schrödinger equation, the relativistically corrected Schrödinger equation, as well as the Dirac equation. Solvers for both the relativistic Schrödinger equation and Dirac equation have been implemented into the EDICo-code.[44] The calculations were performed for unpolarized electrons. We find no significant difference between the relativistic Schrödinger and Dirac result and conclude that effects of spin can be neglected, as also stated in ref. 42. In terms of beam propagation

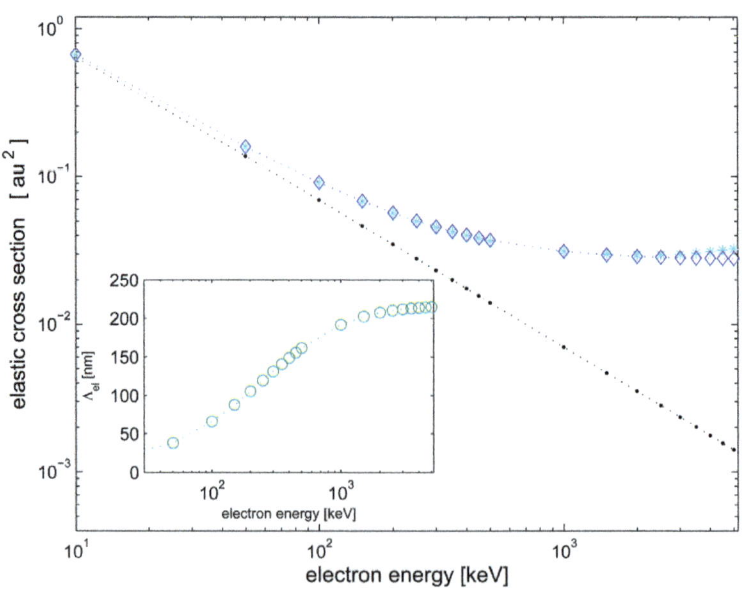


Fig. 1 Comparison of different methods to calculate electron cross-sections in aluminum using the relations given in ref. 42. The inset shows the according mean free path in aluminum for the cross-sections derived from the Dirac equation.

through the sample and effect on imaging, the electron beam broadens its energy distribution in the sample by inelastic collisions. Typically, the total cross-sections for elastic and inelastic scattering show a qualitatively similar behavior with energy. The ratio $\eta = \sigma_{\text{inel}}/\sigma_{\text{el}}$ was experimentally found to be $\eta = 20/Z$[45] and theoretically $\eta = 26/Z$.[43] Since the cross-section for elastic and inelastic scattering follow similar behavior with energy, $\eta$ is almost constant in the keV to MeV range. One can assume that multiple scattering will be reduced by the same factor for both elastic and inelastic scattering. Damage due to ionization will also be reduced by the same argument, but knock-on damage will be more severe than for lower energy devices.

When the sample thickness exceeds the mean free path in the material, one faces two problems in transmission electron microscopy that stem from multiple scattering: multiple elastic scattering increases the minimal spot size of a point like object; and inelastic scattering similarly leads to aberrations in the reconstructed real space image. Thus, the resolution will be reduced and distorted. Such effects are reduced in electron microscopy with relativistic electrons, and allow in the case of light, low $Z$, organic materials, the study of micron thick samples with nanometer resolution. High resolution microscopy with thin samples in which atomic resolution is desired may better be performed in conventional electron microscopes employing aberration correctors[46] that are not currently available for relativistic electron imaging systems.

We studied the feasibility of dynamic real space imaging with pulse propagation simulations in ASTRA. The simulation is launched from the sample position on. A pulse of $10^4$ particles with a Gaussian density profile and an energy of 3 MeV is tracked through the magnetic field of a realistic electromagnetic solenoid, which is a likely implementation for the objective lens. The solenoid field was calculated using the CST software suite.[47] A schematic of the imaging system implemented in the existing REGAE setup is shown in Fig. 2(a). With the pulse, 500 test particles were used to probe the mean field space charge that travel along the propagation direction. For a chosen spot size of 10 μm at the sample, the test particles travel off-center at a distance of 1 μm. The relative energy spread for the pulse is chosen to be $2 \times 10^{-4}$. The transverse momentum spread of the test particles was chosen to match scattered electrons up to typical angles of 5 mrad. We performed the simulation for three different focal lengths of 1.1 cm, 5 cm, and 10 cm. The results depending on the bunch charge are presented in Fig. 2(b). We tracked the test particles until they reached the image plane, and then computed the RMS width to obtain a point spread function. We find higher resolution for lower focal lengths, which is to be expected from the respective spherical aberration coefficients. Furthermore, we observe a decrease in resolution depending on charge, which we interpret as the result of the inhomogeneous space charge field in the bunch, an effect resembling spherical aberrations.[48] The simulation has been repeated for different pulse lengths and spot sizes. We find a similar scaling for all pulse shapes, as visible in Fig. 2(b). In order to reach 1 nm resolution the charge must be kept in the fC range. Longer pulses would certainly relax the issue of space charge induced aberration, as proposed in,[49,50] but then additional, higher harmonic cavities need to be employed to decrease the energy spread that comes with the generation of a long pulse in an RF cavity, which has not been treated in these works.

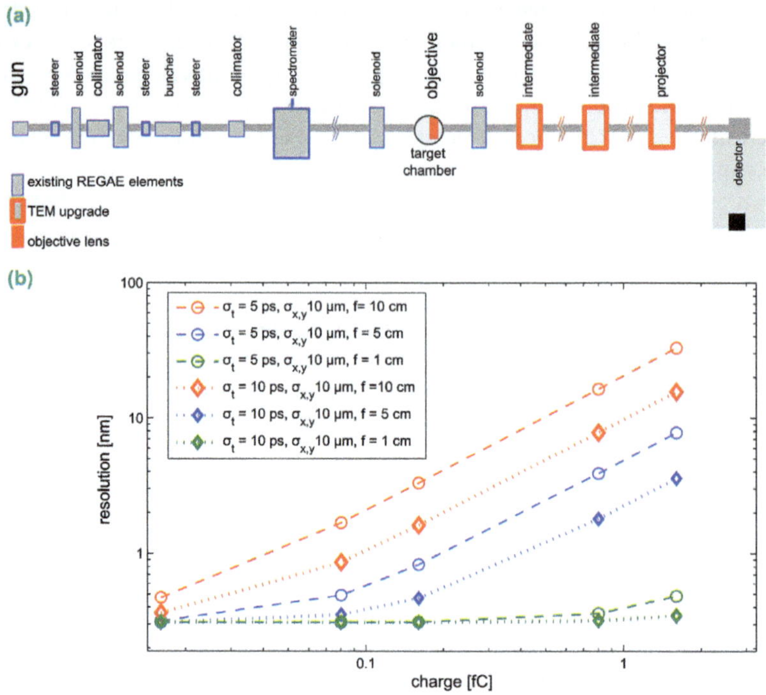


Fig. 2 (a) Schematic of the implementation of a TEM mode within the existing REGAE setup. (b) Effect of charge and objective focal length on resolution for a pulse length of 5 and 10 ps. The beam shape is Gaussian for all cases.

### 2.3 Electron diffraction with REGAE

**2.3.1 The REGAE apparatus.** The relativistic electron gun for atomic exploration, REGAE, is a radio frequency (RF) accelerator based gun design operating in a pulsed mode with 50 bursts per second. It is mainly dedicated to the study of structural dynamics in solid, liquid and gas phase of organic and inorganic samples. As elucidated in the introduction, such a device has to fulfill the requirements of short pulses, high brightness, and high coherence at the same time. Electron pulses are created *via* photoemission by 266 nm, 500 fs pulses of a frequency-tripled Ti:Sa laser impinging the cathode within a 1.5 cell S-band cavity. For the studies presented here, we use either molybdenum, gold, or $Cs_2Te$ as the cathode material. By means of a coaxial on-axis coupler, a travelling 3 GHz S-band wave of some micro-second length launched by a klystron is fed in resonance into the gun cavity, where field gradients of up to 100 MV $m^{-1}$ can be obtained leading to electron energies up to 5 MeV. Running the cavity at high fields has, apart from the beam energy itself, the major advantage that high bunch charge densities can be extracted from the cathode overcoming conventional space charge limitations. Depending on the cathode material and its quantum efficiency, bunch charges in the range from pC for gold and other metal cathodes and 10 pC for $Cs_2Te$ can be extracted. For typical diffraction experiments, we aim for a bunch charge of 100–200 fC (about 1 million electrons) at the target position, as a compromise between beam quality and signal strength. The elements of the beamline are depicted in

 

Fig. 3. The gun section is followed by electro-magnetic solenoids and collimators to shape the beam. The collimators are a critical feature and require a high degree of engineering to ensure proper alignment under the ultrahigh vacuum conditions needed to run the RF cavity. These elements are used for collimation and beam clean up as well as to reduce dark current travelling in the beam direction. One of the most important distinctions of this relativistic electron gun is the low dark current[51] that is primarily due to the very low base pressure of $10^{-10}$ mbar employed in this design concept. Additional beam diagnostic stations allow for determination of transverse electron bunch parameters, *e.g.* beam width, energy, and energy spread. The beam charge can be detected without disturbing the beam by means of a resonator, where the electron pulse induces a TM01 mode at 1.3 GHz. The bunch charge can be measured with fC resolution.[52] The other key feature to this design is the introduction of a rebunching cavity to obtain the shortest possible pulses at the sample position. The design employs the same klystron amplifier with an RF splitter to drive both cavities, which requires finding unique phase solutions for which the pulse is optimally compressed for a given amplification. The rebunching cavity is designed to temporally refocus the electron pulse at a distance of 5.5 m from the photocathode position. For detection, a scintillator on a fiber optics plate in the beam line converts the electrons' density distribution to photons, which are coupled to an EMCCD chip outside of vacuum. The high dynamic range of the EMCCD combined with its high sensitivity down to single electron detection allows for simultaneous detection of the undiffracted beam and diffraction patterns without the necessity for a beam block (Fig. 4).

**2.3.2 Beam dynamics and time resolution at REGAE.** The beam dynamics after the gun are controlled by solenoid magnets, indicated by arrows in Fig. 3. Depending on the demands on beam size at the sample position and extent of the diffraction pattern on the detector screen, we either focus the beam through the target on the screen with solenoid lenses (see Fig. 3), or – if a higher intensity at the target is called for – have a small spot at the sample and refocus on the detector position. Ideal beam dynamics for diffraction in terms of a phase advance[53] approaching 90° has been found only without an intermediate focus.

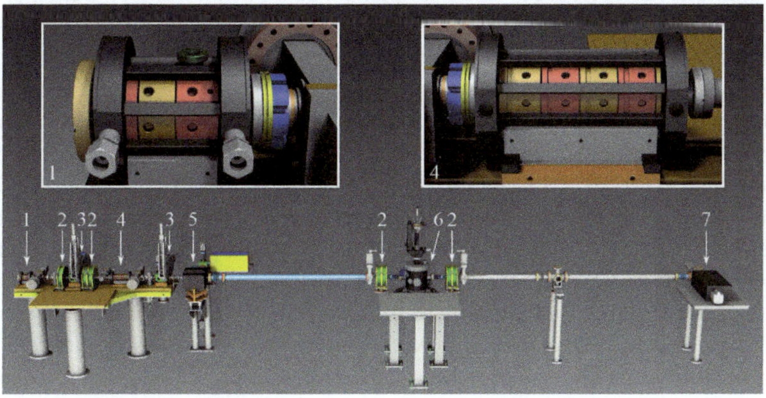


Fig. 3 REGAE scheme. Indicated with numbers are: (1) gun cavity, (2) solenoids, (3) collimators, (4) rebunching cavity, (5) dipole for electron energy measurements, (6) target interaction chamber, and (7) detector.

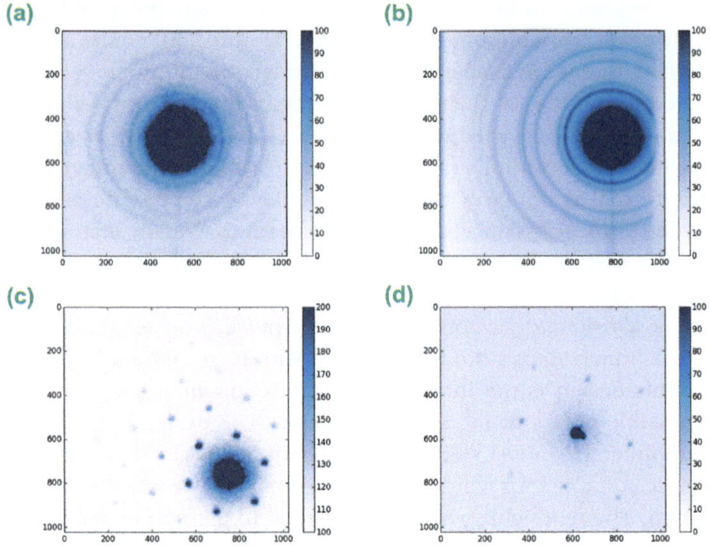

Fig. 4 Example diffraction patterns from (a, b): polycrystalline gold, (c): MoS2, (d): Si; the quality of the diffraction is particularly notable for the single crystal samples MoS2 and Si.

Beam size, divergence, and coherence are controlled by the photocathode and gun cavity, and transverse beam shaping elements, as solenoids, collimators and steerers. At a distance of 1.3 m from the accelerator, a rebunching cavity is placed in the beam line. It consists of four cells and is operated off-crest compared to the arrival time of the electron bunch. The rebunching cavity is driven by a fraction of the RF field that also supplies the gun. The relative phase between the cavities is adjusted with phase shifters. While the parameter space of this coupled cavity system is finite, efforts have to be made to overcome complications due to reflections and unwanted feedback between the two coupled cavities.[54] The electron bunch can, if higher acceleration is required, also be injected on-crest into the rebunching cavity. Potential beam parameters along the propagation axis are summarized in Fig. 5. For this example, the beam dynamics for a bunch of 100 fC is tracked by the ASTRA[55] simulation software, which includes space charge fields. For the initial beam parameters, we use an RMS spot size of the laser at the cathode of 10 μm, a pulse length of 0.5 ps and a charge of 100 fC. For this charge, the beam size at the target reaches 500 μm with a transverse coherence length above 10 nm. With the buncher tuned to the appropriate phase, a bunch length of 10 fs can be expected at the longitudinal focus at the sample position. We then use a solenoid close to the sample position to focus the beam and diffraction pattern to the detector. It needs to be emphasized that a pulse duration of 10 fs is sufficiently short to capture even the fastest possible nuclear motions involved in chemical processes. The achievable transverse coherence is also notable in that the design objectives were to provide sufficient transverse coherence to enable the study of systems as large as proteins. The unit cell of protein systems capable of being phototriggered to execute their biological functions such as heme proteins and the family of rhodopsin photoactive systems are on the order of 6 nm such that the transverse coherence is sufficient for this purpose.

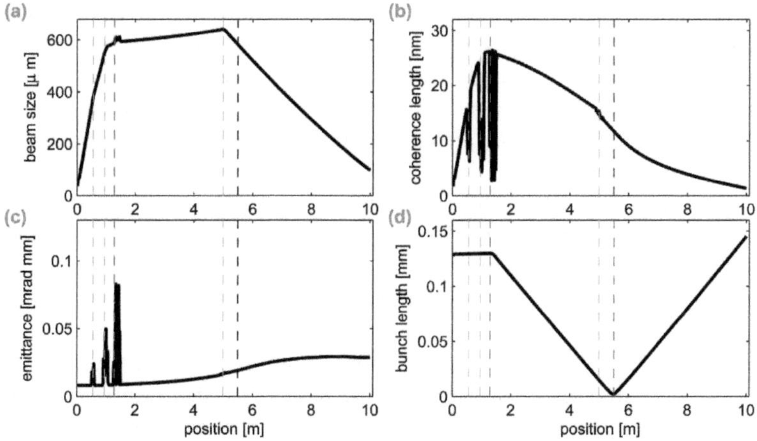


Fig. 5 ASTRA simulations for diffraction with REGAE: (a) RMS beam size showing the beam dynamics, (b) transverse coherence length, (c) emittance, and (d) bunch length in mm, corresponding to 10 fs at the longitudinal focus. The dashed lines indicate the position of solenoids (green), the rebunching cavity (red) and the sample position (black).

## 2.4 Static diffraction from aluminum: thickness dependence

One of the main limitations in the use of electron structural probes is the effect of multiple scattering due to the intrinsically high scattering cross-section of electrons. The increased electron penetration depth and reduced scattering angles for relativistic electrons promise the feasibility of ultra-fast electron diffraction (UED) with thick samples. In conventional electron microscopes and electron guns at intermediate energies (DC guns) typical sample thicknesses range from few tens of nanometers to 200 nm, depending on the density and atomic weight of the material. However, many interesting systems, such as protein crystals, water soluble crystalline systems, and solution phase chemistry call for a probe with a higher penetration depth. From the point of view of material science, a higher penetration depth supports the study of bulk properties with minimal effects from interfaces or surface effects. The relativistic electron energy regime has not been extensively studied and, as discussed above, only approximate relationships in terms of scaling relations for elastic and inelastic scattering are available. The effect of multiple electron scattering on image resolution needs to be explicitly determined. We explored these possibilities using films of polycrystalline aluminum of increasing thickness up to 800 nm. The calculated mean free path according to the inset in Fig. 1 is just above 200 nm. The films are supported by a 30 nm thin SiN membrane. Diffraction patterns were recorded by integrating 300 shots of 180 fC pulses at 4 MeV. Before azimuthal integration of the diffraction pattern a background image is subtracted. The incoherent contribution and the diffuse diffraction ring from the SiN membrane are modeled with a multi-component function and removed as well. These components are shown as gray lines in Fig. 6(a). The off-center peak at about 0.78 $Å^{-1}$ corresponds to diffuse diffraction from the supporting SiN membrane. We need to determine the effect of multiple electron scattering as a function of thickness for a given system and use the $Z$ dependence of the scattering to provide general guidelines for limiting this effect on reciprocal (diffraction) or real space resolution. Coherently scattered electrons interfere after multiple elastic scattering, thus, the diffracted intensities are altered compared to the single scattering regime. Inelastic scattering leads to a broadening of both probe beam and diffracted electrons and can affect the resolution. The addition of both effects is visible as an increase of diffraction ring width, as shown in Fig. 7(c). The question whether multi-scattering effects are necessary to describe such diffraction patterns in theoretical treatments or the degree to which structural information is lost are still open. Modulations in diffraction intensity might be averaged out in polycrystalline samples, as expected from multi-slice calculations. To aid in resolving this issue, we compare the ratio of diffraction intensities for different rings in Fig. 7(a) with the expected result for a completely random powder sample, obtained from the software CrystalMaker.[56] The rather surprising result is that the multiple electron scattering effects, while occurring, are not significantly altering the relative ratios of the different diffraction orders as needed for dynamic structure determination. The small deviation from ideal results is within signal to noise limits for some of the diffraction orders. To be fully conclusive, the effect of substructure, grain size and texture within the polycrystalline sample would need to be studied by a different method.

This study gives a direct determination of the effect of multiple electron scattering on a well-defined system. In the present case, Al is close in $Z$ to organic systems. We conclude that time-resolved electron diffraction studies of systems, at least for light materials, is possible up to micrometer thickness in the relativistic energy range. Here is especially important to note that these experiments always deal with systems of known structure. The experimental challenge is to be able to discern small changes in the relative intensities of the different diffraction

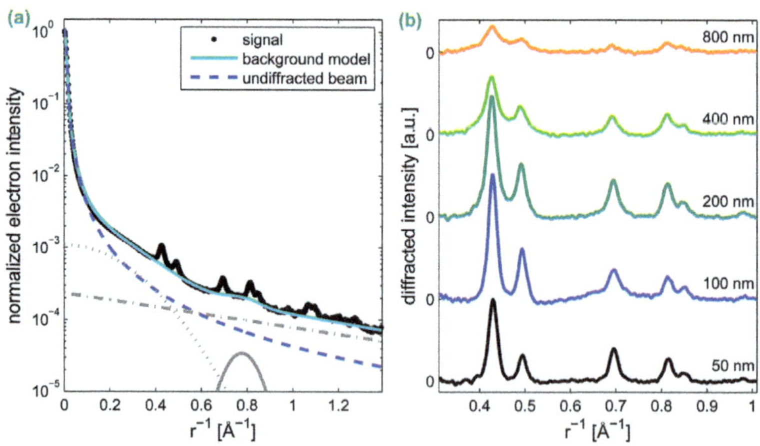


Fig. 6 (a) Example for electron diffraction from poly-crystalline aluminum film after azimuthal integration. The signal (black) has been normalized to the transmitted peak intensity. A reference image including dark current has been subtracted before integration. Further subtraction removes the un-diffracted beam (dashed blue line) and incoherent background (solid blue line) from the coherently diffracted signal. Incoherent contributions are shown in grey. (b) Diffracted intensity after background subtraction. The diffraction patterns are stacked for better visibility.

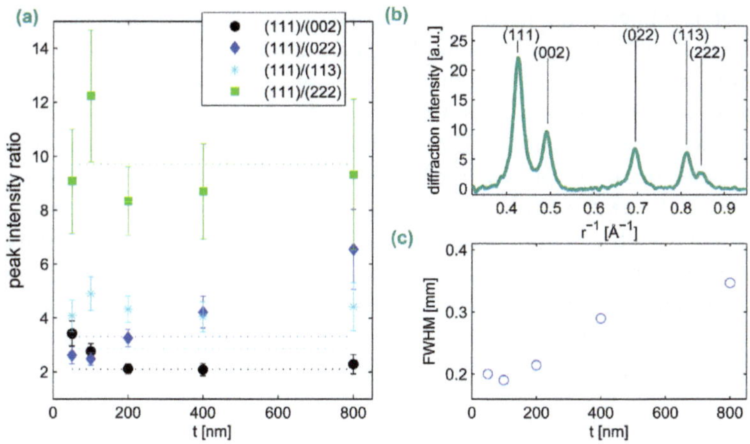


Fig. 7 (a) Ratio of peak intensities in diffraction pattern. The dashed lines gives the scattering ratio for a perfect powder sample in the single scattering regime for comparison. An example diffraction pattern for 200 nm thickness is shown in (b), where the integrated diffraction rings are labeled with Miller indices. (c) Width of rings in diffraction pattern with increasing thickness.

orders. This control study shows that it is possible to extract structural dynamics from samples as thick as several hundred nanometers to micron thickness at this energy range. This feature of relativistic electrons was one of the primary motivations for developing this source as it dramatically increases the ease of sample preparation and opens up solution phase chemistry to atomic inspection, which is much easier to realize with micron scale path lengths as opposed to 100 nm needed for nonrelativistic electron studies.

# 3 Ultrafast dynamics with REGAE

## 3.1 Finding $t = 0$

In order to perform time-resolved electron diffraction studies, one needs to characterize the instrument's time resolution. Before such studies can be performed in detail however, the temporal overlap ($t = 0$) between pump and probe, in our case excitation laser and electron pulse, needs to be found. Here, a process that is reversible under photoexcitation is favorable. We used the laser induced plasma formation as a means to determine the electron-laser timing using a copper mesh at the target position with an incident fluence of 14.9 mJ cm$^{-2}$. Images of the mesh are acquired on the detector in transmission mode. The magnification is close to one, so that the lines with a spacing of 24.5 μm are not resolved in these images. Similar experiments have been presented in ref. 57. Examples of images prior to and after laser irradiation are shown in Fig. 8(a) and (b). The excitation laser with a width of 500 μm, a wavelength of 800 nm and a pulse width of ≈80 fs induces a plasma in the center of the mesh, from where a charged plume is expelled. The probe electron beam is partially deflected by this spatial gradient in charge, which can be observed as the depleted area in the electron density distribution, as depicted in Fig. 8(b). The electron beam seems to be affected on a smaller area than expected from the laser beam size, which we interpret as higher order absorption processes in the high intensity center of the laser beam. The higher order multi photon absorption processes lead in turn to a reduction in the effective beam diameter giving rise to the observed effect. The excitation arrival time was varied using a delay stage and a series of shadow images was recorded scanning through the time delay between the laser excitation and electron probe pulses. In order to retrieve $t = 0$ from the shadow images, we analyzed the RMS intensity changes in the central area of the beam. As plotted in Fig. 8(c), we observed a step in the signal, the onset of which is assumed to indicate the zero – time point. Furthermore, we evaluated the width of the differential change in intensity and find a FWHM of 0.9 ps where the changes are occurring fastest in the center of the imaged profile. The result indicates that the electron bunch length is in the sub picosecond range.

### 3.2 Laser induced structural dynamics of gold

The first structural dynamics at REGAE were observed in a similar geometry to that used for the plasma studies, but under single shot conditions. Free standing membranes of polycrystalline gold were used to observe a laser induced structural change. The membranes with a size of 200 μm by 200 μm were excited one by one with a laser pulse, and the resultant diffraction images were recorded. At a fluence of 150 mJ cm$^{-2}$ and 100 fs pulse duration, the excitation leads to melting and therefore damage of the sample, and is not reversible. For the data presented here, 300 of these membranes fabricated on a chip were used. Again, example images before and after laser irradiation are shown in Fig. 9(a) and (b).

The diffraction image after laser irradiation still exhibits diffraction rings, but the amplitude of the outer rings is diminished. This change indicates a loss of long range order due to melting of the sample. The timescale of this change from ordered to a disordered state is shown in Fig. 9(c). Here, the diffraction rings' amplitudes are plotted *versus* delay time between laser and electron bunch. Each time step was repeated three times. Within five picoseconds after excitation, the outer rings are fading, while the ring indexed with (111) gains in intensity. This gain is attributed to an overlap of the (111) diffraction ring and diffraction from a liquid state. At longer delay times we find the sample completely destroyed. The above results are the first time-resolved studies of REGAE that clearly show its capabilities for single shot structure determination on the 100 fs time scale. The

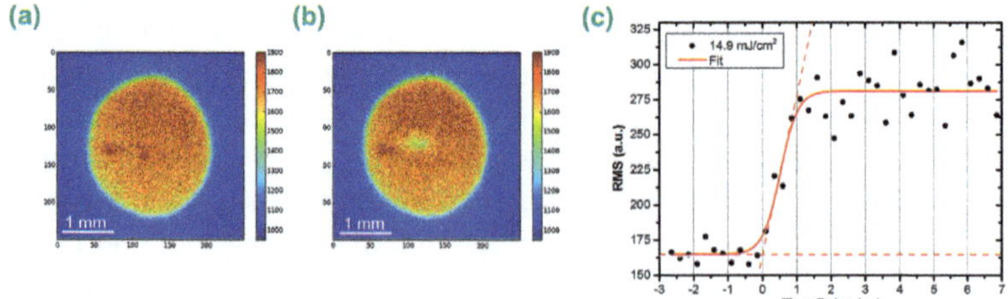


Fig. 8 Shadow images of a copper mesh before (a) and after (b) laser irradiation. The deflection of the electron bunch can be observed as a depleted region in the electron beam profile. The analysis of the signal is shown in (c): the signal RMS in the central region is evaluated and shows a step with a time constant of 0.9 ps.

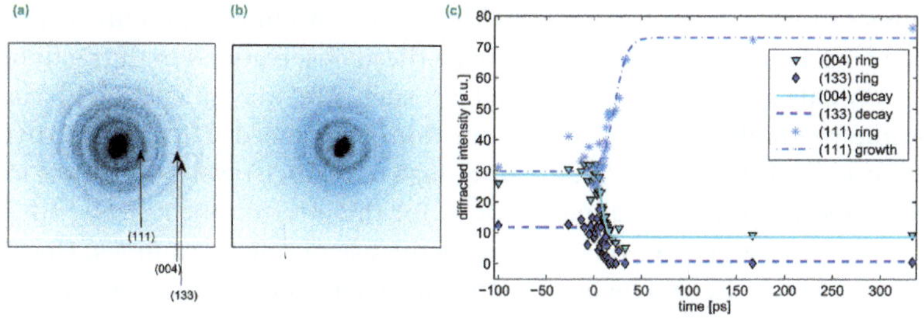


Fig. 9 Single shot diffraction patterns before (a) and after (b) laser excitation. (c) Intensity of the diffraction rings as a function of time delay between pump and probe pulse.

observed dynamics for the particular fluences match very well those found earlier.[58] The present time resolution is limited by RF timing issues and it must be stated that this limit is primarily due to problems in finding a suitably fast reference system to optimize the phase of the rebunching cavity for optimal compression. The best solution will be to introduce an RF streak camera for enabling optimization of the pulse compression process.[62] This diagnostic tracks the phase jitter of the RF so it is not suitable for jitter compensation between the RF and the laser excitation. However, by simultaneously recording the central beam position with the high dynamic range detector, it will be possible to achieve an overall time resolution in the 10 fs range, as demonstrated using photo-triggered streak cameras for time stamping and correction of the jitter.[26–29]

# 4 Concluding remarks

The present space-time limitations have been increased to the 10 femtosecond time scale with changes in diffraction intensities corresponding to net atomic displacements of less than 0.05 Å with single shot capabilities. This space-time resolution is sufficient now to capture the fastest nuclear motions involved in chemistry. In this respect, the single most restrictive condition on space-time resolution, not discussed within the context of the electron source physics, is the sample constraints. The use of electron structural probes to provide table top high brightness sources by its nature requires very thin samples, on the order of few 10 s of nanometers to 100 nm scale for nonrelativistic electrons. It is the sample limitations that impose the brightness condition in the first place. These are precious samples in which excitation induced damage and thermal effects limiting sampling rate have dictated the need for ultrabright sources. There are some samples however that are just not amenable to such short path lengths. Two prime examples of great interest are solution phase chemistry and protein dynamics. Although, there have been major advances in the development of nanofluidics for this expressed purpose,[59,60] it is much simpler to use micron scale flow conditions to achieve the desired flow rates for sample exchange. Similarly for protein systems, the highest spatial resolution is achieved in diffraction and the functionally relevant motions involving the protein are spatially distributed leading to very small net RMS atomic motions, yet are central to the proteins role in transducing stored chemical potential into function.[4] If we are ever to make the key connections between correlated motions inherent to particular protein structures or motifs as part of a generalized understanding of protein structure–function correlations, we will need to make these observations in diffraction. All protein crystals are water soluble and as such are not amenable to classic microtome methods to fabricate 100 nm thin slices needed for nonrelativistic electron probes. Herein is the major advantage of relativistic electrons over lower energy sources. There are scaling relations for the electron mean free path that indicate higher energy electrons will enable the use of thicker samples. However, the difference inelastic and inelastic scattering need to be determined as there are approximations as is evident from the noted differences between experimental measurements and calculated scattering cross-sections. Most important, sample quality has an enormous influence on spatial resolution (*vide infra*). In the present work, we have conducted a systematic study of the effect of sample thickness on the apparent structure resolution. By comparing both experiment and theoretical calculations based on multi-slice methods for dealing with multiple electron scattering, we find that it is possible to use samples up to thicknesses of nearly 1 micron for materials of low $Z$ corresponding to the range of most organic and biological systems of interest. This finding also means that it should be possible to dramatically relax the engineering requirements for introducing flowing liquid cells with electron transparent SiN windows. In terms of spatial resolution limits for discerning structural changes, it is important to emphasize further that the measurement does not have the same requirements as needed to resolve a previously unknown structure. The experiment relies on knowing the initial and in most cases final structure and observing the differential displacements in atomic positions that connect one structure to another. This observation gives us directly information on the key modes coupled to the reaction coordinate and enables observation of the reduction in dimensionality that occurs in curve cross regions between reactant and product surfaces. This feature is central to chemistry and it is this enormous reduction in dimensionality in barrier or curve crossing regions that ultimately makes chemistry a transferable concept. These relatively small but highly correlated motions are observable as long as the relative ratios of the various diffraction orders are conserved and within the dynamic range of the structurally induced changes in intensity. We therefore conclude that it will be possible to pull out the structural changes for many classes of time-resolved diffraction experiments, involving samples that were previously thought intractable due to either their liquid nature or sample thickness issues. This finding that multiple electron scattering is not as big an effect as sample quality and attenuation with sample thickness has enormous implications for improving sample preparation methods. This study also reaffirms the primary motivation for the development of relativistic electron sources for atomically resolved structural dynamics as this source greatly relaxes sample requirements for all classes of study. Future investigations comparing data to calculated diffraction patters will give more insight into the sensitivity that is required to further confirm our observation.

In terms of space-time resolution, relativistic sources also offer the highest possible source brightness for a given photocathode. The high-Q RF cavities provide the highest possible field gradients to accelerate the electron bunch up to the relativistic regime prior to any excessive space-charge broadening effects can spoil the space time correlation of the electrons in the accelerating field. This condition is central to the use of rebunching cavities to correct the small velocity spread from the field gradient. As shown in the present work, based largely on well tested simulations, pulses as short as 10 fs with electron bunch charges of $10^6$ to $10^7$ electrons per pulse for typical matching of electron beam to laser excitation beam parameters are achievable. It should be noted that there is a great deal of effort going into further increasing field gradients for next generation electron accelerators and light sources. The use of extremely high peak power few cycle THz pulses has recently been demonstrated to give uniform acceleration and deceleration of nonrelativistic seed electrons with estimated field gradients in excess of 1 GV $m^{-1}$ or more than an order of magnitude larger than the best RF cavities.[61] This development would lead to correspondingly shorter electron pulses for a given bunch density. With respect to improving time resolution, concepts similar to streaking with RF deflecting structures might be implemented in the future as well.[62]

The most important advances in electron source brightness will deal with the transverse coherence limitations. The present photocathode typically has initial transverse energy spreads of 0.2 to 0.6 eV. This initial energy spread is the greatest limiting factor in the transverse coherence and ability to image complex molecular systems such as proteins. Even within this range of transverse momentum spread, it has been possible to now track atomic motions for large molecular systems with unit cells of >3 nm that are comparable to unit cells of many important protein systems.[63] Further decreases in transverse momentum spread will make higher resolution of even minute motions as well as scaling to larger and more complex systems. In this respect, the use of ultracold atom sources using threshold ionization of Rydberg states have given results close to the theoretical quantum limit for fermions with respect to the minimum transverse velocity spread, yielding effective electron temperatures of 10 K or less than 0.6 meV.[41] This approach, however, is limited in electron density by the relatively low atom density achievable in ultracold atom traps. In order to get sufficient electron numbers for near single shot conditions, the initial beam size approaches mm dimensions that cancel the net reduction in emittance. The electron beam is indeed very cold but the angular distribution does not lead to much improvement in the transverse coherence at the sample position due to the very large source size at the photocathode. We contend that the brightest photocathode will still be based on solid state cathodes. Significant progress has been made lately in understanding the scattering issues and boundary conditions for photoemission from solid state systems. Recent studies of trialkali photocathodes have given evidence for thermally limited photoemission with energy spreads of 160 meV at room temperature,[64] and factors of 10 reduction are clearly possible with excitation closer to threshold. These photocathodes also enjoy relatively high quantum yields as found for other semiconductor photocathodes such as the routinely used $Cs_2Te$ materials. Although it is not straightforward, it should be possible to use cryogenically cooled solid state photocathodes. The challenge is to thermally isolate the cathode section of the electron gun while still having sufficient conduction to avoid excessive charge build up and break down at the high fields used for extraction. It should nevertheless be possible to go to 10 K for suitable source sizes that would give an increase in source brightness of approximately 2 orders of magnitude over the commonly used photocathodes in femtosecond electron diffraction studies. In this event, it will be possible to scale up the experiments to the study of effectively any molecular system and potentially even biological systems.

The above technical achievements could be made in the relatively short term. As with all experiments aimed at structure determination, the quality of the sample ultimately dictates the limits. Future efforts will need to focus on new methods of introducing samples into the vacuum environment of electron diffraction or real space imaging systems. With the relatively recent advent of nanotechnology into this domain, there is reason to believe that it will soon be possible to introduce large arrays of crystals, with the potential for M-pixel arrays of nanocrystals into the sample chambers. High throughput methods developed for XFEL experiments could also be adopted for electron use with appropriate design of environmental enclosures around the sample viewing area. Ion milling and *in situ* growth of 2D crystals and nanocrystals will need to be explored.[19] In all cases, the lessons learned in increasing source brightness will pay enormous dividends for increasing the spatial resolution using electrons for static structure determinations. In the end, it will be the sample that will dictate the space-time resolution limits – as it should be.

## Acknowledgements

We gratefully acknowledge the support received for this collaborative effort by DESY (member of the Helmholtz Association (HGF)), the Max Planck Society, and the Hamburg Centre for Ultrafast Imaging, CUI (DFG-EXC1074). We are indebted to R. Brinkmann and H. Dosch of DESY for supporting this project. Also, we thank Hamburg University (UHH) for supporting the setup of the experiment as well as the MCS, MIN, MKK, MSK, MVS groups of DESY for countless support during setup and experiment. In this respect we particularly like to thank M. Hüning, J. Herrmann, R. Jonas, R. Kammering, T. Kozak, A. Labudda, U. Laustoer, S. Lederer, L. Lilje, D. Lipka, J. Müller, K. Müller, U. Mavric, I. Peperkorn, S. Pfeiffer, H. Schlarb, C. Schmidt, M. Seebach, M. Titberidze, M. Walla, S. Weisse, and K. Witt for great support. T. Hasegawa thanks J. K. Dewhurst for providing EDiCo electron diffraction code and his technical support. We are also grateful for technical support from D. Gitaric and J. Gonschior.

## References

1 J. C. Polanyi and A. H. Zewail, *Acc. Chem. Res.*, 1995, **28**, 119–132.
2 J. R. Dwyer, C. T. Hebeisen, R. Ernstorfer, M. Harb, V. B. Deyirmenjian, R. E. Jordan and R. J. D. Miller, *Philos. Trans. R. Soc., A*, 2006, **364**, 741–778.
3 B. J. Siwick, J. R. Dwyer, R. E. Jordan and R. J. D. Miller, *Science*, 2003, **302**, 1382–1385.
4 R. J. D. Miller, *Annu. Rev. Phys. Chem.*, 2014, **65**, 583–604.
5 M. Gao, C. Lu, H. Jean-Ruel, L. C. Liu, A. Marx, K. Onda, S.-y. Koshihara, Y. Nakano, X. Shao, T. Hiramatsu, G. Saito, H. Yamochi, R. R. Cooney, G. Moriena, G. Sciaini and R. J. D. Miller, *Nature*, 2013, **496**, 343–346.
6 M. J. J. Vrakking and T. Elsaesser, *Nat. Photonics*, 2012, **6**, 645–647.
7 P. Emma, R. Akre, J. Arthur, R. Bionta, C. Bostedt, J. Bozek, A. Brachmann, P. Bucksbaum, R. Coffee, F.-J. Decker, Y. Ding, D. Dowell, S. Edstrom, A. Fisher, J. Frisch, S. Gilevich, J. Hastings, G. Hays, P. Hering, Z. Huang,

 

R. Iverson, H. Loos, M. Messerschmidt, A. Miahnahri, S. Moeller, H.-D. Nuhn, G. Pile, D. Ratner, J. Rzepiela, D. Schultz, T. Smith, P. Stefan, H. Tompkins, J. Turner, J. Welch, W. White, J. Wu, G. Yocky and J. Galayda, *Nat. Photonics*, 2010, **4**, 641–647.

8 A. Barty, J. Küpper and H. N. Chapman, *Annu. Rev. Phys. Chem.*, 2013, **64**, 415–435.

9 P. Hänggi and M. Borkovec, *Rev. Mod. Phys.*, 1990, **62**, 251–341.

10 D. D. Dlott, *J. Opt. Soc. Am. B*, 1990, **7**, 1638–1652.

11 R. J. D. Miller, *Annu. Rev. Phys. Chem.*, 1991, **42**, 581–614.

12 P. Altoè, A. Cembran, M. Olivucci and M. Garavelli, *PNAS*, 2010, **107**, 20172–20177.

13 V. I. Prokhorenko, A. Halpin, P. J. M. Johnson, R. J. D. Miller and L. S. Brown, *J. Chem. Phys.*, 2011, **134**, 085105.

14 R. J. D. Miller, *Science*, 2014, **343**, 1108–1116.

15 M. Harmand, R. Coffee, M. R. Bionta, M. Chollet, D. French, D. Zhu, D. M. Fritz, H. T. Lemke, N. Medvedev, B. Ziaja, S. Toleikis and M. Cammarata, *Nat. Photonics*, 2013, **7**, 215–218.

16 A. Aquila, M. S. Hunter, R. B. Doak, R. a. Kirian, P. Fromme, T. a. White, J. Andreasson, D. Arnlund, S. Bajt, T. R. M. Barends, M. Barthelmess, M. J. Bogan, C. Bostedt, H. Bottin, J. D. Bozek, C. Caleman, N. Coppola, J. Davidsson, D. P. DePonte, V. Elser, S. W. Epp, B. Erk, H. Fleckenstein, L. Foucar, M. Frank, R. Fromme, H. Graafsma, I. Grotjohann, L. Gumprecht, J. Hajdu, C. Y. Hampton, A. Hartmann, R. Hartmann, S. Hau-Riege, G. Hauser, H. Hirsemann, P. Holl, J. M. Holton, A. Hömke, L. Johansson, N. Kimmel, S. Kassemeyer, F. Krasniqi, K.-U. Kühnel, M. Liang, L. Lomb, E. Malmerberg, S. Marchesini, A. V. Martin, F. R. N. C. Maia, M. Messerschmidt, K. Nass, C. Reich, R. Neutze, D. Rolles, B. Rudek, A. Rudenko, I. Schlichting, C. Schmidt, K. E. Schmidt, J. Schulz, M. M. Seibert, R. L. Shoeman, R. Sierra, H. Soltau, D. Starodub, F. Stellato, S. Stern, L. Strüder, N. Timneanu, J. Ullrich, X. Wang, G. J. Williams, G. Weidenspointner, U. Weierstall, C. Wunderer, A. Barty, J. C. H. Spence and H. N. Chapman, *Opt. Express*, 2012, **20**, 2706–2716.

17 E. Möhr-Vorobeva, S. L. Johnson, P. Beaud, U. Staub, R. De Souza, C. Milne, G. Ingold, J. Demsar, H. Schaefer and A. Titov, *Phys. Rev. Lett.*, 2011, **107**, 036403.

18 A. Zarrine-Afsar, C. Müller, F. O. Talbot and R. J. D. Miller, *Anal. Chem.*, 2011, **83**, 767–773.

19 A. Zarrine-Afsar, T. R. M. Barends, C. Müller, M. R. Fuchs, L. Lomb, I. Schlichting and R. J. D. Miller, *Acta Crystallogr., Sect. D: Biol. Crystallogr.*, 2012, **68**, 321–323.

20 J. C. H. Spence, U. Weierstall and H. N. Chapman, Reports on progress in physics, *Phys. Soc.*, 2012, **75**, 102601.

21 A. H. Zewail, *Annu. Rev. Phys. Chem.*, 2006, **57**, 65–103.

22 P. Baum and A. H. Zewail, *Proc. Natl. Acad. Sci. U. S. A.*, 2006, **103**, 16105–16110.

23 B. J. Siwick, J. R. Dwyer, R. E. Jordan and R. J. D. Miller, *J. Appl. Phys.*, 2002, **92**, 1643.

24 T. van Oudheusden, E. F. de Jong, S. B. van der Geer, W. P. E. M. O. t. Root, O. J. Luiten and B. J. Siwick, *J. Appl. Phys.*, 2007, **102**, 093501.

25 L. Veisz, G. Kurkin, K. Chernov, V. Tarnetsky, a. Apolonski, F. Krausz and E. Fill, *New J. Phys.*, 2007, **9**, 451–451.

26 M. Gao, H. Jean-Ruel, R. R. Cooney, J. Stampe, M. D. Jong, M. Harb, G. Sciaini, G. Moriena and R. J. D. Miller, *Opt. Express*, 2012, **20**, 799–802.
27 R. P. Chatelain, V. R. Morrison, C. Godbout and B. J. Siwick, *Appl. Phys. Lett.*, 2012, **101**, 081901.
28 R. P. Chatelain, V. Morrison, C. Godbout, B. van der Geer, M. de Loos and B. J. Siwick, *Ultramicroscopy*, 2012, **116**, 86–94.
29 M. Gao, Y. Jiang, G. H. Kassier and R. J. Dwayne Miller, *Appl. Phys. Lett.*, 2013, **103**, 033503.
30 T. Elsaesser and M. Woerner, *Acta Crystallogr., Sect. A: Found. Crystallogr.*, 2010, **66**, 168–178.
31 V. Juvé, M. Holtz, F. Zamponi, M. Woerner, T. Elsaesser and a. Borgschulte, *Phys. Rev. Lett.*, 2013, **111**, 217401.
32 M. Woerner, F. Zamponi, Z. Ansari, J. Dreyer, B. Freyer, M. Prémont-Schwarz and T. Elsaesser, *J. Chem. Phys.*, 2010, **133**, 064509.
33 J. B. Hastings, F. M. Rudakov, D. H. Dowell, J. F. Schmerge, J. D. Cardoza, J. M. Castro, S. M. Gierman, H. Loos and P. M. Weber, *Appl. Phys. Lett.*, 2006, **89**, 184109.
34 P. Musumeci, J. T. Moody, C. M. Scoby, M. S. Gutierrez and M. Westfall, *Appl. Phys. Lett.*, 2010, **97**, 063502.
35 R. Li, W. Huang, Y. Du, L. Yan, Q. Du, J. Shi, J. Hua, H. Chen, T. Du, H. Xu and C. Tang, *Rev. Sci. Instrum.*, 2010, **81**, 036110.
36 P. Musumeci, J. T. Moody, C. M. Scoby, M. S. Gutierrez, M. Westfall and R. K. Li, *J. Appl. Phys.*, 2010, **108**, 114513.
37 C. M. Scoby, P. Musumeci, J. T. Moody and M. S. Gutierrez, *Phys. Rev. Spec. Top.–Accel. Beams*, 2010, **13**, 022801.
38 Y. Murooka, N. Naruse, S. Sakakihara, M. Ishimaru, J. Yang and K. Tanimura, *Appl. Phys. Lett.*, 2011, **98**, 251903.
39 A. M. Michalik, E. Y. Sherman and J. E. Sipe, *J. Appl. Phys.*, 2008, **104**, 054905.
40 G. Sciaini and R. J. D. Miller, *Rep. Prog. Phys.*, 2011, **74**, 096101.
41 W. J. Engelen, M. a. van der Heijden, D. J. Bakker, E. J. D. Vredenbregt and O. J. Luiten, *Nat. Commun.*, 2013, **4**, 1693.
42 A. Rother and K. Scheerschmidt, *Ultramicroscopy*, 2009, **109**, 154–160.
43 L. Reimer and H. Kohl, *Transmission Electron Microscopy*, Springer, 2008.
44 K. Dewenhurst and D. Rankin, *EDICo – Electron diffraction code*, http://edico.sourceforge.net/.
45 R. F. Egerton, *Phys. Status Solidi A*, 1976, **663**, 663–668.
46 B. Kabius, P. Hartel, M. Haider, H. Müller, S. Uhlemann, U. Loebau, J. Zach and H. Rose, *J. Electron Microsc.*, 2009, **58**, 147–155.
47 CST STUDIO SUITE, CST Computer Simulation Technology AG, http://www.cst.com.
48 M. Reiser, *Theory and Design of Charged Particle Beams*, Wiley, 2nd edn, 2008.
49 R. K. Li and P. Musumeci, *Phys. Rev. Applied*, 2014, **2**, 024003.
50 D. Xiang, F. Fu, J. Zhang, X. Huang, L. Wang, X. Wang and W. Wan, *Nucl. Instrum. Methods Phys. Res., Sect. A*, 2014, **759**, 74–82.
51 H. Delsim-Hashemi and K. Flöttmann, *Proceedings of IPAC2014*, Dresden, Germany, 2014.
52 D. Lipka and M. Seebach, *Proceedings of IBIC2013*, 2013, pp. 872–875.
53 S. Bayesteh and K. Floettmann, *Proceedings of IPAC2014*, Dresden, Germany, 2014, pp. 3515–3517.

 

54 M. Hoffmann, H. Kay, U. Mavric, H. Schlarb, C. Schmidt, W. Jalmuzna, T. Kozak and A. Piotrowski, *Proceedings of IPAC2013*, Shanghai, China, 2013, pp. 2938–2940.

55 K. Floettmann, *ASTRA - A Space Charge Tracking Algorithm*, http://www.desy.de/mpyflo/, 1997–2014.

56 CrystalMaker Software Ltd, http//www.crystalmaker.com.

57 C. M. Scoby, R. K. Li and P. Musumeci, *Ultramicroscopy*, 2013, **127**, 14–18.

58 R. Ernstorfer, M. Harb, C. T. Hebeisen, G. Sciaini, T. Dartigalongue and R. J. D. Miller, *Science*, 2009, **323**, 1033–1037.

59 J. M. Grogan, N. M. Schneider, F. M. Ross and H. H. Bau, *J. Indian Inst. Sci.*, 2012, **92**, 295.

60 C. Mueller, M. Harb, J. R. Dwyer and R. J. D. Miller, *J. Phys. Chem. Lett.*, 2013, **4**, 2339–2347.

61 S. Carbajo, L. Wong, E. Nanni, R. Miller and F. Kärtner, in *Frontiers in Optics*, 2014, OSA Technical Digest, http://dx.doi.org/10.1364/FIO.2014.FTh2A.2.

62 K. Floettmann and V. V. Paramonov, *Phys. Rev. ST Accel. Beams*, 2014, **17**, 024001.

63 S. Hayes, 2014, in preparation.

64 L. Cultrera, I. Bazarov, A. Bartnik, B. Dunham, S. Karkare, R. Merluzzi and M. Nichols, *Appl. Phys. Lett.*, 2011, **99**, 152110.

Faraday Discussions

PAPER

# Two-dimensional sum-frequency generation (2D SFG) spectroscopy: summary of principles and its application to amyloid fiber monolayers

Ayanjeet Ghosh, Jia-Jung Ho, Arnaldo L. Serrano, David R. Skoff, Tianqi Zhang and Martin T. Zanni*

Received 12th September 2014, Accepted 23rd October 2014
DOI: 10.1039/c4fd00173g

By adding a mid-infrared pulse shaper to a sum-frequency generation (SFG) spectrometer, we have built a 2D SFG spectrometer capable of measuring spectra analogous to 2D IR spectra but with monolayer sensitivity and SFG selection rules. In this paper, we describe the experimental apparatus and provide an introduction to 2D SFG spectroscopy to help the reader interpret 2D SFG spectra. The main aim of this manuscript is to report 2D SFG spectra of the amyloid forming peptide FGAIL. FGAIL is a critical segment of the human islet amyloid polypeptide (hIAPP or amylin) that aggregates in people with type 2 diabetes. FGAIL is catalyzed into amyloid fibers by many types of surfaces. Here, we study the structure of FGAIL upon deposition onto a gold surface covered with a self-assembled monolayer of methyl-4-mercaptobenzoate (MMB) that produces an ester coating. FGAIL deposited on bare gold does not form ordered layers. The measured 2D SFG spectrum is consistent with amyloid fiber formation, exhibiting both the parallel ($a^+$) and perpendicular ($a^-$) symmetry modes associated with amyloid β-sheets. Cross peaks are observed between the ester stretches of the coating and the FGAIL peptides. Simulations are presented for two possible structures of FGAIL amyloid β-sheets that illustrate the sensitivity of the 2D SFG spectra to structure and orientation. These results provide some of the first molecular insights into surface catalyzed amyloid fiber structure.

## Introduction

Humans and animals suffer from diseases caused by aggregated proteins, such as cataracts and type 2 diabetes.[1] The protein involved in type 2 diabetes, called the islet amyloid polypeptide (IAPP or amylin) is particularly interesting because its sequence is conserved across species except for the middle segment spanning residues 20–29.[2] This region includes the FGAIL segment in humans, which forms

*Department of Chemistry, University of Wisconsin, 1101 University Avenue, Madison, WI 53706, USA. E-mail: zanni@chem.wisc.edu; Tel: +1 608 262 4783*

well-ordered amyloid fibers under many conditions.[2–6] In rats, mice, and other species that do not contract type 2 diabetes, the corresponding FGAIL sequence contains 1 or more mutations that prevent it from aggregating.[2] Thus, there is a standing hypothesis that the FGAIL and nearby amino acids dictate amyloid fiber formation and thus the propensity for contracting type 2 diabetes.[2,4,6,7] That hypothesis is now being questioned and refined by more recent experiments on the full-length protein using solid-state NMR and by 2D IR spectroscopy.[8,9] Nonetheless, it is clear that the FGAIL sequence plays an important role in dictating amyloid fiber formation.

It is also established that amyloid fiber formation is catalyzed by many types of surfaces.[10–12] IAPP and other amyloid proteins are charged and so are attracted to charged surfaces. Human IAPP is amphipathic. Surfaces, whether or not they are charged, may also serve to effectively increase the protein concentration since diffusion in two rather than three dimensions will more quickly produce protein–protein contacts from which aggregation can commence. Indeed, there are many images collected using atomic force microscopy of amyloid fibers formed on mica and other surfaces.[13,14] However, the atomic structures of these amyloid fibers are not known and it is not known if these fibers have the same structure as when formed spontaneously without surfaces present. Indeed, polymorphs, which are alternate structures of amyloid fibers formed from different conformations of the same protein, are now recognized as a common phenomena,[15] probably because amyloid fibers are kinetically trapped structures that do not ever reach the structure that would be dictated by the thermodynamic minima.

There have been a number of experiments and simulations of the FGAIL and NFGAIL segments of IAPP.[4,16–18] In solution, they form very flat and long fibers, as determined by TEM.[4] Additionally, the β-sheet structure was confirmed by FTIR.[19] *N*-Methylated derivatives of these peptides have even been shown to inhibit fibrillation of full length hIAPP.[20] ssNMR studies on the longer 20–29 segment found the aggregates form into anti-parallel sheets, and X-ray crystallography has shown NNFGAIL to form parallel sheets.[21] While no such experiments have been carried out on FGAIL or NFGAIL, various simulation studies have alternatively predicted either the parallel or anti-parallel structure. Interestingly, a recent study found that CFGAILSS forms anti-parallel amyloid fibrils in solution but when constrained to gold nanoparticle surfaces through thiol tethering, it forms parallel sheets.[22] This result exemplifies the importance of surfaces to the detailed structural dynamics of amyloids. In this case, the specific thiol–gold interaction is critical to the morphology. Evidence for other specific interactions have been shown to be critical to amyloid morphology: MD simulations on the related NFGAIL fragment and its mutant NAGAIL highlighted the importance of Phe–Phe T-stacking interactions in the stability of fibrils, as the latter could not form stable oligomers.[23] Such interactions could serve as targets for forming non-covalent monolayers and nanostructures out of naturally self-assembling peptides or for tuning the morphology of a given peptides amyloid structure.

In this article, we collected heterodyne-detected 2D SFG spectra of the FGAIL peptide on a gold surface coated with an ester functional group *via* the self-assembly of a monolayer of methyl-4-mercaptobenzoate. The goal is to probe the structure of the assembled peptides on these surfaces and to understand the role of hydrophobicity, hydrogen bonding, and ring stacking contacts that could play a role in structure formation. The 2D SFG spectrometer is built by adding a mid-IR pulse shaper to a standard broadband SFG spectrometer. The pulse shaper is used to generate the additional pulse pair from which the second dimension is generated. Phase cycling of the pulse pair is used to set the rotating frame and subtract background noise without having to lose signal from chopping. The spectra are simulated using an excitonic coupling model and selection rules calculated for 2D SFG spectra. The comparison of simulation with experiment provides a preliminary model for the assembly of FGAIL on the gold and ester coated surface.

## Introduction to 2D SFG spectroscopy

Sum-frequency generation spectroscopy is one of the most common techniques for studying surfaces and interfaces.[24–30] SFG spectroscopy is a second-order technique in which a resonant mid-IR pulse combines with a non-resonant visible pulse to produce a signal at the sum of the two frequencies.[25,27] Fig. 1A shows a SFG pulse sequence in which a femtosecond laser pulse is followed by a picosecond visible pulse, as is typical for a broadband SFG spectrometer.[31] (The envelope of the visible pulse is shown as a decaying exponential as would be generated by a filter. Other methods for creating the narrowband visible pulse would have different shapes in the time-domain.) When the signal is scaled back to the mid-IR by subtracting the visible pulse frequency, one obtains a vibrational spectrum. The frequencies in the spectrum would match those of IR and Raman

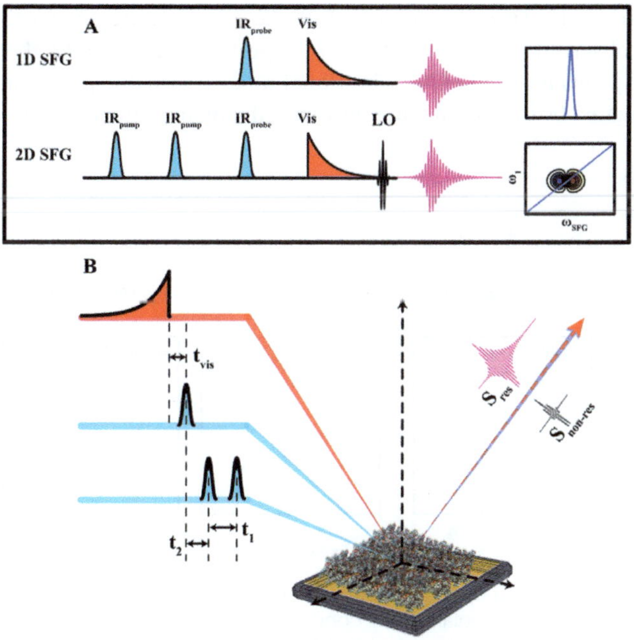


Fig. 1 (A) Outline of the pulse sequence used in the 1D and 2D SFG experiments. (B) Schematic of the beam geometry in 2D SFG experiments. All pulses are p-polarized. $t_1$, $t_2$ and $t_{vis}$ denote the delays between the two mid-IR pump pulses, between the pump and probe mid-IR pulses and the probe and visible pulses respectively.

spectra, if they could be measured, with intensities that scale as a combination of the two since both the vibrational transition dipole and the polarizability contribute. In addition, some modes may not contribute at all if they are isotropically distributed, which is what makes SFG spectroscopy surface selective.

Two modifications are needed to generate a 2D SFG spectrum. First, a 2D spectrum requires a second vibrational frequency-axis to be measured, which necessitates additional infrared laser pulse(s). One can use a picosecond mid-IR laser pulse whose center frequency is scanned, but that limits the time-resolution of the experiment and causes spectral distortions.[32] Early 2D IR experiments used frequency scanning methods, but that has now largely given way to impulsive experiments that only use femtosecond laser pulses.[33–35] Thus, rather than use a single picosecond mid-IR pulse, we use two femtosecond laser pulses, as shown in Fig. 1A. The time-delay, $t_1$, between the pulses is scanned and the data Fourier transformed to give the second frequency axis but with no loss in time-resolution and improved lineshapes. Second, the signal must be heterodyne detected, otherwise phase distortions are present that make it difficult to interpret the spectra or to compare to either FTIR or 2D IR spectra which are always heterodyne detected. For the samples studied here, heterodyne detection is automatically accomplished with the non-resonant signal from the gold interface. For non-metallic surfaces, one needs to add an additional pulse commonly called a local oscillator. Because our pulse pairs are generated by a pulse shaper, as described below, they are collinear, resulting in the beam geometry shown in Fig. 1B.

Shown in Fig. 2 are simulated heterodyne-detected 2D SFG spectra for two coupled oscillators, such as would be measured for a small organic molecule, a dipeptide, a DNA base, or any other molecule with two coupled vibrational modes like two carbonyls. Three variations of 2D SFG spectra are shown to illustrate commonly observed features.[36] In the first 2 spectra (Fig. 2A and B), there is a pair of diagonal peaks for each of the vibrational modes, with one peak exactly on the diagonal created by transitions that only include the fundamentals ($\nu = 0 \rightarrow 1$) of

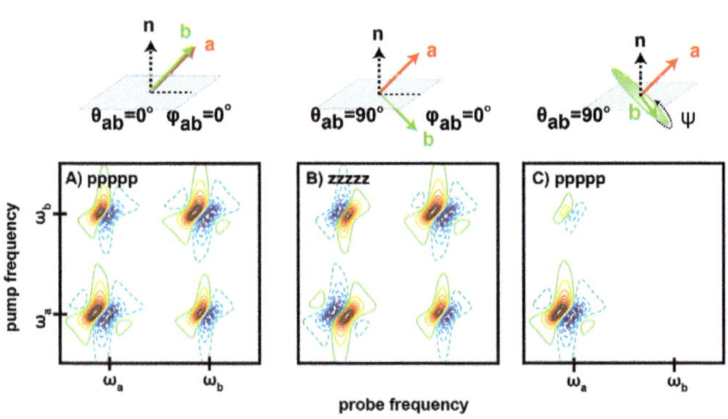


Fig. 2 Simulated 2D SFG spectra for a model system of two coupled oscillators for different orientations and polarization conditions. (A) The relative angle between the modes is set to zero. Spectra are calculated for the ppppp polarization condition. (B) The relative angle between the modes is zero, but they have different tilt angles ($\theta_{ab} = 90°$) with respect to the surface normal. Spectra are calculated for zzzzz polarization. (C) 2D SFG spectra for a $\phi$- and $\psi$-averaged system. The mode a is assumed to lie along the rotation axis. Spectra are calculated using all p-polarized pulses (adapted from ref. 36).

each mode and an out-of-phase peak that includes a transition to the overtone ($\nu = 1 \rightarrow 2$). Because of coupling, pairs of cross peaks are also observed that correspond to transitions involving the vibrational modes from both oscillators. The phases of the peak pairs will depend on the directions of the transition dipole and polarizability tensors with respect to the plane of the surface, thereby providing information on the orientation of the molecule (compare Fig. 2A and B). If a particular mode is isotropically distributed across the interface, as shown in Fig. 2C, then it will not be SFG active and so will not appear in either 1D SFG or along the diagonal in 2D SFG spectra. However, cross peaks can still appear to SFG forbidden modes. That occurs because an SFG *inactive* mode is still pumped by the mid-IR pulses, and so can emit a signal if it is coupled to an SFG *active* mode. This property is proving very useful for interpreting 2D SFG signals.[37,38]

## Experimental implementation

To implement 2D SFG spectroscopy, we add a mid-IR pulse shaper to a broadband SFG spectrometer. The pulse shaper is used to generate the pair of pump pulses, scan their time delays, and increment their phases for background subtraction and shifting data collection in the rotating frame. A schematic of our apparatus is shown in Fig. 3. It consists of a regeneratively-amplified Ti:Sapphire laser that outputs femtosecond pulses (800 nm, ~50 fs FWHM, ~4 mJ per pulse) at a 1 kHz repetition rate. Three quarters of the 800 nm output (3 mJ per pulse) pumps an optical parametric amplifier with $AgGaS_2$ difference frequency generation to produce about 30 μJ per pulse of mid-IR light centered at 6 μm. A $CaF_2$ window is used to split the mid-IR pulse into a pump and probe beam path. About 95% of the light is sent through the mid-IR pulse shaper using a transverse Ge acousto-optic modulator to create the pulse pair. We refer to the two pulses as the $E_1$ and $E_2$ pump beams. The remaining 5% is used as the mid-IR probe pulse, $E_3$. The 800 nm light not used for mid-IR generation (1 mJ per pulse) is frequency-narrowed using a 1 nm FWHM interference filter centered at 805 nm. After filtering, the visible pulse power is 15 μJ per pulse. The visible pulse and mid-IR pulses are focused using 30 cm and 20 cm lenses respectively and overlapped at the sample. The angle of incidence for the visible pulse is $\theta_{vis} = 65°$, and for the mid-IR pulses $\theta_{IR} = 75°$ from normal. The visible pulse and all mid-IR pulses are p polarized to collect a ppppp 2D SFG signal. The signal beam is re-collimated and a 735 nm short pass filter is used to remove unwanted 800 nm light. This signal is sent into

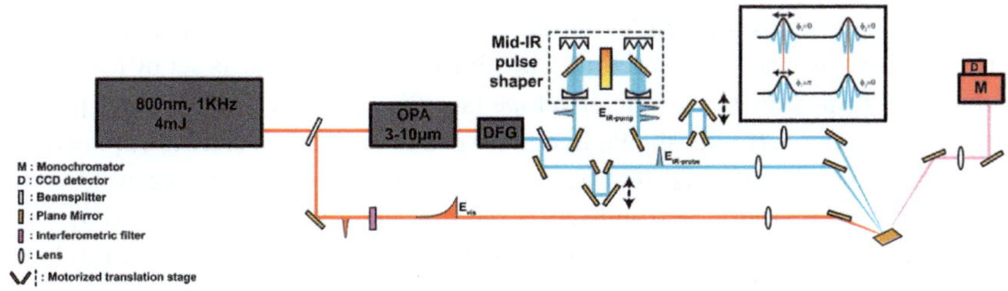


Fig. 3 Schematic layout of the 2D SFG experimental setup. The phase cycling procedure adopted in the experiments is illustrated in the inset.

a monochromator and frequency dispersed onto a CCD detector. To prevent sample degradation during the experiment, the sample was scanned in the *xy*-plane such that a spot on the sample was only probed for 5 minutes (*i.e.*, 5000 laser shots) before changing to a new spot.

## Phase cycling, background subtraction, and step sizes

Because the $E_1$ and $E_2$ pump beams are collinear, the $4^{th}$-order SFG electric field, $E_{res}^{(4)}(\omega_3, t_1, t_2)$ that we want to measure is emitted in the same direction as the $2^{nd}$-order SFG field, $E_{res}^{(4)}(\omega_3, t_1, t_2)$, created by $E_3$ and the visible pump pulse. Both will cover the same frequency range and so the unwanted $2^{nd}$-order SFG field cannot be removed by filtering. In addition, there is the second-order non-resonant field $E_{nonres}^{(2)}(\omega_3, t_1, t_2)$, which in our experiments we use as our local oscillator pulse. On the detector, all three of these fields will interfere, giving the measured signal:

$$\begin{aligned} S(\omega_3, t_1, t_2) &= \left|E_{res}^{(2)} + E_{nonres}^{(2)} + E_{res}^{(4)}\right|^2 \equiv \left|E_{res}^{(2)} + E_{LO}^{(2)} + E_{res}^{(4)}\right|^2 \\ &= \left|E_{res}^{(2)}\right|^2 + \left|E_{LO}^{(2)}\right|^2 + \left|E_{res}^{(4)}\right|^2 + 2E_{res}^{(2)}E_{LO} + 2E_{res}^{(4)}E_{res}^{(2)} + 2E_{res}^{(4)}E_{LO} \end{aligned} \tag{1}$$

The cross terms $E_iE_j$ in eqn (1) depend on the phase of the incident laser pulses, which is:

$$\begin{aligned} E_{res}^{(2)}E_{LO} &\sim \cos\left[\omega t + \left(\phi_3 + \phi_{vis} + \Delta\phi_{res,\,nonres}\right) - \left(\phi_3 + \phi_{vis}\right)\right] \\ &\sim \cos\left(\omega t + \Delta\phi_{res,\,nonres}\right) \\ E_{res}^{(4)}E_{LO} &\sim \cos\left[\omega t + \left(\phi_3 + \phi_{vis} \pm \Delta\phi_{1,2} + \Delta\phi_{res,\,nonres}\right) - \left(\phi_3 + \phi_{vis}\right)\right] \\ &\sim \cos\left(\omega t \pm \Delta\phi_{1,2} + \Delta\phi_{res,\,nonres}\right) \\ E_{res}^{(4)}E_{res}^{(2)} &\sim \cos\left[\omega t + \left(\phi_3 + \phi_{vis} \pm \Delta\phi_{1,2}\right) - \left(\phi_3 + \phi_{vis}\right)\right] \\ &\sim \cos\left(\omega t \pm \Delta\phi_{1,2}\right) \end{aligned} \tag{2}$$

Of these terms, we only want the cross term $E_{res}^{(4)}E_{LO}$, which most faithfully measures the $4^{th}$-order electric field because the nonresonant $E_{LO}$ from gold has a well-defined phase $\Delta\phi_{res,nonres}$ with respect to the resonant contribution to the signal and is temporally short.[39] In previous reports, the term $E_{res}^{(4)}E_{res}^{(2)}$ was used to generate a 2D SFG spectrum,[40] but that causes problems in spectral interpretation because the $2^{nd}$-order signal is a complicated function of amplitude and phase that is not necessarily well-understood.[41] To obtain our desired term and eliminate the others, we use phase cycling. The phase of a laser pulse is set by the delay of the carrier wave relative to the envelope (see Fig. 3 inset). The different phase dependence above terms as shown in eqn (2) can be utilized to extract our desired signal from the others (inset). For each $t_1$ delay, the signal is first measured with $E_1$ and $E_2$ having phases $\phi_1 = 0$ and$\phi_2 = 0$, respectively, and then $\phi_1 = \pi$ and $\phi_2 = 0$. The two signals are then subtracted. Mathematically, one substitutes $\phi_1$ and $\phi_2$ into the equations above, which gives

$$S(\Delta\phi_{1,2} = 0) - S(\Delta\phi_{1,2} = \pi) = 4E_{res}^{(4)}E_{LO} + 4E_{res}^{(2)}E_{res}^{(4)} \tag{3}$$

 

and contains the term that we want. The second term cannot be removed, but since $E_{\mathrm{LO}} \gg E^{(2)}_{\mathrm{res}}$, it is much smaller than $E^{(4)}_{\mathrm{res}}E_{\mathrm{LO}}$, and so we ignore it. Instead of phase cycling, one could also chop the $E_1$ and $E_2$ beams, like often done in transient absorption spectroscopy.[42] Phase cycling is preferable, because signal is collected with every laser shot, and thus data collection is effectively twice the repetition rate as chopping. There are several experimental methods for setting the carrier phase, but pulse shaping makes it very easy because it is done by computer programming.

Phase cycling is also used to shift the apparent frequency of the measured 2D SFG signal.[28,43] For condensed phase peptides, sufficient resolution is achieved by scanning to maximum delays of 1500 to 2500 fs. The Nyquist theorem says that two points per period must be collected to fully sample a sine wave. The peptide vibrations measured here absorb at about 6 microns, which has a period of about 20 fs. Thus, step sizes cannot be larger than 9 fs to follow the Nyquist theorem, requiring ~170 measurements to reach a maximum delay of 1500 fs. Instead, we take 50 fs steps, but still sample at the Nyquist frequency by shifting the apparent frequency of the measure field to 50 microns, which has a period of 167 fs. This is done by using a different phase for $E_1$ at each delay (in addition to the phase flips described above) that partially counteracts the phase shift caused by the delay itself. Data collection in the rotating frame is commonly used in NMR spectroscopy. Phase cycling makes its application in optical spectroscopies very straightforward to implement.

## Coupled vibrations and simulations of the 2D SFG spectra

To explore possible structures that might explain the experimental 2D SFG spectra presented below, we have simulated the 2D SFG spectra using an excitonic coupling model and the formalism for heterodyne detected 2D SFG spectroscopy. The ester monolayer is built by arranging MMB into a close packed array. Each ester is oriented such that the C3–S13 bond aligns with surface normal and the benzene ring lies on the *X*–*Z* plane. The structure, transition dipole and Raman tensor are calculated by Gaussian 09. The amyloid fibers are built assuming standard dihedral angles for a β-strand and then arranging them into parallel β-sheets. Transition dipole coupling is used to describe all coupling elements. The nearest-neighbor interactions are modeled separately by a coupling constant of 0.8 cm$^{-1}$, since a transitional dipole couplings model is known to overestimate the couplings of nearest neighbors in peptides.[44] For the peptide, each amide bond is assigned a transition dipole that is oriented 20° from the carbonyl bond.[45] The amide transition dipole is approximated to be equal in magnitude to the ester dipole for simplicity. A similar procedure is used for the polarizability tensors. The fundamental frequency for the amides is set to 1700 cm$^{-1}$ and for esters at 1745 cm$^{-1}$. The anharmonic shift of all stretches is approximated as 20 cm$^{-1}$. From these parameters, 1-quantum and 2-quantum Hamiltonians are built and diagonalized to obtain the transition dipoles and directions of the excitons. These quantities are then input into the Feynman path formalism form which the 2D SFG spectra are generated.[36]

## Preparation of the monolayer and peptide samples

### Peptide synthesis

The FGAIL pentapeptide was obtained from a 0.1 mmol scale solid-phase synthesis using standard FMOC chemistry,[46] with double 2 h couplings of each amino acid. PAL-PEG-PS resin was employed in order to generate an amidated C-terminus. The resin was cleaved for 4 h in 95% (v/v) TFA in $H_2O$. Crude peptide was then purified by reverse phase HPLC and verified by mass using electrospray ionization mass-spectrometry.

### Preparation of MMB monolayers on gold

Methyl-4-mercaptobenzoate (MMB) was obtained from Toronto Research Chemicals Inc. and used without further purification. The MMB monolayers on gold were deposited using established protocols.[47,48] Fresh gold substrates (Platypus Technologies – 100 nm Au with 5 nm Ti adhesion layer on silicon ⟨1,0,0⟩ wafers, ~1 cm × 1 cm dimension) were soaked in 1 mM MMB solutions in methanol (Sigma Aldrich, HPLC grade, 99.9%) for 24 hours. The substrates were then removed from solution, rinsed thoroughly with pure methanol and dried under nitrogen. The samples were then used for the 2D-SFG measurements. For the deposition of FGAIL, ester coated gold substrates were prepared as described above, following which a 400 μM solution of FGAIL was added on the substrate and dried.

## Results

Shown in Fig. 4A is the heterodyne detected 2D SFG spectrum of methyl-4-mercaptobenzoate (MMB) on gold for $t_2 = 0$. Three strong peaks are observed, labeled

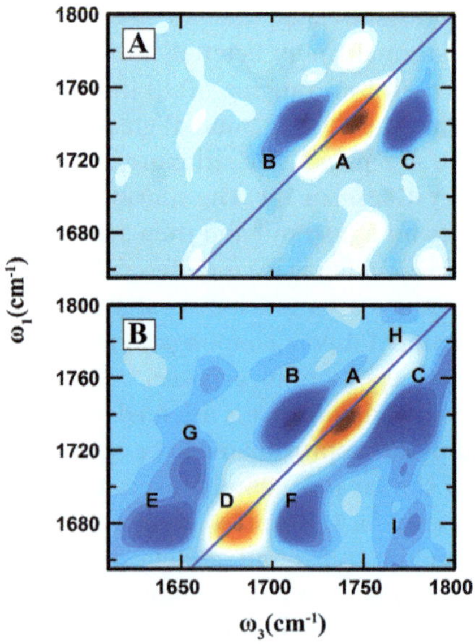


Fig. 4 Experimental 2D SFG spectra at $t_2 = 0$ of (A) MMB monolayers on gold substrates and (B) FGAIL on MMB coated gold substrates.

in Fig. 4A as "A", "B" and "C". Peaks A and B appear at $\{\omega_{pump}, \omega_{probe}\}$ = $\{1740\ cm^{-1}, 1742\ cm^{-1}\}$ and $\{1740\ cm^{-1}, 1714\ cm^{-1}\}$. These peaks are caused by transitions to the fundamental and overtone states, as described above, and so their difference in frequency is caused by the anharmonic shift (the actual anharmonic shift is smaller than the peak separation because the picosecond pulse broadens the spectra along the $x$-axis).[37] Peak C has not been observed in previous 2D SFG spectra. It may arise from a distribution of eigenstates caused by coupling between esters, as suggested by simulations presented below. The noise level in this spectrum is about one contour level, and so only peaks A, B and C lie appreciably above the noise.

The 2D SFG spectrum of the monolayer soaked in a solution of FGAIL peptide is shown in Fig. 4B. It exhibits peaks A, B and C, as observed in Fig. 4A for the ester stretches of the MMB monolayers. In addition, there are peaks that we label D, E and F, as well as other features above the noise level. The 2D SFG spectrum of the MMB monolayer did not have any features in the region of these three peaks. Thus, we conclude that peaks A, B and C arises from the amide vibrations of adsorbed peptide on MMB. That is an interesting observation because samples made by soaking gold substrates in 400 μM aqueous FGAIL solution without the MMB monolayer produced no measureable signal (data not shown) even though the surface was clearly coated with peptide (by visual inspection). Thus, the MMB monolayer is necessary to create ordered protein conformations, which is a point that we come back to in the discussion section.

In addition to peaks D, E and F, we also mark 3 other features that are above the noise, labeled G, H and I. Peak G also lies below the frequency range observed for the MMB esters, and so we assign it to a cross peak associated with FGAIL. Peak H lies above A, B or C. We believe it is diagonal peak of the MMB mode causing peak C. Finally, peak I appears to correlate peak H with peak E, making it a cross peak between the ester stretches and the FGAIL peptide.

## Simulations of MMB and MMB-FGAIL 2D SFG spectra

Shown in Fig. 5 are simulated spectra of the MMB monolayer, and the monolayer with a 4-stranded parallel β sheet in two different orientations. The MMB monolayer exhibits features similar to peaks A, B and C. Examination of the eigenvectors that produce the peaks reveals that peak C is caused by coupling between bright and dark eigenstates, although it is much weaker in the simulations than experiment.

The peptide features all appear at lower frequencies than MMB. In the perpendicular orientation (Fig. 5B), the β-strands of the amyloid fiber stand up straight from the monolayer so that the fiber axis lies along the surface whereas in the parallel orientation the strands lie on the MMB monolayer so that the fiber axis points away from the surface. The 2D SFG spectra for the peptide are very different in these two orientations. Parallel β-sheets have an $a^-$ mode that runs across the β-strands and an $a^+$ mode that lies (somewhat) along the strands.[37,49] The $a^-$ mode absorbs 50–60 $cm^{-1}$ lower than the $a^+$ mode and usually dominates an IR spectrum because its transition dipole is much larger. That is not the case in the perpendicular orientation simulated here. Since the $a^-$ mode lies along the

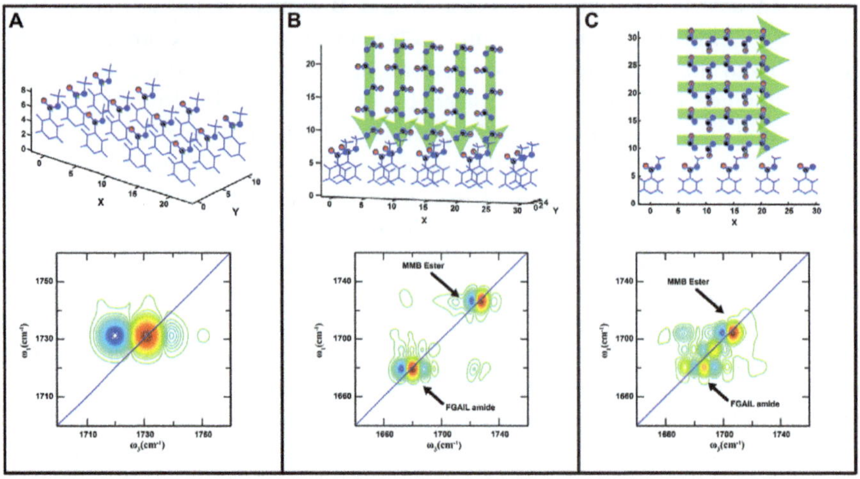


Fig. 5 Simulated 2D SFG spectra for MMB monolayers with and without adsorbed FGAIL. (A) MMB monolayer. (B) FGIAL-MMB, with the fibril axis parallel to the monolayer. (C) FGAIL-MMB, with the fibril axis perpendicular to the monolayer.

surface, it is only weakly SFG active whereas the $a^+$ mode lies in an orientation that produces the strongest SFG signal. Thus, both modes have above equal intensity in this orientation and produce strong cross peaks between them. In the parallel orientation (Fig. 5A), it is the $a^+$ mode that is SFG inactive and the $a^-$ mode pointing perpendicular to the surface, making its intensity dominate.

Both simulated spectra predict cross peaks between the ester stretches and the peptides. Thus, the vibrational coupling is large enough to become delocalized across the interface. It is also interesting to note that the peptides alter the features of the MMB itself, producing a peak similar to peak C from the experiments.

## Discussion

The data presented in this study make it clear that the membranes are playing an active role in fiber orientation, if not their growth. No SFG signal is observed for bare gold surfaces, even though dried peptide deposits are observed. Thus, the gold surface does not orient the molecules and any amyloid fibers formed in solution deposit onto the surface randomly. With MMB present, very strong SFG signals are observed, indicating ordering. Control experiments will have to be performed to determine if the peptides are aggregating in solution and then ordering upon binding to MMB or if the MMB causes amyloid fibers to grow on the surface. Based on literature reports, we believe the latter. Many interfaces cause amyloid fiber aggregation. Lipid interfaces typically cause amyloid aggregation many times faster than observed in solution, making surface aggregation the dominant process.[10,50] Thus, we believe that the MMB surface either binds the FGAIL monomers, thereby increasing their concentration and accelerating aggregation, or serving to nucleate a particular structure that can seed fiber formation. Either way, the mechanism must be uniform, since the process results in well-ordered β-sheets.

 

Neither of the simulations presented here match the experimental spectrum well-enough to be representative of the true molecular structure, but the simulations do help their interpretation. First, the simulations show that the ratio of the $a^-$ and $a^+$ modes will depend on the β-sheet orientation. The experiments have a stronger $a^-$ mode (peak D) and the phase of the $a^-$ mode and the ester stretch (B) are both positive, suggesting that the β-strands are largely parallel to the surface so that the amyloid β-sheets grow away from the surface. With that interpretation, peak G would be a cross peak between the $a^-$ and $a^+$ modes, albeit more intense than predicted in the simulations. Qualitatively, it appears that the FGAIL peptides bind in a linear conformation to the surface, like shown in Fig. 5B.

The simulations also seem to support the assignment of peak I to a cross peak between the ester and the peptide. In the simulations the coupling is large enough to create cross peaks between the two. However, cross peaks appear in both FGAIL orientations, and so it is not clear how to interpret the magnitude of the peak I. The fact that peak I is correlating the $a^-$ mode of the β-sheet to an MMB mode that is normally forbidden, may be a consequence of the peptide disordering the MMB monolayer. Disorder will break the 2D symmetry and redistribute oscillator strength, thereby making forbidden MMB modes more intense. Amyloid fiber formation is known to disrupt lipid monolayers.[51,52] The cross peaks certainly indicate hydrogen bonding between the esters of MMB and the peptide backbone.

Many other factors need to be taken into account for more reliable simulations. For example, the coupling model used to calculate these numbers has not been calibrated for ester stretches, and so the magnitude of these couplings may be wrong. The monolayer structure is assumed to be well-ordered, whereas structural disorder is probably present. Presumably, the ester stretches will not all point in the same direction like they do in our simulations. Moreover, the relative surface coverages of MMB and FGAIL also need to be evaluated, as the surface density dictates the SFG signal intensity. Nevertheless, simulations like these are able to reproduce experimental 2D IR and 2D SFG spectra reasonably well and discriminate between possible structures. We expect the same to be ultimately true here as well.

## Conclusions

Surfaces play an important role in catalyzing amyloid fiber formation, yet the mechanism and resulting structures are largely unknown. In this paper, we show that we can collect 2D SFG spectra of amyloid peptides formed by an ester monolayer. The spectra clearly resolve β-sheets that are consistent with amyloid fiber formation and a comparison with simulations suggests that the fiber axis is perpendicular to the surface. Although these results are still preliminary, if this conclusion holds, then it may indicate that the surface promotes seed formation by aligning the FGAIL peptides into strands parallel to the surface that then nucleate fiber growth. Further experiments are needed to test this hypothesis. 2D IR spectroscopy has become an important tool for studying peptide and amyloid structures. Extension of these techniques to surface monolayers opens up a new realm of experiments.

## Acknowledgements

The authors thank the National Science Foundation (NSF) for funding through Grant CHE-1266422. Funding for A.L.S. was provided by a Diversity Supplement to NIH DK079895. D.R.S. was supported by the University of Wisconsin Materials Research Science and Engineering Center (UW-MRSEC, DMR-1121288).

## References

1 D. Eisenberg and M. Jucker, *Cell*, 2012, **148**, 1188–1203.
2 P. Westermark, U. Engström, K. H. Johnson, G. T. Westermark and C. Betsholtz, *Proc. Natl. Acad. Sci. U. S. A.*, 1990, **87**, 5036–5040.
3 C. Betsholtz, L. Christmansson, U. Engström, F. Rorsman, V. Svensson, K. H. Johnson and P. Westermark, *FEBS Lett.*, 1989, **251**, 261–264.
4 K. Tenidis, M. Waldner and J. Bernhagen, *J. Mol. Biol.*, 2000, **295**, 1055–1071.
5 T. T. Ashburn, M. Auger and P. T. Lansbury, *J. Am. Chem. Soc.*, 1992, **114**, 790–791.
6 G. G. Glenner, E. David Eanes and C. A. Wiley, *Biochem. Biophys. Res. Commun.*, 1988, **155**, 608–614.
7 Y. Porat, Y. Mazor, S. Efrat and E. Gazit, *Biochemistry*, 2004, **43**, 14454–14462.
8 S. Luca, W.-M. Yau, R. Leapman and R. Tycko, *Biochemistry*, 2007, **46**, 13505–13522.
9 L. E. Buchanan, E. B. Dunkelberger, H. Q. Tran, P.-N. Cheng, C.-C. Chiu, P. Cao, D. P. Raleigh, J. J. de Pablo, J. S. Nowick and M. T. Zanni, *Proc. Natl. Acad. Sci. U. S. A.*, 2013, **110**, 19285–19290.
10 J. D. Knight and A. D. Miranker, *J. Mol. Biol.*, 2004, **341**, 1175–1187.
11 S. A. Jayasinghe and R. Langen, *Biochemistry*, 2005, **44**, 12113–12119.
12 J. D. Knight, J. A. Hebda and A. D. Miranker, *Biochemistry*, 2006, **45**, 9496–9508.
13 J. Adamcik, J.-M. Jung, J. Flakowski, P. De Los Rios, G. Dietler and R. Mezzenga, *Nat. Nanotechnol.*, 2010, **5**, 423–428.
14 T. Kowalewski and D. M. Holtzman, *Proc. Natl. Acad. Sci. U. S. A.*, 1999, **96**, 3688–3693.
15 W. Qiang, K. Kelley and R. Tycko, *J. Am. Chem. Soc.*, 2013, **135**, 6860–6871.
16 C. Wu, H. Lei and Y. Duan, *J. Am. Chem. Soc.*, 2005, **127**, 13530–13537.
17 A. Melquiond, J.-C. Gelly, N. Mousseau and P. Derreumaux, *J. Chem. Phys.*, 2007, **126**, 065101.
18 D. Zanuy, Y. Porat, E. Gazit and R. Nussinov, *Structure*, 2004, **12**, 439–455.
19 A. Kapurniotu, A. Schmauder and K. Tenidis, *J. Mol. Biol.*, 2002, **315**, 339–350.
20 M. Tatarek-Nossol, L.-M. Yan, A. Schmauder, K. Tenidis, G. Westermark and A. Kapurniotu, *Chem. Biol.*, 2005, **12**, 797–809.
21 J. J. W. Wiltzius, S. A. Sievers, M. R. Sawaya, D. Cascio, D. Popov, C. Riekel and D. Eisenberg, *Protein Sci.*, 2008, **17**, 1467–1474.
22 C. P. Shaw, D. A. Middleton, M. Volk and R. Lévy, *ACS Nano*, 2012, **6**, 1416–1426.
23 D. Zanuy, B. Ma and R. Nussinov, *Biophys. J.*, 2003, **84**, 1884–1894.
24 Y. R. Shen, *J. Phys. Chem. C*, 2013, **117**, 11884.
25 Y. R. Shen, *The principles of nonlinear optics*, Wiley-Interscience, Hoboken, N.J., Wiley classics library edn, 2003.
26 Y. R. Shen and V. Ostroverkhov, *Chem. Rev.*, 2006, **106**, 1140–1154.

 

27 A. G. Lambert, P. B. Davies and D. J. Neivandt, *Appl. Spectrosc. Rev.*, 2005, **40**, 103–145.
28 W. Xiong, J. E. Laaser, R. D. Mehlenbacher and M. T. Zanni, *Proc. Natl. Acad. Sci. U. S. A.*, 2011, **108**, 20902–20907.
29 A. M. Jubb, W. Hua and H. C. Allen, *Acc. Chem. Res.*, 2012, **45**, 110–119.
30 L. Fu, G. Ma and E. C. Y. Yan, *J. Am. Chem. Soc.*, 2010, **132**, 5405–5412.
31 I. V. Stiopkin, H. D. Jayathilake, A. N. Bordenyuk and A. V. Benderskii, *J. Am. Chem. Soc.*, 2008, **130**, 2271–2275.
32 S. H. Shim, D. B. Strasfeld, Y. L. Ling and M. T. Zanni, *Proc. Natl. Acad. Sci. U. S. A.*, 2007, **104**, 14197–14202.
33 P. Hamm, M. H. Lim and R. M. Hochstrasser, *J. Phys. Chem. B*, 1998, **102**, 6123–6138.
34 M. T. Zanni, N. H. Ge, Y. S. Kim and R. M. Hochstrasser, *Proc. Natl. Acad. Sci. U. S. A.*, 2001, **98**, 11265–11270.
35 M. Khalil, N. Demirdoven and A. Tokmakoff, *Phys. Rev. Lett.*, 2003, **90**, 047401.
36 J. E. Laaser and M. T. Zanni, *J. Phys. Chem. A*, 2013, **117**, 5875–5890.
37 J. E. Laaser, D. R. Skoff, J. J. Ho, Y. Joo, A. L. Serrano, J. D. Steinkruger, P. Gopalan, S. H. Gellman and M. T. Zanni, *J. Am. Chem. Soc.*, 2014, **136**, 956–962.
38 D. R. Skoff, J. J. Ho, A. Ghosh and M. T. Zanni, in preparation.
39 A. Lagutchev, S. A. Hambir and D. D. Dlott, *J. Phys. Chem. C*, 2007, **111**, 13645–13647.
40 S. Nihonyanagi, J. A. Mondal, S. Yamaguchi and T. Tahara, *Annu. Rev. Phys. Chem.*, 2013, **64**, 579–603.
41 M. Cho, *Two-dimensional optical spectroscopy*, CRC Press, Boca Raton, 2009.
42 P. Hamm and M. T. Zanni, *Concepts and methods of 2d infrared spectroscopy*, Cambridge University Press, Cambridge, New York, 2011.
43 S.-H. Shim and M. T. Zanni, *Phys. Chem. Chem. Phys.*, 2009, **11**, 748–761.
44 C. Lee and M. H. Cho, *J. Phys. Chem. B*, 2004, **108**, 20397–20407.
45 W. H. Moore and S. Krimm, *Biopolymers*, 1976, **15**, 2439–2464.
46 P. Marek, A. M. Woys, K. Sutton, M. T. Zanni and D. P. Raleigh, *Org. Lett.*, 2010, **12**, 4848–4851.
47 A. Michota and J. Bukowska, *J. Raman Spectrosc.*, 2003, **34**, 21–25.
48 J. C. Love, L. A. Estroff, J. K. Kriebel, R. G. Nuzzo and G. M. Whitesides, *Chem. Rev.*, 2005, **105**, 1103–1170.
49 C. M. Cheatum, A. Tokmakoff and J. Knoester, *J. Chem. Phys.*, 2004, **120**, 8201–8215.
50 Y. L. Ling, D. B. Strasfeld, S. H. Shim, D. P. Raleigh and M. T. Zanni, *J. Phys. Chem. B*, 2009, **113**, 2498–2505.
51 E. Sparr, M. F. M. Engel, D. V. Sakharov, M. Sprong, J. Jacobs, B. de Kruijff, J. W. M. Höppener and J. Antoinette Killian, *FEBS Lett.*, 2004, **577**, 117–120.
52 M. Anguiano, R. J. Nowak and P. T. Lansbury, *Biochemistry*, 2002, **41**, 11338–11343.

Faraday Discussions

PAPER

# Phase modulation nanoscopy: a simple approach to enhanced optical resolution

Robert Pal*

Received 15th August 2014, Accepted 5th November 2014
DOI: 10.1039/c4fd00158c

A new modular super-resolution technique called Phase Modulation Nanoscopy (PhMoNa) has been developed in order to break the optical diffraction barrier in Confocal Laser Scanning Microscopy (LSCM). This technique is based on using spatially modulated illumination intensity, whilst harnessing the fluorophore's non-linear emission response. It allows experimental resolution in both lateral and axial domains to be improved by at least a factor of 2. The work is in its initial phase, but by using a custom built Electro Optical Modulator (EOM) in conjunction with functionalised Ln(III) complexes as probes, a sub-diffraction resolution of ~60 nm was achieved of selected cellular organelles in long term live cell imaging experiments.

## Introduction

In the field of optical fluorescence microscopy, many researchers have dedicated their entire scientific career to developing better and improved cellular stains or associated microscopy techniques and equipment, pushing the boundaries of both signal detection and resolution. However, the intrinsic resolution of fluorescence microscopy is limited by diffraction, which determines the extension of the focused light emitted by a point source object, in other words the point spread function (PSF). In recent years, renewed efforts have been made in optical microscopy and associated life sciences to break through the optical diffraction barrier and visualize even smaller parts of the 'living' cell in 'super-resolution'. Governed by Abbé's law (1873) the highest achievable theoretical spatial resolution ($d$) of a given experimental setup is dictated by the lowest applied excitation light ($\lambda_{exc.}$) $d = \lambda_{exc.}/2NA$ (NA: numerical aperture of objective).[1,2] Almost all commercially available cellular stains and subsequently the majority of 'off-the-shelf' microscopes are equipped with light sources possessing excitation wavelengths $\geq$405 nm through the visible and NIR promoting maximal achievable resolutions of ~200 nm.[3] However, much of the fundamental biology of the cell occurs below this threshold; hence breaking this limit will allow the wide multidisciplinary scientific community to look deeper in higher resolution.

*Department of Chemistry, Durham University, South Road, DH1 3LE Durham, UK. E-mail: robert.pal@dur.ac.uk*

As noted by Köhler in the beginning of the 20th century, one of the simplest experimental modifications to improve lateral resolution is to illuminate samples using shorter wavelength ultraviolet (UV) light.[4] This technique was fully exploited in the early days of microscopy and has always been associated with unwanted auto-fluorescence from endogenous chromophores and more importantly disruptive UV induced processes, such as DNA damage leading to high phototoxicity.[5] Thus, the approach of improved resolution, ever since the step-changing invention of Confocal Microscopy by Minsky in 1953,[6,7] has turned towards the development of new optical (hardware) and computer (software) based methodologies,[8,9] revolutionising the way in which we investigate nanostructures. The current key technologies are: SIM, where structured patterns are introduced into the illumination pathway ($d_{\text{lateral}}$ = 100–120 nm),[10–12] and RESOLFT, which utilises quantum effects to spatially inhibit the applied stain fluorescence (*e.g.* STED which is currently the most sought after and applied commercially available technique, 20–100 nm).[13–15] These techniques dominate alongside the near-field nano-scanning technique utilising a propagating evanescent wave NSOM (20–120 nm),[16,17] and the pointillistic localisation methods PALM/STORM (20–50 nm)[18,19] using photo-switchable fluorophores. Recent years have seen many refined and evolved variants of the above mentioned techniques by improvements applied either or both to the illumination or detection process leading to improvement in both precision and resolution. But these new, still fundamentally diffraction limited techniques all have drawbacks too, such as limitation of suitable stains, disruptively high laser powers, slow/limited imaging speed, poor SNR due to loss of information and specialized expensive and bulky instrumentation. Taking the above into account, the obvious conclusion is to revisit the use of near-UV excitation with tailored bright and non-invasive molecular probes to investigate cellular organelles in finer detail. The primary future research aim renders itself to introduce a novel user friendly resolution enhancing method into any LSCM microscope harnessing the already superior resolution[20] of the confocal detection principle.

## Methods and results

The instrumental technique detailed herein, is currently still in the development phase and is predominantly used in conjunction with bright luminescent lanthanide probes recently developed in Durham.[21,22] In the future, it aims to be used with any currently existing LSCM microscope with high speed acquisition and improved resolution of ~60 nm (achievable lateral resolution when coupled to a suitable 355 nm light source). The use of well-designed lanthanide complexes as probes for spectroscopy or microscopy can be advantageous due to their unique properties.[23,24] They exhibit long-lived emission (μs–ms), typically many orders of magnitude longer than that of background bio-auto-fluorescence (ns, similar lifetime as commercial organic cell stains), allowing for facile time-gated detection.[25] They have large pseudo-Stokes' shifts of hundreds of nanometers and possess well-defined, narrow peaks in the emission spectrum that can be exploited for more complex analysis.[26] Several examples exhibit dynamic responses to local conditions such as pH,[27–29] citrate[30] or bicarbonate,[31] allowing the measurement of these species within particular organelles of mammalian[32] or plant cells.[33]

 

Fluorescence confocal microscopy is characterised by rather low SNR. However this can be boosted by pixel or line average scanning or multiple line or frame accumulation. This is an obvious trade off with acquisition speed as it prolongs scan intervals and subsequent, often disruptive specimen light exposure. However, with the new and higher sensitivity detectors, scan speed can be increased whilst excitation power can be minimised to trigger a good SNR probe fluorescence response, leading to exponentially lower levels of excitation exposure.

The experimental setup detailed herein is based on a novel combined structured illumination technique, where the region of support of the Optical Transfer Function (OTF) has been doubled leading to reciprocal sharpening of the Point Spread Function (PSF).[11] The approach uses a sequential frequency matched 3D oscillation of a carefully chosen optical grating (OG, paired with a corresponding first order filter) in the illumination beam path, inducing synchronized spatial periodicity (spatially modulated illumination intensity fluctuation, in other words, disruption in coherence) into the raster scanned (flying spot) excitation laser beam (Fig. 1 upper).[34] Images recorded under such experimental conditions due to the frequency mixing of the sub-diffraction limited structure of the sample and the scanning patterned excitation cluster display low frequency Moiré

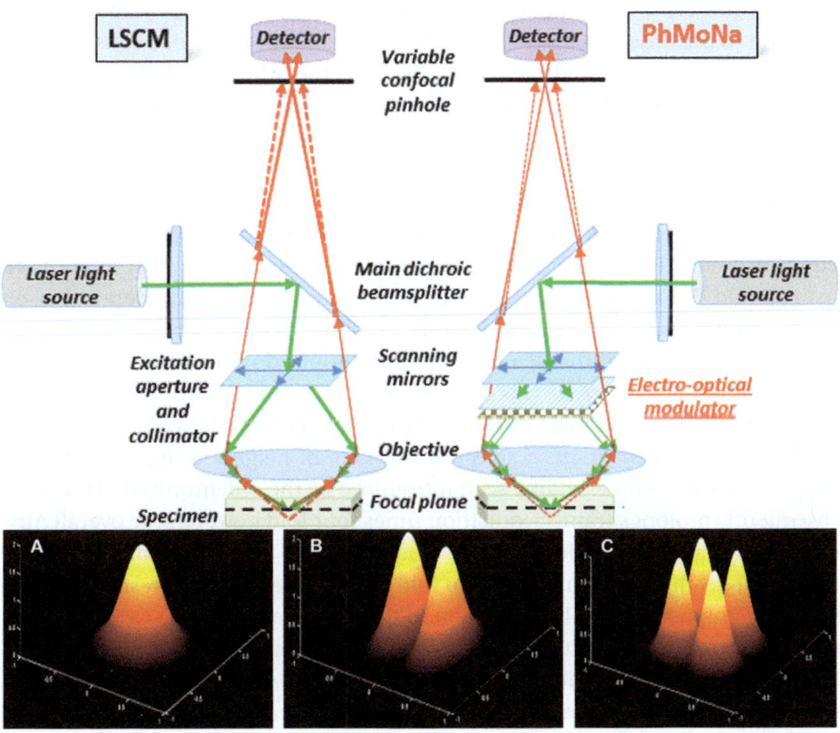


Fig. 1 (Upper) schematic diagram highlighting the key differences between the LSCM and PhMoNa microscope hardware and light path. (Lower) theoretical evolution of the excitation beam profile: (A) standard CLSM, (B) 1D-EOM oscillation and (C) 2D-EOM oscillation PhMoNa at identical scan speeds (100 Hz) $x$ and $y$ in Airy units (AU, using $\lambda_{exc.}$ 355 nm and 1.4 NA). PSF calculations were based on the findings of Cremer and Heintzmann.[10]

fringes. Simultaneously, this can easily be resolved and sub diffraction resolution images can be reconstructed *in situ* through automated synchronised image averaging.

In essence, Phase Modulation Nanoscopy (PhMoNa) operates by utilizing the optical grid pattern projected by the scanned first order diffracted excitation beam projected onto the back focal plane of the objective. This structured light modulation introduces high spatial frequencies in the illuminating light, whilst also acting as a finite detection mask, leading to an advantageous shift of the observed finite object's spatial frequencies to generate improved experimental resolution.

Gaussian beam profile calculations (Fig. 1 lower) revealed that this scan frequency matched synchronised grid oscillation provides the best and hence fastest results, *e.g.* least number of averaged iteration is needed per scan to result in an evenly illuminated flat field of view (FOV), when the generated excitation pattern resembles a cluster of four identically symmetric beams ($n = 4$). Theoretically, subsequent $n^2$ splitting of the beam could improve the achievable resolution even further. However, it needs to be considered that to achieve the same unified beam height, in more simplistic terms excitation power delivery and subsequent fluorescent response, the spatial merge of neighbouring Gaussian beams ($\rho$, bleeding-in factor) would be substantial ($\rho^n = \sqrt{\pi} = 0.18$ when $n = 4$). This undesired spatial light saturation would lead to the recombination of the induced finite excitation pattern, consequently degrading the experimental resolution to the orders of a standard CLSM. Another advantageous consequence of forming such a structured excitation pattern is the inherent sharpening of the individual excitation PSFs. Preliminary results obtained, using our modified Leica SP5 II LSCM system, revealed that phase modulation induced transversal momentum disruption of the excitation laser beam can sharpen the experimental PSF by a factor of $[2\pi/\sqrt{(2\pi)}] \times (1 - \rho^n) = 2(0.05)$ ($n = 4$), eliminating spectral focal blur (*e.g.* pixel-cross-talk and saturation) by completely suppressing spatial frequencies outside the OTF, and subsequently improving lateral resolution at least two-fold.[11] It needs to be emphasized that in order to regain identical laser density, whilst recording identical levels of fluorescence probe response (*i.e.* similar even image brightness) a carefully synchronised individual frame or line Nyquist sampling rate of $n^x = 4$ averaging ($n = 2$, and as 2D diagonal EOM oscillation $x = 2$) needed to be introduced to compensate for loss of fluorescence. This loss is attributed to offset selective masking of neighbouring pixels in the detection pathway with each line scan introduced by the synchronised EOM. This subsequently prolongs frame acquisition times four fold however, the overall Airy disk (AD) light exposure levels in 3D-Fourier space are maintained constant throughout, leading to an inherent two fold improvement in axial resolution and SNR. This simplistic optical modulation to the excitation path of a CLSM using a previously reported Europium(III)–9N3 complex[21] revealed that improved lateral (~60 nm) and axial resolution (~200 nm both at 1.40NA) can be achieved (Fig. 2) using functionalised lanthanide complexes and a linearly polarised CW-NdYAG laser (3$^{rd}$ harmonic at 355 nm).

An overall 8 fold reduction in experimental voxel size was achieved, as the optical sectioning pinhole diameter, determining axial resolution, was kept at 0.5 AU throughout. The value was automatically calculated using the FWHM of the individual Gaussian beam generated in the scanning excitation cluster.[35,36]

 

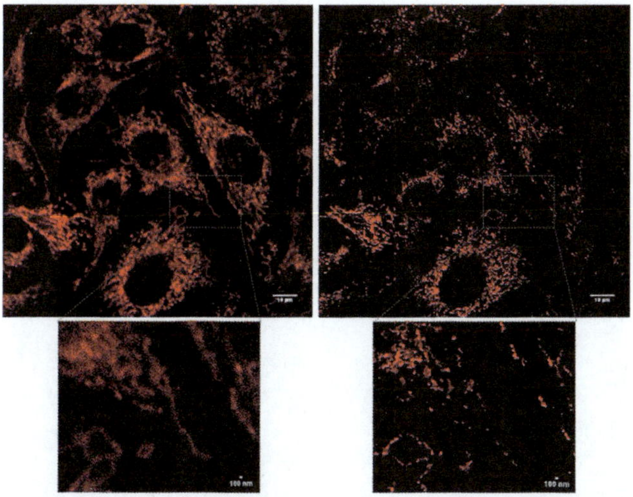

Fig. 2 Comparison of CLSM (top left) and PhMoNa enhanced LSCM (top right) live cell images of the same FOV using a Eu(III)dye[21] ($C_{loading}$ = 10 μM, $\lambda_{ex}$ = 355 nm) revealing enhanced 62 nm lateral resolution (100 Hz, 2 line accumulation and 4 frame averaging sequence, 2048 × 2048 pixel scan on the Leica SP2 II) of the mitochondrial network of NIH 3T3 cells. Total time of scan: LSCM 40 s, PhMoNa 160 s. Voxel size: CLSM 126 × 126 × 390 nm, PhMoNa: 63 × 63 1 × 195 nm; scalebars are 10 μm. (Bottom) Digital magnification of highlighted sections as indicated for LSCM (left) and PhMoNa (right); scalebars are 100 nm.

Further reduction of the confocal pinhole diameter could result in achieving a unified voxel of $a^3$ ($a$ = 62 nm). However, due to reduction of the studied optical section thickness, loss in fluorescence emission could only be compensated by an increase in applied laser power that could lead to undesired photo-induced probe and cell damage resulting in shortened live cell experiments.

Functionalised lanthanide(III) complexes possess unique bio- and spectrophysical properties, such as long luminescence lifetimes (orders of ms), large pseudo-Stokes'-shift (Eu(III): 200 nm) and broad range (150 nm) yet finite structured emission profiles.[26,37–39] Due to the high brightness and cellular accumulation of such probes, we minimize the exposure to UV excitation below non photo-damaging level during live cell imaging experiments. Subsequent modular instrumental development also facilitated time-resolved imaging, spectral and lifetime profiling of the internalised dyes and probes.[32,40] The drawback, in this case, is that lanthanide probes possess low photo emission rates. However, this has proven to be an advantage in this early developmental stage, by allowing slow scan speeds to be used (10 Hz, 100 ms per pixel laser dwell time) while precise synchronisation of the structured illumination process was established. Once this sequence was experimentally incorporated and validated, as a result of probe brightness, the scan speed and subsequent excitation patterning could be increased (up to 400 Hz was used), resulting in faster overall image acquisition times. It is also important to emphasize that by using luminescent lanthanide complexes, it is possible to achieve time-resolved (TR-)PhMoNa. Such a technique will not only eliminate the spatial focal blur contribution of UV induced short

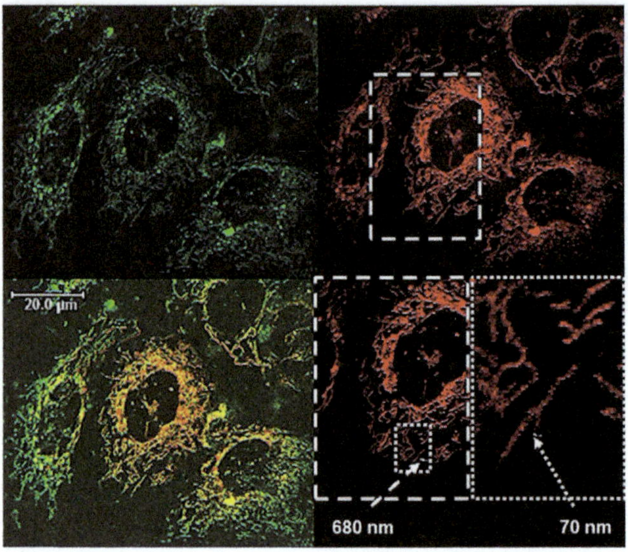


Fig. 3 PhMoNa enhanced live-cell LSCM images of the Eu(III)dye[21] used above (Fig. 2) (*red*) ($C_{loading} = 1$ μM, 40 s image acquisition, $\lambda_{ex} = 355$ nm) co-stained with MitoTracker Green (*green*) ($C_{loading} = 200$ nM, 40 s image acquisition, $\lambda_{ex} = 488$ nm) revealing enhanced 62 nm (*red channel*) and 87 nm (*green channel*) lateral resolution respectively (400 Hz, 2 pixel by pixel accumulation and 4 line-averaging sequence, smallest detectable individual pixel size using a 2048 × 2048 scan on the Leica SP2 II) of the mitochondrial network of NIH 3T3 cells. (*Orange*) RGB merge to confirm co-localisation, (bottom right) digital magnification sequence (with respective scalebar values) to reveal 2-fold improved experimental lateral resolution.

lived (ns) auto-fluorescence from endogenous chromophores and subsequently improve SNR, but also will drastically reduce photo-exposure of the cell due to pulsed excitation. These experimental aspects will also further prolong the observed cells' natural homeostatic state and minimise long term probe photobleaching whilst retaining optimal scanning speed. Another advantage to be explored in the future could be to facilitate biologically favourable multi photon excitation (range 730–770 nm) with these bright lanthanide stains that also possess improved two photon cross-section.[32] This is an important point to make that as with the application of tuneable (380–2200 nm broad spectral range) White Laser Light (WLL) sources the need of expensive quartz optical elements ($\lambda_{exc.} < 355$ nm) in the illumination path will be eliminated. Most importantly WLL-PhMoNa and its 8 fold overall experimental resolution improvement can be used with any commercially available cellular stain, at any given excitation wavelength on existing confocal microscopes, making it an attractive and affordable live-cell imaging technique to be used by the broad microscopy community.

Owing to its simplistic modular approach, PhMoNa can theoretically also be readily adapted to any preferred wavelength of excitation to achieve synchronised illumination–emission reduction, with a subsequent confocal voxel size reduction factor of $2^3 = 8$ ($d_{lateral} \sim \lambda_{exc.}/4$). Thus, its versatility can be extended to any commercially available polarised laser excitation source. Since our test Leica SP5

 

II CLSM framework is equipped with 8 different laser lines (ranging from 355–633) attention has focused on utilising our custom EOM to structure a more widely used laser line. Using the commercially available MitoTracker Green probe alongside the Eu(III) dye the EOM device was incorporated into the optical path by placing it below the microscope objective. In this manner, both 355 and 488 nm identically oriented polarised laser lines were used simultaneously in a frame by frame recording sequence. Superior resolution images were captured, maintaining both scan speed and light exposure adequately in each channel (Fig. 3).

One of the most important user friendly aspects to be considered with the above detailed experimental modification is in fact an inherent feature of the donor LSCM setup. In detail, due to the software determined line averaging sequence combined with the synchronised frequency matched nature of excitation cluster generation (*i.e.* spatially structured illuminating periodicity) during raster scanning, this technique is rendered free from any unnecessary time consuming post image processing deconvolution algorithm.

## Conclusion and future work

The primary aim was set to develop a low-cost super-resolution microscope creating a major and formidable advance over conventional LSCM. Initially the setup was tested with functionalised Ln(III) complexes, synthesized parallel to instrumental development in Durham. By employing a custom built Electro Optical Modulator (EOM) spatio-temporal evolution of a sinusoidal phase modulated laser beam has been carried out. The resulting modular attachment can be used *in situ*, as part of an existing LSCM instrument. The modified system allowed even more detailed information of cellular substructures to be extracted at favourable high scanning speeds. Overall, identical laser power was employed during PhMoNa recordings as was used with the standard LSCM images used for comparison. The only experimental difference was prolonging the scan time by a factor of four (up to 400 Hz scan speed), but strictly not the excitation exposure time (*i.e.* overall dwell time per pixel). It is extremely important to emphasize that this form of excitation PSF sharpening technique is only in its early phase and introducing an optical element oscillating at high frequencies into any optical path can lead to unwelcome vibration which can impair resolution. This was considered as being one major drawback of this technique, the other being the introduced four times increase in overall image acquisition. Hence, several novel approaches using a cleverly constructed new generation of EOMs are being currently developed to obviate this. Our aim still remains the modular attachment approach and not introducing a special cavity modulated laser to the setup to generate such finite excitation patterns. Once further developed, with a suitable EOM device to fulfil all our critical design criteria, PhMoNa may be employed as a cost effective 'add-on' resolution enhancing module to all LSCMs.

In the future PhMoNa and its 8 fold overall experimental resolution improvement could be used with any commercially available or novel emerging (including both transition-metal and lanthanide complexes) cellular stain, at any given excitation wavelength on any existing confocal microscope, making it an attractive and affordable live-cell imaging technique, to be used by the broad microscopy community Thus, saturation eliminated high spatial resolution 3D reconstructions can be created maintaining fast scan speed, low laser power and exposure time inherently minimising photo-bleaching. Another important advantage to emphasize is that PhMoNa promises to be one of the few applicable experimental nanoscopy techniques that can be used safely for live-cell imaging, eliminating the need for tedious after image processing algorithms.

## Acknowledgements

I would like to express my sincere gratitude to Professor David Parker for his expert help and guidance on lanthanide(III) complexes, Prof. Andrew Beeby for his time and devotion on optical detection optimisation, Mr Kelvin Appleby (Electrical Workshop) for all his valuable expertise and work on electric components, and EPSRC, ERC and Royal Society URF for the funding.

## References

1 E. Abbe, *Arch. Mikr. Anat.*, 1873, **9**, 413–418.
2 J. W. Strutt (Lord Rayleigh), Philos. Mag., 1896, **42**, pp. 167–195.
3 D. Spector and R. D. Goldman, *Basic Methods in Microscopy Protocols and Concepts from Cells: A Laboratory Manual*, Book News Inc, 2006.
4 A. Köhler and W. Loos, *Naturwissenschaften*, 1941, **29**, 49–61.
5 J. C. F. Wong and A. V. Parisi, *2nd Internet Conference on Photochemistry and Photobiology: Protection Against the Hazards of UVR*, 1999.
6 M. Minsky, *Scanning*, 1988, **10**, 128–138.
7 C. Cremer and T. Cremer, *Microsc. Acta*, 1974, 31–44.
8 L. Schermelleh, R. Heintzmann and H. Leonhardt, *J. Cell Biol.*, 2010, **190**, 165–175.
9 S. Habuchi, *Front. Bioeng. Biotechnol.*, 2014, 2.
10 R. Heintzmann and C. G. Cremer, *Proc. SPIE 3568*, 1999, vol. 3568, pp. 185–196.
11 M. G. L. Gustafsson, *J. Microsc.*, 2000, **198**, 82–87.
12 R. Heintzmann, *Micron*, 2003, **34**, 283–291.
13 S. W. Hell and J. Wichmann, *Opt. Lett.*, 1994, **19**, 780–782.
14 K. Willig, S. Rizzoli, V. Westphal, R. Jahn and S. Hell, *Nature*, 2006, **440**, 935–939.
15 T. A. Klar, S. Jakobs, M. Dyba, A. Egner and S. W. Hell, *Proc. Natl. Acad. Sci.*, 2000, **97**, 8206–8210.
16 E. Betzig and J. K. Trautman, *Science*, 1992, **257**, 189–195.
17 F. de Lange, A. Cambi, R. Huijbens, B. de Bakker, W. Rensen, M. Garcia-Parajo, N. van Hulst and C. G. Figdor, *J. Cell Sci.*, 2001, **114**, 4153–4160.
18 K. Lidke, B. Rieger, T. Jovin and R. Heintzmann, *Opt. Express*, 2005, **13**, 7052–7062.
19 S. T. Hess, T. P. K. Girirajan and M. D. Mason, *Biophys. J.*, 2006, **91**, 4258–4272.
20 O. Haeberlé and B. Simon, *Opt. Commun.*, 2009, **282**, 3657–3664.
21 S. J. Butler, L. Lamarque, R. Pal and D. Parker, *Chem. Sci.*, 2014, **5**, 1750–1756.
22 S. J. Butler, M. Delbianco, L. Lamarque, B. K. McMahon, E. R. Neil, J. W. Walton, R. Pal, D. Parker and J. M. Zwier, *Dalton Trans.*, 2014, Advance Article.
23 S. J. Butler and D. Parker, *Chem. Soc. Rev.*, 2013, **42**, 1652–1666.
24 S. V. Eliseeva and J.-C. G. Bunzli, *Chem. Soc. Rev.*, 2010, **39**, 189–227.

 

25 D. Parker and J. A. G. Williams, *J. Chem. Soc., Dalton Trans.*, 1996, 3613–3628.
26 C. P. Montgomery, B. S. Murray, E. J. New, R. Pal and D. Parker, *Acc. Chem. Res.*, 2009, **42**, 925–937.
27 B. K. McMahon, R. Pal and D. Parker, *Chem. Commun.*, 2013, **49**, 5363–5365.
28 R. Pal and D. Parker, *Org. Biomol. Chem.*, 2008, **6**, 1020–1033.
29 D. G. Smith, B. K. McMahon, R. Pal and D. Parker, *Chem. Commun.*, 2012, **48**, 8520–8522.
30 R. Pal, D. Parker and L. C. Costello, *Org. Biomol. Chem.*, 2009, **7**, 1525–1528.
31 D. G. Smith, G. Law, B. S. Murray, R. Pal, D. Parker and K.-L. Wong, *Chem. Commun.*, 2011, **47**, 7347–7349.
32 S. J. Butler, B. K. McMahon, R. Pal, D. Parker and J. W. Walton, *Chem.–Eur. J.*, 2013, **19**, 9511–9517.
33 A. Palmer, S. H. Ford, S. J. Butler, T. J. Hawkins, P. J. Hussey, P. Robert, J. W. Walton and D. Parker, *RSC Adv.*, 2014, **4**, 9356.
34 R. Heintzmann, T. M. Jovin and C. Cremer, *J. Opt. Soc. Am. A*, 2002, **19**, 1599–1609.
35 T. Wilson, *Confocal Microscopy*, London UK, 1990.
36 C. J. R. Sheppard, Scanning optical microscopy, in *Advances in Optical and Electron Microscopy 10*, ed. Barer R. and Cosslett V. E., Academic press, London UK, 1987.
37 J. W. Walton, A. Bourdolle, S. J. Butler, M. Soulie, M. Delbianco, B. K. McMahon, R. Pal, H. Puschmann, J. M. Zwier, L. Lamarque, O. Maury, C. Andraud and D. Parker, *Chem. Commun.*, 2013, **49**, 1600–1602.
38 E. J. New, D. Parker, D. G. Smith and J. W. Walton, *Curr. Opin. Chem. Biol.*, 2010, **14**, 238–246.
39 M. C. Heffern, L. M. Matosziuk and T. J. Meade, *Chem. Rev.*, 2014, **114**, 4496–4539.
40 R. Pal and A. Beeby, *Methods Appl. Fluoresc.*, 2014, **2**, 037001.

Faraday Discussions

DISCUSSIONS

# Future challenges: general discussion

E. D. Jemmis, Sankarampadi Aravamudhan, Elangannan Arunan, Abhishek Shahi, Neil Hunt, Christoph Schnedermann, John R. Helliwell, Mike Ashfold, Himangshu Prabal Goswami, Artur Nenov, Volker Deckert, Priyadarshi Roy Chowdhury, Kenneth Ghiggino, R. J. Dwayne Miller, Debabrata Goswami, Wolfgang Junge, Judith Howard, Keisuke Tominaga, Tim Brandt van Driel, Martin Zanni, Siva Umapathy, Martin Meedom Nielsen, R. Pal and Shaul Mukamel

DOI: 10.1039/C5FD90019K

**Martin Zanni** opened a general discussion of the paper by **Shaul Mukamel:** Shaul – the experiments that you are proposing are fascinating. Your idea of doing a "Raman" style experiment with X-ray pulses strikes me as a terrific idea because it will be much simpler to implement experimentally yet yield very informative data. I had a few questions. A Raman pulse will presumably be on-resonance to improve signal strengths. Are there differences in the electronic states of the atoms that would make one excited state preferable over another?

**Shaul Mukamel** answered: In addition to being stronger, Resonance stimulated X-ray Raman signals (SXRS), have another notable advantage: spatial selectivity. A core resonant Raman process creates a wavepacket of valence excitations localized in the vicinity of the selected atom. Other Raman pulses, including the detection pulse, can similarly watch valence excitations around other selected atoms. Thus the signals are very sensitive to the selected resonances and offer an unusual spatial resolution. Stated differently: in an analogous manner to the way resonant optical/UV Raman selects vibrations belonging to a specific chromophore, Resonant X-ray Raman selects valence excitations localized near the atom where the core hole resides.[1]

1 J. Biggs, D. Healion, Y. Zhang, and S. Mukamel, Multidimensional Attosecond Resonant X-ray Spectroscopy of Molecules; Lessons from the Optical Regime, *Ann. Rev. Phys. Chem.*, 2013, **64**, 101–127.

**Elangannan Arunan** followed up by asking: Could you comment on the difference between X-ray Raman scattering and Compton scattering? Raman scattering was considered the 'visible analog' of Compton scattering observed in X-ray.

**Shaul Mukamel** replied: Fundamentally there are two types of scattering of radiation by matter: elastic (Rayleigh) and inelastic (Raman). The former is coherent and scales as $N^2$ with the number of scatterers, and the latter is incoherent and $N$ scaling. Historically the elastic (inelastic) scattering of electromagnetic radiation by a free charged particle has been named Thompson (Compton). Compton scattering was an important demonstration that X-ray photons can be treated as particles. Taking a broader viewpoint, it is an example of Raman scattering. Similarly, Brillouin scattering is simply a Raman process involving acoustic phonons .

**Kenneth Ghiggino** remarked: You mentioned briefly in your presentation about results for a Zn and Ni porphyrin dimer. Can you expand briefly on the information that was obtained for this system using the approach outlined?

**Shaul Mukamel** responded: Using the REW-TDDFT computational protocol,[1] simulated SXRS signals of various Zn–Ni porphyrin heterodimers with different linkers and linking conformations were obtained.[2,3] The time-domain signals can be directly connected to the motion of the excited state wavepacket created by the pump pulse. Both the one-color (pump and probe at the same energy edge) and two-color (pump and probe at different edges) SXRS signals show a real time image of a back-and-forth excitation energy transfer (EET) in the system. This is not available from time-resolved fluorescence anisotropy decay measurements. This study demonstrates that SXRS could be a powerful tool for revealing EET mechanisms in molecular aggregates and could help the design of highly efficient solar energy devices.

1 Y. Zhang, J. D. Biggs, D. Healion, N. Govind and S. Mukamel, Core and Valence Excitations in Resonant X-ray Spectroscopy using Restricted Excitation Window Time-dependent Density Functional Theory, *J. Chem. Phys.*, 2012, **137**, 194306.
2 J. D. Biggs, Y. Zhang, D. Healion and S. Mukamel, Watching Energy Transfer in Metalloporphyrin Heterodimers with Stimulated X-ray Raman Spectroscopy in Real Time, *Proc. Natl. Acad. Sci.*, 2013, **110**, 15597.
3 Y. Zhang, J. D. Biggs and S. Mukamel, Understanding Excitation Energy Transfer in Metalloporphyrin Heterodimers with Different Linkers, Bonding Structures and Geometries through Stimulated X-Ray Raman Spectroscopy, *J. Mod. Opt.*, 2014, **61**, 558.

**Wolfgang Junge** asked: Shaul, how does your approach compare in signal-to-noise with the UV-vis experiments on a similar topic by Harry Gray[1–3]?

1 J. N. Onuchic, D. B. Beratan, J. R. Winkler and H. B. Gray, *Annu. Rev. Biophys. Biomol. Struct.*, 1992, **21**, 349–377.
2 H. B. Gray and J. R. Winkler, *Annu. Rev. Biochem.*, 1996, **65**, 537–561.
3 J. J. Regan, B. E. Ramirez, J. R. Winkler, H. B. Gray and B. G. Malmstrom, *J. Bioenerg. Biomembr.*, 1998, **30**, 35–39.

**Shaul Mukamel** answered: The X-ray measurements are more difficult. The signal-to-noise depends on the pulse characteristics and detection modes, which are being improved continuously, so it is hard to give a definite answer. The important point is that they provide qualitatively different information on many electronic states which are not available from narrower band UV/VIS experiments.

**Debabrata Goswami** questioned: When you have an extreme ultrafast pulse interaction that has a huge bandwidth, it may no longer allow us to use the Born–Oppenheimer approximation, nor the Frank–Condon principle. How do you get the separation of variables? With these kinds of transitions how do you reconcile them with your theory? When you are no longer in that regime, how do you separate them?

**Shaul Mukamel** replied: I would state the problem differently. A broadband pulse creates a superposition of many electronic states. The standard expansion in adiabatic states given by products of electronic and nuclear states does not necessarily fail but becomes tedious and impractical. In our applications we looked at very short times where nuclear motions may be neglected and only considered the purely electronic response. Treating nuclear dynamics classically pauses no major difficulty. However, developing efficient simulation protocols that treat both electrons and nuclei quantum mechanically is an open challenge. Cederbaum has addressed this issue in recent work.

**Himangshu Prabal Goswami** commented: Can we not capture the Auger transitions by starting from the original TRPES Hamiltonian in your eqn 1 and going to a higher order in perturbation in the $H_{\mathbf{p}}$, $H_v(t)$, $H_x(t)$ or a combination of the interacting Hamiltonians, instead of introducing the operators $H_{\mathbf{k}}$ and $A$, as introduced in eqn 44 of your paper, separately in the Hamiltonian.

**Shaul Mukamel** responded: The interactions represented by $H_x$ and $H_v$ are field-matter interactions and while a suitable combination could accomplish the same orbital re-arrangement as the Auger process, it would alter the field (the transition would be radiative). The Auger process, in contrast, is a non-radiative transition mediated by Coulomb matrix elements (as opposed to transition dipoles or combinations thereof).

**Artur Nenov** asked: How is the dephasing during $t_1$ and $t_3$ described for the TRPES signal (eqn 30 and 37 of your paper)? If a constant dephasing was used, what values were used for generating the TRPES spectrum?

**Shaul Mukamel** answered: The spectrum in our Fig. 3 is created by using eqn 37 in our paper, which averages over a set of semi-classical trajectories on a signal level. The nuclear degrees of freedom are treated classically, implying that the dephasing due to the vibrational motion is included in the model. By applying the surface hopping method and ionizing from a specific state the dephasing is implicitly included. Coherence between electronic states is encoded in the fluctuating gap coordinate. In Fig. 3, both initial pumping and photoionization were taken to be impulsive and thus the dephasing was taken into account only for time period ($t_0$). Equation 30 is more general and uses the fluctuating gap coordinates to recast the coherences into populations during the time periods.

**Artur Nenov** remarked: The expression for the TRPES signal (your eqn 30) is obtained under the assumption of a coherent superposition during the coherence time $t_1$ since one follows the fluctuation of the energy gap along each trajectory. Does this assumption hold for ultrafast dynamics where the conical intersection is reached from the excited state within few tens of fs, while it is inaccessible from the ground state? Does one need to correct for the decoherence of the wavepackets propagating on the ground and excited states?

**Shaul Mukamel** replied: If the conical intersection (CI) can be reached during the pumping pulse, then it is possible for the ket (or bra) to reach the CI before the bra (or ket) can interact with the field to generate the presumed population at time $t_2$. Relaxing the assumption of a population-Green's function for the time period $t_2$ would include the processes indicated by the question. This could be done by introducing a fluctuating gap coordinate as was done for $t_1$ and $t_3$ .

**E. D. Jemmis** opened the discussion of the paper by **John R. Helliwell:** Your results clearly indicate that the Pt–Br bond-breaking process in the protein–$PtBr_6$ complex is faster than the Pt–I bond-breaking process. Is this in tune with the bond strengths involved? Are there any computational or experimental data available on bond strengths? Are the observed differences a result of an unusual coincidence of the cavity size available to the protein fitting well for the $PtI_6$ complex and rather loosely for the $PtBr_6$ complex?

**John R. Helliwell** responded: As you indicate the expectation would be that the Pt–I bond breaking would be quicker than the Pt–Br as the former bond, I expect, is weaker than the latter, *i.e.* the Pt to bromide and Pt to iodide bond distances are respectively 2.4 *versus* 2.7 Angstrom (see our PDB codes 4owh and 4owc described in ref. 1). Secondly, I like your observation that the observed differences in X-ray radiation sensitivity might be a result of an unusual coincidence of the protein surface cavity size available to the $PtI_6$ complex matching it better than for the $PtBr_6$ complex.

1 S. W. M. Tanley, L.-V. Starkey, L. Lamplough, S. Kaenket and John R. Helliwell, *Acta Cryst.*, 2014, **F70**, 1132–1134.

**Mike Ashfold** asked: I was very struck by the sequence of images showing the progressive X-ray induced loss of Br atoms from the room temperature hen egg white lysozyme bound $PtBr_6$ species in your Fig. 4. Given that these are held in a non-spherical pocket, is it fanciful to imagine that one might be able to recognise site specific Br atom loss – now or in the future?

**John R. Helliwell** answered: The bromines obviously go somewhere as they depart the platinum hexabromide. Since this crystal form of hen egg white lysozyme is grown from NaCl as precipitant, the bromines would compete with the chlorides already bound to the protein, so most likely the bromines will go into the crystal solvent channels and thereby will not be visible to X-ray crystal structure analysis. The other aspect is that we expect the bromines to be steadily replaced with waters as per our Fig. 5. These, at fractional occupancy, are not obviously visible in the electron density of our Fig. 4.

**Neil Hunt** commented: The abstract of your article refers to lattice-free single molecule diffraction. What are the challenges facing the practical achievement of this?

**John R. Helliwell** replied: The macromolecular structure determination studies at the Stanford LINAC Coherent Light Source, as I explained in my Science Perspective article (reference 7 of our article), are currently working with a microcrystal size range of samples. The next step would be nanocrystals, then nanoclusters and then finally single molecules (or complexes such as my own research sample alpha crustacyanin with a 320 kDa molecular weight). To reach these successive stages of sample volume capability is the challenge, one that our article discusses in detail. I would add though that it was openly discussed at The Royal Society Discussion Meeting held in October 2013 (ref. 5 of our paper is a collection of most of those articles; from these see my book review at *J. Synchrotron Rad.*, 2015, **22**, 191–192) that a higher peak X-ray pulse flux would help the objective of measuring X-ray diffraction data from smaller samples. However a counter argument was made by Dr Abbas Ourmazd that ultra weak X-ray diffraction patterns should also lead to interpretable electron density maps from samples smaller than microcrystals (see especially his abstract at https://royalsociety.org/events/2013/xray-lasers-satellite/).

**Elangannan Arunan** addressed **John R. Helliwell:** In his talk Prof. Helliwell mentioned finding 'electron density within a molecule' by using X-ray methods. Two slides on the recent Atomic Force Microscope images of chemical bonds were shown at the end of this session. One of them had a beautiful image of a 'hydrogen bond'.[1] This image has been questioned recently in another study by Swart and coworkers[2] who have titled their paper 'Intermolecular contact in AFM images without intermolecular bonds'. I welcome any comments about whether this question can be unambiguously answered by any of the techniques discussed during this meeting. X-ray, attosecond, ...?

1 J. Zhang *et al.*, *Science*, 2013, **342**(6158), 611–614.
2 S. K. Hämäläinen, N. van der Heijden, J. van der Lit, S. den Hartog, P. Liljeroth and I. Swart, *Phys. Rev. Lett.*, 2014, **113**, 186102.

**John R. Helliwell** answered: Thank you for drawing my attention to these two articles on AFM results. In the short time available to me set by the Faraday Discussion deadline I have looked to understand how AFM might allow quantitative structural chemistry studies (bond distances and angles having standard deviations that allow bond order discrimination for example). The two articles you quote seem not to provide such details. Nevertheless, from the reference list of the *Phys. Rev. Lett.* article, I found the associated article: *Science*, 2012, **337**(6100), 1326–1329 entitled "Bond-Order Discrimination by Atomic Force Microscopy" by L. Gross *et al.*. Their abstract states "The greater electron density in bonds of higher bond order led to a stronger Pauli repulsion, which enhanced the brightness of these bonds in high-resolution AFM images. The apparent bond length in the AFM images decreased with increasing bond order because of tilting of the CO molecule at the tip apex." These are certainly impressive studies by AFM. The AFM approach seems to allow a semi-quantitative scrutiny of the structural chemistry unlike crystal structure analysis by X-rays (or neutrons) where detailed molecular crystal structure refinement is undertaken and standard deviations on bond distances and angles in a molecule can be derived. [*N.B.* on a point of nomenclature standard deviations are now referred to as standard uncertainties.]

**Neil Hunt** continued the general discussion of the papers by **John R. Helliwell**, **Shaul Mukamel** and **Martin Meedom Nielsen**: In light of the session being one that addresses future directions, can you comment on how you envision your respective research areas evolving over the next decade?

**John R. Helliwell** commented: I have no budget for the methods development and structural biology (on the 320 kDa alpha crustacyanin multi-macromolecular complex) research that I have outlined in my article, however the fact that the UK is now a formal participant in the Euro XFEL encourages me to think that it should be possible to undertake the work, *i.e.* it might be funded.

**Shaul Mukamel** responded: Single molecule diffraction is an exciting future development made possible by recently developed ultrafast and intense free electron laser pulses. The connection of diffraction to spectroscopy should be clarified. Classically, diffraction patterns are thought of as the interference of elastically scattered X-ray wave-fronts emanating from different particles.[1] A fundamental difficulty in extending diffraction to single molecules is that X-ray scattering (as any light scattering) from single molecules may not be simply described as "diffraction". There are two basic microscopic mechanisms for light scattering from an assembly of $N$ particles in the ground state. The first is incoherent, where the entire process occurs with a single particle. The second is coherent, and involves a pair of particles each generating a scattering amplitude.[2] These two components of the X-ray scattering signal come with different form factors, unlike in the classical theory.[3] Additionally, they scale as $N$ and $N^2$, respectively, with the number of particles. Coherent scattering from the ground state is elastic, whereas incoherent scattering contains both elastic and inelastic components. This classification is based on the fundamental way in which the field and matter interact (how the detected field mode is populated from the vacuum) and holds regardless of the initial state of the system (be it stationary, pure or mixed). In optical spectroscopy, elastic and inelastic scattering is known as Rayleigh and Raman, respectively, whereas elastic and inelastic scattering of X-rays from a free charged particle are historically known as Thompson and Compton, respectively. In the spectroscopic language, X-ray diffraction is thus an example of Rayleigh scattering. For a single molecule , $N = 1$, elastic and inelastic signals may be comparable and the scattering process is better labeled as Rayleigh and Raman rather than diffraction.

1 J. Als-Nielsen and D. McMorrow, Elements of Modern X-Ray Physics, Wiley, Hoboken, 2011.
2 K. E. Dorfman, K. Bennett, Y. Zhang and S. Mukamel, Nonlinear Light Scattering in Molecules Triggered by an Impulsive X-Ray Raman Process, *Phys. Rev. A*, 2013, **87**, 0853826.
3 K. Bennett, J. D. Biggs, Y. Zhang, K. E. Dorfman and S. Mukamel, Time-, Frequency-, and Wavevector-Resolved X-Ray Diffraction from Single-Molecules, *J. Chem. Phys.*, 2014, **140**, 204311.

**Martin Meedom Nielsen** answered: For an ultrafast time-resolved molecular structure, I see a very interesting development towards using combinations of different spectroscopic and scattering techniques, simultaneously in the same experiment. This has the potential of providing not only complementary information about spin and structural dynamics of the system, but interestingly, as the measurements are truly simultaneous, it will enable to cross correlate findings between the methods and enhance the ability to extract quantitative results. For the sample systems, I see a move towards ever more complex systems. Also systems that are close to 'real' applications in chemistry, materials science, and electronics. This is a very important development for the field, which has for a long time been limited to a set of model systems.

**Judith Howard** asked: Where do you think the subject will be in another 10–20 years with respect to membrane proteins?

**John R. Helliwell** replied: As an objective measure the website http://blanco.biomol.uci.edu/MP_Structure_Progress.html monitors and analyses the PDB deposition statistics for membrane protein structures in detail and the growth rate of this category of depositions in the PDB suggests that "we can expect the number of new membrane protein structures by 2025 to be over 2000". In January 2015 the number of membrane protein structures in the PDB, mainly due to X-ray crystallography, logged by the same website were: "unique proteins in database = 522; coordinate files in database = 1592; published reports of membrane protein structures in database = 931 (These counts do not include pre-publication structures)." Clearly the progress in macromolecular crystallisation techniques for this previously challenging category of protein structures for crystal structure analysis has therefore been substantial. It would also be likely that the new X-ray lasers and ultra-bright upgraded synchrotron radiation X-ray sources, such as the world leading ESRF and initiatives like the MASSIF (massively automated sample selection integrated facility) for the automatic screening of huge numbers of samples to assess their single crystal diffraction characteristics, are likely to accelerate the rate of membrane protein crystal structure determinations further. Also most major research infrastructure synchrotron sites in Europe have, or are developing, multidisciplinary centres, like the Partnership for Structural Biology (Grenoble), the Centre for Structural Systems Biology (Hamburg), the Membrane Protein Laboratory and the Oxford Protein Production Facility (RAL). In the USA, the APS is building its Advanced Protein Crystallization Facility, which will allow the production, characterization and crystallization of proteins. All these, often multi-facility/multi-laboratory partnerships, stretch facility impact beyond the supply of X-rays and/or neutrons to helping users prepare proteins, perform quality control and crystallize them, and refine conditions.[1]

1 J. R. Helliwell and E. Mitchell, Synchrotron radiation macromolecular crystallography: science and spin-offs, *IUCrJ*, 2015, **2**, 283–291.

**R. J. Dwayne Miller** continued the discussion of the paper by **Martin Meedom Nielsen**: I am sure you did not go into these experiments wanting to get into the details of the detector response. Characterizing the detector was really heroic work. Your method of correcting for the nonlinear response of the detector was key to pulling out the dynamics of interest. However, the detector is still intrinsically nonlinear. The noise characteristics are not the same as a linear detector under which the detector can operate within the shot noise limit. For most experiments, the biggest source of noise is the laser excitation and probe contributions to the overall signal detection, as well as sample issues. At XFELs, one is using a SASE source in which the shot to shot variations are nearly chaotic with variances over 100%. In a time-resolved measure, one is doing a differential measurement for time-resolved diffraction or diffuse scattering experiments and must detect signal changes on the level of 1% or less. Even if the normalization for shot to shot variations, timing jitter, spectrum fluctuations are handled perfectly, there are noise contributions in the detection used for normalization, in addition to the nonlinear detector response and associated increased sensitivity to noise in the low intensity detection range. The very significant difference in noise amplitude in the low intensity relative to the nonlinear higher intensity regimes of the detector response may lead to problems in trying to detect very small changes in signal. For just simple femtosecond pump–probe spectroscopy, one needs around the order of 0.5% rms noise, from all sources (laser, detectors) within the detection bandwidth. In this case, the laser excitation/probe are the largest contributor to the noise and laser operation around 0.5% rms is needed to pull out changes in signal of the order of 0.1%. Can you comment on the effect of noise on the measurement, notably the contributions from the X-ray probe noise (after all the normalizations) and the nonlinear noise contributions to the detector? Given the noise limits, can you give an approximate estimate of the concentration and fraction excited that is needed to clearly resolve structure changes above background scatter from the solvent? This estimate will of course depend on the $Z$ contrast of the photoactive system under study. If you could give a range for different $Z$ values it would be appreciated to help guide future experiments.

**Tim Brandt van Driel** replied: The challenges at XFELs, as opposed to synchrotron sources, have been the systematic errors and nonlinear effects, as well as artifact-like fluctuations, that could not be corrected for or averaged out.

 

This is an area of constant development both with regards to detector development and data processing. After the presented corrections have been applied, the CSPAD detector still exhibits common mode fluctuations that do not average out with the number of images usually measured for X-ray diffuse scattering. After corrections, the remaining fluctuations are common mode and Poisson distributed counting noise similar to CCD cameras.

The scattering power of the sample depends on $Z$, but the difference scattering varies in the measured $Q$-range and depends on the Fourier transform of the changes in the electron density distribution. The background noise depends on the Fourier transform of the full electron density. Therefore the resulting difference signal is given by the absolute number of excited molecules and the noise is given by the total number of molecules in the beam path. From recent experiments a difference signal of at least 0.1% is needed for the easy identification of a difference scattering signal. For solute molecules containing heavy metal centers (Ir or Pt), or even polypyridyl metal-centered complexes (Fe, Co and Ru), these signal levels are currently achievable with 10 mM concentrations and excitation fractions around 10%, but it depends on the experiment.

**Martin Meedom Nielsen** added: Regarding the nonlinearity of the detector, referring to Fig. 8b and 8c in our paper, the detector is linear to better than 1% over a quite wide range of beam intensities. Even so, some of our analysis has been enabled by grouping measurements according to relatively narrow intensity ranges allowing a less than 10% variation in measured intensity in the groups. It is true that the shot-to-shot variations of the SASE source may lead to problems in detecting small changes in the signal. These problems can be checked for by comparing the difference signals obtained over a range of variations in incoming intensity. The difference signal is observed to converge as this range is systematically reduced. In this way, we further reduce the effect of non-linearities and can significantly enhance the ability to observe small changes in the signal. Referring to Fig. 14 in our paper, we can reliably detect changes in signal intensity well below 0.1% after applying the corrections described in this paper.

Comparing our Fig. 14 with Fig. 4 in reference 1, a sensitivity of order 0.1% is more than adequate for distinguishing between solvent and solute contributions to the data. We have not performed quantitative simulations of this, but based on recent experience with 100 μm thick liquid jets we find that around 10 mM concentration is fully adequate for Fe and Co-based compounds such as $Fe(bpy)_3$ and $Co(terpy)_2$. Typical excitation fractions under beamline experimental conditions have been in the range of 20–40% for these compounds.

1 K. Haldrup, G. Vankó, W. Gawelda, A. Galler, G. Doumy, A. M. March, E. Kanter, A. Bordage, A. Dohn, T. van Driel, K. Kjær, H. Lemke, S. Canton, J. Uhlig, V. Sundstrom, L. Young, S. Southworth, M. Meedom Nielsen, C. Bressler, Guest-Host Interactions Investigated by Time-Resolved X-ray Spectroscopies and Scattering at MHz Rates: Solvation Dynamics and Photoinduced Spin Transition in Aqueous $Fe(bipy)_3^{2+}$, *J. Phys. Chem. A*, 2012, **116**, 9878–9887.

**John R. Helliwell** said: On page 20 of your article you make the point that the SVD type of correction method you have harnessed can be used for any fluctuations relating to (variations in the) detector electronics. Once you move away from the detector prototype that you used thus far, *i.e.* to more mature devices, such detector instabilities are unlikely. The effects on the X-ray diffraction intensities of changes in air scattering will be small in any case. Can you please comment on both these two points?

**Martin Meedom Nielsen** replied: Ideally, one should never be forced to use correction schemes, such as the ones we have developed. However, the area detectors used since the start of operation of the XFEL facilities have all been suffering from electronic fluctuations and non-linearities, making the interpretation of difference scattering signals very challenging. It is certainly a hope, and often also true, that fluctuations in detector electronics become less and less important as the technology matures. In the latest generation of the CSPAD detector, we can directly observe that the relative magnitude of the correction terms become smaller compared to the older versions. Regarding changes in air scattering, it is a concern when measuring the difference in scattering signals. Often the sample is held in a He environment, but some amount of atmospheric air is always present, and if this amount changes, *e.g.* due to leakage, it changes the difference scattering signal. Also without He enclosures, differences due to air scattering are readily observed if the air surrounding the sample is moving, for example if air is blown over the sample from an air conditioning unit. Using difference scattering, we are extremely sensitive to changes in the scattering signals. This is a great benefit for analysing changes in molecular systems, but also a challenge when phenomena not related to the molecular structure are fluctuating and causing changes in the scattering signal. Hence we need tools to identify different sources of such changes, and here we have had good experience using the tools described in the paper.

**Neil Hunt** queried: How do you see access to large facilities, such as XFELs evolving as technology advances?

**Martin Meedom Nielsen** responded: Access to making experiments at XFELs will presumably always be more limited than access to making experiments at synchrotron sources, as long as there are fewer XFEL facilities and as long as only a limited number of their experimental stations are capable of operating simultaneously. The latter will improve. LCLS has developed and is already operating a 'beam sharing' mode, and the European XFEL is from the design already envisaged to be able to operate more than one experiment at the same time. Additionally, I see a big potential in development of software for accelerating the output of results from XFELs and making the experiments more efficient. Virtual experiments will help users design and optimize their experiments, and standardised software solutions for data reduction and analysis will help the users convert their data into results much more rapidly and reliably. This will undoubtedly also be an important parameter in opening XFEL science to a wider community.

 

**Wolfgang Junge** opened a general discussion of the paper by **R. J. Dwayne Miller** by communicating: Dwayne, how do you overcome a possible limitation of time-resolution by Coulomb spread of the electron pulse?

**R. J. Dwayne Miller** communicated in reply: Thanks for this question as it allows me to expand on this key enabling development. This question is related to the one on linear chirp on the electron pulse. We discovered that it is possible to minimize Coulombic repulsion effects on the spatial-temporal resolution by properly managing the longitudinal Coulombic or space charge effects on pulse broadening, without affecting the transverse beam divergence significantly. The longitudinal broadening (defined by the pulse propagation direction) leads to loss in time-resolution. The transverse broadening to this direction leads to increased beam divergence and loss of spatial resolution. The epiphany moment in how to optimize electron pulses for femtosecond structural studies came through an effectively exact solution to the coupled equations of motion for up to 10 000 electrons.[1] It should be noted that with the very high scattering cross section of electrons, this number of electrons is sufficient for single shot atomic structure determination of many solid state systems. For a 100 micron beam parameters needed to match typical laser excitation conditions, the space-charge broadening was confined primarily to longitudinal pulse broadening. This detail needs to be fully appreciated. Moreover, there were two solutions that jumped out of these calculations that were completely missed in previous work using solely analytical methods with overly simplified approximations. The first most obvious solution was that if the electron pulse broadens with propagation, then simply don't let it propagate very far. In this work, we could determine exactly the electron gun dimensions needed to achieve 100 femtosecond electron pulses without sacrificing the spatial resolution. This point was obviously understood qualitatively but we could not determine precisely the spatial scale for electron acceleration. Now we can. This new insight led to the development of the "compact electron gun", which is capable of over $10^5$ electrons with sub-100 fs pulse durations, and over 5 nm transverse coherence, in our present designs. These systems are by far the simplest and most robust sources for structural dynamics. With new developments in photocathodes giving lower emittance, or higher transverse spatial coherence, these sources will soon be capable of studying systems as large as proteins. The second solution is more involved but it was also immediately apparent from these calculations. We discovered that the longitudinal space charge effects still conserve the space-time correlation. The electrons at the front of the electron pulse experience the electron charge distribution at the back of the pulse and are accelerated; whereas the electrons at the back of the pulse are decelerated. The electrons at the front of the pulse stay at the front and the electrons at the back stay at the back. For nonrelativistic electrons, the higher energy electrons from this process move forward in time relative to the retarded electrons. The big surprise was that this "chirp" or variation in space-time is effectively perfectly linear. This effect was completely unexpected and certainly missed by analytical methods prior to this work. When one sees such a linear chirp, it is clear that simple dispersive elements can be used to temporally re-focus the electron pulse down to near its initial conditions at birth and place a sample at the re-focusing point. Basically any dispersive element that makes the higher energy electrons travel a longer path than the lower energy electrons will work, with the condition that the spatial overlap should be reconstituted at the sample position. These elements are well known in the accelerator community. The best dispersive element for this application has been developed by the Luiten group using a so-called pill lens for rf pulse compression. This device creates a close to linear chirp of opposite sign, without affecting the emittance or spatial properties of the pulse, to recompress the pulse at the sample position. Pulses with up to $10^6$ electrons and pulse durations as short as 30 fs can be generated with this approach. There are technical challenges associated with timing jitter between the laser excitation and the rf that limit the time-resolution to approximately 200–300 fs, the exact same problem as incurred with XFELs for the same reason. Time stamping methods have been developed that enable correction of this jitter to 30 fs.[2]

Of the two technical achievements in optimizing electron pulses for structural dynamics, the compact electron gun is the simplest to use and by far the most robust. It reduces the real time observation of atomic motions to experiments no more difficult than conventional pump–probe experiments that are routinely done in the femtosecond laser community. I would only recommend using rf pulse compression or other pulse compression methods if the higher time-resolution (<100 fs) is warranted to solve the problem of interest.

1 B. J. Siwick, J. R. Dwyer, R. E. Jordan and R. J. D. Miller, *J. Appl. Phys.*, 2002, **92**, 1643–1648.
2 M. Gao, Y. Jiang, G. H. Kassier *et al.*, *Appl. Phys. Lett.*, 2013, **103**(3), 033503.

**John R. Helliwell** commented: The $10^6$ gain of electrons as a probe over X-rays in their scattering efficiency by matter is indeed very attractive, but traditionally this strong interaction has made interpretation difficult (due to dynamical scattering effects) and in particular standard deviations on atomic positions are either not quoted or are quite a lot larger than those derived from X-rays or neutrons as probes of the structure of matter. Can you please comment on this as it will underpin how well electron diffraction can be harnessed in quantitative structural chemistry or structural biology studies?

**R. J. Dwayne Miller** answered: The high scattering cross section and the very short de Broglie wavelengths are major distinguishing features of electron probes in relation to X-ray probes of structure. The much shorter wavelength actually transforms to higher spatial resolution (in the limit the samples are of sufficient quality and the diffraction occurs in the kinematic scattering regime). This much higher spatial resolution and scattering cross section is explicitly exploited in gas phase diffraction from which most of the bond length information is determined. I make this distinction to point out that intrinsically electrons have a much higher information content per scattering event and a much higher spatial resolution. The real problem as you point out is the effect of multiple electron scattering for samples that are too thick relative to the electron elastic mean free path and/or too thick relative to the inelastic electron mean free path that limits electron transmission and creates background scatter. In the past, it has been very difficult to attain samples sufficiently thin to avoid this problem. There have been only a few classic examples of solving protein structures using electrons for systems such as bacteriorhodopsin that naturally form 2D crystals. However, with the latest developments in nanofabrication methods, this problem is no longer restricting the use of electron probes. Moreover, it is becoming apparent that some of the most important protein targets, such as G-protein coupled receptors, intrinsically do not grow large crystals of sufficient quality for structure determination, but it is possible to attain submicron or nanocrystals of these and related systems. The ability to do protein nano-crystallography has been one of the main scientific motivations for XFELS. If protein nano-crystallography is indeed as important a new direction for structural biology as currently thought, electrons have a decided advantage. This length scale is perfect for electrons to avoid multiple electron scattering. The real challenge for both X-rays and electron approaches is sample delivery.

To specifically address your point, we deliberately did a detailed study of the effect of multiple electron scattering on the problem. In our Fig. 1, we did a first principles calculation of the electron scattering cross section using Al as a well-defined test case. In Fig. 6 and 7, we show the effect of different sample thicknesses on the observed diffraction pattern and compare to the ideal single scatter limit. It is clear that we can still resolve diffraction intensities well correlated to structures up to 3–4 times the electron mean free path. For relativistic electron regimes, this means samples close to micron thickness can be studied. This finding may seem surprising. However, issues raised over multiple electron scatterings do not consider the diffuse scatter of the crystal itself, even within the single scatter limit. It is only for perfect crystals that one could state that multiple electron scattering is limiting the resolution. In fact, it is usually crystal quality and not multiple scattering that will limit the ultimate spatial resolution for a given sample. This statement will be especially true for nanoprotein crystals where it is unclear if such systems will be inherently poorly diffracting systems due to the surface strain and disorder effects. The spatial resolution one can attain for protein nanocrystals remains an open question. There is no fundamental limitation in the resolving power of electrons in this case. High resolution structure determination will be possible. There is also the prospect that the new high coherence electron sources could even provide atomic resolution for poorly diffracting nanocrystals as there is additional information in the scattering process that has yet to be exploited.

I should also respond to your question with respect to extracting transient structures for photoinduced structural transitions and photochemical processes, which is the main emphasis of this work. We are not trying to resolve unknown static structures. We know the starting structure and in most cases the final fully relaxed structure. We can use *ab initio* phasing methods such as charge flipping methods, but given the enormous number of diffraction orders and constraints on the structural changes, it is simplest to model the structural changes and do a correlation analysis.[1] We have found a robust convergence in tracking the structural changes. The variances to the fits are obtained from the coefficients in the Pearson correlation analysis. For cases where there is a singularly important parameter such as the bond elongation in undergoing a spin transition or metal–ligand charge transfer band, your point is well taken. We have not reported on such systems as yet. It would be helpful to explicitly state the error bars in regard to specific displacements, especially for comparison to theoretical predictions.

1 M. Gao, C. Lu, H. Jean-Ruel, L. C. Liu, A. Marx, K. Onda, S.-ya Koshihara, Y. Nakano, X. Shao, T. Hiramatsu, G. Saito, H. Yamochi, R. R. Cooney, G. Moriena, G. Sciaini and R. J. Dwayne Miller, *Nature*, 2013, **496**, 343–346.

**Siva Umapathy** asked: When you observe the motion of particles, if there is a solvent effect (such as inertial solvation) on the dynamics, which would be at the same time scale as pulse evolution, can you get information of the coordinates?

**R. J. Dwayne Miller** replied: We have developed nanofluidic cells to enable the study of solution phase chemistry. We have demonstrated path-length control to achieve a stable 100 nm flow, which is sufficient for even non-relativistic electrons to be used to probe solution phase processes and even real space imaging within conventional TEMs. As discussed the development of REGAE was in fact motivated to take advantage of the much higher penetration depth of relativistic electrons to enable direct observation of atomic motions in liquid environments. This feature is important as most chemistry occurs in the solution phase. However, to date we have not studied solution phase chemical processes as we are further refining the nanocell concept. I can however comment that the time-resolution is sufficient to follow even the fastest solvation dynamics as exemplified in water. The solvation dynamics have an inertial contribution as you point out. For water the fastest motions involve hindered librational motions with an approximate density of states weighted relaxation time of 70 fs. This time scale can be resolved. The specific motions may be very difficult as this is a collective response involving small net displacements of some 80 water molecules within a typical solvation volume. The effect is larger in terms of the relaxation energetics but the driving force is spread out over a large number of molecules. There are similar relaxation processes within the solid state in which adjacent moieties or unit cells undergo a concerted relaxation around photoproduct states and here one can observe the motions. The motions of the $PF_6$-counterion in EDO–TTF is a case in point in which we were able to well resolve its motions.

**Martin Meedom Nielsen** commented: You describe multiple scatterings of electrons as a problem for understanding the scattering signal. However for Low Energy Electron Diffraction (LEED), multiple scattering has been used to enhance the sensitivity of the method, by measuring the scattered intensity as a function of the acceleration voltage of the electron beam (the so called Intensity–Voltage (IV) curves), see *e.g.* the books by J. B. Pendry, *Low Energy Electron Diffraction*, 1974, and M. A. van Hove and S. Y. Tong, *Surface Crystallography by LEED*, 1979. Would it be possible to change the energy of the electrons in order to measure IV curves and retain the time-resolution in your ultrashort electron pulses? Would it limit the time-resolution when using electrons at low energies?

**R. J. Dwayne Miller** answered: To be clear, my main point is that multiple electron scattering can be well handled by proper care in preparing the samples. For time-resolved studies, one knows the starting, and in most cases, the end point structures, so that less information is needed for determining the changes in structures relative to the static structure determination. This feature is due to the differential detection with and without laser excitation that removes much of the inelastic scattering contributions that are more or less constant for chemical reactions, especially on picosecond to subpicosecond time scales for which there is negligible non-radiative relaxation into lattice heating. From our studies of the thickness dependence on the diffraction intensities for different diffraction orders, samples with thicknesses at least 2–3 times the inelastic mean free path will be sufficient. We generally try to keep the sample closer to the single scatter limit as we did for the organic systems we have studied. I quite like your comment about scanning the electron energy to exploit multiple scattering to enhance sensitivity and by that structural resolution. This approach can easily be done as previously with LEED. The time-resolution is challenging as it is difficult to keep the electron propagation path short enough for low energy electrons to in turn keep the pulses subpicosecond with sufficient electrons. I expect it should be possible to attain picosecond time-resolution with much averaging of the low electron bunch charge to keep the pulses as short as possible. If a reversible surface process can be found then 100 femtosecond time-resolutions will be achievable. I am most interested in surface chemistry, which is generally non-reversible. Studies of surface chemistry will require low excitation and sampling rates with sufficient times for the regeneration of the surface between excitation events – all under UHV conditions. It is challenging but possible.

**Martin Meedom Nielsen** added: Is it possible to tune the energy of the electron bunches, and if yes, could one use this to make spectroscopic studies?

**R. J. Dwayne Miller** responded: It is rather simple to tune the voltage in a programmable way. The only issue is the field gradient. We have developed a tunable gap to keep the gradient constant as one scans the electron energies. This question really represents an excellent suggestion. The answer is yes.

**Debabrata Goswami** asked: You have used linearly compressed pulses, do you know what happens if you do a systematic study as a function of chirp on the results? Are the results only dependent on the pulse width?

**R. J. Dwayne Miller** communicated in reply: There are both the pulse widths, or time-resolution, and the spatial resolving power that require attention when using compressed electron pulses. The chirp that naturally develops for non-relativistic electrons is exceptionally linear, amazingly so. For relativistic electrons, it is very close to linear within the field gradient used. Small nonlinear terms can be corrected in the rebunching cavity. The only effect is that one can broaden, linearly, the pulse to stretch it and enable more electrons per pulse without incurring a space charge time broadening of the pulse. In principle, there is no limit to the pulse compression in the linear chirp regime. The only problem is that the path-length at the sample position over which the pulse is compressed.

If you inspect the inset to our Fig. 5 (panel d), you will see that the pulse is compressed at the sample position and then increases in pulse duration after this point. For REGAE parameters, the length over which the pulse is optimally compressed is in the order of a few centimeters so it is relatively easy to temporally focus at the sample position. However, if we increase the electron bunch charge more, the compression results in space charge effects that lead to a loss in transverse spatial coherence and the optimally compressed region becomes smaller than practical. We see this effect as a loss of spatial resolution or diffraction quality. The limit to electron bunch density is very easy to determine. The numbers we have presented are for optimal conditions with the maximum electrons per pulse without sacrificing spatial resolving power for following molecular processes for systems as large as proteins.

**Tim Brandt van Driel** communicated: When using electron diffraction to record 10 fs resolution movies in a single shot, is this only doable for large structural changes such as sample melting? Are the 10 fs the full instrument response function and what are the main contributions such as laser and electron pulse duration? Additionally, do you have similar problems as X-rays with sample damage, sample charging or Coulomb explosion?

**R. J. Dwayne Miller** communicated in reply: The concept of using a streak camera was not discussed in our paper but is mentioned here to give some perspective on future directions. This idea of obtaining full movies in a single shot has been discussed previously. This prospect reduces sample requirements by nearly two orders of magnitude in terms of having a sufficient surface area to collect the atomically resolved dynamics relative to stroboscopic, single time point sampling. In this regard, I would like to particularly point out the work of Heinrich Schwoerer's group where they have shown significantly improved signal-to-noise using a streaked time base to collect all the dynamics at once.[1] They show full movies captured for the photoinduced modulation of the structure order parameter for the charge density waves in TaS2. This is a rather complex diffraction pattern and the motions involved are very small (0.1 Å or less depending on the excitation level). The time-resolution they achieved was approximately 200 fs, limited by the field gradient they could apply. Similar demonstrations of the concept have been given by Musumeci's group where they gave a proof of principle study of melting using relativistic electrons in which a 400 fs time-resolution was demonstrated.[2] The latter study did not take into account the nonlinear features of the rf field used for streaking. We have modelled the parameters for REGAE and have developed a more linear sweep field. We expect to achieve a 10 fs time-resolution based on the streak velocity and pixel resolution of our detector. We have also developed analytical methods that take advantage of the known initial structure to extract the dynamics from the 2D streaked diffraction image. It is clear we can use this approach for molecular systems as large as EDO-TTF with unit cells up to several nm.[3] The question is what is the minimum amount of information one needs to construct an image? It is not inconceivable that this approach could scale to systems as large as proteins given the known starting structures and fully relaxed structure or end points to constrain the fits. The streak camera approach provides sufficient time-resolutions to capture the fastest possible nuclear motions while enabling the study of precious samples, *i.e.* irreversible systems that cannot be obtained in sufficiently high quality crystals for multiple sampling methods. It effectively increases the source brightness by another factor of 100 or more depending on the dynamic range sampled by the streak field. One can keep the electron bunch charge per unit time constant and simply stretch the pulse by a factor of 100 or more and increase the total number of electrons accordingly to sample the dynamic range of interest in a single shot. This is truly a unique feature of electron sources that should be fully exploited.

As per the electron induced damage of the samples, we have never observed this effect. In a conventional TEM, the areal beam current is many orders of magnitude higher than the average currents used for femtosecond electron diffraction. We have observed sample damage in seconds within a TEM yet we have collected data for months on the same sample without any noticeable changes using our femtosecond electron sources (EDO-TTF as a case in point). We often inspect samples for diffraction quality first in a TEM before committing beam time for femtosecond diffraction studies. We can attest that electron induced damage is not an issue as it is for X-rays or TEM focusing conditions. Basically, the peak current for the femtosecond electron pulses is high but the average current is many orders of magnitude less than conventional TEMs. There is essentially no change in the lattice temperature from the inelastic scattering of the electrons at the average currents used, which completely eliminates thermally induced lattice damage. In terms of electron induced ionization effects, we are many orders of magnitude below the 1–10 $e^-$/atom flux where this effect dominates. The sole source of sample damage is due to the optical excitation, which has been the driving force for developing the brightest electron sources possible to collect atomically resolved structures for even complex systems in a single shot – and single shot atomically resolved full movies are now possible.

1 M. Eichberger, N. Erasmus, K. Haupt, G. Kassier, A. von Flotow, J. Demsar and H. Schwoerer, *Appl. Phys. Lett.*, 2013, **102**(12), 121106.
2 P. Musumeci, J. T. Moody, C. M. Scoby, M. S. Gutierrez, M. Westfall and R. K. Li, *J. Appl. Phys.*, 2010, **108**(11), 114513.
3 M. Gao, C. Lu, H. Jean-Ruel, L. C. Liu, A. Marx, K. Onda, S.-ya Koshihara, Y. Nakano, X. Shao, T. Hiramatsu, G. Saito, H. Yamochi, R. R. Cooney, G. Moriena, G. Sciaini and R. J. Dwayne Miller, *Nature*, 2013, **496**, 343–346.

**Priyadarshi Roy Chowdhury** asked: What are the basic factors that should be taken into consideration during the mapping motions with ultra bright electrons in the cases of different types of molecules in solid, semi-solid or volatile samples?

**R. J. Dwayne Miller** responded: Let me first address solid state samples for femtosecond electron diffraction studies where one has the highest space-time-resolution. The machine physics for attaining a sufficiently bright electron source has been solved in that it is no longer limiting problem selection. The real challenge now is the samples. Generally speaking, one would like to probe different aspects of fundamental issues related to structural dynamics and material properties. I am personally mostly interested in chemical reaction dynamics as this methodology allows the observation of the relative atomic motions that propagate the system from one distinct molecular structure to another, *i.e.* a direct observation of the defining moments that lead to chemistry. It also allows us to observe how a complex many body system reduces to a few key motions at the far from equilibrium points involved in barrier crossing. In this respect, the relative figures of merit for samples to probe these issues are critical considerations. You need to find a system that probes the fundamental issue in question and this process must be capable of being photoinitiated with femtosecond laser pulses, otherwise it is not possible to address the relevant timescales for the atomic motions. Basically, it must be possible to prepare a system optically on an excited state potential energy surface that is effectively barrierless to the process of interest and the quantum yield must be close to unity. As a rule of thumb, it is necessary to have approximately 10% of the lattice undergoing the structural change of interest to get above the background diffraction and diffuse scatter. In the femtosecond time domain, there are peak power limitations that prevent exciting 100% of the sample as this necessarily leads to multiphoton ionization artifacts. It is for this reason that the system should exhibit a high quantum yield for the photoinduced structural process of interest, and certainly should be above 10% as a threshold for consideration. In addition, the sample must form high quality crystals that can be prepared as thin as 20–200 nm thick, depending on $Z$ of the dominant lattice host atom and associated electron scattering cross section. We either grow crystals directly to meet this condition or microtome them to the desired thickness. The biggest problem is with water soluble crystals where microtoming is not possible as the high surface tension of water is used to pick up a sample after microtoming thin sections. The real frontier of this field now is in designing crystals to probe different aspects of chemistry (and biology) with the above constraints.

With respect to gas phase systems, the major limitation is the number density of gas molecules one can deliver, with the above caveats for quantum yield for the process of interest fully respected. For any given system there will be a limit to the number density and this further stipulates a sample path-length to achieve sufficient electron diffraction. This aspect to the problem is well known in the gas electron diffraction community where the emphasis is on high resolution structures. However for time-resolved studies, the temporal resolution is limited by the velocity mismatch between the speed of the laser excitation pulse to induce the structure changes and the speed of the electron pulse probing the structural changes. For typical molecular beam dimensions and electron energies of less than 100 keV, as typically used for gas phase studies, the time-resolution is limited to a few picoseconds, with 10 picoseconds being more typical when electron broadening effects in propagation to the molecular beam are taken into account. Basically the time-resolution is limited by the difference in propagation time for the electrons to traverse the excited volume relative to the laser excitation. It is possible to use tilted phase fronts for the laser excitation to minimize the velocity mismatch but this is rather difficult to implement and has not been done as yet for gas phase samples. It is precisely in this domain where relativistic electron sources will open up a 10 femtosecond time-resolution to chemical reaction dynamics. There is effectively no velocity mismatch, *i.e.* the electrons travel very near the speed of light to the point that path length limitations are no longer a concern. REGAE was designed to meet this goal for the gas phase to take advantage of the higher velocity and to also open up solution phase studies for atomically resolving reaction dynamics to take advantage of the longer electron mean free paths in the relativistic regime. We are now in the process of introducing different gas and liquid cell concepts based on our work using nanofluidics[1] to tune the desired sample path-length for a given problem from 100 nm (liquid) to 100 microns (gas). It would be a dream come true if we could compare the reaction dynamics for the same system under isolated collision free conditions of the gas phase to that in the solution phase where most chemistry occurs to truly separate the role of the solvent in directing chemical processes.

1 C. Mueller, M. Harb, J. R. Dwyer and R. J. Dwayne Miller, *J. Phys. Chem. Lett.*, 2013, **4**(14), 2339–2347.

**Priyadarshi Roy Chowdhury** queried: How does the bucket size increase during the snapshots of your molecules?

**R. J. Dwayne Miller** answered: Depending on the dynamics of interest, we control the bunch charge density, or bucket as you term it, to give us the most electrons possible without losing the required time-resolution. Control over the "bucket size" is readily achieved by controlling the intensity of the laser excitation used for electron photoinjection. For 100 femtosecond dynamics, we typically use between $10^5$ to $10^6$ electrons per pulse in an approximately 100 micron spot. For the sample thicknesses used, this source intensity is comparable to $10^{11}$ to $10^{12}$ X-ray photons from an XFEL in terms of detected signal counts. For this comparison as a calibration, these are very bright sources. The main limitation is in the transverse coherence where we are currently limited to transverse coherence lengths of a few nanometers, which limits the size of the unit cell one can study with atomic resolution to the ensuing dynamics. To date, we have achieved very high space-time-resolutions for unit cells of approximately 4–5 nm, which is approaching protein crystal unit cell dimensions. REGAE in this respect holds promise to go to unit cells larger than 10 nm.

Again, the bunch charge density must be adjusted to attain the required space-time-resolution. Higher time-resolution generally means lower electron numbers per pulse that are compensated for by averaging more laser-electron pump–probe shots to achieve the desired signal-to-noise in the diffraction pattern. Since the laser excitation generally damages the crystal (not the electron pulse as in the X-ray case), this requirement puts great demands on the samples. In the case of our studies on the ring closing reaction for diarylethene we used literally hundreds of crystals to acquire full atomic movies of the key features in the reaction dynamics.[1] Each shot may give a sufficient diffraction intensity to correlate to the structure, however, to distinguish different types of relatively small motions with respect to the interatomic separation requires acquiring a higher signal-to-noise ratio to better constrain the fits.

1 H. Jean-Ruel, M. Gao, M. A. Kochman, C. Lu, L. C. Liu, R. R. Cooney, C. A. Morrison and R. J. Dwayne Miller, *J. Phys. Chem. B.*, 2013, **117**(49), 15894–15902.

**Priyadarshi Roy Chowdhury** communicated: What are the precautions that must be taken during mapping motions with ultra bright electrons? What is the maximum resolution that has been achieved to date in case of mapping motions with ultra bright electrons?

**R. J. Dwayne Miller** replied: With regard to the first question, I am going to assume that one is using optimized electron pulses for the dynamics of interest and the sample issues discussed are under control. In terms of execution of the experiment, the main precaution is that the laser excitation should be below 100 GW $cm^{-2}$ to avoid excessively high peak powers that lead to multiphoton ionization artifacts. There is always a tendency to go to higher excitations to get further above the background to observe a structural change. Ionization will ensure you observe structural changes but they will not be related to electronic surfaces of interest that can be connected to the chemistry of interest, as opposed to Coulombic repulsion or trivial electrostatic charging effects on the lattice. There is an equally important issue that photoionization leads to photoemission with electrons moving away from the surface region creating large surface fields that also modify the diffraction pattern and can be confused with diffraction from lattice motion.[1] There are both lattice charging effects and surface field artifacts at the high power that must be avoided. This peak power issue is a major concern in which control experiments as a function of peak power are needed to be sure this effect does not contribute to the observed dynamics. Such controls are often not conducted due to limited signal-to-noise in the use of low brightness or unstable sources for the structural probe. It is in this respect that the use of ultrabright electron sources has significant advantages. The higher brightness and robust, stable nature of these sources enables the attainment of a higher signal-to-noise ratio for any given excitation level relative to other sources at the present time. These features enable studies at sufficiently low peak powers/intensities to avoid this problem.

With respect to your second question regarding maximum resolution, there are both spatial and temporal aspects to consider. The maximum time-resolution achieved to date to a structural transition is in the order of 100 fs. For examples important to condensed matter physics, see the works of Sciaini *et al.*[2] on the study of the electronically induced melting of Bi, and Morrison *et al.*[3] on the photoinduced phase transition of VO2 as representative examples. For molecular processes, see Gao *et al.*[4] on a formally photoinduced charge transfer process in a charge ordered organic system and Jean-Ruel *et al.* for an example of a ring closing reaction.[5] There are faster nuclear motions possible but it must be born in mind that structural transitions are dominated by the lowest frequency, mostly anharmonic, motions coupled to the reaction or structural transition, that undergo the largest net displacement along the reaction coordinate. The relaxation or dissipation processes to propagate the system to the new structural minimum dominate the dynamics and these modes tend to have half periods in the order of 100 fs. To date, the spatial-temporal resolution has achieved the necessary limits to follow the primary dynamics involved in the structural transitions studied. There are structural transitions such as the primary processes in vision that will necessitate a 10–20 fs time-resolution with better than 0.1 Å spatial resolution, but here there are sample issues. In principle, the current generation of electron sources are up to this task. It is really the sample that is most limiting in terms of what dynamics can be probed. In this respect, it is interesting to also note that the spatial resolution of the structural changes is in the order of 0.01 Å or less. This estimate can be gleaned from gas phase electron diffraction studies and from signal-to-noise analyses of solid state diffraction patterns. We typically see diffraction out to better than 0.1 Å (for *e.g.* charged ordered molecular crystals such as EDO-TTF)[4] with good signal-to-noise, relative to 1–2 Å for X-ray diffraction. (We generally cut the camera off at 0.2 Å but will extend the reciprocal space for higher resolution in the future.) This difference is a reflection of the much shorter wavelength of electrons relative to X-rays typically used in diffraction studies. Again, the sample quality must be good enough to support this higher intrinsic spatial resolution. It is interesting to note that the spatial resolution is in principle high enough to literally observe nuclear vibrations associated with say IR transitions. The root mean square (rms) motion in going from $\nu = 0$ to $\nu = 1$ is in this order . The problem is that for a harmonic oscillator there is no net motion; the mean remains the same. The anharmonicity is the key factor and far from equilibrium motions must explore higher degrees of anharmonicity in the interatomic potential to observe these net displacements along the potential. It will be interesting to use high power IR excitation for strong field control of atomic motions to probe this region and directly follow the light matter interactions with femtosecond electron diffraction probes. We are just in the process of such studies and I expect we will be able to significantly push the spatial resolution limits to the fundamental space-time limits of the source technology in this class of experiments.

1 H. Park and J.-M. Zou, *Phys. Rev. Lett.*, 2010, **105**, 059603.
2 G. Sciaini, M. Harb, S. G. Kruglik, T. Payer, C. T. Hebeisen, F.-J. Meyer zu Heringdorf, M. Yamaguchi, M. Horn-von Hoegen, R. Ernstorfer and R. J. Dwayne Miller, *Nature*, 2009, **458**, 56–59.
3 V. R. Morrison, R. P. Chatelain, K. L. Tiwari, A. Hendaoui, A. Bruhács, M. Chaker, B. J. Siwick, *Science*, 2014, **346**(6208), 445–448.
4 M. Gao, C. Lu, H. Jean-Ruel, L. C. Liu, A. Marx, K. Onda, S.-ya Koshihara, Y. Nakano, X. Shao, T. Hiramatsu, G. Saito, H. Yamochi, R. R. Cooney, G. Moriena, G. Sciaini and R. J. Dwayne Miller, *Nature*, 2013, **496**, 343–346.
5 H. Jean-Ruel, M. Gao, M. A. Kochman, C. Lu, L. C. Liu, R. R. Cooney, C. A. Morrison and R. J. Dwayne Miller, *J. Phys. Chem. B.*, 2013, **117**(49), 15894–15902.

**Elangannan Arunan** opened the discussion of **Martin Zanni**'s paper: The discussion on your Fig. 4, in particular the assignments of the A, B and C peaks, is confusing to me. On page 9, line 8, it is mentioned that peak C is caused by coupling between esters. On the same page, in line 40, it is mentioned that peak C is caused by coupling between bright and dark eigen states (in general 'dark and bright states' are used to refer to the states in one isolated molecule while discussing IVR). Are these two statements contradicting?

**Martin Zanni** responded: We believe that the C peak is caused by coupling between the bright and dark eigenstates of the esters, in analogy to the $a^+$ and $a^-$ peaks of a beta-sheet. The "dark" eigenstate may still carry a little oscillator strength, but not very much since it is not prominent along the diagonal.

**John R. Helliwell** asked: In Fig. 5 of your article you give calculated spectra for different monolayer structures. Have you considered using these to interpret the experimental spectrum of Fig. 4, which could perhaps be explained by the fractional occupancies of Fig. 5A, 5B and 5C, *i.e.* Fig. 4 could be due to a mixture of those monolayer structural states?

**Martin Zanni** answered: The simulations I presented in my paper are very rudimentary. Not only did we not consider fractional occupancies, which is a good suggestion, but we have also only simulated a small range of possible structures and used quite course models. So, the comparison to the experiments can definitely be improved and the structures refined.

**Abhishek Shahi** communicated: Your article is written very clearly. I have four questions:

(i) Can you help us in understanding the $a^-$ and $a^+$ modes, are they normal modes of vibration?

(ii) If we can see the cross peak G in the high intensity, why can we not see both the parent peaks ($a^-$ and $a^+$) from where the peak G originates?

(iii) Why is there inconsistency in the intensity of predicted and observed spectra (particularly for peak G)? Is there another factor which determines the intensity in experiments? I guess in the model, only the change in dipole moment is accounted for.

(iv) How is the FGAIL oriented in the absence of MMB ? Is it clearly random or can the $a^-/a^+$ ratio give some information about the orientation?

**Martin Zanni** replied:

(i) Yes, the $a^+$ and $a^-$ are normal modes. See my book "Concepts and methods of 2D infrared spectroscopy".[1]

(ii) The cross peaks can sometimes be much stronger than one of the two diagonal peaks. The intensity of a cross peak is given by the mu1*mu1*mu2*alpha2, where mu is the infrared transition dipole and alpha is the polarizability tensor for a given mode 1 or 2, respectively. The diagonal peaks are given by mu*mu*mu*alpha. One way that the diagonal peak can be very weak but the cross peak strong, is if alpha1 is large but alpha2 is small.

(iii) Please keep in mind that our work is preliminary. Our simulations are quite rudimentary and we have not tested a wide range of possible monolayer or peptide structures. Thus, our conclusions may change. If you are thinking of modeling these spectra, please feel free to come to your own conclusions about our assignments!

(iv) In the absence of the membrane we did not measure the SFG signal for FGAIL. The peptides definitely deposited, but there was no signal, meaning either that they did not form fibers or that the fibers were isotropically distributed. In principle the $a^+$ and $a^-$ features can give information about the orientation, but their relative intensities are also influenced by the size of the sheet and the structural disorder, which are two factors that are difficult to characterize accurately.

 

1 P. Hamm and M. Zanni, *Concepts and Methods of 2D Infrared Spectroscopy*, Cambridge University Press, Cambridge, UK, 2011.

**Debabrata Goswami** followed the discussion by asking: Could you please comment on how your Femtosecond Pulse Shaped 2D-IR distinctively stands out compared to the more conventional interferometer based time delayed 2D-IR spectrometers?

**Martin Zanni** answered: 2D-IR *via* pulse shaping is much more versatile and has faster data collection than conventional interferometers. A conventional interferometer moves a translation stage to increment the time delay between two laser pulses. Physically moving the stage is slow and most stages are not precise enough to make accurate time delays. The simplest use of a pulse shaper is to replace the delay stage, because the shaper can increment the delays instantaneously up to laser repetition rates of 100 kHz. Updating the delay shot-to-shot is advantageous not only because data collection is faster, but also because the signal-to-noise is improved by shifting the signal frequency further from the noise spectrum. That is why commercial FTIR spectrometers dither their interferometers as fast as possible, but no mechanical device can compare to the speed of an acousto-optic modulator based pulse shaper. There are additional benefits with a pulse shaper as well. The phase of the pulses can also be modified in order to phase cycle. That enables data collection to be collected in the rotating frame, as Goswami and Warren showed many years ago, as well as removing background scatter without using chopping. In my own research laboratory, we now have 5 pulse shaping 2D-IR spectrometers in operation and not a single interferometric system.

**Sankarampadi Aravamudhan** commented: In your paper you state that "Thus, rather than use a single picoseconds mid-IR pulse, we use two femtosecond laser pulses, as shown in Fig. 1A. The time-delay, $t_1$, between the pulses is scanned and the data Fourier transformed to give the second frequency axis but with no loss in time-resolution and improved line shapes." Are the pulses specified in the excerpt mid IR and which are femtosecond visible region pulses? Clearly specify the above with respect to the $t_1$ parameter. You also mention "Second, the signal must be heterodyne detected, otherwise phase distortions are present that make it difficult to interpret the spectra or to compare to either FTIR or 2D IR spectra which are always heterodyne detected. For the samples studied here, heterodyne detection is automatically accomplished with the non-resonant signal from the gold interface." With the specification in the previous extract, a more vivid explanation/description of the content of lines is highly desirable if the real advantage is to become well appreciated by one and all among the interested readers.

**Martin Zanni** communicated in reply: Generally speaking, the local oscillator should be the same frequency as the signal that is emitted from the sample. For

2D-IR spectroscopy, that means that the local oscillator is identical to any of the three pulses preceding it and so can be easily generated with a beam splitter. For SFG, it is a little more difficult, because the emitted signal is at a shorter wavelength than any of the excitation pulses. Thus, the local oscillator needs to be generated by summing the visible and infrared excitation pulses, similar to the SFG process of the sample itself, but using some other material with a non-resonant response so that the local oscillator does not contain molecular signals but is just a short and smoothly varying envelope. One can use a non-linear crystal or even a piece of quartz. In the experiment that I presented in this Faraday Discussion, we used the gold substrate under the sample to generate the local oscillator. Not only does it provide a nice pulse at the correct frequency, but the local oscillator is also aligned with the signal.

**Volker Deckert** asked: What are the limitations of your depth resolution, with respect to the surface sensitivity?

**Martin Zanni** communicated in reply: The "depth resolution" will be set by the system itself. If the interface is a single monolayer, then it will be detected, even if it is buried between two large masses. For example, the interface between two liquids or the interface between a liquid and solid. That is because the SFG signal only arises from non-centro symmetric places. As a result, the signal will arise for whatever portion of the sample is non-centro symmetric. So, if the interface is many layers thick, they will all give a signal just so long as they are all oriented the same. The extreme case would be a non-linear crystal like BBO that is designed to sum frequencies together very efficiently. Thus, the "depth resolution" for SFG cannot be answered generally or compared to microscopies without more specifics.

**Keisuke Tominaga** queried: Is it possible to study the liquid gas interface? It is far away from the metal surface, and the IR pulse is absorbed by the liquid.

**Martin Zanni** answered: Yes, you can study liquid gas interfaces. As you point out, the IR pulse is absorbed by the liquid, and so the IR beam should impinge on the interface from the gas side. 2D-SFG experiments on gas–liquid interfaces have been performed by Tahei Tahara and Mischa Bonn.

**Keisuke Tominaga** followed up by asking: Could you explain the selection rules for your 2D-SFG measurements?

**Martin Zanni** communicated in reply: The selection rules for 2D-SFG are quite interesting. In 1D-SFG, the modes have to be both IR and Raman active, which is also true for the diagonal peaks in 2D-SFG, but not so for the cross peaks. This fact is briefly alluded to in our Faraday Discussion paper. It was first presented in ref. 1. In other words, a cross peak in a 2D-SFG spectrum can appear even if one of the two diagonal peaks that it connects is completely SFG inactive. The reason is the following. The two photon IR pump pulse will excite a mode even if it is SFG inactive, because they are absorbed due to the infrared transition dipole, even if it does not emit. If that SFG inactive mode is then coupled to a SFG active mode, then a cross peak will appear because the active mode will emit. Thus, the cross peaks will map out modes that are invisible in standard SFG. There are other unintuitive situations that arise in 2D SFG, which are explained in ref. 2.

1 J. E. Laaser, D. R. Skoff, J.-J. Ho, Y. Joo, A. L. Serrano, J. D. Steinkruger, P. Gopalan, S. H. Gellman and M. T. Zanni, *J. Am. Chem. Soc.*, 2013, **136**(3), 956–962.
2 J. E. Laaser and M. T. Zanni, *J. Phys. Chem. A*, 2013, **117**, 5875–5890.

**Siva Umapathy** communicated: Could you please comment on a comparative aspect of vibrational SFG *vs.* 2D-IR-SFG. In the experiment 2D coupling that you monitor, would you observe similar coupling in vibrational 2D-SFG?

**Martin Zanni** responded: Our paper "Extracting Structural Information from the Polarization Dependence of One- and Two-Dimensional Sum Frequency Generation Spectra" provides a good comparison of SFG, 2D-SFG and 2D-IR spectroscopy.[1] 2D-SFG is to SFG like 2D-IR is to FTIR spectroscopy. The 2D versions visualize the couplings through the cross peaks. In addition, 2D-IR and 2D-SFG will both measure the same couplings for a given sample. The spectra might look a little different because the 2D-SFG spectra are weighted by a Raman tensor, but if both the 2D-SFG and 2D-IR spectra are collected with heterodyne detection, like was first done by my group in 2011,[2] then the comparison of the 2D-IR and 2D-SFG spectra is very informative. A 2D-SFG spectrum provides information that cannot be obtained by standard vibrational SFG. It can resolve relative molecular orientations and SFG forbidden peaks, for example. Please see our paper above for more details.

1 J. E. Laaser and M. T. Zanni, *J. Phys. Chem. A*, 2013, **117**, 5875–5890.
2 W. Xiong, J. E. Laaser, R. D. Mehlenbacher and M. T. Zanni, *PNAS*, 2011, **108**(52), 20902–20907.

**Kenneth Ghiggino** opened a general discussion of the paper by **Robert Pal**: You have reported results using a 355 nm excitation light source. Can you expand on the reasons for using this excitation wavelength in your work considering potential issues with optics transmission and increased light scattering? Can multiphoton excitation of the sample be used with this technique and would it offer any advantages?

**Robert Pal** answered: Initially PhMoNa and its associated methodology was established using functionalised lanthanide(III) complexes. The reason behind this is that these Ln(III) complexes upon excitation at 355 nm possess long lived sensitized emissive lifetimes (orders of ms). In this case the initial electro-optical modulation of the excitation beam and galvanometric sample raster scan speeds could be synchronized in time scales as slow as 1Hz. After successful hardware configuration it was facile to perform enhanced resolution scans using commercial cellular stains (*e.g.* MitoTracker Green) at 400Hz using its corresponding (488 nm for MTG) excitation laser. It is well known that commercial optics could display lower optical performance below 400 nm, however, in our case LambaBlue (Leica) objectives with excellent >350nm optical performance were initially used. Light scattering and UV induced sample bio-auto-fluorescence (BAF) was eliminated by carefully adjusting the scanning parameters *via* unstained control experiments and time-gated dye detections in case applied Ln(III) complexes could be facilitated too (ns BAF *vs.* ms Ln(III) lifetimes). Sample photobleaching was not observed as we maintained live cell sustainable laser parameters throughout (Ln(III) dye photobleaching has been extensively studied in ref. 1). This technique could be transferred to a multiphoton excitation setup, by simply employing a tunable laser and suitable 2PE dye combination, however, it could not overcome the inherent resolution limitations of multiphoton excitation using low NA objectives to achieve deeper sample penetrations.

1 S. J. Butler *et al.*, *Chem. Sci.*, 2014, 5, 1750.

**Christoph Schnedermann** asked: Optical microscopy is often employed to track nanoscopic objects in time. What are the limitations of your technique in this regard and how does it compare to more routinely used techniques?

**Robert Pal** replied: The limitation of tracking nanoscopic objects is limited in this set up by the mechanical vibration of the applied electro-optical modulator, generating standing waves. This has a maximum limiting value of 400 Hz per line (minimum 1024 pixel per line). Combine this with the required automatic reconstruction of a 4 line averaging scan, due to the raster scanning of the *in situ* formed improved PSF excitation cluster, this line scan speed is technically 100 Hz per line, which could give the best scenario of 9.76 ns/pixel excitation and readout/cycle. Further proposed improvements in and the development of a novel non-mechanical electro-optical modulator could foresee a 10 fold improvement in the above limiting scan speed.

**Sankarampadi Aravamudhan** commented: In Fig. 1c of your paper (see Fig. 1 below), slice-level 1 (the lower trace) has to a greater extent the possibility for the circles to overlap than what appears to be possible at slice-level 2 (the upper trace). Hence the resolution must be defined with an appropriate convention, so that it becomes applicable without ambiguity in all contexts encountered. In the above two levels one must specify the spectrometer settings unambiguously for a comparison of resolution.

 

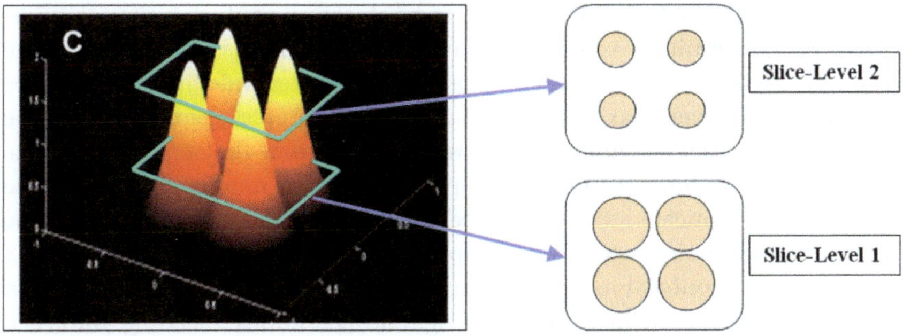


Fig. 1 Qualitative Resolution depending on the image slicing-level.

**Robert Pal** communicated in reply: In this form of structurally modulated enhanced confocal microscopy the achieved 2 fold improvement in the *x*,*y* (lateral) resolution allows the user to harness the advantages presented by the variable confocal pinhole of the system and set it to the corresponding airy disk size which will result in the 2 fold improvement in optical sectioning and subsequent *z* (axial) resolution. Any observed *z* saturation could also be eliminated if needed by a set 10% overlap in *z* sections applying a cross section eliminating algorithm, although due to the default line averaging nature of the set up this has not been needed. The answer to this question in my response herein is referring to axial resolution, for the lateral resolution element of this question my explanation is detailed in my response to the question asked by Siva Umapathy below.

**Siva Umapathy** asked: What is your prediction of the best resolution you could get spatially?

**Robert Pal** responded: Currently this version of PhMoNa provides half of the diffraction limited resolution (at a given excitation wavelength and objective NA, ~60nm lateral at 355 nm and 1.40 NA; axial resolution is governed by the applied pinhole size which is a function of lateral resolution), providing an eight times reduced detectable voxel size. This is a result of the 2 by 2 cluster generated in the excitation beam profile. Further splitting can be achieved, however, this will not improve the resolution at this point any further as the generated peaks will overlap at their origin more than 20%, which will produce destructive wave-patterns prohibiting good signal-to-noise and Nyquist sampled emission detections.

**Debabrata Goswami** queried: Is it true that as you make the confocal point bigger, you have less *z* resolution? However, can you still do slicing along the *z*-axis to get the 3D image?

**Robert Pal** communicated in reply: It is in design principle true that in confocal microscopy with increased detection pinhole size the studied optical thickness is increased, but the set-up (NA of objective, applied excitation wavelength, *etc.*) determined axial resolution will determine the *xyz* 'clarity' of the image. 3D slicing is possible even without a pinhole (*e.g.* wide-field fluorescence microscopy), however, maximum image clarity and *z* profile accuracy can only be achieved if this axial resolution limit is used to determine the applied *z* step size (and keeping the Rayleigh criterion, Sparrow limit in mind for efficient true contrast). If the optical sections have been set to be 'thinner' than the achievable axial resolution would permit, over-sampling (bigger axial PSF component than *z* parameter of the confocal voxel) will occur. This will actually reduce the resolution in the reconstructed 3D image due to neighboring sections 'bleeding' into each other saturating the voxels as the image is being formed along the *z* axis. This, for the same principle, could subsequently also reduce the observed lateral resolution.

**Volker Deckert** queried: Is there a difference in lateral resolution if you compare your combination of SIM and confocal with bright field SIM?

**Robert Pal** answered: The achievable lateral resolution is at the same levels as with wide-field SIM, however in this case it is done *via* a variable raster cluster scanning pattern harnessing the advantageous superior optical sectioning (axial resolution) of a confocal setup. Theoretically the lateral resolution could be further improved (by at least a further factor of 2) and surpass the performance of a WF-SIM setup, but due to prolonged scanning times, complications in optical components compatibility and reduced signal-to-noise ratio this has not been pursued further.

**Debabrata Goswami** commented: In context to the work presented in the paper by Dr Pal, the use of an idea like the PALM super-resolution technique introduced by Dr Eric Betzig (based on the switchable fluorophores of Dr W. E. Moerner)[1,2] would be more appropriate compared to the STED microscopy technique of Dr Stefen Hell.[3] All these have been recognized with the 2014 Chemistry Nobel prize. In the particular case of Dr Pal, the multiple image idea of PALM would work very well.

1 E. Betzig, G. H. Patterson, R. Sougrat, O. W. Lindwasser, S. Olenych, Imaging Intracellular Fluorescent Proteins at Nanometer Resolution, *Science*, 2006, **313**, 1642–1645.
2 R. M. Dickson, A. B. Cubitt, R. Y. Tsien, W. E. Moerner, On/off blinking and switching behavior of single molecules of green fluorescent protein, *Nature*, 1997, **388**, 355–358.
3 S. W. Hell and J. Wichman, Breaking the diffraction resolution limit by stimulated emission depletion microscopy, *Opt. Lett.*, 1994, **19**, 780–782.

**Robert Pal** responded: Thank you for the valuable comment, yes indeed the 2014 Nobel prize has been given for the 'physicochemically approached' (fluorophore modulation and localisation) resolution enhancement to achieve subdiffraction limit (predominantly 2D) images. 3D reconstruction in all these pointilistic techniques is already achieved. In the case of PALM/STORM this has been done *via* the methodology of interferometric determination of the tracked fluorophores' axial position. These methods could be detailed further, however the vital point of PhMoNa is not to compete with these superior techniques, but to combine the advantageous properties on LSCM and SIM. Therefore by doing so this auxiliary technique can literally half the current diffraction limited resolution of a given confocal microscope without the need of applying a range of novel compatible fluorophores, expensive new equipment and often disruptive experimental parameters (such as high laser powers). The achieved aim was to provide the broad imaging community with a 'way' of performing live cell experiments with higher resolutions by keeping both existing LSCM instrumentation and applied experimental techniques in place.

**Priyadarshi Roy Chowdhury** asked: What precautions should be followed during slicing?

**Robert Pal** communicated in reply: There are no precautions to address during optical slicing since it is experimentally identical to a standard confocal microscope.

Faraday Discussions

PAPER

# Spatio-temporal resolution of primary processes of photosynthesis

Wolfgang Junge*

Received 24th March 2015, Accepted 25th March 2015
DOI: 10.1039/c5fd90015h

Technical progress in laser-sources and detectors has allowed the temporal and spatial resolution of chemical reactions down to femtoseconds and Å-units. In photon-excitable systems the key to chemical kinetics, trajectories across the vibrational saddle landscape, are experimentally accessible. Simple and thus well-defined chemical compounds are preferred objects for calibrating new methodologies and carving out paradigms of chemical dynamics, as shown in several contributions to this *Faraday Discussion*. Aerobic life on earth is powered by solar energy, which is captured by microorganisms and plants. Oxygenic photosynthesis relies on a three billion year old molecular machinery which is as well defined as simpler chemical constructs. It has been analysed to a very high precision. The transfer of excitation between pigments in antennae proteins, of electrons between redox-cofactors in reaction centres, and the oxidation of water by a $Mn_4Ca$-cluster are solid state reactions. ATP, the general energy currency of the cell, is synthesized by a most agile, rotary molecular machine. While the efficiency of photosynthesis competes well with photovoltaics at the time scale of nanoseconds, it is lower by an order of magnitude for crops and again lower for bio-fuels. The enormous energy demand of mankind calls for engineered (bio-mimetic or bio-inspired) solar-electric and solar-fuel devices.

## 1 Introduction

When the founders of atomic and molecular physics convened in 1927 at the $5^{th}$ Solvay Conference on "Electrons and Photons" in Brussels (see photograph in ref. 1) they had established quantum mechanics, relativity and the interplay of stochastic and conservative forces in the nano-world. Despite their immense insight it was probably all but evident to them how molecular events could ever be resolved down to molecular scales of time (femto-second) and space (1 Å-unit = 0.1 nanometer). Neither was it evident that such a high resolution might be applicable not only to small molecules but also to complex systems such as oligomers, interfaces, proteins, membranes, cells and organisms. High resolution became feasible by dramatic innovations in physical instrumentation. In the

*Dept. Biology & Chemistry, University of Osnabrück, R. 35/E42 Barbarastrasse 11, 49076 Osnabrück, Germany. E-mail: junge@uos.de; Tel: +49-15112180743*

1960s pulse lasers opened up the time domain of nanoseconds, soon followed by pico- to femtoseconds (ps–fs) in the seventies. The whole spectral range from the infrared (IR) over the visible to the ultraviolet (UV) has been covered by pulsed photon sources. It culminated in the free electron "laser" (rather a coherent super-radiation source) in the seventies, which provides femtosecond X-ray or electron pulses. Pulse sources together with greatly improved detectors and data handling capacities have allowed the high resolution of structure and kinetics, particularly in those physical, chemical and biological systems that can be stimulated by light.

Recent technical progress was covered in the *Faraday Discussion* on "Emerging Photon Technologies for Chemical Dynamics" (FD 171). The present *Faraday Discussion* (FD 177) expands this theme into "Temporally and Spatially Resolved Molecular Science". It aims at a rigorous exploration of the energy landscape and the path of extremely rapid photophysical and photochemical reactions involving a manifold of vibronic and torsional states. Two- and more-dimensional photon-echo spectroscopies have been essential to overcome the lack of spectral resolution when using photon-pulses of fs-duration. Prominent topics of FD 177 have been quantum coherence, failure of the micro-canonical approximation, solvent effects on vibrational spectra and relaxations, and time-resolved structure analysis by fs-pulses of X-rays or electrons. A common quest in most contributions to FD 177 has been the perfection of data and the outmost rigor in their theoretical interpretation. It has called for simple and well defined systems, mostly small inorganic or organic molecules. The paradigms derived from the behaviour of simple systems – physical chemistry at its best – have provided firm grounds for bio-physical-chemists and biophysicists to address their more complex research objects. Although being complex in design some of nature's essential constructs, like the proteins that supply cells with energy, originated about three billions years ago and have been largely conserved during evolution. They are thus standardized and often as well defined as simple chemical compounds.

When dealing with these fundamental biological systems the canon of questions to be asked is expanded. The key questions of physical chemistry – *what, where* and *how fast* – are to be supplemented by *what for* and *why*. The former three address the object, its molecular structure and evolution in time. The question *what for* addresses the physiological role of a given construct and the question *why* whether a particular construct is exacted by laws of physics or merely reveals nature's fancy as a legacy of evolution. In the following the relation between pure and biologically oriented physical chemistry is illustrated by taking oxygenic photosynthesis as an example.

## 2 Oxygenic photosynthesis

Oxygenic photosynthesis by cyanobacteria and plants uses sunlight and produces oxygen and biomass. Biomass serves man as food, fuel, fibre and platform chemical. Early attempts to understand this process go as far back as the 18$^{th}$ century. Jan Ingen-Housz in his study on vegetables noticed their "Great Power of Purifying the Common Air in the Sunshine and of Injuring it in the Shade and at Night".[2] It was a first appreciation of the production and re-consumption – in the reaction cycle between photosynthesis and cell respiration – of the gases that were later coined oxygen and carbon dioxide. A rigorous spectroscopic analysis of photosynthesis started half a century ago using microsecond flashes of light for excitation. In 1961 Duysens,[3] Kok,[4] and Witt[5] independently arrived at the conclusion that green plant photosynthesis is powered by two photosystems in series. They drive electrons from water to $NADP^+$. Water is oxidized by photosystem II (PSII) to yield oxygen and protons. PSII, in turn, reduces photosystem I (PSI) which reduces $NADP^+$ to NADPH, as illustrated in Fig. 1.[6] In that same year Mitchell (Nobel Prize 1978) postulated that ATP is synthesized from ADP and $P_i$ at the expense of a proton-motive force across the respective coupling membrane.[7] The energy conserved in the chemical difference of the products $NADPH/O_2$ (by PSII and PSI) and $ATP/ADP \cdot P_i$ (by ATP synthase ($F_OF_1$)) drives the reduction of $CO_2$ to biomass.

A decade later the time resolution of the primary electron transfer was advanced into the picosecond range.[8] But it took more than another decade until Deisenhofer, Huber and Michel (Nobel Prize 1988) published the first structural model of a photosynthetic reaction centre (resolution 3 Å).[9] It was the first ever structural model of any membrane protein. Today, structural models are available for all the proteins central to photosynthesis. PSI of a cyanobacterium is resolved at a 2.5 Å resolution,[10] and even the larger PSI of green plants with a molecular mass of 660 000 a.u. is resolved at 3.4 Å.[11] The latter hosts almost 200 chlorophyll molecules, mostly with antennae function, plus the redox-active core with four chlorophyll- and two pheophytin molecules.[12] Cyanobacterial PSII, the water–quinone oxidoreductase, has been resolved starting at 3.8 Å[13] and now at a 1.9 Å resolution.[14] The structure of ATP synthase ($F_OF_1$) has only partially been determined at the atomic resolution.[15,16] However, its full structure is evident from low resolution data of the holoenzyme (see ref. 6 and 17 and references therein). PSII and PSI are solid-state devices. $F_OF_1$ is a most agile rotary machine composed of two rotary motor generators. $F_O$, the ion-driven motor[18,19] is mechanically coupled[20] to $F_1$, the chemical generator.[15,21]

The key machinery of bioenergetics (photosynthesis and respiration) dates about three billion years ago. Its design has not changed much since then. It justifies casting into one coherent scheme the kinetic and structural insights which have been elaborated on enzymes from different organisms (microorganisms, plants and animals).

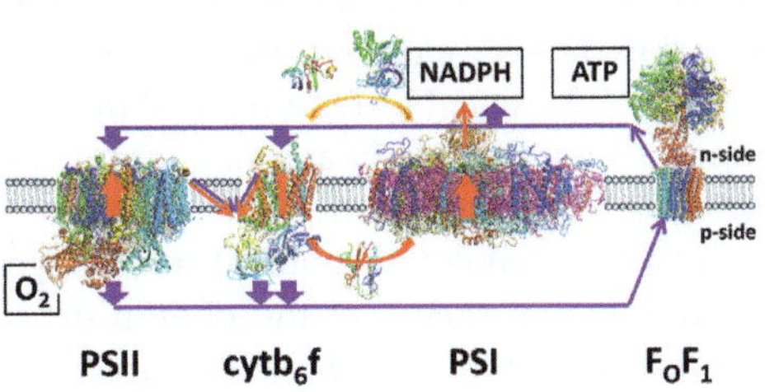


Fig. 1 Electron and proton transfer in the coupling membrane of oxygenic photosynthesis. (a) Electron transfer (*red arrows*) and proton transfer (*purple arrows*) involving photosystems I and II (PSI and PSII), cytochrome $b_6f$ (cyt $b_6f$), and $F_OF_1$. It produces $O_2$, NADPH and ATP. Modified with permission from the *Annual Review of Biochemistry*, Volume 84. © 2015, Annual Reviews, http://www.annualreviews.org.[6]

## 3 Excitation energy transfer

Photosynthetic reaction centres are surrounded by pigments with antenna function, in PSII between 200 and 300 chlorophyll-molecules. Being embedded in membrane proteins they enhance the absorption cross-section of the photo-reactive core. In chlorophyll-based reaction centres from all organisms the core hosts four plus two tetra-pyrroles, the former four being bacterio-chlorophyll and the latter two either bacterio-pheophytin or bacterio-chlorophyll (see Fig. 2). These pigments are very closely spaced and thus strongly coupled. The core was first isolated in the pure form from bacteria. Starting in the eighties femtosecond spectroscopy revealed that the first electron transfer step from bacterio-chlorophyll to bacterio-pheophytin relaxes in picoseconds.[22,23] Soon thereafter vibrational coherence between these pigments became evident.[24] As in several contributions to this *Faraday Discussion* femtosecond excitation and multidimensional photon-echo spectroscopy became the clue to investigate processes where the micro-canonical approximation fails. An impressive series of studies on photosynthesis has revealed quantum beats in antennae proteins (for reviews and references see ref. 25–27, for a recent survey of quantum biology see ref. 28).

Taking up the above biophysical question of physical necessity *versus* legacy of evolution one wonders whether coherent excitation transfer is a necessary prerequisite for the high quantum yield of photosynthesis. In a theoretical study on a structurally well characterized bacterial antenna complex it is argued that it is rather unecessary.[29] If the reorganisation time of the environment is not much faster than the one of the system coherence is prolonged. However, it does not increase the efficiency of energy transfer, except when excitation decay competes with trapping.[29] An experimental study on the same antenna protein was presented at this meeting in a poster by Battacharyya and Sebastian. Mechanisms for the dephasing of coherence by the environment were discussed in posters by Kayal *et al.* and Roy *et al.*, and the mechanism of photon-induced intra-molecular excimer formation and charge transfer was treated by Mishra *et al.*[30]

In photosynthetic antennae the excitation decay does not to compete with trapping. This is has been corroborated by the following recent experiments. Van Grondelle's group analysed the excitation transfer between six pigments (4 chlorophyll a, 2 pheophytin a) in the purified core of PSII.[31] Quantum beats persisted for several 100 femtoseconds even at ambient temperature. Several coherent excimer-/charge-transfer states coexisted before the creation of a metastable radical pair. The authors suggested that quantum design of the core is pivotal for the high quantum yield.[31] In these experiments the core of PSII is isolated from its antennae complement and it is excited by femtosecond pulses. *In vivo* the core is surrounded by antennae. Quanta trickle in at a rate of much less than 1000 per second. Holzwarth and his colleagues studied this situation in oxygen evolving PSII particles containing a complement of about 80 chlorophyll molecules.[32,33] To avoid singlet–singlet annihilation they kept the photon density of the exciting pulse low. The total time for exciton trapping and charge separation was about 100 ps.[32] The calculated intrinsic time of charge separation in the core, on the other hand, was much shorter, 2.7 ps.[33] This situation is paralleled in purple photosynthetic bacteria. Excitation transfer in the B850-ring of the antenna LH2 as well as in the B875-ring of LH1 is very fast (about 100 fs) and it involves

 

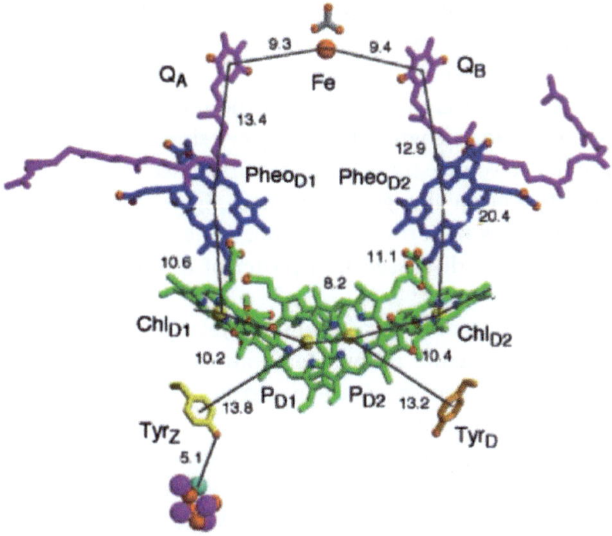


Fig. 2 Structural model of the redox cofactors in the core of cyanobacterial PSII (modified with permission of *Science*).[114] The electron transfer uses the D1-branch of the hetero-dimer. Chorophyll a molecules are given in green and blue, pheophytin a in blue and the primary ($Q_A$) and secondary ($Q_B$) quinones in purple. The catalytic $CaMn_4$-cluster is linked *via* a tyrosine ($Tyr_Z$) to the chlorophyll ($P_{D1}$). Centre-to-centre distances in Å-units.

coherence. The transfer from LH2 to LH1 takes 3 ps, and further on from LH1 into the trap 35 ps.[25] The long and trap-limited lifetime of excitation greatly exceeds the decay time of coherence. Independent of whether coherent transfer is involved or not, the only condition for high quantum yield is that the transfer beats the dissipative loss of excitation in the whole system. This physical constraint is rather mild, and it provides a leeway for various transfer mechanisms and structural designs. In fact, photosynthetic organisms use a wide palette of mechanisms for excitation transfer, ranging from hopping by Förster-resonance over Frenkel-excitons to more delocalized modes involving the coherent states of many pigment molecules. Accordingly, the antennae structure varies considerably between organisms. Chlorophyll molecules are either enwrapped in proteins as in the LHC2 of plants,[34] and in the LH1 and LH2 of purple bacteria,[35,36] or self-organized as in the chlorosomes of green bacteria.[37] The energy transfer may be directed towards a deep trap (low energy trap), as in PSI, or undirected if the trap is shallow, as in PSII.[33] In all cases the quantum yield of trapping is close to one.

# 4 Electron transfer

Excitation transfer into the very core of any reaction centre results in a very rapid charge separation. The first metastable product is a radical pair which is formed in picoseconds. In PSII this pair is $pheophytin_{D1}{}^{+}$–$chlorophyll_{D1}{}^{-}$ (see Fig. 2). Herein the subscript D1 refers to the subunit to which the respective pigment is attached. While the positive charge is localized on this particular chlorophyll in PSII ($P_{D1}$ in Fig. 2), it is delocalized over two bacterio-chlorophylls, the "special pair", in the reaction centre of purple bacteria.[38] The delocalization of the unpaired electron over this "special pair" hence is not pivotal for a high quantum yield, in contrast with previous beliefs.[39]

The charge separation is directed across the membrane. The membrane is charged by 50 mV (field strength $10^7$ V m$^{-1}$) per single turnover of each photosystem. Light induced voltage transients were first detected in plant chloroplasts. Electrochromic absorption changes of intrinsic pigments served as a molecular voltmeter[40,41] with a very high time-resolution.[42] The vibrational Stark-effect as a molecular voltmeter was discussed in a poster of this *Faraday Discussion* by S. Bagchi *et al.*

Structural pseudo-$C_2$-symmetry is a common feature of most reaction centres. Their hetero-dimeric structure is considered an evolutionary legacy of a homo-dimeric and truly $C_2$-symmetrical common ancestor.[43–46] Anaerobic green bacteria still host a homo-dimeric reaction centre.[43]

Independent of the type of reaction centre and organism, the positions and orientations of the six pigments in the core are very much the same. There are three pigments on each branch of the dimer. While only one branch is redox-active in type II centres (of purple bacteria and in PSII of plants and cyanobacteria), both are similarly active in a type I centre (PSI). The functional asymmetry in the former has been attributed (by Stark-effect spectroscopy) to different dielectric screenings in the two branches.[47] The asymmetric activity of type II centres is linked to its function as a one-electron to two-electron gate, whereas PSI, with its longer line of secondary acceptors, functions mono-electronically (for a thorough discussion and references see ref. 48).

The mechanism of electron transfer between cofactors that are embedded in the protein is adequately described by the Marcus theory.[49,50] It models tunnelling within the limits of the Born–Oppenheimer approximation. The rate is a function of the (edge-to-edge) distance between donor and acceptor ($r$), and it depends on three parameters: the standard free energy difference between the donor and the acceptor ($\Delta G^0$), the reorganization energy of their environment ($\lambda$) and the decay constant of wave-function-overlap ($\beta$):

$$\text{rate} \sim \exp\left[\frac{(\Delta G^0 + \lambda)^2}{4\lambda kT}\right]\exp(-\beta r) \tag{1}$$

This simple relation in terms of only three parameters has been experimentally very well met. In synthetic donor–spacer–acceptor triads it holds over three orders of magnitude.[51] As for the electron transfer between cofactors in a protein, various pathways were screened in cytochrome c. Its heme served as the electron acceptor for a photosensitizer which was covalently attached to various positions at the surface. The transfer rate decreased depending on whether the electron passed mainly along single bonds, hydrogen bonds or through a void volume in the protein.[52,53] In a related study, Dutton and colleagues[54] analysed the wealth of kinetic and structural data on the electron transfer in photosynthesis and cell respiration. They corrected the observed rates for the matching driving force (determined!) and reorganization energy (assumed!), namely $-\Delta G^0 = \lambda$. Over 13 orders of magnitude the normalized rates decline mono-exponentially with the edge-to-edge distance ($r$) between donor and acceptor. Deviations by one order of magnitude can be attributed to local variations of $\lambda$ and $\beta$. The donor–acceptor distance is hence the major determinant of the rate of electron transfer.[55] The reorganization energy ($\lambda$) and the decay constant of the electronic wave functions ($\beta$) in proteins do not vary too much. Nature's engineering of structures for rapid electron transport has thus relied mainly on the distance between redox-cofactors ($r$) and on the driving force ($\Delta G^0$).[55]

Once the core of the reaction centre is hit by an excitation the primary electron transfer is very fast. In PSII the radical pair (chlorophyll$^+$–pheophytin$^-$) appears in picoseconds. The electron is then transferred to a first quinone ($Q_A$) in about 400 picoseconds and from there on to a second quinone ($Q_B$) in some 100 microseconds.[56,57] The electron hole, on the other hand, is rapidly transferred in some ten nanoseconds to a tyrosine[58] and further in microseconds to the $Mn_4Ca$-cluster. All of these reactions outrun by orders of magnitude the respective wasteful back-reactions of excited states and radical pairs. Rapid forward reactions and slow back-reactions guarantee the high quantum yield of photosynthesis. On the other hand they require losses of free energy. In other words, photosynthesis sacrifices energy efficiency for directionality.[48]

# 5 Water oxidation

PSII is a water–quinone oxidoreductase. The structure of cyanobacterial PSII is resolved at 3.8 Å,[13] 3.5 Å,[59] and 1.9 Å.[14] Water (-derivatives) are oxidized by the catalytic $Mn_4Ca$-cluster and oxygen plus protons are liberated. When dark adapted PSII is excited by a series of short light flashes the $Mn_4Ca$-cluster is clocked through sequentially higher and metastable oxidation states, until reaching the highest state (see ref. 60 and references therein). Only then the reaction with bound water proceeds (in one millisecond) to yield dioxygen (see ref. 61). The stepwise accumulation and pooling of four oxidizing equivalents, before initiating what still appears as the reaction of four-electrons-at-once, serves two purposes. It homogenizes the energy demand of successive electron transfer steps, and controls hazardous intermediates (*e.g.* hydroxyl radical and superoxide) on the way from water to dioxygen. There is a wealth of detailed kinetic studies on water oxidase, addressing the chlorophyll-moiety, an intermediate tyrosine, the $Mn_4Ca$-cluster, electrons, protons and oxygen. A wide range of optical (X-ray, UV, VIS and IR), magnetic and mass spectroscopic techniques have been utilized to resolve electron transfer, proton transfer, conformational changes of the protein, substrate binding and product release in the time-domain from nano- to milliseconds (for reviews and references see ref. 62–65).

The precise structure of the manganese cluster is still under contention. During standard X-ray exposure of PSII-crystals $Mn_4$ is the prime target of radiation damage[66,67] (for radiation damage of metal markers for protein structural studies see Helliwell *et al.*).[68] The reduction of Mn may influence its ligation state. The lack of unequivocal structural models of the $Mn_4Ca$-cluster in its sequential oxidation states is a major obstacle for a thorough understanding of the final reaction cascade. For the time being, two non-invasive approaches, namely by magnetic resonance spectroscopy (Endor[69]) and by theoretical chemistry (density functional theory[71]), have converged towards one particular structural model of the $CaMn_4$-cluster and its ligands including water (derivatives). X-ray crystal structural analysis will eventually take up, either challenging or corroborating the present structural concept by a probe-before-destroy-approach (for principles and perspectives of this technique see Miller *et al.*).[70] PSII-crystals have been exposed to an ultra-short and intense X-ray pulse (100 fs) of a free-electron-laser.[72,73] The feasibility of analysing the cluster-structure in its sequential oxidation states has been clearly demonstrated, although at a limited resolution so far (5–5.5 Å).[73] Structural details on the $Mn_4Ca$-moiety, bound water (derivatives), and amino acid ligands is the essential complement of kinetic data when aiming to understand the detailed reaction mechanism of this "holy grail" of photosynthesis.

# 6 ATP synthesis

Mitchell's daring hypothesis on proton-driven ATP synthesis[7] roused the unprepared bioenergetics community. Soon thereafter essential elements of his hypothesis were detected in plant chloroplasts. Illumination generates transmembrane voltage[40] and a pH-difference,[74] and the pH-difference[75] and/or the voltage[76] can drive the synthesis of ATP. The molecular structure of ATP synthase was then still unknown. Only its bipartite construction with a membrane portion, $F_O$, which conducts protons, and a peripheral portion, $F_1$, which interacts with nucleotides, was apparent. Boyer (Nobel Prize 1996) and his colleagues found that $F_1$ hosts two, if not three identical catalytic sites, that operate alternately,[77] if not in a rotary mode.[78] Conformational energy stored in the enzyme might drive the formation of ATP from ADP and $P_1$.[79] In 1994 Walker (Nobel Prize 1996) and his colleagues presented the first structural model of its chemical generator at a 2.3 Å resolution,[15] now at 1.9 Å,[80] which strongly favoured a rotary function of $F_1$. $F_1$ is a pseudo-hexagon of two types of subunits arranged as $(\alpha\beta)_3$ with a cranked shaft (subunit $\gamma$) in its centre. It suggests that the rotation of the central shaft drives ATP synthesis when progressing from one catalytic site on $\beta$ to the next. Soon thereafter the rotation of the central shaft was experimentally established, starting without time resolution by the biochemical crosslink technique,[81] and then time resolved either by polarized photobleaching and recovery[82,63] or fluorescence microscopy.[83] The latter technique became the gold standard in the field (for original video recordings see ref. 119). Its principle is illustrated in Fig. 3a. The hexagonal body of the enzyme is fixed on a solid support so that the hydrolysis of ATP drives the shaft around. In a pioneering experiment, a fluorescent actin filament was attached to the foot of the central stalk of $F_1$ and the rotation of the filament was video-recorded in a fluorescence microscope.[83] In the author's lab the fluorescent probe was attached to the holo-enzyme, $F_OF_1$, as shown in Fig. 3a. Notably the procession of this enzyme over its reaction coordinate can be recorded in real-time. By certain tricks it has been feasible to correlate the dynamic jumps and transient dwells of the live enzyme with the still pictures of the inhibited enzyme as a crystal structure.[84–86] Moreover, the rotor position has been externally manipulated by an attached hyper-paramagnetic bead.[87,88] Ongoing studies in this line will likely provide more precise data on the energy landscape of $F_1$, a prerequisite for thorough theoretical descriptions (see ref. 89–91 and references therein). The wealth of structural and kinetic data qualifies ATP synthase as a hydrogen atom of nano-motoring. The sophisticated, yet simple electro-mechano-chemical operation of ATP synthase has been described in several reviews.[6,17–20,92,93] For animations of the enzyme and its two motors, $F_O$ and $F_1$, see ref. 121. A few salient features shall be emphasized in the following.

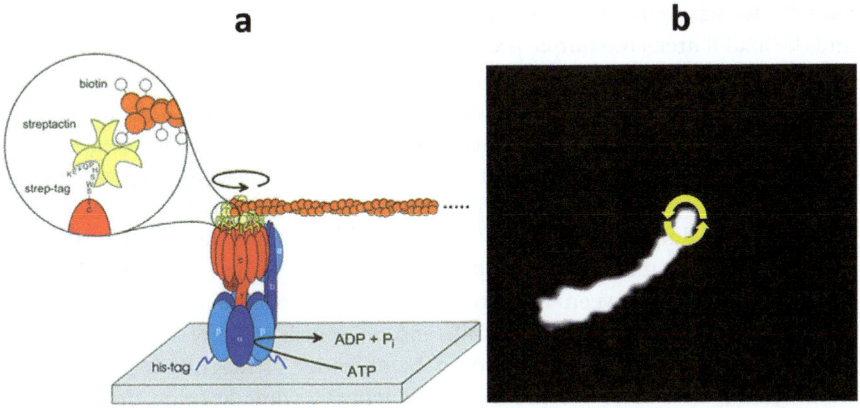


Fig. 3 The micro-videographic assay for the rotary activity of $F_OF_1$.[115,116] (a) The holo-enzyme is solubilized in detergent and the crown of $F_1$ is attached by His-tags to a solid support. A fluorescent actin filament is attached to the rotor ring of $F_O$ (modified with permission from *FEBS Lett.*[116]). (b) When ATP hydrolysis drives the rotor ring round, the actin filament, subject to viscous drag on the filament, is curved. The elastic parameters of the filament are determined by fluctuation analysis (modified figures from ref. 115 with permission). For the original video-recordings see ref. 120.

Both $F_O$ and $F_1$ are stepping motors. When driven by ATP hydrolysis $F_1$ rotates with a period of 120°, with substeps of 40° and 80°.[94–96] This stepping reflects the presence of three equivalent reaction sites on $F_1$ and torque production during substrate binding, catalysis and product release. On the other hand $F_O$ steps by 36°,[97,98] reflecting the $C_{10}$ symmetry of the particular bacterial $F_O$. The mechanism and the structure of the rotary ion-motor, $F_O$, is not discussed here. For the principle of operation see ref. 18, for the kinetic properties ref. 99, for the structure ref. 16 and for simulations of its reaction dynamics ref. 89,99,100. Here we focus on the cooperation of $F_O$ and $F_1$ in the holo-enzyme.

If the proton-motive force is larger than the thermodynamic force of ATP hydrolysis the electrochemical motor of ATP synthase, $F_O$, drives the chemical generator, and $F_1$ synthesizes ATP. If the force-relation is reversed, motor and generator change roles. ATP hydrolysis by $F_1$ drives $F_O$ to generate the proton-motive force. If the forces match the enzyme rests. Thermodynamic quasi-equilibrium between the two motors was used to determine the torque generated by ATP hydrolysis.[101] As illustrated in Fig. 3a a long filament ($\sim$3 μm) was attached to the rotary electromotor of the holo-enzyme. It slows the turnover rate of the enzyme by several orders of magnitude into a state of quasi-equilibrium. The torque was calculated from the elastic deformation of the filament (documented in Fig. 3b), which reflects the counteraction of the drive and the viscous drag. The mean torque, 56 pN nm, conformed with the calculated driving force of ATP-hydrolysis (70 kJ mol$^{-1}$),[101] which implies that the two motors do not slip against each other.

Although torque production by $F_1$ was expected to progress in steps (under the given conditions), the torque output to the actin filament on $F_O$ was almost constant. This surprising observation has been explained by solving the Fokker–Planck equation of a stepping nanomotor when elastically coupled to a heavy load.[101] Decreasing the stiffness of the elastic buffer between the stepping drive and the load flattens the torque profile at the output. It increases the turnover rate by orders of magnitude over the one with a stiff transmission (see Fig. 4a). In other words, an elastic force–torque transmission is pivotal for high *kinetic* efficiency of the stepping nanomotor that drives a heavy load.[102] This benefit of a compliant transmission applies to all of nature's intrinsically stepping nanomotors. The distribution of elastically compliant and stiff domains over $F_OF_1$ was determined by fluctuation analysis.[20,103,104] The result is illustrated in Fig. 4b. The most compliant domain (stiffness 70 pN nm) of the holo-enzyme is located on the central rotor and between the sites where the respective power strokes are generated (see ref. 6 and 20 for details and references). The rest of the enzyme, and in particular the stator, are stiffer by an order of magnitude. The elastic transmission decouples the two motors kinetically while keeping them strictly coupled, both thermodynamically and, in the time average, also under a steady turnover. It explains why this enzyme can run with different gears in different organisms. The gear ratio of $F_O : F_1$, and likewise the expected $H^+ : ATP$ ratio, ranges from 8 : 3 in mammalian mitochondria[105] to 10 : 3 in yeast mitochondria[105] and 14 : 3 in chloroplasts,[106] and it varies between 10 : 3 and 15 : 3 in different bacteria (see ref. 107 and references therein). Organisms (organelles) thriving at a large and constant ion-motive force, like mammalian mitochondria, run at low gear (speedsters), and those at a low and/or fluctuating force, like chloroplasts and in alkaliphilic bacteria, at high gear (tractors).[6] Because the two motors are kinetically decoupled the enzyme can operate with a strictly proton specific ion motor in some organisms,[99] and optionally on sodium or proton[108,109]

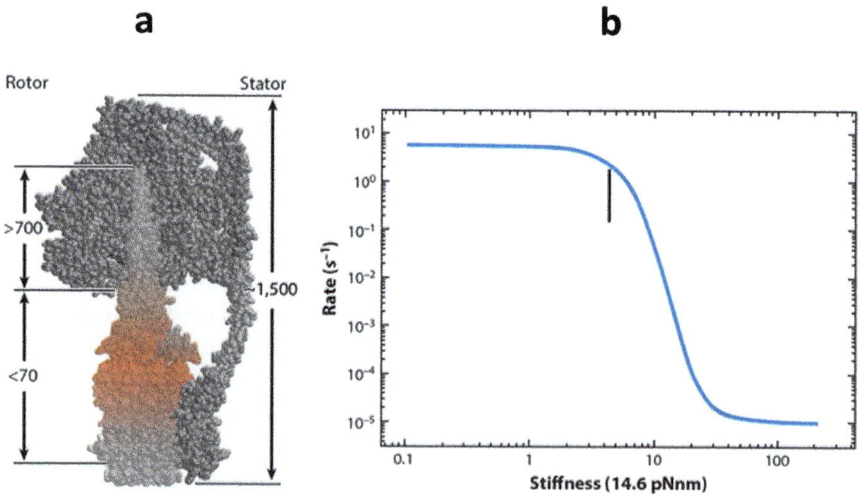


Fig. 4 On the elastic torque transmission between the ion-driven motor, $F_O$, and the chemical generator, $F_1$, of rotary ATP synthase. (a) The distribution of elastically compliant (red) and stiff domains (grey) over the enzyme. Numbers give the torsion stiffness in pN nm.[6,117,118] (b) Calculated dependence of the turnover rate of a stepping and rotary nanomotor, which is coupled to a heavy load, on the torsion stiffness of the transmission to the load.[115] With permission from the *Annual Review of Biochemistry*, Volume 84. © 2015, Annual Reviews, http://www.annualreviews.org.

in others, depending on the ambient ion concentration. The elastically compliant transmission accounts for considerable freedom of design for both rotor and stator in the similarly bipartite relatives of $F_OF_1$, namely the archeal A-ATPase and the V-ATPase of eukaryotes.[6,110] They all share with other nucleotide triphosphatases the (pseudo-)hexagonal design of the catalytic headpiece, namely with helicases and bacterial DNA-translocases which rotate and/or translocate proteins, RNA or DNA, in their central cavity and probably share a common ancestry.[111]

# 7 Outlook

Oxygenic photosynthesis in its present and fossil forms provides man with food, fiber and fuel. The quantum yield of photosynthesis is almost perfect. As to primary energy efficiency in relation to the solar spectrum (about 20%), oxygenic photosynthesis compares well with single-band-gap photovoltaic cells.[112] For the yearly average of crops in the field however the efficiency drops down to 2%, and for bio-ethanol production to 0.2%, if not being energy negative (as were biofuels of the first generation). The total useful energy of biomass which is globally produced in one year (on land) is only about 5-times greater than man's yearly energy consumption. It is obvious that present day biomass, and even more so biofuels, cannot satisfy man's ever growing energy demand. Although bio-ethanol is produced and used on a large scale in Brazil, it is no option for countries with higher population densities and/or industrialization, namely China, the EU and India. The products of photosynthesis should thus rather be reserved for food, feed, fibre and platform chemicals. A sustainable energy supply calls for technical utilization of solar energy. Wind-power, photo-thermal and photo-voltaic devices are technically established, economically competitive, and increasingly installed worldwide. It is a major scientific challenge to develop new techniques for the production of solar fuels, artificial photosynthesis with greater efficiency, and based on low cost catalysts.[113]

# References

1 http://en.wikipedia.org/wiki/Solvay_Conference accessed on the 25th of March 2015.
2 J. Ingen-Housz, *Experiments Upon Vegetables*, P. Elmsly and H. Payne, London, 1779.
3 L. N. Duysens, J. Amesz and B. M. Kamp, *Nature*, 1961, **190**, 510–511.
4 B. Kok, *Biochim. Biophys. Acta*, 1961, **48**, 527–533.
5 H. T. Witt, A. Mueller and B. Rumberg, *Nature*, 1961, **192**, 967–969.
6 W. Junge and N. Nelson, *Annu. Rev. Biochem.*, 2015, **84**, 631–657.
7 P. Mitchell, *Nature*, 1961, **191**, 144–148.
8 T. L. Netzel, P. M. Rentzepis and J. Leigh, *Science*, 1973, **182**, 238–241.
9 J. Deisenhofer, O. Epp, K. Miki, R. Huber and H. Michel, *J. Mol. Biol.*, 1984, **180**, 385–398.
10 P. Jordan, P. Fromme, H. T. Witt, O. Klukas, W. Saenger and N. Krauss, *Nature*, 2001, **411**, 909–917.
11 A. Amunts, O. Drory and N. Nelson, *Nature*, 2007, **447**, 58–63.
12 N. Nelson and W. Junge, *Annu. Rev. Biochem.*, 2015, **84**, 659–683.

13 A. Zouni, H. T. Witt, J. Kern, P. Fromme, N. Krauss, W. Saenger and P. Orth, *Nature*, 2001, **409**, 739–743.

14 Y. Umena, K. Kawakami, J. R. Shen and N. Kamiya, *Nature*, 2011, **473**, 55–60.

15 J. P. Abrahams, A. G. W. Leslie, R. Lutter and J. E. Walker, *Nature*, 1994, **370**, 621–628.

16 T. Meier, P. Polzer, K. Diederichs, W. Welte and P. Dimroth, *Science*, 2005, **308**, 659–662.

17 J. E. Walker, *Biochem. Soc. Trans.*, 2013, **41**, 1–16.

18 W. Junge, H. Lill and S. Engelbrecht, *Trends Biochem. Sci.*, 1997, **22**, 420–423.

19 C. von Ballmoos, A. Wiedenmann and P. Dimroth, *Annu. Rev. Biochem.*, 2009, **78**, 649–672.

20 W. Junge, H. Sielaff and S. Engelbrecht, *Nature*, 2009, **459**, 364–370.

21 P. D. Boyer, *Annu. Rev. Biochem.*, 1997, **66**, 717–749.

22 J. M. Breton, J. L. Martin, A. Migus, A. Antonetti and A. Orszag, *Proc. Natl. Acad. Sci. U. S. A.*, 1986, **83**, 957–961.

23 W. Holzapfel, U. Finkele, W. Kaiser, D. Oesterhelt, H. Scheer, H. U. Stilz and W. Zinth, *Proc. Natl. Acad. Sci. U. S. A.*, 1990, **87**, 5168–5172.

24 M. H. Vos, J. C. Lambry, S. J. Robles, D. C. Youvan, J. Breton and J. L. Martin, *Proc. Natl. Acad. Sci. U. S. A.*, 1991, **88**, 8885–8889.

25 G. R. Fleming and R. van Grondelle, *Curr. Opin. Struct. Biol.*, 1997, **7**, 738–748.

26 G. D. Scholes, G. R. Fleming, A. Olaya-Castro and R. van Grondelle, *Nat. Chem.*, 2011, **3**, 763–774.

27 G. D. Scholes and C. Smyth, *J. Chem. Phys.*, 2014, 140.

28 *Quantum Effects in Biology*, ed. M. Mohseni, Y. Omar and G. Engel and M. B. Plenio, Cambridge University Press, Cambridge, UK, 2014.

29 A. G. Dijkstra and Y. Tanimura, *New J. Phys.*, 2012, 14.

30 A. K. Pati, S. J. Gharpure and A. K. Mishra, *Faraday Discuss.*, 2015, DOI: 10.1039/C4FD00170B.

31 E. Romero, R. Augulis, V. I. Novoderezhkin, M. Ferretti, J. Thieme, D. Zigmantas and R. Van Grondelle, *Nat. Phys.*, 2014, **10**, 676–682.

32 G. H. Schatz, H. Brock and A. R. Holzwarth, *Proc. Natl. Acad. Sci. U. S. A.*, 1987, **84**, 8414–8418.

33 G. H. Schatz, H. Brock and A. R. Holzwarth, *Biophys. J.*, 1988, **54**, 397–405.

34 W. Kuhlbrandt, D. N. Wang and Y. Fujiyoshi, *Nature*, 1994, **367**, 614–621.

35 M. Z. Papiz, S. M. Prince, T. Howard, R. J. Cogdell and N. W. Isaacs, *J. Mol. Biol.*, 2003, **326**, 1523–1538.

36 R. J. Cogdell, N. W. Isaacs, A. A. Freer, J. Arrelano, T. D. Howard, M. Z. Papiz, A. M. Hawthornthwaite-Lawless and S. Prince, *Prog. Biophys. Mol. Biol.*, 1997, **68**, 1–27.

37 A. R. Holzwarth, K. Griebenow and K. Schaffner, *J. Photochem. Photobiol., A*, 1992, **65**, 61–71.

38 J. J. Katz, J. R. Norris, L. L. Shipman, M. C. Thurnauer and M. R. Wasielewski, *Annu. Rev. Biophys. Bioeng.*, 1978, **7**, 393–434.

39 L. L. Shipman, T. M. Cotton, J. R. Norris and J. J. Katz, *Proc. Natl. Acad. Sci. U. S. A.*, 1976, **73**, 1791–1794.

40 W. Junge and H. T. Witt, *Zeitschrift für Naturforschung*, 1968, **23b**, 244–254.

41 W. Junge, *Annu. Rev. Plant Physiol.*, 1977, **28**, 503–536.

42 C. Wolff, H. E. Buchwald, H. Ruppel, H. Witt and H. T. Witt, *Zeitschrift für Naturforschung*, 1969, **B 24**, 1038–1042.


 

43 M. Buttner, D. L. Xie, H. Nelson, W. Pinther, G. Hauska and N. Nelson, *Biochim. Biophys. Acta, Bioenerg.*, 1992, **1101**, 154–156.
44 G. Hauska, T. Schoedl, H. Remigy and G. Tsiotis, *Biochim. Biophys. Acta, Bioenerg.*, 2001, **1507**, 260–277.
45 A. Ben-Shem, F. Frolow and N. Nelson, *FEBS Lett.*, 2004, **564**, 274–280.
46 N. Nelson and A. Ben-Shem, *BioEssays*, 2005, **27**, 914–922.
47 M. A. Steffen, K. Q. Lao and S. G. Boxer, *Science*, 1994, **264**, 810–816.
48 A. W. Rutherford, A. Osyczka and F. Rappaport, *FEBS Lett.*, 2012, **586**, 603–616.
49 R. A. Marcus, *J. Chem. Phys.*, 1956, **24**, 966–978.
50 R. A. Marcus and N. Sutin, *Biochim. Biophys. Acta, Rev. Bioenerg.*, 1985, **811**, 265–322.
51 G. L. Closs and J. R. Miller, *Science*, 1988, **240**, 440–447.
52 H. B. Gray and J. R. Winkler, *Annu. Rev. Biochem.*, 1996, **65**, 537–561.
53 J. N. Onuchic, D. B. Beratan, J. R. Winkler and H. B. Gray, *Annu. Rev. Biophys. Biomol. Struct.*, 1992, **21**, 349–377.
54 C. C. Page, C. C. Moser, X. Chen and P. L. Dutton, *Nature*, 1999, **402**, 47–52.
55 D. Noy, C. C. Moser and P. L. Dutton, *Biochim. Biophys. Acta, Bioenerg.*, 2006, **1757**, 90–105.
56 F. J. E. van Mieghem, K. Brettel, B. Hillmann, A. Kamlowski, A. W. Rutherford and E. Schlodder, *Biochemistry*, 1995, **34**, 4798–4813.
57 B. Hillmann, K. Brettel, F. J. E. van Mieghem, A. Kamlowski, A. W. Rutherford and E. Schlodder, *Biochemistry*, 1995, **34**, 4814–4827.
58 M. Haumann, A. Mulkidjanian and W. Junge, *Biochemistry*, 1999, **38**, 1258–1267.
59 K. N. Ferreira, T. M. Iverson, K. Maghlaoui, J. Barber and S. Iwata, *Science*, 2004, **303**, 1831–1838.
60 P. Joliot and B. Kok, in *Bioenergetics of photosynthesis*, ed. R. A. Govindjee, Academic Press., New York, 1975, pp. 387–412.
61 A. Klauss, M. Haumann and H. Dau, *J. Phys. Chem. B*, 2015, **119**, 2677–2689.
62 J. Messinger and G. Renger, in *Primary processes of photosynthesis – part 2*, ed. G. Renger, The Royal Society of Chemistry, Cambridge, 2008, vol. 1, pp. 291–349.
63 D. Sabbert, S. Engelbrecht and W. Junge, *Proc. Natl. Acad. Sci*, 1997, **94**, 4401–4405.
64 J. Yano and V. Yachandra, *Chem. Rev.*, 2014, **114**, 4175–4205.
65 W. Junge, M. Haumann, R. Ahlbrink, A. Mulkidjanian and J. Clausen, *Philos. Trans. R. Soc. London, Ser. B*, 2002, **357**, 1407–1417.
66 J. Yano, J. Kern, K. D. Irrgang, M. J. Latimer, U. Bergmann, P. Glatzel, Y. Pushkar, J. Biesiadka, B. Loll, K. Sauer, J. Messinger, A. Zouni and V. K. Yachandra, *Proc. Natl. Acad. Sci. U. S. A.*, 2005, **102**, 12047–12052.
67 M. H. Grabolle, C. Müller, P. Liebisch and H. Dau, *J. Biol. Chem.*, 2006, **281**, 4580–4588.
68 J. R. Helliwell, A. Brink, S. Kaenket, V. L. Starkey and S. W. M. Turner, *Faraday Discuss.*, 2015, DOI: 10.1039/C4FD00166D.
69 L. Rapatskiy, N. Cox, A. Savitsky, W. M. Ames, J. Sander, M. M. Nowaczyk, M. Rogner, A. Boussac, F. Neese, J. Messinger and W. Lubitz, *J. Am. Chem. Soc.*, 2012, **134**, 16619–16634.

70 Stephanie Manz, Albert Casandruc, Dongfang Zhang, Yinpeng Zhong, Rolf A. Loch, Alexander Marx, Taisuke Hasegawa, Lai Chung Liu, Shima Bayesteh, Hossein Delsim-Hashemi, Matthias Hoffmann, Matthias Felber, Max Hachmann, Frank Mayet, Julian Hirscht, Sercan Keskin, Masaki Hada, Sascha W. Epp, Klaus Flöttmann and R. J. Dwayne Miller, *Faraday Discuss.*, 2015, DOI: 10.1039/C4FD00204K.

71 P. E. Siegbahn, *J. Photochem. Photobiol., B*, 2011, **104**, 94–99.

72 J. Kern, R. Alonso-Mori, R. Tran, J. Hattne, R. J. Gildea, N. Echols, C. Glockner, J. Hellmich, H. Laksmono, R. G. Sierra, B. Lassalle-Kaiser, S. Koroidov, A. Lampe, G. Han, S. Gul, D. Difiore, D. Milathianaki, A. R. Fry, A. Miahnahri, D. W. Schafer, M. Messerschmidt, M. M. Seibert, J. E. Koglin, D. Sokaras, T. C. Weng, J. Sellberg, M. J. Latimer, R. W. Grosse-Kunstleve, P. H. Zwart, W. E. White, P. Glatzel, P. D. Adams, M. J. Bogan, G. J. Williams, S. Boutet, J. Messinger, A. Zouni, N. K. Sauter, V. K. Yachandra, U. Bergmann and J. Yano, *Science*, 2013, **340**, 491.

73 C. Kupitz, S. Basu, I. Grotjohann, R. Fromme, N. A. Zatsepin, K. N. Rendek, M. S. Hunter, R. L. Shoeman, T. A. White, D. Wang, D. James, J. H. Yang, D. E. Cobb, B. Reeder, R. G. Sierra, H. Liu, A. Barty, A. L. Aquila, D. Deponte, R. A. Kirian, S. Bari, J. J. Bergkamp, K. R. Beyerlein, M. J. Bogan, C. Caleman, T. C. Chao, C. E. Conrad, K. M. Davis, H. Fleckenstein, L. Galli, S. P. Hau-Riege, S. Kassemeyer, H. Laksmono, M. Liang, L. Lomb, S. Marchesini, A. V. Martin, M. Messerschmidt, D. Milathianaki, K. Nass, A. Ros, S. Roy-Chowdhury, K. Schmidt, M. Seibert, J. Steinbrener, F. Stellato, L. Yan, C. Yoon, T. A. Moore, A. L. Moore, Y. Pushkar, G. J. Williams, S. Boutet, R. B. Doak, U. Weierstall, M. Frank, H. N. Chapman, J. C. Spence and P. Fromme, *Nature*, 2014, **513**, 261–265.

74 W. Schliephake, W. Junge and H. T. Witt, *Zeitschrift für Naturforschung*, 1968, **23**, 1571–1578.

75 A. T. Jagendorf and E. Uribe, *Proc. Natl. Acad. Sci. U. S. A.*, 1966, **55**, 170–177.

76 W. Junge, B. Rumberg and H. Schroeder, *Eur. J. Biochem.*, 1970, **14**, 575–581.

77 C. Kayalar, J. Rosing and P. D. Boyer, *J. Biol. Chem.*, 1977, **252**, 2486–2491.

78 P. D. Boyer and W. E. Kohlbrenner, in *Energy coupling in photosynthesis*, eds. B. R. Selman and S. Selman-Reimer, Elsevier, Amsterdam, 1981, pp. 231–241.

79 P. D. Boyer, *Trends Biochem. Sci.*, 1977, **2**, 38–41.

80 M. W. Bowler, M. G. Montgomery, A. G. Leslie and J. E. Walker, *J. Biol. Chem.*, 2007, **282**, 14238–14242.

81 T. M. Duncan, V. V. Bulygin, Y. Zhou, M. L. Hutcheon and R. L. Cross, *Proc. Natl. Acad. Sci. U. S. A.*, 1995, **92**, 10964–10968.

82 D. Sabbert, S. Engelbrecht and W. Junge, *Nature*, 1996, **381**, 623–625.

83 H. Noji, R. Yasuda, M. Yoshida and K. Kinosita, *Nature*, 1997, **386**, 299–302.

84 H. Sielaff, H. Rennekamp, S. Engelbrecht and W. Junge, *Biophys. J.*, 2008, **95**, 4979–4987.

85 D. Okuno, R. Fujisawa, R. Iino, Y. Hirono-Hara, H. Imamura and H. Noji, *Proc. Natl. Acad. Sci. U. S. A.*, 2008, **105**, 20722–20727.

86 T. Masaike, F. Koyama-Horibe, K. Oiwa, M. Yoshida and T. Nishizaka, *Nat. Struct. Mol. Biol.*, 2008, **15**, 1326–1333.

87 H. Itoh, A. Takahashi, K. Adachi, H. Noji, R. Yasuda, M. Yoshida and K. Kinosita, *Nature*, 2004, **427**, 465–468.


 

88 R. Watanabe and H. Noji, *Sci. Rep.*, 2014, **4**, 4962.
89 S. Mukherjee and A. Warshel, *Proc. Natl. Acad. Sci. U. S. A.*, 2012, **109**, 14876–14881.
90 S. Mukherjee and A. Warshel, *Proc. Natl. Acad. Sci. U. S. A.*, 2011, **108**, 20550–20555.
91 K.-i. Okazaki and G. Hummer, *Proc. Natl. Acad. Sci. U. S. A.*, 2013, **110**, 16468–16473.
92 D. Stock, A. G. Leslie and J. E. Walker, *Science*, 1999, **286**, 1700–1705.
93 K. Kinosita Jr, *Adv. Exp. Med. Biol.*, 2012, **726**, 5–16.
94 R. Yasuda, H. Noji, M. Yoshida, K. Kinosita Jr and H. Itoh, *Nature*, 2001, **410**, 898–904.
95 K. Shimabukuro, R. Yasuda, E. Muneyuki, K. Y. Hara, K. Kinosita Jr and M. Yoshida, *Proc. Natl. Acad. Sci. U. S. A.*, 2003, **100**, 14731–14736.
96 H. Ueno, T. Suzuki, K. Kinosita Jr and M. Yoshida, *Proc. Natl. Acad. Sci. U. S. A.*, 2005, **102**, 1333–1338.
97 M. G. Düser, N. Zarrabi, D. J. Cipriano, S. Ernst, G. D. Glick, S. D. Dunn and M. Börsch, *EMBO J.*, 2009, **28**, 2689–2696.
98 R. Ishmukhametov, T. Hornung, D. Spetzler and W. D. Frasch, *EMBO J.*, 2010, **29**, 3911–3923.
99 B. A. Feniouk, M. A. Kozlova, D. A. Knorre, D. Cherepanov, A. Mulkidjanian and W. Junge, *Biophys. J.*, 2004, **86**, 4094–4109.
100 D. Pogoryelov, A. Krah, J. D. Langer, O. Yildiz, J. D. Faraldo-Gomez and T. Meier, *Nat. Chem. Biol.*, 2010, **6**, 891–899.
101 O. Pänke, D. A. Cherepanov, K. Gumbiowski, S. Engelbrecht and W. Junge, *Biophys. J.*, 2001, **81**, 1220–1233.
102 W. Junge, O. Pänke, D. A. Cherepanov, K. Gumbiowski, M. Muller and S. Engelbrecht, *FEBS Lett.*, 2001, **504**, 152–160.
103 A. Wächter, Y. Bi, S. D. Dunn, B. D. Cain, H. Sielaff, F. Wintermann, S. Engelbrecht and W. Junge, *Proc. Natl. Acad. Sci. U. S. A.*, 2011, **108**, 3924–3929.
104 H. Sielaff, H. Rennekamp, A. Wächter, H. Xie, F. Hilbers, K. Feldbauer, S. D. Dunn, S. Engelbrecht and W. Junge, *Proc. Natl. Acad. Sci. U. S. A.*, 2008, **105**, 17760–17765.
105 I. N. Watt, M. G. Montgomery, M. J. Runswick, A. G. Leslie and J. E. Walker, *Proc. Natl. Acad. Sci. U. S. A.*, 2010, **107**, 16823–16827.
106 H. Seelert, A. Poetsch, N. A. Dencher, A. Engel, H. Stahlberg and D. J. Mueller, *Nature*, 2000, **405**, 418–419.
107 D. Pogoryelov, A. L. Klyszejko, G. O. Krasnoselska, E. M. Heller, V. Leone, J. D. Langer, J. Vonck, D. J. Muller, J. D. Faraldo-Gomez and T. Meier, *Proc. Natl. Acad. Sci. U. S. A.*, 2012, **109**, E1599–E1608.
108 W. Laubinger and P. Dimroth, *Eur. J. Biochem.*, 1987, **168**, 475–480.
109 A. Krah, D. Pogoryelov, J. D. Langer, P. J. Bond, T. Meier and J. D. Faraldo-Gomez, *Biochim. Biophys. Acta, Bioenerg.*, 2010, **1797**, 763–772.
110 N. Nelson, *Biochim. Biophys. Acta, Bioenerg.*, 1992, **1100**, 109–124.
111 A. Y. Mulkidjanian, M. Y. Galperin, K. S. Makarova, Y. I. Wolf and E. V. Koonin, *Biol. Direct*, 2008, **3**, 13.
112 R. E. Blankenship, D. M. Tiede, J. Barber, G. W. Brudvig, G. Fleming, M. Ghirardi, M. R. Gunner, W. Junge, D. M. Kramer, A. Melis, T. A. Moore,

C. C. Moser, D. G. Nocera, A. J. Nozik, D. R. Ort, W. W. Parson, R. C. Prince and R. T. Sayre, *Science*, 2011, **332**, 805–809.

113 T. A. Faunce, W. Lubitz, A. W. B. Rutherford, D. R. MacFarlane, G. F. Moore, P. Yang, D. G. Nocera, T. A. Moore, D. H. Gregory, S. Fukuzumi, K. B. Yoon, F. A. Armstrong, M. R. Wasielewski and S. Styring, *Energy Environ. Sci.*, 2013, **6**, 695–698.

114 K. N. Ferreira, T. M. Iverson, K. Maghlaoui, J. Barber and S. Iwata, *Science*, 2004, **303**, 1831–1838.

115 O. Pänke, D. A. Cherepanov, K. Gumbiowski, S. Engelbrecht and W. Junge, *Biophys. J.*, 2001, **81**, 1220–1233.

116 O. Pänke, K. Gumbiowski, W. Junge and S. Engelbrecht, *FEBS Lett.*, 2000, **472**, 34–38.

117 A. Wächter, Y. Bi, S. D. Dunn, B. D. Cain, H. Sielaff, F. Wintermann, S. Engelbrecht and W. Junge, *Proc. Natl. Acad. Sci. U. S. A.*, 2011, **108**, 3924–3929.

118 H. Sielaff, H. Rennekamp, A. Wächter, H. Xie, F. Hilbers, K. Feldbauer, S. D. Dunn, S. Engelbrecht and W. Junge, *Proc. Natl. Acad. Sci. U. S. A.*, 2008, **105**, 17760–17765.

119 Original video recordings of rotary motion of $F_1$: http://www.k2.phys.waseda.ac.jp/Movies.html.

120 Original video recordings of $F_OF_1$: http://www.home.uni-osnabrueck.de/wjunge/Media.html.

121 Animations of the rotary activity of $F_1$, $F_O$, and $F_OF_1$: http://www.mrc-mbu.cam.ac.uk/research/atp-synthase/molecular-animations-atp-synthase http://www.home.uni-osnabrueck.de/wjunge/Media.html.

 

# Poster titles

Hydrogen-bond dynamics and ferroelectricity in planar monocyclic oxoacids, **S. Mukhopadhyay, M. Gutmann, M. Krzystyniak and F. Fernandez-Alonso**, *ISIS Pulsed Neutron and Muon Source, UK*

Spectrally resolved photon echo spectroscopy of quantum dots at far-from-resonance excitation, **A. Kumar, S. Singhal, D. Kumar and D. Goswami**, *Indian Institute of Technology Kanpur, India*

Curcumin encapsulated pluronic block copolymer micelles synthesis as a promising therapeutic agent, **J. George, Nibisha P. and F. John**, *Sacred Heart College Thevara, India*

Explaining the novel underlying interference structures in the supercontinuum spectrum, **S. Dinda and D. Goswami**, *Indian Institute of Technology Kanpur, India*

Anomalous density function fluctuations in super-critical water: Timescales of dynamical heterogeneity measured by spectral anisotropy in 2D-IR spectroscopy, **T. Samanta, R. Biswas and B. Bagchi**, *IISC, India*

Outer Membrane channel CymA from *K. oxytoca*: Selective filter for bulky cyclodextrin molecules, **S. P. Bhamidimarri, J. P. Dahyabhai, B. van den Berg, U. Kleinekathoefer and M. Winterhalter**, *Jacobs University Bremen, Germany*

Diffusion on a rugged energy landscape with spatial correlations, **S. Banerjee, R. Biswas, K. Seki and B. Bagchi**, *Indian Institute of Science, India*

Effect of dephasing on persistent charge current in a molecular junction, **H. K. Yadalam and U. Harbola**, *Indian Institute of Science, India*

Effect of quantum nuclear motion on hydrogen bonding, **B. Athokpam, R. H. McKenzie, C. Bekker and S. G. Ramesh**, *Indian Institute of Science, India*

Electroluminescence in molecular junctions, **H. Prabal Goswami and U. Harbola**, *Indian Institute of Science, India*

ESA relationship and its impacts in interstellar chemistry, **E. Etim and E. Arunan**, *Indian Institute of Science Bangalore, India*

Reaction dynamics of UV excited phenol in aqueous solution, **G. Kumar, A. Roy and S. Bradforth**, *University of Southern California, U.S.A.*

Tuning the solid state optical properties of pyrenyl using C—H···H—C, C—H···O and π—π interactions, **S. K. Rajagopal, A. M. Philip, K. Nagarajan**

**and M. Hariharan**, *Indian Institute of Science Education and Research Thiruvananthapuram, India*

Controlling light absorption in semiconductor nanocrystals, **R. Mahadevu and A. Pandey**, *Solid State and Structural Chemistry Unit, India*

Intermolecular hydrogen bonding in the ground and excited electronic states: Steady state and time resolved absorption, resonance raman studies, **R. K. Venkatraman and S. Umapathy**, *Indian Institute of Science, India*

Light harvesting donor—acceptor vesicular scaffold limits the rate of charge recombination in the presence of an electron donor, **A. R. Mallia, R. T. Cheriya and M. Hariharan**, *IISER-Thiruvananthapuram, India*

Modeling vibrational relaxation in 2D-IR spectra, **A. G. Dijkstra, T. la Cour Jansen and J. Knoester**, *Max Plank Institute for the Structure and Dynamics of Matter, Germany*

Neutron backscattering spectroscopy at ANSTO, Australia, **N. de Souza, A. J. Edwards, G. N. Iles and A. Klappworth**, *Australian Nuclear Science and Technology Organisation, Australia*

Photophysics of diphenylbutadiynes, **A. K. Pati, S. J. Gharpure and A. K. Mishra**, *Indian Institute of Technology Madras, India*

Preparation of organomontmorillonite to study the effect of chain length on adsorption of Cu(II) ion in aqueous medium, **H. Medhi and K. G. Bhattacharyya**, *Gauhati University, India*

Exciton dynamics in lowband-gap conjugated polymers: implications for photoinduced charge generation, **A. Jha, P. Roy, V. Benyamin, Hadiya V. M., C Kanimozhi, S. Patil and J. Dasgupta**, *Tata Institute of Fundamental Research, India*

Real-time tracking of structural rearrangements after photoexcitation, **C. Schedermann, T. Wende, G. Bassolino, M. Liebel and P. Kukura**, *Oxford University, UK*

Revealing single molecule translocation through biological nanopores, **H. Bajaj, M. Ceccarelli, R. Benz and M. Winterhalter**, *University Bremen, Germany*

Sensitivity of polarization fluctuations to the nature of protein water interactions: Study of biological water in four different protein—water systems, **R. Ghosh, S. Banerjee, M. Hazra and S. Roy**, *Solid State and Structural Chemistry Unit, India*

Spatial and temporal evolution of composition fluctuation during spinodal decomposition in a binary mixture, **S. Sarkar and B. Bagchi**, *Indian Institute of Science, India*

Spectroscopic investigation for ground state proton transfer in benzaldehyde—water clusters, $[BZ\text{-}(H_2O)_n]H^+$ for $n \geq 3$, **O. Dopfer, A. Patzer, S. Chakraborty, I. Alata, R. Omidyan, M. Broquier, C. Dedonder and C. Jouvet**, *Birla Institute of Technology and Science (BITS), India*

Supramolecular assemblies of some organic entities with –COOH functionality, **V. R. Pedireddi, L. Giri and A. Chand**, *Indian Institute of Technology Bhubaneswar, India*

 

Synthesis and characterisation of novel Mn/Ti layered double hydroxide as efficient photocatalyst for degredation of azo dyes in water, **P. Roy Chowdhury and K. G. Bhattacharyya**, *Gauthati University, India*

Effect of solvent on ultrafast coherent nuclear dynamics of photo-excited Tetraphenylethylene (TPE) from ultrafast RAMA loss and transient absorption spectroscopy, **S. Kayal, K. Roy and S. Umapathy**, *Indian Institute of Science, India*

Understanding the stability and transformations of pharmaceutical polymorphs using *ab initio* lattice dynamics, **A. Pallipurath, J. Skelton, N. Kamali, P. McArdle and A. Erxleben**, *National University of Ireland, Galway, Ireland*

The X-ray radiation stability of $K_2PtI_6$, $K_2PtBr_6$ and $Ta_6Br_{12}$ bound to lysozyme, **S. Kaenket, L. Lamplough, V. L. Starkey, J. Habash, R. Raftery, J. R. Helliwell, C. Levy, P. Bryant, A. Fitch, I. Margiolaki, Y. Waiter, J. Wright and A. Bell**, *University of Manchester, UK*

Excited state dynamics of cylindrical Pi conjugated system, **K. Roy, S. Kayal, S. Juna and S. Umapathy**, *Indian Institute of Science, India*

Effect of methyl substitution in the complexes of indole with methylated benzenes: Interplay among electrostatic, dispersion interaction and steric repulsion, **S. Kumar, S. K. Singh, J. Vaishnav, G. Hill and A. Das**, *Indian Institute of Science Education and Research Pune, India*

Structural synthon approach for aluminum hydroxide layers using density functional theory, **S. Nagendran, G. Periyasamy and V. Kamath**, *Bangalore University, India*

Spectroscopic evidences of the electrostatic dynamic and non-electrostatic contribution to Hydrogen boning interactions, **P. Deb, S. M. Kashid, T. Haldar and S. Bagchi**, *CSIR - National Chemical Laboratory India, India*

The Skinner Prize for the best poster was awarded to Mr Christoph Schnerdermann of Oxford University for his poster on real-time tracking of structural rearrangements after photoexcitation. The joint runner up prize was awarded to Mr Saikat Banerjee, for his poster on the diffusion on a rugged energy landscape with spatial correlations, and to Mr Ajay Jha for his poster on exciton dynamics in low band-gap conjugated polymers: implications for photoinduced charge generation.

# List of participants

Professor Dr Vincenzo Aquilanti, *Dipartimento di Chimica,Università di Perugia, Italy*
Professor Sankarampadi Aravamudhan, *Department of Chemistry, North Eastern Hill University, India*
Dr Freek Ariese, *Indian Institute of Science, India*
Professor Elangannan Arunan, *Indian Institute of Science, India*
Professor Michael Ashfold, *University of Bristol, United Kingdom*
Miss Bijyalaxmi Athokpam, *Indian Institute of Science, India*
Professor Biman Bagchi, *Indian Institute of Science, India*
Miss Harsha Bajaj, *Jacobs University Bremen, Germany*
Mr Saikat Banerjee, *Indian Institute of Science, India*
Miss Satya Prathyusha Bhamidimarri, *Jacobs University Bremen, Germany*
Dr Atanu Bhattacharya, *Indian Institute of Science, India*
Professor Marylise Buron-Le Cointe, *University Rennes 1, France*
Dr Shamik Chakraborty, *Birla Institute of Technology and Science (BITS), Pilani. Pilani Campus, India*
Miss Apramita Chand, *Indian Institute of Technology, Bhubaneswar, India*
Dr Emma Cooper, *Royal Society of Chemistry, United Kingdom*
Mr Pranab Deb, *NCL, India*
Professor Dr Volker Deckert, *University of Jena, Germany*
Mr Kunal Dhoke, *Indian Institute of Science, India*
Dr Arend G. Dijkstra, *Max Planck Institute for the Structure and Dynamics of Matter, Germany*
Professor E. D. Jemmis, *Indian Institute of Science-Bangalore, India*
Dr Alison Edwards, *Bragg Institute, ANSTO, Australia*
Dr Rainer Eichberger, *Helmholtz Zentrum Berlin, Germany*
Mr Emmanuel Etim, *Indian Institute of Science, India*
Dr Aparna Ganguly, *Royal Society of Chemistry, India*
Dr Jinu George, *S.H college, M.G University, India*
Professor Ken Ghiggino, *University of Melbourne, Australia*
Ms Rikhia Ghosh, *Solid State and Strucutral Chemistry Unit, India*
Miss Lopamudra Giri, *IIT Bhubaneswar, India*
Miss Sharon Gnanasekar, *Indian Institute of Science, India*
Professor Dr Debabrata Goswami, *Indian Institute of Technology Kanpur, India*
Miss Nikita Gupta, *Royal Society of Chemistry, India*
Professor Tayur Guru Row, *Indian Institute of Science, India*
Mr Prasenjit Halder, *Indian Institute of Technology Madras, India*
Mr Sudeeksha Hanehalli, *Indian Institute of Science, India*
Dr Upendra Harbola, *Indian Institute of Science, India*
Dr Mahesh Hariharan, *Indian Institute of Science Education and Research Thiruvananthapuram, India*
Professor Dr John Helliwell, *University of Manchester, United Kingdom*
Professor Jonathan Hirst, *University of Nottingham, United Kingdom*

Professor Judith Howard FRS, *Durham University*, *United Kingdom*
Dr Neil Hunt, *University of Strathclyde*, *United Kingdom*
Miss Veena Iyer, *Indian Institute of Science*, *India*
Mr Sanjib Jana, *Indian Institute of Science*, *India*
Mr Ajay Jha, *Tata Institute of Fundamental Research*, *India*
Dr Franklin John, *S.H college, M.G University*, *India*
Mr Amit Kumar Joshi, *Indian Institute of Science*, *India*
Mr Jyothish Joy, *Indian Institute of Science Education and Research-Thiruvananthapuram*, *India*
Professor Dr Wolfgang Junge, *University of Osnabrück*, *Germany*
Mr Surajit Kayal, *Indian Institute of Science*, *India*
Professor Srihari Keshavamurthy, *Indian Institute of Technology Kanpur*, *India*
Mr Gaurav Kumar, *University of Southern California*, *U.S.A.*
Mr Kabir Kumbhar, *Indian Institute of Science, Bangalore*, *India*
Dr Adithya Lakshmanna, *Indian Institute of Science Education and Research Thiruvananthapuram*, *India*
Professor Dr King-Chuen Lin, *National Taiwan Univ.*, *Chinese Taipei*
Ms Rekha M, *IISC*, *India*
Mr Ajith R. Mallia, *IISER-Thiruvananthapuram*, *India*
Miss Himani Medhi, *Department of Chemistry, Gauhati University, Guwahati, Assam - 781014*, *India*
Professor Stephen Meech, *University of East Anglia*, *United Kingdom*
Professor Martin Meedom Nielsen, *DTU Physics, Technical University of Denmark*, *Denmark*
Professor Dr R J Dwayne Miller, *MPSD/University of Toronto*, *Germany*
Professor Dr Ashok Kumar Mishra, *Indian Institute of Technology Madras*, *India*
Mr Madhu Mohan, *Indian Institute of Technology Madras*, *India*
Mr Kiran Moirangthem, *Indian Institute of Science*, *India*
Dr Heather Montgomery, *Royal Society of Chemistry*, *United Kingdom*
Professor Shaul Mukamel, *University of California, Irvine*, *USA*
Dr Sanghamitra Mukhopadhyay, *ISIS Neutron and Muon Facility*, *UK*
Mr Supreeth Nagendran, *PhD Student*, *India*
Mr Andres Neira, *King's College London*, *United Kingdom*
Dr Artur Nenov, *Università di Bologna*, *Italy*
Dr Robert Pal, *Durham University*, *United Kingdom*
Dr Anuradha Pallipurath, *National University of Ireland, Galway*, *Ireland*
Mr Rajesh Parishwad, *Royal Society of Chemistry*, *India*
Mr Avik Kumar Pati, *Indian Institute of Technology Madras*, *India*
Professor V. R. Pedireddi, *Indian Institute of Technology Bhubaneswar*, *India*
Dr Carla Pegoraro, *Royal Society of Chemistry*, *United Kingdom*
Dr Ganga Periyasamy, *UGC-Assistant Professor*, *India*
Mr Himangshu Prabal Goswami, *Indian Institute of Science*, *India*
Dr Abhiram Prabhu, *Royal Society of Chemistry*, *India*
Mr Sanjay Prasad, *Indian Institute of Science*, *India*
Dr Sai Ramesh, *IPC Department, IISc-Bangalore*, *India*
Mr Jyotirmoy Ray, *Indian Institute of science (IISc.)*, *India*
Dr Chilla Malla Reddy, *IISER Kolkata*, *India*
Mr Khokan Roy, *Indian Institute of Science*, *India*
Mr Priyadarshi Roy Chowdhury, *Department of Chemistry, Gauhati University, Guwahati-781014, Assam*, *India*
Professor Gopinathan Sankar, *University College London*, *United Kingdom*
Ms Sarmistha Sarkar, *Indian Institute of Science*, *India*
Mr Christoph Schnedermann, *Oxford University*, *United Kingdom*

 

Mr Abhishek Shahi, *Indian Institute of Science, India*
Dr Helena Shepherd, *University of Bath, United Kingdom*
Mr Santosh Kumar Singh, *Indian Institute of Science Education and Research, India*
Mr Sumit Singhal, *Indian Institute of Technology, Kanpur, India*
Dr Jonathan Skelton, *University of Bath, United Kingdom*
Dr Hazel Sparkes, *University of Bristol, United Kingdom*
Professor Keisuke Tominaga, *Molecular Photoscience Research Center, Kobe University, Japan*
Professor Siva Umapathy, *Indian Institute of Science, India*
Mr Raja Mani Kandan V B, *Indian Institute of Technology Madras, India*
Dr Tim Brandt van Driel, *Technical University of Denmark, Denmark*
Mr Ravi Kumar Venkatraman, *Indian Institute of Science, India*
Dr Mark Warren, *Diamond Light Source Ltd, United Kingdom*
Mr Hari Kumar Yadalam, *Indian Institute of Science, India*
Dr Veeranna Yempally, *Texas A&M University at Qatar, Qatar*
Professor Martin Zanni, *University of Wisconsin-Madison, United States*